Selected Titles in This Series

37 **Hershel M. Farkas and Irwin Kra,** Theta constants, Riemann surfaces and the modular group, 2001

36 **Martin Schechter,** Principles of functional analysis, second edition, 2001

35 **James F. Davis and Paul Kirk,** Lecture notes in algebraic topology, 2001

34 **Sigurdur Helgason,** Differential geometry, Lie groups, and symmetric spaces, 2001

33 **Dmitri Burago, Yuri Burago, and Sergei Ivanov,** A course in metric geometry, 2001

32 **Robert G. Bartle,** A modern theory of integration, 2001

31 **Ralf Korn and Elke Korn,** Option pricing and portfolio optimization: Modern methods of financial mathematics, 2001

30 **J. C. McConnell and J. C. Robson,** Noncommutative Noetherian rings, 2001

29 **Javier Duoandikoetxea,** Fourier analysis, 2001

28 **Liviu I. Nicolaescu,** Notes on Seiberg-Witten theory, 2000

27 **Thierry Aubin,** A course in differential geometry, 2001

26 **Rolf Berndt,** An introduction to symplectic geometry, 2001

25 **Thomas Friedrich,** Dirac operators in Riemannian geometry, 2000

24 **Helmut Koch,** Number theory: Algebraic numbers and functions, 2000

23 **Alberto Candel and Lawrence Conlon,** Foliations I, 2000

22 **Günter R. Krause and Thomas H. Lenagan,** Growth of algebras and Gelfand-Kirillov dimension, 2000

21 **John B. Conway,** A course in operator theory, 2000

20 **Robert E. Gompf and András I. Stipsicz,** 4-manifolds and Kirby calculus, 1999

19 **Lawrence C. Evans,** Partial differential equations, 1998

18 **Winfried Just and Martin Weese,** Discovering modern set theory. II: Set-theoretic tools for every mathematician, 1997

17 **Henryk Iwaniec,** Topics in classical automorphic forms, 1997

16 **Richard V. Kadison and John R. Ringrose,** Fundamentals of the theory of operator algebras. Volume II: Advanced theory, 1997

15 **Richard V. Kadison and John R. Ringrose,** Fundamentals of the theory of operator algebras. Volume I: Elementary theory, 1997

14 **Elliott H. Lieb and Michael Loss,** Analysis, 1997

13 **Paul C. Shields,** The ergodic theory of discrete sample paths, 1996

12 **N. V. Krylov,** Lectures on elliptic and parabolic equations in Hölder spaces, 1996

11 **Jacques Dixmier,** Enveloping algebras, 1996 Printing

10 **Barry Simon,** Representations of finite and compact groups, 1996

9 **Dino Lorenzini,** An invitation to arithmetic geometry, 1996

8 **Winfried Just and Martin Weese,** Discovering modern set theory. I: The basics, 1996

7 **Gerald J. Janusz,** Algebraic number fields, second edition, 1996

6 **Jens Carsten Jantzen,** Lectures on quantum groups, 1996

5 **Rick Miranda,** Algebraic curves and Riemann surfaces, 1995

4 **Russell A. Gordon,** The integrals of Lebesgue, Denjoy, Perron, and Henstock, 1994

3 **William W. Adams and Philippe Loustaunau,** An introduction to Gröbner bases, 1994

2 **Jack Graver, Brigitte Servatius, and Herman Servatius,** Combinatorial rigidity, 1993

1 **Ethan Akin,** The general topology of dynamical systems, 1993

Theta Constants, Riemann Surfaces and the Modular Group

Theta Constants, Riemann Surfaces and the Modular Group

An Introduction with Applications to Uniformization Theorems, Partition Identities and Combinatorial Number Theory

Hershel M. Farkas
Irwin Kra

Graduate Studies in Mathematics
Volume 37

American Mathematical Society
Providence, Rhode Island

2000 *Mathematics Subject Classification.* Primary 30F35, 30F30, 11F20, 11F25, 11F30, 11P81, 11P82, 11P83, 14H42, 14H45, 14H55, 20H10.

ABSTRACT. The book is an introduction to the theory of Theta Functions and Theta Constants with rational characteristics showing their close connections to Riemann Surfaces and the Modular Group. Applications to uniformization theorems, partition identities and combinatorial number theory are developed. Projective embeddings of modular curves and Ramanujan partition congruences are explored. Theta identities and their combinatorial contents are studied. Many examples are provided, and directions for future studies are indicated.

Library of Congress Cataloging-in-Publication Data

Farkas, Hershel M.

Theta constants, Riemann surfaces, and the modular group : an introduction with applications to uniformization theorems, partition identities and combinatorial number theory / Hershel M. Farkas, Irwin Kra.

p. cm. — (Graduate studies in mathematics, ISSN 1065-7339 ; v. 37)

Includes bibliographical references and index.

ISBN 0-8218-1392-7 (acid-free paper)

1. Functions, Theta. 2. Riemann surfaces. 3. Modular groups. I. Kra, Irwin. II. Title. III. Series.

QA345.F37 2001
515′.984—dc21 2001035711

∞ The paper used in this book is acid-free and falls within the guidelines
established to ensure permanence and durability.
Visit the AMS home page at URL: `http://www.ams.org/`

10 9 8 7 6 5 4 3 2 1 06 05 04 03 02 01

Contents

Introduction

The theory of compact Riemann surfaces brings together diverse areas of mathematics. Its building blocks include vast areas of analysis (including Lie theory), geometry/topology and algebra. This was our point of view in our book on Riemann surfaces [**6**] and it dictated the material to be included in that volume. In particular, we presented a modern approach to the theory of compact Riemann surfaces based on classical methods that prepared the reader to study the modern theories of moduli of surfaces. In this book we head in a different direction and develop another classical connection: to combinatorial number theory. We do not neglect, however, the connections to the problem of uniformizing surfaces represented by very special Fuchsian groups. Problems in number theory can be reformulated as questions about Riemann surfaces, and many of the answers to some of these questions are obtained using function theory. Even though it is an old idea to use function theory (compact Riemann surfaces and automorphic forms) to study analytic and combinatorial number theory and there are many results in these fields, we found it hard to dig out the underlying function theory in the publications of number theorists. No doubt, this is our failing. But since others may also have a deficiency in this area, we decided to organize the material from this point of view. There is new material in this book that has not previously appeared in print, and part of our aim is to present this material to as wide an audience as possible. Our more important aim, however, is to expose the reader to a beautiful chapter in function theory and its applications.

The main actors in our presentation are genus one theta functions and theta constants[1] (including the classical η-function), the modular group $\Gamma = \text{PSL}(2,\mathbb{Z})$, and some of the Riemann surfaces that arise as quotients of the action of finite index subgroups of Γ on $\mathbb{H}^2$. We are particularly interested in the principal congruence subgroups $\Gamma(k)$ and the related subgroups $\Gamma_o(k)$ for (usually small) primes k. Some very interesting combinatorial identities follow from the function theory on these surfaces.[2]

Theta functions and theta constants with integral characteristics are classical objects intimately connected with the principal congruence subgroup of level 2, $\Gamma(2)$. This theory is well understood and has as one of its consequences the theorem of Picard: every entire function which omits two values is constant. As is well known, the basic ingredients in the proof of Picard's theorem are that the holomorphic universal covering of the sphere punctured at three points is the upper half plane and that its fundamental group is $\Gamma(2)$. We use theta constants with even integral characteristics to construct the universal covering map, and in this way obtain, without using the general uniformization theorem, the hyperbolicity of the three times punctured sphere. The universal covering map is constructed here as a quotient of fourth powers of any two of the three theta constants. We noticed that in this construction the three even characteristics correspond in a natural way to the three punctures on $\mathbb{H}^2/\Gamma(2)$, and we began to wonder about natural generalizations. In this book, we present the answer to these inquiries. We uniformize the Riemann surfaces $\mathbb{H}^2/\Gamma(k)$ using theta constants with special rational characteristics, and establish a one-to-one (almost canonical) correspondence between the punctures on $\mathbb{H}^2/\Gamma(k)$ and certain equivalence classes of characteristics. For example, the four punctures on $\mathbb{H}^2/\Gamma(3)$ correspond to the characteristics

$$\begin{bmatrix} 1 \\ \frac{1}{3} \end{bmatrix}, \begin{bmatrix} \frac{1}{3} \\ \frac{1}{3} \end{bmatrix}, \begin{bmatrix} \frac{1}{3} \\ 1 \end{bmatrix}, \begin{bmatrix} \frac{1}{3} \\ \frac{5}{3} \end{bmatrix}.$$

Furthermore, the Riemann surface is uniformized by a quotient of cubes of any two of the four corresponding theta constants. Similar, but obviously more complicated, expressions uniformize the surfaces represented by the higher level congruence groups. Multiple uniformizations of the same

[1]Thus the θ-functions we study, $\theta[\chi](\zeta,\tau)$, depend on three variables: a characteristic $\chi \in \mathbb{R}^2$; a variable $\zeta \in \mathbb{C}$; and a parameter $\tau \in \mathbb{H}^2$, the upper half plane. Fixing the variable $\zeta = 0$ yields the family of theta constants, an abuse of notation since these are holomorphic functions on $\mathbb{H}^2$; as functions of the local coordinates $q = e^{2\pi\imath\tau}$ these are classically known as q-series. We will use the symbol x for the local variable, since tradition in (parts of complex analysis) reserves the letter q for the weight of an automorphic form.

[2]What is interesting is clearly in the eyes of the beholder. The identities we discuss are obviously interesting to us. The reader must decide whether or not to share our enthusiasm.

Riemann surface lead to theta identities. It is an open problem to determine uniformizations of all four punctured spheres by the methods described above.

The theta constants which appear in our constructions of the uniformizing functions for $\mathbb{H}^2/\Gamma(3)$ are closely related with the formulae used by Euler and Ramanujan in the theory of partitions. Specifically, we note that[3]

$$\theta\left[\begin{array}{c}\frac{1}{3}\\1\end{array}\right](0,\tau)=\exp\left(\frac{\pi\imath}{6}\right)x^{\frac{1}{24}}\sum_{n=-\infty}^{\infty}(-1)^n x^{\frac{3n^2+n}{2}}=\exp\left(\frac{\pi\imath}{6}\right)x^{\frac{1}{24}}\prod_{n=1}^{\infty}(1-x^n),$$

where $x=\exp\left(\frac{2\pi\imath\tau}{3}\right)$. Continuing in this direction, we discover that uniformizations of the Riemann surfaces $\mathbb{H}^2/\Gamma(k)$ involve functions which appear in the Jacobi triple product. We give a function theoretic proof of this famous formula and then generalize it to the quintuple and septuple product identities, explaining along the way why the formulae obtained are natural from the point of view of the theory of N-th order theta functions. The highlights of the book are systematic studies of theta constant identities, uniformizations of surfaces represented by subgroups of the modular group, partition identities and Fourier series coefficients of automorphic functions, and identities involving the σ-function and Fourier series coefficients of automorphic forms. More detailed information on the contents of each of the chapters follows.

In Chapter 1 we explain the genesis of the modular group in our theory. This group appears naturally when one classifies compact Riemann surfaces of genus one (elliptic curves) up to conformal equivalence. We discuss the generators of this group, find all the fixed points of elements of this group and describe some of the subgroups we shall need in the sequel. Almost everything we do in this chapter is well known and covered in a standard course on complex variables. We describe the structure of the Riemann surfaces $\mathbb{H}^2/G$ for subgroups G of $\mathrm{PSL}(2,\mathbb{Z})$. In order to show that this well known and elementary material has nontrivial consequences, we use this theory to show that factors of integers of the form N^2+1 are always the sums of two squares, and we give a geometric criterion for N^2+1 to be a prime number. The result is that N^2+1 is prime if and only if the portion in the upper half plane of the straight line joining the origin to the point $N+\imath$ in the complex plane intersects the orbit of $\imath$ under $\mathrm{PSL}(2,\mathbb{Z})$ in exactly two points, namely $N+\imath$ and $\frac{N+\imath}{N^2+1}$. We have included some of the function theoretic prerequisites in this chapter. However, most of the prerequisites will be described when needed. In general, we provide full definitions of all concepts. We do not repeat proofs or arguments readily available in

[3]The reader may at this point conclude that the η-function is a disguised theta constant with a rational characteristic. It will also become obvious that the prime 3 plays a special role in our drama.

other books, but do reproduce, usually in modified form, proofs from many research papers. The bibliography of relevant books (after the last chapter) is followed by a set of bibliographical notes containing an (incomplete) list of research and expository notes on the material covered by this volume.

In Chapter 2 we define the theta functions and theta constants with characteristics and specialize to rational characteristics of the form $\begin{bmatrix} \frac{m}{k}, \\ \frac{m'}{k} \end{bmatrix}$ with m, m' and k integers[4] of the same parity in order to construct a correspondence between equivalence classes of sets of characteristics and the punctures on the surface $\mathbb{H}^2/\Gamma(k)$. In this chapter we derive a most important property of the theta functions and theta constants, the transformation formula (a significant generalization of known transformation rules) for the action of $\text{PSL}(2,\mathbb{Z})$ on the upper half plane, and we give a function theoretic proof of the Jacobi triple product formula and some generalizations. The transformation formula allows us to use theta functions to construct modular and cusp forms for subgroups of $\text{PSL}(2,\mathbb{Z})$. The function theoretic proof of the Jacobi triple product formula yields new proofs of important identities of Jacobi and Euler that are needed for our presentation of partition theory in Chapter 5. We construct theta constant identities which turn out to agree with discoveries of Ramanujan. Our derivations of these identities are on the one hand quite natural, and on the other hand lead to simpler expressions of the equivalent identities discovered by Ramanujan in the sense that they do not involve irrationalities (extracting roots of single valued functions) until they are artificially introduced. It appears that the theta constants which we use are a lot richer than the ones that Ramanujan had at his disposal.

Chapter 3, in a sense, contains the most important material of the book. In it we construct automorphic forms and functions for the principal congruence subgroups and some related groups. The theory we describe is particularly well suited for the study of $\mathbb{H}^2/\Gamma(k)$, and we obtain holomorphic mappings of these Riemann surfaces into projective spaces of rather low dimensions. Some interesting geometry and topology emerges as we observe connections of the principal congruence subgroups with the Platonic solids. This phenomenon first occurs for $k = 3$, 4 and 5. In these cases, $\Gamma/\Gamma(k) \cong \text{PSL}(2,\mathbb{Z}_k)$ are the symmetry groups of the regular tetrahedron, octahedron and icosahedron, respectively. This suggests a relation between the images of these curves in the projective space and the regular solids and leads to a generalization of the regular solids based on curves of (some) positive genera. While our development is most suited for the groups $\Gamma(k)$, for many of the most important applications we need to construct automorphic

[4]Assume, unless otherwise stated, for these introductory remarks that k is a (positive) prime.

forms for $\Gamma_o(k)$. Part of the extra difficulties involves the presence of torsion in these groups. We need more detailed analysis, in this and subsequent chapters, to handle these richer groups.

Chapter 4 is a systematic study of theta identities. Theta constant identities are interesting for several reasons. One reason is their inherent elegance and symmetry. There is something tantalizingly beautiful about the identities of Jacobi, for example,

$$\begin{aligned}\theta^2\begin{bmatrix}0\\0\end{bmatrix}&(0,\tau)\theta^2\begin{bmatrix}0\\0\end{bmatrix}(z,\tau)\\&=\theta^2\begin{bmatrix}0\\1\end{bmatrix}(0,\tau)\theta^2\begin{bmatrix}0\\1\end{bmatrix}(z,\tau)+\theta^2\begin{bmatrix}1\\0\end{bmatrix}(0,\tau)\theta^2\begin{bmatrix}1\\0\end{bmatrix}(z,\tau)\end{aligned}$$

or its restriction to $z=0$ (known as the Jacobi quartic identity)

$$\theta^4\begin{bmatrix}0\\0\end{bmatrix}(0,\tau)=\theta^4\begin{bmatrix}0\\1\end{bmatrix}(0,\tau)+\theta^4\begin{bmatrix}1\\0\end{bmatrix}(0,\tau).$$

Aside from the inherent beauty of the form there is a combinatorial content to the identity. It relates the number of representations of an integer as a sum of four squares to its representation as a sum of four triangular numbers. This is of course just the beginning of a chapter. As one delves deeper into the theory, one finds more and more beautiful identities with more and more combinatorial content.

We present four distinct ways of constructing such identities. In two of theses methods, we use the classical technique of constructing finite dimensional linear spaces of theta functions or modular forms or functions on certain Riemann surfaces and use linear algebra and the simple notions of independence and dependence. We present another newer technique which makes use of the fact that we can use theta functions to construct elliptic functions and the fact that the sum of the residues of an elliptic function in a period parallelogram vanishes. This technique is very powerful and succeeds in giving a rather large set of identities. The main idea here is to construct the correct elliptic function, which turns out to be more an art than a science, that leads to an interesting identity. The fourth method for constructing identities uses uniformizations of Riemann surfaces. In a very nontrivial sense the next two chapters are also studies of theta identities, this time of a very special sort with a very special purpose.

In Chapter 5 we turn to the congruences discovered by Ramanujan for the partition function and show how they follow in a rather simple way from function theory on the appropriate Riemann surfaces. The main ingredient is the construction of the same function in more than one way. Some of the constructions involve averaging operators. It turns out that the averaging processes produce in some cases constant functions. We study the many implications of these constructions, especially of the appearance of constant functions. This material may not be new to the literature. We present it in a unified way based on function theoretic foundations that most of the time remove and in general isolate the mysteries in many of the research monographs on the subject. This chapter is based almost entirely on the properties of the classical η-function, a very special case of the theory described in the previous chapters. We need to know rather detailed information on its multiplier system. The needed number theoretic arguments are found in [**16**], for example. We would like to avoid dependence on these. We have only partially succeeded in doing so and have hence not included in this volume most of the results of this effort.

In Chapter 6 we begin by reviewing some concepts from covering space theory and show how these ideas lead to beautiful identities among theta constants and their interpretation as identities among infinite products. Here the main tools are the uniformizations of the Riemann surfaces in question. We then continue by showing how many of the ideas used in the congruences related with the partition function and its generalizations carry over to other modular forms. We treat in particular the $\jmath$-function and the congruences satisfied by the coefficients of its Laurent series expansion.

In Chapter 7 we show how statements about partitions are related to other combinatorial quantities such as representations of positive integers as sums of squares or of triangular numbers, and most importantly to the divisors of an integer. In particular, we describe relations to the question of primality of integers depending on statements about partitions. This suggests that while primality is usually thought of as a subject in multiplicative number theory, it can also be viewed as a part of additive number theory.

We give some examples to show what type of results can be expected in this chapter in the expectation that these applications are the main interests of some readers of this text. We emphasize that these results were not the reason for writing this book. The list of examples is by no means exhaustive. Let $\sigma(n)$ denote the classical σ-function; that is, $\sigma(n)$ is the sum of the divisors of the positive integer n. We show that

$$\sum_{j=-\infty}^{\infty} \sigma\left(n - \frac{3j^2+j}{2}\right) = 0$$

whenever n is not of the form $\frac{3m^2+m}{2}$ with $m \in \mathbb{Z}$. A companion related result is

$$\sum_{j=0}^{\infty} \sigma\left(n - \frac{j^2+j}{2}\right) = 4\sum_{j=0}^{\infty} \sigma\left(\frac{n}{2} + \frac{j^2+j}{4}\right)$$

whenever n is not of the form $\frac{m^2+m}{2}$ with $m \in \mathbb{Z}$. We obtain a variant of Jacobi's result on the number of ways a positive integer can be written as a sum of four squares by replacing squares with triangular numbers, obtaining, in our view, a cleaner result. A last example of the type of result we will discuss in this chapter is the following. Let S be a set consisting of the positive integers with an additional copy of those positive integers congruent to zero modulo 7. Decompose S into its even and odd parts, E and O respectively. Denote by $P_E(k)$ and $P_O(k)$ the number of partitions of k with parts taken from the sets E and O respectively. We prove that for all nonnegative integers k,

$$P_E(2k) = P_O(2k+1).$$

The prerequisites for this book are a thorough understanding of material traditionally covered in first year graduate courses – especially the contents of the complex analysis course. We review, however, the most salient points about elliptic function theory portions of this course. Although a knowledge of Riemann surfaces and Fuchsian groups is helpful, it is not needed by the reader who is willing to accept the summaries of the required material (with references to the literature). Although we do not, in general, reproduce material available in other textbooks, we make an exception for material on theta functions and theta constants despite the availability of excellent sources (for example, [**23**]). We do so, not only for the convenience of the reader, but also to emphasize our point of view. We have also ignored, to a great extent, the combinatorial and special functions connection. These are discussed in [**2**], [**7**] and [**10**], for example.

A road map

While we did not intend to write an encyclopedic text, the result has been quite a large book. We take the liberty of offering the readers our suggestions for possible ways of going through this text, which was written with several different types of readers in mind. These range from the beginning graduate mathematics student through the professional mathematician whose interests are either in combinatorial mathematics (partition theory, representation as sums of squares, counting points on conic sections) or function theory (Riemann surfaces, modular forms). Theoretical physicists might be interested in portions of the material we cover.

The reader is expected to have a reasonable knowledge of the theory of functions of a complex variable, through the Riemann mapping theorem, and enough mathematical maturity to follow an argument even though unfamiliar with the proofs of all the tools used. Thus, the book can be used as a text for a topics course in either analysis or analytic number theory, and as such a reasonable approach would be to go through Chapter 1 sections 1-4.5 and sections 6 and 7. The above material is essentially background material to acquaint the reader with the domains on which we will be doing analysis to obtain the combinatorial results. This reader should then continue with Chapter 2, where the theory of the one dimensional theta function is presented. This chapter should be read in its entirety. In Chapter 3 the reader or instructor pressed for time could read quickly the first section for the definitions and then go on to a careful study of sections 2 through 8.4. Chapter 4 should be read in its entirety. The above could constitute a one semester topics course for beginning graduate students.

The above suggestion leaves out Chapters 5, 6 and 7, which occupy much of this book and are an important part of it, since they deal with the theory of congruences for the Ramanujan partition function, the congruences for the $\jmath$-function, and the combinatorial interpretations of many of the identities derived in Chapter 4. In a one-year course the material studied could include Chapter 5 through section 10.8 and section 12,[5] all of Chapters 6 and 7.

The professional mathematician who is interested in Riemann surface theory should study Chapter 1, including sections 4.6 through 5.7, in order to get a picture of where the theory can possibly go. If conversant with the theory of the Riemann theta function, the reader can skip section 1 of Chapter 2 and read the remainder of that chapter. The reader should then proceed to the beginning of Chapter 3. Some of the introductory material of this chapter can be skipped or read quickly to get acquainted with the notation used; the choice of which of the remaining sections to read should be guided by interests; we suggest that this include section 10. Chapter 4 should be read in its entirety and then Chapter 5 and Chapter 6, once again guided by the interests of the reader. Chapter 7 should also be read in its entirety.

The professional mathematician whose interests are in combinatorial mathematics may wish to begin by looking at Chapter 7 and then proceed backwards through the theory. Needless to say, Chapters 2, 3 and 4 will have to be read at some point, and if interested in Ramanujan congruences, Chapter 5 is a must. In any event section 12 of Chapter 5 should be reviewed.

There is lots of flexibility in the way the text can be studied and/or approached. We trust the various readers will find their way through the maze and enjoy the material they stop to study or, as we and others have said, will enjoy this tour of Ramanujan's garden and the flowers they pick there.

[5]In a course where all nonstandard material is included, the instructor might want to spend some time on the multiplicative properties of the η-function. These could be taken from Knopp's book [**16**].

We have included descriptions of many special cases and summarized the results of many calculations. Most of the nontrivial calculations used the symbolic manipulation programs MATHEMATICA and/or MAPLE. In order to excite the reader about the flowers in the beautiful garden we are cultivating here, we start each chapter with (what we regard as) a handsome example of what will follow. We have included a number of accessible exercises and research level problems. The latter may be quite challenging and are at times only conjectures. The reader should also approach the many special cases we have included as challenges to obtain independent solutions. They are presented in the spirit of exercises, with solutions supplied by the authors.

Numbering systems. The book consists of seven chapters and a set of bibliographical notes that will be maintained and updated on the web. Chapters are subdivided into sections; these into subsections. Definitions, lemmas, propositions, theorems, exercises, problems and remarks are labeled consecutively as a single group within each section. A typical item is Theorem section.number; number starts with 1 for the first item in the section. Thus, for example, in Chapter 2, Definition 2.32 (in section 2) in our numbering scheme is followed by Lemma 4.1 (in section 4). Equations that will be referenced subsequently in the text are labeled by a decimal: chapter.number; number starts with 1 for the first numbered equation in the chapter. Tables and figures are numbered consecutively in the book.

Acknowledgments. We thank
1. the mathematicians who have had the patience to listen to our lectures on the subject of this book: whether our students at the Hebrew University in Jerusalem or the State University of New York in Stony Brook, who at times have provided us with constructive criticism, or colleagues at various institutions who have invited us to speak about this subject;
2. our graduate students who read previous drafts of the book and pointed out errors and ambiguities (needless to say, the remaining errors and ambiguities are the authors' responsibilities);
3. our undergraduate students who helped with the MATHEMATICA calculations and the figures;
4. the editorial and technical staff of the American Mathematical Society, especially Deborah Smith and Janet Pecorelli who guided us through the final preparation of the manuscript; and most importantly,

5. our respective wives, who patiently suffered with us through this long project.

Finally we wish to acknowledge and thank the U.S. - Israel Binational Science Foundation, the U.S. National Science Foundation, The Hebrew University and the State University of New York at Stony Brook for their support over the years that helped produce the research for this book; for providing us with the needed time, facilities and equipment; and for making possible visits by each of us to the institution of the other, not only to work on this volume.

Hershel M. Farkas
Jerusalem, Israel

Irwin Kra
Stony Brook, NY, U.S.A.

Chapter 1

The modular group and elliptic function theory

This chapter contains mostly well known material on the modular group $\text{PSL}(2, \mathbb{Z})$ and elliptic curves along with some relatively new material on once punctured tori. We present in §4 a more or less classical treatment of function theory on elliptic curves (a euclidean theory) as a model for subsequent work. It is followed by an outline of the equivalent (noneuclidean) theory on once punctured tori using Poincaré series to illustrate another approach to some of the problems we will encounter in this book. The most important fact for our subsequent chapters from this material is that the sum of the residues of an elliptic function is zero. We introduce and describe the geometry of several families of subgroups of the modular group that will be needed in the sequel. This section (§6) establishes much of the notation that will be used throughout our presentation. The chapter ends with the elementary application to primality testing discussed in the introduction.

Most of the material in this chapter can be found in the references supplied. Much of the material is presented as background, a guide to different approaches, and to establish notation. As pointed out above, an exception is made in subsections of §4, where complete arguments are provided.

1. Möbius transformations

The material in this book deals mostly with compact Riemann surfaces and subgroups of $\mathrm{PSL}(2,\mathbb{Z})$. We start with a discussion of the more general group: $\mathrm{PSL}(2,\mathbb{C})$. Good additional references for this introductory material are [**4**, Ch. 4], [**12**, Ch. I] and [**21**, Ch. I].

An element $A = \begin{bmatrix} a & b \\ c & d \end{bmatrix} \in \mathrm{PSL}(2,\mathbb{C})$ acts on the Riemann sphere $\hat{\mathbb{C}} = \mathbb{C} \cup \{\infty\}$ as the Möbius transformation

$$z \mapsto \frac{az+b}{cz+d}.$$

We think of $\mathrm{PSL}(2,\mathbb{C})$ as a topological group with the following topology. Consider $\mathrm{SL}(2,\mathbb{C})$ as the closed subset of complex euclidean 4-space defined by the equation

$$(a,b,c,d) \in \mathbb{C}^4;\ ad - bc = 1,$$

and give $\mathrm{PSL}(2,\mathbb{C}) = \mathrm{SL}(2,\mathbb{C})/\{\pm I\}$ the quotient topology.

It is convenient to introduce a most important and useful classification for Möbius transformations. Let $A = \begin{bmatrix} a & b \\ c & d \end{bmatrix} \in \mathrm{PSL}(2,\mathbb{C})$, $A \neq I$. Then (recall that $ad - bc = 1$)

$$A \text{ is } parabolic \iff \mathrm{tr}^2(A) = (a+d)^2 = 4,$$

$$A \text{ is } elliptic \iff 0 \leq \mathrm{tr}^2(A) < 4,$$

$$A \text{ is } hyperbolic \iff \mathrm{tr}^2(A) > 4,$$

and

$$A \text{ is } loxodromic \iff \mathrm{tr}^2(A) \notin [0,4].$$

We note that in our definition hyperbolic is a special case of loxodromic. The (square of the) trace of an element of $\mathrm{PSL}(2,\mathbb{C})$ is a conjugacy class invariant; that is,

$$\mathrm{tr}(A) = \mathrm{tr}(B \circ A \circ B^{-1}) \text{ for all } A, B \in \mathrm{SL}(2,\mathbb{C}).$$

Conversely, if $A \neq I$ and $B \neq I$ are two nontrivial Möbius transformations[6] with the same squared trace, then they are conjugate in $\mathrm{PSL}(2,\mathbb{C})$; if the two motions are in $\mathrm{PSL}(2,\mathbb{R})$ and they have the same squared trace, then we can only conclude that A is conjugate in $\mathrm{PSL}(2,\mathbb{R})$ to either B or B^{-1}.

[6] We denote the identity matrix and the identity Möbius transformation by the same symbol I.

The motion A has either one or two fixed points α, β in $\hat{\mathbb{C}}$; they are given by the formulae

$$\{\alpha,\ \beta\} = \left\{\frac{1}{2c}((a-d) \pm \sqrt{\operatorname{tr}^2(A) - 4})\right\} \text{ if } c \neq 0,$$

and

$$\{\alpha,\ \beta\} = \left\{\infty,\ -\frac{b}{a - a^{-1}}\right\} \text{ if } c = 0.$$

It follows from the above formulae[7] that A is parabolic if and only if A has one fixed point in $\hat{\mathbb{C}}$, if and only if A is conjugate in PSL(2,$\mathbb{C}$) to $\begin{bmatrix} 1 & 1 \\ 0 & 1 \end{bmatrix}$. In all other cases $A \neq I$ has two fixed points and is hence conjugate in PSL(2,$\mathbb{C}$) to $\begin{bmatrix} \kappa & 0 \\ 0 & \kappa^{-1} \end{bmatrix}$ with $\kappa \in \mathbb{C} - \{0, 1, -1\}$. In this case A is elliptic if and only if $|\kappa| = 1$, A is hyperbolic if and only if $\kappa \in \mathbb{R}$, and A is loxodromic if and only if $|\kappa| \neq 1$.

The square of $\kappa, \kappa^2(A)$ is called the *multiplier* of A. The multiplier is defined uniquely except that a number and its reciprocal are the two multipliers of the same Möbius transformation.[8] For A parabolic or the identity, we set $\kappa^2(A) = 1$. We note that

$$\kappa^2(A^n) = (\kappa^2(A))^n, \text{ for all } A \in \text{ PSL}(2, \mathbb{C}).$$

Hence only elliptic elements can have finite order. A Möbius transformation is of finite order if and only if its multiplier is a root of unity. The *order* n of such a motion A is the smallest positive integer n such that $\kappa^{2n}(A) = 1$.

The trace and the square root of the multiplier are related by the formula

$$\operatorname{tr} = \kappa + \kappa^{-1}.$$

A parabolic motion A with finite fixed point a and *translation length* α can be written in *normal form* as a motion of the sphere as

$$\frac{1}{A(z) - a} = \frac{1}{z - a} + \alpha$$

(with $\alpha \in \mathbb{C}^*$), or equivalently as an element of PSL(2,$\mathbb{C}$) as

$$A = \begin{bmatrix} 1 + a\alpha & -a^2\alpha \\ \alpha & 1 - a\alpha \end{bmatrix}.$$

[7] Many times hereafter we will derive onlyformulae for finite values of the parameters (or variables) (here fixed points). We will usually leave it to the reader to make the necessary adjustments for the parameter assuming the value ∞.

[8] For loxodromic motions, it is always possible to choose the multiplier so that its magnitude is > 1. For primitive elliptic motions we may choose the multiplier so that its argument lies in $(0, \pi]$.

The case of infinity as a fixed point corresponds to $a = \infty$ and $-a^2\alpha$ finite and nonzero (this latter quantity is the translation length in this case); thus $\alpha = 0 = a\alpha$. We record for future use

$$A'(a) = 1 \text{ and } A''(a) = -2\alpha.$$

A motion A with fixed points a and b and multiplier κ^2 is given in *normal form* by

$$\frac{A(z) - a}{A(z) - b} = \kappa^2 \frac{z - a}{z - b},$$

and equivalently as a matrix (of determinant 1)

$$A = \frac{1}{\kappa(a - b)} \begin{bmatrix} a - \kappa^2 b & ab(\kappa^2 - 1) \\ 1 - \kappa^2 & \kappa^2 a - b \end{bmatrix}.$$

Simple calculations show that

$$A'(a) = \kappa^2 \text{ and } A'(b) = \kappa^{-2}.$$

It follows that, unless A has order two, the selection of a multiplier allows us to distinguish one fixed point from the other. For loxodromic A, we call a the *attractive* (respectively, *repulsive*) fixed point provided $|A'(a)| < 1$ (respectively, $|A'(a)| > 1$).

As an example, we consider a sequence of nonparabolic motions

$$A_n = \begin{bmatrix} a_n & b_n \\ c_n & d_n \end{bmatrix} \in \mathrm{PSL}(2,\mathbb{C})$$

whose elements do not fix ∞ and that converges to a parabolic motion with fixed point α. This means that (we may assume after multiplying the entries of A_n by -1, if necessary)

$$\lim_{n\to\infty} \frac{a_n - d_n}{2c_n} = \alpha \text{ and } \lim_{n\to\infty} a_n + d_n = 2.$$

If, on the other hand, A_n fixes ∞ and α_n and $\alpha = \infty$, then

$$A_n = \begin{bmatrix} \kappa_n & \alpha_n(\kappa_n^{-1} - \kappa_n) \\ 0 & \kappa_n^{-1} \end{bmatrix}$$

with

$$\lim_{n\to\infty} \kappa_n = 1, \ \lim_{n\to\infty} \alpha_n = \infty \text{ and } \lim_{n\to\infty} \alpha_n(\kappa_n^{-1} - \kappa_n) = b \in \mathbb{C}^*.$$

2. Riemann surfaces

More on the material in this section can be found in [**6**, Ch. I].

Definition 2.1. A *Riemann surface* M is a one (complex) dimensional connected complex analytic manifold.

This means that we can find open sets $\{U_\alpha;\ \alpha \in A\}$ that cover M (that is, $M \subset \cup_{\alpha \in A} U_\alpha$) and homeomorphisms called *(local) coordinates*

$$z_\alpha : U_\alpha \to \mathbb{C}$$

whose *transition functions*

$$f_{\alpha\beta} = z_\alpha \circ z_\beta^{-1} : z_\beta(U_\alpha \cap U_\beta) \to z_\alpha(U_\alpha \cap U_\beta)$$

are complex analytic.

Definition 2.2. A compact Riemann surface is *closed*; a noncompact, *open.*

Definition 2.3. A continuous map

$$f : M \to N$$

between Riemann surfaces is *holomorphic* if for every local coordinate z on M and every local coordinate ζ on N, the map $\zeta \circ f \circ z^{-1}$ is holomorphic. A holomorphic equivalence (an injective holomorphic map of one surface onto another) is a *conformal* map.

Theorem 2.4. *Let $f : M \to N$ be a holomorphic map between Riemann surfaces. If M is compact, then f is either constant or surjective.*

Proof. If f is not constant, then it is an open mapping. Thus $f(M)$ is open in N. It is closed in N because it is a compact subset of N. Since N is connected $f(M) = N$.

Definition 2.5. A holomorphic map of a Riemann surface M into the Riemann sphere $\hat{\mathbb{C}}$ other than the constant map with value ∞ is called a *meromorphic function.* The set of meromorphic functions on M forms a field, $\mathcal{K}(M)$.

3. Kleinian groups

See [**4**, Ch. 5] and [**21**, Ch. II] for more details on this subject. We outline several aspects of the theory of Kleinian groups. The reader is warned that NOT all claims made are easily verifiable. We will, however, subsequently prove all that we need for our work and is not easily available in other books.

3.1. Generalities. Let X be a subset of the sphere. Consider an inclusion of groups $G \subset \Gamma \subset \mathrm{PSL}(2, \mathbb{C})$. We say that X is *precisely invariant under G in* Γ provided

$$g(X) = X \text{ for all } g \in G \text{ and } \gamma(X) \cap X = \emptyset \text{ for all } \gamma \in \Gamma - G.$$

We say that Γ acts *discontinuously* at the point z in $\hat{\mathbb{C}}$ provided the stabilizer Γ_z of z in Γ is finite and there exists a neighborhood U of z that is precisely invariant under Γ_z in Γ. The *region of discontinuity* or *ordinary set* $\Omega = \Omega(\Gamma)$ is the largest (automatically open) set on which the group Γ acts discontinuously. Its complement $\Lambda = \Lambda(\Gamma)$ is the *limit set* of Γ. We call Γ *Kleinian* provided the region of discontinuity is nonempty. In this case, the limit set contains zero, one, two or infinitely many points. If Λ contains more than two points, then it is a closed perfect subset of $\hat{\mathbb{C}}$. If Γ is infinite and Kleinian, then the limit set is the set of accumulation points of the orbit under Γ of a single ordinary point of the group.

A group Γ of Möbius transformations is called *elementary* provided every two elements of infinite order in the group have a common fixed point. If Γ is Kleinian, then it is elementary if and only if $\Lambda(\Gamma)$ has two or fewer points. For nonelementary Kleinian groups Γ, the limit set can also be characterized as the smallest nonempty Γ-invariant closed subset of the sphere. The elementary discrete groups are all Kleinian and well known.

Remark 3.1. If Γ is a subgroup of PSL(2,$\mathbb{C}$) and Γ acts discontinuously somewhere on $\hat{\mathbb{C}}$, then Γ is automatically discrete (in the topology for the group PSL(2,$\mathbb{C}$) described above). Hence every Kleinian group is countable. Discreteness does not imply discontinuity. The *Picard group*, $\mathrm{PSL}(2, \mathbb{Z}[\imath])$, is discrete but does not act discontinuously anywhere on the Riemann sphere. The situation with the Picard group is in marked contrast to the case of $\mathrm{PSL}(2, \mathbb{R})$ where discreteness is equivalent to the notion of acting discontinuously on the upper half plane. The reader should not be perplexed by this since even in the case of $\mathrm{PSL}(2, \mathbb{R})$, the action on the real line is not discontinuous. Even though we are viewing PSL(2,$\mathbb{C}$) as acting on the extended complex plane, there is another action on the upper half space. The complex plane is the boundary of that space just as the real line is the boundary of the upper half plane.

Definition 3.2. Let G be a group of self-homeomorphisms of a topological space X. The group introduces an equivalence relation on the space. Two points x and y in X are *equivalent (modulo G)* if there exists a $g \in G$ with $g(x) = y$. The set of equivalence classes is *the quotient space* X/G. Let $\pi : X \to X/G$ be the quotient map that assigns to each $x \in X$ its equivalence class modulo G. We topologize X/G by giving the factor space the weakest topology so that π is continuous.

The factor space Ω/Γ is a union of (at most countably many) Riemann surfaces whenever the group Γ is Kleinian, as a direct consequence of the next exercise.

Exercise 3.3. If Γ is a Kleinian group, then for each $z \in \Omega(\Gamma)$, the *stabilizer*

$$\Gamma_z = \{\gamma \in \Gamma;\ \gamma(z) = z\}$$

is a finite cyclic group. Describe the stabilizer of a limit point of a Kleinian group.

Exercise 3.4. If Γ is a Kleinian group and $A \in \mathrm{PSL}(2,\mathbb{C})$, then the *conjugate* group

$$A\Gamma A^{-1} = \{A \circ \gamma \circ A^{-1};\ \gamma \in \Gamma\}$$

is also Kleinian. Further
(a) $\Omega(A\Gamma A^{-1}) = A(\Omega(\Gamma))$,
(b) $\Lambda(A\Gamma A^{-1}) = A(\Lambda(\Gamma))$ and
(c) $\Omega(A\Gamma A^{-1})/A\Gamma A^{-1} \cong \Omega(\Gamma)/\Gamma$.

Definition 3.5. Let G be a group of Möbius transformations operating on a set $D \subset \hat{\mathbb{C}}$. A *fundamental domain* for the action of G on D is a connected open set $\omega = \omega(G)$ bounded by analytic arcs $\cup_{i\in I}(c_i^+ \cup c_i^-)$ satisfying:
1. For each $x \in D$, there exists a $z \in$ cl ω and a $g \in G$ such that $g(x) = z$.
2. If z_1 and z_2 belong to ω and $g(z_1) = z_2$ for some $g \in G$, then g is the identity.
3. For each $i \in I$, there exists a $g_i \in G$ with $g_i(c_i^+) = c_i^-$.

A *fundamental set* for the action of G on D is a set containing exactly one point for each G-orbit.

Roughly speaking, Ω/Γ is obtained by boundary identifications on the closure of a fundamental domain for the action of Γ on Ω. We let

$$\Delta_1,\ \Delta_2,\ \ldots,\ \Delta_i, \ldots$$

be a maximal set of Γ-inequivalent components of Ω, and we let

$$\Gamma_i = \{\gamma \in \Gamma;\ \gamma(\Delta_i) = \Delta_i\}$$

be the *stabilizer* of Δ_i in Γ. The stabilizer subgroups are finitely generated whenever the group itself is finitely generated. But, this is far from obvious. It is quite obvious, however, that

$$\Omega/\Gamma \cong \bigcup_i \Delta_i/\Gamma_i.$$

It turns out that the above is a finite (disjoint) union whenever Γ is finitely generated. As remarked earlier, it is easy to conclude that we have at most a countable union of surfaces.

A Kleinian group Γ is called a *function* group provided it has an invariant component,[9] that is, provided there is a component Δ of Ω with $\gamma(\Delta) = \Delta$ for all $\gamma \in \Gamma$. A finitely generated Kleinian group can have at most two invariant components.[10] The stabilizers of the components of the ordinary set of a Kleinian group are, of course, function groups. A function group is called a *b-group* (b after boundary or Bers) provided its invariant component

[9] By abuse of language, components of $\Omega(G)$ are also called components of Γ.

[10] As with many claims in this section, this is not obvious but easily proven. We will not need this fact for our presentation.

is simply connected. A Kleinian group is called *Fuchsian* if it is conjugate to a subgroup of PSL(2,$\mathbb{R}$). A b-group is Fuchsian if and only if its invariant component is a disc.

A Kleinian group Γ that leaves invariant a disc Δ is automatically Fuchsian with Δ contained in $\Omega(\Gamma)$. The limit set is, in this case, a subset of $\partial(\Delta)$, the set theoretic boundary of the domain Δ. The group is called *of the first kind* if $\Lambda = \partial(\Delta)$ and *of the second kind* otherwise. A Fuchsian group of the first kind Γ represents on $\Omega = \Omega(\Gamma)$ a Riemann surface and its *mirror* image, while a group of the second kind represents a single Riemann surface Ω/Γ: the *double* of Δ/Γ. In this latter case, the surface Ω/Γ is Δ/Γ glued to its mirror image along their common boundary, $(\partial(\Delta) - \Lambda(\Gamma))/\Gamma$.

3.2. The situation of interest. We consider the special case that we study in this book: a Fuchsian group G operating on a disc Δ and the natural projection $P : \Delta \to \Delta/G$. The map P is locally one-to-one at each $z_o \in \Delta$ whose stabilizer G_{z_o} in G is trivial; if $\mu = |G_{z_o}|$, it is locally μ-to-one at z_o. In the latter case, z_o is called a *branch point of order* μ (it is also called a *ramification point of order* $\mu - 1$). The point $P(z_o) \in \Delta/G$ is called a *branch value* of the map P. A *puncture* x on Δ/G is a region $D^* \subset \Delta/G$ that is conformally equivalent to $\mathbb{D}^* = \{0 < |z| < 1\}$ (say $f : \mathbb{D}^* \to D^*$ is this equivalence) with the property that P is unramified over D^* and for all sequences $\{z_n\}$ in $\mathbb{D}^*$ with $\lim_n z_n = 0$, the sequence $\{P(z_n)\} \subset D^*$ is discrete in Δ/G (this sequence converges to the puncture x). We assign the branch number ∞ to the puncture x. A puncture on Δ/G is the image of a cusp z on the boundary of Δ for the group G. If we define $\bar{\Delta}$ to be the union of Δ and the *cusps* of G (these are the fixed points of the parabolic elements of G), then P extends by continuity to $\bar{\Delta}$ and $P(\bar{\Delta}) = \overline{\Delta/G}$ is the *completion*[11] of Δ/G obtained by *filling in* the punctures. For the puncture x on Δ/G, we can choose a *horodisc* D precisely invariant under a cyclic parabolic subgroup $H \subset G$ with $P(D) = D^*$ (the fixed point z of the parabolic elements in H is on the boundary of D and $P(z) = x$).

The group G and the surfaces Δ/G and $\overline{\Delta/G}$ are of *finite (analytic) type* (p, n) if $\overline{\Delta/G}$ is a compact surface of genus p and Δ/G has a total of n punctures and branch values. If $\{z_1, z_2, \ldots, z_n\}$ is the set of points in $\bar{\Delta}$ that project to the punctures and branch values, and if μ_j is the order of the branch point or cusp z_j, then the $n+2$-tuple

$$(p, n;\ \mu_1,\ \mu_2,\ \ldots,\ \mu_n)$$

[11] In many of the interesting cases, the *compactification*.

is the *signature* of the group G and the orbifolds Δ/G and $\overline{\Delta/G}$; their *(negative) Euler characteristic* is

$$\chi(G) = 2p - 2 + \sum_{j=1}^{\infty} \left(1 - \frac{1}{\mu_j}\right),$$

where we use the usual convention $\frac{1}{\infty} = 0$.

4. The elliptic paradise

The special case of surfaces of genus one[12] is much better understood than the more general case of higher genus. We discuss in this section a classical theory. It is a model for the subsequent work. This is one of the very few cases where complete computations are possible.

4.1. The family of tori. More on material in this section can be found in [**1**, Ch. 7] and [**14**, §1.2].

We begin by constructing a (one could almost say "the") family of tori. This procedure is well known. It is nevertheless a good starting point. Let τ be any point in the upper half plane

$$\mathbb{H}^2 = \{\tau \in \mathbb{C};\ \Im\tau > 0\}.$$

The point τ determines a free abelian group $\tilde{G} = \tilde{G}_\tau$ generated by the two translations

$$\tilde{A}(z) = z + 1,\ \tilde{B}(z) = z + \tau.$$

The complex plane $\mathbb{C}$ factored by the equivalence relation defined by $\tilde{G}_\tau$ is a torus that we will denote by the symbol T_τ:

$$T_\tau = \mathbb{C}/\tilde{G}_\tau.$$

Let $a \in \mathbb{C}$ be arbitrary. A fundamental domain $\tilde{\omega} = \omega(\tilde{G}_\tau)$ for the action of the group $\tilde{G}$ on $\mathbb{C}$ is the interior of the parallelogram $\mathcal{P}_a$ with vertices $a,\ a+1,\ a+1+\tau,\ a+\tau$; the interior of $\mathcal{P}_a$ with the edges from $a+\tau$ to a to $a+1$, excluding the vertices $a+\tau$ and $a+1$, forms a fundamental set for this action.

The choice of a is essentially immaterial, but, as will be seen, an appropriate choice will simplify proofs. Abel's theorem[13] tells us that every

[12]To the best of our knowledge the catchy title that we assigned to this section is due to Lipman Bers.

[13]Not proven in this book. This special uniformization theorem for tori is also a fairly trivial consequence of the general uniformization theorem (that there are only three distinct simply connected Riemann surfaces).

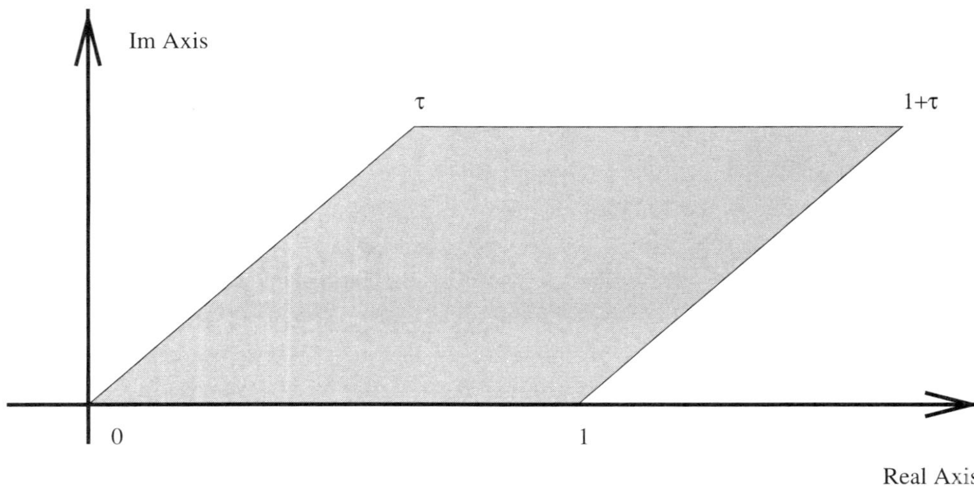

Figure 1. Fundamental polygon. $\mathcal{P}_0$

torus[14] is of the above form (that is, the quotient of the complex plane by a group of translations). As a consequence, the upper half plane can be naturally identified with the Teichmüller space $\mathbf{T}(1,0)$ of compact surfaces of genus 1. A point in this Teichmüller space is a "marked torus"; the marking is provided by the point τ which gives an isomorphism from the fundamental group of the torus to the covering group $\tilde{G}$.

We should point out that we have, in reality, two definitions of the torus. The one we have emphasized so far is a manifold $\mathbb{C}$ factored by the action of a discontinuous group $\tilde{G}$; hence T_τ is a complex analytic manifold. The *lattice* points L of the group $\tilde{G}$ are the points

$$L = L(\tilde{G}) = \{n + m\tau;\ n \text{ and } m \in \mathbb{Z}\}.$$

The lattice points form a (normal) subgroup of the abelian group of additive complex numbers. Hence $T_\tau = \mathbb{C}/L$ is also a commutative group, in fact a one dimensional complex Lie group.

Theorem 4.1. *Two points $\tau_1,\ \tau_2 \in \mathbb{H}^2$ determine conformally equivalent tori if and only if there exists a Möbius transformation $M \in \mathrm{PSL}(2,\mathbb{Z})$ with $\tau_2 = M(\tau_1)$.*

[14]For our purposes, we can define a *torus* as a surface conformally equivalent to T_τ with $\tau \in \mathbb{H}^2$. A more abstract (equivalent, of course) definition of a torus is a Riemann surface whose fundamental group is isomorphic to $\mathbb{Z} \oplus \mathbb{Z}$.

Proof. Assume that f is a conformal map from T_{τ_1} onto T_{τ_2}. Then there exists a conformal selfmap F of $\mathbb{C}$ so that

$$\begin{array}{ccc} \mathbb{C} & \xrightarrow{F} & \mathbb{C} \\ \downarrow & & \downarrow \\ T_{\tau_1} & \xrightarrow{f} & T_{\tau_2} \end{array}$$

commutes, where the vertical arrows are the natural projections. Since F is conformal, it is of the form

$$z \mapsto az + b,\ a \in \mathbb{C}^* = \mathbb{C} - \{0\},\ b \in \mathbb{C}.$$

Translation by b induces a conformal selfmap of T_{τ_2}. Hence we may and do assume that $b = 0$. The affine map F must conjugate the group $\tilde{G}_{\tau_1}$ to $\tilde{G}_{\tau_2}$ (that is, $F\tilde{G}_{\tau_1}F^{-1} = \tilde{G}_{\tau_2}$). Let us call this conjugation θ. Identifying the translation $z \mapsto z + b$ with the complex number b, we see that the $\theta(1) = F(1) = a$ and $\theta(\tau_1) = F(\tau_1) = a\tau_1$ must be generators for $\tilde{G}_{\tau_2}$. Thus there exists a matrix $A = \begin{bmatrix} \alpha & \beta \\ \gamma & \delta \end{bmatrix} \in \mathrm{GL}(2, \mathbb{Z})$ so that

$$a = \alpha + \beta\tau_2 \text{ and } a\tau_1 = \gamma + \delta\tau_2.$$

Hence

$$\tau_1 = \frac{\gamma + \delta\tau_2}{\alpha + \beta\tau_2}.$$

Since both τ_1 and τ_2 belong to $\mathbb{H}^2$, A must belong to $\mathrm{SL}(2, \mathbb{Z})$, and $M = \begin{bmatrix} \alpha & -\gamma \\ -\beta & \delta \end{bmatrix}$.

Conversely, if $\tau_2 = M(\tau_1)$ with M as above, then we define

$$F(z) = (\alpha + \beta\tau_2)z,\ z \in \mathbb{C}.$$

Simple calculations show that

$$F(z + 1) = F(z) + (\alpha + \beta\tau_2)$$

and

$$F(z + \tau_1) = F(z) + (\gamma + \delta\tau_2),\ \text{for all } z \in \mathbb{C}.$$

It follows that F defines a conformal map of T_{τ_1} onto T_{τ_2}.

Exercise 4.2. Let G be a group of Möbius transformations. Assume that G is discontinuous and consists in addition to the identity of only parabolic transformations. Show that G is either cyclic and hence conjugate to the group generated by the selfmap[15] B of $\hat{\mathbb{C}}$

$$B : z \mapsto z + 1,\ z \in \hat{\mathbb{C}},$$

[15]We will reserve the symbol B for this motion in most of the rest of this manuscript, in particular, when dealing with subgroups of the modular group. In our discussion of tori we use the symbol $\tilde{A}$ for this motion.

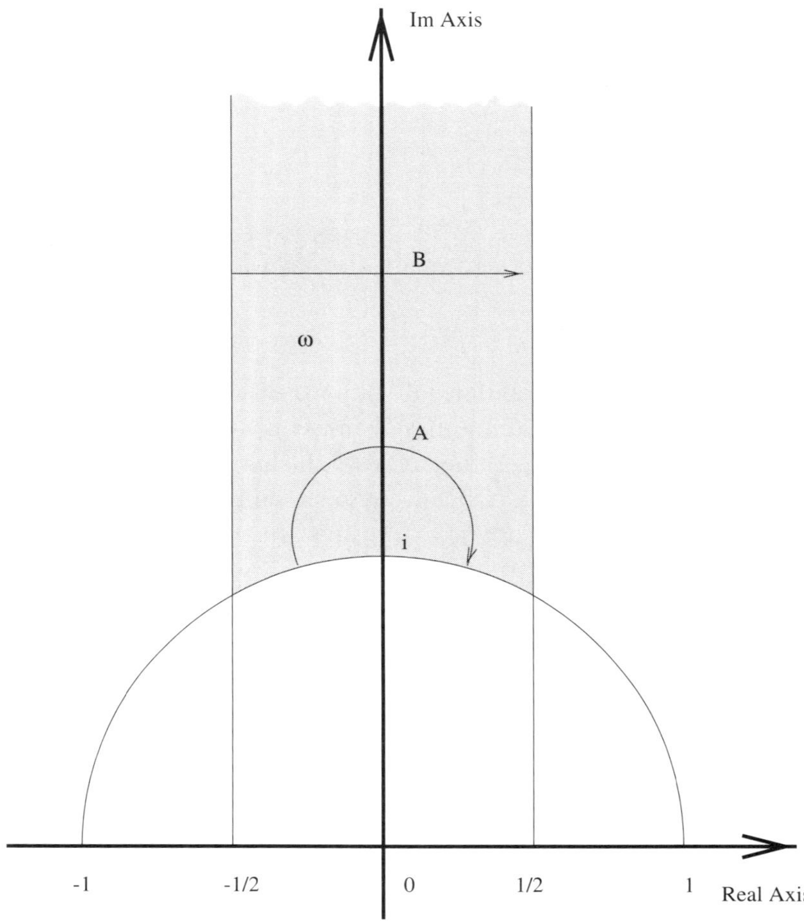

Figure 2. Fundamental domain for the modular group Γ.

or G is conjugate to $\tilde{G}_\tau$ for some $\tau \in \mathbb{H}^2$.

Theorem 4.3. *A fundamental domain for the action of* $\Gamma = \mathrm{PSL}(2,\mathbb{Z})$ *on* $\mathbb{H}^2$ *is the open region* $\omega(\mathrm{PSL}(2,\mathbb{Z}))$ *in the upper half plane bounded by the lines* $\Re\tau = -\frac{1}{2}$, $\Re\tau = \frac{1}{2}$ *and the semi-circle* $|\tau| = 1$. *A fundamental set for this action can be obtained by adjoining the circular arc from* $\imath$ *to* $\frac{1}{2} + \frac{\sqrt{3}}{2}\imath$ *(including both end points) and the vertical line segment from* $\frac{1}{2} + \frac{\sqrt{3}}{2}\imath$ *to* $\imath\infty$.

Proof. Let us denote by ω the set described by the last sentence in the statement of the theorem. Assume that we have two points τ_1 and τ_2 in ω that are equivalent under Γ. Without loss of generality we may assume that

$$\Im\tau_2 \geq \Im\tau_1.$$

Let

$$\tau_2 = \frac{a\tau_1 + b}{c\tau_1 + d} \text{ with } C = \begin{bmatrix} a & b \\ c & d \end{bmatrix} \in \mathrm{SL}(2, \mathbb{Z}).$$

Since

$$\Im \tau_2 = \frac{\Im \tau_1}{|c\tau_1 + d|^2},$$

we must have that $|c\tau_1 + d| \leq 1$. We now exploit the fact that we are dealing with integer matrices.

If $c = 0$, then $ad = 1$ and both a and d must be ± 1. Without loss of generality we may assume that $a = 1 = d$. It follows that $\tau_2 = \tau_1 + b$. Hence $|b| = |\Re\tau_2 - \Re\tau_1| < 1$ and $b = 0$.

Assume now that $c \neq 0$. The condition $|\tau_1 + \frac{d}{c}| \leq \frac{1}{|c|}$ implies that $|c| = 1$. For if $|c| \geq 2$, then τ_1 would be at a distance at most $\frac{1}{2}$ from $\mathbb{R}$, which is impossible since the point in ω nearest to $\mathbb{R}$ is $\frac{1}{2} + \frac{\sqrt{3}}{2}\imath$. Without loss of generality we may now assume that $c = 1$ and hence $|\tau_1 + d| \leq 1$. The geometric shape of ω tells us that this inequality is possible only for $|d| \leq 1$. The inequality $|\tau_1 + 1| \leq 1$ is never satisfied for points $\tau_1 \in \omega$ since $\exp \frac{2\pi\imath}{3} \notin \omega$. The inequality $|\tau_1 - 1| \leq 1$ is satisfied only by the point $\exp \frac{\pi\imath}{3} \in \omega$. For this point $|c\tau_1 + d| = 1$. Hence $\Im\tau_2 = \Im\tau_1$, and by the shape of ω, $\tau_2 = \tau_1$. There remains only the case $c = 1$ and $d = 0$. Hence $b = -1$. Since $|\tau_1| \leq 1$ and $|\tau_1| \geq 1$, we conclude that $|\tau_1| = 1$ and $\tau_2 = a - \frac{1}{\tau_1} = a - \bar{\tau}_1$. Hence $\Re(\tau_1 + \tau_2) = a$, and this is possible only when $a = 0$. Thus $\tau_2 = -\frac{1}{\tau_1}$, which tells us that $\tau_1 = \imath = \tau_2$.

It remains to show that every $\tau \in \mathbb{H}^2$ is Γ-equivalent to at least one point in the closure of ω. It is convenient to show first that the region $\omega(\mathrm{PSL}(2, \mathbb{Z}))$ is the same as

$$\omega_1 = \left\{ z \in \mathbb{H}^2;\ |\Re z| < \frac{1}{2} \text{ and } |cz + d| > 1 \text{ for all } (c, d) \in \mathbb{Z}^2 - \{(0, 0)\} \right\}.$$

It is obvious that $\omega_1 \subset \omega(\mathrm{PSL}(2, \mathbb{Z}))$. For the reverse inclusion, if $z \in \omega(\mathrm{PSL}(2, \mathbb{Z}))$, and $(c, d) \in \mathbb{Z}^2 - \{(0, 0)\}$, then

$$|cz + d|^2 = (c\Re z + d)^2 + (c\Im z)^2 =$$

$$(c^2)(\Re z^2 + \Im z^2) + 2cd\Re z + d^2 > c^2 - cd + d^2 \geq 1.$$

Every $\tau \in \mathbb{H}^2$ is certainly equivalent to a $\tau_1 \in \mathbb{H}^2$ with maximal height[16] (that is, to a $\tau_1 \in \mathbb{H}^2$ with $|c\tau_1 + d| \geq 1$, for all integers $(c, d) \in \mathbb{Z}^2 - \{(0, 0)\}$). Translating by powers of B we can certainly get an equivalent point to land on the closure of ω_1 (since the height of a point is unaffected by the translation).

[16] If $c \neq 0$, then C maps the upper half plane $\{\Im\tau \geq 1\}$ to the closed disc in $\mathbb{H}^2$ of diameter $\frac{1}{c^2}$ tangent to $\mathbb{R}$ at $\frac{a}{c}$.

A point τ in the interior of this fundamental domain $\omega(\mathrm{PSL}(2,\mathbb{Z}))$ represents a very particular kind of marking on the surface. The condition that τ is in (the interior of) the fundamental domain can be restated as

$$|\tau| > 1 \text{ and } |\tau| < |\tau \pm 1|.$$

Thus on the torus T_τ we are choosing a basis for the fundamental group with $\tilde{A}$ representing the shortest closed geodesic[17] and $\tilde{B}$ representing the shortest curve that is not homotopic to a power of $\tilde{A}$. Almost all tori (except those corresponding to boundary points of the fundamental domain) have a unique marking of this type. The factor space $\mathbb{H}^2/\mathrm{PSL}(2,\mathbb{Z})$ is the Riemann or moduli space of surfaces of genus 1, $\mathbf{R}(1,0)$. It is the complex sphere $\mathbb{C}\cup\{\infty\}$ with three distinguished points, a Riemann surface (orbifold) of signature $(0,3;\ 2,3,\infty)$. The distinguished points correspond to the fixed points $\tau = \imath$ and $\tau = \frac{1+\sqrt{3}\imath}{2}$ of the elliptic motions $\begin{bmatrix} 0 & -1 \\ 1 & 0 \end{bmatrix}$ and $\begin{bmatrix} 0 & -1 \\ 1 & -1 \end{bmatrix}$ of order 2 and 3 (respectively) and the fixed point $\tau = \infty$ of the parabolic motion $\begin{bmatrix} 1 & 1 \\ 0 & 1 \end{bmatrix}$. We will see later that these fixed points determine special tori.

4.2. The algebraic curve associated to a torus. The material on tori in this section is included in most books on basic function theory, for example, in [**1**, Ch. 7]. The material on abstract surfaces comes from [**6**, Chs. I and III]. Fix a point $\tau \in \mathbb{H}^2$.

Definition 4.4. A meromorphic function f on $\mathbb{C}$ is *elliptic* or *(doubly) periodic* (with periods 1 and τ) if

$$f(z+1) = f(z) = f(z+\tau), \text{ for all } z \in \mathbb{C}.$$

It is clear that the space of periodic functions can be naturally identified with $\mathcal{K}(T_\tau)$. If $f \neq 0$ is a meromorphic function on a Riemann surface X and z is a local coordinate on X vanishing at x, then f has a Laurent series expansion

$$f(z) = \sum_{i=n}^{\infty} a_i z^i,\ a_n \neq 0.$$

We call n the *order* of f at x, in symbols, $\mathrm{ord}_x f$. If the periodic function f assumes the value c at $z \in \mathbb{C}$, then it assumes the same value c at every point in $\mathbb{C}$ that is $\tilde{G}_\tau$-equivalent to z. Hence in counting the number of solutions x (usually with multiplicity) of $f(x) = c$, we include only inequivalent points x. We are, in effect, counting solutions on the torus T_τ. A periodic function f also has a well defined residue, $\mathrm{res}_x f$, at each point $x \in \mathbb{C}$. Again, the concept of residues should be viewed on the torus (since f has the same residue at $\tilde{G}_\tau$-equivalent points).

[17]In any flat (zero curvature) metric, for example, $|dz|$.

Remark 4.5. Let X be a Riemann surface and $0 \neq f \in \mathcal{K}(X)$. It makes sense to define $\mathrm{ord}_x f$ for each $x \in X$. However, the residue of a function is not a well defined concept on an arbitrary Riemann surface. On an arbitrary surface one can define meromorphic one forms φ. These are assignments of a meromorphic function f to each local coordinate z on the surface such that $f(z)dz$ is invariant under coordinate changes. It then makes sense to define $\mathrm{res}_x \varphi$. There is a canonical isomorphism on every torus between the space of meromorphic functions and the space of meromorphic one forms, because of the presence of a regular (holomorphic) global one form without zeros (namely, the projection of dz to T_τ). For each $q \in \mathbb{Z}$, one defines a meromorphic q-form as an assignment of a meromorphic function φ to each local coordinate z such that

$$\varphi(z)dz^q$$

is invariant under coordinate changes. These spaces are useful tools in the study of function theory on surfaces.

Theorem 4.6. *The sum of the residues (over a maximal set of inequivalent points) of an elliptic function is* 0.

Proof. Let f be a periodic function. Since the poles of a meromorphic function are isolated, we can choose $a \in \mathbb{C}$ so that f is regular on the boundary $\partial\mathcal{P}_a$ of the period parallelogram $\mathcal{P}_a$. It follows that

$$\sum_{x \in \mathcal{P}_a} \mathrm{res}_x f = \frac{1}{2\pi\imath} \int_{\partial\mathcal{P}_a} f(z)dz$$

$$= \frac{1}{2\pi\imath}\left(\int_a^{a+1} f(z)dz + \int_{a+1}^{a+1+\tau} f(z)dz + \int_{a+1+\tau}^{a+\tau} f(z)dz + \int_{a+\tau}^{a} f(z)dz\right).$$

Using the periodicity of the function f and a change of variable, we conclude the sum of the first and third integrals, as well as the sum of the second and fourth integrals, on the last line must be zero.

Corollary 4.7. *Let f be a nonconstant elliptic function and let $c \in \hat{\mathbb{C}}$. The number of solutions of $f(z) = c$ is independent of c and will be called the degree,* $\deg f$, *of the function f.*

Proof. Assume that c is finite. Then $g = \frac{f'}{f-c} \neq 0$ is an elliptic function. Since $\mathrm{res}_x g = \mathrm{ord}_x(f-c)$, the corollary for f follows from the theorem for g.

Corollary 4.8. *A nonconstant elliptic function must have poles.*

Remark 4.9. The genus $p = p(X)$ of a compact Riemann surface X is defined as $\frac{1}{2}\mathrm{rank}\ H_1(X, \mathbb{Z})$. It is a complete topological invariant for compact surfaces: two compact Riemann surfaces are homeomorphic if and only if they have the same genus. Let $f : X \to Y$ be a nonconstant map between compact Riemann surfaces. We have already seen that f is surjective. Let $x \in X$. We can find local coordinates z on X vanishing at x and ζ on $Y = f(X)$ vanishing at $f(x)$ such that in terms of these coordinates

$$\zeta = z^{b_f(x)+1},$$

where $b_f(x)$ is a nonnegative integer (which equals zero at all but finitely many $x \in X$) known as the *branch number* of f at x. We define the *degree* of f, deg f, as

$$\sum_{x\in f^{-1}(y)} (b_f(x)+1);\ y \in Y.$$

It is independent of the choice of y, and the **Riemann Hurwitz** relation

$$2p(X) - 2 = \deg f\ (2p(Y) - 2) + \sum_{x\in X} b_f(x)$$

holds.

Corollary 4.10. *A torus does not carry any meromorphic functions of degree one.*

Proof. (Topological) Such a function would establish an isomorphism (in particular, a topological equivalence) between the torus (whose fundamental group is $\mathbb{Z}^2$) and the Riemann sphere (which is simply connected).
(Analytic) The sum of the residues of such a function would have to be nonzero.

Theorem 4.11. *Let f be a nonconstant elliptic function of degree n. Let a_1, ..., a_n be the zeros of f (listed according to their multiplicities) and b_1, ..., b_n its poles. Then*

$$\sum_{i=1}^{n}(a_i - b_i) = 0$$

(in the group T_τ).

Proof. By properly choosing $a \in \mathbb{C}$ (close to 0), we may assume that f is regular and nonzero on the boundary of $\mathcal{P}_a$. By the residue theorem, the sum to be evaluated is

$$\frac{1}{2\pi\imath}\int_{\partial\mathcal{P}_a} z\frac{f'(z)}{f(z)}dz$$

$$= \frac{1}{2\pi\imath}\left(\int_a^{a+1} + \int_{a+1}^{a+1+\tau} - \int_{a+\tau}^{a+\tau+1} - \int_a^{a+\tau}\right) z\frac{f'(z)}{f(z)}dz.$$

We now group the first and third integral, the second and fourth integral, and use the periodicity of the elliptic function $\frac{f'}{f}$ to continue the evaluation as

$$\left(\frac{-\tau}{2\pi\imath}\int_a^{a+1} + \int_a^{a+\tau}\right)\frac{f'(z)}{f(z)}dz$$

$$= \frac{1}{2\pi\imath}\left(-\tau(\log\ f(a+1) - \log\ f(a)) + (\log\ f(a+\tau) - \log\ f(a))\right) = n+m\tau,$$

for some integers n and m.

Remark 4.12. There is a converse to the above theorem. It will be established in Chapter 2; see Theorem 5.1 of that chapter. The theorem and its converse are part of a more general theory. On an arbitrary compact Riemann surface X of genus $p > 0$, one constructs a *canonical homology basis* (for $H_1(X, \mathbb{Z})$)

$$a_1,\ ...,\ a_p,\ b_1,\ ...,\ b_p$$

(this means that the intersection numbers of the cycles (curves) satisfy

$$a_i \times a_j = 0 = b_i \times b_j,\ a_i \times b_j = \delta_{ij},\ 1 \le i,j \le p,$$

where δ_{ij} is the Kronecker δ-symbol). Then one proves that there exists a *dual* basis

$$\omega_1,\ ...,\ \omega_p$$

for the space of holomorphic one forms on X. This means that

$$\int_{a_i} \omega_j = \delta_{ij},\ i,j = 1,\ ...,\ p.$$

The *period matrix* π defined by

$$\pi_{ij} = \int_{b_i} \omega_j,\ i,j = 1,\ ...,\ p,$$

is a symmetric matrix whose imaginary part is positive definite. It follows that the p columns of the $p \times p$ identity matrix together with the p columns of π generate a rank $2p$ lattice $L(X) \subset \mathbb{C}^p$. The quotient $J(X) = \mathbb{C}^p/L(X)$ is a p-dimensional torus known as the *Jacobi variety* of the Riemann surface X. There is a holomorphic mapping

$$\varphi : X \to J(X)$$

defined by sending the point $x \in X$ to the projection to $J(X)$ of the vector $\int_{x_o}^{x} \omega$, where x_o is a fixed point in X (the *base point* for the embedding) and ω is the vector valued (column) differential on X whose components are $\omega_1,\ ...,\ \omega_p$. The map φ is injective and of maximal rank everywhere.

Exercise 4.13. Show that for a surface T_τ of genus one, $J(T_\tau) = T_\tau$ and φ is the identity map.

Exercise 4.14. The map $C : z \mapsto \frac{z}{c\tau+d}$ is (induces) a conformal map from T_τ to $T_{\frac{a\tau+b}{c\tau+d}}$, and hence the map $f \mapsto f \circ C$ is an isomorphism of $\mathcal{K}(T_{\frac{a\tau+b}{c\tau+d}})$ onto $\mathcal{K}(T_\tau)$.

The divisor group of the surface X is the free (multiplicative) group on the points of X; thus a *divisor* D is a formal product $D = x_1^{\alpha_1} \dots x_n^{\alpha_n}$ with $n \in \mathbb{Z}^+ \cup \{0\}$, $\alpha_i \in \mathbb{Z}$, and $x_i \in X$, $1 \leq i \leq n$. The *degree* of the divisor D is then $\sum_{i=1}^{n} \alpha_i$. If $0 \neq f \in \mathcal{K}(X)$, then the zeros and poles of f form a *principal* divisor

$$(f) = \prod_{x \in X} x^{\mathrm{ord}_x f}$$

of degree 0. Since the divisors on X and its Jacobi variety are (abelian) groups, the map φ extends in an obvious way to divisors: the divisor D is sent by this extension φ to $\sum_{i=1}^{p} \alpha_i \varphi(x_i)$. The generalization of the last result now reads as follows.

Theorem 4.15 (Abel). *A divisor D of degree 0 on the compact Riemann surface X of positive genus is principal if and only if $\varphi(D) = 0$.*

The theorem has as an immediate consequence: the

Corollary 4.16. *If $p(X) = 1$, then $\varphi : X \to J(X)$ is an isomorphism.*

Every torus is also an algebraic curve of a particularly simple form. To obtain the curve associated to a Riemann surface, one must produce "good" meromorphic functions on the surface. In the case under study, we must produce periodic functions. We know that a torus cannot carry a meromorphic function of degree one. Thus we seek functions of degree two. The Riemann Hurwitz formula tells us that such a function must have 4 branch points. It involves no loss of generality to assume that infinity is a branch value (that is, the function we seek should have a single double pole). Finally, the location of the pole is arbitrary since the group of conformal automorphisms of a torus is transitive (the translation $z \mapsto z + a$ maps the origin to the equivalence class of the arbitrary point $a \in \mathbb{C}$). Thus we may as well place the pole at the origin of the torus. The Laurent series expansion of such a function f about the origin is

$$f(z) = \frac{a_{-2}}{z^2} + a_0 + a_1 z + \dots \tag{1.1}$$

since a_{-1}, the residue of f at the origin, must vanish. We can normalize so that $a_{-2} = 1$ and $a_0 = 0$ by choosing

$$g(z) = \frac{1}{a_{-2}}(f(z) - a_0) = \frac{1}{z^2} + A_1 z + \dots \quad .$$

Since the function h defined by

$$h(z) = g(z) - g(-z), \; z \in \mathbb{C},$$

is also elliptic and holomorphic (because the single pole has been killed), it follows that g is an even function. We conclude that if an elliptic function of degree 2 with a single pole at the origin exists, it must be an even function. It remains to construct this function.

We introduce the simplest doubly periodic function, the *Weierstrass* $\wp$-*function*:

$$\wp(z)=\frac{1}{z^2}+\sum_{(n,m)\in\mathbb{Z}\oplus\mathbb{Z}-\{(0,0)\}}\left\{\frac{1}{(z-n-m\tau)^2}-\frac{1}{(n+m\tau)^2}\right\},\quad z\in\mathbb{C}.$$

Theorem 4.17. *The infinite series defining $\wp$ converges uniformly and absolutely (as a series of meromorphic functions) on compact subsets of $\mathbb{C}$ and defines a doubly periodic function (with periods 1 and τ).*

Proof. We write $\omega=n+m\tau$, with n and $m\in\mathbb{Z}$. For $z\in\mathbb{C}$, with $|\omega|>2|z|$, we have

$$\left|\frac{1}{(z-\omega)^2}-\frac{1}{\omega^2}\right|=\left|\frac{z(2\omega-z)}{\omega^2(z-\omega)^2}\right|\le\frac{10|z|}{|\omega|^3}.$$

Thus the series defining $\wp$ converges uniformly and absolutely on compact subsets of $\mathbb{C}$ provided

$$\sum_{\omega\neq 0}\frac{1}{|\omega|^3}<\infty.$$

We leave the reader the task of showing that this last series converges. It is quite obvious that

$$\begin{aligned}\wp'(z)&=-\tfrac{2}{z^3}-\textstyle\sum_{(n,m)\in\mathbb{Z}\oplus\mathbb{Z}-\{(0,0)\}}\frac{2}{(z-n-m\tau)^3}\\&=-2\textstyle\sum_{(n,m)\in\mathbb{Z}\oplus\mathbb{Z}}\frac{1}{(z-n-m\tau)^3},\quad z\in\mathbb{C}\end{aligned},$$

is an elliptic function. Thus each of the entire functions

$$z\mapsto\wp(z+1)-\wp(z)\text{ and }z\mapsto\wp(z+\tau)-\wp(z)$$

is constant. Since $\wp$ is an even function, evaluation of these last two functions at $-\frac{1}{2}$ and $-\frac{\tau}{2}$, respectively, shows that the constants are 0.

It is quite easy to show that (a formal calculation will do since all the series involved converge)

$$\wp(z)=\frac{1}{z^2}+\sum_{k=1}^{\infty}(2k+1)\sum_{(n,m)\in\mathbb{Z}\oplus\mathbb{Z}-\{(0,0)\}}\frac{1}{(n+m\tau)^{2k+2}}z^{2k},\quad z\in\mathbb{C}.\tag{1.2}$$

It is obvious that

$$G_k=G_k(\tau)=\sum_{(n,m)\in\mathbb{Z}\oplus\mathbb{Z}-\{(0,0)\}}\frac{1}{(n+m\tau)^k}$$

converges for every integer $k\ge 3$. It is also obvious that $G_k=0$ for all odd k since both $n+m\tau$ and $-n-m\tau$ appear in the sum. We thus once again see that $\wp$ is an even function.

The function $\wp$ has a double pole at the lattice points and is regular elsewhere on the plane. The derivative, $\wp'$, of the $\wp$-function is again periodic with a triple pole at the origin.

Theorem 4.18. *For every doubly periodic function f (with periods 1 and τ), we can find rational functions a and b so that*

$$f = a(\wp)\wp' + b(\wp).$$

Proof. Let $z_1, \ldots, z_k$ be a maximal list of $\tilde{G}$-inequivalent points where f has poles. We assume that the equivalence class of the origin has been temporarily removed from this list (if f has a pole there). By choosing sufficiently large integers $\mu_1, \ldots, \mu_k$,

$$f_1 = f \prod_{i=1}^{k} (\wp - \wp(z_i))^{\mu_i}$$

is a doubly periodic function which is holomorphic except at the lattice points. For the positive integer k, $\wp^k$ has a pole of order $2k$ at the origin, and $\wp^{k-1}\wp'$ has a pole of order $2k+1$ at the origin. Thus we can find polynomials c and d such that

$$f_2 = f_1 - c(\wp) - d(\wp)\wp'$$

is holomorphic except for perhaps a simple pole at the lattice points. If this doubly periodic function had a simple pole at the origin, then it would define a degree 1 function from the torus T_τ to the sphere. Since this is impossible for topological reasons, f_2 is constant. This gives the desired representation for the function f.

The elliptic functions $\wp$ and $\wp'$ are not (algebraically) independent. To derive "the" relation between them, we compute the first few terms of the Laurent series expansion about 0 for

$$\wp'(z)^2 - 4\wp(z)^3 + 60G_4\wp(z) = -140G_6 + \ldots .$$

The left hand side is an elliptic function. The right hand side shows that this function is free from singularities, hence constant. We conclude that

$$\wp'(z)^2 = 4\wp(z)^3 - 60G_4\wp(z) - 140G_6, \text{ for all } z \in \mathbb{C}. \tag{1.3}$$

We now let $y = \wp'(z)$, $x = \wp(z)$ and rewrite the above equation as

$$y^2 = 4(x-e_1)(x-e_2)(x-e_3), \tag{1.4}$$

which is obviously irreducible. Thus the field of meromorphic functions on the torus T_τ is isomorphic to the function field

$$\frac{\mathbb{C}(x)[y]}{(y^2 - 4(x-e_1)(x-e_2)(x-e_3))}.$$

It is most important to note that the passage from the covering group of the torus to its algebraic curve involved the transcendental step of constructing the Weierstrass $\wp$-function. It is also customary to set $g_2 = 60G_4$ and $g_3 = 140G_6$ and translate (1.3) to

$$\wp'(z)^2 = 4\wp(z)^3 - g_2\wp(z) - g_3.$$

Remark 4.19. Conversely, every curve determines an algebraic function field in one variable and a covering group (via the general uniformization theorem). The situation in genus 1 is particularly simple since every torus can be defined by essentially the same equation (1.4) with variable parameters (e_1, e_2 and e_3).

For an arbitrary compact surface X of genus ≥ 2, one can find two meromorphic functions z and w on the surface that generate the function field $\mathcal{K}(X)$. Further, these two functions satisfy an irreducible polynomial identity of the form

$$\sum_{ij} a_{ij} z^i w^j = 0, \ a_{ij} \in \mathbb{C}.$$

For the special case under consideration, it remains to identify the constants e_i, $i = 1, 2, 3$. The elliptic function $\wp'$ is odd. Hence

$$\wp'(1-z) = \wp'(\tau - z) = \wp'(1+\tau-z) = -\wp'(z).$$

By setting $z = \frac{1}{2}$, $\frac{\tau}{2}$ and $\frac{1+\tau}{2}$, respectively, we see that $\wp'$ vanishes at the three half periods. (We have computed the divisor $(\wp')$.) We conclude that we may take

$$e_1 = \wp\left(\frac{1}{2}\right), \ e_2 = \wp\left(\frac{\tau}{2}\right), \ e_3 = \wp\left(\frac{1+\tau}{2}\right).$$

We also observe that since $\wp$ is of degree 2, the three values e_i are distinct.

The Weierstrass $\wp$-function depends on two variables z and τ. It is a meromorphic function of these two variables; its value at $(z, \tau) \in \mathbb{C} \times \mathbb{H}^2$ should be denoted by the symbol $\wp(z, \tau)$. The semidirect product of $\mathbb{Z}^2$ and $\text{PSL}(2, \mathbb{Z})$ acts on the function $\wp$. We have for all $(z, \tau) \in \mathbb{C} \times \mathbb{H}^2$,

$$\wp(z + n + m\tau, \tau) = \wp(z, \tau) \text{ for } n, \ m \in \mathbb{Z},$$

and

$$\wp\left(\frac{z}{c\tau + d}, \frac{a\tau + b}{c\tau + d}\right) = (c\tau + d)^2 \ \wp(z, \tau), \tag{1.5}$$

for all $\gamma = \begin{bmatrix} a & b \\ c & d \end{bmatrix} \in \text{PSL}(2,\mathbb{Z})$. To verify the second formula, let $\hat{z} = \frac{z}{c\tau+d}$ and $\hat{\tau} = \frac{a\tau+b}{c\tau+d}$. Then[18]

$$\frac{1}{c\tau+d} = a - c\hat{\tau} \text{ and } \frac{\tau}{c\tau+d} = -b + d\hat{\tau}.$$

It follows that $z \mapsto \wp(\hat{z},\hat{\tau})$ defines a function in $\mathcal{K}(T_\tau)$. This function is holomorphic except at the lattice points, where it has a double pole. Since its Laurent series about the origin is

$$\left(\frac{z}{c\tau+d}\right)^{-2} + a_2\left(\frac{z}{c\tau+d}\right)^{2} + \ldots,$$

we obtain (1.5) by comparison with (1.1).

It also follows from (1.2) and (1.5) that

$$\frac{1}{\hat{z}^2} + \sum_{k=1}^{\infty}(2k+1)G_{2k+2}(\hat{\tau})\hat{z}^{2k} = \gamma'(\tau)^{-1}\left(\frac{1}{z^2} + \sum_{k=1}^{\infty}(2k+1)G_{2k+2}(\tau)z^{2k}\right).$$

Comparison of Laurent series coefficients yields

$$G_{2k+2}(\gamma(\tau))\gamma'(\tau)^{k+1} = G_{2k+2}(\tau), \text{ for all } k \in \mathbb{Z}^+,\ \gamma \in \Gamma,\ \tau \in \mathbb{H}^2.$$

Thus, in language to be introduced subsequently (in §1 of Chapter 3), for each positive integer k, G_{2k+2} is an automorphic $k+1$-form for the modular group Γ. It has a limit as τ approaches $\imath\infty$; as a matter of fact, there is a connection to the Riemann ζ-function

$$\lim_{\tau\to\imath\infty} G_k(\tau) = \sum_{n\in\mathbb{Z},n\neq 0} n^{-k} = \begin{cases} 0 \text{ for odd } k \geq 3 \\ 2\zeta(k) \text{ for even } k \end{cases}.$$

Hence, using symbols to be introduced in Chapter 3, $G_{2k+2} \in \mathbb{A}^+_{k+1}(\mathbb{H}^2,\Gamma)$. It follows, using the same notation and known values of the Riemann ζ-function (see, for example, [**26**, pg. 91]), that

$$G_4^3 - \frac{2\zeta(4)^3}{\zeta(6)^2}G_6^2 \in \mathbb{A}_6(\mathbb{H}^2,\Gamma).$$

[18] It is useful to record that if we write

$$z = \alpha + \beta\tau,\ \alpha,\ \beta \in \mathbb{R},$$

then

$$\hat{z} = (\alpha a - \beta b) + (-\alpha c + \beta d)\hat{\tau}.$$

4.3. Invariants for tori. For more information see [**1**, Ch. 7]. The theory for tori we have discussed so far can be described as the transcendental theory (involving the parameter τ). To begin to study the algebraic theory (parameters λ and $\jmath$), we proceed to describe the covering of the Riemann space $\mathbf{R}(1,0)$ by the Teichmüller space $\mathbf{T}(1,0)$. First of all, the modular group PSL(2,$\mathbb{Z}$) has a big torsion free normal subgroup of finite index, the level 2 principal congruence subgroup

$$\Gamma(2) = \left\{ \begin{bmatrix} a & b \\ c & d \end{bmatrix} \in \mathrm{PSL}(2,\mathbb{Z});\ a \equiv d \equiv 1 (\mathrm{mod}\ 2).\ b \equiv c \equiv 0 (\mathrm{mod}\ 2) \right\}.$$

This group is freely generated by the parabolic motions $B^2 = \begin{bmatrix} 1 & 2 \\ 0 & 1 \end{bmatrix}$ and $C = \begin{bmatrix} 1 & 0 \\ -2 & 1 \end{bmatrix}$.

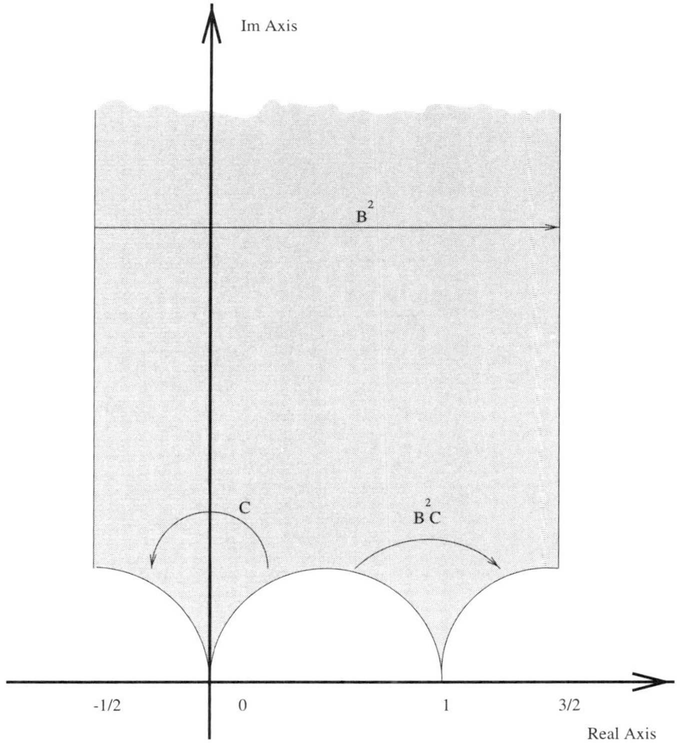

Figure 3. Fundamental domain for the level 2 principal congruence subgroup $\Gamma(2)$.

The groups $\Gamma(2)$ and $\mathrm{PSL}(2,\mathbb{Z})$ fit into a short exact sequence

$$1 \to \Gamma(2) \to \mathrm{PSL}(2,\mathbb{Z}) \xrightarrow{\theta} \mathrm{Perm}(3) \to 1,$$

where Perm(3) is the permutation group on three letters. The group $\Gamma(2)$ is the holomorphic universal covering group of the thrice punctured sphere $\mathbb{C} - \{0, 1\}$. The permutation group Perm(3) acts on the thrice punctured sphere to produce the moduli space $\mathbf{R}(1, 0)$. The covering map of the thrice punctured sphere can be described in terms of elliptic functions.

Theorem 4.20. *The function*

$$\lambda(\tau) = \frac{e_3 - e_2}{e_1 - e_2} = \frac{\wp\left(\frac{1+\tau}{2}\right) - \wp\left(\frac{\tau}{2}\right)}{\wp\left(\frac{1}{2}\right) - \wp\left(\frac{\tau}{2}\right)}, \quad \tau \in \mathbb{H}^2,$$

is a holomorphic universal covering map of the plane punctured at 0 *and* 1 *whose group of deck transformations is* $\Gamma(2)$.

Proof. We only outline the proof since we will subsequently present different approaches to the issues that arise in the proof. The motion $z \mapsto \frac{z-e_2}{e_1-e_2}$ is a holomorphic map of the complex sphere taking e_2 to the origin, e_1 to the point 1, that preserves the point ∞. $\lambda(\tau)$ is the image of the point e_3 under this map, which may be viewed as a change of parameters defining conformally equivalent tori. We use some of the notation introduced in the previous subsection. The function λ is holomorphic and $\Gamma(2)$-invariant; that is,

$$\lambda \circ \gamma = \lambda, \text{ for all } \gamma \in \Gamma(2).$$

To verify the claim about invariance, we observe that Γ essentially permutes the half periods as a consequence of (1.5): For $\gamma = \begin{bmatrix} a & b \\ c & d \end{bmatrix} \in \mathrm{PSL}(2, \mathbb{Z})$,

$$(c\tau + d)^2 e_1(\tau) = (c\tau + d)^2 \wp\left(\frac{1}{2}, \tau\right) = \wp\left(\frac{\hat{1}}{2}, \hat{\tau}\right) = \wp\left(\frac{1}{2}(a - c\hat{\tau}), \hat{\tau}\right),$$

which equals

$$\begin{cases} \wp\left(\frac{\hat{\tau}}{2}, \hat{\tau}\right) = e_2(\hat{\tau}) \text{ if } a \text{ is even and } c \text{ is odd} \\ \wp\left(\frac{1}{2}, \hat{\tau}\right) = e_1(\hat{\tau}) \text{ if } a \text{ is odd and } c \text{ is even} \\ \wp\left(\frac{1+\hat{\tau}}{2}, \hat{\tau}\right) = e_3(\hat{\tau}) \text{ if } a \text{ is odd and } c \text{ is odd} \end{cases}.$$

Note that it is impossible for both a and c to be even. Similarly,

$$(c\tau + d)^2 e_2(\tau) = (c\tau + d)^2 \wp\left(\frac{\tau}{2}, \tau\right) = \wp\left(\frac{\hat{\tau}}{2}, \hat{\tau}\right) = \wp\left(\frac{1}{2}(-b + d\hat{\tau}), \hat{\tau}\right),$$

which equals

$$\begin{cases} \wp\left(\frac{\hat\tau}{2},\hat\tau\right)=e_2(\hat\tau) \text{ if } b \text{ is even and } d \text{ is odd} \\ \wp\left(\frac{1}{2},\hat\tau\right)=e_1(\hat\tau) \text{ if } b \text{ is odd and } d \text{ is even} \\ \wp\left(\frac{1+\hat\tau}{2},\hat\tau\right)=e_3(\hat\tau) \text{ if } b \text{ is odd and } d \text{ is odd} \end{cases},$$

and

$$\begin{aligned}(c\tau+d)^2e_3(\tau) = & \ (c\tau+d)^2\wp\left(\tfrac{1+\tau}{2},\tau\right)=\wp\left(\tfrac{\hat{1}+\hat\tau}{2},\hat\tau\right) \\ = & \ \wp\left(\tfrac{1}{2}(a-b-(c-d)\hat\tau),\hat\tau\right)\end{aligned},$$

which equals

$$\begin{cases} \wp\left(\frac{1+\hat\tau}{2},\hat\tau\right)=e_3(\hat\tau) \text{ if } a-b \text{ is odd and } c-d \text{ is odd} \\ \wp\left(\frac{1}{2},\hat\tau\right)=e_1(\hat\tau) \text{ if } a-b \text{ is odd and } c-d \text{ is even} \\ \wp\left(\frac{\hat\tau}{2},\hat\tau\right)=e_2(\hat\tau) \text{ if } a-b \text{ is even and } c-d \text{ is odd} \end{cases}.$$

An element $\gamma\in\Gamma(2)$ (thus a and d are odd, while b and c are even) fixes each half period. It is not too hard to construct a fundamental domain for the action of $\Gamma(2)$ on $\mathbb{H}^2$ (it is described by the last figure; see also §6.4) and show that λ maps this fundamental domain injectively into $\mathbb{C}-\{0,1\}$. We can avoid this argument by observing (we already know that λ is holomorphic and misses 0 and 1) that $\Gamma(2)$ has three equivalence classes of parabolic fixed points, which can be taken to be ∞, 0, 1. Further, the function λ extends by continuity to the fixed points of the parabolic elements of the group $\Gamma(2)$, that is, to the rational points and the ideal point at infinity. One can now show that

$$\lambda(\infty)=0,\ \lambda(0)=1,\ \lambda(1)=\infty;$$

as a matter of fact

$$\lambda(\tau)\exp\{-\pi\imath\tau\}\to 16 \text{ as } \Im\tau\to\infty.$$

(We will produce at least two alternate formulae for the function λ: one involving Poincaré series (we will only outline the theory needed for this approach), and the second derived from the Riemann θ-function (based on the main ideas discussed in this book).) From the last formula it will be easy to see that in terms of the local coordinate $x=\exp(\pi\imath\tau)$,

$$\lambda(\tau)=16x(1-48x+O(x^2)).$$

Finally $(\mathbb{H}^2\cup\mathbb{Q}\cup\{\infty\})/\Gamma(2)$ is compact. Since λ is a nonconstant holomorphic map from this compact surface onto $\hat{\mathbb{C}}$ with a single simple pole, it is surjective and a holomorphic universal covering of its image which must be $\mathbb{C} - \{0, 1\}$.

To obtain the map from the punctured sphere to the moduli space $\mathbf{R}(1, 0)$, we must find out how the permutation group acts on the thrice punctured sphere. One can show that Perm(3) is faithfully represented by the six Möbius transformations that permute the points 0, 1 and ∞. This is, in fact, the full automorphism group of the thrice punctured sphere $\hat{\mathbb{C}} - \{0,\ 1,\ \infty\}$. The maps

$$\tau \mapsto -\frac{1}{\tau} \text{ and } \tau \mapsto 1 - \frac{1}{\tau}$$

generate $\mathrm{PSL}(2, \mathbb{Z})$ (thus the corresponding $\Gamma(2)$-cosets generate the quotient $\mathrm{PSL}(2, \mathbb{Z})/\Gamma(2)$). The first of these maps interchanges zero and infinity and sends one to minus one (which is equivalent modulo $\Gamma(2)$ to plus one). Hence its image in Perm(3) must be the map

$$\lambda \mapsto 1 - \lambda;$$

that is,

$$\lambda\left(-\frac{1}{\tau}\right) = 1 - \lambda(\tau).$$

Similarly, the image of the second map must be

$$\lambda \mapsto 1 - \frac{1}{\lambda}.$$

Since fixed points of maps in the τ-plane project to fixed points of maps in the λ-plane, we have

$$\lambda(\imath) = \frac{1}{2} \text{ and } \lambda\left(\frac{1}{2}(1 + \sqrt{3}\imath)\right) = \frac{1}{2}(1 \pm \sqrt{3}\imath).$$

(We do not determine which choice of sign is correct; it can be done easily.) In the above calculations we have used the fact that $\mathrm{PSL}(2, \mathbb{Z})$ is the normalizer of $\Gamma(2)$, that Perm(3) is isomorphic to $\mathrm{Aut}(\mathbb{H}^2/\Gamma(2))$, and that we have for each $\gamma \in \mathrm{PSL}(2, \mathbb{Z})$ the commutative diagram

$$\begin{array}{ccc} \mathbb{H}^2 & \xrightarrow{\gamma} & \mathbb{H}^2 \\ \downarrow & & \downarrow \\ \mathbb{H}^2/\Gamma(2) & \xrightarrow{\theta(\gamma)} & \mathbb{H}^2/\Gamma(2) \end{array}$$

To get a function invariant under the full modular group, we take any rational function of λ with a single simple pole at a point that is not fixed by any element in the group Perm(3) (that is, a rational function of degree one with a pole at a point not fixed by the group of Möbius transformations that permute the three points $0, 1, \infty$) (for example, we can use the function $\lambda \mapsto$

$\frac{1}{\lambda-3}$) and average this function over the group Perm(3).[19] This averaging process leads to the rational function

$$\left(\frac{1}{\lambda-3}+\frac{\lambda}{1-3\lambda}-\frac{1}{\lambda+2}+\frac{1-\lambda}{3\lambda-2}+\frac{\lambda-1}{3-2\lambda}-\frac{\lambda}{2\lambda+1}\right).$$

To normalize, we postcompose the resulting average with a Möbius transformation C, so that the composite map sends $\lambda=\infty$ to the point ∞, $\lambda=\frac{1+\sqrt{3}\imath}{2}$ to 0 and $\lambda=2$ to 1. The Möbius transformation C is computed to be $\left(\frac{7}{9}\right)^2\frac{7z+15}{3z+5}$, and we obtain the famous $\jmath$-invariant

$$\jmath(\lambda)=\frac{4}{27}\frac{(1-\lambda+\lambda^2)^3}{\lambda^2(1-\lambda)^2}.$$

The above may appear a bit mysterious and magical since we began with an arbitrary rational function with a simple pole at a nonfixed point of Perm(3) (at 3 in our case), averaged it over this finite group of order 6 to obtain a rational function of degree 6 with double poles at the orbit of 3 under the group. We then find the Möbius transformation taking the poles to $0,1,\infty$. It is perhaps the last part which is most mysterious, so we present an additional procedure which may be a little less mysterious.

In order to obtain a Γ-invariant function, we average a $\Gamma(2)$-invariant function over the group $\Gamma(2)\backslash\Gamma$. In place of the function we used above we can average the function λ itself. This produces a constant function. So we try to use the next simplest case and average the function λ^2. This leads us to the rational function

$$F(\lambda)=\frac{2\lambda^6-6\lambda^5+9\lambda^4-8\lambda^3+9\lambda^2-6\lambda+2}{\lambda^2(1-\lambda)^2}.$$

The function F has the virtue of having the correct poles. The reader can now check that

$$\frac{F(\lambda)+3}{2}=\frac{(1-\lambda-\lambda^2)^3}{\lambda^2(1-\lambda)^2}.$$

This still does not explain the constant $\frac{4}{27}$. We leave it to the reader to consult books on elliptic functions to discover the reasons for this normalization. Based on our work on θ-constants we will conclude that in terms of the local coordinate $x=\exp(2\pi\imath\tau)$,

$$\jmath(\tau)=\frac{1}{4\cdot 27x}(1-4^3 7x-O(x^2)).$$

---

[19] Realized, as before, as the group of fractional linear transformations that permute the three points ∞, 0 and 1. If we identify the permutation $\sigma\in$ Perm(3) with the motion in PSL(2,$\mathbb{C}$) that sends λ to $\sigma(\lambda)$ for $\lambda=0$, 1 and ∞, then we summarize this identification by the entries $\sigma=\sigma(\lambda)$ in

$$(-)=\lambda,\ (0,1)=1-\lambda,\ (0,\infty)=\frac{1}{\lambda},\ (0,1,\infty)=\frac{1}{1-\lambda},\ (0,\infty,1)=\frac{\lambda-1}{\lambda},\ (1,\infty)=\frac{\lambda}{\lambda-1}.$$

We are using here standard permutation notation; see, for example, [**9**, Ch. 5].

More important than the formulae produced is an understanding of the maps involved. We are studying the coverings

$$\mathbb{H}^2 \xrightarrow{\lambda} \mathbb{C} - \{0, 1\} \cong \mathbb{H}^2/\Gamma(2) \xrightarrow{\jmath} \mathbb{C} \cong \mathbb{H}^2/\Gamma.$$

The group $\Gamma(2)$ is torsion free (isomorphic to the fundamental group of the thrice punctured sphere $\mathbb{C} - \{0, 1\}$), and λ is a holomorphic universal covering map. The map $\jmath \circ \lambda$ is the branched universal covering map of the orbifold $\mathbb{H}^2/\Gamma$. Since Γ has only 2 and 3 torsion, the map $\jmath$ must be branched. Viewing $\jmath$ as a selfmap of the sphere, we see that $\jmath$ has branch number 2 at ∞ as well as -1, $\frac{1}{2}$ and 2; it has branch number 3 at $e^{\frac{\pi\imath}{3}}$ and $e^{\frac{5\pi\imath}{3}}$. We conclude that $\jmath$ sends the three punctures ($\{0, 1, \infty\}$) on $\mathbb{H}^2/\Gamma(2)$ to ∞. The punctures on $\mathbb{H}^2/\Gamma(2)$ are the image of $\mathbb{Q} \cup \{\infty\}$ under λ; the Γ-orbit of $\imath$ is sent by λ to $\{-1,\ \frac{1}{2},\ 2\}$ and by $\jmath$ to 1; λ sends the Γ-orbit of $\frac{1}{2} + \frac{\sqrt{3}}{2}\imath = e^{\frac{\pi\imath}{3}}$ to $\left\{\frac{1}{2} \pm \frac{\sqrt{3}}{2}\right\}$ and by $\jmath$ to 0.

Let us summarize what we have accomplished. For a point $\tau \in \mathbb{H}^2$, we constructed a torus T_τ and its $\jmath$-invariant $\jmath(\lambda(\tau))$ which is a complete conformal invariant for the family of tori in the following sense. Given two points τ_1 and τ_2 in $\mathbb{H}^2$, then T_{τ_1} is conformally equivalent to T_{τ_2} if and only if $\jmath(\lambda(\tau_1)) = \jmath(\lambda(\tau_2))$ and for each point $\alpha \in \mathbb{C}$, there is a $\tau \in \mathbb{H}^2$ such that $\alpha = \jmath(\lambda(\tau))$. For almost all $\alpha \in \mathbb{C}$ (we shall see shortly that we are only excluding α in the segment $(-\infty, 1] \subset \mathbb{R}$), we can find a unique (also canonical) τ in the interior of $\omega(\mathrm{PSL}(2,\mathbb{Z}))$ with $j(\lambda(\tau)) = \alpha$. The complex plane is, of course, not compact. We would like to have a natural compactification of the space of tori. We need a point τ such that $j(\lambda(\tau)) = \infty$. Such a point τ must hence be sent by λ to 0, 1 or ∞. Hence τ cannot belong to $\mathbb{H}^2$; it necessarily will belong to $\mathbb{Q} \cup \{\infty\}$. We will see in a little while (§4.6) how to add the point at infinity to our space of moduli of surfaces of genus one.

Exercise 4.21. Show that for each $\gamma \in \Gamma$, there is a permutation σ_γ of the integers 1, 2 and 3 such that

$$(e_i \circ \gamma)\gamma' = e_{\sigma_\gamma(i)},\ i = 1, 2, 3.$$

Further, the map $\gamma \mapsto \sigma_\gamma$ is a surjective homomorphism from Γ to the permutation group on three letters with kernel $\Gamma(2)$.

4.4. Tori with symmetries. The group of conformal automorphisms of the torus T_τ, Aut T_τ, contains the subgroup of translations by elements of T_τ. Hence we should consider instead the factor group (Aut $T_\tau)/T_\tau$ which can be canonically identified with the (extra) *symmetries* of T_τ

$$\mathrm{Aut}_o T_\tau = \{C \in \mathrm{Aut}\ T_\tau;\ C(0) = 0\}.$$

It is best to encode the above information in a short exact sequence that splits

$$0 \to \mathrm{Aut}_o T_\tau \to \mathrm{Aut}\ T_\tau \to T_\tau \to 0;$$

the second arrow is the inclusion map and the third arrow sends the automorphism C to the point $C(0)$. The splitting map sends the point on T_τ determined by $a \in \mathbb{C}$ to the translation of the torus induced by the affine map $z \mapsto z + a$. We observe that every torus has at least one symmetry. It is given by the conformal involution $\tilde{E}(z) = -z + 1 + \tau$ with fixed point the half period $\frac{1+\tau}{2}$. We are choosing this particular involution because it preserves the fundamental parallelogram $\tilde{\omega}$ for $\tilde{G}_\tau$. It is equivalent modulo T_τ to the motion $z \mapsto -z \in \mathrm{Aut}_o T_\tau$. The involution has the property that it conjugates $\tilde{A}$ and $\tilde{B}$ to their respective inverses. The generic torus has no other symmetries. The "square" torus (the one corresponding to $\tau = \imath$) has the additional symmetry corresponding to the motion of order four, $z \mapsto \imath z$. The "rhombic" torus ($\tau = \frac{1+\imath\sqrt{3}}{2}$) has the extra symmetry of order six, $z \mapsto \tau z$. We have accounted for two of the three distinguished points in the compactification of the moduli space $\mathbf{R}(1,0)$. The third point is distinguished because it is missing from the moduli space $\mathbf{R}(1,0)$. It is also distinguished for other reasons as we shall see shortly.

It is useful to describe tori with additional symmetries. The tori with purely imaginary moduli ($\Re(\tau) = 0$) have the anti-conformal involution $\tilde{J}$ given by

$$\tilde{J}(z) = \bar{z} + \tau.$$

This involution fixes the point $\frac{1+\tau}{2}$ and conjugates $\tilde{A}$ to itself and $\tilde{B}$ to its inverse.

Tori with moduli of absolute value 1 ($|\tau| = 1$) have a different anti-conformal involution $\tilde{J}_1$ defined by

$$\tilde{J}_1(z) = \tau\bar{z}.$$

This involution also fixes $\frac{1+\tau}{2}$ and conjugates $\tilde{A}$ to $\tilde{B}$ and $\tilde{B}$ to $\tilde{A}$.

We will distinguish two more families of tori: those with $\Re\tau = -\frac{1}{2}$ and those with $\Re\tau = \frac{1}{2}$. Each member of the first (second) family has the involution $\tilde{J}_3$ ($\tilde{J}_2$) $z \mapsto -\bar{z}$ ($z \mapsto -\bar{z} + 1$) that conjugates $\tilde{A}$ to its inverse and $\tilde{B}$ to $\tilde{A} \circ \tilde{B}$ ($\tilde{A}$ to its inverse and $\tilde{B}$ to $\tilde{A}^{-1} \circ \tilde{B}$).

The families of tori discussed above lie on lines in $\mathbb{H}^2$ determined by segments on the boundary of $\omega(\mathrm{PSL}(2,\mathbb{Z}))$ and on the axis of symmetry of the fundamental domain we have chosen for $\mathrm{PSL}(2,\mathbb{Z})$).[20] The one symmetry that is present on all tori can be described more intrinsically. Given a torus T and a point x on it, there exists a unique conformal involution $\tilde{E}$ of T that

[20]The fundamental domain $\omega(\mathrm{PSL}(2,\mathbb{Z}))$ is invariant under the motion $z \mapsto -\bar{z}$.

fixes x. If we now identify x with the origin of T, then $\tilde{E}$ fixes, in addition to 0, the other three half periods (points of order 2) of T.

We can use symmetries to obtain information about the $\jmath$-invariant. The Weierstrass $\wp$-function is the unique doubly periodic function of degree two with Laurent series expansion at the origin of the form

$$\wp(z) = \frac{1}{z^2} + \sum_{n=1}^{\infty} a_n z^{2n}.$$

The function $\wp$ is an even function of z. However, the coefficients appearing in the Laurent series expansion are not in general real. If for τ in the upper half plane we have $\tau = -\bar{\tau}$, then the torus T_τ is invariant under the involution $z \mapsto \bar{z}$, and we conclude from the uniqueness of the Laurent series expansion (or the original formula) for $\wp$ that $\overline{\wp(\bar{z})} = \wp(z)$ for all $z \in \mathbb{C}$. It follows that $e_1 \in \mathbb{R}$. Also

$$e_2 = \wp\left(\frac{\tau}{2}\right) = \wp\left(\frac{\bar{\tau}}{2}\right) = \overline{\wp\left(\frac{\tau}{2}\right)} = \overline{e_2}.$$

Similarly we see that e_3 is also real. Hence (if τ is imaginary) $\lambda(\tau) \in \mathbb{R}$ and $\jmath(\tau) \in \mathbb{R}$. Thus we see that

$$\tau = -\bar{\tau} \Longrightarrow \lambda = \bar{\lambda} \text{ and } \jmath = \bar{\jmath}.$$

The segment $\{\imath y \in \mathbb{C};\ y > 1\}$ gets mapped injectively by $j \circ \lambda$ to the interval $[1, \infty) \subset \mathbb{R}$. We consider $\tau \in \mathbb{H}^2$ with $|\tau| = 1$. The torus T_τ is invariant under the map $z \mapsto \tau\bar{z}$. The Laurent series for the $\wp$-function now shows that

$$\wp(\tau\bar{z}) = \tau^{-2}\overline{\wp(z)},\ z \in \mathbb{C}.$$

We conclude that

$$\wp\left(\frac{\tau}{2}\right) = \tau^{-2}\overline{\wp\left(\frac{1}{2}\right)},$$

and

$$\tau^{-2}\overline{\wp\left(\frac{1}{2}(1+\tau)\right)} = \wp\left(\frac{1}{2}(1+\tau)\right).$$

One easily computes from these identities that

$$|\tau| = 1 \Longrightarrow \lambda + \bar{\lambda} = 1.$$

We conclude once again that $\lambda = \frac{1}{2}$ for $\tau = \imath$ and that $\jmath$ is real for τ of absolute value one. The function $j \circ \lambda$ maps the circular segment $\{\exp \imath\theta\colon \frac{\pi}{3} \leq \theta \leq \frac{\pi}{2}\} \subset \mathbb{C}$ injectively to $[-\frac{4}{27}, 1] \subset \mathbb{R}$. It is also easy to see that λ takes on real values on the lines $\Re\tau = \pm\frac{1}{2}$, and that $j \circ \lambda$ maps the segment $\{\frac{1}{2} + \imath y;\ y > \frac{\sqrt{3}}{2}\}$ in $\mathbb{C}$ injectively to $(-\infty, -\frac{4}{27}] \subset \mathbb{R}$.

4.5. Congruent numbers. See [**17**, Ch. I]. We briefly illustrate a connection to number theory that will not be pursued in this book.

Definition 4.22. An element $r \in \mathbb{Q}^+$ is a *congruent number* provided it is the area of a right triangle with rational sides. It involves no loss of generality to also assume that r is a square free positive integer (that is, no square of an integer divides r).

Fix a square free $n \in \mathbb{Z}^+$, and consider positive rationals X, Y and Z with $X < Y < Z$. The map

$$(X, Y, Z) \mapsto x = \left(\frac{Z}{2}\right)^2$$

establishes a one-to-one correspondence between right triangles with legs X and Y, hypotenuse Z and area n and rational numbers x for which x and $x \pm n$ are each square free; the inverse map is

$$x \mapsto (X, Y, Z) = (\sqrt{x+n} - \sqrt{x-n},\ \sqrt{x+n} + \sqrt{x-n},\ 2\sqrt{x}).$$

One can show that if (x, y) is a point with rational coordinates on the curve

$$y^2 = x^3 - n^2 x$$

and if x is the square of a rational number with even denominator whose numerator is relatively prime to n, then the last map produces from x a right triangle that shows that n is a congruent number.

4.6. The plumbing construction. We now describe an alternate construction of the torus. This alternate construction can be generalized (from the euclidean case to the hyperbolic case) considerably to yield all Riemann surfaces of finite analytic type. It will also give us an intermediate moduli space between the Teichmüller space and the Riemann space. Let us look at the most complicated (topologically) euclidean (that is, with zero curvature) surface without moduli: the infinite cylinder, $\mathcal{C}$, which we view as the complex plane $\mathbb{C}$ factored by the cyclic subgroup generated by the motion $\tilde{A}$. Perfectly reasonable (global) coordinates each vanishing at one of the punctures on $\mathcal{C}$ are $z = e^{-2\pi\imath\zeta}$ and $w = e^{2\pi\imath\zeta}$, $\zeta \in \mathbb{C}$. For each $t \in \mathbb{C}$ with $0 < |t| < 1$, we construct a torus S_t by a *plumbing procedure*. Remove from $\mathcal{C}$ two punctured discs[21] $\{0 < |z| \leq |t|\}$ and $\{0 < |w| \leq |t\ \}$ to obtain a finite cylinder $\mathcal{C}'$. Identify two points P and Q on $\mathcal{C}'$ if and only if $z(P)w(Q) = t$, and thus obtain a torus S_t together with a *central* curve on it described in our local coordinates by the equations $\{|z| = |t|^{\frac{1}{2}}\} = \{|w| = |t|^{\frac{1}{2}}\}$. This

[21] It turns out that we have a lot of freedom in choosing which punctured discs to remove – the same "plumbed" surface would be obtained with many other choices. One takes advantage of this fact in the hyperbolic plumbing construction.

central curve is freely homotopic to $\{|z| = 1\} = \{|w| = 1\}$. We proceed to analyze this *plumbing construction.* There are two disjoint annuli on $\mathcal{C}'$:

$$\mathcal{A}_1 = \{|t| < |z| < 1\} \text{ and } \mathcal{A}_2 = \{|t| < |w| < 1\}.$$

There is no identification of points in $\mathcal{C}' - (\mathcal{A}_1 \cup \mathcal{A}_2)$ (which consists of the circle $\{|z| = 1\} = \{|w| = 1\}$). Each point of $\mathcal{A}_1$ is identified by the plumbing construction with precisely one point of $\mathcal{A}_2$ to form an annulus $\mathcal{A}$ on S_t. The construction can, of course be completely described in terms of the single global coordinate z on $\mathcal{C}'$. In the z coordinate, a point z_1 is identified with z_2 if and only if $\frac{z_1}{z_2} = t$. The plumbing construction is equivalent to a boundary identification on the cylinder

$$\mathcal{C}'' = \mathcal{C} - (\{0 < |z| < |t|^{\frac{1}{2}}\} \cup \{0 < |w| < |t|^{\frac{1}{2}}\}).$$

The cylinder $\mathcal{C}''$ has two boundary components (circles). On the first $|z| = |t|^{\frac{1}{2}}$, and on the second $|w| = |t|^{\frac{1}{2}}$. We identify each point P on the first boundary circle with the unique point Q on the second that satisfies $z(P)w(Q) = t$.

It is easy to see that our two constructions (the classical elliptic function construction on $\mathbb{C}$ and the plumbing construction on $\mathcal{C}$) are related:

$$S_{e^{2\pi\imath\tau}} = T_\tau \text{ for all } \tau \in \mathbb{H}^2.$$

Thus it is clear that every torus can be constructed by this plumbing operation. The limiting case ($t = 0$) makes perfectly good sense and corresponds to a torus with a node. It can be viewed as the cylinder $\mathcal{C}$ with its two punctures (filled in and then) identified to form this node.

The torus S_t carries three conformal involutions (the Klein 4-group acts on S_t) . One can be defined by the map $z \mapsto -z$, $w \mapsto -w$ (in terms of the local coordinate ζ on $\mathbb{C}$, the holomorphic universal coordinate it is given by $\zeta \mapsto \zeta + \frac{1}{2}$). It has no fixed points if $t \neq 0$ (for $t = 0$, the involution fixes the node). This involution fixes (as a set) the central curve on S_t (as well as any curve parallel to it). A second, more interesting, involution is defined by the map $z \mapsto \frac{1}{z}$, $w \mapsto \frac{1}{w}$ (in terms of ζ, $\zeta \mapsto -\zeta$). The third involution is the composite of the first two. This second involution has four fixed points, $z = \pm 1$ and $w = \pm 1$. Thus it may be identified with the map $\tilde{E}$ for any of these fixed points.

Exercise 4.23. a. Describe the group of conformal automorphisms of the torus T_τ, $\mathrm{Aut}(T_\tau)$.
b. Find the formula for each element of $\mathrm{Aut}(S_t)$ in terms of the plumbing parameters on S_t.

The punctured unit disc (the set of nonzero t-coordinates) is an intermediate moduli space for tori; it corresponds to $\mathbf{T}(1,0)/\mathbb{Z}$, where $\mathbb{Z}$ acts on

$\mathbb{H}^2 \cong \mathbf{T}(1,0)$ by translation and the generator of $\mathbb{Z}$ is the Dehn twist about the central curve (changing τ to $1+\tau$). The addition of the origin gives a partial compactification of $\mathbf{T}(1,0)/\mathbb{Z}$. The extra point in the compactified moduli space $\overline{\mathbf{R}(1,0)}$ corresponding to $t=0$ is the point at infinity. It corresponds to the image of fixed points, for example, $\tau=\infty$, of the parabolic elements of the modular group. The unit disc is an example of a deformation space of a graph; the general case is studied extensively in the literature.

If we normalize the metric (of zero curvature) so that S_t has area one, then the length of the central curve on this surface is $\sqrt{\frac{-2\pi}{\log|t|}}$. Thus the limiting surface corresponding to $t=0$ is obtained by shrinking the central curve on S_t to obtain the node. On the surface S_t, there is also a geodesic *transverse* to the central curve of length $\frac{|\log t|}{2\pi}\sqrt{\frac{-2\pi}{\log|t|}}$.[22] On the limiting surface S_0, the transverse curve is infinitely long. Creating a node involves some (mild) catastrophic behavior. Infinities are forced to appear.

4.7. Teichmüller and moduli spaces for tori. We have encountered several moduli spaces of tori. Most important are the Teichmüller space $\mathbf{T}(1,0) \cong \mathbb{H}^2$ and the moduli space $\mathbf{R}(1,0) \cong \mathbb{H}^2/\Gamma$. The compactification $\overline{\mathbf{R}(1,0)}$ of $\mathbf{R}(1,0)$ can be naturally identified with the orbifold of signature $(0,3;\ 2,3,\infty)$. The plumbing parameters yield the moduli space $\mathbb{H}^2/<B>$, where $B=\begin{bmatrix} 1 & 1 \\ 0 & 1 \end{bmatrix}$. This space is analytically equivalent to the punctured unit disc. The origin can be added; it represents the unique singular surface needed to compactify $\mathbf{R}(1,0)$.

In the study of tori we also encounter the moduli space $\mathbb{H}^2/\Gamma(k)$. The groups involved will be defined later; their study constitutes a central portion of this book. It suffices, for the moment, to say that this moduli space represents marked level k structures on tori.

4.8. Fiber spaces – the Teichmüller curve. We present one more interpretation of our constructions. Let us look at the product space $\mathbb{C}\times\mathbb{H}^2$. The group $\mathbb{Z}\oplus\mathbb{Z}$ acts on the product space. Let (m,n) be a point in the group; the induced automorphism on the product space sends (z,τ) to $(z+n+m\tau,\tau)$. It is easy to see that the factor space $\mathbf{V}(1,0)=(\mathbb{C}\times\mathbb{H}^2)/(\mathbb{Z}\oplus\mathbb{Z})$ is a complex manifold (of dimension two) and a fiber space over the Teichmüller space $\mathbf{T}(1,0)$ (the projection onto the second coordinate induces the map from the total space to the base space of the fibration). Using this construction and the explicit formulae for the functions $\wp$ and $\wp'$, we can simultaneously

[22]The multivaluedness of this function reflects the fact that there are countably many geodesics transverse to the central curve.

uniformize all tori. The map

$$\wp : (z, \tau) \mapsto (1, \wp(z, \tau), \wp'(z, \tau))$$

defines a holomorphic function from $\mathbf{V}(1, 0)$ to complex projective two-space $\mathbf{P}\mathbb{C}^2$. It is injective on the fiber T_τ over each $\tau \in \mathbb{H}^2$. These facts are easily verifiable. We note that for z sufficiently close to zero,

$$\wp(z, \tau) = \left(1, \frac{1}{z^2} + o(1), \frac{-2}{z^3} + o(1)\right) = (z^3, z + o(|z|^3), -2 + o(|z|^3)).$$

Thus the origin on T_τ gets mapped to the coordinate vector $(0, 0, 1)$ and no other point gets mapped to this vector. Hence if $\wp(z_1, \tau) = \wp(z_2, \tau) \neq (0, 0, 1)$, then $z_1 \neq n + m\tau \neq z_2$ (all n and $m \in \mathbb{Z}$), $\wp(z_1) = \wp(z_2)$ and $\wp'(z_1) = \wp'(z_2)$. The inequalities and first equality show that $z_2 = \pm z_1$ up to periods. If the minus sign were to hold, then we would conclude from the second equality that

$$\wp'(z_1) = \wp'(z_2) = \wp'(-z_1) = -\wp'(z_1) = 0,$$

from which it follows that z_1 and z_2 are equivalent half periods.

Remark 4.24. It was already well known in the last century that a compact Riemann surface X of genus $p \geq 2$ can be mapped holomorphically into $\mathbf{P}\mathbb{C}^{p-1}$ using holomorphic 1-forms. This map is an injection of maximal rank, except in case X is hyperelliptic ($\mathcal{K}(X)$ contains a function of degree 2). The corresponding map using holomorphic q-forms, $q \geq 2$, has $\mathbf{P}\mathbb{C}^{(2q-1)(p-1)}$ as target and is an injection of maximal rank, except in case $p = 2 = q$. (See [**6**, §III.10].) The fact that one can choose these maps to vary holomorphically with moduli was not established rigorously until the middle of the twentieth century.

To analyze the action of the modular group on the fiber space introduced above, we need different descriptions of the Teichmüller space and the modular group. As a principal tool, we shall use orientation preserving affine (not necessarily analytic) self-maps of the complex plane. These are mappings of the form

$$w : z \mapsto az + b\bar{z} + c,$$

where a, b, c are complex numbers with $|b| < |a|$.[23] For many purposes it will suffice to consider only such affine mappings that also fix the origin (that is, with $c = 0$). The inverse of the map w is the affine mapping

$$z \mapsto \frac{\bar{a}(z - c) - b(\bar{z} - \bar{c})}{|a|^2 - |b|^2}.$$

[23]This condition is necessary for the mapping w to preserve orientation. It is also sufficient for w to be invertible.

If we let A_α denote translation by α (that is, $A_\alpha(z) = z + \alpha$), then

$$w \circ A_\alpha \circ w^{-1} = A_{a\alpha + b\bar{\alpha}}.$$

We start with a fixed group

$$\tilde{G} = \tilde{G}_{\tau_o} \text{ for some } \tau_o \in \mathbb{H}^2.$$

(For our illustrations below, we will take $\tau_o = \imath$.) An isomorphism θ of $\tilde{G}$ into Aut $\mathbb{C}$ is called *geometric* if there exists an affine mapping w such that

$$\theta(g) = w \circ g \circ w^{-1} \text{ for all } g \in \tilde{G}.$$

Two geometric isomorphisms θ_1 and θ_2 are *equivalent* provided there exists a conformal affine mapping C (that is, an element of Aut $\mathbb{C}$) such that

$$\theta_2(g) = C \circ \theta_1(g) \circ C^{-1} \text{ for all } g \in \tilde{G}.$$

The *Teichmüller space* of $\tilde{G}$, $\mathbf{T}(\tilde{G})$, may be defined as the set of equivalence classes of geometric isomorphisms. The equivalence class of the geometric isomorphism induced by the map w will be denoted by $[w]$. Since the group of conformal automorphisms of the complex plane is doubly transitive, each equivalence class contains a unique isomorphism that is induced by a *normalized* affine map (a map that fixes 0 and 1), that is, by a map w of the form

$$w(z) = az + (1-a)\bar{z} \text{ with } |1-a| < |a|.$$

We have thus produced the same set of coordinates for all the $\mathbf{T}(\tilde{G}_\tau)$ consisting of $\{a \in \mathbb{C};\ \Re a > \frac{1}{2}\}$. The above map w commutes with the motion $z \mapsto z+1$ and conjugates the motion $z \mapsto z + \tau$ to the motion $z \mapsto z + a\tau + (1-a)\bar{\tau}$. We conclude that the map which sends the normalized affine mapping w to $\tau = w(\imath) = (2a-1)\imath$ establishes a complex analytic isomorphism between $\mathbf{T}(\tilde{G}_\imath)$ and $\mathbb{H}^2$. We can view this Teichmüller space as the values at $z = \imath$ of normalized affine mappings.

The (lower case) *modular group* of $\tilde{G}$, mod $\tilde{G}$, may be defined as the set of affine maps that fix the lattice points of the group $\tilde{G}$. The (upper case) *Modular group* of $\tilde{G}$, Mod $\tilde{G}$, is defined as the modular group of $\tilde{G}$ factored by the group $\tilde{G}$. It is hence isomorphic to the stabilizer of the origin in mod $\tilde{G}$. We thus have a split exact sequence

$$0 \to \tilde{G} \to \text{mod } \tilde{G} \to \text{Mod } \tilde{G} \to 0.$$

If $\omega(z) = dz + e\bar{z}$, then the condition that it preserve the lattice of the group $\tilde{G}_\imath$ is

$$d + e = \gamma\imath + \delta,\ d\imath - e\imath = \alpha\imath + \beta,$$

where

$$\alpha, \beta, \gamma, \delta \in \mathbb{Z},\ \alpha\delta - \beta\gamma = \epsilon = \pm 1,$$

from which it follows that

$$2d = (\alpha + \delta) + \imath(\gamma - \beta), \ 2e = (\delta - \alpha) + \imath(\gamma + \beta).$$

The condition that ω be orientation preserving guarantees that $\epsilon = |d|^2 - |e|^2 = 1$. The above remarks make it clear that the modular group Mod $\tilde{G}_\imath$ can be identified with the group SL(2,$\mathbb{Z}$). We emphasize that the identification is with the special linear and not the projective special linear group as we might have expected.

The Modular group Mod $\tilde{G}$ acts on the Teichmüller space $\mathbf{T}(\tilde{G})$. The element of the modular group induced by the mapping ω sends the isomorphism class $[w]$ to the isomorphism class $[w \circ \omega^{-1}]$. If w is normalized, then this automorphism of the Teichmüller space $\mathbf{T}(\tilde{G}_\imath)$ must send $\tau = w(\imath)$ onto $\hat{\tau} = \hat{w}(\imath)$, where $\hat{w} = C \circ w \circ \omega^{-1}$ and the conformal affine map C is chosen in order to normalize $\hat{w}$. Routine calculations show that this map sends the point a to the point

$$\hat{a} = \frac{a(\bar{d} + \bar{e}) - \bar{e}}{a(\bar{d} + \bar{e} - d - e) + d - \bar{e}}.$$

Translating to the τ-coordinates, we see that τ gets mapped to

$$\hat{\tau} = \frac{\delta\tau - \beta}{-\gamma\tau + \alpha} = M^{-1}(\tau),$$

where $M = \begin{bmatrix} \alpha & \beta \\ \gamma & \delta \end{bmatrix}$ is the obvious element of SL(2,$\mathbb{Z}$). It should be noted that SL(2,$\mathbb{Z}$) does not act effectively on the Teichmüller space; the matrix $-I$ acts as the identity.

The group mod $\tilde{G}$ acts on the fiber space $\mathbf{T}(\tilde{G}) \times \mathbb{C}$ by sending the point $([w], z)$ (here the map w is normalized) to the point $([\hat{w}], \hat{z})$, where

$$\hat{z} = (\hat{w} \circ \omega \circ w^{-1})(z) = C(z).$$

To compute (note that we may no longer assume that ω fixes the origin) the value of $\hat{z}$, we make a number of preliminary observations. We work with $\tau_o = \imath$, as before. The normalized map w takes the lattice point $n + m\imath$ (here n and m are integers) for the group $\tilde{G}_\imath$ to the lattice point $n + m\tau$ for the group $\tilde{G}_\tau$ and induces an "affine" map from $T_\imath$ onto T_τ. The most general ω is of the form

$$\omega(z) = dz + e\bar{z} + k + l\imath,$$

where d and e satisfy the integrality conditions previously discussed and k and l are integers. It follows that

$$\omega(0) = k + l\imath, \ \omega(1) = \gamma\imath + \delta + k + l\imath, \text{ and } \omega(\imath) = \alpha\imath + \beta + k + l\imath.$$

From these observations it is easy to compute that

$$C(z) = \frac{z - (\beta l - \alpha k) - \tau(\gamma k - \delta l)}{\alpha - \gamma\tau}.$$

We note that the map C sends $0, 1$ and τ to $k + l\hat{\tau}$, $\frac{1}{\alpha-\gamma\tau} + k + l\hat{\tau}$ and $\frac{\tau}{\alpha-\gamma\tau} + k + l\hat{\tau}$, respectively. Since $\frac{1}{\alpha-\gamma\tau} = \delta + \gamma\hat{\tau}$ and $\frac{\tau}{\alpha-\gamma\tau} = \beta + \alpha\hat{\tau}$, we see that C induces a conformal map from the torus T_τ to the torus $T_{\hat{\tau}}$.

We leave it to the reader to verify that mod $\tilde{G}$ is a semi-direct product of SL(2,$\mathbb{Z}$) with $\mathbb{Z} \oplus \mathbb{Z}$. We note that the matrix $-I$ belongs to SL(2,$\mathbb{Z}$) and acts on the fiber space as the map $(z, \tau) \mapsto (-z, \tau)$. An element of SL(2,$\mathbb{Z}$) with a fixed point induces a conformal automorphism of the surface corresponding to the fixed point. For example, the element of SL(2,$\mathbb{Z}$) of order two, corresponding to $\alpha = 0 = \delta$, $\beta = 1 = -\gamma$, yields the selfmap of order four of $T_\imath$ that sends z to $-\imath z$. It is also easy to show that the action of mod $\tilde{G}$ on $\mathbf{T}(\tilde{G}) \times \mathbb{C}$ is effective. The quotient of this action is a fiber space $\mathbf{Y}(1,0)$ over the moduli space $\mathbf{R}(1,0)$; the fiber over a point τ is the torus T_τ factored by its symmetries. Thus the generic fiber is a surface (orbifold) of type $(0, 4;\ 2, 2, 2, 2)$.

We have seen that the moduli space $\mathbf{R}(1,0)$ can be compactified to produce $\overline{\mathbf{R}(1,0)}$. Similarly, the space $\mathbf{Y}(1,0)$ can be compactified to obtain a space $\overline{\mathbf{Y}(1,0)}$; the fiber over the point at infinity in $\overline{\mathbf{R}(1,0)}$ is the torus with a node (corresponding to $t = 0$ in the plumbing construction) factored by its involution $\tilde{E}$. The involution fixes the node and two nonsingular points on the noded surface. Two of the fixed points of the generic fiber (over points in $\mathbf{R}(1,0)$) have coalesced to form the node on the singular fiber.

Problem 4.25. Can one construct a complex analytic fiber space over the moduli space $\mathbf{R}(1,0)$ such that the fiber over each point is the Riemann surface represented by the point?

5. Hyperbolic version of elliptic function theory

The calculations of the last section involved mostly euclidean geometry. Generally, it is more difficult to do calculations in hyperbolic geometry. Only for surfaces of low genera and few punctures can one more or less compute explicitly some of the interesting quantities associated with uniformizations. We proceed to illustrate the simplest hyperbolic case. Most of the material in this section is based on recent research papers and many details are omitted. It is included to orient the reader; it is independent of the rest of this book. We present it here as the motivation for the remainder of the book in the sense that we would like to use some of the ideas we present in this book in the study of more general surfaces than the ones appearing in this section and in subsequent chapters of this book.

5.1. Fuchsian representation. For some background material, consult [**4**, Chs. 9 and 10] and [**6**, Ch. IV]. Given a torus T_τ and any two points on it, we can certainly find a conformal selfmap of the torus that takes one of these points to the other. It follows from this observation that the study of tori is in some sense equivalent to the study of once punctured tori. Of course, a punctured torus has a non-abelian fundamental group. The fundamental group of the torus is the free abelian group on two generators $\mathbb{Z} \oplus \mathbb{Z}$; the fundamental group of the punctured torus is the free group on two generators $\mathbb{Z} * \mathbb{Z}$. Fix a point τ in the upper half plane. Let us consider the complex plane punctured at the lattice points:

$$\mathbb{C}_\tau = \mathbb{C} - \{n + m\tau;\ n \text{ and } m \in \mathbb{Z}\} = \mathbb{C} - L(\tilde{G}_\tau).$$

Let us choose a holomorphic universal covering map

$$h : \mathbb{H}^2 \to \mathbb{C}_\tau \tag{1.6}$$

and a lift G for the action of $\tilde{G} = \tilde{G}_\tau$ on $\mathbb{C}_\tau$; that is,

$$G = \{g \in \text{ PSL}(2, \mathbb{R});\ h \circ g = \tilde{g} \circ h \text{ for some } \tilde{g} \in \tilde{G}\}.$$

We note that $\mathbb{C}_\tau$ is invariant under the action of $\tilde{G}$ and that $\mathbb{C}_\tau/\tilde{G}$ is the punctured torus $T_\tau - \{0\}$. Since $\mathbb{H}^2/G$ is conformally equivalent to $\mathbb{C}_\tau/\tilde{G}$, we have constructed a Fuchsian group representing the torus T_τ punctured at the origin (the same point that we distinguished to get a group structure for T_τ). We choose a base point $z_o \in \mathbb{H}^2$ with $h(z_o) = \frac{1+\tau}{2}$ and a lift[24] $\omega = \omega(G)$ of $\tilde{\omega} = \omega(\tilde{G})$ that contains the point z_o. See Figure 4. Then ω is a fundamental domain for the action of G on $\mathbb{H}^2$. It is a topological rectangle with vertices on $\mathbb{R} \cup \{\infty\}$, the "boundary of hyperbolic space" $\mathbb{H}^2$. The vertices are fixed points of parabolic elements of the group G and are mapped by h onto the four vertices of $\tilde{\omega}$. Since the map h is continuous on the union of the upper half plane with the parabolic fixed points of G (these parabolic fixed points are dense in $\hat{\mathbb{R}} = \mathbb{R} \cup \{\infty\}$), we may replace G by a conjugate group (conjugation takes place in PSL(2,$\mathbb{R}$)) and hence replace h by $h \circ C$ for some $C \in$ PSL$(2, \mathbb{R})$, and then assume that we have

$$h(0) = 0,\ h(1) = 1,\ h(1 + a) = 1 + \tau,\ h(\infty) = \tau,$$

for some positive $a \in \mathbb{R}$. The sides of ω are pairwise identified by hyperbolic motions[25] A and B in G that satisfy

$$A(\infty) = 1 + a,\ A(0) = 1,\ B(0) = \infty,\ B(1) = 1 + a$$

and thus also

$$h \circ A = \tilde{A} \circ h,\ h \circ B = \tilde{B} \circ h.$$

[24] The set $h^{-1}(\tilde{\omega})$ contains infinitely many components that are permuted by the group H, defined below. We are choosing one component of this set.

[25] The standard usage in this book of the symbol B is different. Current usage is limited to this section only.

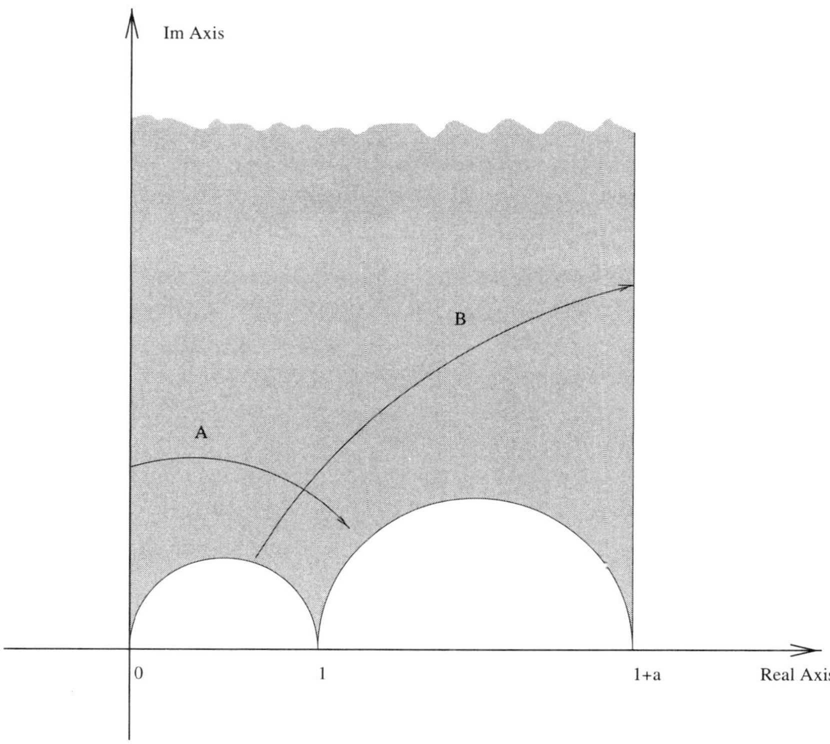

Figure 4. Fundamental domain for $G(a, b)$.

It is convenient, for our current purposes, to represent Möbius transformations fixing the upper half plane $\mathbb{H}^2$ by matrices in $\mathrm{PGL}(2,\mathbb{R})^+$, that is, by 2 by 2 real matrices with positive determinant. In this context the matrix $M = \begin{bmatrix} a & b \\ c & d \end{bmatrix}$ in $\mathrm{PGL}(2,\mathbb{R})^-$, with negative determinant, will be viewed as either the orientation reversing automorphism

$$M(z) = \frac{a\bar{z} + b}{c\bar{z} + d}$$

of $\mathbb{H}^2$, or the orientation preserving map

$$M(z) = \frac{az + b}{cz + d}$$

sending $\mathbb{H}^2$ onto the lower half plane $\mathbb{H}^2_*$. The context should make it clear which meaning is intended. With these conventions in mind it is easy to see that

$$A = \begin{bmatrix} (1+a)b & 1 \\ b & 1 \end{bmatrix}, \quad B = \begin{bmatrix} 1+(1+a)c & -1 \\ c & 0 \end{bmatrix}$$

for positive real numbers b, c. The element $P = B^{-1} \circ A^{-1} \circ B \circ A$ belongs to the covering group of h (it is mapped onto the identity by the natural homomorphism from G onto $\tilde{G}$) and fixes zero. It must be parabolic (because the curve in $\mathbb{C}_\tau$ corresponding to the motion P is simple and has winding number one about the origin) and hence has trace (as an element of $\mathrm{SL}(2, \mathbb{R})$) minus two (rather than plus two). Thus we see that $c = b$ (which will be assumed until further notice) and hence

$$P = \begin{bmatrix} -1 & 0 \\ -2(1 + b + ab) & -1 \end{bmatrix}.$$

The group G we have constructed above depends on two real positive parameters a and b. We will hence denote it by the symbol $G(a, b)$. Let us make a number of observations.

Since G is the covering group of a punctured torus which is isomorphic to the fundamental group of a punctured torus, we see that G is isomorphic to $\mathbb{Z} * \mathbb{Z}$; the normalizer H (smallest normal subgroup of G containing all of the parabolic elements of G, which can also be described as

$$H = \{g \in \ \mathrm{PSL}(2, \mathbb{R});\ h \circ g = h\},$$

is infinitely generated and isomorphic to the fundamental group of $\mathbb{C}_\tau$, and we have a short exact sequence

$$1 \to H \to G \to \tilde{G} \to 1.$$

The parallelogram ω (because of the way it has been constructed) is not in general bounded by geodesics. It can be replaced (we assume from now on that this substitution has been made) by a parallelogram with the same vertices (∞, 0, 1, and $1 + a$) but bounded by geodesics. Now any four points on the extended real axis determine a geodesic quadrilateral. If we introduce motions A and B that (correctly with regard to orientation)[26] identify the opposite sides of the quadrilateral, then these motions always generate a Fuchsian group; the group represents a punctured torus if and only if the commutator of A and B is parabolic. In this case, the quadrilateral is a fundamental domain for the group generated by the two motions. We conclude from these remarks that if we start with any pair of positive numbers a and b and if we use the above formulae to define fractional linear transformations A and B, then the group $G(a, b)$ generated by these motions represents a punctured torus and that every punctured torus is so

[26]For a and b points on $\mathbb{R} \cup \{\infty\}$, we let $\overline{a, b}$ denote the oriented geodesic in $\mathbb{H}^2$ from a to b; it is the half circle passing through a and b orthogonal to the real axis. A circle through ∞ is a straight line. The correct orientation requirement reads

$$A(\overline{\infty, 0}) = \overline{1 + a, 1} \text{ and } B(\overline{0, 1}) = \overline{\infty, 1 + a}.$$

represented. Hence the ordered pair (a, b) gives global parameters for the space of marked punctured tori.

It is important to observe that the parameters a and b that describe the group $G(a, b)$ can be defined in terms of (fixed points of words in) the generators of the group in a conjugacy invariant form. For a parabolic motion C, we let $f(C)$ denote the fixed point of C. It follows that

$$0 = f(P), \ \infty = B(0) = f(A^{-1} \circ B \circ A \circ B^{-1}),$$

$$1 = A(0) = f(A \circ B^{-1} \circ A^{-1} \circ B), \ 1 + a = A(\infty) = f(A \circ B^{-1} \circ A^{-1} \circ B).$$

Let us also introduce the cross ratio function[27]

$$\operatorname{cr}(z, u, v, x) = \frac{z - v}{z - u} \frac{x - u}{x - v}.$$

Then

$$1 + a = \operatorname{cr}(A(B(f(P))), B(f(P)), f(P), A(f(P))).$$

Since in our calculations

$$A^{-1}(\infty) = -\frac{1}{b},$$

we conclude that

$$-\frac{1}{b} = \operatorname{cr}(A^{-1}(B(f(P))), B(f(P)), f(P), A(f(P))).$$

5.2. Symmetries of once punctured tori. We would like to determine an explicit relation (formula)[28] between the parameter τ for the Teichmüller space $\mathbf{T}(1, 0)$ of tori and the parameters a and b for the Teichmüller space $\mathbf{T}(1, 1)$ of once punctured tori. We use the various symmetries of the torus T_τ to obtain information on the corresponding punctured torus $\mathbb{H}^2/G(a, b)$. We observe that each of the symmetries of a torus $(\tilde{E}, \tilde{J}, \tilde{J}_i, \ i = 1, 2, 3)$ discussed in the previous section is also a symmetry of the corresponding punctured torus. Let us begin by lifting $\tilde{E}$ to obtain E. Since the involution $\tilde{E}$ fixes $\frac{1+\tau}{2}$, the involution E must fix z_o, and from the fact that

$$\tilde{E}(0) = 1 + \tau, \ \tilde{E}(1) = \tau,$$

we conclude that

$$E(0) = 1 + a, \ E(1) = \infty.$$

This is enough to determine the involution uniquely:

$$E = \begin{bmatrix} 1 & -1 - a \\ 1 & -1 \end{bmatrix}.$$

[27] We fix once and for all this particular cross ratio; thus $\operatorname{cr}(z, u, v, x)$ is the value $A(z)$ at z of the unique Möbius transformation A that sends u to ∞, v to 0 and x to 1. The other cross ratios are obtained by following this one by an element of Perm(3).

[28] It is not possible to get an algebraic formula.

It is now trivial to check that

$$E \circ A \circ E = A^{-1}, \; E \circ B \circ E = B^{-1}.$$

Since the (unique) fixed point of E in the upper half plane must project under h onto $\frac{1+\tau}{2}$, it follows that

$$z_o = 1 + \imath\sqrt{a}.$$

The above point projects under h onto one of the three (nonlattice point) half-periods of the torus T_τ. It is easy to locate pre-images of the other two half-periods. The fixed points

$$\frac{\imath}{\sqrt{b}} \text{ and } \frac{1+\imath\sqrt{ab}}{1+ab}$$

of the conformal involutions $E \circ A$ and $E \circ B$ are mapped by h onto $\frac{\tau}{2}$ and $\frac{1}{2}$, respectively.

We study the family of tori defined by $\Re\tau = 0$. The lift J of the involution $\tilde{J}$ that fixes the base point z_o, must be an involution that satisfies

$$J(0) = \infty, \; J(1) = 1 + a.$$

It must hence be given by

$$J = \begin{bmatrix} 0 & 1+a \\ 1 & 0 \end{bmatrix}.$$

Since J must commute with A, we conclude that

$$b = \frac{1}{1+a}.$$

It is now easy to check that under these conditions J conjugates B to its inverse, as expected. The second family of tori to study is defined by the equation $|\tau| = 1$. The involution J_1 must satisfy

$$J_1(0) = 0, \; J_1(1) = \infty, \; J_1(1+a) = 1 + a.$$

The first two conditions force

$$J_1 = \begin{bmatrix} 1 & 0 \\ 1 & -1 \end{bmatrix}.$$

The last condition forces the defining relation $a = 1$. It now trivial to check that J_1 conjugates A to B as required. The lifts of the family of tori with $\Re\tau = \frac{1}{2}$ have involutions J_2 that satisfy

$$J_2(0) = 1, \; J_2(\infty) = \infty, \; J_2(1+a) = A^{-1}(\infty) = -\frac{1}{b}.$$

The first two conditions force

$$J_2 = \begin{bmatrix} -1 & 1 \\ 0 & 1 \end{bmatrix}.$$

The last condition forces the defining relation $ab = 1$. We now check that

$$J_2 \circ A \circ J_2 = A^{-1},\ J_2 \circ B \circ J_2 = A^{-1} \circ B$$

as expected. The last family of tori that we will consider is defined by $\Re\tau = -\frac{1}{2}$. The tori in this family have involutions J_3 that satisfy

$$J_3(0) = 0,\ J_3(\infty) = 1 + a,\ J_3(1) = A^{-1}(0) = -\frac{1}{(1+a)b}.$$

The first two conditions force

$$J_3 = \begin{bmatrix} 1+a & 0 \\ 1 & -1-a \end{bmatrix}.$$

The last condition forces the defining relation

$$b = \frac{a}{(1+a)^2}.$$

It is easy to check that

$$J_3 \circ A \circ J_3 = A^{-1},\ J_3 \circ B \circ J_3 = B \circ A$$

as expected.

The above calculations are best summarized in a table.

τ − region	involution on torus	(a,b) − region	involution on punctured torus
$\Re\tau = 0$	$z \mapsto \bar{z} + \tau$	$b = \frac{1}{1+a}$	$z \mapsto \frac{1+a}{\bar{z}}$
$\lvert\tau\rvert = 1$	$z \mapsto \tau\bar{z}$	$a = 1$	$z \mapsto \frac{\bar{z}}{\bar{z}-1}$
$\Re\tau = \frac{1}{2}$	$z \mapsto -\bar{z} + 1$	$ab = 1$	$z \mapsto -\bar{z} + 1$
$\Re\tau = -\frac{1}{2}$	$z \mapsto -\bar{z}$	$b = \frac{a}{(1+a)^2}$	$z \mapsto \frac{(1+a)\bar{z}}{\bar{z}-1-a}$

The involutions of the tori used above are characterized by the fact that they fix the fundamental domain $\tilde{\omega}$, and equivalently by the fact that they fix the half-period $\frac{1+\tau}{2}$.

5.3. The modular group. We now have enough information to determine a fundamental domain for the action on $\mathbf{T}(1,1)$ of the modular group Mod(1,1). This modular group identifies two points in $\mathbf{T}(1,1)$ if and only if they represent conformally equivalent punctured tori. The fundamental region, in our coordinates, is described by

$$\left\{(x,y) \in \mathbb{R}^2;\ x > 1,\ \frac{x}{(1+x)^2} < y < \frac{1}{x}\right\}.$$

We can also describe the action of generators of the modular group Mod(1,1) on the Teichmüller space $\mathbf{T}(1,1)$. Translation by 1 in the τ-plane is generated by the (group-) automorphism of $\tilde{G}$ that fixes $\tilde{A}$ and sends $\tilde{B}$ to $\tilde{A} \circ \tilde{B}$. We consider the map that sends A to itself and B to $A \circ B$. It defines an *allowable* (in the sense that it sends parabolic elements of the group onto

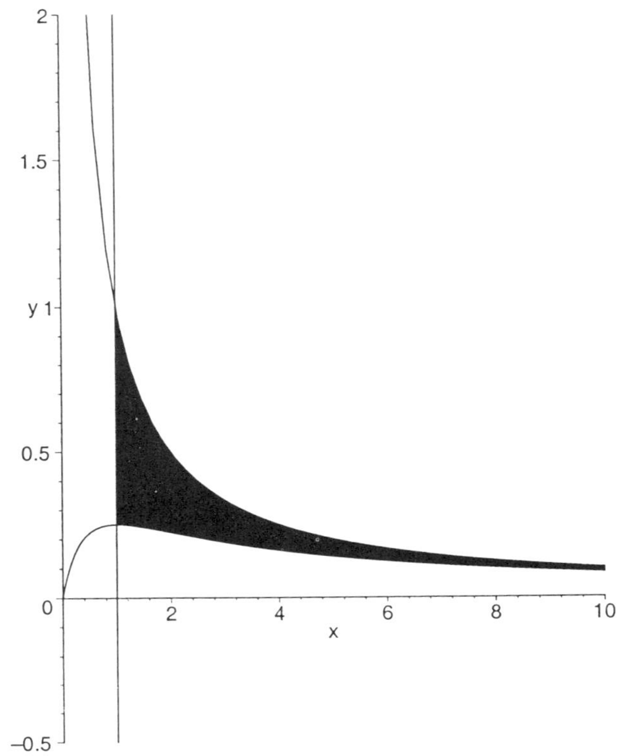

Figure 5. Fundamental domain for action of Mod(1, 1) **on** $\mathbf{T}(1,1)$.

parabolic elements) automorphism of the group $G(a,b)$. Let $(\hat{a},\hat{b})$ be the coordinates of the new (normalized) group. Since the parabolic element P is preserved by the automorphism, we have

$$1+\hat{a} = \mathrm{cr}(A((A\circ B)(0)),(A\circ B)(0),0,1)$$

$$= \mathrm{cr}(A(1+a),1+a,0,1) = 1+(1+a)^2 b$$

and

$$-\frac{1}{\hat{b}} = \mathrm{cr}(A^{-1}((A\circ B)(0)),1+a,0,1) = -a.$$

Thus

$$(\hat{a},\hat{b}) = \left((1+a)^2 b, \frac{1}{a}\right).$$

It should be observed that this automorphism does send the curve $b=\frac{a}{(1+a)^2}$ to the curve $ab=1$, avoiding, once again, an internal contradiction in mathematics. The automorphism of the τ-plane given by $\tau\mapsto -\frac{1}{\tau}$ corresponds to the element of Mod(1,1) induced by the automorphism of $G(a, b)$ that sends A to B and B to A^{-1}. The induced action on $\mathbf{T}(1, 1)$ is given by

$$(a, b) \mapsto \left(\frac{1}{a}, \frac{a}{b(1+a)^2}\right).$$

We note that this map has period two, leaves the line $a = 1$ invariant and fixes the point $(1, \frac{1}{2})$. The composite of the last two maps, the order two map followed by the translation, is the map

$$(a, b) \mapsto \left(\frac{1}{ab}, a\right);$$

it has period three and fixes the point (1,1). The "square" torus is thus represented by the point $\left(1, \frac{1}{2}\right)$ and the "rhombic " torus by $(1, 1)$. Thus we have a rather complete picture of the action of the modular group on the Teichmüller space for surfaces of finite analytic type $(1, 1)$.

5.4. Geometric interpretations. We have constructed intrinsic (conjugation invariant) real-analytic coordinates for the Teichmüller space $\mathbf{T}(1, 1)$. What is the geometric significance of these coordinates? Perhaps the most important invariant attached to an element of PSL(2,$\mathbb{C}$) is its trace (see §1), tr, which is defined up to a plus or minus sign. For elements of PSL(2,$\mathbb{R}$), we can always choose the trace to be non-negative. A second invariant is the square root of the multiplier, κ. For hyperbolic transformations, we can always choose the multiplier to be bigger than one. In general, we have the relation

$$\kappa + \frac{1}{\kappa} = \text{tr}.$$

The above formula serves to define the multiplier for parabolic elements to be one. If C is a hyperbolic element of a Fuchsian group G acting on $\mathbb{H}^2$, then the length of the geodesic in the free homotopy class on $\mathbb{H}^2/G$ corresponding to the motion C is $2 \log\ \kappa(C)$. In the case under investigation here

$$\text{tr}(A) = \sqrt{ab} + \sqrt{\frac{b}{a}} + \frac{1}{\sqrt{ab}} \text{ and } \text{tr}(B) = \sqrt{b} + \sqrt{\frac{1}{b}} + a\sqrt{b}.$$

Thus the only way for the geodesic corresponding to A to shrink is for ab to approach one and for $\frac{b}{a}$ to approach zero. This means that b must approach zero and hence a must approach infinity. The matrices in PSL(2, $\mathbb{R}$) corresponding to A converge to the translation by one and those corresponding to B diverge. Under these conditions $\text{tr}(B)$ approaches infinity. Thus we see that as the curve corresponding to A shrinks, the transverse curve (corresponding to B) becomes infinitely long. Note that both curves can become infinitely long simultaneously (for example, let a go to one and b go to zero). We note also that $\text{tr}(A) = \text{tr}(B)$ if and only if $a = 1$. Hence the marked tori whose canonical curves have equal length correspond to marked punctured tori with geodesics arising from the generators also having equal length. From these remarks it would appear that we could get a good map from $\mathbf{T}(1,1)$ to $\mathbb{R}^2$ by sending the pair (a,b) to the pair $(\mathrm{tr}(A), \mathrm{tr}(B))$. Simple calculations show that the Jacobian of this map vanishes if and only if $(1+a)b = 1$. Thus the two geodesic lengths do not provide even local coordinates in any neighborhood of a marked punctured torus corresponding to a purely imaginary τ. Another straightforward calculation shows that for any pair of Möbius transformations A and B, we have

$$\mathrm{tr}(A)^2 + \mathrm{tr}(B)^2 + \mathrm{tr}(B \circ A)^2 - \mathrm{tr}(A)\mathrm{tr}(B)\mathrm{tr}(B \circ A) - 2 = \mathrm{tr}([B, A]),$$

where $[B, A]$ is the commutator

$$[B, A] = B^{-1} \circ A^{-1} \circ B \circ A.$$

In our case, the commutator is parabolic and hence has trace minus two, so that

$$x = \mathrm{tr}(A), \;\; y = \mathrm{tr}(B), \;\; z = \mathrm{tr}(B \circ A) \tag{1.7}$$

satisfy the *trace equation*

$$x^2 + y^2 + z^2 = xyz.$$

Lengthy calculations show that the matrices A and B with invariants defined by (1.7) that solve the trace equation may be defined as elements of $\mathrm{PSL}(2, \mathbb{R})$ by the formulae

$$A = \begin{bmatrix} \frac{z}{y} & x - \frac{z}{y} \\ \frac{x^2}{y(xy-z)} & x - \frac{z}{y} \end{bmatrix} \text{ and } B = \begin{bmatrix} y & -y + \frac{z}{x} \\ \frac{x}{xy-z} & 0 \end{bmatrix}.$$

We note that the old parameters are given by

$$a = \left(\frac{y}{x}\right)^2 \text{ and } b = \frac{x^2}{(xy-z)^2}.$$

Thus we see that the traces of the three motions A, B and $B \circ A$ are conjugacy class invariants for the group $G(a, b)$. Hence an alternate description of the Teichmüller space $\mathbf{T}(1,1)$ is provided by the "semi-algebraic" set

$$\{(x, y, z) \in \mathbb{R}^3; \; x > 2, \; y > 2, \; z > 2 \text{ and } x^2 + y^2 + z^2 = xyz\}.$$

It is of interest to investigate the behavior of the groups and surfaces as one or more of the components of (x, y, z) approach 2.

5.5. The period of a punctured torus. Some of the concepts encountered in this and the next two subsections will be defined in later chapters. Almost all nontrivial claims made are very difficult to verify without a great deal of work.

The square of the derivative of a universal covering map h, equation (1.6) of §5.1, of the plane punctured at the lattice points is a cusp form for the group $G = G(a,b)$; that is, it projects to a quadratic differential on the surface $\mathbb{H}^2/G(a,b) \cong T_\tau - \{0\}$, without zeros or poles on T_τ. The derivative h' projects to a holomorphic one form, without zeros, on the torus. To recover the period τ from the group G, one can proceed as follows. Form the Poincaré series (see 1.2 of Chapter 3)

$$\Phi(z) = \sum_{\gamma \in G} \varphi(\gamma(z))\gamma'(z)^2,$$

where

$$\varphi(z) = \frac{a(1+a)}{z(z-1)(z-1-a)}, \quad z \in \mathbb{H}^2.$$

The function Φ is defined because the series converges uniformly and absolutely on compact subsets of $\mathbb{H}^2$. From

$$\Phi(\gamma(z))\gamma'(z)^2 = \Phi(z), \text{ for all } z \in \mathbb{H}^2 \text{ and all } \gamma \in G,$$

and

$$\int\int_{\mathbb{H}^2/G} \left| \Phi(z) \frac{dz\, \overline{dz}}{2} \right| \leq \int\int_{\mathbb{H}^2} \left| \varphi(z) \frac{dz\, \overline{dz}}{2} \right| < \infty$$

one concludes that the projection of $\Phi(z)dz^2$ to $\mathbb{H}^2/G$ is regular on the compactified surface $T_\tau = \mathbb{H}^2/G \cup \{\text{puncture}\}$. The only issue is to show that Φ is not identically zero. This can be shown using either the variational theory of moduli or Eichler cohomology. It follows that $\sqrt{\Phi(z)}dz$ projects to a holomorphic one form on T_τ.[29] Choose a holomorphic function Ψ so that

$$\Psi' = \sqrt{\Phi}.$$

Then there exist constants c_B and c_A such that for all $z \in \mathbb{H}^2$,

$$\Psi(B(z)) - \Psi(z) = c_B \text{ and } \Psi(A(z)) - \Psi(z) = c_A.$$

It follows that

$$h(z) = \frac{\Psi(z)}{c_A}, \text{ for all } z \in \mathbb{H}^2, \text{ and hence } \tau = \frac{c_B}{c_A}.$$

[29]It is trivial that every function η with $\eta^2 = \Phi$ satisfies

$$\eta \circ g = c_g \eta \text{ for all } g \in G, \text{ with } c_g = \pm 1;$$

only slightly more work is needed to prove that $c_g = 1$ for all $g \in G$. Thus $\eta(z)dz$ projects to an abelian differential of the first kind (rather than a Prym differential) on T_τ.

5.6. The function of degree two on the once punctured torus. In order to find the algebraic curve associated to a punctured torus, we may use the value of τ produced in the previous subsection to compute the Weierstrass $\wp$-function or determine directly a function of degree two on $\mathbb{H}^2/G$. Towards this latter end, we need to produce a holomorphic quadratic differential for G whose projection to $\mathbb{H}^2/G$ has a double pole at the puncture. Form the relative Poincaré series

$$\Phi_1(z) = \sum_{\gamma \in <P> \backslash G} \frac{\gamma'(z)^2}{\gamma(z)^4}, \; z \in \mathbb{H}^2.$$

The automorphic form Φ_1 behaves differently from the form Φ of the previous subsection. The function Φ is a cusp form, and Φ_1 is only a modular form.[30] By computing the divisors of the projections of Φ and Φ_1 to the torus, we conclude that $\mathcal{P} = \frac{\Phi_1}{\Phi}$ defines a function of degree 2 on T_τ with a double pole at the origin. Since $\mathcal{P}$ must be an affine function of the Weierstrass $\wp$-function, the λ invariant for the torus is recovered as

$$\lambda = \frac{\mathcal{P}(x_3) - \mathcal{P}(x_2)}{\mathcal{P}(x_1) - \mathcal{P}(x_2)},$$

where x_1, x_2, x_3 are the fixed points in $\mathbb{H}^2$ of the conformal involutions $E \circ B$, $E \circ A$ and E, respectively (thus $x_1 = \frac{1+\imath\sqrt{ab}}{1+ab}$, $x_2 = \frac{\imath}{\sqrt{b}}$, and $x_3 = 1 + \imath\sqrt{a}$). Thus also the $\jmath$ invariant for the torus can be computed from the covering group of the corresponding punctured torus.

5.7. The quasi-Fuchsian representation. We have seen that for each ordered pair of positive real numbers a and b, we can construct a unique marked Fuchsian group $G(a, b)$ that represents a punctured torus. An obvious question appears as a result of this construction. What happens if we try to extend the construction to the complex case? This will lead us to quasi-Fuchsian (obtained from Fuchsian groups by conjugation by a quasiconformal selfmap of the sphere) groups that represent two arbitrary punctured tori rather than a punctured torus and its mirror image as in the case with Fuchsian groups. It may appear surprising that this case already presents many new aspects on uniformization theory. For example, the exact region in $\mathbb{C}^2$ corresponding to quasi-Fuchsian groups representing punctured tori is not known. See the bibliographical notes for a discussion of some related recent developments.

Our development in this chapter leads us to many other interesting questions. For example, what is the picture of the most general (torsion free) group of Möbius transformations that represents a punctured torus? What

[30]The difference is NOT accounted for by the fact that one is a Poincaré series and the other a relative Poincaré series. Different methods are developed to handle the convergence of the series involved in the definition of these two functions.

does the space of two generator discrete groups look like? The answers to these two questions are beyond the scope of this enterprise.

6. Subgroups of the modular group

We begin a discussion of the basic objects in our study: subgroups of $\Gamma = \mathrm{PSL}(2,\mathbb{Z})$. The material we present is part of a well known and long story. Details may be found in the early chapters of [**18**] and [**27**]. The information is also presented to establish notation for subsequent chapters.

6.1. Basic properties. By definition, a Fuchsian group is a group of Möbius transformations G that leaves invariant a proper projective disc $\Delta \subset \hat{\mathbb{C}}$. Replacing G by a conjugate group, if necessary, we can always assume that Δ is one of the following two standard discs: either the upper half plane

$$\mathbb{H}^2 = \{z \in \mathbb{C} : \Im z > 0\},$$

in which case $G \subset \mathrm{PSL}(2,\mathbb{R})$, or the unit disc

$$\mathbb{D} = \{z \in \mathbb{C} : |z| < 1\},$$

in which case the elements of G are of the form $\begin{bmatrix} a & b \\ \bar{b} & \bar{a} \end{bmatrix}$ with a and $b \in \mathbb{C}$, $|a|^2 - |b|^2 = 1$.

Proposition 6.1. *Let $G \subset \mathrm{PSL}(2,\mathbb{C})$ leave invariant the proper disc $\Delta \subset \hat{\mathbb{C}}$. The following conditions are equivalent:*
(a) *The group G is discrete.*
(b) *The group G is Fuchsian.*
(c) *The set $\Delta \subset \Omega(G)$.*
(d) *The group G acts discontinuously at some point $z_o \in \Delta$.*

Proof. See, for example, [**18**, Ch. I, Th. 1.1].

Theorem 6.2. *Let G be a Fuchsian group with invariant component Δ. Then G is finitely generated of the first kind if and only if the orbifold Δ/G is of finite conformal type; this means that Δ/G is a finitely punctured compact Riemann surface and the natural projection $\pi : \Delta \to \Delta/G$ is branched over finitely many points.*

Proof. See, for example, [**18**, Ch. II, Th. 3.2].

Definition 6.3. Assume that G is a finitely generated Fuchsian group of the first kind acting on the disc Δ. Let $\overline{\Delta/G}$ be the compactification of Δ/G, p its genus, and n' the number of punctures on Δ/G. Let x_1, ..., $x_{n''}$ be a maximal set of G-inequivalent branch points of π (these are the fixed points of nontrivial elements of G). Set $\mu_j = |G_{x_j}|$. Without loss of generality, we may assume that

$$\mu_1 \leq \ \dots, \ \leq \mu_{x_{n''}}.$$

Let $n = n' + n''$. The $(n+2)$-tuple

$$(p, n; \ \mu_1, \ \dots, \ \mu_{x_{n''}}, \ \infty, \ \dots, \ \infty)$$

is the *signature* of G and (p, n) is its *type*. The *distinguished* points on Δ/G or $\overline{\Delta/G}$ are its punctures and the branch values of π.

6.2. Poincaré metric on simply connected domains. For this case and generalizations, see [**18**, Ch. II]. Let Ω be a proper simply connected domain in $\mathbb{C}$. We are interested in defining a metric, the *Poincaré metric* on this domain that is invariant under its full automorphism group. We start by describing this (essentially unique) metric on $\mathbb{H}^2$. The metric $ds = \lambda(z)|dz| = \lambda_\Omega(z)|dz|$ on Ω is defined by requiring that it be a conformal invariant. We proceed to describe the construction. We set

$$\lambda_{\mathbb{H}^2}(z) = \frac{1}{\Im z}, \ z \in \mathbb{H}^2.$$

It follows by direct calculation that for all $A \in \mathrm{PSL}(2, \mathbb{R})$,

$$\lambda_{\mathbb{H}^2}(z) = \lambda_{\mathbb{H}^2}(A(z))|A'(z)|, \ z \in \mathbb{H}^2. \tag{1.8}$$

Choose a Riemann map

$$h : \mathbb{H}^2 \to \Omega. \tag{1.9}$$

The map h is not unique; it may be replaced by $h \circ A$ with $A \in \mathrm{PSL}(2, \mathbb{R})$. However, as a consequence of (1.8),

$$\lambda_\Omega(h(z))|h'(z)| = \lambda_{\mathbb{H}^2}(z), \ z \in \mathbb{H}^2, \tag{1.10}$$

yields a well defined function[31] λ_Δ on Δ that is invariant under $\mathrm{Aut}(\Omega)$ in the sense that

$$\lambda_\Omega(\gamma(z))|\gamma'(z)| = \lambda_\Omega(z), \text{ for all } z \in \Omega \text{ and all } \gamma \in \ \mathrm{Aut}(\Omega).$$

Let G be a Kleinian group that leaves invariant a simply connected region Ω. Let

$$\pi : \Omega \to \Omega/G$$

be the natural projection. The invariance properties of λ_Ω show that we can also define a metric with singularities at the distinguished points on Ω/G by projecting down the metric from Ω. Thus we have well defined concepts of

[31]It follows that

$$\lambda_{\mathbb{D}}(z) = \frac{2}{1 - |z|^2}, \ z \in \mathbb{D}.$$

length and area not just on Ω but also on Ω/G. If $c : [0,1] \to \Omega$ is a curve such that π is injective on c, then the *length* of $\pi(c)$ is defined by

$$l(\pi(c)) = \int_0^1 \lambda(c(t))|c'(t)|dt.$$

Often one considers a curve c in Ω precisely invariant under a cyclic subgroup (say with generator A). Then on the surface, we have

$$l(c/ < A >) = \int_{c/<A>} \lambda(z)|dz|.$$

If $D \subset \Omega$ is G-invariant, then the *area* of $\pi(D) = D/G$ is given by

$$\text{Area}(D/G) = \int\int_{D/G} \lambda_\Omega^2(z)\frac{|dz\ d\bar{z}|}{2}.$$

In particular, the area of the orbifold Ω/G is

$$\text{Area}(\Omega/G) = \int\int_{\Omega/G} \lambda_\Omega^2(z)\frac{|dz\ d\bar{z}|}{2}.$$

Equation (1.10) defines a metric on Ω as soon as (1.9) is a covering map; Ω need not be simply connected. It is important, for many applications,[32] to obtain various estimates for λ. Let δ be the distance to the boundary; that is,

$$\delta(z) = \delta_\Omega(z) = \ \inf\ \{|z-\zeta|; \zeta \in \mathbb{C} - \Omega\},\ z \in \Omega.$$

If $\Omega \subset \mathbb{C}$, then

$$\frac{1}{2} \leq \lambda\delta \leq 2;$$

the first inequality is a consequence of the Koebe $\frac{1}{4}$-thecrem, and the second inequality is a consequence of the monotonicity property of λ. If we have an inclusion of simply connected domains on the Riemann sphere

$$\Omega_1 \subset \Omega_2 \subset \hat{\mathbb{C}},$$

then Schwarz's lemma yields

$$\lambda_{\Omega_2}(z) \leq \lambda_{\Omega_1}(z),\ z \in \Omega_1.$$

For $\mathbb{H}^2$ there is a nice interplay between the geometry of its Poincaré metric and its automorphism group $\text{PSL}(2,\mathbb{R})$. Each point $\xi \in \hat{\mathbb{R}}$ determines two foliations of $\mathbb{H}^2$. The first is the *horocyclic foliation.* The *horocycles* are the family of circles (*horocircles*) in $\mathbb{H}^2$ tangent to $\hat{\mathbb{R}}$ at ξ. Each horocircle bounds two *horodiscs.* Each horocircle and each horodisc is invariant under the one parameter group G of parabolic motions in $\text{PSL}(2,\mathbb{R})$ fixing ξ. For

[32]Not for our purposes.

example: If $\xi = \infty$, then the family of horocircles $\{C_y\}$ is parametrized by $y \in \mathbb{R}^+$, and

$$C_y = \{z \in \mathbb{H}^2;\ \Im z = y\};$$

the corresponding group consists of the motions

$$G = \{z \mapsto z + b;\ b \in \mathbb{R}\}.$$

The second is the *geodesic foliation* consisting of maximal geodesics in $\mathbb{H}^2$ terminating at ξ. Each geodesic and each of the two half planes it determines is invariant under the group of motions in $\mathrm{PSL}(2,\mathbb{R})$ fixing or interchanging the end points of the geodesic. (This is a $\mathbb{Z}_2$ extension of the one parameter group of hyperbolic motions fixing the end points.) For example: If $\xi = \infty$, then the family of geodesics $\{C_x\}$ is parametrized by $x \in \mathbb{R}$, where

$$C_x = \{z \in \mathbb{H}^2;\ \Re z = x\}.$$

The corresponding family of groups is

$$G_x = \left\{z \mapsto \lambda(z-x) + x \text{ and } z \mapsto \frac{1}{\lambda(z-x)} + x;\ \lambda \in \mathbb{R}^+ - \{1\}\right\}.$$

6.3. Fundamental domains. There are many constructions of fundamental domains for Fuchsian groups. We will need the following

Theorem 6.4. *Let G be a Fuchsian group operating on the disc Δ. The following are equivalent:*
(a) *G is finitely generated of the first kind.*
(b) *There exists a fundamental domain for G that is finite sided with no sides on the boundary of Δ.*
(c) $Area(\Delta/G) < \infty$.

We are studying, in this section, subgroups G of the modular group $\Gamma = \mathrm{PSL}(2,\mathbb{Z})$ acting as Möbius transformations on the upper half plane $\mathbb{H}^2$. The generators of $\mathrm{SL}(2,\mathbb{Z})$ (we use the same symbols for a matrix and the Möbius transformation it induces)[33] are

$$A = \begin{bmatrix} 0 & -1 \\ 1 & 0 \end{bmatrix} \text{ and } B = \begin{bmatrix} 1 & 1 \\ 0 & 1 \end{bmatrix}.$$

Let G be a subgroup of finite index $\mu = \mu(G)$ of Γ. Alternately, we may describe G as a finitely generated Fuchsian group of the first kind contained in Γ. The index $\mu = [\Gamma : G]$ of G in Γ is given by the formula

$$\mu = \frac{\mathrm{Area}(\mathbb{H}^2/G)}{\mathrm{Area}(\mathbb{H}^2/\Gamma)} = 6\left(2p - 2 + \sum_{j=1}^{n}\left(1 - \frac{1}{\mu_j}\right)\right) = 6\chi(G),$$

[33] An element of even order ν in Γ lifts to two elements of order 2ν in $\mathrm{SL}(2,\mathbb{Z})$; for the lifts of all other (excluding I) elements, orders are preserved.

where $(p, n;\ \mu_1,\ \ldots,\ \mu_n)$ is the signature of G. The above formula defines the *(negative) Euler characteristic* $\chi(G)$ of G.

Corollary 6.5. *If G satisfies the conditions of the last theorem, then (using the notation introduced above)*

$$\text{Area}(\Delta/G) = 2\pi\left(2p - 2 + \sum_{j=1}^{n}\left(1 - \frac{1}{\mu_j}\right)\right) = 2\pi\chi(G).$$

Since Γ contains elliptic elements of orders 2 and 3 only, it make sense to let:

$\nu_2 = \nu_2(G)$ be the number of G-inequivalent elliptic fixed points (in $\mathbb{H}^2$) of order 2,
$\nu_3 = \nu_3(G)$ be the number of G-inequivalent elliptic fixed points (in $\mathbb{H}^2$) of order 3 and
$\nu_\infty = \nu_\infty(G)$ be the number of G-inequivalent parabolic fixed points (in $\mathbb{Q} \cup \{\infty\}$).

Obviously, ν_i is the number of conjugacy classes in G of elliptic subgroups of order $i = 2,\ 3$, and ν_∞ is the number of conjugacy classes in G of maximal (cyclic) parabolic subgroups. While ν_2 and ν_3 are only nonnegative, ν_∞ must be positive (because of the finite index assumption).

Slightly at variance with [**12**], we will call the quadruple $\{\mu;\ \nu_2,\ \nu_3,\ \nu_\infty\}$ the *branch schema* for G. This quadruple is a topological invariant for the orbifold $\mathbb{H}^2/G$; its genus $p = p(G)$ is computed as (see, for example, [**27**, Prop. 1.40])

$$p = 1 + \frac{\mu}{12} - \frac{\nu_2}{4} - \frac{\nu_3}{3} - \frac{\nu_\infty}{2}.$$

The above formula can be obtained from Riemann-Hurwitz applied to the natural covering $\mathbb{H}^2/G \to \mathbb{H}^2/\Gamma$. It can also be obtained immediately from the formula for the index given at the beginning of this subsection. It should also be observed that the signature of G is

$$(p,\ \nu_2 + \nu_3 + \nu_\infty;\ 2, \ldots, 2,\ 3, \ldots, 3,\ \infty, \ldots, \infty)$$

with i listed ν_i times for $i = 2,\ 3$, and ∞. The canonical projection

$$P = P(G) : \mathbb{H}^2 \to \mathbb{H}^2/G$$

extends to the set of fixed points $\mathbb{Q} \cup \{\infty\}$ of parabolic elements of G; we will use the same symbol P to denote the extension, and we will often write P_x for $P(x)$ with $x \in \mathbb{H}^2 \cup \mathbb{Q} \cup \{\infty\}$. The compactification of $\mathbb{H}^2/G$ obtained by filling in the punctures will be denoted by $\overline{\mathbb{H}^2/G}$.

6.4. The principal congruence subgroups $\Gamma(k)$. Throughout this section, k is a positive integer; $k \in \mathbb{Z}^+$. We let $\Gamma(k)$ denote the *level k principal congruence subgroup*[34] of Γ; it consists of Möbius transformations represented by matrices $\begin{bmatrix} a & b \\ c & d \end{bmatrix}$ with

$$a \in \mathbb{Z},\ a \equiv 1 \mod k, \quad b \in \mathbb{Z},\ b \equiv 0 \mod k,$$
$$c \in \mathbb{Z},\ c \equiv 0 \mod k, \quad d \in \mathbb{Z},\ d \equiv 1 \mod k,$$

and $ad - bc = 1$.

Let red denote the canonical reduction (mod k) map

$$\text{red} : \mathbb{Z} \to \mathbb{Z}_k = \mathbb{Z}/k\mathbb{Z}.$$

The map red induces a surjective homomorphism

$$\text{red} : \ \text{SL}(2,\mathbb{Z}) \to \ \text{SL}(2,\mathbb{Z}_k).$$

For $k > 2$, $\Gamma(k)$ is isomorphic to the kernel of this map, while $\Gamma(2)$ is isomorphic to the kernel of this map factored by $\{\pm I\}$. In each case,

$$\Gamma/\Gamma(k) \cong \ \text{PSL}(2,\mathbb{Z}_k).$$

The orbifold $\mathbb{H}^2/\Gamma$ has signature $(0,3;\ 2,3,\infty)$; while for $k \geq 2$, $\Gamma(k)$ is torsion free, and the Riemann surface $\mathbb{H}^2/\Gamma(k)$ is of type $(p(k), n(k))$, where

$$p(2) = 0$$

and

$$p(k) = 1 + \frac{k^2(k-6)}{24} \prod_{\mathfrak{P} \text{ prime},\ \mathfrak{P}|k} \left(1 - \frac{1}{\mathfrak{P}^2}\right) \text{ for } k > 2,$$

while

$$n(2) = 3$$

and

$$n(k) = \frac{k^2}{2} \prod_{\mathfrak{P} \text{ prime},\ \mathfrak{P}|k} \left(1 - \frac{1}{\mathfrak{P}^2}\right) \text{ for } k > 2. \tag{1.11}$$

If k is an odd prime, then the above formulae simplify to

$$p(k) = 1 + \frac{(k-6)(k^2-1)}{24} \text{ and } n(k) = \frac{k^2-1}{2}.$$

It should be observed that for $k \geq 2$, we have

$$(k-6)n(k) = 12(p(k)-1). \tag{1.12}$$

For many applications we need to know generators for the cyclic groups stabilizing the parabolic fixed points of $\Gamma(k)$. It is obvious that B^k is a generator for the stabilizer of ∞. The fact that $\Gamma(k)$ is normal in Γ allows

[34]Thus $\Gamma(1) = \Gamma$.

k	$p(k)$	$n(k)$	$\mu(k)$
2	0	3	6
3	0	4	12
4	0	6	24
5	0	12	60
6	1	12	72
7	3	24	168
8	5	24	192
9	10	36	324
10	13	36	360
11	26	60	660
12	25	48	576
13	50	84	1092

Table 1. INVARIANTS FOR $\Gamma(k)$.

us to easily compute the stabilizer of an arbitrary $x \in \mathbb{Q} \cup \{\infty\}$. Choose any $C \in \Gamma$ with $C(x) = \infty$; then $C^{-1} \circ B^k \circ C$ is a generator of the stabilizer of x in $C^{-1}\Gamma(k)C = \Gamma(k)$.

It is easily seen that for $k \geq 2$,

$$\mu(k) = \mu(\Gamma(k)) = 6(2p(k) - 2 + n(k)) = kn(k) = \frac{12k}{k-6}(p(k) - 1).$$

We tabulate, for the convenience of the reader, the type of $\mathbb{H}^2/\Gamma(k)$ and $\mu(k)$ for low values of k.

For later use, we define[35]

$$n(1) = 1.$$

Let $k = \mathfrak{P}^\alpha k_1$, with α and k_1 positive integers, $\mathfrak{P}$ a positive prime, and $\mathfrak{P} \nmid k_1$. Then

$$n(k) = n(\mathfrak{P}^\alpha k_1) = \mathfrak{P}^{2\alpha}\left(1 - \frac{1}{\mathfrak{P}^2}\right) n(k_1) = 2n(\mathfrak{P}^\alpha)n(k_1) \text{ if } k_1 \neq 2,$$

and

$$n(k) = n(2\mathfrak{P}^\alpha) = \frac{3}{2}\mathfrak{P}^{2\alpha}\left(1 - \frac{1}{\mathfrak{P}^2}\right) = n(\mathfrak{P}^\alpha)n(2) \text{ if } k_1 = 2.$$

In the last formula we have of course assumed that $\mathfrak{P} \neq 2$.

Fundamental domains for $\Gamma(k)$. In general, if ω is a fundamental set for a Kleinian group G and we write ω as a disjoint union $\omega = \omega_1 \cup \omega_2$, then for any $g \in G$, $\omega_* = \omega_1 \cup g(\omega_2)$ is again a fundamental set for G; and

[35]Thus for all $k \in \mathbb{Z}^+$, $n(k) = \nu_\infty(\Gamma(k))$.

if we know some side pairing transformations for ω, then we can determine the corresponding side pairing transformations for ω_*. We illustrate this construction for the modular group. For the rest of this section we consider subgroups of the modular group and their action on the upper half plane. Although the exact shapes of the fundamental domains for subgroups of Γ are not needed for our development, they are presented as a tool for the reader to enable him/her to visualize some examples. We have constructed the fundamental domain ω for Γ, the region in $\mathbb{H}^2$ bounded by the curves

$$C_1 = \left\{\Re z = -\frac{1}{2},\ \Im z \geq \frac{\sqrt{3}}{2}\right\},\ C_2 = \left\{\Re z = \frac{1}{2},\ \Im z \geq \frac{\sqrt{3}}{2}\right\},$$

$$C_3 = \left\{|z| = 1,\ -\frac{1}{2} \leq \Re z \leq 0,\ \Im z > 0\right\}$$

and

$$C_4 = \left\{|z| = 1,\ 0 \leq \Re z \leq \frac{1}{2},\ \Im z > 0\right\}.$$

The side pairing transformations for ω are

$$B(C_1) = C_2 \text{ and } A(C_3) = C_4.$$

We form a new fundamental domain ω_* for Γ as

$$\omega_* = \left\{z \in \omega; -\frac{1}{2} \leq \Re z \leq 0\right\} \cup B^{-1}\left(\left\{z \in \omega; 0 \leq \Re z \leq \frac{1}{2}\right\}\right).$$

Its boundary consists of

$$C_1 = \{\Re z = -1,\ \Im z \geq 1\},\ C_2 = \{\Re z = 0,\ \Im z \geq 1\},$$

$$C_3 = \left\{|z+1| = 1, -1 \leq \Re z \leq -\frac{1}{2}\right\} \text{ and } C_4 = \left\{|z| = 1, -\frac{1}{2} \leq \Re z \leq 0\right\}.$$

The corresponding new side pairing transformations for ω_* are

$$B(C_1) = C_2 \text{ and } (AB)(C_3) = C_4.$$

A calculation tells us that $AB = \begin{bmatrix} 0 & -1 \\ 1 & 1 \end{bmatrix}$.

If ω is any fundamental domain for Γ and we let $\gamma_1, \ldots, \gamma_{kn(k)}$ be a set of representatives for the coset space $\Gamma/\Gamma(k)$, then $\cup_{i=1}^{kn(k)} \gamma_i(\omega)$ is a fundamental domain for the principal congruence subgroup $\Gamma(k)$.

The above two simple observations may be used for the construction of

Fundamental domains for $\Gamma(k)$, $k = 2,\ 3$ and 4.

The construction of the fundamental regions that we describe also gives us the motions which identify the sides and a set of inequivalent cusps for the groups we study.

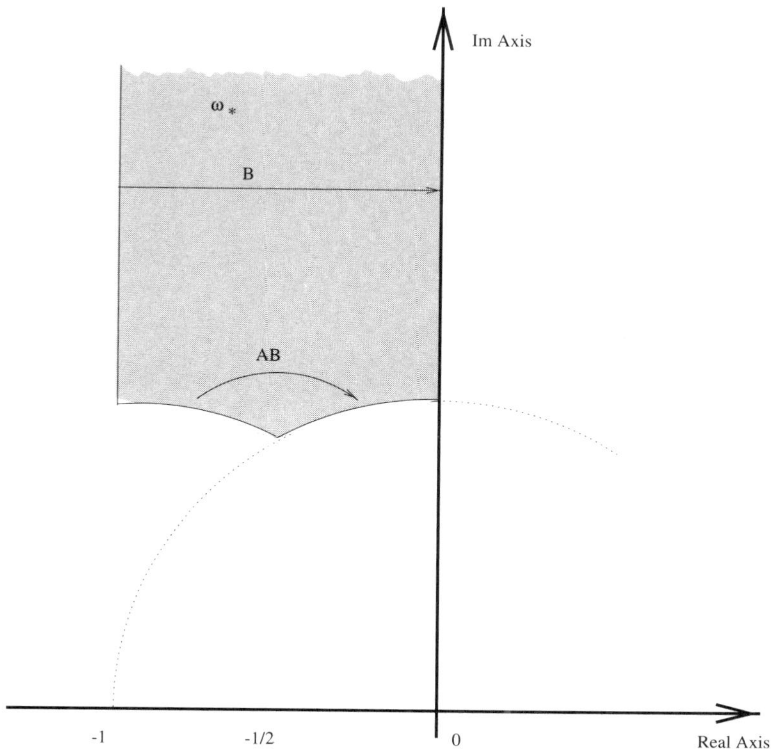

Figure 6. A second fundamental domain for the modular group Γ.

The six motions

$$I,\ B,\ A,\ A\circ B,\ B^{-1}\circ A,\ B\circ A\circ B$$

generate Γ over $\Gamma(2)$. A fundamental region for $\Gamma(2)$ may be taken as the region bounded by the following semi-circles:

$$C_1=\left\{\left|z+\frac{1}{2}\right|=\frac{1}{2},\ \Im z>0\right\},\ C_2=\left\{\left|z-\frac{1}{2}\right|=\frac{1}{2},\ \Im z>0\right\},$$

$$C_3=\{\Re z=-1,\ \Im z>0\},\ C_4=\{\Re z=1,\ \Im z>0\}.$$

One easily checks that the above region is

$$\omega_*\cup B(\omega_*)\cup A(\omega_*)\cup(A\circ B)(\omega_*)\cup\left(B^{-1}\circ A\right)(\omega_*)\cup(B\circ A\circ B)(\omega_*).$$

The sides of this region are identified by the motions B^2 which maps C_3 onto C_4 and $T_2=A\circ B^{-2}\circ A=\begin{bmatrix}1&0\\2&1\end{bmatrix}$ which pairs the sides C_1 and C_2.

We see explicitly the three cusps that project to the punctures on $\mathbb{H}^2/\Gamma(2)$ at the points -1, 0 and ∞. The cusp at 1 is, of course, $\Gamma(2)$-equivalent to

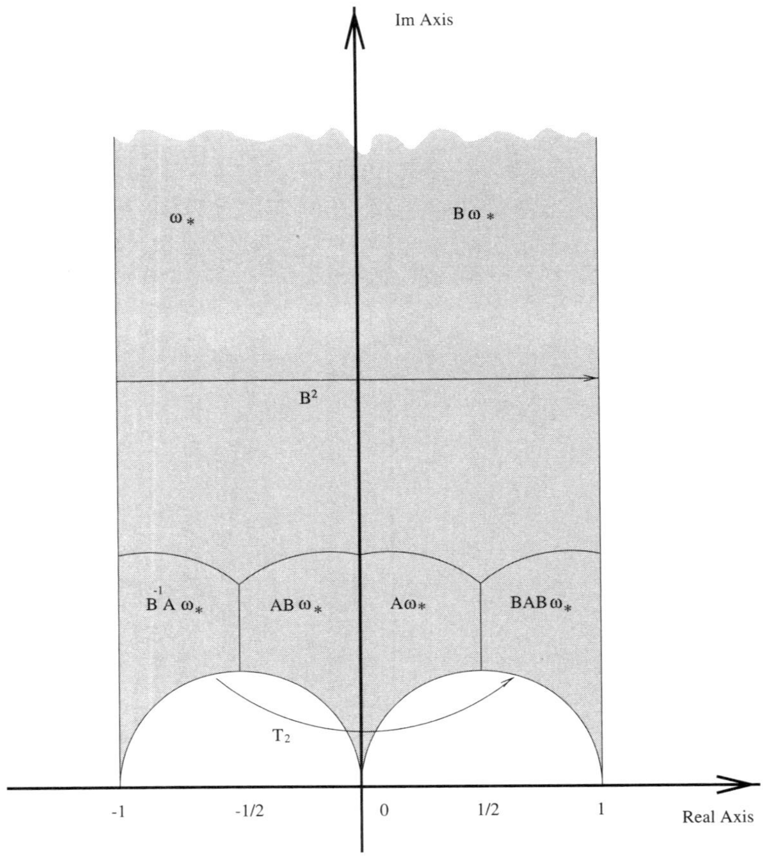

Figure 7. A second fundamental domain for the level 2 principal congruence subgroup $\Gamma(2)$.

the one at -1 via the motion B^2. The reader should now determine the relation of Figure 7 to Figure 3 in §4.3.

An alternative technique to construct fundamental regions for groups of Möbius transformations is by the use of isometric circles. We give the basic definitions of the objects and refer to [**8**, §11] for more details.

Let $C : z \mapsto \frac{az+b}{cz+d}$ be a Möbius transformation with $c \neq 0, ad - bc = 1$. The *isometric circle* I_C of the transformation C is the locus of points $\{|cz + d| = 1\}$. Alternately, it is the set $\{|C'(z)| = 1\}$. It is the circle with center $z = \frac{-d}{c}$ and radius $r = \frac{1}{|c|}$. It is easy to see from the geometric meaning of the isometric circle that a transformation carries its isometric circle onto the isometric circle of its inverse ($C(I_C) = I_{C^{-1}}$) with the interior of the former mapping to the exterior of the latter. For a euclidean isometry $D : z \mapsto e^{\imath\theta} z + \alpha$, we have $D(I_C) = I_{D \circ C}$; for an arbitrary Möbius transformation

D, we do not have such a simple relation between corresponding isometric circles.

The set of isometric circles of a group $G \subset \mathrm{PSL}(2, \mathbb{C})$ can be used to construct a fundamental region for the group. Let R be the region in the extended complex plane $\hat{\mathbb{C}}$ exterior to all the isometric circles of the (nontranslations in the) group. It is quite clear that no two points of R can be identified by an element $\begin{bmatrix} a & b \\ c & d \end{bmatrix}$ of the group with $c \neq 0$. For the moment, let us assume that the group does not contain any translations (that is, there are no transformations in the group other than the identity which have $c = 0$). It is a fundamental fact, proved in Ford's book, that the region R constitutes the interior of a fundamental region for the group (that is every point in $\Omega(G)$ is G-equivalent to at least one point in the closure of R in $\hat{\mathbb{C}}$).

We now explain how our fundamental region for $\Gamma(2)$ can be constructed using isometric circles. The group $\Gamma(2)$ contains the translation $z \mapsto z+2$; the construction will, of course, have to take this complication into account

Let $\gamma = \begin{bmatrix} a & b \\ c & d \end{bmatrix}$, $c \neq 0$, $\in \Gamma(2)$, so that c is an even integer $|c| \geq 2$. The simplest, in some sense, such transformation is $\gamma_1(z) = \frac{z}{2z+1}$ or $\gamma_2(z) = \frac{z}{-2z+1} = \gamma_1^{-1}(z)$. Clearly,

$$C_{\frac{1}{2}} = I_{\gamma_1} = \left\{ \left| z + \frac{1}{2} \right| = \frac{1}{2} \right\},\ C_{\frac{-1}{2}} = I_{\gamma_2} = \left\{ \left| z - \frac{1}{2} \right| = \frac{1}{2} \right\}$$

$$\text{and } \gamma_1 \left(I_{\gamma_1} \right) = I_{\gamma_2}.$$

The subgroup of $\Gamma(2)$ generated by $z \mapsto z+2$ maps these isometric circles onto the set of circles $C_{\frac{d}{2}} = \left\{ \left| z + \frac{d}{2} \right| = \frac{1}{2} \right\}$ with d an arbitrary odd integer. An arbitrary isometric circle I_C for the group $\Gamma(2)$ is $C_{\frac{d}{c}}$ (with center at $\frac{d}{c}$ and radius $\frac{1}{|c|}$ with c an even integer and d an odd integer relatively prime to c). Since

$$C_{\frac{d}{c}} \subset C_{\frac{|d|}{2}} \text{ for } \frac{d}{c} > 0 \text{ and } C_{\frac{d}{c}} \subset C_{\frac{-|d|}{2}} \text{ for } \frac{d}{c} < 0,$$

we conclude that the exteriors of circles $C_{\frac{d}{2}}$ with d odd coincides with the exteriors of circles $C_{\frac{d}{c}}$ with c even and $(c, d) = 1$. The set constructed is not a fundamental region for the group $\Gamma(2)$ since we have neglected the translations in the group. If we consider the intersection of the constructed set with the strip bounded by the vertical lines $\Re z = \pm 1$, we obtain a fundamental domain for the group $\Gamma(2)$ acting on $\mathbb{C} - \mathbb{R}$. If we now intersect this set with the upper half plane, we have reconstructed a fundamental domain for the action of $\Gamma(2)$ on $\mathbb{H}^2$.

The case $k = 3$ is a bit more interesting and a bit more complex. The 12 motions

$$I, B, A, B^2, B \circ A, A \circ B$$

$$B^2 \circ A, A \circ B \circ A, B \circ A \circ B^2, B^2 \circ A \circ B, A \circ B^2, A \circ B^2 \circ A$$

generate Γ over $\Gamma(3)$. A fundamental region for $\Gamma(3)$ may be taken as the region in $\mathbb{H}^2$ bounded by the circular arcs

$$C_1 = \left\{\left|z + \frac{4}{3}\right| = \frac{1}{3},\ \frac{-3}{2} \leq \Re z \leq -1,\ \Im z > 0\right\},$$

$$C_2 = \left\{\left|z + \frac{2}{3}\right| = \frac{1}{3}, -1 \leq \Re z \leq \frac{-1}{2},\ \Im z > 0\right\},$$

$$C_3 = \left\{\left|z + \frac{1}{3}\right| = \frac{1}{3},\ \frac{-1}{2} \leq \Re z \leq 0,\ \Im z > 0\right\},$$

$$C_4 = \left\{\left|z - \frac{1}{3}\right| = \frac{1}{3},\ 0 \leq \Re z \leq \frac{1}{2},\ \Im z > 0\right\},$$

$$C_5 = \left\{\left|z - \frac{2}{3}\right| = \frac{1}{3},\ \frac{1}{2} \leq \Re z \leq 1\right\},\ C_6 = \left\{\left|z - \frac{4}{3}\right| = \frac{1}{3},\ 1 \leq \Re z \leq \frac{3}{2}\right\},$$

$$C_7 = \left\{\Re z = \frac{-3}{2},\ \Im z \geq \frac{\sqrt{3}}{6}\right\},\ C_8 = \left\{\Re z = \frac{3}{2},\ \Im z \geq \frac{\sqrt{3}}{6}\right\}.$$

We define

$$T_2 : z \mapsto \frac{z}{3z+1},\ T_3 : z \mapsto \frac{-2z-3}{3z+4} \text{ and } T_4 : z \mapsto \frac{4z-3}{3z-2}.$$

The side pairings in this case are

$$B^3(C_7) = C_8,\ T_2(C_3) = C_4,\ T_3(C_1) = C_2 \text{ and } T_4(C_5) = C_6.$$

It is easily checked that the four side pairing transformations are not independent; a presentation of $\Gamma(3)$ is provided by these generators and the relation $T_4 \circ T_2 \circ T_3 = B^3$. We again see explicitly the four cusps that project to the punctures on $\mathbb{H}^2/\Gamma(3)$ at the points -1, 0, 1 and ∞ . Another point worth observing here is that one also sees explicitly from the picture of the fundamental domain that the fixed point of the automorphism of the Riemann surface $\mathbb{H}^2/\Gamma(3)$ induced by B is the projection of the intersection of C_7 with C_1.

As in the previous case the exteriors of the isometric circles with the smallest radii ($\frac{1}{3}$ in this case) give rise to the fundamental domain. These are parabolic transformations with fixed points at ± 1 and 0.

The computations for the case $k = 4$ can also be done quite explicitly as can those for any particular $k > 4$, with enough perseverance. We base our construction on the fundamental domain $\omega' = B(\omega_*)$ for the action of Γ on

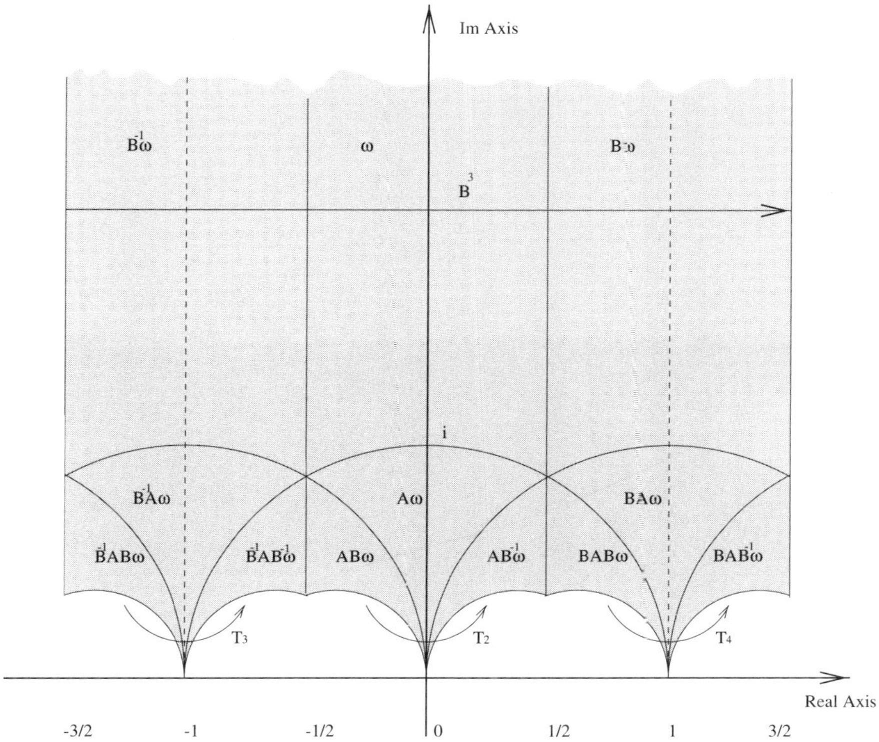

Figure 8. A fundamental domain for the level 3 principal congruence subgroup $\Gamma(3)$.

$\mathbb{H}^2$. A fundamental region for $\Gamma(4)$ may be taken as the region bounded by the half circles:

$$C_0 = \{\Re z = -1,\ \Im z > 0\}\,,\ C_9 = \{\Re z = 3,\ \Im z > 0\}\,,$$

$$C_i = \left\{\left|z + \frac{2i-5}{4}\right| = \frac{1}{4},\ \Im z > 0\right\},\ i = 1,\ 2,\ ...,\ 8.$$

The side pairing motions are

$$B^4(C_0) = C_9,$$

$$T_2 : z \mapsto \frac{z+1}{-4z-3} \text{ that maps } C_1 \text{ to } C_2,$$

$$T_3 : z \mapsto \frac{-3z+1}{-4z+1} \text{ that maps } C_3 \text{ to } C_4,$$

$$T_4 : z \mapsto \frac{-7z+9}{-4z+5} \text{ that maps } C_5 \text{ to } C_6$$

$$\text{and } T_5 : z \mapsto \frac{-11z+25}{-4z+9} \text{ that maps } C_7 \text{ to } C_8.$$

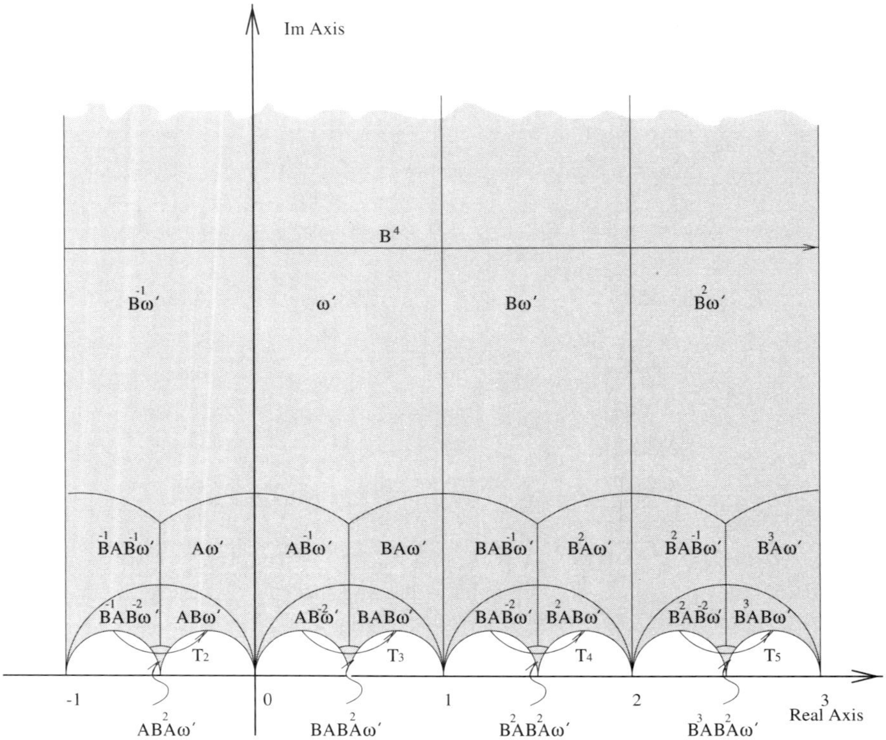

Figure 9. A fundamental domain for the level 4 principal congruence subgroup $\Gamma(4)$.

In Figure 9 one again sees explicitly the distinct cusps that project to punctures; these are at the points -1, 0, 1, 2, ∞ and $\frac{1}{2}$. The cusp at 3 is equivalent to the one at -1 and the cusps $-\frac{1}{2}$, $\frac{3}{2}$, $\frac{5}{2}$ are all equivalent to the cusp at $\frac{1}{2}$. It is also clear that the motions we have described above are all parabolic with fixed points at the cusps.

6.5. Adjoining translations: The subgroups $G(k)$. Let $G(k)$ be the group[36] generated by the elements of $\Gamma(k)$ and the translation B. Our first claim is that (see also Lemma 2.15 of Chapter 2)

$$G(k) = \{g \in \Gamma;\ g^{-1}(\infty) = \infty \mod \Gamma(k)\}.$$

It is clear that $G(k)$ is contained in the group defined by the right hand side of the last displayed equation. Conversely, if for $g \in \Gamma$ there is a $\gamma \in \Gamma(k)$ with $g^{-1}(\infty) = \gamma^{-1}(\infty)$, then $g \circ \gamma^{-1}$ belongs to Γ and fixes ∞; hence it is a power of B. We have also shown that every element $g \in G(k)$ can be

[36]This group is usually denoted as $\Gamma_1(k)$ in the number theory literature.

k	μ	ν_2	ν_3	ν_∞	p
1	1	1	1	1	0
2	3	1	0	2	0
3	4	0	1	2	0
4	6	0	0	3	0
5	6	2	0	2	0
6	12	0	0	4	0
7	8	0	2	2	0
8	12	0	0	4	0
9	12	0	0	4	0
10	18	2	0	4	0
11	12	0	0	2	1
12	24	0	0	6	0
13	14	2	2	2	0
14	24	0	0	4	1
15	24	0	0	4	1
17	18	2	0	2	1
19	20	0	2	2	1
21	32	0	2	4	1
23	24	0	0	2	2
35	48	0	0	4	3

Table 2. THE BRANCH SCHEMA FOR $\Gamma_o(k)$.

decomposed uniquely as

$$g = B^l \circ \gamma, \tag{1.13}$$

with $\gamma \in \Gamma(k)$ and $l \in \{0,\ 1,\ \ldots,\ k-1\}$.

More information on these groups is found in the next subsection for $k \leq 4$ and $k = 6$, and for the general case in §4.2 of Chapter 2.

6.6. The Hecke subgroups $\Gamma_o(k)$. For each positive integer k, we define the *Hecke* group[37] $\Gamma_o(k)$ to consist of those Möbius transformations that can be represented by elements $\gamma = \begin{bmatrix} a & b \\ c & d \end{bmatrix} \in \mathrm{SL}(2, \mathbb{Z})$ with

$$c \equiv 0 \mod k.$$

The branch schema for $\Gamma_o(k)$ is described by [**27**, Prop. 1.43]. We record for selected low values of k this information and the genus p of $\mathbb{H}^2/\Gamma_o(k)$ in Table 2.

[37]This nomenclature is not universal.

It is useful to observe that

$$\mu = k \prod_{\mathfrak{P} \text{ prime}, \ \mathfrak{P}|k} \left(1 + \frac{1}{\mathfrak{P}}\right).$$

For k an odd prime, the branch schema for $\Gamma_o(k)$ is quite simple. We have

$$\mu = k + 1 \text{ and } \nu_\infty = 2,$$

$$\nu_2 = 2 \text{ if } k \equiv 1 \mod 4 \text{ and } \nu_2 = 0 \text{ if } k \equiv 3 \mod 4,$$

and

$$\nu_3 = 1 \text{ if } k = 3, \ \nu_3 = 2 \text{ if } k \equiv 1 \mod 3, \text{ and } \nu_3 = 0 \text{ if } k \equiv 2 \mod 3.$$

Further, for the genus, we have (again, only for primes k)

$$p = \begin{cases} 0 \text{ for } k = 2 \text{ and } 3, \\ \frac{k-13}{12} \text{ if } k \equiv 1 \mod 12, \\ \frac{k-n}{12} \text{ if } k \equiv n \mod 12, \text{ for } n = 5 \text{ and } 7, \\ \frac{k+1}{12} \text{ if } k \equiv 11 \mod 12 \end{cases}$$

and the negative Euler characteristic is given by $\chi = \frac{1}{6}(k+1)$. The cusps ∞ and 0 determine the two punctures on $\mathbb{H}^2/\Gamma_o(k)$. The first cusp is stabilized by $< B >$; the second by $\begin{bmatrix} 1 & 0 \\ k & 1 \end{bmatrix}$. It is of interest to speculate about the connections, if any, between the branch schema for $\Gamma_o(k)$ and the set of classes of characteristics fixed by $\Gamma_o(k)$ (see Lemma 4.18 in Chapter 2).

It is obvious that $G(k) \subset \Gamma_o(k)$ (in fact $\Gamma_o(k)$ is the normalizer of $G(k)$ in Γ). It follows that $G(k) = \Gamma_o(k)$ for $k \le 4$ and $k = 6$ because both groups have the same index in Γ.

It is also convenient to introduce the groups $\Gamma^o(k)$ consisting of Möbius transformations represented by elements $\gamma = \begin{bmatrix} a & b \\ c & d \end{bmatrix} \in \text{SL}(2,\mathbb{Z})$ with $b \equiv 0 \mod k$. Since

$$A \begin{bmatrix} a & b \\ c & d \end{bmatrix} A^{-1} = \begin{bmatrix} d & -c \\ -b & a \end{bmatrix},$$

$A\Gamma_o(k)A^{-1} = \Gamma^o(k)$. A group conjugate to $\Gamma^o(2)$ appears prominently in those parts, not studied in depth in this book, of the theory of theta functions that develop applications to representations of integers as sums of squares. We define

$$\Gamma_\theta =< A, B^2 >;$$

it is easily seen that $B\Gamma_\theta B^{-1} = \Gamma^o(2)$. See Proposition 2.9 of Chapter 5 for a prominent use of the group Γ_θ.

k	μ	ν_2	ν_3	ν_∞	p
1	1	1	1	1	0
2	6	0	0	3	0
3	12	0	0	4	0
4	24	0	0	6	0
5	30	2	0	6	0
6	72	0	0	10	2
7	56	0	2	8	1
8	96	0	0	12	3
9	108	0	0	12	4
10	180	0	0	18	6
11	132	0	0	12	6
12	288	0	0	28	10
13	182	2	2	14	8

Table 3. THE BRANCH SCHEMA FOR $\Gamma(k,k)$.

6.7. Structure of $\Gamma(k,k)$. Define $\Gamma(k,k) = \Gamma_o(k) \cap \Gamma^o(k)$. We show in the next subsections that $\Gamma(k,k) \cong \Gamma_o(k^2)$. Hence it is easy to compute the branch schema and the genus of $\Gamma(k,k)$. We record the results for low values of k Table 3.

We note that

$$\Gamma(k,k) = \Gamma(k) \text{ for } k = 1,\ 2,\ 3 \text{ and } \Gamma(1) = \Gamma.$$

For the remainder of this subsection let us assume that k is an odd prime. For such $k \geq 5$,

$$p(\Gamma(k,k)) = \begin{cases} \frac{k^2-5k-8}{12} & \text{if } k \equiv 1 \mod 12 \\ \frac{k(k-5)}{12} & \text{if } k \equiv 5 \mod 12 \\ \frac{k^2-5k-2}{12} & \text{if } k \equiv 7 \mod 12 \\ \frac{(k-2)(k-3)}{12} & \text{if } k \equiv 11 \mod 12 \end{cases};$$

in particular,

$$p(\Gamma(k,k)) \geq 8 \text{ for } k \geq 13.$$

The hyperbolic Möbius transformation

$$C_k = \begin{bmatrix} 2 & k \\ k & \frac{k^2+1}{2} \end{bmatrix}$$

generates $\Gamma(k,k)$ over $\Gamma(k)$ (also $\Gamma_o(k)$ over $G(k)$), and B generates $\Gamma_o(k)$ over $\Gamma(k,k)$ (also $G(k)$ over $\Gamma(k)$). The last assertion also holds for $k=2$ and nonprimes.

If $G \subset \Gamma$ and $\gamma \in N(G)$, the normalizer of G in Γ, then $\tilde{\gamma}$ will always denote the automorphism of $\mathbb{H}^2/G$ induced by γ. The group of order $\frac{k-1}{2}$ generated by $\tilde{C}_k$ acts fixed point freely on the $\frac{k^2-1}{2}$ punctures of $\mathbb{H}^2/\Gamma(k)$ to produce the $k+1$ punctures

$$P_\infty,\ P_0,\ P_1,\ \ldots,\ P_{k-1}$$

on $\mathbb{H}^2/\Gamma(k,k)$, and the group of order k generated by $\tilde{B}$ acts on the $k+1$ punctures of $\mathbb{H}^2/\Gamma(k,k)$ to produce the 2 punctures P_∞ and P_0 on $\mathbb{H}^2/\Gamma_o(k)$. On $\mathbb{H}^2/\Gamma(k,k)$, $\tilde{B}$ fixes P_∞ and permutes cyclically the other k punctures.

The surface $\mathbb{H}^2/\Gamma(k,k)$ has a conformal involution $J=\tilde{A}$ generated by the motion $A \in N(\Gamma(k,k))$ that conjugates $\begin{bmatrix} a & b \\ c & d \end{bmatrix}$ to $\begin{bmatrix} -d & c \\ b & -a \end{bmatrix}$. We note, for future use, that the surface $\mathbb{H}^2/\Gamma_o(k)$ has a conformal involution J_o generated by the motion[38] $\begin{bmatrix} 0 & -1 \\ k & 0 \end{bmatrix} \in N(\Gamma_o(k))$ that conjugates $\begin{bmatrix} a & b \\ c & d \end{bmatrix}$ to $\begin{bmatrix} -d & \frac{c}{k} \\ bk & -a \end{bmatrix}$.

6.8. A two parameter family of groups. The groups $\Gamma(k,k)$ are a one parameter subfamily of an important two parameter family of subgroups of the modular group. Let p and q be two positive integers.[39] We define

$$\Gamma(p,q) = \left\{ \begin{bmatrix} a & b \\ c & d \end{bmatrix} \in \Gamma;\ p|b \text{ and } q|c \right\} = \Gamma^o(p) \cap \Gamma_o(q).$$

For $c \in \mathbb{C}^*$, we let A_c and M_c be the Möbius transformations induced by the matrices $A_c = \begin{bmatrix} 0 & -1 \\ c & 0 \end{bmatrix}$ and $M_c = \begin{bmatrix} c & 0 \\ 0 & 1 \end{bmatrix} \in \mathrm{GL}(2,\mathbb{C})$, respectively.[40] Our first observation is that for every Möbius transformation

$$\gamma \in \Gamma(p,q),\ A_q \circ \gamma \circ A_q^{-1} \in \Gamma_o(pq)$$

[38] We identify 2×2 nonsingular matrices with the fractional linear transformations they define. Also for subgroups G of Γ, we often (when there can be no confusion) identify an element C in $N(G)$ with the motion $\tilde{C}$ of $\mathbb{H}^2/G$ that it induces.

[39] The use of the symbol p for an index here and the genus of a surface elsewhere should not cause confusion.

[40] Although these matrices are not in $\mathrm{SL}(2,\mathbb{Z})$, for many $c \in \mathbb{Q}^+$ they conjugate subgroups of Γ onto other such subgroups.

as is easily seen from the fact that A_q conjugates $\begin{bmatrix} a & b \\ c & d \end{bmatrix}$ to $\begin{bmatrix} d & -\frac{c}{q} \\ -qb & a \end{bmatrix}$. Hence $\Gamma(p,q) \cong \Gamma_o(pq)$. It follows that

$$[\Gamma_o(q) : \Gamma(p,q)] = \frac{\mu(\Gamma_o(pq))}{\mu(\Gamma_o(q))} = p \prod_{\mathfrak{P}\text{, prime, } \mathfrak{P}|p, \mathfrak{P}\not| q} \left(1 + \frac{1}{\mathfrak{P}}\right).$$

Fix a prime k, and let m and $n \in \mathbb{Z}^+ \cup \{0\}$. Then

$$[\Gamma_o(k^n) : \Gamma(k^m, k^n)] = \begin{cases} k^m \text{ if } m = 0 \text{ or } n > 0 \\ k^{m-1}(1+k) \text{ if } m \neq 0 \text{ and } n = 0 \end{cases}.$$

In particular, the k^m motions

$$B^l;\ l = 0,\ \ldots,\ k^m - 1,$$

provide representatives for a complete set of left or right $\Gamma(k^m, k^n)$-cosets for $\Gamma_o(k^n)$ if $n > 0$. Thus, if $n > 0$ and

$$P_\infty,\ P_0,\ \ldots,\ P_r,$$

is a list of punctures on $\mathbb{H}^2/\Gamma_o(k^n)$, then

$$P_\infty,\ P_{B^l(0)},\ \ldots,\ P_{B^l(r)},\ l = 0,\ 1,\ \ldots,\ k^n - 1,$$

is the corresponding list on $\mathbb{H}^2/\Gamma(k^m, k^n)$. In particular,

$$P_\infty,\ P_0,\ \ldots,\ P_{k^m - 1}$$

are the punctures on $\mathbb{H}^2/\Gamma(k^m, k)$. In the last two lists, there are duplications, illustrated below for $k = 3$, $m = 2$, $n = 1$. We know that

$$\nu_\infty(\Gamma_o(k^n)) = \begin{cases} 2k^{\frac{n-1}{2}} \text{ if } n \text{ is odd} \\ k^{\frac{n}{2}-1}(k+1) \text{ if } n \text{ is even} \end{cases}.$$

Thus $\mathbb{H}^2/\Gamma(9,3)$ has 6 punctures; they are

$$P_\infty,\ P_0,\ P_3,\ P_6,\ P_1 = P_4 = P_7,\ P_2 = P_5 = P_8.$$

The motion $C = \begin{bmatrix} 13 & -9 \\ 3 & -2 \end{bmatrix} \in \Gamma(9,3)$ maps 1 to 4, for example.

If $n = 0$, then P_∞ is the only puncture on $\mathbb{H}^2/\Gamma$; the punctures on $\mathbb{H}^2/\Gamma^o(k^m)$ are described in [**27**, Ch. I]. If $m = 1$, we can take

$$B^l,\ l = 0,\ \ldots,\ k-1, \text{ and } \begin{bmatrix} k & -1 \\ 1 & 0 \end{bmatrix}$$

as right coset representatives for $\Gamma/\Gamma^o(k)$.

A second case of interest is $(p,q) = 1$. For this case,

$$[\Gamma_o(q) : \Gamma(p,q)] = p \prod_{\mathfrak{P}\text{ prime, } \mathfrak{P}|p} \left(1 + \frac{1}{\mathfrak{P}}\right) = \mu(\Gamma_o(p)).$$

In general, the motion $A_{\frac{q}{p}} = \begin{bmatrix} 0 & -1 \\ \frac{q}{p} & 0 \end{bmatrix} = \begin{bmatrix} 0 & -p \\ q & 0 \end{bmatrix}$ (as an element of $\mathrm{PGL}(2,\mathbb{C})$) induces an involution of $\Gamma(p,q)$.

We record for future use that for all $N \in \mathbb{Z}^+$,

$$A_N\Gamma(p,Nq)A_N = \Gamma(q,Np) \text{ and } M_N\Gamma(p,Nq)M_{\frac{1}{N}} = \Gamma(Np,q)$$

(the special case of the first equality (with N replaced by q and q by 1) was considered previously).

7. A geometric test for primality

In this short section, we discuss an example, the polynomial N^2+1, to hint at the connections of the modular group to number theory, and show that very simple geometric properties of Γ can be translated to nontrivial number theoretic observations.

The motion $A = \begin{bmatrix} 0 & -1 \\ 1 & 0 \end{bmatrix}$ has its unique fixed point in $\mathbb{H}^2$ at $\imath$. Every matrix in Γ with trace 0 is conjugate in Γ to A. If $C = \begin{bmatrix} a & b \\ c & -a \end{bmatrix}$ (without loss of generality $c > 0$) is such an involution, then its fixed point in $\mathbb{H}^2$ is $\frac{a+\imath}{c}$. If $C' = \begin{bmatrix} a' & b' \\ c' & d' \end{bmatrix} \in \Gamma$ is chosen so that $C = C'A(C')^{-1}$, then $C'(\imath)$ is the fixed point of C in $\mathbb{H}^2$, and

$$C'(\imath) = \frac{a'\imath + b'}{c'\imath + d'} = \frac{a'c' + b'd' + \imath}{c'^2 + d'^2}.$$

It follows that $c = c'^2 + d'^2$ is the sum of two relatively prime squares.

Using these observations, we establish

Theorem 7.1. *Let N be a positive integer.*
(a) *Every positive factor of N^2+1 can be written as the sum of two relatively prime squares.*
(b) *The number of positive factors of N^2+1 is precisely the number of times the orbit of $\imath$ under Γ meets the ray in $\mathbb{H}^2$ joining $N+\imath$ to the origin.*

Proof. To prove (a), write

$$N^2 + 1 = a\,b, \text{ with } a \text{ and } b \in \mathbb{Z}^+.$$

The matrix

$$\begin{bmatrix} N & -b \\ a & -N \end{bmatrix} \in \mathrm{SL}(2,\mathbb{Z})$$

has trace 0, determinant 1, and fixed point $\frac{N+\imath}{a}$. Hence a is representable as the sum of two relatively prime squares.

To establish (b), we first observe that as a result of part (a), each factor a of N^2+1 determines a fixed point of an involution in Γ: a point, $\frac{N+\imath}{a}$, on the ray joining $N+i$ and 0. We must now show that no other point on the orbit of $\imath$ meets this line. Let $C = \begin{bmatrix} a & b \\ c & d \end{bmatrix} \in \Gamma$ and assume that $C(\imath) = \frac{ac+bd+\imath}{c^2+d^2}$ is on this line. It follows that

$$ac + bd = N.$$

Together with $ad - bc = 1$, this gives us that

$$N^2 + 1 = (ac+bd)^2 + (ad-bc)^2 = (a^2+b^2)(c^2+d^2);$$

that is, c^2+d^2 is a factor of N^2+1, and the point $C(\imath)$ already arose in part (a) as the point produced by the factor c^2+d^2 of N^2+1. This establishes part (b).

We note as a consequence of the proof of part (b) the following:

Corollary 7.2. *Let c and d be relatively prime integers. Then the orbit of $\imath$ under Γ meets the line $y = \frac{1}{c^2+d^2}$ precisely in those points*

$$\frac{r+\imath}{c^2+d^2}$$

for which c^2+d^2 divides r^2+1. In particular, every integer of the form c^2+d^2 divides some number of the form N^2+1.

Chapter 2

Theta functions with characteristics

We begin with the basic definitions of theta functions with characteristics via their expansions as a series – thus we view θ as a function of three variables: a characteristic $\chi = \begin{bmatrix} \epsilon \\ \epsilon' \end{bmatrix} \in \mathbb{R}^2$, a variable $z \in \mathbb{C}$ and a parameter $\tau \in \mathbb{H}^2$. We then derive the basic properties of these functions as functions of the complex variable z and show how the theory of elliptic functions can be derived from theta functions.

We define the space of N-th order theta functions with characteristics and show how this theory begins to lead to interesting identities which have number theoretical content. We postpone the actual derivation of most of the identities to Chapters 4 and 5, but lay here the foundation for their construction. Here we also see how theta functions depend on the characteristics, thus realizing that the characteristics can be partitioned into equivalence classes in such a way that equivalence of characteristics corresponds to θ-functions that differ at most by a multiplicative constant of absolute value 1.

We use properties of theta functions to construct elliptic functions and then once again deduce identities from the residue theorem. The material described above is classical complex function theory. More interesting for our development, however, is to set the variable z in the definition of the theta function equal to 0 and view the resulting function, known classically as a theta constant, as a function of the parameter τ in the upper half plane. At this point function theory on Riemann surfaces begins to dominate the discussion.

A highlight of this chapter is the derivation of the transformation theory of theta functions. The proof of the transformation formula depends on the heat equation, the residue theorem and lots of calculations. This formula allows us to use (in the next chapter) theta constants with appropriate choices of characteristics to define functions automorphic with respect to certain subgroups of the modular group. We explain how the set of punctures of the Riemann surface $\mathbb{H}^2/\Gamma(k)$ can be put into a rather natural correspondence with a set of equivalence classes of characteristics. We obtain in this way a nice geometric picture of the set of punctures of this Riemann surface as a set of towers. The chapter ends with a derivation, in a natural way from the point of view of the theory of theta functions, of the Jacobi triple product formula, thus connecting the infinite sums defining the theta functions with infinite products. We also indicate how our approach leads naturally to generalizations of the Jacobi triple product, for example, to the quintuple product identity. Among the classical results we recover are *Jacobi's identity*,

$$\prod_{n=1}^{\infty}(1-x^n)^3 = \sum_{n=0}^{\infty}(-1)^n(2n+1)x^{\frac{n(n+1)}{2}}, \tag{2.1}$$

and *Euler's identity*,

$$\prod_{n=1}^{\infty}(1-x^n) = \sum_{n=-\infty}^{\infty}(-1)^n x^{\frac{n(3n+1)}{2}} = 1+\sum_{n=1}^{\infty}(-1)^n\left[x^{\frac{n(3n-1)}{2}}+x^{\frac{n(3n+1)}{2}}\right], \tag{2.2}$$

described in the introduction. We end the chapter with a discussion of those properties of the classical η-function that will be needed for the study in Chapter 5 of partition identities.

1. Theta functions and theta constants

1.1. Definitions and basic properties. In this section we define the theta functions and theta constants and describe or derive the properties we shall be using. The results here are quite classical and can be found, in even more generality, for example in Chapter VI of [**6**] as well as in many other places. For the convenience of the reader, though, we shall in most cases provide proofs of results in order to describe the general flavor of the approach we are following and to make the treatment of θ-functions self contained. General references for this section are Chapters I and II of [**24**] and Chapter VI of [**6**]. We begin with

Definition 1.1. The *theta function with characteristic* $\begin{bmatrix} \epsilon \\ \epsilon' \end{bmatrix} \in \mathbb{R}^2$ is defined by the following series which converges uniformly and absolutely on compact subsets of $\mathbb{C} \times \mathbb{H}^2$:

$$\theta\begin{bmatrix} \epsilon \\ \epsilon' \end{bmatrix}(z,\tau) = \sum_{n\in\mathbb{Z}} \exp 2\pi\imath \left\{ \frac{1}{2}\left(n+\frac{\epsilon}{2}\right)^2 \tau + \left(n+\frac{\epsilon}{2}\right)\left(z+\frac{\epsilon'}{2}\right) \right\}.$$

We begin with the observation that there really is only one theta function

$$\theta\begin{bmatrix} 0 \\ 0 \end{bmatrix} = \theta,$$

and the theta functions with characteristic are expressed in terms of this function by the formula

$$\theta\begin{bmatrix} \epsilon \\ \epsilon' \end{bmatrix}(z,\tau) = \exp 2\pi\imath \left\{ \frac{1}{8}\epsilon^2\tau + \frac{1}{2}\epsilon z + \frac{1}{4}\epsilon\epsilon' \right\} \theta\left(z + \frac{\epsilon'}{2} + \tau\frac{\epsilon}{2}, \tau\right),$$

which also implies that we may work[41] with complex (not just real) characteristics. The derivation of the above formula is straightforward. From the definition

(2.3)

$$\theta\begin{bmatrix} \epsilon \\ \epsilon' \end{bmatrix}(z,\tau) = \sum_{n\in\mathbb{Z}} \exp 2\pi\imath \left\{ \frac{1}{2}\left(n^2 + n\epsilon + \frac{\epsilon^2}{4}\right)\tau + nz + n\frac{\epsilon'}{2} + \frac{\epsilon}{2}z + \frac{\epsilon\epsilon'}{4} \right\}$$

$$= \exp 2\pi\imath \left\{ \frac{1}{8}\epsilon^2\tau + \frac{1}{2}\epsilon z + \frac{1}{4}\epsilon\epsilon' \right\} \sum_{n\in\mathbb{Z}} \exp 2\pi\imath \left\{ \frac{1}{2}n^2\tau + n\left(\frac{1}{2}\epsilon\tau + z + \frac{\epsilon'}{2}\right) \right\}$$

$$= \exp 2\pi\imath \left\{ \frac{1}{8}\epsilon^2\tau + \frac{1}{2}\epsilon z + \frac{1}{4}\epsilon\epsilon' \right\} \theta\begin{bmatrix} 0 \\ 0 \end{bmatrix}\left(z + \frac{\epsilon'}{2} + \frac{\epsilon}{2}\tau, \tau\right).$$

We proceed to establish the convergence of the series defining $\theta\begin{bmatrix} 0 \\ 0 \end{bmatrix}(z,\tau)$. We estimate for $\tau \in \mathbb{H}^2$ with $\Im\tau \geq a > 0$ and $z \in \mathbb{C}$ with $|z| \leq b$, $b > 0$,

$$|\theta(z,\tau)| \leq \sum_{n\in\mathbb{Z}} \exp \pi\{-n^2\Im\tau + 2|n||z|\} \leq \sum_{n\in\mathbb{Z}} \exp \pi\{-n^2 a + 2|n|b\}.$$

For only finitely many $n \in \mathbb{Z}$, we have $2|n|b > \frac{1}{2}n^2 a$. Hence we have to show that $\sum_{n\in\mathbb{Z}} \exp \pi \left\{\frac{-n^2 a}{2}\right\} < \infty$, which follows by the Cauchy root test.

Term by term differentiation of the defining series leads to the *heat equation* for theta functions:

$$\frac{\partial^2\theta\begin{bmatrix} \epsilon \\ \epsilon' \end{bmatrix}(z,\tau)}{\partial z^2} = 4\pi\imath \frac{\partial\theta\begin{bmatrix} \epsilon \\ \epsilon' \end{bmatrix}(z,\tau)}{\partial\tau}.$$

We will abbreviate

$$' = \frac{\partial}{\partial z} \text{ and } \cdot = \frac{\partial}{\partial\tau}.$$

[41]We will not do so.

The theta functions satisfy the following easily derivable properties. For all integers m and n,

(2.4)
$$\theta\begin{bmatrix}\epsilon\\ \epsilon'\end{bmatrix}(z+n+m\tau,\tau) = \exp 2\pi\imath\left\{\frac{n\epsilon - m\epsilon'}{2} - mz - \frac{m^2}{2}\tau\right\}\theta\begin{bmatrix}\epsilon\\ \epsilon'\end{bmatrix}(z,\tau);$$

in particular,

$$\theta\begin{bmatrix}\epsilon\\ \epsilon'\end{bmatrix}(z+1,\tau) = \exp(\pi\imath\epsilon)\,\theta\begin{bmatrix}\epsilon\\ \epsilon'\end{bmatrix}(z,\tau) \tag{2.5}$$

and

$$\theta\begin{bmatrix}\epsilon\\ \epsilon'\end{bmatrix}(z+\tau,\tau) = \exp(-\pi\imath\left\{\epsilon' + 2z + \tau\right\})\,\theta\begin{bmatrix}\epsilon\\ \epsilon'\end{bmatrix}(z,\tau). \tag{2.6}$$

Exercise 1.2. Derive formula (2.4).

Remark 1.3. As a consequence of (2.5), the entire function

$$z \mapsto \exp(-\pi\imath\epsilon z)\,\theta\begin{bmatrix}\epsilon\\ \epsilon'\end{bmatrix}(z,\tau)$$

is periodic with period 1. Hence we have the Fourier series expansion (for fixed τ as a function of $z \in \mathbb{C}$)

(2.7)
$$\theta\begin{bmatrix}\epsilon\\ \epsilon'\end{bmatrix}(z,\tau) = \sum_{n=-\infty}^{\infty} a_n \exp\pi\imath z(2n+\epsilon) = \exp\{\pi\imath\epsilon z\}\sum_{n=-\infty}^{\infty} a_n\exp\{2\pi\imath n z\},$$

where for each $n \in \mathbb{Z}$,

$$\begin{aligned} a_n &= \int_0^1 \exp\{-\pi\imath(2n+\epsilon)(\zeta+t)\}\,\theta\begin{bmatrix}\epsilon\\ \epsilon'\end{bmatrix}(\zeta+t,\tau)\,dt \\ &= \exp 2\pi\imath\left\{\tfrac{1}{2}\left(n+\tfrac{\epsilon}{2}\right)^2\tau + \left(n+\tfrac{\epsilon}{2}\right)\tfrac{\epsilon'}{2}\right\} \end{aligned}$$

In the above, the point $\zeta \in \mathbb{C}$ is arbitrary. It is important to observe that $a_n \neq 0$ for all $n \in \mathbb{Z}$.

A formula slightly more general than (2.4) is: for arbitrary real numbers m and n,

$$\theta\begin{bmatrix}\epsilon\\ \epsilon'\end{bmatrix}\left(z+\tau\frac{m}{2}+\frac{n}{2},\tau\right) \tag{2.8}$$
$$= \exp 2\pi\imath\left\{-\frac{1}{2}mz - \frac{1}{8}m^2\tau - \frac{1}{4}m(\epsilon'+n)\right\}\theta\begin{bmatrix}\epsilon+m\\ \epsilon'+n\end{bmatrix}(z,\tau).$$

In fact, a double application of this formula with n and m integers (or a single application with integers n and m replaced by $2n$ and $2m$) gives (2.4) because of the next identity. For integers m and n,

(2.9) $$\theta\left[\begin{array}{c}\epsilon+2m\\ \epsilon'+2n\end{array}\right](z,\tau)=\exp(\pi\imath\epsilon n)\,\theta\left[\begin{array}{c}\epsilon\\ \epsilon'\end{array}\right](z,\tau).$$

Further,

(2.10)
$$\theta\left[\begin{array}{c}-\epsilon\\ -\epsilon'\end{array}\right](z,\tau)=\theta\left[\begin{array}{c}\epsilon\\ \epsilon'\end{array}\right](-z,\tau)\text{ and }\theta'\left[\begin{array}{c}-\epsilon\\ -\epsilon'\end{array}\right](z,\tau)=-\theta'\left[\begin{array}{c}\epsilon\\ \epsilon'\end{array}\right](-z,\tau).$$

The *theta constants* are defined by setting the variable $z=0$ in the definition of the theta functions. We shall abbreviate

$$\theta\left[\begin{array}{c}\epsilon\\ \epsilon'\end{array}\right](0,\tau)=\theta\left[\begin{array}{c}\epsilon\\ \epsilon'\end{array}\right]\text{ and }\theta'\left[\begin{array}{c}\epsilon\\ \epsilon'\end{array}\right](0,\tau)=\theta'\left[\begin{array}{c}\epsilon\\ \epsilon'\end{array}\right]$$

when there can be no confusion.[42] In general,

(2.11) $$\theta\left[\begin{array}{c}\pm\epsilon+2m\\ \pm\epsilon'+2n\end{array}\right]=(\exp\pm\pi\imath\epsilon n)\,\theta\left[\begin{array}{c}\epsilon\\ \epsilon'\end{array}\right]$$

(in the last equation we use the same sign in each of its three appearances), and thus

(2.12) $$\theta\left[\begin{array}{c}1\\ 2-\epsilon'\end{array}\right]=-\theta\left[\begin{array}{c}1\\ \epsilon'\end{array}\right]\text{ and }\theta\left[\begin{array}{c}0\\ 2-\epsilon'\end{array}\right]=\theta\left[\begin{array}{c}0\\ \epsilon'\end{array}\right]$$

as well as

(2.13) $$\theta\left[\begin{array}{c}2-\epsilon\\ 1\end{array}\right]=\exp(-\pi\imath\epsilon)\,\theta\left[\begin{array}{c}\epsilon\\ 1\end{array}\right]\text{ and }\theta\left[\begin{array}{c}2-\epsilon\\ 0\end{array}\right]=\theta\left[\begin{array}{c}\epsilon\\ 0\end{array}\right].$$

Similarly,

$$\theta'\left[\begin{array}{c}-\epsilon\\ -\epsilon'\end{array}\right]=-\theta'\left[\begin{array}{c}\epsilon\\ \epsilon'\end{array}\right],$$

$$\theta'\left[\begin{array}{c}2-\epsilon\\ 1\end{array}\right]=-\exp(-\pi\imath\epsilon)\,\theta'\left[\begin{array}{c}\epsilon\\ 1\end{array}\right]\text{ and }\theta'\left[\begin{array}{c}2-\epsilon\\ 0\end{array}\right]=\theta'\left[\begin{array}{c}\epsilon\\ 0\end{array}\right].$$

There is little difference between theta functions and theta constants. The various expressions of the same objects merely facilitate the bookkeeping. Since 1 and τ are linearly independent over $\mathbb{R}$, an arbitrary $z\in\mathbb{C}$ can be written as $z=\frac{n}{2}+\frac{m}{2}\tau$, with n and m in $\mathbb{R}$. It thus follows from (2.8) that

(2.14) $$\theta\left[\begin{array}{c}\epsilon\\ \epsilon'\end{array}\right]\left(\tau\frac{m}{2}+\frac{n}{2},\tau\right)$$

---

[42]It should be clear from the context when the symbol $\theta\left[\begin{array}{c}\epsilon\\ \epsilon'\end{array}\right]$ stands for the theta function (of two variables), for the theta constant (a function of one variable), and for the value of the theta constant at a point.

$$= \exp\left(-\pi\imath\left\{\frac{1}{4}m^2\tau+\frac{1}{2}m(\epsilon'+n)\right\}\right)\theta\left[\begin{array}{c}\epsilon+m\\ \epsilon'+n\end{array}\right](0,\tau),$$

for all m and $n \in \mathbb{R}$.

Each of the above formulae follows easily from the definitions. We illustrate a typical argument by establishing (2.8).

Proof. We compute

$$\theta\left[\begin{array}{c}\epsilon\\ \epsilon'\end{array}\right]\left(z+\frac{\tau m}{2}+\frac{n}{2},\tau\right)$$

$$= \sum_{k\in\mathbb{Z}}\exp 2\pi\imath\left(\frac{1}{2}\left(k+\frac{\epsilon}{2}\right)^2\tau+\left(k+\frac{\epsilon}{2}\right)\left(z+\frac{\tau m}{2}+\frac{n}{2}+\frac{\epsilon'}{2}\right)\right)$$

$$= \sum_{k}\exp 2\pi\imath\left(\frac{1}{2}\left(k+\frac{\epsilon}{2}\right)^2\tau+\left(k+\frac{\epsilon}{2}\right)\tau\frac{m}{2}+\left(k+\frac{\epsilon}{2}\right)\left(z+\frac{n+\epsilon'}{2}\right)\right)$$

$$= \exp 2\pi\imath\left(-\frac{m^2\tau}{8}-\frac{m(\epsilon'+n)}{4}-\frac{mz}{2}\right)\theta\left[\begin{array}{c}\epsilon+m\\ \epsilon'+n\end{array}\right](z,\tau).$$

The last formula follows from the preceding one by expanding the expression in parentheses to

$$\frac{1}{2}\left(k+\frac{\epsilon+m}{2}\right)^2\tau+\left(k+\frac{\epsilon+m}{2}\right)\left(z+\frac{n+\epsilon'}{2}\right)-\frac{m^2\tau}{8}-\frac{m}{2}\left(z+\frac{n+\epsilon'}{2}\right).$$

It is an immediate consequence of (2.10) and (2.9) (along with the cheap trick that for all characteristics χ, $-\chi = \chi - 2\chi$) that if $\left[\begin{array}{c}\epsilon\\ \epsilon'\end{array}\right] \in \mathbb{Z}^2$, then

$$\theta\left[\begin{array}{c}\epsilon\\ \epsilon'\end{array}\right](-z,\tau) = \exp(\pi\imath\epsilon\epsilon')\,\theta\left[\begin{array}{c}\epsilon\\ \epsilon'\end{array}\right](z,\tau).$$

We conclude that if we restrict to integral characteristics $\left[\begin{array}{c}\epsilon\\ \epsilon'\end{array}\right] \in \mathbb{Z}^2$, the θ-functions decompose into even and odd functions and up to multiplicative constants ± 1 there are only four such functions. Three of these, $\theta\left[\begin{array}{c}0\\ 0\end{array}\right]$, $\theta\left[\begin{array}{c}0\\ 1\end{array}\right]$ and $\theta\left[\begin{array}{c}1\\ 0\end{array}\right]$, are even and one, $\theta\left[\begin{array}{c}1\\ 1\end{array}\right]$, is odd.

We record one more formula for arbitrary characteristics.

Lemma 1.4. *For each* $N \in \mathbb{Z}^+$,

$$\theta\left[\begin{array}{c}\epsilon\\ \epsilon'\end{array}\right](z,\tau) = \sum_{l=0}^{N-1}\theta\left[\begin{array}{c}\frac{\epsilon+2l}{N}\\ N\epsilon'\end{array}\right](Nz,N^2\tau).$$

Proof. The proof is a straightforward calculation.

$$\theta\begin{bmatrix}\epsilon\\ \epsilon'\end{bmatrix}(z,\tau)=\sum_{n\in\mathbb{Z}}\exp 2\pi\imath\left(\frac{1}{2}\left(n+\frac{\epsilon}{2}\right)^2\tau+\left(n+\frac{\epsilon}{2}\right)\left(z+\frac{\epsilon'}{2}\right)\right)$$

$$=\sum_{m\in\mathbb{Z}}\sum_{l=0}^{N-1}e^{2\pi\imath\left(\frac{1}{2}\left(Nm+l+\frac{\epsilon}{2}\right)^2\tau+\left(Nm+l+\frac{\epsilon}{2}\right)\left(z+\frac{\epsilon'}{2}\right)\right)}$$

$$=\sum_{m\in\mathbb{Z}}\sum_{l=0}^{N-1}e^{2\pi\imath\left(\frac{1}{2}\left(m+\frac{l}{N}+\frac{\epsilon}{2N}\right)^2N^2\tau+\left(m+\frac{l}{N}+\frac{\epsilon}{2N}\right)N\left(z+\frac{\epsilon'}{2}\right)\right)}$$

$$=\sum_{l=0}^{N-1}\theta\begin{bmatrix}\frac{\epsilon+2l}{N}\\ N\epsilon'\end{bmatrix}(Nz,N^2\tau).$$

Corollary 1.5. *For each $N\in\mathbb{Z}^+$,*

$$\theta'\begin{bmatrix}\epsilon\\ \epsilon'\end{bmatrix}(z,\tau)=N\sum_{l=0}^{N-1}\theta'\begin{bmatrix}\frac{\epsilon+2l}{N}\\ N\epsilon'\end{bmatrix}(Nz,N^2\tau).$$

Although the lemma and its corollary are formal consequences of the definitions, they have many interesting consequences as we shall see later. Among the immediate consequences are the following identities:

$$\theta\begin{bmatrix}\epsilon\\ 0\end{bmatrix}(0,\tau)=\theta\begin{bmatrix}\frac{\epsilon}{2}\\ 0\end{bmatrix}(0,4\tau)+\theta\begin{bmatrix}1-\frac{\epsilon}{2}\\ 0\end{bmatrix}(0,4\tau),$$

$$\theta\begin{bmatrix}\epsilon\\ 1\end{bmatrix}(0,\tau)=\exp\left(\frac{\pi\imath\epsilon}{2}\right)\left(\theta\begin{bmatrix}\frac{\epsilon}{2}\\ 0\end{bmatrix}(0,4\tau)-\theta\begin{bmatrix}1-\frac{\epsilon}{2}\\ 0\end{bmatrix}(0,4\tau)\right),$$

$$\theta'\begin{bmatrix}\epsilon\\ 0\end{bmatrix}(0,\tau)=2\left(\theta'\begin{bmatrix}\frac{\epsilon}{2}\\ 0\end{bmatrix}(0,4\tau)-\theta\begin{bmatrix}1-\frac{\epsilon}{2}\\ 0\end{bmatrix}(0,4\tau)\right)$$

and

$$\theta'\begin{bmatrix}\epsilon\\ 1\end{bmatrix}(0,\tau)=2\exp\left(\frac{\pi\imath\epsilon}{2}\right)\left(\theta'\begin{bmatrix}\frac{\epsilon}{2}\\ 0\end{bmatrix}(0,4\tau)+\theta'\begin{bmatrix}1-\frac{\epsilon}{2}\\ 0\end{bmatrix}(0,4\tau)\right),$$

which imply in particular that

$$\theta\begin{bmatrix}0\\ 0\end{bmatrix}(0,\tau)=\theta\begin{bmatrix}0\\ 0\end{bmatrix}(0,4\tau)+\theta\begin{bmatrix}1\\ 0\end{bmatrix}(0,4\tau),$$

$$\theta\begin{bmatrix}1\\ 0\end{bmatrix}(0,\tau)=2\,\theta\begin{bmatrix}\frac{1}{2}\\ 0\end{bmatrix}(0,4\tau),$$

$$\theta\begin{bmatrix}0\\ 1\end{bmatrix}(0,\tau)=\theta\begin{bmatrix}0\\ 0\end{bmatrix}(0,4\tau)-\theta\begin{bmatrix}1\\ 0\end{bmatrix}(0,4\tau)$$

and

$$\theta'\begin{bmatrix}1\\ 1\end{bmatrix}(0,\tau)=4\imath\left(\theta'\begin{bmatrix}\frac{1}{2}\\ 0\end{bmatrix}(0,4\tau)\right).$$

More rearrangement of summation formulae will be useful. In particular, we will need the following "doubling the parameter" formulae. The formula we give here holds in a larger setting. It is true for theta functions and not only theta constants. Furthermore, it is valid in the several variable case as well.

Lemma 1.6. *For all characteristics* $\begin{bmatrix} \epsilon \\ \epsilon' \end{bmatrix}$, $\begin{bmatrix} \delta \\ \delta' \end{bmatrix}$ *and all* $\tau \in \mathbb{H}^2$, *we have*

$$\theta\begin{bmatrix} \epsilon \\ \epsilon' \end{bmatrix}(0,\tau)\theta\begin{bmatrix} \delta \\ \delta' \end{bmatrix}(0,\tau) \tag{2.15}$$

$$= \theta\begin{bmatrix} \frac{\epsilon+\delta}{2} \\ \epsilon'+\delta' \end{bmatrix}(0,2\tau)\theta\begin{bmatrix} \frac{\epsilon-\delta}{2} \\ \epsilon'-\delta' \end{bmatrix}(0,2\tau)$$

$$+\theta\begin{bmatrix} \frac{\epsilon+\delta}{2}+1 \\ \epsilon'+\delta' \end{bmatrix}(0,2\tau)\theta\begin{bmatrix} \frac{\epsilon-\delta}{2}+1 \\ \epsilon'-\delta' \end{bmatrix}(0,2\tau).$$

Proof. It follows from the definitions that $\theta\begin{bmatrix} \epsilon \\ \epsilon' \end{bmatrix}(0,\tau)\theta\begin{bmatrix} \delta \\ \delta' \end{bmatrix}(0,\tau)$ is the sum over all integers n and m of

$$\exp(2\pi\imath)\left[\frac{1}{2}\left(n+\frac{\epsilon}{2}\right)^2\tau+\left(n+\frac{\epsilon}{2}\right)\frac{\epsilon'}{2}+\frac{1}{2}\left(m+\frac{\delta}{2}\right)^2\tau+\left(m+\frac{\delta}{2}\right)\frac{\delta'}{2}\right],$$

which can be rewritten as

$$\exp(2\pi\imath)\left[\frac{1}{2}\left(n+m+\frac{\epsilon+\delta}{2}\right)^2\frac{\tau}{2}+\frac{1}{2}\left(n-m+\frac{\epsilon-\delta}{2}\right)^2\frac{\tau}{2}\right.$$
$$\left.+\frac{1}{2}\left(n+m+\frac{\epsilon+\delta}{2}\right)\left(\frac{\epsilon'+\delta'}{2}\right)+\frac{1}{2}\left(n-m+\frac{\epsilon-\delta}{2}\right)\left(\frac{\epsilon'-\delta'}{2}\right)\right].$$

We now make the change of index $n-m=p$, and conclude that the preceding quantity equals

$$\exp(2\pi\imath)\left[\frac{1}{2}\left(2m+p+\frac{\epsilon+\delta}{2}\right)^2\frac{\tau}{2}+\frac{1}{2}\left(p+\frac{\epsilon-\delta}{2}\right)^2\frac{\tau}{2}\right.$$
$$\left.+\frac{1}{2}\left(2m+p+\frac{\epsilon+\delta}{2}\right)\left(\frac{\epsilon'+\delta'}{2}\right)+\frac{1}{2}\left(p+\frac{\epsilon-\delta}{2}\right)\left(\frac{\epsilon'-\delta'}{2}\right)\right];$$

this expression is to be summed over all integers m and p. The crucial step is to sum separately over the even and odd integers p; in other words the above expression breaks into a sum (write $p=2k$ and $p=2k+1$, respectively)

$$\exp(2\pi\imath)\left\{\frac{1}{2}\left(2m+2k+\frac{\epsilon+\delta}{2}\right)^2\frac{\tau}{2}+\frac{1}{2}\left(2k+\frac{\epsilon-\delta}{2}\right)^2\frac{\tau}{2}\right.$$
$$\left.+\frac{1}{2}\left(2m+2k+\frac{\epsilon+\delta}{2}\right)\left(\frac{\epsilon'+\delta'}{2}\right)+\frac{1}{2}\left(2k+\frac{\epsilon-\delta}{2}\right)\left(\frac{\epsilon'-\delta'}{2}\right)\right\}$$

$$+\exp(2\pi\imath)\left[\frac{1}{2}\left(2m+2k+1+\frac{\epsilon+\delta}{2}\right)^2\frac{\tau}{2}+\frac{1}{2}\left(2k+1+\frac{\epsilon-\delta}{2}\right)^2\frac{\tau}{2}\right.$$
$$\left.+\frac{1}{2}\left(2m+2k+1+\frac{\epsilon+\delta}{2}\right)\left(\frac{\epsilon'+\delta'}{2}\right)+\frac{1}{2}\left(2k+1+\frac{\epsilon-\delta}{2}\right)\left(\frac{\epsilon'-\delta'}{2}\right)\right].$$

We sum the above expression over all m and k. If we let $m+k=l$, the above reduces to

$$\exp(2\pi\imath)\left[\frac{1}{2}\left(l+\frac{\epsilon+\delta}{4}\right)^2 2\tau+\frac{1}{2}\left(k+\frac{\epsilon-\delta}{4}\right)^2 2\tau\right.$$
$$\left.+\left(l+\frac{\epsilon+\delta}{4}\right)\left(\frac{\epsilon'+\delta'}{2}\right)+\left(k+\frac{\epsilon-\delta}{4}\right)\left(\frac{\epsilon'-\delta'}{2}\right)\right]$$
$$+\exp(2\pi\imath)\left[\frac{1}{2}\left(l+\frac{\frac{\epsilon+\delta}{2}+1}{2}\right)^2 2\tau+\frac{1}{2}\left(k+\frac{\frac{\epsilon-\delta}{2}+1}{2}\right)^2 2\tau\right.$$
$$\left.+\left(l+\frac{\frac{\epsilon+\delta}{2}+1}{2}\right)\left(\frac{\epsilon'+\delta'}{2}\right)+\left(k+\frac{\frac{\epsilon-\delta}{2}+1}{2}\right)\left(\frac{\epsilon'-\delta'}{2}\right)\right].$$

Summing the above expression over all integers l and k gives the right hand side of (2.15). Another formula which we will find convenient in the sequel and which is essentially (2.8) after setting $z=0$ is given by:

Lemma 1.7. *Let* $\begin{bmatrix}\epsilon\\ \epsilon'\end{bmatrix}\in\mathbb{R}^2$.

(a) *For all* δ *and* $\delta'\in\mathbb{R}$,

$$\theta\begin{bmatrix}\epsilon\\ \epsilon'\end{bmatrix}\left(\frac{1-\delta'}{2}+\frac{1-\delta}{2}\tau,\tau\right)$$
$$=\left(\exp\left(-\pi\imath\left\{\frac{1}{4}(1-\delta)^2\tau+\frac{1}{2}(1-\delta)(1+\epsilon'-\delta')\right\}\right)\right)\theta\begin{bmatrix}\epsilon+1-\delta\\ \epsilon'+1-\delta'\end{bmatrix}(0,\tau).$$

(b)

$$\theta'\begin{bmatrix}\epsilon\\ \epsilon'\end{bmatrix}\left(\frac{1-\epsilon'}{2}+\frac{1-\epsilon}{2}\tau,\tau\right)$$
$$=\exp\left(-\pi\imath\left\{\frac{1}{4}(1-\epsilon)^2\tau+\frac{1}{2}(1-\epsilon)\right\}\right)\theta'\begin{bmatrix}1\\ 1\end{bmatrix}(0,\tau).$$

The next theorem describes the divisor of the theta function. For $\tau\in\mathbb{H}^2$, we let as before $\mathcal{P}_a$ be the period parallelogram with vertices a, $a+1$, $a+1+\tau$ and $a+\tau$.

Theorem 1.8. *Let* $\chi=\begin{bmatrix}\epsilon\\ \epsilon'\end{bmatrix}$ *be a characteristic. The function* $\theta[\chi](\cdot,\tau)$ *has a single simple zero in any period parallelogram. The zero is at the point* $\frac{1-\epsilon'}{2}+\frac{1-\epsilon}{2}\tau$ *(and its translates by lattice points).*

Proof. As a consequence of (2.3), it suffices to prove the theorem for the trivial characteristic $\chi = 0$. Since the zeros of a holomorphic function are isolated, by properly choosing the origin a of $\mathcal{P} = \mathcal{P}_a$, we may assume that $\theta = \theta(\cdot, \tau)$ does not vanish on the boundary $\partial\mathcal{P}$ of $\mathcal{P}$. Thus the number of zeros of $\theta(\cdot, \tau)$ equals

$$\frac{1}{2\pi\imath}\int_{\partial\mathcal{P}}\frac{\theta'}{\theta}dz = \frac{1}{2\pi\imath}\left(\int_a^{a+1} - \int_{a+\tau}^{a+1+\tau} - \int_a^{a+\tau} + \int_{a+1}^{a+\tau+1}\right)\frac{\theta'}{\theta}dz,$$

where the last four line integrals are over line segments joining the limits of integration. Using (2.4), we conclude that the first two terms of the expression enclosed by parentheses add up to $2\pi\imath$; the last two to 0. We have shown that θ has a single simple zero. Its location is found by evaluating, as above, the integral (see also (2.31), below)

$$\frac{1}{2\pi\imath}\int_{\partial\mathcal{P}} z\frac{\theta'}{\theta}dz = \frac{1}{2\pi\imath}\int_a^{a+1} z\left(\frac{\theta'(z)}{\theta(z)} - \frac{\theta'(z+\tau)}{\theta(z+\tau)}\right)dz - \frac{\tau}{2\pi\imath}\int_a^{a+1}\frac{\theta'(z+\tau)}{\theta(z+\tau)}dz$$

$$-\frac{1}{2\pi\imath}\int_a^{a+\tau} z\left(\frac{\theta'(z)}{\theta(z)} - \frac{\theta'(z+1)}{\theta(z+1)}\right)dz + \frac{1}{2\pi\imath}\int_a^{a+\tau}\frac{\theta'(z+1)}{\theta(z+1)}dz$$

$$= \int_a^{a+1} zdz - \frac{\tau}{2\pi\imath}(\log\theta(a+1+\tau) - \log\theta(a+\tau))$$

$$+\frac{1}{2\pi\imath}(\log\theta(a+\tau+1) - \log\theta(a+1))$$

$$= \left(a+\frac{1}{2}\right) + \left(-a-1-\frac{\tau}{2}\right) = -\frac{1}{2} - \frac{\tau}{2} + \text{ a period.}$$

A period enters the picture because of the choices of the branches of the logarithm of the exponential and θ-function that we need to use. Alternatively, the result folllows from simple parity cosiderations.

Corollary 1.9. *Let f be an entire function that satisfies the functional equations (2.5) and (2.6) for some ϵ and ϵ' in $\mathbb{R}$; that is, for all $z \in \mathbb{C}$,*

$$f(z+1) = \exp(\pi\imath\epsilon)\ f(z) \text{ and } f(z+\tau) = \exp(-\pi\imath\{\epsilon' + 2z + \tau\})\ f(z).$$

Then there exists $c \in \mathbb{C}$ such that

$$f = c\,\theta\begin{bmatrix}\epsilon\\ \epsilon'\end{bmatrix}(\cdot, \tau).$$

Proof. The function

$$\frac{f}{\theta\begin{bmatrix}\epsilon\\ \epsilon'\end{bmatrix}(\cdot, \tau)}$$

is elliptic with at most one simple pole in a period parallelogram, hence constant.

Corollary 1.10. *Let $\chi \in \mathbb{R}^2$. The theta constant $\theta[\chi]$ vanishes (identically) if and only if χ is equivalent to the characteristic $\begin{bmatrix} 1 \\ 1 \end{bmatrix}$. If χ is not equivalent to $\begin{bmatrix} 1 \\ 1 \end{bmatrix}$, then the theta constant $\theta[\chi]$ never vanishes (viewed as a (holomorphic) function on $\mathbb{H}^2$).*

1.2. The transformation formula. A property of theta functions deeper than the previous identities and which is central to the development in this text is contained in the next formula. For any characteristic $\chi = \begin{bmatrix} \epsilon \\ \epsilon' \end{bmatrix} \in \mathbb{R}^2$, and any element $\gamma = \begin{bmatrix} a & b \\ c & d \end{bmatrix}$ of $\mathrm{SL}(2,\mathbb{Z})$, we have

$$\text{(2.16)} \qquad \frac{\exp \pi \imath \left\{\frac{-cz^2}{c\tau+d}\right\} \theta \begin{bmatrix} \epsilon \\ \epsilon' \end{bmatrix} \left(\frac{z}{c\tau+d}, \frac{a\tau+b}{c\tau+d}\right)}{\theta \begin{bmatrix} a\epsilon + c\epsilon' - ac \\ b\epsilon + d\epsilon' + bd \end{bmatrix} (z,\tau)} = \kappa\left(\begin{bmatrix} \epsilon \\ \epsilon' \end{bmatrix}, \gamma\right) (c\tau+d)^{\frac{1}{2}},$$

for all $z \in \mathbb{C}$, $\tau \in \mathbb{H}^2$, where[43] $\kappa\left(\begin{bmatrix} \epsilon \\ \epsilon' \end{bmatrix}, \gamma\right)$ is a constant depending on the characteristic $\begin{bmatrix} \epsilon \\ \epsilon' \end{bmatrix}$ and the matrix γ. We will show that

$$\text{(2.17)} \qquad \kappa\left(\begin{bmatrix} \epsilon \\ \epsilon' \end{bmatrix}, \gamma\right)$$

$$= \exp 2\pi\imath \left\{-\frac{1}{4}(a\epsilon + c\epsilon')bd - \frac{1}{8}(ab\epsilon^2 + cd\epsilon'^2 + 2bc\epsilon\epsilon')\right\} \kappa\left(\begin{bmatrix} 0 \\ 0 \end{bmatrix}, \gamma\right),$$

with $\kappa\left(\begin{bmatrix} 0 \\ 0 \end{bmatrix}, \gamma\right)$ an eighth root of unity.

Theorem 1.11. *Let $\chi = \begin{bmatrix} \epsilon \\ \epsilon' \end{bmatrix} \in \mathbb{R}^2$ and $\gamma = \begin{bmatrix} a & b \\ c & d \end{bmatrix} \in \mathrm{SL}(2,\mathbb{Z})$. The theta constants satisfy the transformation rule*

$$\text{(2.18)} \qquad \theta \begin{bmatrix} \epsilon \\ \epsilon' \end{bmatrix} (0, \gamma(\tau)) = \kappa (c\tau+d)^{\frac{1}{2}} \theta \begin{bmatrix} a\epsilon + c\epsilon' - ac \\ b\epsilon + d\epsilon' + bd \end{bmatrix} (0,\tau),$$

for all $\tau \in \mathbb{H}^2$, where $\kappa = \kappa(\chi,\gamma)$ is the constant depending on the characteristic χ and the matrix γ defined by the above formula (2.17).

---

[43] To fix κ, we must make a choice for the square root of $(c\tau + d)$; we will always choose it with argument in $[0, \pi)$.

Proof. We begin by establishing (2.16). Fix τ and let g be the function defined by the left side of the equality. It is obviously a meromorphic function of z on $\mathbb{C}$. Our first task is to show that g is a doubly periodic function for the lattice generated by 1 and τ. Towards this end we begin with an examination of the periodicity of the function

$$z \mapsto \theta\begin{bmatrix}\epsilon\\ \epsilon'\end{bmatrix}\left(\frac{z}{c\tau+d}, \frac{a\tau+b}{c\tau+d}\right).$$

We write (as in Chapter 1)

$$\frac{1}{c\tau+d} = a - c\frac{a\tau+b}{c\tau+d} \text{ and } \frac{\tau}{c\tau+d} = -b + d\frac{a\tau+b}{c\tau+d}, \tag{2.19}$$

and conclude from these identities that

$$\theta\begin{bmatrix}\epsilon\\ \epsilon'\end{bmatrix}\left(\frac{z+1}{c\tau+d}, \frac{a\tau+b}{c\tau+d}\right)$$

$$= \exp 2\pi\imath\left\{\frac{a\epsilon+c\epsilon'}{2} + c\frac{z}{c\tau+d} - \frac{c^2}{2}\frac{a\tau+b}{c\tau+d}\right\}\theta\begin{bmatrix}\epsilon\\ \epsilon'\end{bmatrix}\left(\frac{z}{c\tau+d}, \frac{a\tau+b}{c\tau+d}\right)$$

and

$$\theta\begin{bmatrix}\epsilon\\ \epsilon'\end{bmatrix}\left(\frac{z+\tau}{c\tau+d}, \frac{a\tau+b}{c\tau+d}\right)$$

$$= \exp 2\pi\imath\left\{\frac{-b\epsilon-d\epsilon'}{2} - d\frac{z}{c\tau+d} - \frac{d^2}{2}\frac{a\tau+b}{c\tau+d}\right\}\theta\begin{bmatrix}\epsilon\\ \epsilon'\end{bmatrix}\left(\frac{z}{c\tau+d}, \frac{a\tau+b}{c\tau+d}\right).$$

It is immediate that

$$\exp\pi\imath\left\{\frac{-c(z+1)^2}{c\tau+d}\right\} = \exp\pi\imath\left\{\frac{-2cz}{c\tau+d} - \frac{c}{c\tau+d}\right\}\exp\pi\imath\left\{\frac{-cz^2}{c\tau+d}\right\}$$

and

$$\exp\pi\imath\left\{\frac{-c(z+\tau)^2}{c\tau+d}\right\} = \exp\pi\imath\left\{\frac{-2c\tau z}{c\tau+d} - \frac{c\tau^2}{c\tau+d}\right\}\exp\pi\imath\left\{\frac{-cz^2}{c\tau+d}\right\}.$$

Finally

$$\theta\begin{bmatrix}a\epsilon+c\epsilon'-ac\\ b\epsilon+d\epsilon'+bd\end{bmatrix}(z+1,\tau)$$

$$= \exp\pi\imath\{a\epsilon+c\epsilon'-ac\}\,\theta\begin{bmatrix}a\epsilon+c\epsilon'-ac\\ b\epsilon+d\epsilon'+bd\end{bmatrix}(z,\tau)$$

and

$$\theta\begin{bmatrix}a\epsilon+c\epsilon'-ac\\ b\epsilon+d\epsilon'+bd\end{bmatrix}(z+\tau,\tau)$$

$$= \exp 2\pi\imath\left\{-z-\frac{\tau}{2} - \frac{b\epsilon+d\epsilon'+bd}{2}\right\}\theta\begin{bmatrix}a\epsilon+c\epsilon'-ac\\ b\epsilon+d\epsilon'+bd\end{bmatrix}(z,\tau).$$

It is easy to see that we have arranged our functions to have cancellation; that is,

$$g(z+1) = g(z) = g(z+\tau), \text{ for all } z\in\mathbb{C}.$$

The function g has (at most) a single simple pole at a single point (in a period parallelogram). See Theorem 1.8. Hence it is a constant which depends on the parameter τ, the characteristic $\begin{bmatrix} \epsilon \\ \epsilon' \end{bmatrix}$ and the matrix γ. We fix for the moment the characteristic and matrix, and denote the constant (as a function of z) by $\kappa(\tau)$. To obtain the formula for $\kappa(\tau)$, we rewrite our basic identity as

$$\exp \pi\imath \left\{\frac{-cz^2}{c\tau+d}\right\} \theta \begin{bmatrix} \epsilon \\ \epsilon' \end{bmatrix} \left(\frac{z}{c\tau+d}, \frac{a\tau+b}{c\tau+d}\right) = \kappa(\tau)\, \theta \begin{bmatrix} a\epsilon + c\epsilon' - ac \\ b\epsilon + d\epsilon' + bd \end{bmatrix} (z,\tau)$$

and expand both sides in a power series (in z) about the origin. We begin with

$$\exp \pi\imath \left\{\frac{-cz^2}{c\tau+d}\right\} = 1 - \frac{\pi\imath c z^2}{c\tau+d} + \dots.$$

$$\theta \begin{bmatrix} \epsilon \\ \epsilon' \end{bmatrix} \left(\frac{z}{c\tau+d}, \frac{a\tau+b}{c\tau+d}\right) = \theta \begin{bmatrix} \epsilon \\ \epsilon' \end{bmatrix} \left(0, \frac{a\tau+b}{c\tau+d}\right)$$

$$+ \frac{\theta' \begin{bmatrix} \epsilon \\ \epsilon' \end{bmatrix} (0, \frac{a\tau+b}{c\tau+d})}{c\tau+d} z + \frac{\theta'' \begin{bmatrix} \epsilon \\ \epsilon' \end{bmatrix} (0, \frac{a\tau+b}{c\tau+d})}{2(c\tau+d)^2} z^2 + \frac{\theta''' \begin{bmatrix} \epsilon \\ \epsilon' \end{bmatrix} (0, \frac{a\tau+b}{c\tau+d})}{6(c\tau+d)^3} z^3 + \dots$$

and

$$\theta \begin{bmatrix} a\epsilon + c\epsilon' - ac \\ b\epsilon + d\epsilon' + bd \end{bmatrix} (z,\tau)$$

$$= \theta \begin{bmatrix} a\epsilon + c\epsilon' - ac \\ b\epsilon + d\epsilon' + bd \end{bmatrix} (0,\tau) + \theta' \begin{bmatrix} a\epsilon + c\epsilon' - ac \\ b\epsilon + d\epsilon' + bd \end{bmatrix} (0,\tau) z$$

$$+ \frac{1}{2}\theta'' \begin{bmatrix} a\epsilon + c\epsilon' - ac \\ b\epsilon + d\epsilon' + bd \end{bmatrix} (0,\tau) z^2 + \frac{1}{6}\theta''' \begin{bmatrix} a\epsilon + c\epsilon' - ac \\ b\epsilon + d\epsilon' + bd \end{bmatrix} (0,\tau) z^3 + \dots,$$

where the prime ($'$) after the theta function denotes (as usual) differentiation with respect to z. Equating coefficients of powers of z leads to (four equations that we need)

$$\theta \begin{bmatrix} \epsilon \\ \epsilon' \end{bmatrix} \left(0, \frac{a\tau+b}{c\tau+d}\right) = \kappa(\tau) \theta \begin{bmatrix} a\epsilon + c\epsilon' - ac \\ b\epsilon + d\epsilon' + bd \end{bmatrix} (0,\tau),$$

$$\frac{1}{c\tau+d}\theta' \begin{bmatrix} \epsilon \\ \epsilon' \end{bmatrix} \left(0, \frac{a\tau+b}{c\tau+d}\right) = \kappa(\tau) \theta' \begin{bmatrix} a\epsilon + c\epsilon' - ac \\ b\epsilon + d\epsilon' + bd \end{bmatrix} (0,\tau),$$

$$\frac{1}{2(c\tau+d)^2}\theta'' \begin{bmatrix} \epsilon \\ \epsilon' \end{bmatrix} \left(0, \frac{a\tau+b}{c\tau+d}\right) - \frac{\pi\imath c}{c\tau+d}\theta \begin{bmatrix} \epsilon \\ \epsilon' \end{bmatrix} \left(0, \frac{a\tau+b}{c\tau+d}\right)$$
$$= \frac{\kappa(\tau)}{2}\theta'' \begin{bmatrix} a\epsilon + c\epsilon' - ac \\ b\epsilon + d\epsilon' + bd \end{bmatrix} (0,\tau)$$

and

$$\frac{1}{6(c\tau+d)^3}\theta''' \begin{bmatrix} \epsilon \\ \epsilon' \end{bmatrix} \left(0, \frac{a\tau+b}{c\tau+d}\right) - \frac{\pi\imath c}{(c\tau+d)^2}\theta' \begin{bmatrix} \epsilon \\ \epsilon' \end{bmatrix} \left(0, \frac{a\tau+b}{c\tau+d}\right)$$

$$= \frac{\kappa(\tau)}{6}\theta''' \begin{bmatrix} a\epsilon + c\epsilon' - ac \\ b\epsilon + d\epsilon' + bd \end{bmatrix} (0, \tau).$$

Dividing the third of the above equations by the first, we obtain

$$\frac{\theta'' \begin{bmatrix} \epsilon \\ \epsilon' \end{bmatrix} (0, \gamma(\tau))}{\theta \begin{bmatrix} \epsilon \\ \epsilon' \end{bmatrix} (0, \gamma(\tau))} \gamma'(\tau) = \frac{2\pi\imath c}{c\tau + d} + \frac{\theta'' \begin{bmatrix} a\epsilon + c\epsilon' - ac \\ b\epsilon + d\epsilon' + bd \end{bmatrix} (0, \tau)}{\theta \begin{bmatrix} a\epsilon + c\epsilon' - ac \\ b\epsilon + d\epsilon' + bd \end{bmatrix} (0, \tau)}. \tag{2.20}$$

Similarly, from the other two equations

$$\frac{\theta''' \begin{bmatrix} \epsilon \\ \epsilon' \end{bmatrix} (0, \gamma(\tau))}{\theta' \begin{bmatrix} \epsilon \\ \epsilon' \end{bmatrix} (0, \gamma(\tau))} \gamma'(\tau) = \frac{6\pi\imath c}{c\tau + d} + \frac{\theta''' \begin{bmatrix} a\epsilon + c\epsilon' - ac \\ b\epsilon + d\epsilon' + bd \end{bmatrix} (0, \tau)}{\theta' \begin{bmatrix} a\epsilon + c\epsilon' - ac \\ b\epsilon + d\epsilon' + bd \end{bmatrix} (0, \tau)}.$$

Differentiating the first of our four equations with respect to τ and using the heat equation yield

$$\frac{1}{4\pi\imath} \frac{\theta'' \begin{bmatrix} \epsilon \\ \epsilon' \end{bmatrix} (0, \gamma(\tau))}{(c\tau + d)^2}$$

$$= \frac{\kappa(\tau)}{4\pi\imath}\theta'' \begin{bmatrix} a\epsilon + c\epsilon' - ac \\ b\epsilon + d\epsilon' + bd \end{bmatrix} (0, \tau) + \kappa'(\tau)\theta \begin{bmatrix} a\epsilon + c\epsilon' - ac \\ b\epsilon + d\epsilon' + bd \end{bmatrix} (0, \tau)$$

(the prime denotes differentiation with respect to z for the θ-function (as before) and with respect to τ for κ). "Dividing by the first equation" results in

$$\frac{\theta'' \begin{bmatrix} \epsilon \\ \epsilon' \end{bmatrix} (0, \gamma(\tau))}{\theta \begin{bmatrix} \epsilon \\ \epsilon' \end{bmatrix} (0, \gamma(\tau))} \gamma'(\tau) = \frac{\theta'' \begin{bmatrix} a\epsilon + c\epsilon' - ac \\ b\epsilon + d\epsilon' + bd \end{bmatrix} (0, \tau)}{\theta \begin{bmatrix} a\epsilon + c\epsilon' - ac \\ b\epsilon + d\epsilon' + bd \end{bmatrix} (0, \tau)} + 4\pi\imath \frac{\kappa'(\tau)}{\kappa(\tau)}.$$

Combining this last equation with (2.20) leads to the ordinary differential equation

$$\frac{\kappa'(\tau)}{\kappa(\tau)} = \frac{c}{2(c\tau + d)}$$

satisfied by our unknown function κ. From this it easily follows that

$$\kappa(\tau) = \kappa_o(c\tau + d)^{\frac{1}{2}},$$

where the constant κ_o depends on the characteristic $\begin{bmatrix} \epsilon \\ \epsilon' \end{bmatrix}$ and the matrix γ.

It is thus a consequence of what we have already proved (for the characteristic $\begin{bmatrix} 0 \\ 0 \end{bmatrix}$) that

$$\exp \pi \imath \left\{ \frac{-cz^2}{c\tau+d} \right\} \theta \begin{bmatrix} 0 \\ 0 \end{bmatrix} \left(\frac{z}{c\tau+d}, \frac{a\tau+b}{c\tau+d} \right)$$
$$= \kappa \left(\begin{bmatrix} 0 \\ 0 \end{bmatrix}, \gamma \right) (c\tau+d)^{\frac{1}{2}} \theta \begin{bmatrix} -ac \\ bd \end{bmatrix} (z, \tau).$$

Set $z = \zeta + \tau \frac{a\epsilon + c\epsilon'}{2} + \frac{b\epsilon + d\epsilon'}{2}$ with ϵ, ϵ' arbitrary real numbers and obtain

(2.21)

$$\exp \pi \imath \left\{ \frac{-c(\zeta + \tau \frac{a\epsilon+c\epsilon'}{2} + \frac{b\epsilon+d\epsilon'}{2})^2}{c\tau+d} \right\} \theta \begin{bmatrix} 0 \\ 0 \end{bmatrix} \left(\frac{\zeta + \tau \frac{a\epsilon+c\epsilon'}{2} + \frac{b\epsilon+d\epsilon'}{2}}{c\tau+d}, \gamma(\tau) \right)$$
$$= \kappa \left(\begin{bmatrix} 0 \\ 0 \end{bmatrix}, \gamma \right) (c\tau+d)^{\frac{1}{2}} \theta \begin{bmatrix} -ac \\ bd \end{bmatrix} \left(\zeta + \tau \frac{a\epsilon + c\epsilon'}{2} + \frac{b\epsilon + d\epsilon'}{2}, \tau \right).$$

We now note that from (2.14), it follows that

(2.22)
$$\theta \begin{bmatrix} -ac \\ bd \end{bmatrix} \left(\zeta + \tau \frac{a\epsilon + c\epsilon'}{2} + \frac{b\epsilon + d\epsilon'}{2}, \tau \right)$$
$$= \exp 2\pi \imath \left\{ -\frac{1}{2}(a\epsilon + c\epsilon')\zeta - \frac{1}{8}(a\epsilon + c\epsilon')^2 \tau - \frac{1}{4}(a\epsilon + c\epsilon')(bd + b\epsilon + d\epsilon') \right\}$$
$$\times \theta \begin{bmatrix} a\epsilon + c\epsilon' - ac \\ b\epsilon + d\epsilon' + bd \end{bmatrix} (\zeta, \tau).$$

From (2.8), using (2.19), we see that

(2.23)
$$\theta \begin{bmatrix} 0 \\ 0 \end{bmatrix} \left(\frac{\zeta + \tau \frac{a\epsilon+c\epsilon'}{2} + \frac{b\epsilon+d\epsilon'}{2}}{c\tau+d}, \gamma(\tau) \right)$$
$$= \exp 2\pi \imath \left\{ -\frac{1}{2}\epsilon \frac{\zeta}{c\tau+d} - \frac{1}{8}\epsilon^2 \frac{a\tau+b}{c\tau+d} - \frac{1}{4}\epsilon\epsilon' \right\} \theta \begin{bmatrix} \epsilon \\ \epsilon' \end{bmatrix} \left(\frac{\zeta}{c\tau+d}, \tau \right).$$

From (2.21), using (2.22) and (2.23), we obtain a formula which has the form

$$\exp \pi \imath \left\{ -\frac{c\zeta^2}{c\tau+d} \right\} \theta \begin{bmatrix} \epsilon \\ \epsilon' \end{bmatrix} \left(\frac{\zeta}{c\tau+d}, \gamma(\tau) \right)$$
$$= \kappa \left(\begin{bmatrix} 0 \\ 0 \end{bmatrix}, \gamma \right) (c\tau+d)^{\frac{1}{2}} \theta \begin{bmatrix} a\epsilon + c\epsilon' - ac \\ b\epsilon + d\epsilon' + bd \end{bmatrix} (\zeta, \tau) E,$$

with $E = E(\chi, \gamma)$ a complicated exponential expression involving all the variables (depending on the characteristic χ and the matrix γ) we have used. With a little bit of perseverance, however, one can simplify E to be

$$\exp 2\pi \imath \left\{ -\frac{1}{4}(a\epsilon + c\epsilon')bd - \frac{1}{8}(ab\epsilon^2 + cd\epsilon'^2 + 2bc\epsilon\epsilon') \right\}$$

from which we obtain (2.17), our almost final formula for the constant κ. There now only remains the problem of computing $\kappa\left(\begin{bmatrix}0\\0\end{bmatrix},\gamma\right)$ for arbitrary $\gamma \in \mathrm{SL}(2,\mathbb{Z})$. We do this for our three[44] favorite generators for the group of unimodular matrices: $-I, A$, and B. It is transparent from the transformation formula that for all characteristics $\begin{bmatrix}\epsilon\\\epsilon'\end{bmatrix}$,

$$\kappa\left(\begin{bmatrix}\epsilon\\\epsilon'\end{bmatrix},-I\right) = -\imath.$$

It follows from the definition of theta functions that

$$\theta\begin{bmatrix}0\\0\end{bmatrix}(0,B(\tau)) = \sum_{n=-\infty}^{\infty} \exp 2\pi\imath\left\{\frac{1}{2}n^2(\tau+1)\right\}$$

$$= \sum_{n=-\infty}^{\infty} \exp 2\pi\imath\left\{\frac{1}{2}n^2\tau+\frac{n}{2}\right\} = \theta\begin{bmatrix}0\\1\end{bmatrix}(0,\tau)$$

(we have used in the penultimate of the above equalities the fact that for integers N, $\exp\pi\imath N$ depends only on the parity of N and hence we can replace $\exp\pi\imath n^2$ by $\exp\pi\imath n$), so that

$$\kappa\left(\begin{bmatrix}0\\0\end{bmatrix},B\right) = 1.$$

We now treat the case of the motion A. Here we observe that $\tau=\imath$ is a fixed point of the motion, and since κ is independent of the point τ, we compute at that point. We have

$$\theta\begin{bmatrix}0\\0\end{bmatrix}(0,A(\imath)) = \kappa\left(\begin{bmatrix}0\\0\end{bmatrix},A\right)\sqrt{\imath}\,\theta\begin{bmatrix}0\\0\end{bmatrix}(0,\imath).$$

Since $A(\imath)=\imath$, we get

$$\kappa\left(\begin{bmatrix}0\\0\end{bmatrix},A\right) = \frac{1}{\sqrt{\imath}} = \exp\left(-\frac{\pi\imath}{4}\right).$$

It remains to derive information about $\kappa\left(\begin{bmatrix}0\\0\end{bmatrix},\gamma\right)$ for arbitrary $\gamma \in \mathrm{SL}(2,\mathbb{Z})$, from the information already obtained (the formula for this mysterious quantity for the generators $-I$, A, and B of $\mathrm{SL}(2,\mathbb{Z})$) which will complete the proof of the theorem.

We write our basic identity (with m and m' integers) as

$$\theta^2\begin{bmatrix}m\\m'\end{bmatrix}(0,\gamma(\tau)) = \kappa^2\left(\begin{bmatrix}m\\m'\end{bmatrix},\gamma\right)(c\tau+d)\theta^2\begin{bmatrix}am+cm'-ac\\bm+dm'+bd\end{bmatrix}(0,\tau),$$

[44]Since in $\mathrm{SL}(2,\mathbb{Z})$, $A^2=-I$, we can dispense with the calculation for $-I$.

and conclude that

$$\kappa^2\left(\begin{bmatrix} m \\ m' \end{bmatrix}, \gamma_1\gamma_2\right) = \kappa^2\left(\begin{bmatrix} m \\ m' \end{bmatrix}, \gamma_1\right)\ \kappa^2\left(\begin{bmatrix} a_1m + c_1m' - a_1c_1 \\ b_1m + d_1m' + b_1d_1 \end{bmatrix}, \gamma_2\right)$$

for all $\gamma_1 = \begin{bmatrix} a_1 & b_1 \\ c_1 & d_1 \end{bmatrix}$ and $\gamma_2 = \begin{bmatrix} a_2 & b_2 \\ c_2 & d_2 \end{bmatrix}$ in $\mathrm{SL}(2,\mathbb{Z})$ and all integers m, m'. The last formula together with (2.17) allows us to conclude that

$$\kappa^2\left(\begin{bmatrix} 0 \\ 0 \end{bmatrix}, \gamma_1\gamma_2\right) = \kappa^2\left(\begin{bmatrix} 0 \\ 0 \end{bmatrix}, \gamma_1\right)\kappa^2\left(\begin{bmatrix} 0 \\ 0 \end{bmatrix}, \gamma_2\right) \tag{2.24}$$

$$\times e^{2\pi\imath\left\{\frac{1}{2}(a_1c_1a_2 - b_1d_1c_2)b_2d_2 - \frac{1}{4}(a_1^2c_1^2a_2b_2 + b_1^2d_1^2c_2d_2 - 2a_1b_1c_1d_1b_2c_2)\right\}}.$$

The above formula and the fact that $\kappa\left(\begin{bmatrix} 0 \\ 0 \end{bmatrix}, \gamma\right)$ is an 8-th root of unity for $\gamma = -I$, A, and B allow us to conclude (because A and B generate $\mathrm{SL}(2,\mathbb{Z})$) that $\kappa\left(\begin{bmatrix} 0 \\ 0 \end{bmatrix}, \gamma\right)$ is an 8-th root of unity for all $\gamma \in \mathrm{SL}(2,\mathbb{Z})$.

Our next claim is that $\kappa\left(\begin{bmatrix} \frac{m}{k} \\ \frac{m'}{k} \end{bmatrix}, \gamma\right)$ is an $8k$-th root of unity for every motion γ in the preimage[45] in $\mathrm{SL}(2,\mathbb{Z})$ of $\Gamma(k)$, for all primes k and all integers m and m'. Also, for integral characteristics $\begin{bmatrix} m \\ m' \end{bmatrix}$, $\kappa\left(\begin{bmatrix} m \\ m' \end{bmatrix}, \gamma\right)$ is an 8-th root of unity for all $\gamma \in \mathrm{SL}(2,\mathbb{Z})$. Both cf these claims follow directly from (2.17) and (2.24).

Let, for the time being, $Z(k)$ denote the set of characteristics of the form $\begin{bmatrix} \frac{m}{k} \\ \frac{m'}{k} \end{bmatrix}$ with $m, m', k \in Z$, and $X(k)$ denote the subset of $Z(k)$ with the property that the integers m, m' have the same parity as k. We summarize the basic facts about κ in

Proposition 1.12. *Let $\chi \in Z(k)$ and $\gamma \in \Gamma(k) \subset \mathrm{SL}(2,\mathbb{Z})$. Then*
(a) *$\kappa(\chi,\gamma)$ is an $8k$-th root of unity,*
(b) *$\kappa(\chi,\gamma)$ is a $4k$-th root of unity if $\chi \in X(k)$ and k is even, and*
(c) *$\kappa(\chi,\gamma)$ is a $2k$-th root of unity if $\chi \in X(k)$ and $4|k$ or $k = 2$. Further,*
(d) *$\kappa(\chi,\gamma)$ is an 8-th root of unity, if $\chi \in X(2)$ and $\gamma \in \Gamma$.*

An equivalence relation on characteristics as well as sets of distinguished classes of characteristics are introduced in the next section.

1.3. More transformation formulae.

We rewrite (2.16) as

$$f(z,\tau)\ \theta[\chi](z\gamma'(\tau)^{\frac{1}{2}}, \gamma(\tau)) = \kappa(\chi,\gamma)\gamma'(\tau)^{-\frac{1}{4}}\ \theta[\chi\gamma](z,\tau),$$

[45] When there can be little confusion, we will identify $\Gamma(k)$ with its preimage in $\mathrm{SL}(2,\mathbb{Z})$.

where

$$f(z,\tau) = \exp \pi\imath \left\{ \frac{-cz^2}{c\tau+d} \right\} = \exp \pi\imath\{-cz^2\gamma'(\tau)^{\frac{1}{2}}\}.$$

Differentiating the above formula n times with respect to z, we obtain

$$\sum_{k=0}^{n} \binom{n}{k} f^{(k)}(z,\tau) \;\; \gamma'(\tau)^{\frac{1}{2}(n-k)} \; \theta^{(n-k)}[\chi] \left(\frac{z}{c\tau+d}, \gamma(\tau) \right)$$

$$= \kappa(\chi,\gamma)\gamma'(\tau)^{-\frac{1}{4}} \; \theta^{(n)}[\chi\gamma](z,\tau).$$

For $z=0$, the above formula simplifies to

$$\sum_{k=0}^{n} \binom{n}{k} f^{(k)}(0,\tau) \;\; \gamma'(\tau)^{\frac{1}{2}(n-k)} \; \theta^{(n-k)}[\chi](0,\gamma(\tau))$$

$$= \kappa(\chi,\gamma)\gamma'(\tau)^{-\frac{1}{4}} \; \theta^{(n)}[\chi\gamma](0,\tau).$$

From the fact that f is an even function (of z) and $f(0)=1$, we conclude (the case $n=1$) that

$$\theta'[\chi](0,\gamma(\tau)) = \kappa(\chi,\gamma)\gamma'(\tau)^{-\frac{3}{4}} \; \theta'[\chi\gamma](0,\tau). \tag{2.25}$$

It is a direct consequence of (2.16) that

$$f(z,\tau)\frac{\theta[\chi](z\gamma'(\tau)^{\frac{1}{2}},\gamma(\tau))}{\theta[\chi](0,\gamma(\tau))} = \frac{\theta[\chi\gamma](z,\tau)}{\theta[\chi\gamma](0,\tau)}, \tag{2.26}$$

whenever the denominators do not vanish (equivalently, whenever the characteristic χ is not equivalent (as defined in the next section) to $\begin{bmatrix} 1 \\ 1 \end{bmatrix}$), and

$$f(z,\tau)\frac{\theta[\chi](z\gamma'(\tau)^{\frac{1}{2}},\gamma(\tau))}{\theta'[\chi](0,\gamma(\tau))} = \gamma'(\tau)^{\frac{1}{2}} \frac{\theta[\chi\gamma](z,\tau)}{\theta'[\chi\gamma](0,\tau)}, \tag{2.27}$$

whenever one of the characteristics appearing in the above formulae is (hence both are) equivalent to $\begin{bmatrix} 1 \\ 1 \end{bmatrix}$. It also follows that

$$\gamma'(\tau)^{\frac{1}{2}} \left(-2\pi\imath cz + \frac{\theta'[\chi](z\gamma'(\tau)^{\frac{1}{2}},\gamma(\tau))}{\theta[\chi](z\gamma'(\tau)^{\frac{1}{2}},\gamma(\tau))} \right) = \frac{\theta'[\chi\gamma](z,\tau)}{\theta[\chi\gamma](z,\tau)}. \tag{2.28}$$

We will be studying theta functions for characteristics $\chi = \begin{bmatrix} \frac{m}{k} \\ \frac{m'}{k} \end{bmatrix}$, with m, m' and $k \in \mathbb{Z}$ of the same parity, and their transformations under motions $\gamma = \begin{bmatrix} 1+\tilde{a}k & \tilde{b}k \\ \tilde{c}k & 1+\tilde{d}k \end{bmatrix} \in \Gamma(k)$. Even though for such characteristics it is in general not true that $\chi\gamma = \chi$, the fact that (see Lemma 2.7) $\chi\gamma - \chi \in 2\mathbb{Z}^2$ simplifies the transformation formula considerably.

Remark 1.13. We shall use often (2.4) and (2.16). They will often be referred to simply as the *translation and transformation formula for theta functions*. Similarly, we shall refer to (2.18) and (2.25) as the *transformation formula for theta constants* and *theta constant derivatives*, respectively.

2. Characteristics

2.1. Classes of characteristics. The formulae developed until now, especially (2.9), suggest that it is a good idea to work with equivalence classes of characteristics, rather than characteristics. We define an important equivalence relation on $\mathbb{R}^2$, viewed as *theta characteristics*.

Definition 2.1. Two characteristics $\left[\begin{array}{c} x_1 \\ y_1 \end{array}\right], \left[\begin{array}{c} x_2 \\ y_2 \end{array}\right] \in \mathbb{R}^2$ are *equivalent* provided their sum or difference belongs to $2\mathbb{Z}^2$.

We denote this equivalence by the symbol $\equiv$.[46] The quotient of $\mathbb{R}^2$ by this equivalence $\equiv$ (the space of classes of characteristics) can also be viewed as the quotient space of $\mathbb{R}^2$ by the group of motions generated by the three transformations $\left[\begin{array}{c} x \\ y \end{array}\right] \mapsto \left[\begin{array}{c} -x \\ -y \end{array}\right]$, $\left[\begin{array}{c} x \\ y \end{array}\right] \mapsto \left[\begin{array}{c} x+2 \\ y \end{array}\right]$, and $\left[\begin{array}{c} x \\ y \end{array}\right] \mapsto \left[\begin{array}{c} x \\ y+2 \end{array}\right]$. Using the usual identification of $\mathbb{R}^2$ with $\mathbb{C}$, we see that the space of equivalence classes can be identified with the Riemann surface (orbifold) of signature $(0,4;\ 2,2,2,2)$ corresponding to the elementary group generated by $z \mapsto -z$, $z \mapsto z+2$ and $z \mapsto z+2\imath$. See [**6**, pg. 227].

We define a right action of $\mathrm{SL}(2,\mathbb{Z})$ on $\mathbb{R}^2$. For the unimodular matrix $\gamma = \left[\begin{array}{cc} a & b \\ c & d \end{array}\right] \in \mathrm{SL}(2,\mathbb{Z})$ and the characteristic $\chi = \left[\begin{array}{c} \epsilon \\ \epsilon' \end{array}\right] \in \mathbb{R}^2$, we define the characteristic $\chi\gamma$ by the formula $\left[\begin{array}{c} a\epsilon + c\epsilon' - ac \\ b\epsilon + d\epsilon' + bd \end{array}\right]$. This does not define a group action of $\mathrm{SL}(2,\mathbb{Z})$ on characteristics. It does define a (right) group action of $\mathrm{PSL}(2,\mathbb{Z})$ on equivalence classes (see Proposition 2.2). We will also need to select finite sets of equivalence classes that are fixed pointwise and/or permuted by the modular group and its principal congruence subgroups.

We have defined the space of equivalence classes as the quotient of $\mathbb{R}^2$ by a group of rigid motions. A convenient fundamental set for this group action is

$$\{(x,y) \in \mathbb{R}^2;\ x>0,\ y>0 \text{ and } x+y<2\} \cup \{(0,y) \in \mathbb{R}^2;\ 0 \le y \le 1\}$$
$$\cup\{(x,0) \in \mathbb{R}^2;\ 0<x \le 1\} \cup \{(x,2-x) \in \mathbb{R}^2;\ 0<x \le 1\}.$$

[46]We will also use this symbol for the usual congruences involving integers. This abuse of symbols should not cause any confusion.

Another convenient fundamental domain (which will hereafter be denoted by $\mathcal{P}$) for this group action is

$$\{(x,y)\in\mathbb{R}^2;\ 0<x<1,\ 0<y<2\}\cup\{(0,y)\in\mathbb{R}^2;\ 0\le y\ \le 1\}$$
$$\cup\{(x,0)\in\mathbb{R}^2;\ 0<x\ \le 1\}\cup\{(1,y)\in\mathbb{R}^2;\ 0<y\ \le 1\}.$$

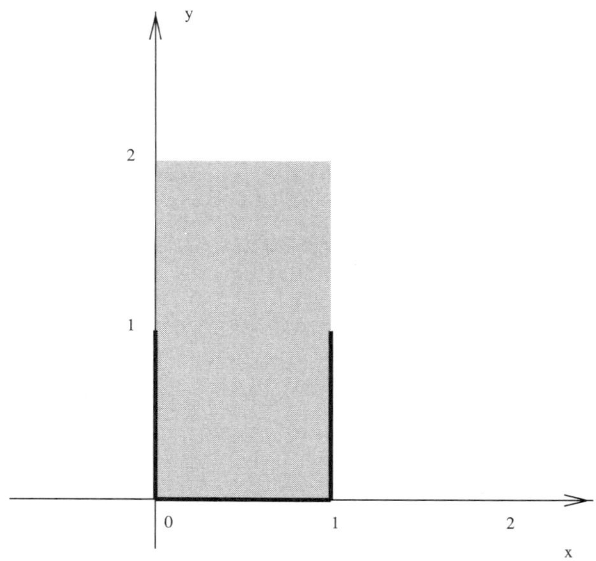

Figure 10. Fundamental polygon $\mathcal{P}$ for equivalence classes of characteristics.

Fix an odd prime k. We are interested in equivalence classes represented by vectors of the form $\begin{bmatrix}\frac{m}{k}\\ \frac{m'}{k}\end{bmatrix}$ with m and m' odd integers. We usually eliminate from our list the (equivalence class of the) characteristic $\begin{bmatrix}1\\1\end{bmatrix}$ (since the corresponding theta constant vanishes identically and this (equivalence class) characteristic is invariant under the entire modular group). Up to equivalence there are $\frac{k^2-1}{2}$ such characteristics. We may choose as representatives for these classes, the characteristics

$$\begin{bmatrix}1\\ \frac{1}{k}\end{bmatrix},\ \begin{bmatrix}1\\ \frac{3}{k}\end{bmatrix},\ \dots,\ \begin{bmatrix}1\\ \frac{k-2}{k}\end{bmatrix},$$
$$\begin{bmatrix}\frac{1}{k}\\ 1\end{bmatrix},\ \begin{bmatrix}\frac{3}{k}\\ 1\end{bmatrix},\ \dots,\ \begin{bmatrix}\frac{k-2}{k}\\ 1\end{bmatrix},$$
$$\begin{bmatrix}\frac{1}{k}\\ \\ \frac{1}{k}\end{bmatrix},\ \begin{bmatrix}\frac{3}{k}\\ \\ \frac{1}{k}\end{bmatrix},\ \dots,\ \begin{bmatrix}\frac{k-2}{k}\\ \\ \frac{1}{k}\end{bmatrix},$$

$$\begin{bmatrix} \frac{1}{k} \\ \\ \frac{3}{k} \end{bmatrix}, \begin{bmatrix} \frac{3}{k} \\ \\ \frac{3}{k} \end{bmatrix}, \ldots, \begin{bmatrix} \frac{k-2}{k} \\ \\ \frac{3}{k} \end{bmatrix},$$

$$\cdots\cdots\cdots$$

$$\begin{bmatrix} \frac{1}{k} \\ \\ \frac{k-2}{k} \end{bmatrix}, \begin{bmatrix} \frac{3}{k} \\ \\ \frac{k-2}{k} \end{bmatrix}, \ldots, \begin{bmatrix} \frac{k-2}{k} \\ \\ \frac{k-2}{k} \end{bmatrix},$$

$$\begin{bmatrix} \frac{1}{k} \\ \\ \frac{k+2}{k} \end{bmatrix}, \begin{bmatrix} \frac{3}{k} \\ \\ \frac{k+2}{k} \end{bmatrix}, \ldots, \begin{bmatrix} \frac{k-2}{k} \\ \\ \frac{k+2}{k} \end{bmatrix},$$

$$\cdots\cdots\cdots$$

$$\begin{bmatrix} \frac{1}{k} \\ \\ \frac{2k-1}{k} \end{bmatrix}, \begin{bmatrix} \frac{3}{k} \\ \\ \frac{2k-1}{k} \end{bmatrix}, \ldots, \begin{bmatrix} \frac{k-2}{k} \\ \\ \frac{2k-1}{k} \end{bmatrix}.$$

In terms of the second fundamental region $\mathcal{P}$ described above, the first row of the above list consists of characteristics on the vertical line $x = 1$ and the remaining rows are on the horizontal lines $y = 1$, $y = \frac{1}{k}$, $y = \frac{3}{k}$, ..., $y = \frac{k-2}{k}$, $y = \frac{k+2}{k}$, ..., $y = \frac{2k-1}{k}$. They also have the following interesting and important algebraic interpretation: The first row is the set of representatives of those classes of characteristics that are invariant (kept fixed) under the element B of Γ (see §4.11). The second row is the image of the first row under the map A, and the collection of elements in the subsequent rows are the images of the second row under the group generated by the motion B. As a matter of fact, this group permutes the columns of the array obtained by excluding the first row of the above table.

It is easy to obtain other useful arrays by rearranging the one described above. As usual for $x \in \mathbb{R} \cup \{\infty\}$, let Γ_x denote the stabilizer of x in Γ. (This group is infinite cyclic for rational x and trivial otherwise.) Then our new array consists, as above, of $k+1$ rows (each with $\frac{k-1}{2}$ elements). The first row consists of those classes of characteristics that are fixed pointwise by Γ_∞, while the elements in the j-th row are fixed by Γ_{j-2} for $j = 2$, ..., $k+1$. We call this array the *list of equivalence classes of characteristics.*

Proposition 2.2. *The action of the unimodular matrices on characteristics defines a group action of* Γ *on classes of characteristics.*

Proof. The result is known and can be found in textbooks (see, for example, [**24**, Ch. III]). A proof is included here for the convenience of the reader. We write the (affine) action of $\gamma \in \mathrm{SL}(2, \mathbb{Z})$ on the characteristic $v \in \mathbb{R}^2$ as a linear transformation followed by a translation

$$v\gamma = \gamma^t v + T_\gamma.$$

(As usual γ^t is the transpose of the matrix γ and for $\gamma = \begin{bmatrix} a & b \\ c & d \end{bmatrix}$ with $ad - bc = 1$, $T_\gamma = \begin{bmatrix} -ac \\ bd \end{bmatrix}$.) We first show that the class of $v\gamma$ depends only on the class of the characteristic v and the Möbius transformation determined by the matrix γ. We note that for all real ϵ and ϵ', we have

$$\begin{bmatrix} \epsilon+2 \\ \epsilon' \end{bmatrix}\gamma = \begin{bmatrix} \epsilon \\ \epsilon' \end{bmatrix}\gamma + 2\begin{bmatrix} a \\ b \end{bmatrix}, \quad \begin{bmatrix} \epsilon \\ \epsilon'+2 \end{bmatrix}\gamma = \begin{bmatrix} \epsilon \\ \epsilon' \end{bmatrix}\gamma + 2\begin{bmatrix} c \\ d \end{bmatrix},$$

$$(-v)\gamma = -v\gamma + 2\begin{bmatrix} -ac \\ bd \end{bmatrix}$$

and

$$v(-\gamma) = -v\gamma + 2\begin{bmatrix} -ac \\ bd \end{bmatrix}.$$

We show next that we have a group action of $\mathrm{PSL}(2,\mathbb{Z})$ on the set of equivalence classes of characteristics. For γ_1 and γ_2 in $\mathrm{SL}(2,\mathbb{Z})$, we have

$$v(\gamma_1\gamma_2) = (\gamma_1\gamma_2)^t v + T_{\gamma_1\gamma_2}$$

and

$$(v\gamma_1)\gamma_2 = \gamma_2^t(\gamma_1^t v + T_{\gamma_1}) + T_{\gamma_2} = \gamma_2^t\gamma_1^t v + \gamma_2^t T_{\gamma_1} + T_{\gamma_2}.$$

Thus to have a (right) group action it suffices to show that

$$T_{\gamma_1\gamma_2} - (\gamma_2^t T_{\gamma_1} + T_{\gamma_2}) \in 2\mathbb{Z}^2.$$

Writing $\gamma_i = \begin{bmatrix} a_i & b_i \\ c_i & d_i \end{bmatrix}$, $i = 1,\ 2$, we need to study the difference between

$$\begin{aligned}&\begin{bmatrix} -(a_1a_2+b_1c_2)(c_1a_2+d_1c_2) \\ (a_1b_2+b_1d_2)(c_1b_2+d_1d_2) \end{bmatrix} \\ &= \begin{bmatrix} -(a_1c_1a_2^2 + a_1d_1a_2c_2 + b_1c_1a_2c_2 + b_1d_1c_2^2) \\ a_1c_1b_2^2 + a_1d_1b_2d_2 + b_1c_1b_2d_2 + b_1d_1d_2^2 \end{bmatrix}\end{aligned}$$

and

$$\begin{bmatrix} a_2 & c_2 \\ b_2 & d_2 \end{bmatrix}\begin{bmatrix} -a_1c_1 \\ b_1d_1 \end{bmatrix} + \begin{bmatrix} -a_2c_2 \\ b_2d_2 \end{bmatrix} = \begin{bmatrix} -a_1c_1a_2 + b_1d_1c_2 - a_2c_2 \\ -a_1c_1b_2 + b_1d_1d_2 + b_2d_2 \end{bmatrix}.$$

We will show that the last two characteristics differ by twice an integral characteristic. Since signs are irrelevant mod 2, it suffices to show that

$$a_1c_1a_2 + b_1d_1c_2 + a_2c_2 \equiv a_1c_1a_2^2 + a_1d_1a_2c_2 + b_1c_1a_2c_2 + b_1d_1c_2^2 \pmod 2$$

and

$$a_1c_1b_2 + b_1d_1d_2 + b_2d_2 \equiv a_1c_1b_2^2 + a_1d_1b_2d_2 + b_1c_1b_2d_2 + b_1d_1d_2^2 \pmod 2.$$

But

$$a_1c_1a_2^2 + a_1d_1a_2c_2 + b_1c_1a_2c_2 + b_1d_1c_2^2 = a_1c_1a_2^2 + a_2c_2(a_1d_1 + b_1c_1) + b_1d_1c_2^2$$

$$\equiv a_1c_1a_2^2 + a_2c_2(a_1d_1 - b_1c_1) + b_1d_1c_2^2 \pmod 2 = a_1c_1a_2^2 + a_2c_2 + b_1d_1c_2^2.$$

We are hence reduced to showing that

$$a_1c_1a_2 + b_1d_1c_2 \equiv a_1c_1a_2^2 + b_1d_1c_2^2 \pmod 2.$$

This last equation is obvious since

$$a_1c_1a_2(1 - a_2) + b_1d_1c_2(1 - c_2)$$

is the sum of two even integers ($a_2(1 - a_2)$ and $c_2(1 - c_2)$ are both even). Similarly, for the second entries of our theta characteristics.

Remark 2.3. The above proof relies on the fact that we are dealing with matrices in $\mathrm{SL}(2,\mathbb{Z})$ and of course does not work for matrices in $\mathrm{SL}(2,\mathbb{R})$.

Problem 2.4. Is there an action of $\mathrm{PSL}(2,\mathbb{R})$ on equivalence classes of characteristics that is an extension of the action of $\mathrm{PSL}(2,\mathbb{Z})$?

2.2. Integral classes of characteristics. Let $v_1 = \begin{bmatrix} 1 \\ 0 \end{bmatrix}$ and $v_2 = \begin{bmatrix} 0 \\ 1 \end{bmatrix}$ be the usual basis for $\mathbb{R}^2$. There are only four equivalence classes of *integral* characteristics (that is, characteristics $\begin{bmatrix} \epsilon \\ \epsilon' \end{bmatrix} \in \mathbb{R}^2$ with both ϵ and $\epsilon' \in \mathbb{Z}$). They are the equivalence classes of the characteristics 0, v_1, v_2 and $v_1 + v_2$. The first three are *even*;[47] the fourth is *odd*. The modular group permutes the even integral equivalence classes and fixes the odd integral class. It is easily seen that (as characteristics)

$$0\ A = 0,\ v_1\ A = -v_2,\ v_2\ A = v_1,\ (v_1 + v_2)\ A = v_1 - v_2$$

and

$$0\ B = v_2,\ v_1\ B = v_1 + 2v_2,\ v_2\ B = 2v_2,\ (v_1 + v_2)\ B = v_1 + 3v_2.$$

2.3. Rational classes of characteristics. A characteristic $\begin{bmatrix} \epsilon \\ \epsilon' \end{bmatrix} \in \mathbb{R}^2$ is called *rational* if both ϵ and ϵ' are. For the positive integer k, we let $Z(k)$[48] denote the equivalence classes of characteristics represented by vectors in $\mathbb{R}^2$ of the form $\begin{bmatrix} \frac{m}{k} \\ \frac{m'}{k} \end{bmatrix}$ with m and m' in $\mathbb{Z}$. (All our definitions are in terms of characteristics, but apply to equivalence classes of characteristics. We will normally leave it to the reader to verify this.)

[47] We are using classical terminology here. The parity of a characteristic $\begin{bmatrix} m \\ n \end{bmatrix} \in \mathbb{Z}^2$ is defined traditionally as the parity of $mn \in \mathbb{Z}$. For the purposes of this text, it is convenient to have a different definition of the parity of a characteristic. See below.

[48] Previously $Z(k)$ represented the set of characteristics. It now represents the set of equivalence classes of characteristics. Similarly for $X(k)$ as redefined in this subsection.

Definition 2.5. The characteristic $\begin{bmatrix} \frac{m}{k}, \\ \frac{m'}{k} \end{bmatrix} \in Z(k)$ is said to be *odd* or *of odd parity* if mm' is an odd integer, to be *even* if both m and m' are even integers, and to be *of mixed parity* if precisely one of m and m' is even. This concept of parity for a characteristic that belongs to $Z(k)$ is well defined on the level of classes, but does depend on k. For example, consider $\mathbb{Z}^2 \subset \mathbb{R}^2$. For $k = 1$, 0 is even, v_1 and v_2 are mixed, and $v_1 + v_2$ is odd; while for $k = 2$, all four are even characteristics.

We introduce a decomposition of $Z(k)$ by defining $X(k)$ to consist of the classes in $Z(k)$ of the same parity as k, and letting $Y(k)$ be the complement of $X(k)$ in $Z(k)$. We will denote the cardinality of the set S, by the symbol $|S|$. It is easily seen that

$$|Z(k)| = 2(k^2 + 1),$$

$$|X(k)| = \frac{k^2+1}{2} \text{ for } k \text{ odd and } |X(k)| = \frac{k^2}{2} + 2 \text{ for } k \text{ even.}$$

It follows that

$$|Y(k)| = \frac{3}{2}(k^2 + 1) \text{ for } k \text{ odd and } |Y(k)| = \frac{3}{2}k^2 \text{ for } k \text{ even.}$$

It is easily seen that

$$Z(k) = X(2k).$$

Definition 2.6. The *primitive* k characteristic is

$$\chi_o = \chi_o(k) = \begin{bmatrix} 1 \\ \frac{\alpha}{k} \end{bmatrix},$$

where
$\alpha = 1$ if $k \equiv 1 \mod 2$,
$\alpha = 2$ if $k \equiv 0 \mod 4$ and
$\alpha = 4$ if $k \equiv 2 \mod 4$ and $k > 2$. We set

$$\chi_o(2) = v_1 \text{ and } \chi_o(1) = v_1 + v_2.$$

It is also convenient to define

$$\chi_1 = \chi_1(k) = \begin{bmatrix} \frac{\alpha}{k} \\ 1 \end{bmatrix}, \quad \chi_2 = \chi_2(k) = \begin{bmatrix} \frac{\alpha}{k} \\ \frac{\alpha}{k} \end{bmatrix}, \quad k > 2,$$

$$\chi_1(2) = v_2 \text{ and } \chi_2(2) = v_1 + v_2.$$

Lemma 2.7. *For every integer $k \geq 1$,*
(a) *Γ permutes the classes in $Z(k)$ as well as those in $X(k)$,*
(b) *the elements of $\Gamma(k)$ fix pointwise the characteristic classes in $X(k)$, and thus*
(c) *the finite group $\Gamma/\Gamma(k)$ operates on the finite set $X(k)$.*

Proof. It is obvious that Γ permutes the classes in $Z(k)$. To prove the rest of (a), we use the formula for the action of motions on characteristics to observe that the image of $\begin{bmatrix} \frac{m}{k} \\ \frac{m'}{k} \end{bmatrix} \in X(k)$ under the action of $\begin{bmatrix} a & b \\ c & d \end{bmatrix} \in \Gamma$ is $\begin{bmatrix} \frac{am+cm'}{k} - ac \\ \frac{bm+dm'}{k} + bd \end{bmatrix}$. This characteristic certainly represents a class in $X(k)$ if k is even. If k is odd, it represents a class in $X(k)$ as long as either a or c is odd and either b or d is odd (the excluded cases are impossible since they would contradict the fact that $ad - bc = 1$). This completes the proof of part (a).

As a consequence of Proposition 2.2 and (a), we have a well defined homomorphism

$$\eta : \Gamma \to \operatorname{Perm}(X(k)),$$

defined by sending the motion $\gamma \in \Gamma$ to the permutation of the set $X(k)$ that it induces. For $\gamma \in \Gamma$, the permutation $\eta(\gamma)$ sends the class of the characteristic χ to the class of the characteristic $\chi\gamma$.

We show next that the kernel of this homomorphism is $\Gamma(k)$. Assume that k is odd and that $k > 2$ ($k = 1$ is left to the reader). Consider an arbitrary element of our space of characteristics $\chi = \begin{bmatrix} \frac{m}{k} \\ \frac{m'}{k} \end{bmatrix}$ with m and m' odd. The image of this characteristic under the motion $\begin{bmatrix} kr+1 & ks \\ kt & ku+1 \end{bmatrix} \in \Gamma(k)$ is the characteristic $\begin{bmatrix} \frac{m}{k} + mr + m't - kt(kr+1) \\ \frac{m'}{k} + ms + m'u + ks(ku+1) \end{bmatrix}$. This characteristic is equivalent to χ provided $mr+m't-kt(kr+1)$ and $ms+m'u+ks(ku+1)$ are even integers. This fails to happen only if r is odd and t is even or u is odd and s is even, in which case we contradict that $(kr+1)(ku+1)-k^2st = 1$. The argument just given also works for even k. We have hence completed the proof of (b). Part (c) is a direct consequence of (a) and (b).

We continue to study the kernel of the homomorphism η. The above argument has shown that $\Gamma(k)$ is contained in the kernel of η. We show next that if $\gamma \in \Gamma$ is not an element of $\Gamma(k)$, $k > 1$ odd, then there is a characteristic which is not fixed by it. As usual, let $\gamma = \begin{bmatrix} a & b \\ c & d \end{bmatrix}$ and assume that b is not congruent to zero mod k. Then we see that the (class of the) characteristic $\chi_1 = \begin{bmatrix} \frac{1}{k} \\ 1 \end{bmatrix}$ is not fixed by γ. Similarly, if c is not congruent to zero mod k, then the characteristic $\chi_0 = \begin{bmatrix} 1 \\ \frac{1}{k} \end{bmatrix}$ is not fixed by γ. It thus follows that we may assume that b and c are congruent to zero mod k. It follows that ad is congruent to one mod k. Consider now the characteristic $\chi_2 = \begin{bmatrix} \frac{1}{k} \\ \\ \frac{1}{k} \end{bmatrix}$. It is mapped by γ to $\begin{bmatrix} \frac{a}{k} + m \\ \\ \frac{d}{k} + n \end{bmatrix}$ (for some integers m and n) which can be equivalent to its preimage only when a and d are both congruent to either ± 1 mod k. This means that γ is in $\Gamma(k)$.

For $k = 2$, if γ fixes the class of v_1, then a is odd and c is even. If it fixes the class of v_2, then d is odd and b is even. Thus if it fixes both classes, $\gamma \in \Gamma(2)$. Assume next that $k \equiv 0 \bmod 4$. If γ fixes (the class of) $\chi_0 = \begin{bmatrix} 1 \\ \frac{2}{k} \end{bmatrix}$, then $1 + c\frac{2}{k} - ac$ is an odd integer. Hence $c \equiv 0 \bmod \frac{k}{2}$. In particular, $c \in 2\mathbb{Z}$, and further c must be an even multiple of $\frac{k}{2}$. Thus $c \equiv 0 \bmod k$. Similarly, if γ fixes χ_1, then $b \equiv 0 \bmod k$. If γ fixes these two classes, then, in addition, $ad \equiv 1 \bmod k$. If γ also fixes χ_2, then both a and d are congruent to ± 1 mod k. We have shown that only elements in $\Gamma(k)$ are in the kernel of η. The argument for the case $k > 2$, $k \equiv 2 \bmod 4$ is similar. We have established

Theorem 2.8. *The group $\Gamma/\Gamma(k)$ is isomorphic to a subgroup of the permutation group of $X(k)$.*

We will prove a stronger result (Theorem 2.28).

Definition 2.9. Let $\chi = \begin{bmatrix} \frac{m}{k} \\ \frac{m'}{k} \end{bmatrix} \in Z(k)$, and write

$$m = 2kr + \alpha \text{ and } m' = 2kr' + \alpha'$$

with r, r', $\alpha, \alpha' \in \mathbb{Z}$ with $0 \le \alpha < 2k$ and $0 \le \alpha' < 2k$. We define

$$m(\chi) = \min\{\alpha,\ 2k - \alpha\} \text{ and } m'(\chi) = \min\{\alpha',\ 2k - \alpha'\}.$$

We observe that the integers $m(\chi)$ and $m'(\chi)$ depend only on k and the class of χ and are positive integers between 0 and k (inclusive).

Remark 2.10. If we assume that the equivalence class of the characteristic $\chi = \begin{bmatrix} \frac{m}{k} \\ \frac{m'}{k} \end{bmatrix} \in Z(k)$ is represented by a point in the fundamental domain $\mathcal{P}$ of §2.1, then $m(\chi) = m$ and $m'(\chi) = \min\{m', 2k - m'\}$. Each characteristic χ whose class lands in $Z(k)$ has a representation as $\begin{bmatrix} \frac{m(\chi)}{k} \\ \\ \frac{m'(\chi)}{k} \end{bmatrix}$. Each such characteristic represents precisely one class if $m(\chi) = 0$ or k, and precisely two classes otherwise. We can view $\chi \mapsto \begin{bmatrix} \frac{m(\chi)}{k} \\ \\ \frac{m'(\chi)}{k} \end{bmatrix}$ as a selfmap of $Z(k)$. It is generically two-to-one. It is useful to view this map as acting on the array listing equivalence classes of characteristics introduced (see §2.8) below. In the terminology to be defined subsequently, the map is one-to-one on the towers over 0 and 1; two-to-one on each of the other towers.

Remark 2.11. In our discussion of theta functions and characteristics we have developed thus far many formulae. We shall develop many more and we need to distinguish formulae that are valid for theta characteristics from those that hold, more generally, for equivalence classes of characteristics. In this regard, we observe that (2.16), (2.17) and (2.24) are identities for characteristics and are in general not correct on the level of equivalence classes. This sometimes makes it difficult to apply a formula such as (2.16) to a situation of interest. In addition, the formulae involving motions $\gamma \in \mathrm{PSL}(2,\mathbb{Z})$ may depend on the choice of matrix in $\mathrm{SL}(2,\mathbb{Z})$ that represents the motion.

2.4. Invariant classes for $\Gamma(k)$.

Definition 2.12. For a subgroup G of Γ, let $X(G)$ be the set of equivalence classes of characteristics fixed pointwise by all elements of G. We will call this set the *invariants* or *invariant classes (of characteristics)* for the group G.

The set $X(G)$ is never empty since it always contains the class of the vector $v_1 + v_2$; we will say that $X(G)$ is *trivial* if it contains nothing else. If $G_1 \subset G_2 \subset \Gamma$, then

$$X(\Gamma) \subset X(G_2) \subset X(G_1).$$

Lemma 2.13. *For each positive integer k,*

$$X(\Gamma(k)) = X(k).$$

Proof. We have shown that the classes in $X(k)$ are fixed by $\Gamma(k)$. To establish the converse, assume that $\chi = \begin{bmatrix} \epsilon \\ \epsilon' \end{bmatrix} \in \mathcal{P}$ is fixed by B^k (which, of course, belongs to $\Gamma(k)$). If $\epsilon = 0$, we conclude that for even k, ϵ' is arbitrary and for odd k, $\epsilon' = \frac{1}{2}$. If $\epsilon \neq 0$, then $\epsilon = \frac{m}{k}$ with $m \in \mathbb{Z}$ of the same parity as k and ϵ' is arbitrary. Next we apply the same argument to the motion corresponding to the matrix $\begin{bmatrix} 1 & 0 \\ k & 1 \end{bmatrix}$ and conclude as follows. If $\epsilon' = 0$, then for even k, ϵ is arbitrary and for odd k, $\epsilon = \frac{1}{2}$. If $\epsilon' \neq 0$, then ϵ is arbitrary and $\epsilon' = \frac{m'}{k}$ with $m' \in \mathbb{Z}$ of the same parity as k. These observations suffice to conclude that $\chi \in X(k)$.

Remark 2.14. Our arguments also show that $X(k)$ is the set of invariants for the group generated by the two parabolic matrices $\begin{bmatrix} 1 & k \\ 0 & 1 \end{bmatrix}$ and $\begin{bmatrix} 1 & 0 \\ k & 1 \end{bmatrix}$.

2.5. Punctures on $\mathbb{H}^2/\Gamma(k)$ and the classes $X_o(k)$. We proceed to establish a correspondence of the punctures on $\mathbb{H}^2/\Gamma(k)$ with certain equivalence classes of characteristics in $X(k)$. There is a degree of arbitrariness in choosing an "origin" for this correspondence. We identify the puncture determined by ∞ with the class of the *primitive* k-characteristic χ_o. Note that the characteristic $\chi_o = \begin{bmatrix} 1 \\ \frac{\alpha}{k} \end{bmatrix}$ is characterized by α is the smallest positive integer so that $\chi_o \in X(k)$ but $\chi_o \notin X(k')$ for all positive integers $k' < k$ with $k'|k$.

Lemma 2.15. *Let* $\gamma = \begin{bmatrix} a & b \\ c & d \end{bmatrix} \in \Gamma$*; then*

$$\chi_o\gamma \equiv \chi_o \text{ if and only if } \gamma^{-1}(\infty) = \infty \mod \Gamma(k);$$

that is, $\chi_o\gamma \equiv \chi_o$ *if and only if* $\gamma \in G(k)$.

Proof. We leave to the readers the cases $k = 1$ and 2. So we assume that $k > 2$. We start with the formula

$$\chi_o\gamma = \begin{bmatrix} 1 \\ \frac{\alpha}{k} \end{bmatrix} \begin{bmatrix} a & b \\ c & d \end{bmatrix} = \begin{bmatrix} a + \frac{\alpha c}{k} - ac \\ b + \frac{\alpha d}{k} + bd \end{bmatrix}.$$

Assume that $\chi_o\gamma \equiv \chi_o$. If $c = 0$, then (since $d \neq 0$) $\gamma^{-1}(\infty) = -\frac{d}{c} = \infty$. Assume that $c \neq 0$. We consider cases.
1. Assume k is odd. Then $\alpha = 1$ and we conclude that

$$c \equiv 0 \mod k \text{ and } d \equiv \pm 1 \mod k. \tag{2.29}$$

Then $\gamma^{-1}(\infty)$ is not equal to ∞, but is $\Gamma(k)$-equivalent to ∞.
2. Assume that k and $\frac{k}{2}$ are even. Then $\alpha = 2$ and we conclude that

$$c \equiv 0 \mod \frac{k}{2} \text{ and } d \equiv \pm 1 \mod k.$$

We write $c = \tilde{c}\frac{k}{2}$ (thus c is even and a is odd) and use the fact that $a(1-c)+\tilde{c}$ is odd to conclude that $\tilde{c}$ is even. Thus $c \equiv 0 \mod k$, and $\gamma^{-1}(\infty)$ is $\Gamma(k)$-equivalent to ∞.
3. The last possibility is that k is even and $\frac{k}{2}$ is odd. Then $\alpha = 4$ and

$$c \equiv 0 \mod \frac{k}{2} \text{ and } d \equiv \pm 1 \mod k.$$

As above we write $c = \tilde{c}\frac{k}{2}$ and use the fact that $a(1-c)+2\tilde{c}$ is odd to conclude that c is even. Then $\tilde{c}$ must also be even and $c \equiv 0 \mod k$.

To establish the converse, note that if $\gamma^{-1}(\infty) = \infty$, then $c = 0$ and $a = d = \pm 1$ and it follows trivially that $\chi_o\gamma$ is equivalent to χ_o. Assume hence that $\gamma^{-1}(\infty) \neq \infty$, but $\gamma^{-1}(\infty)$ is $\Gamma(k)$-equivalent to ∞. Then $c \neq 0$, but (2.29) holds. Thus for some integers $\tilde{c}$ and $\tilde{d}$, we have

$$c = \tilde{c}k \text{ and } d = \pm 1 + \tilde{d}k.$$

The first reduction involves replacing every entry in γ by its negative (if necessary) in order to assume that the plus sign holds in the last equality. Again, it is easiest to consider the three cases:
1. k is odd. Then

$$\chi_o\gamma = \begin{bmatrix} a(1-\tilde{c}k)+\tilde{c} \\ 2b+\frac{1}{k}+\tilde{d}(1+bk) \end{bmatrix}.$$

The first entry in the above matrix is an odd integer (it is most convenient to check this assertion separately for odd and even $\tilde{c}$). It now suffices to check that $\tilde{d}(1+bk)$ is an even integer. If b is even, the d must be odd and hence $\tilde{d}$ must be even. If b is odd, then $1+bk$ is even.
2. k and $\frac{k}{2}$ are even. As before,

$$\chi_o\gamma = \begin{bmatrix} a(1-\tilde{c}k)+2\tilde{c} \\ 2b+\frac{2}{k}+\tilde{d}(2+bk) \end{bmatrix}.$$

Now $a(1-\tilde{c}k)$ is odd because c is even (and hence a is odd) and $\tilde{d}(2+bk)$ is even (because k is).
3. k is even and $\frac{k}{2}$ is odd. In this case

$$\chi_o\gamma = \begin{bmatrix} a(1-\tilde{c}k)+4\tilde{c} \\ 2b+\frac{4}{k}+\tilde{d}(4+bk) \end{bmatrix},$$

which is easily seen to be equivalent to χ_o.

We have seen that the finite group

$$\Gamma/\Gamma(k) \cong \text{PSL}(2, \mathbb{Z}_k)$$

operates on the finite set $X(k)$. It is clear that if $\chi \in X(k')$ for some positive integer $k'|k$, then the $\Gamma/\Gamma(k)$-orbit of χ (which coincides, of course, with the $\Gamma/\Gamma(k')$ orbit of χ) is contained in $X(k')$ since $\Gamma(k) \subset \Gamma(k')$. In particular, the orbit of the class of the characteristic $\begin{bmatrix} 1 \\ 1 \end{bmatrix}$ is contained in $X(1)$ – which consists, of course, of only this point.

The $\Gamma/\Gamma(k)$-orbit of ∞ (which coincides with the $G(k)\backslash\Gamma$-orbit of ∞, but, $G(k)\backslash\Gamma$ is no longer a group since $G(k)$ is NOT normal in Γ) is naturally identified with the punctures on $\mathbb{H}^2/\Gamma(k)$, and [**27**, Lemmas 1.41 and 1.42] gives a good alternate description of this orbit. We need a similar description of the $(\Gamma/\Gamma(k))$-orbit of χ_o.

Definition 2.16. For each $k \in \mathbb{Z}^+$, define a subset of $X(k)$ by

$X_o(k) = \{\chi \in X(k); \chi \notin X(k') \text{ for all } k' \in \mathbb{Z} \text{ such that } 0 < k' < k \text{ and } k'|k\}.$

Remark 2.17. The characteristics $\chi_o(k)$ and $\chi_i(k)$, $i = 1, 2$, belong to $X_o(k)$.

2.6. The classes in $X_o(k)$.

Lemma 2.18. *Let m and m' be nonnegative integers such that the class of $\chi = \begin{bmatrix} \frac{m}{k} \\ \frac{m'}{k} \end{bmatrix} \in X(k)$ and let $d = \gcd(m, m', k)$.*[49]
(a) *For $k \equiv 1 \mod 2$, the class of $\chi \in X_o(k)$ if and only if $d = 1$.*
(b) *For $k \equiv 2 \mod 4$, the class of $\chi \in X_o(k)$ if and only if at least one of the integers $\frac{m}{2}$ or $\frac{m'}{2}$ is even and $d = 2$.*
(c) *For $k \equiv 0 \mod 4$, the class of $\chi \in X_o(k)$ if and only if $d = 2$ (and hence not both $\frac{m}{2}$ and $\frac{m'}{2}$ can be even).*

Proof. We consider the three cases separately.
(a) Assume that k is odd and $d > 1$. Then $\chi \in X\left(\frac{k}{d}\right)$. Conversely, if $\chi \in X(k')$ with $1 \le k' < k$ and $k'|k$, then $d \ge \frac{k}{k'} > 1$.
(b) Assume that k is even and $\frac{k}{2}$ is odd. In this case $2|d$, but $4 \not| d$. If $d > 2$, then $\chi \in Z(\frac{k}{d}) = X(\frac{2k}{d})$. Conversely, if $\chi \in X(k')$ with $k' \le k$ and $k'|k$, then

$$\chi = \begin{bmatrix} \frac{m}{k} \\ \frac{m'}{k} \end{bmatrix} = \begin{bmatrix} \frac{m_1}{k'} \\ \frac{m_1'}{k'} \end{bmatrix},$$

for some nonnegative integers m_1 and m_1'. Now if $k' = \frac{k}{2}$, then both $m_1 = \frac{m}{2}$ and $m_1' = \frac{m'}{2}$ would be odd. If $k' < \frac{k}{2}$, then $\gcd(m, m', k) \ge \frac{k}{k'} > 2$.
(c) Finally assume that k and $\frac{k}{2}$ are even. Again $2|d$. If $d > 2$, then as before $\chi \in Z\left(\frac{k}{d}\right) = X(\frac{2k}{d})$. If both $\frac{m}{2}$ and $\frac{m'}{2}$ were even, then $\chi \in X(\frac{k}{2})$. Conversely, if $\chi \in X(k')$ with $1 \le k' < k$, then $d \ge \frac{k}{k'} \ge 2$. Thus $k' = \frac{k}{2}$ and both $\frac{m}{2}$ and $\frac{m'}{2}$ must be even.

Corollary 2.19. *If k is a prime, then*

$$X_o(k) = X(k) - \left\{\begin{bmatrix} 1 \\ 1 \end{bmatrix}\right\}.$$

[49] If $\alpha_1, \ldots, \alpha_l$ is a finite collection of integers, then we shall also write

$$(\alpha_1, \ldots, \alpha_l) = \gcd(\alpha_1, \ldots, \alpha_l).$$

Let $\chi \in X(k)$. There is a unique smallest k' such that $1 \leq k' \leq k$, $k'|k$, and $\chi \in X_o(k')$. We will call k' the *level* of the equivalence class of the characteristic χ and denote it by the symbol $l(\chi)$. It is clear that

$$X(k) = \bigcup_{k' \in \mathbb{Z},\ 1 \leq k' \leq k,\ k'|k} X_o(k')$$

is a decomposition of $X(k)$ into a disjoint union of Γ-invariant subsets. It is a consequence of the next two lemmas that Γ acts transitively on each of these subsets.

Lemma 2.20. *For each integer $k > 1$, $2n(k) > |X(k)|$.*

Recall that $n(k)$ is the number of punctures of $\mathbb{H}^2/\Gamma(k)$ (defined in Chapter 1).

Proof. Factor k into a product of powers of distinct (positive) primes: $k = \prod_{i=1}^{j} \mathfrak{P}_i^{\alpha_i}$. We prove first that for sufficiently large k, $2n(k) > \frac{k^2}{2} + 2 \geq |X(k)|$. Now, $2n(k) = k^2 \prod_{i=1}^{j} \left(1 - \frac{1}{\mathfrak{P}_i^2}\right)$. Thus $2n(k) > \frac{k^2}{2} + 2$ if and only if

$$\prod_{i=1}^{j} \left(1 - \frac{1}{\mathfrak{P}_i^2}\right) > \frac{1}{2}\left(1 + \frac{4}{k^2}\right).$$

Since (see, for example, [**3**, pp. 219 and 129])

$$\frac{6}{\pi^2} = \prod_{\mathfrak{P}>0,\ \text{prime}} \left(1 - \frac{1}{\mathfrak{P}^2}\right) < \prod_{i=1}^{j} \left(1 - \frac{1}{\mathfrak{P}_i^2}\right),$$

the lemma is proven for $k \geq 5$. For $k = 2,\ 3$, or 4, direct calculations show that the inequality $2n(k) > |X(k)|$ holds.

The last lemma shows that the Γ-orbit of χ_o is the unique largest orbit in the action of Γ on $X(k)$. Lemma 2.15 showed that this orbit has $n(k)$ elements. Since $\chi_o \in X_o(k)$, the Γ- orbit of χ_o is contained in $X_o(k)$. To obtain a more complete picture concerning the decomposition of $X(k)$ into Γ-orbits, we establish

Lemma 2.21. *For each integer $k \geq 1$,*
(a) $|X(k)| = \sum_{k'|k} n(k')$ *and*
(b) $|X_o(k)| = n(k)$.

Proof. In the statement of the lemma and in this argument, all indices of summation (k', for example) are assumed to be positive integers. The lemma is certainly true for $k = 1$ and for prime k. We prove the general case by induction on k. We may and do assume that $k \geq 3$.

Assume that $k = \mathfrak{P}^\alpha k_1$, where $\mathfrak{P}$ is prime, α is a positive integer, and $\mathfrak{P} \nmid k_1$. An arbitrary divisor of k can be written uniquely as $\mathfrak{P}^\beta k'$, where $0 \le \beta \le \alpha$ and $k'|k_1$. Assume that $\mathfrak{P}$ is odd and k_1 is even. Then

$$\begin{aligned}
\sum_{k'|k} n(k') &= n(1) + n(2) + \sum_{k'|k, k' \neq 1, k' \neq 2} n(k') \\
&= \sum_{\beta=0}^{\alpha} n(\mathfrak{P}^\beta) + \sum_{\beta=0}^{\alpha} n(2\mathfrak{P}^\beta) + \sum_{k'|k_1, k'\neq 1, k' \neq 2} \sum_{\beta=0}^{\alpha} n(k'\mathfrak{P}^\beta) \\
&= n(1) + \frac{1}{2}\left(1 - \frac{1}{\mathfrak{P}^2}\right) \sum_{\beta=1}^{\alpha} \mathfrak{P}^{2\beta} + n(2) + \frac{3}{2}\left(1 - \frac{1}{\mathfrak{P}^2}\right) \sum_{\beta=1}^{\alpha} \mathfrak{P}^{2\beta} \\
&\quad + \left(\sum_{k'|k_1, k' \neq 1, k' \neq 2} n(k')\right)\left(1 + \left(1 - \frac{1}{\mathfrak{P}^2}\right)\sum_{\beta=1}^{\alpha} \mathfrak{P}^{2\beta}\right) \\
&= 4 + 2(\mathfrak{P}^{2\alpha} - 1) + \left(\frac{k_1^2}{2} - 2\right)\mathfrak{P}^{2\alpha} = \frac{k^2}{2} + 2.
\end{aligned}$$

The next to last equality follows from the induction hypothesis, that is, by the hypothesis that the result is true for k_1. The argument for $\mathfrak{P}$ even and k_1 odd is similar. This proves part (a) at least for the case of even k. To establish (b), note that for each odd k' that divides k, the class of the characteristic $\begin{bmatrix} 1 \\ \frac{1}{k'} \end{bmatrix}$ belongs to $X(k)$, while for each even k' that divides k, the classes of the characteristics $\begin{bmatrix} 1 \\ 0 \end{bmatrix}$, $\begin{bmatrix} 1 \\ \frac{2}{k'} \end{bmatrix}$, and $\begin{bmatrix} 1 \\ \frac{4}{k'} \end{bmatrix}$ belong to $X(k)$. Hence for each positive integer $k'|k$ an orbit of size $n(k')$ is contained in $X_o(k') \subset X(k)$. Since these orbits are disjoint, $X(k)$ can contain no other points by part (a). Hence (b) is also proven. We leave the case of odd k to the reader.

Corollary 2.22. *For each $\chi \in X_o(k)$, there exists a $\gamma \in \Gamma$ with $\chi_o\gamma = \chi$.*

Corollary 2.23. *The group Γ (equivalently, $\Gamma/\Gamma(k)$) acts transitively on $X_o(k)$.*

Remark 2.24. The first of the above corollaries shows that $X_o(k)$ is in one-to-one correspondence with the punctures on $\mathbb{H}^2/\Gamma(k)$. The collection of ideas discussed above shows that whereas $X_o(k)$ carries information about the curve $\mathbb{H}^2/\Gamma(k)$, the larger set $X(k)$ encodes some information about all the curves $\mathbb{H}^2/\Gamma(k')$ with $0 < k'|k$. They also show that the choice of χ_o was somewhat arbitrary. We could have replaced it by any vector in $X_o(k)$. The circle of ideas developed above permits many explicit computations. For example, if $\mathfrak{P}$ is an odd positive prime, then

$$X(2\mathfrak{P}) = X_o(2\mathfrak{P}) \cup X_o(\mathfrak{P}) \cup X_o(2) \cup X_o(1) \text{ and } X_o(2\mathfrak{P}) = Y(\mathfrak{P}) - X_o(2).$$

Definition 2.25. We define a map $\mathcal{I}$ from classes of characteristics to punctures on $\mathbb{H}^2/\Gamma(k)$ by sending for each $\gamma \in \Gamma$ the class of the characteristic $\chi_o\gamma$ to the puncture determined by the cusp $\gamma^{-1}(\infty)$.

Theorem 2.26. *The image in* $\text{Perm}(X(k))$ *of* Γ *under the homomorphism* η *is isomorphic to the group of conformal selfmaps of* $\mathbb{H}^2/\Gamma(k)$. *The action of* Γ *on the set of equivalence classes of characteristics* $X(k)$ *corresponds to the right permutation of the punctures; that is, for each equivalence class* $\chi \in X(k)$ *and all* $\gamma \in \Gamma$, *we have*

$$\mathcal{I}(\chi\gamma) = \gamma^{-1}(\mathcal{I}(\chi)).$$

Proof. The image of the homomorphism η is isomorphic to $\Gamma/\Gamma(k)$ which is well known to be the group of conformal selfmaps of the Riemann surface $\mathbb{H}^2/\Gamma(k)$. The remark concerning the permutation of the punctures is a consequence of our identification of the characteristics with the punctures via the map $\mathcal{I}$; specifically, the characteristic χ is given as $\chi_o\gamma_o$ for some $\gamma_o \in \Gamma$. Hence

$$\begin{aligned} \mathcal{I}(\chi\gamma) &= \mathcal{I}((\chi_o\gamma_o)\gamma) \\ = (\gamma_o \circ \gamma)^{-1}(\infty) &= \gamma^{-1}(\gamma_o^{-1}(\infty)) = \gamma^{-1}(\mathcal{I}(\chi_o\gamma_o)) = \gamma^{-1}(\mathcal{I}(\chi)). \end{aligned}$$

Remark 2.27. The punctures on $\mathbb{H}^2/\Gamma(k)$ are the projections under

$$P : \mathbb{H}^2 \cup \mathbb{Q} \cup \{\infty\} \to (\mathbb{H}^2 \cup \mathbb{Q} \cup \{\infty\})/\Gamma(k)$$

of the parabolic fixed points $(\mathbb{Q} \cup \{\infty\})$ of $\Gamma(k)$. These punctures are in one to one correspondence with the $G(k)\backslash\Gamma$-orbit of ∞, while the equivalence classes of characteristics in $X(k)$ contain the $G(k)\backslash\Gamma$-orbit of χ_o. We will show below that this last set has useful alternate descriptions; it is precisely the set we called $X_o(k)$. If we take a set of representatives

$$\gamma_1^{-1}, \ \ldots, \ \gamma_{n(k)}^{-1}$$

for the cosets $G(k)\backslash\Gamma$, then

$$P_{\gamma_1(\infty)}, \ \ldots, \ P_{\gamma_{n(k)}(\infty)}$$

lists the punctures on $\mathbb{H}^2/\Gamma(k)$.

Theorem 2.28. *The group* $\Gamma/\Gamma(k)$ *is isomorphic to a transitive subgroup of the permutation group of* $X_o(k)$.

Proof. We have already shown that we have a homomorphism η from $\Gamma/\Gamma(k)$ to a transitive subgroup of the permutation group of $X_o(k)$. We only have to verify that η is injective. Let $\gamma = \begin{bmatrix} a & b \\ c & d \end{bmatrix} \in \Gamma - \Gamma(k)$. We will show that there is a characteristic (an equivalence class of characteristics)

$\chi \in X_o(k)$ not fixed γ. We consider 3 cases:

1. $k \equiv 1 \mod 2$. This case was treated in detail in the proof of Theorem 2.8, where we showed that either $\begin{bmatrix} \frac{1}{k} \\ 1 \end{bmatrix}$, $\begin{bmatrix} 1 \\ \frac{1}{k} \end{bmatrix}$, or $\begin{bmatrix} \frac{1}{k} \\ \\ \frac{1}{k} \end{bmatrix}$ is not a fixed point of γ.

2. $k \equiv 0 \mod 4$. If $b \not\equiv 0 \mod \frac{k}{2}$ $(c \not\equiv 0 \mod \frac{k}{2})$, then $\begin{bmatrix} \frac{2}{k} \\ 0 \end{bmatrix}$ $\left(\begin{bmatrix} 0 \\ \frac{2}{k} \end{bmatrix}\right)$ is not a fixed point of γ. Thus we may assume that

$$b \equiv 0 \equiv c \mod \frac{k}{2}$$

and in addition that

$$a \equiv d \equiv \pm 1 \mod k.$$

We can assume without loss of generality that the value is $+1$. In other words, in order that the equivalence class of $\begin{bmatrix} \frac{2}{k} \\ 0 \end{bmatrix}$ $\left(\begin{bmatrix} 0 \\ \frac{2}{k} \end{bmatrix}\right)$ be fixed by γ, it must be the case that

$$\gamma = \begin{bmatrix} sk+1 & \frac{lk}{2} \\ \frac{mk}{2} & rk+1 \end{bmatrix}.$$

Note that we may assume that at least one of the integers l or m must be odd. Otherwise, $\gamma \in \Gamma(k)$.

Consider now

$$\begin{bmatrix} \frac{\epsilon}{k} \\ \frac{\epsilon'}{k} \end{bmatrix} \gamma = \begin{bmatrix} \frac{\epsilon}{k} + s\epsilon + m\frac{\epsilon'}{2} - \frac{mk}{2} \\ \frac{\epsilon'}{k} + r\epsilon + l\frac{\epsilon}{2} + \frac{lk}{2} \end{bmatrix} \equiv \begin{bmatrix} \frac{\epsilon}{k} + m\frac{\epsilon'}{2} \\ \frac{\epsilon'}{k} + l\frac{\epsilon}{2} \end{bmatrix}.$$

Since ϵ, ϵ' are both even, in order for $\begin{bmatrix} \frac{\epsilon}{k} \\ \frac{\epsilon'}{k} \end{bmatrix}$ to be fixed by γ it is necessary that

$$m\frac{\epsilon'}{2} - \frac{mk}{2} \equiv 0 \mod 2 \text{ and } l\frac{\epsilon}{2} + \frac{lk}{2} \equiv 0 \mod 2.$$

Since $\frac{k}{2}$ is even, in order for the characteristic to be in $X_o(k)$, at least one of $\frac{\epsilon}{2}, \frac{\epsilon'}{2}$ must be odd. Furthermore one of l and m is odd. It follows that both equations cannot be satisfied simultaneously. In other words there is always a choice of $\begin{bmatrix} \frac{\epsilon}{k} \\ \frac{\epsilon'}{k} \end{bmatrix}$ which will not be fixed unless $\gamma \in \Gamma(k)$.

3. $k \equiv 2 \mod 4$. The proof for this case is identical with the proof of the previous case until the last step. Then, since $\frac{k}{2}$ is odd at least one of $\frac{\epsilon}{2}, \frac{\epsilon'}{2}$ must be even and the result follows.

2.7. Invariant quadruples. The following lemma is based on an important observation of Y. Kopeliovich.

Lemma 2.29. *For all integers m and m', the classes of the four characteristics*

$$\begin{bmatrix} \frac{m}{k} \\ \frac{m'}{k} \end{bmatrix}, \begin{bmatrix} \frac{m}{k} \\ \frac{m'}{k}+1 \end{bmatrix}, \begin{bmatrix} \frac{m}{k}+1 \\ \frac{m'}{k} \end{bmatrix}, \begin{bmatrix} \frac{m}{k}+1 \\ \frac{m'}{k}+1 \end{bmatrix}$$

are invariant (as a set) under the action of $\Gamma(k)$. Further:
(a) *For odd k, one of these classes is in $X(k)$ (hence fixed by $\Gamma(k)$), and the other three are permuted by this group.*
(b) *For even k and $\gamma \in \Gamma(k)$, let $\chi = \begin{bmatrix} \frac{m}{k} \\ \frac{m'}{k} \end{bmatrix}$ and $\chi\gamma = \begin{bmatrix} \frac{m_1}{k} \\ \frac{m_1'}{k} \end{bmatrix}$. Then each of the pairs m, m_1 and m', m_1' has the same parity. Thus for m and m' even, each of the above four classes of characteristic is in $X(k)$ (and hence fixed by $\Gamma(k)$). For m and m' not both even, the above classes form a single $\Gamma(k)$-orbit.*
(c) *Each element of Γ maps the above quadruple onto a similar quadruple.*

Proof. Write an arbitrary element $\gamma \in \Gamma(k)$ as a matrix

$$\begin{bmatrix} 1+kr & ks \\ kt & 1+ku \end{bmatrix},$$

with r, s, t, and $u \in \mathbb{Z}$ such that $(1+kr)(1+ku) - k^2ts = 1$. Then for $\chi = \begin{bmatrix} \frac{m}{k} \\ \frac{m'}{k} \end{bmatrix}$, we have

$$\chi\gamma = \begin{bmatrix} \frac{m}{k} + mr + m't - kt(1+kr) \\ ms + \frac{m'}{k} + m'u + ks(1+ku) \end{bmatrix},$$

from which all the statements of parts (a) and (b) of the lemma follow (sometimes by examining cases). The proof of (c) is left to the reader.

Remark 2.30. The four classes listed above need not be distinct. For example, if k is even, $m = k$, and $m' = \frac{k}{2}$.

Definition 2.31. We shall call the quadruple appearing in the statement of the previous lemma an *adherent* quadruple of classes of characteristics.

2.8. Towers. It is convenient to represent the classes of characteristics in $Z(k) \cap \mathcal{P}$ as a set of towers (columns) in $\mathbb{R}^2$. The *tower* over $\frac{m}{k}$ with $m \in \mathbb{Z}$, $0 \le m \le k$, consists of the characteristics $\begin{bmatrix} \frac{m}{k} \\ \frac{m'}{k} \end{bmatrix}$ with $m' \in \mathbb{Z}$, $0 \le m' \le k$ for $m = 0$ and $m = k$, and $0 \le m' < 2k$ for $0 < m < k$. We observe that each tower is invariant under B. The punctures on $\mathbb{H}^2/\Gamma(k)$ are in one to one correspondence with the points in $X_o(k) \cap \mathcal{P}$. If k and m have the same parity, then the tower over $\frac{m}{k}$ must contain at least one point of $X_o(k)$. If $k > 2$, this tower then contains the characteristic $\begin{bmatrix} \frac{m}{k} \\ \\ \frac{1}{k} \end{bmatrix}$ if k is odd; the characteristic $\begin{bmatrix} \frac{m}{k} \\ \\ \frac{2}{k} \end{bmatrix}$ if k is even, $\frac{k}{2}$ is odd, and $4|m$, or k and $\frac{k}{2}$ are even; and the characteristic $\begin{bmatrix} \frac{m}{k} \\ \\ \frac{4}{k} \end{bmatrix}$ if k is even, $\frac{k}{2}$ is odd, and $4 \not| \, m$.

Definition 2.32. We denote by $\mathcal{T}_{\frac{m}{k}}$ the tower in $X_o(k)$ over $\frac{m}{k}$.

3. Punctures and characteristics

3.1. A correspondence. Let G be a subgroup of Γ that contains a $\Gamma(k)$ for some positive integer k: $\Gamma(k) \subset G \subset \Gamma$. The natural projection $\mathbb{H}^2 \to \mathbb{H}^2/G$ is a composition of the two projections: $\mathbb{H}^2 \to \mathbb{H}^2/\Gamma(k)$ and $\mathbb{H}^2/\Gamma(k) \to \mathbb{H}^2/G$. The last of these three projections is a finite sheeted Galois covering with Galois group $G/\Gamma(k)$. The covering $\mathbb{H}^2/G \to \mathbb{H}^2/\Gamma$ is also finite sheeted, but Galois if and only if G is a normal subgroup of Γ. However, it can always be viewed as the quotient of the surface $\mathbb{H}^2/G$ by the right finite coset space $G\backslash\Gamma$. Theorem 2.26 and Corollary 2.23 state that for all $\gamma \in \Gamma$, the following diagram commutes:

$$\begin{array}{ccc} X_o(k) & \xrightarrow{\mathcal{I}} & \text{Punctures on } \mathbb{H}^2/\Gamma(k) \\ \gamma\big\downarrow & & \big\downarrow \qquad\qquad \gamma^{-1}. \\ X_o(k) & \xrightarrow{\mathcal{I}} & \text{Punctures on } \mathbb{H}^2/\Gamma(k) \end{array}$$

The group Γ acts on the set of equivalence classes of characteristics on the right and on punctures on the left (accounting for the presence of the inverse in the above diagram). It is a consequence of this commutative diagram (and the fact that the punctures on $\mathbb{H}^2/\Gamma(k)$ are in one-to-one correspondence with the classes of characteristics in $X_o(k)$) that the punctures on $\mathbb{H}^2/G$ are in one-to-one correspondence with the points in the quotient of $X_o(k)$ by $\Gamma(k)\backslash G$.

3.2. Branching. For any Fuchsian group G acting on $\mathbb{H}^2$ and any $x \in \mathbb{H}^2 \cup \hat{\mathbb{R}}$, we let G_x be the stabilizer of x in G. If x is a fixed point of an elliptic (parabolic) element of G, then G_x is a finite (infinite) cyclic group. We are interested in studying the map (notation as in the previous paragraph)

$$\pi : \overline{\mathbb{H}^2/\Gamma(k)} \to \overline{\mathbb{H}^2/G}.$$

The branch number of the map π at $y \in \overline{\mathbb{H}^2/\Gamma(k)}$ is given by

$$b_\pi(y) = |G_x|$$

if y is the image of $x \in \mathbb{H}^2$ under the natural projection $P(\Gamma(k))$, and by

$$b_\pi(y) = [G_x : \Gamma(k)]$$

if $y = P(\Gamma(k))_x$ for a parabolic fixed point x of $\Gamma(k)$.

4. More invariant classes

4.1. Invariant classes for $G(k)$.

Lemma 4.1. *Let* $\chi_i = \begin{bmatrix} \frac{m_i}{k} \\ \frac{m_i'}{k} \end{bmatrix}$, $i = 1, 2$, *be two characteristics in* $Z(k)$. *Then* χ_2 *is* $< B >$*-equivalent to* χ_1 *if and only if*

$$m_1 + (-1)^\epsilon m_2 \equiv 0 \mod 2k,$$

and there exists an $l \in \mathbb{Z}$ *such that*

$$m_1' + (-1)^\epsilon m_2' + l(m_1 + k) \equiv 0 \mod 2k,$$

with $\epsilon = 0$ *or* 1.

Proof. The proof is straightforward and left to the reader.

Remark 4.2. If k is an odd prime, $\chi_i \in X_o(k)$ for $i = 1, 2$, and $0 < m_1 = m_2 < k$, then we can choose $\epsilon = 1$ and always find an l as above. Thus $< B >$ acts transitively on all but the last (the one over 1) tower of $X_o(k)$. The action on the tower over 1 is certainly not transitive since each characteristic in this tower is fixed by $< B >$.

Lemma 4.3. *An equivalence class of characteristics is fixed by the group* $G(k)$ *if and only if it is in* $X(k)$ *and equivalent to either*

(a) $\begin{bmatrix} 1 \\ \frac{m'}{k} \end{bmatrix}$ *with* $m' \in \mathbb{Z}$ *or*

(b) *(if* k *and* $\frac{k}{2}$ *are even)* $\begin{bmatrix} 0 \\ \frac{1}{2} \end{bmatrix}$.

Proof. Left to reader.

Remark 4.4. The last lemma shows that $X(G(k))$ consists of the tower in $X(k)$ above 1 and, in addition, the class of the characteristic $\frac{1}{2}v_2$ if $k \equiv 0 \mod 4$.

As a consequence of the last lemma, we are able to construct, as in parts (a) and (b) of Theorem 2.8 of Chapter 3, automorphic forms for $G(k)$. See §2.5 of Chapter 3.

The surface $\mathbb{H}^2/G(k)$ is isomorphic to $\mathbb{H}^2/\Gamma(k)/ < \tilde{B} >$. The automorphism $\tilde{B}$ of $\overline{\mathbb{H}^2/\Gamma(k)}$ has period k. The stabilizers of points in $\overline{\mathbb{H}^2/\Gamma(k)}$ have orders that divide k. Hence unless $2|k$ or $3|k$, the group $< \tilde{B} >$ acts fixed point freely on $\mathbb{H}^2/\Gamma(k)$. Every element g of $G(k)$ can be written in the form $B^l \circ \gamma$, where $l \in \mathbb{Z}$, $0 \leq l \leq k-1$, and $\gamma \in \Gamma(k)$. It is a simple consequence of this decomposition that $g \in G(k)$ has a fixed point in $\mathbb{H}^2$ if and only if $k \leq 3$. We already saw that in these cases $G(k) = \Gamma_o(k)$. In particular, $G(k)$ is torsion free for $k \geq 4$.

The punctures on $\mathbb{H}^2/G(k)$ are the images under the natural projection $\mathbb{H}^2/\Gamma(k) \to \mathbb{H}^2/G(k)$ of the punctures of $\mathbb{H}^2/\Gamma(k)$; thus they correspond to the finitely many cosets $X_o(k)/ < \tilde{B} >$. A good way to study this coset space is in terms of the towers in $\mathcal{P}$ discussed previously.

Assume that $k \geq 2$. Consider the tower in $X_o(k)$ over the point 1. In particular, χ_o is in this tower. It is a consequence of Lemma 4.3 that each element of this tower is fixed by B, and hence that the cover $\overline{\mathbb{H}^2/\Gamma(k)} \to \overline{\mathbb{H}^2/G(k)}$ is completely ramified at the punctures corresponding to these classes. If $k \neq 4$, then these are the only fixed points of $\tilde{B}$ on $\overline{\mathbb{H}^2/\Gamma(k)}$. The class of $\chi_1 = \begin{bmatrix} 0 \\ \frac{1}{2} \end{bmatrix} \in X_o(k)$ if and only if $k = 4$; in this case the puncture corresponding to χ_1 is also a fixed point of $\tilde{B}$. It is easy to see that B operates as a permutation of the elements of each tower in $X_o(k)$. We need to determine the number of $< B >$-orbits in each tower and also the number of elements in each orbit.

Fix an integer m of the same parity as k with $0 \leq m \leq k$. Let $N(k, m)$ be the number of elements in the tower above $\frac{m}{k}$ in $X_o(k)$; that is,

$$N(k, m) = |\mathcal{T}_{\frac{m}{k}}|.$$

As an immediate consequence of Lemma 2.18 we compute $N(k, m)$.

Lemma 4.5. *Let $k \in \mathbb{Z}^+$, $k \geq 2$. Let m be a nonnegative integer of the same parity as k. Let $d = (k, m)$ for odd k and $d = \left(\frac{k}{2}, \frac{m}{2}\right)$ for even k. We have:*
(a) *For $k \equiv 1 \mod 2$, $N(k, m)$ equals the number of positive odd integers $< 2k$ relatively prime to d if $1 \leq m \leq k-2$, and*
(a') *$N(k, k)$ equals the number of positive odd integers $< k$ relatively prime to k.*
(b) *$N(2, 0) = 2$ and $N(2, 2) = 1$.*
(c) *For $k \geq 6$ and $\equiv 2 \mod 4$, $N(k, 0)$ equals the number of nonnegative integers $\leq \frac{k}{2}$ relatively prime to $d = \frac{k}{2}$,*
(c') *$N(k, m)$ equals the number of nonnegative integers $\leq k-1$ relatively prime to d if $0 < m < k$ and $\frac{m}{2}$ is even,*

(c") *$N(k,m)$ equals the number of nonnegative even integers $\leq k-2$ relatively prime to d if $0 < m < k$ and $\frac{m}{2}$ is odd, and*
(c"') *$N(k,k)$ equals the number of nonnegative even integers $\leq \frac{k}{2}-1$ relatively prime to $d = \frac{k}{2}$.*
(d) *For $k \equiv 0 \mod 4$, $N(k,0)$ equals the number of positive odd integers $< \frac{k}{2}$ relatively prime to $d = \frac{k}{2}$,*
(d') *$N(k,m)$ equals the number of nonnegative integers $\leq k-1$ relatively prime to d if $1 < m < k-1$ and $\frac{m}{2}$ is odd,*
(d") *$N(k,m)$ equals the number of positive odd integers $\leq k-1$ relatively prime to d if $1 < m < k-1$ and $\frac{m}{2}$ is even, and*
(d"') *$N(k,k)$ equals the number of positive odd integers $\leq \frac{k}{2}-1$ relatively prime to $d = \frac{k}{2}$.*

Corollary 4.6. *For all integers $k \geq 2$, we have*

$$\sum_{l=0}^{\lfloor \frac{k}{2} \rfloor} N\left(k, 2\left(l + \frac{k}{2} - \left\lfloor \frac{k}{2} \right\rfloor\right)\right) = n(k),$$

where $\lfloor x \rfloor$ denotes the greatest integer less than or equal to $x \in \mathbb{R}$.

Lemma 4.7. *Let $k \in \mathbb{Z}^+$, $k \geq 2$. Let m be an integer of the same parity as k with $0 \leq m \leq k$ and $d = (k,m)$. There exists a unique smallest positive integer $j(k,m)$ so that B^j fixes each class in the tower over $\frac{m}{k}$ in $X_o(k)$. We have for $m < k$:*
(a) *$j(k,m) = \frac{k}{d}$ if k is odd,*
(b) *$j(k,m) = \frac{k}{d}$ if k and $\frac{k}{d} + \frac{m}{d}$ are even and $m \neq 0$,*
(c) *$j(k,m) = 2\frac{k}{d}$ if k is even, $\frac{k}{d} + \frac{m}{d}$ is odd and $m \neq 0$,*
(d) *$j(k,0) = 2$ if $k \neq 4$ and*
(e) *$j(4,0) = 1$. Further,*
(f) *$j(k,k) = 1$.*

Proof. For every pair of integers m' and j,

$$\begin{bmatrix} \frac{m}{k} \\ \\ \frac{m'}{k} \end{bmatrix} B^j = \begin{bmatrix} \frac{m}{k} \\ \\ \frac{j(m+k)}{k} + \frac{m'}{k} \end{bmatrix}. \tag{2.30}$$

It follows from this that the smallest positive integer j that fixes the class of $\begin{bmatrix} \frac{m}{k}, \\ \frac{m'}{k} \end{bmatrix}$ is independent of m'. The rest of the lemma also follows because m and k have the same parity.

Many of our claims are simplified by the following simple proposition that is almost self evident.

Proposition 4.8. *Let k be a positive integer and m and m' nonnegative integers of the same parity as k so that* $\begin{bmatrix} \frac{m}{k} \\ \\ \frac{m'}{k} \end{bmatrix} \in X(k)$ *and let* $d = (k, m, m')$. *Then*

$$\begin{bmatrix} \frac{\frac{m}{d}}{\frac{k}{d}} \\ \\ \frac{\frac{m'}{d}}{\frac{k}{d}} \end{bmatrix} \in \begin{cases} X_o\left(\frac{k}{d}\right) & \text{if } \frac{k}{d}\frac{m}{d}\frac{m'}{d} \text{ is odd} \\ X_o\left(2\frac{k}{d}\right) & \text{if } \frac{k}{d}\frac{m}{d}\frac{m'}{d} \text{ is even} \end{cases}.$$

Remark 4.9. If k is an odd prime, then we have considerable simplification in the computations:

$$N(k, m) = k \text{ for } m = 1,\ 3,\ \dots,\ k-2,\ N(k,k) = \frac{k-1}{2},$$

and

$$j(k, m) = k \text{ for } k = 1,\ 3,\ \dots,\ k-2.$$

Remark 4.10. We proceed to express the numbers $N(k, m)$ in terms of the Euler φ function.

k **odd.** In this case m is also odd and we assume for the moment that $0 < m < k$. We let $(m, k) = d$. We want to compute $N(k, m)$, the number of odd positive integers $\leq (2k-1)$ that are relatively prime to d. Now the even positive integer $j \leq (k-1)$ is relatively prime to d if and only if the odd positive integer $j + k$, $k + 1 \leq j + k \leq 2k - 1$, is relatively prime to d. Hence $N(k, m)$ is the number of positive integers $\leq k$ that are relatively prime to d. It follows that $N(k, m) = \frac{k}{(k,m)}\varphi((k, m))$. The same reasoning gives $N(1, 1) = 1$ and $N(k, k) = \frac{1}{2}\varphi(k)$ for $k > 1$.

k **even.** The calculations proceed via an analysis of cases and are very similar to the ones for odd k. The table in this subsection summarizes the results.

4.2. Characterization of $G(k)$. Consider the subgroup $\mathcal{G}(k)$ of $\Gamma_o(k)$ defined as those motions represented by matrices in

$$\left\{ \begin{bmatrix} a & b \\ c & d \end{bmatrix} \in \mathrm{SL}(2, \mathbb{Z});\ a \equiv 1 \equiv d \mod k,\ c \equiv 0 \mod k \right\}.$$

Clearly, $G(k) \subset \mathcal{G}(k) \subset \Gamma_o(k)$. We shall show that $G(k) = \mathcal{G}(k)$. An easy calculation shows that $\mathcal{G}(k)$ is normal in $\Gamma_o(k)$. We need the following

Lemma 4.11. *If* $\begin{bmatrix} 1 \\ \frac{m'}{k} \end{bmatrix} \in Z(k)$ *is fixed by an element* $\gamma \in \Gamma$, *then* $\gamma \in \mathcal{G}(k)$.

Conditions on k	Conditions on m	$N(k,m)$
$k=1$	$m=1$	1
$k=2$	$m=0$	2
	$m=2$	1
$k>1$ and $k\equiv 1 \mod 2$	$m<k$	$\frac{k}{(k,m)}\varphi((k,m))$
	k	$\frac{1}{2}\varphi(k)$
$k\equiv 0 \mod 4$	$m=0$ or $m=k$	$\varphi\left(\frac{k}{2}\right)$
	$0<m<k$	$\frac{2k}{(k,m)}\varphi\left(\frac{(k,m)}{2}\right)$
$k>2$ and $k\equiv 2 \mod 4$	$m=0$ or $m=k$	$\varphi\left(\frac{k}{2}\right)$
	$0<m\le k$	$(2,\frac{m}{2})\frac{k}{(k,m)}\varphi\left(\frac{(k,m)}{2}\right)$

Table 4. FORMULAE FOR $N(k,m)$ IN TERMS OF THE EULER φ-FUNCTION. The integers k and m are assumed to have the same parity for the entries in the table.

Proof. If the class of $\begin{bmatrix} 1 \\ \frac{m'}{k} \end{bmatrix}$ is fixed by $\gamma = \begin{bmatrix} a & b \\ c & d \end{bmatrix}$, then $\begin{bmatrix} a + c\frac{m'}{k} - ac \\ b + d\frac{m'}{k} + bd \end{bmatrix}$ is equivalent to $\begin{bmatrix} 1 \\ \frac{m'}{k} \end{bmatrix}$. This implies that c is congruent to zero modulo k and that d is congruent to either plus or minus 1 modulo k (without loss of generality we may assume plus 1). This shows that $\gamma \in \mathcal{G}(k)$.

Lemma 4.12. *The group $\mathcal{G}(k)$ maps each tower of $X(k)$ onto itself.*

Proof. We observe that (all entries in the next equation are integers)

$$\begin{bmatrix} \frac{m}{k} \\ \frac{m'}{k} \end{bmatrix} \begin{bmatrix} 1+kt & b \\ kr & 1+ks \end{bmatrix} = \begin{bmatrix} \frac{m}{k} + mt + m'r - kr(1+kt) \\ \frac{m'}{k} + b\frac{m}{k} + m's + b(1+ks) \end{bmatrix}.$$

If k is even, then it is immediate that a class of characteristics and its image are in the same tower. If k is odd, we argue as follows. If r is even, then t must be even (for otherwise $1+kt$ would be even). Therefore we see once again that each tower is preserved. If r is odd, then the result follows provided $mt - kr(1+kt)$ is odd. This is so independently of whether t is even or odd.

Proposition 4.13. *We have $G(k) = \mathcal{G}(k)$.*

Proof. The class of $\chi_o = \begin{bmatrix} 1 \\ \frac{\alpha}{k} \end{bmatrix}$ is fixed by each $g \in \mathcal{G}(k)$ so that ∞ is fixed mod $\Gamma(k)$ by g^{-1}. This means that $\mathcal{G}(k) \subset G(k)$. Since the reverse inclusion is trivial, the proposition follows.

k	$\mu = n(k)$	ν_∞	p
4	6	3	0
5	12	4	0
6	12	4	0
7	24	6	0
8	24	6	0
9	36	8	0
10	36	8	0
11	60	10	1
12	48	10	0
13	84	12	2

Table 5. THE SIGNATURE OF $G(k)$.

4.3. The surface $\mathbb{H}^2/G(k)$. The orbifold $\mathbb{H}^2/G(k)$ for $k \leq 4$ is described by the table in Chapter 1, §6.6. So assume $k \geq 4$. The group $G(k)$ is torsion free of index $n(k)$ in Γ. The quotient map

$$\overline{\mathbb{H}^2/\Gamma(k)} \to \overline{\mathbb{H}^2/G(k)}$$

is k-sheeted and ramified only over punctures. We conclude that $\mathbb{H}^2/G(k)$ is of type

$$\left(1 + \frac{n(k)}{12} - \frac{1}{2}\sum_{*m} \frac{N(k,m)}{j(k,m)}, \sum_{*m} \frac{N(k,m)}{j(k,m)}\right),$$

where $\sum_{*m}$ stands for the sum over integers m between 0 and k of the same parity as k. If k is an odd prime, then the above reduces to

$$\left(\frac{(k-5)(k-7)}{24}, k-1\right).$$

In terms of the Euler φ-function description

$$\nu_\infty(G(k)) = \sum_{*m} \frac{N(k,m)}{j(k,m)}$$

$$= \begin{cases} \sum_{l=0}^{\frac{k-3}{2}} \varphi((2l+1,k)) + \frac{1}{2}\varphi(k) & \text{if } k \equiv 0 \mod 2 \\ \sum_{l=1}^{\frac{k}{2}-1} \varphi((l,\frac{k}{2})) + \frac{3}{2}\varphi(\frac{k}{2}) & \text{if } k \equiv 0 \mod 4 \\ \sum_{l=1}^{\frac{k}{2}-1} \varphi((l,\frac{k}{2})) + \varphi(\frac{k}{2}) & \text{if } k \equiv 2 \mod 4 \end{cases}.$$

We have tabulated the relevant numbers for the group $G(k)$ for low values of k. To study automorphic forms for the group $G(k)$, we need to know good local coordinates at the punctures of $\mathbb{H}^2/G(k)$. The map

$$\pi : \overline{\mathbb{H}^2/\Gamma(k)} \to \overline{\mathbb{H}^2/G(k)}$$

can be used to conclude that if ζ is a local coordinate at the puncture x on $\mathbb{H}^2/\Gamma(k)$, then $\zeta^{b_\pi(x)+1}$ is a good local coordinate at the puncture $\pi(x)$ on $\mathbb{H}^2/G(k)$. Further, if the puncture x is identified with the characteristic $\chi = \begin{bmatrix} \frac{m}{k} \\ \frac{m'}{k} \end{bmatrix} \in X_o(k)$ (that is, $\mathcal{I}(\chi) = x$), then

$$b_\pi(x) + 1 = \frac{k}{j(k,m)}.$$

As an application, we describe for k an odd prime, the structure of the $k-1$ punctures on $\mathbb{H}^2/G(k)$ as a consequence of our work on towers (for $\Gamma(k)$). We can take as a set of representatives for the action of $\Gamma(k)\backslash G(k)$ on $X_o(k)$ the two lists of characteristics

$$\begin{bmatrix} \frac{2l+1}{k} \\ 1 \end{bmatrix}, \begin{bmatrix} 1 \\ \frac{2l+1}{k} \end{bmatrix}, \; l = 0,\ 1,\ \ldots,\ \frac{k-3}{2}.$$

Let $\alpha \in \mathbb{Z}$, $1 \le \alpha \le \frac{k-1}{2}$. Choose b and $d \in \mathbb{Z}$ such that $d\alpha - bk = 1$. Then

$$\begin{bmatrix} 1 \\ \frac{1}{k} \end{bmatrix} \begin{bmatrix} d & -b \\ -k & \alpha \end{bmatrix} \equiv \begin{bmatrix} 1 \\ \frac{\alpha}{k} \end{bmatrix} \text{ or } \begin{bmatrix} 1 \\ \frac{k-\alpha}{k} \end{bmatrix},$$

depending on whether α is even or odd. We conclude that the punctures corresponding to the second list of characteristics are the images under P of the rational numbers

$$\frac{\alpha}{k};\ \alpha = 1,\ 2,\ \ldots,\ \frac{k-1}{2}.$$

It is also not hard to see that these punctures on $\mathbb{H}^2/G(k)$ are $\Gamma_o(k)/G(k)$ equivalent to P_∞. Similarly, the image under P of

$$\frac{k}{\alpha};\ \alpha = 1,\ 2,\ \ldots, \frac{k-1}{2},$$

are the punctures corresponding to the first list of characteristics; these punctures are $\Gamma_o(k)/G(k)$ equivalent to P_0.

4.4. Invariant classes for $\Gamma_o(k)$. New methods must be developed to treat the function theory for $\Gamma_o(k)$ since in general there are few invariant nontrivial classes of characteristics for this group as is shown by the next few lemmas.

Lemma 4.14. *The group $\Gamma_o(k)$ permutes the elements in the tower $\mathcal{T}_{\frac{k}{k}}$ above 1 in $X_o(k)$.*

Proof. It suffices to show that for each $m' \in \mathbb{Z}$ of the same parity as k and each $\gamma \in \Gamma_o(k)$, $\begin{bmatrix} 1 \\ \frac{m'}{k} \end{bmatrix} \gamma$ is again of the form $\begin{bmatrix} 1 \\ \frac{m''}{k} \end{bmatrix}$ with $m'' \in \mathbb{Z}$ (it will automatically have the same parity as k). If, as usual, $\gamma = \begin{bmatrix} a & b \\ \tilde{c}k & d \end{bmatrix}$, then we must show that $a + \tilde{c}m' - a\tilde{c}k$ is odd. An examination of cases yields this easily.

The above lemma allows us to define a homomorphism of $\Gamma_o(k)$ into the permutation group on $N(k,k)$ letters whose kernel is $G(k)$. We show next that $\Gamma_o(k)$ acts transitively on the $N(k,k)$ letters.

Lemma 4.15. *The group $\Gamma_o(k)$ acts transitively on the elements in $X_o(k)$ in the tower over 1. If k is even, the group $\Gamma_o(k)$ acts transitively on the elements in $X_o(k)$ in the tower over 0. Hence the finite group $\Gamma_o(k)/G(k)$ acts transitively on these towers.*

Proof. We consider three cases.
(a) Assume that k is odd. Then an arbitrary element in the tower over 1 is $\chi = \begin{bmatrix} 1 \\ \frac{m'}{k} \end{bmatrix}$ with m' a positive odd integer, $< k$, and relatively prime to k. Choose integers a and b so that $am' - bk = 1$. The element χ is then the image of χ_o under $\begin{bmatrix} a & b \\ k & m' \end{bmatrix} \in \Gamma_o(k)$.
(b) Assume that k is even and $\frac{k}{2}$ is odd. An arbitrary element in the tower is $\chi = \begin{bmatrix} 1 \\ \frac{2m'}{k} \end{bmatrix}$ with m' a nonnegative even integer, $< \frac{k}{2}$, and relatively prime to $\frac{k}{2}$. In this case, $(\frac{m'}{2}, k)$ is 1 or 2. If $(\frac{m'}{2}, k) = 1$, choose integers a and b with $\frac{am'}{2} - bk = 1$. Then $\begin{bmatrix} a & b \\ k & \frac{m'}{2} \end{bmatrix} \in \Gamma_o(k)$ maps χ_o onto χ. If $(\frac{m'}{2}, k) = 2$, then $(\frac{m'}{2} + \frac{k}{2}, k) = 1$, and we choose the integers a and b so that $a(\frac{m'}{2} + \frac{k}{2}) - bk = 1$. Then $\begin{bmatrix} a & b \\ k & \frac{m'}{2} + \frac{k}{2} \end{bmatrix} \in \Gamma_o(k)$ maps χ_o onto χ.
(c) Assume that k and $\frac{k}{2}$ are even. An arbitrary element in the tower is $\chi = \begin{bmatrix} 1 \\ \frac{2m'}{k} \end{bmatrix}$ with m' a positive odd integer, $< \frac{k}{2}$, and relatively prime to $\frac{k}{2}$. It follows that $(m', k) = 1$. Choosing integers a and b with $am' - bk = 1$, it follows that $\begin{bmatrix} a & b \\ k & m' \end{bmatrix} \in \Gamma_o(k)$ maps χ_o onto χ.

To prove the second statement (about even k) repeat the argument of parts (b) and (c) with the characteristic χ_o replaced by $\begin{bmatrix} 0 \\ \frac{\alpha}{k} \end{bmatrix}$.

Corollary 4.16. *We have*

$$[\Gamma_o(k) : G(k)] = N(k,k).$$

Lemma 4.17. *The tower in $X_o(k)$ over 1 is never empty; it contains more than one element if and only if $k = 5$ or $k > 6$.*

Proof. For all k, $\chi_o \in X_o(k)$. If $k \equiv 1 \mod 2$, $k > 3$, then $\chi_o = \begin{bmatrix} 1 \\ \frac{1}{k} \end{bmatrix}$ and $\begin{bmatrix} 1 \\ \frac{k-2}{k} \end{bmatrix}$ are distinct and belong to $X_o(k)$. If $k \equiv 0 \mod 4$, $k \neq 4$, then $\chi_o = \begin{bmatrix} 1 \\ \frac{2}{k} \end{bmatrix}$ and $\begin{bmatrix} 1 \\ \frac{k-2}{k} \end{bmatrix}$ are distinct and belong to $X_o(k)$. Finally, if $k \equiv 2 \mod 4$, $k > 6$, then $\chi_o = \begin{bmatrix} 1 \\ \frac{4}{k} \end{bmatrix}$ and $\begin{bmatrix} 1 \\ \frac{k-2}{k} \end{bmatrix}$ are distinct and belong to $X_o(k)$.

Lemma 4.18. *The invariant classes of characteristic $X(\Gamma_o(k))$ are as follows:*

(a) *For $k \equiv 1,\ 5,\ 7,$ or $11 \mod 12$, $X(\Gamma_o(k))$ is trivial, that is, consists only of the class of*

$$\begin{bmatrix} 1 \\ 1 \end{bmatrix}.$$

(b) *For $k \equiv 2$ or $10 \mod 12$, $X(\Gamma_o(k)) =$*

$$X(\Gamma_o(2)) = \left\{ \begin{bmatrix} 1 \\ 0 \end{bmatrix} \text{ and } \begin{bmatrix} 1 \\ 1 \end{bmatrix} \right\}.$$

(c) *For $k \equiv 3$ or $9 \mod 12$, $X(\Gamma_o(k)) =$*

$$X(\Gamma_o(3)) = \left\{ \begin{bmatrix} 1 \\ \frac{1}{3} \end{bmatrix} \text{ and } \begin{bmatrix} 1 \\ 1 \end{bmatrix} \right\}.$$

(d) *For $k \equiv 4$ or $8 \mod 12$, $X(\Gamma_o(k)) =$*

$$X(\Gamma_o(4)) = \left\{ \begin{bmatrix} 1 \\ 0 \end{bmatrix}, \begin{bmatrix} 1 \\ \frac{1}{2} \end{bmatrix}, \begin{bmatrix} 1 \\ 1 \end{bmatrix}, \text{ and } \begin{bmatrix} 0 \\ \frac{1}{2} \end{bmatrix} \right\}.$$

(e) *For $k \equiv 6 \mod 12$, $X(\Gamma_o(k)) =$*

$$X(\Gamma_o(6)) = \left\{ \begin{bmatrix} 1 \\ 0 \end{bmatrix}, \begin{bmatrix} 1 \\ \frac{1}{3} \end{bmatrix}, \begin{bmatrix} 1 \\ \frac{2}{3} \end{bmatrix}, \text{ and } \begin{bmatrix} 1 \\ 1 \end{bmatrix} \right\}.$$

(f) *For $k \equiv 0 \mod 12$, $X(\Gamma_o(k)) =$*

$$X(\Gamma_o(12)) = \left\{ \begin{bmatrix} 1 \\ 0 \end{bmatrix}, \begin{bmatrix} 1 \\ \frac{1}{3} \end{bmatrix}, \begin{bmatrix} 1 \\ \frac{1}{2} \end{bmatrix}, \begin{bmatrix} 1 \\ \frac{2}{3} \end{bmatrix}, \begin{bmatrix} 1 \\ 1 \end{bmatrix}, \text{ and } \begin{bmatrix} 0 \\ \frac{1}{2} \end{bmatrix} \right\}.$$

Proof. Only the classes of characteristics fixed by $G(k)$ are candidates for $\Gamma_o(k)$-invariants. Thus we are left to study the classes in the tower in $X(k)$ over 1 and the class of the characteristic $\chi_1 = \begin{bmatrix} 0 \\ \frac{1}{2} \end{bmatrix}$ (if $4|k$). If $1 < k'|k$, then $X(\Gamma_o(k)) \subset X(\Gamma_o(k'))$. The tower over 1 in $X(k)$ decomposes into sections lying in the various $X_o(k')$ with $1 \leq k' \leq k$ and $k'|k$. Each of these sections is $\Gamma_o(k')$ and thus also $\Gamma_o(k)$-invariant. Hence as a consequence of the last two lemmas, $X(\Gamma_o(k))$ is trivial unless $2|k$ or $3|k$. We have proven part (a). Assume now that $2|k$. Then $X(\Gamma_o(k))$ contains the tower over 1 in $X(2)$ which consists of the classes of the two vectors v_1 and $v_1 + v_2$. If $3 \not| k$ and $4 \not| k$, then $X(\Gamma_o(k))$ contains nothing else, and we have hence established part (b). If $3|k$, then $X(\Gamma_o(k))$ contains the tower over 1 in $X(3)$ which consists of the classes of the two vectors listed in part (c). If $2 \not| k$, then $X(\Gamma_o(k))$ contains nothing else, and we have hence established part (c). If $4|k$, then $X(\Gamma_o(k))$ contains the tower over 1 in $X(4)$ which has two parts: classes in $X_o(4)$ and classes in $X(2)$. In addition the vector χ_1 is fixed by $\Gamma_o(k)$. If $3 \not| k$, then $X(\Gamma_o(k))$ contains nothing else. Part (d) follows. If $6|k$, then $X(\Gamma_o(k))$ contains $X(\Gamma_o(3)) \cup X(\Gamma_o(2))$ and the one class in the tower in $X_o(6)$ over 1. Part (e) is an immediate consequence. Finally, if $12|k$, then $X(\Gamma_o(k))$ contains $X(\Gamma_o(6)) \cup X(\Gamma_o(4))$. This suffices to prove part (f).

4.5. More homomorphisms. As a result of the lemmas, we have established a homomorphism η from $\Gamma_o(k)$ to the permutation group of $\mathcal{T}_{\frac{k}{k}}$ with kernel $G(k)$. Thus, in particular, $G(k)$ is a normal subgroup of $\Gamma_o(k)$. Further, the finite group $\Gamma_o(k)/G(k)$ acts transitively on the tower in $X_o(k)$ over 1; this action is fixed point free for $k = 5$ and $k > 6$.

It is easy to see without the use of the homomorphism η that $G(k)$ is a normal subgroup of $\Gamma_o(k)$, as already observed in §4.2. This serves as a reminder that $\Gamma_o(k)/G(k)$ acts as a subgroup of the automorphism group of $\mathbb{H}^2/G(k)$. As such it acts as a permutation group on the punctures of $\mathbb{H}^2/G(k)$. Our last result shows that this action is transitive on the set of punctures corresponding to the characteristics in $\mathcal{T}_1$.

Lemma 4.19. *Let $0 < m_1,\ m_2 < k$ be integers with the same parity as k. For k odd, a necessary and sufficient condition that an element of $\Gamma_o(k)$ map $\mathcal{T}_{\frac{m_1}{k}}$ to $\mathcal{T}_{\frac{m_2}{k}}$ is that $(m_1,k) = (m_2,k) \neq k$. For k even, we must require in addition that the parity of $\frac{m_1}{2}$ be equal to the parity of $\frac{m_2}{2}$.*

Proof. Let $\chi = \begin{bmatrix} \frac{m}{k} \\ \frac{m'}{k} \end{bmatrix} \in X(k)$ and $\gamma = \begin{bmatrix} a & b \\ \tilde{c}k & d \end{bmatrix} \in \Gamma_o(k)$. Then

$$\chi\gamma = \begin{bmatrix} a\frac{m}{k} + \tilde{c}(m' - ak) \\ * \end{bmatrix}.$$

If k is even, so is $m' - ak$. If k is odd, then $\tilde{c}(m' - ak)$ is odd (even) if and only if a is even (odd). We conclude that γ maps towers to towers. For $m \in \mathbb{Z}$ with the same parity as k and $0 < m < k$, let $(m,k) = d$. If $(m_1,k) = d$, then $m_1 = dl$ with $(k,l) = 1$, so that l is invertible mod k. This l is necessarily odd (no matter what the parity of k), and we can choose an integer l' so that $ll' = 1 + \tilde{c}k$ for some $\tilde{c} \in \mathbb{Z}$. Hence $\gamma = \begin{bmatrix} l & 1 \\ \tilde{c}k & l' \end{bmatrix} \in \Gamma_o(k)$ and $\mathcal{T}_{\frac{d}{k}}$ gets mapped by γ to $\mathcal{T}_{\frac{m_1}{k}}$. The argument for even k is similar.

It is also clear that $\mathcal{T}_0$ is mapped to itself by $\Gamma_o(k)$, and we have already proved that $\mathcal{T}_1$ is mapped to itself by this group.

It is easy to determine the subgroup of $\Gamma_o(k)$ that stabilizes a given tower $\mathcal{T}_{\frac{m}{k}}$. It suffices for our purposes to observe that $G(k)$ stabilizes each of the towers. Hence we have a homomorphism

$$\eta_2 : \Gamma_o(k)/G(k) \to \mathcal{S}_{\lfloor \frac{k-1}{2} \rfloor},$$

where $\mathcal{S}_{\lfloor \frac{k-1}{2} \rfloor}$ is viewed as the group of permutations of the integers

$$\begin{matrix} 1,\ 3,\ \ldots,\ k-2 \text{ for } k \text{ odd} \\ 2,\ 4,\ \ldots,\ k-2 \text{ for } k \text{ even} \end{matrix}.$$

For more on homomorphisms from $\Gamma_o(k)/G(k)$ to permutation groups see Chapter 3.

5. Elliptic function theory revisited

The approach to elliptic function theory developed here may also be found in [**25**] as well as some of the research papers discussed in the notes at the end of the book. We fix a point $\tau \in \mathbb{H}^2$ and a point $a \in \mathbb{C}$. We let $\tilde{G}_\tau$ be the group generated by translations by 1 and τ; $\mathcal{P}_a$ be the period parallelogram with vertices a, $a+1$, $a+1+\tau$ and $a+\tau$; and T_τ denote the corresponding torus. The θ-functions are not $\tilde{G}_\tau$-equivariant; as a consequence of the quasiperiodicity of these functions, their zero sets are.

5.1. Function theory on a torus. Actually the first result of this section should have been Theorem 1.8 and the two corollaries which followed it. We needed these results previously so we placed them earlier. A consequence of these results is

Theorem 5.1. *Fix $n \in \mathbb{Z}^+$. Let $a_1, \ldots, a_n$ and $b_1, \ldots, b_n$ be points on T_τ with $a_i \neq b_j$, $i, j = 1, \ldots, n$. There exists an elliptic function with divisor*[50]

$$D = \sum_{i=1}^{n} (a_i - b_i)$$

if and only if $D = 0$ (in the group T_τ).

[50]On tori, it is more natural to use additive notation for divisors.

Proof. Necessity of the condition on D has already been established (Theorem 4.11 of Chapter 1). We prove sufficiency. For $i = 1, \ldots, n$, let us choose $(\epsilon_i, \epsilon_i')$ and (δ_i, δ_i') in $\mathbb{R}^2$ such that

$$a_i = \frac{1-\epsilon_i'}{2} + \frac{1-\epsilon_i}{2}\tau \text{ and } b_i = \frac{1-\delta_i'}{2} + \frac{1-\delta_i}{2}\tau.$$

Define

$$f(z) = \frac{\prod_{i=1}^n \theta \begin{bmatrix} \epsilon_i \\ \epsilon_i' \end{bmatrix} (z,\tau)}{\prod_{i=1}^n \theta \begin{bmatrix} \delta_i \\ \delta_i' \end{bmatrix} (z,\tau)}, \quad z \in \mathbb{C}.$$

The function f is meromorphic on $\mathbb{C}$, vanishes at each a_i, and has a pole at each b_j (with required multiplicity). The periodicity of f is a consequence of the quasiperiodicity of θ and the identities

$$\epsilon_i = 1 - 2\frac{\Im a_i}{\Im \tau}, \quad \epsilon_i' = 1 - 2\Re a_i + 2\Re\tau\frac{\Im a_i}{\Im\tau},$$

$$\delta_i = 1 - 2\frac{\Im b_i}{\Im \tau} \text{ and } \delta_i' = 1 - 2\Re b_i + 2\Re\tau\frac{\Im b_i}{\Im\tau}.$$

Lemma 5.2. *Let $\chi = \chi_1$, χ_2, χ_3, and χ_4 be characteristics. Assume that $(\chi_1+\chi_2)-(\chi_3+\chi_4) \in 2\mathbb{Z}^2$. Then*

$$f_1 = \frac{\theta[\chi_1](\cdot,\tau)\theta[\chi_2](\cdot,\tau)}{\theta[\chi_3](\cdot,\tau)\theta[\chi_4](\cdot,\tau)}, \quad f_2 = \left(\frac{\theta'[\chi](\cdot,\tau)}{\theta[\chi](\cdot,\tau)}\right)'$$

and

$$f_3 = \frac{\theta'[\chi_1](\cdot,\tau)}{\theta[\chi_1](\cdot,\tau)} - \frac{\theta'[\chi_2](\cdot,\tau)}{\theta[\chi_2](\cdot,\tau)}$$

are periodic functions (with periods 1 and τ), where $'$ denotes the partial derivative with respect to the first variable.

Proof. Let us write $\chi_i = \begin{bmatrix} \epsilon_i \\ \epsilon_i' \end{bmatrix}$ for $i = 1,2,3,4$. For $z \in \mathbb{C}$ and n and $m \in \mathbb{Z}$, we have as a consequence of (2.4)

$$f_1(z+n+m\tau) = \exp 2\pi\imath \left\{\frac{n(\epsilon_1+\epsilon_2-\epsilon_3-\epsilon_4) - m(\epsilon_1'+\epsilon_2'-\epsilon_3'-\epsilon_4')}{2}\right\} f_1(z);$$

hence f_1 is doubly periodic. From (2.4), it also follows that

$$\theta'[\chi](z+n+m\tau,\tau) = \exp 2\pi\imath\left\{\frac{n\epsilon - m\epsilon'}{2} - mz - \frac{m^2}{2}\tau\right\}\theta'[\chi](z,\tau) \tag{2.31}$$

$$-2\pi\imath m \exp 2\pi\imath\left\{\frac{n\epsilon - m\epsilon'}{2} - mz - \frac{m^2}{2}\tau\right\}\theta[\chi](z,\tau).$$

It follows from this identity that f_2 and f_3 are also periodic.

Despite the fact that we shall have a chapter for theta function and theta constant identities, we pause here for some identities which are immediate from the theory of elliptic functions. An additional reason for deriving them at this stage is that we shall soon see how to use the theory developed so far in order to solve a problem in conformal mapping. We shall describe, in a very concrete way, the conformal map of a rectangle onto the unit disc. The identity of the next theorem is a first step on the road towards better identities. It will be improved significantly.

Theorem 5.3. *We have the following identities among theta constants and their derivatives:*

$$\left(\frac{\theta'\begin{bmatrix}1\\1\end{bmatrix}}{\theta\begin{bmatrix}0\\0\end{bmatrix}\theta\begin{bmatrix}0\\1\end{bmatrix}\theta\begin{bmatrix}1\\0\end{bmatrix}}\right)^2$$

$$= \frac{\dfrac{\theta''\begin{bmatrix}0\\1\end{bmatrix}}{\theta\begin{bmatrix}0\\1\end{bmatrix}} - \dfrac{\theta''\begin{bmatrix}1\\0\end{bmatrix}}{\theta\begin{bmatrix}1\\0\end{bmatrix}}}{\theta^4\begin{bmatrix}0\\0\end{bmatrix}} = \frac{\dfrac{\theta''\begin{bmatrix}0\\0\end{bmatrix}}{\theta\begin{bmatrix}0\\0\end{bmatrix}} - \dfrac{\theta''\begin{bmatrix}1\\0\end{bmatrix}}{\theta\begin{bmatrix}1\\0\end{bmatrix}}}{\theta^4\begin{bmatrix}0\\1\end{bmatrix}} = \frac{\dfrac{\theta''\begin{bmatrix}0\\1\end{bmatrix}}{\theta\begin{bmatrix}0\\1\end{bmatrix}} - \dfrac{\theta''\begin{bmatrix}0\\0\end{bmatrix}}{\theta\begin{bmatrix}0\\0\end{bmatrix}}}{\theta^4\begin{bmatrix}1\\0\end{bmatrix}}.$$

Proof. Fix $\tau \in \mathbb{H}^2$. The functions

$$f(z) = \frac{\theta^2\begin{bmatrix}0\\1\end{bmatrix}(z,\tau)}{\theta^2\begin{bmatrix}0\\0\end{bmatrix}(z,\tau)} \text{ and } g(z) = \frac{\theta^2\begin{bmatrix}1\\1\end{bmatrix}(z,\tau)}{\theta^2\begin{bmatrix}0\\0\end{bmatrix}(z,\tau)}, \quad z \in \mathbb{C},$$

are both periodic with a single double pole at $\frac{1}{2} + \frac{\tau}{2}$. It follows that 1, f, and g are linearly dependent; that is, there exist constants c_1, c_2, and c_3 such that

$$c_1\,\theta^2\begin{bmatrix}0\\1\end{bmatrix}(z,\tau) - c_2\,\theta^2\begin{bmatrix}1\\1\end{bmatrix}(z,\tau) - c_3\,\theta^2\begin{bmatrix}0\\0\end{bmatrix}(z,\tau) = 0 \text{ for all } z \in \mathbb{C}.$$

Setting $z = \frac{\tau}{2}$, 0, $\frac{1}{2} + \frac{\tau}{2}$, respectively, we get

$$-c_2\,\theta^2\begin{bmatrix}1\\1\end{bmatrix}\left(\frac{\tau}{2},\tau\right) - c_3\,\theta^2\begin{bmatrix}0\\0\end{bmatrix}\left(\frac{\tau}{2},\tau\right) = 0,$$

$$c_1\,\theta^2\begin{bmatrix}0\\1\end{bmatrix}(0,\tau) - c_3\,\theta^2\begin{bmatrix}0\\0\end{bmatrix}(0,\tau) = 0,$$

and

$$c_1\,\theta^2\begin{bmatrix}0\\1\end{bmatrix}\left(\frac{1}{2}+\frac{\tau}{2},\tau\right) - c_2\,\theta^2\begin{bmatrix}1\\1\end{bmatrix}\left(\frac{1}{2}+\frac{\tau}{2},\tau\right) = 0.$$

Using (2.4), the above equations can be translated to ones in terms of theta constants yielding that 0 equals

$$c_2\,\theta^2\begin{bmatrix}0\\1\end{bmatrix} - c_3\,\theta^2\begin{bmatrix}1\\0\end{bmatrix} = c_1\,\theta^2\begin{bmatrix}0\\1\end{bmatrix} - c_3\,\theta^2\begin{bmatrix}0\\0\end{bmatrix}$$

$$= c_1\,\theta^2\begin{bmatrix}1\\0\end{bmatrix} - c_2\,\theta^2\begin{bmatrix}0\\0\end{bmatrix}.$$

A basis for the one dimensional space of solutions is easily seen to be

$$c_1 = \theta^2\begin{bmatrix}0\\0\end{bmatrix},\ c_2 = \theta^2\begin{bmatrix}1\\0\end{bmatrix},\ c_3 = \theta^2\begin{bmatrix}0\\1\end{bmatrix},$$

leading to the identity that 0 equals

(2.32)

$$\theta^2\begin{bmatrix}0\\0\end{bmatrix}\theta^2\begin{bmatrix}0\\1\end{bmatrix}(z,\tau) - \theta^2\begin{bmatrix}1\\0\end{bmatrix}\theta^2\begin{bmatrix}1\\1\end{bmatrix}(z,\tau) - \theta^2\begin{bmatrix}0\\1\end{bmatrix}\theta^2\begin{bmatrix}0\\0\end{bmatrix}(z,\tau).$$

Exercise 5.4. Derive the identities that

$$\theta^2\begin{bmatrix}0\\0\end{bmatrix}\theta^2\begin{bmatrix}1\\0\end{bmatrix}(z,\tau) + \theta^2\begin{bmatrix}0\\1\end{bmatrix}\theta^2\begin{bmatrix}1\\1\end{bmatrix}(z,\tau) - \theta^2\begin{bmatrix}1\\0\end{bmatrix}\theta^2\begin{bmatrix}0\\0\end{bmatrix}(z,\tau)$$

and

$$\theta^2\begin{bmatrix}0\\1\end{bmatrix}\theta^2\begin{bmatrix}1\\0\end{bmatrix}(z,\tau) + \theta^2\begin{bmatrix}0\\0\end{bmatrix}\theta^2\begin{bmatrix}1\\1\end{bmatrix}(z,\tau) - \theta^2\begin{bmatrix}1\\0\end{bmatrix}\theta^2\begin{bmatrix}0\\1\end{bmatrix}(z,\tau)$$

equal 0.

We now expand each of the functions defined by the left hand side of the last three identities in a Taylor series about the origin, and equate the coefficient of z^2 to zero to obtain the identity in the theorem.

Corollary 5.5 (Quartic identity). *We have*

$$\theta^4\begin{bmatrix}0\\0\end{bmatrix} = \theta^4\begin{bmatrix}0\\1\end{bmatrix} + \theta^4\begin{bmatrix}1\\0\end{bmatrix}.$$

Proof. This identity is a simple algebraic consequence of the last three equalities in the statement of the theorem.

It is however possible to obtain a much stronger result which will imply the previous corollary. In the proof of the previous theorem we derived equation (2.32). Set $z = \zeta + \frac{1}{2}$ in that identity and obtain

$$\theta^2\begin{bmatrix}0\\0\end{bmatrix}\theta^2\begin{bmatrix}0\\1\end{bmatrix}\left(\zeta+\frac{1}{2},\tau\right) - \theta^2\begin{bmatrix}1\\0\end{bmatrix}\theta^2\begin{bmatrix}1\\1\end{bmatrix}\left(\zeta+\frac{1}{2},\tau\right)$$

$$-\theta^2\begin{bmatrix}0\\1\end{bmatrix}\theta^2\begin{bmatrix}0\\0\end{bmatrix}\left(\zeta+\frac{1}{2},\tau\right) = 0,$$

which by (2.8) yields that

(2.33)

$$\theta^2\begin{bmatrix}0\\0\end{bmatrix}\theta^2\begin{bmatrix}0\\0\end{bmatrix}(\zeta,\tau)-\theta^2\begin{bmatrix}1\\0\end{bmatrix}\theta^2\begin{bmatrix}1\\0\end{bmatrix}(\zeta,\tau)-\theta^2\begin{bmatrix}0\\1\end{bmatrix}\theta^2\begin{bmatrix}0\\1\end{bmatrix}(\zeta,\tau)$$

equals 0.

The quartic identity for theta constants is now the special case $\zeta = 0$. We shall give other proofs of this quartic identity in the sequel as well as of the above quadratic identity for theta functions.

Lemma 5.6. *If we define for $z \in \mathbb{C}$,*

$$f_1(z,\tau) = f_1(z) = \frac{\theta'\begin{bmatrix}1\\1\end{bmatrix}(z,\tau)}{\theta\begin{bmatrix}1\\1\end{bmatrix}(z,\tau)} - \frac{\theta'\begin{bmatrix}0\\0\end{bmatrix}(z,\tau)}{\theta\begin{bmatrix}0\\0\end{bmatrix}(z,\tau)}$$

and

$$f_2(z,\tau) = f_2(z) = \frac{\theta'\begin{bmatrix}1\\1\end{bmatrix}\theta\begin{bmatrix}0\\0\end{bmatrix}\theta\begin{bmatrix}0\\1\end{bmatrix}(z,\tau)\,\theta\begin{bmatrix}1\\0\end{bmatrix}(z,\tau)}{\theta\begin{bmatrix}0\\1\end{bmatrix}\theta\begin{bmatrix}1\\0\end{bmatrix}\theta\begin{bmatrix}0\\0\end{bmatrix}(z,\tau)\,\theta\begin{bmatrix}1\\1\end{bmatrix}(z,\tau)},$$

then $f_1 = f_2$.

Proof. The previous lemma tells us that both functions are periodic. Both functions have simple poles at 0 and $\frac{1+\tau}{2}$. The residue of these functions at 0 is 1. Hence $f_1 - f_2$ has at most one simple pole and must therefore be constant. The conclusion of the lemma follows from the observation that

$$f_1(z) - f_2(z) = O(z),\ z \to 0.$$

It follows from (2.26), (2.27), (2.28) and the fact that Γ fixes the odd integral class and permutes the even integral classes that for all $\gamma \in \Gamma$,

$$\gamma'(\tau)^{\frac{1}{2}} f_1(z\gamma'(\tau)^{\frac{1}{2}}, \gamma(\tau)) = \frac{\theta'[v_1+v_2](z,\tau)}{\theta[v_1+v_2](z,\tau)} - \frac{\theta'[0\ \gamma](z,\tau)}{\theta[0\ \gamma](z,\tau)}$$

and

$$\gamma'(\tau)^{\frac{1}{2}} f_2(z\gamma'(\tau)^{\frac{1}{2}}, \gamma(\tau))$$

$$= \frac{\theta[0\ \gamma]}{\theta[0\ \gamma](z,\tau)}\ \frac{\theta[v_1\ \gamma](z,\tau)}{\theta[v_1\ \gamma]}\ \frac{\theta[v_2\ \gamma](z,\tau)}{\theta[v_2\ \gamma]}\ \frac{\theta'[v_1+v_2]}{\theta[v_1+v_2](z,\tau)}.$$

Theorem 5.7. *We have the identity*

$$\frac{\theta'''\begin{bmatrix}1\\1\end{bmatrix}}{\theta'\begin{bmatrix}1\\1\end{bmatrix}} = \frac{\theta''\begin{bmatrix}0\\0\end{bmatrix}}{\theta\begin{bmatrix}0\\0\end{bmatrix}} + \frac{\theta''\begin{bmatrix}0\\1\end{bmatrix}}{\theta\begin{bmatrix}0\\1\end{bmatrix}} + \frac{\theta''\begin{bmatrix}1\\0\end{bmatrix}}{\theta\begin{bmatrix}1\\0\end{bmatrix}}. \tag{2.34}$$

Proof. The previous lemma tells us that if we define for $z \in \mathbb{C}$,

$$f(z) = \frac{\theta\begin{bmatrix}0\\0\end{bmatrix}(z,\tau)\,\theta'\begin{bmatrix}1\\1\end{bmatrix}(z,\tau)}{\theta\begin{bmatrix}0\\0\end{bmatrix}\theta'\begin{bmatrix}1\\1\end{bmatrix}} - \frac{\theta\begin{bmatrix}1\\1\end{bmatrix}(z,\tau)\,\theta'\begin{bmatrix}0\\0\end{bmatrix}(z,\tau)}{\theta\begin{bmatrix}0\\0\end{bmatrix}\theta'\begin{bmatrix}1\\1\end{bmatrix}}$$
$$- \frac{\theta\begin{bmatrix}0\\1\end{bmatrix}(z,\tau)\,\theta\begin{bmatrix}1\\0\end{bmatrix}(z,\tau)}{\theta\begin{bmatrix}0\\1\end{bmatrix}\theta\begin{bmatrix}1\\0\end{bmatrix}},$$

then the function f is identically zero. For each $n \in \mathbb{Z}^+ \cup \{0\}$, the coefficient of z^n in the Taylor series expansion of f about the origin is

$$\frac{1}{n!}\frac{\sum_{k=0}^n \binom{n}{k}\left(\theta^{(k)}\begin{bmatrix}0\\0\end{bmatrix}\theta^{(n-k+1)}\begin{bmatrix}1\\1\end{bmatrix} - \theta^{(k)}\begin{bmatrix}1\\1\end{bmatrix}\theta^{(n-k+1)}\begin{bmatrix}0\\0\end{bmatrix}\right)}{\theta\begin{bmatrix}0\\0\end{bmatrix}\theta'\begin{bmatrix}1\\1\end{bmatrix}}$$
$$-\frac{1}{n!}\frac{\sum_{k=0}^n \binom{n}{k}\theta^{(k)}\begin{bmatrix}0\\1\end{bmatrix}\theta^{(n-k)}\begin{bmatrix}1\\0\end{bmatrix}}{\theta\begin{bmatrix}0\\1\end{bmatrix}\theta\begin{bmatrix}1\\0\end{bmatrix}}.$$

For n odd, each of the above sums vanishes. For $n = 0$, we obtain a trivial identity. For $n = 2$ we obtain the identity of the theorem.

Corollary 5.8 (Jacobi's derivative formula)**.** *We have the identity*

$$\theta'\begin{bmatrix}1\\1\end{bmatrix} = -\pi\,\theta\begin{bmatrix}0\\0\end{bmatrix}\theta\begin{bmatrix}0\\1\end{bmatrix}\theta\begin{bmatrix}1\\0\end{bmatrix}.$$

Proof. Using the heat equation, (2.34) can be rewritten as

$$\frac{d}{d\tau}\log\,\theta'\begin{bmatrix}1\\1\end{bmatrix} = \frac{d}{d\tau}\log\,\theta\begin{bmatrix}0\\0\end{bmatrix} + \frac{d}{d\tau}\log\,\theta\begin{bmatrix}0\\1\end{bmatrix} + \frac{d}{d\tau}\log\,\theta\begin{bmatrix}1\\0\end{bmatrix}.$$

Integrating with respect to τ yields

$$\theta'\begin{bmatrix}1\\1\end{bmatrix} = c\,\theta\begin{bmatrix}0\\0\end{bmatrix}\theta\begin{bmatrix}0\\1\end{bmatrix}\theta\begin{bmatrix}1\\0\end{bmatrix},$$

for some $c \in \mathbb{C}$.

We now record for current and future use some formulae in terms of the variable $x = \exp(\pi \imath \tau)$, $\tau \in \mathbb{H}^2$. For the theta constants with integral characteristic,

$$\theta\begin{bmatrix} 0 \\ 0 \end{bmatrix}(0,\tau) = \sum_{n\in\mathbb{Z}} x^{n^2} = 1 + 2\sum_{n=1}^{\infty} x^{n^2}, \tag{2.35}$$

$$\theta\begin{bmatrix} 0 \\ 1 \end{bmatrix}(0,\tau) = \sum_{n\in\mathbb{Z}} (-1)^n x^{n^2} = 1 + 2\sum_{n=1}^{\infty} (-1)^n x^{n^2} \tag{2.36}$$

and

$$\theta\begin{bmatrix} 1 \\ 0 \end{bmatrix}(0,\tau) = x^{\frac{1}{4}} \sum_{n\in\mathbb{Z}} x^{n(n+1)} = 2x^{\frac{1}{4}} \sum_{n=0}^{\infty} x^{n(n+1)}. \tag{2.37}$$

Some useful formulae for z-derivatives are

$$\begin{aligned} \theta'\begin{bmatrix} 1 \\ 1 \end{bmatrix}(0,\tau) &= -\pi x^{\frac{1}{4}} \sum_{n\in\mathbb{Z}} (2n+1)(-1)^n x^{n(n+1)} \\ &= -2\pi x^{\frac{1}{4}} \sum_{n=0}^{\infty} (2n+1)(-1)^n x^{n(n+1)} \end{aligned} \tag{2.38}$$

and

$$0 = \theta'\begin{bmatrix} 0 \\ 0 \end{bmatrix} = \theta'\begin{bmatrix} 1 \\ 0 \end{bmatrix} = \theta'\begin{bmatrix} 0 \\ 1 \end{bmatrix}.$$

Finally, a formula involving third integer characteristics that we will need later:

$$\begin{aligned} \theta\begin{bmatrix} \frac{1}{3} \\ 1 \end{bmatrix}(0,3\tau) &= \exp\left\{\frac{\pi\imath}{6}\right\} x^{\frac{1}{12}} \sum_{n\in\mathbb{Z}} (-1)^n x^{3n^2+n} \\ &= \exp\left\{\frac{\pi\imath}{6}\right\} x^{\frac{1}{12}} \left(\sum_{n=0}^{\infty} (-1)^n x^{n(3n+1)} + \sum_{n=1}^{\infty} (-1)^n x^{n(3n-1)}\right). \end{aligned} \tag{2.39}$$

Each of the above five formulae defines a series that converges for $|x| < 1$.

The value of c is easily obtained from the first four of the five expansions ((2.35), (2.36), (2.37) and (2.38)) given above.

The purpose of the current section is to give the reader a feeling for the usefulness of the theory of theta functions and theta constants. In particular, we already indicated how theta function theory is used as a basis for the theory of elliptic functions. We take two more steps in this direction. We derive a formula for the Weierstrass $\wp$-function in terms of θ-functions and study projective embeddings of families of tori. Then we take a different direction and give an application to conformal mappings.

For fixed τ,

$$f(z,\tau) = \frac{\theta^2\begin{bmatrix}0\\0\end{bmatrix}(z,\tau)}{\theta^2\begin{bmatrix}1\\1\end{bmatrix}(z,\tau)}$$

is a doubly periodic function with periods 1 and τ with a double zero at $\frac{1+\tau}{2}$ and a double pole at the origin. It follows that there exist constants $a(\tau) \in \mathbb{C}^*$ and $b(\tau) \in \mathbb{C}$ so that

$$f(z,\tau) = a(\tau)\wp(z,\tau) + b(\tau), \text{ for all } (z,\tau) \in \mathbb{C} \times \mathbb{H}^2.$$

Hence we have obtained

Theorem 5.9. *An alternate formula for the λ-function is given by*

$$\lambda(\tau) = \frac{f\left(\frac{\tau}{2},\tau\right)}{f\left(\frac{\tau}{2},\tau\right) - f\left(\frac{1}{2},\tau\right)}, \quad \tau \in \mathbb{H}^2.$$

The exact relation between the $\wp$-function and θ-functions is provided by

Theorem 5.10. *We have*

$$\wp(z,\tau) = -\left(\frac{\theta'\begin{bmatrix}1\\1\end{bmatrix}(z,\tau)}{\theta\begin{bmatrix}1\\1\end{bmatrix}(z,\tau)}\right)' + \frac{1}{3}\,\frac{\theta'''\begin{bmatrix}1\\1\end{bmatrix}(0,\tau)}{\theta'\begin{bmatrix}1\\1\end{bmatrix}(0,\tau)}.$$

Proof. Lemma 5.2 tells us at once that the right hand side of the equality defines an even periodic function. Its only singularity is at the origin. It is immediate that the Laurent series about the origin starts with $\frac{1}{z^2} + a_0 + a_2z^2 + \ldots$. A little more work shows that this series has the form

$$\frac{1}{z^2} + \left[\frac{1}{6}\left(\frac{\theta'''\begin{bmatrix}1\\1\end{bmatrix}}{\theta'\begin{bmatrix}1\\1\end{bmatrix}}\right)^2 - \frac{1}{10}\,\frac{\theta^{(5)}\begin{bmatrix}1\\1\end{bmatrix}}{\theta'\begin{bmatrix}1\\1\end{bmatrix}}\right]z^2$$

$$+\left[\frac{1}{24}\,\frac{\theta'''\begin{bmatrix}1\\1\end{bmatrix}\theta^{(5)}\begin{bmatrix}1\\1\end{bmatrix}}{\theta'^2\begin{bmatrix}1\\1\end{bmatrix}} - \frac{5}{108}\left(\frac{\theta'''\begin{bmatrix}1\\1\end{bmatrix}}{\theta'\begin{bmatrix}1\\1\end{bmatrix}}\right)^3 - \frac{1}{168}\,\frac{\theta^{(7)}\begin{bmatrix}1\\1\end{bmatrix}}{\theta'\begin{bmatrix}1\\1\end{bmatrix}}\right]z^4 + \ldots.$$

The new formula for the Weierstrass $\wp$-function follows immediately.

5.2. Projective embeddings of the family of tori. Define

$$f = \frac{\theta^2\begin{bmatrix}0\\0\end{bmatrix}(\cdot,\tau)}{\theta^2\begin{bmatrix}1\\1\end{bmatrix}(\cdot,\tau)} \text{ and } g = \frac{\theta\begin{bmatrix}0\\0\end{bmatrix}(\cdot,\tau)\,\theta\begin{bmatrix}0\\1\end{bmatrix}(\cdot,\tau)\,\theta\begin{bmatrix}1\\0\end{bmatrix}(\cdot,\tau)}{\theta^3\begin{bmatrix}1\\1\end{bmatrix}(\cdot,\tau)}.$$

It is easily seen that there are constants $a \neq 0 \neq c$ and b such that (the first fact has been previously used)

$$\wp' = cg \text{ and } \wp = af + b.$$

It follows that for some $\alpha \in \mathbb{C}^*$,

$$\alpha g^2 = \left(f - f\left(\frac{1}{2}\right)\right)\left(f - f\left(\frac{\tau}{2}\right)\right)\left(f - f\left(\frac{1+\tau}{2}\right)\right).$$

We hence uniformize all tori simultaneously by producing a map from the factor space $\mathbf{V}(1,0) = (\mathbb{C}\times\mathbb{H}^2)/(\mathbb{Z}\oplus\mathbb{Z})$ to $\mathbf{P}\mathbb{C}^2$ by sending $(z,\tau)\in\mathbb{C}\times\mathbb{H}^2$ to the projective equivalence of the point $(Z_1, Z_2, Z_3)\in\mathbb{C}^3 - \{0\}$, where

$$Z_1 = \theta^3\begin{bmatrix}1\\1\end{bmatrix}(z,\tau),\ Z_2 = \theta^2\begin{bmatrix}0\\0\end{bmatrix}(z,\tau)\,\theta\begin{bmatrix}1\\1\end{bmatrix}(z,\tau)$$

and

$$Z_3 = \theta\begin{bmatrix}0\\0\end{bmatrix}(z,\tau)\,\theta\begin{bmatrix}0\\1\end{bmatrix}(z,\tau)\,\theta\begin{bmatrix}1\\0\end{bmatrix}(z,\tau).$$

This map is injective on each fiber T_τ over $\tau\in\mathbb{H}^2$, and the relation between the functions f and g leads to the theta identity

$$\theta^4\begin{bmatrix}0\\0\end{bmatrix}(0,\tau)\,\theta^2\begin{bmatrix}0\\1\end{bmatrix}(z,\tau)\,\theta^2\begin{bmatrix}1\\0\end{bmatrix}(z,\tau)$$

$$= \left(\theta^2\begin{bmatrix}1\\0\end{bmatrix}(0,\tau)\,\theta^2\begin{bmatrix}0\\0\end{bmatrix}(z,\tau) - \theta^2\begin{bmatrix}0\\1\end{bmatrix}(0,\tau)\,\theta^2\begin{bmatrix}1\\1\end{bmatrix}(z,\tau)\right)$$

$$\times\left(\theta^2\begin{bmatrix}0\\1\end{bmatrix}(0,\tau)\,\theta^2\begin{bmatrix}0\\0\end{bmatrix}(z,\tau) + \theta^2\begin{bmatrix}1\\0\end{bmatrix}(0,\tau)\,\theta^2\begin{bmatrix}1\\1\end{bmatrix}(z,\tau)\right).$$

The theta identity identifies the torus T_τ with the projective algebraic curve

$$\theta^4\begin{bmatrix}0\\0\end{bmatrix}Y^2Z = X\left(\theta^2\begin{bmatrix}1\\0\end{bmatrix}X - \theta^2\begin{bmatrix}0\\1\end{bmatrix}Z\right)\left(\theta^2\begin{bmatrix}0\\1\end{bmatrix}X + \theta^2\begin{bmatrix}1\\0\end{bmatrix}Z\right).$$

We note that the limiting case of this equation (as $\tau\to\imath\infty$) is

$$Y^2 + X^2 = 0.$$

6. Conformal mappings of rectangles and Picard's theorem

The Riemann mapping theorem guarantees the existence of a conformal map between any two simply connected regions of the extended plane which have at least two boundary points. More generally there exists a conformal map of any simply connected region of the extended plane onto either the extended plane, the plane or the unit disc. It is generally a difficult problem to determine this map explicitly; however in certain cases it is possible to do so. One of these cases is the map of a rectangle onto the disc. This is a problem solved in elementary courses in complex function theory, and the solution is generally given as a Schwarz-Christoffel map. It is usually commented that the integral involved is an elliptic integral, and the matter is usually left there. We shall here derive an explicit formula for this mapping using the theory of theta functions and the identity we derived in equation (2.33).

A region in $\hat{\mathbb{C}}$ is called *hyperbolic* if it has the unit disc as its holomorphic universal covering manifold. One of the pretty proofs of Picard's theorem which says that an entire function which omits two values is constant is based on the fact that the sphere minus three points is hyperbolic. Using the transformation formula we shall prove that this is so by exhibiting the universal covering map of the sphere minus three points as the fourth power of a quotient of theta constants with integer characteristics. We establish the following two results:

Theorem 6.1. *Let R denote the rectangle with vertices at the points*

$$(0, 1, 1 + \imath t, \imath t),$$

where t is a positive real number. Define for $\zeta \in \mathbb{C}$,

$$H(\zeta) = \frac{\theta\begin{bmatrix} 1 \\ 0 \end{bmatrix}(0, \imath t)\theta\begin{bmatrix} 1 \\ 0 \end{bmatrix}(\zeta, \imath t)}{\theta\begin{bmatrix} 0 \\ 0 \end{bmatrix}(0, \imath t)\theta\begin{bmatrix} 0 \\ 0 \end{bmatrix}(\zeta, \imath t)}, \quad J(\zeta) = \frac{\theta\begin{bmatrix} 0 \\ 1 \end{bmatrix}(0, \imath t)\theta\begin{bmatrix} 0 \\ 1 \end{bmatrix}(\zeta, \imath t)}{\theta\begin{bmatrix} 0 \\ 0 \end{bmatrix}(0, \imath t)\theta\begin{bmatrix} 0 \\ 0 \end{bmatrix}(\zeta, \imath t)}.$$

Then, $H \pm \imath J$ maps the boundary of the rectangle R onto the unit circle. The function $H + \imath J$ maps the interior of the R onto the interior of the disc, and $H - \imath J$ maps the interior of R onto the exterior of the disc (in $\hat{\mathbb{C}}$). In both cases the mapping is an injection of the closed rectangle R.

Theorem 6.2. *The Riemann sphere punctured at three points is hyperbolic.*

The proof of Theorem 6.1 will require some elementary observations.

Exercise 6.3. Prove Theorem 6.2 based on the action of $\mathrm{PSL}(2, \mathbb{Z})$ on classes of characteristics.

6.1. Reality conditions. Let $t \in \mathbb{R}^+$. Note that:

1) The three functions (of ζ) $\theta\begin{bmatrix}0\\0\end{bmatrix}(\zeta, it)$, $\theta\begin{bmatrix}1\\0\end{bmatrix}(\zeta, it)$ and $\theta\begin{bmatrix}0\\1\end{bmatrix}(\zeta, it)$ are real for ζ on the real axis and for ζ pure imaginary.

2) The functions $\dfrac{\theta\begin{bmatrix}1\\0\end{bmatrix}(\zeta,it)}{\theta\begin{bmatrix}0\\0\end{bmatrix}(\zeta,it)}$, $\dfrac{\theta\begin{bmatrix}0\\1\end{bmatrix}(\zeta,it)}{\theta\begin{bmatrix}0\\0\end{bmatrix}(\zeta,it)}$ are real on the real axis, the imaginary axis, the line $\Im\zeta = 1$ and the line $\Re\zeta = 1$.

The first of the above observations is verified directly from the definition of the functions. The second uses the first observation and the fact that

$$\frac{\theta\begin{bmatrix}1\\0\end{bmatrix}(\zeta+1,\tau)}{\theta\begin{bmatrix}0\\0\end{bmatrix}(\zeta+1,\tau)} = -\frac{\theta\begin{bmatrix}1\\0\end{bmatrix}(\zeta,\tau)}{\theta\begin{bmatrix}0\\0\end{bmatrix}(\zeta,\tau)}, \quad \frac{\theta\begin{bmatrix}1\\0\end{bmatrix}(\zeta+\tau,\tau)}{\theta\begin{bmatrix}0\\0\end{bmatrix}(\zeta+\tau,\tau)} = \frac{\theta\begin{bmatrix}1\\0\end{bmatrix}(\zeta,\tau)}{\theta\begin{bmatrix}0\\0\end{bmatrix}(\zeta,\tau)}$$

and

$$\frac{\theta\begin{bmatrix}0\\1\end{bmatrix}(\zeta+1,\tau)}{\theta\begin{bmatrix}0\\0\end{bmatrix}(\zeta+1,\tau)} = \frac{\theta\begin{bmatrix}0\\1\end{bmatrix}(\zeta,\tau)}{\theta\begin{bmatrix}0\\0\end{bmatrix}(\zeta,\tau)}, \quad \frac{\theta\begin{bmatrix}0\\1\end{bmatrix}(\zeta+\tau,\tau)}{\theta\begin{bmatrix}0\\0\end{bmatrix}(\zeta+\tau,\tau)} = -\frac{\theta\begin{bmatrix}0\\1\end{bmatrix}(\zeta,\tau)}{\theta\begin{bmatrix}0\\0\end{bmatrix}(\zeta,\tau)}.$$

It is clear from the preceding that

$$(H + \imath J)(\zeta + 1) = (-H + \imath J)(\zeta) \text{ and } (H + \imath J)(\zeta + \imath t) = (H - \imath J)(\zeta).$$

Furthermore, the function $H + \imath J$ is an elliptic function with periods 2 and $2\imath t$. We also see that

$$(H + \imath J)(\zeta + 1 + \imath t) = -(H + \imath J)(\zeta).$$

As a consequence of the above observations, we show that $H + iJ$ maps the rectangle with vertices 0, 2, $2 + 2\imath t$ and $2\imath t$ onto the sphere and that the map is two-to-one. It suffices to show that the function has precisely two poles in the above rectangle. The only possible poles of the function in the rectangle are at the points

$$\frac{1 + \imath t}{2}, \ \frac{3 + \imath t}{2}, \ \frac{1 + 3\imath t}{2} \text{ and } \frac{3 + 3\imath t}{2}.$$

Checking these values we find that the second and third are indeed poles, in fact simple poles, but that the first and consequently the last are zeros of this function.

We now prove Theorem 6.1. Consider the functions $H \pm \imath J$ defined on the boundary of the rectangle with vertices 0, 1, $1 + \imath t$ and $\imath t$. Both H and J are real on the boundary, and hence the square of the modulus of $H \pm \imath J$

on the boundary is $H^2 + J^2$, which by our theta identity (2.33) is equal to 1. We conclude that on the boundary of the rectangle $(H + \imath J)(H - \imath J) = 1$. It thus follows that $H \pm \imath J$ maps the interior of the rectangle with the above vertices into either the interior or the exterior of the unit disc and maps the boundary of the rectangle into the boundary of the unit disc. We claim that the rectangle is mapped into the interior of the disc by $H + \imath J$. It suffices to show that at least one point in the interior of the rectangle is mapped to the interior of the unit disc. A straightforward calculation shows that the point $\frac{1+\imath t}{2}$ is mapped by $H + \imath J$ to the origin. Since the functions $H \pm \imath J$ are of modulus 1 on the boundary of the rectangle, they assume every value in the interior of the disc the same number of times (in our case, once). The fact that $(H + iJ)(\zeta + 1 + \imath t) = -(H + \imath J)(\zeta)$ now assures us that the origin has been assumed twice in the big rectangle with vertices 0, 2, $2 + 2\imath t$ and $2\imath t$ and that is the number of times it can be assumed so that the origin is assumed only once in each of the smaller rectangles.

Another example of a conformal mapping problem solved by theta functions is provided by the next result, whose proof is left to the reader.

Theorem 6.4. *Let $y \in \mathbb{R}$, $y > 0$, and let* $\mathbf{R}$ *be the closed rectangle with vertices at* $(0,0)$, $(1,0)$, $(1,y)$ *and* $(0,y)$. *Then*

$$z \mapsto \frac{\theta \begin{bmatrix} 1 \\ 0 \end{bmatrix} (z, 2\imath y)}{\theta \begin{bmatrix} 0 \\ 0 \end{bmatrix} (z, 2\imath y)}$$

maps the rectangle $\mathbf{R}$ *conformally onto the closed upper half plane with the vertices sent to the four* μ, $-\mu$, $-\frac{1}{\mu}$ *and* $\frac{1}{\mu}$, *where* $\mu = \dfrac{\theta \begin{bmatrix} 1 \\ 0 \end{bmatrix} (0, 2\imath y)}{\theta \begin{bmatrix} 0 \\ 0 \end{bmatrix} (0, 2\imath y)} > 1$.

6.2. Hyperbolicity and Picard's theorem. The ideas involved in the proof of the hyperbolicity of the thrice punctured sphere all follow from the material in the section on the transformation theory of the theta function and, in particular, from the formula for the action of $\mathrm{SL}(2, \mathbb{Z})$ on the classes of characteristics. Rather that refer to the previous propositions we shall pretend that we have at our disposal only the transformation theory (for classes of integral characteristics), that $\Gamma(2)$ acts as the identity on these classes, and that $\kappa(\chi, \gamma)$ is an 8-th root of unity for χ an integral characteristic and γ an element of $\mathrm{SL}(2, \mathbb{Z})$. With this information at our disposal we consider the functions defined on the upper half plane

$$\lambda(\tau) = \frac{\theta^4 \begin{bmatrix} 0 \\ 0 \end{bmatrix} (0, \tau)}{\theta^4 \begin{bmatrix} 1 \\ 0 \end{bmatrix} (0, \tau)}, \quad \Lambda = \lambda^2.$$

It follows from the above remarks that Λ defines a holomorphic function $\tilde{\Lambda}$ on the Riemann surface $\mathbb{H}^2/\Gamma(2)$. Observe that $\tilde{\Lambda}$ can be extended to be a meromorphic function on the compactified Riemann surface $\overline{\mathbb{H}^2/\Gamma(2)}$ obtained by filling in the three punctures (P_0, P_1 and P_∞) on $\mathbb{H}^2/\Gamma(2)$. We observe next that we have obtained a meromorphic function on the Riemann sphere, a compact surface, with a double zero at one of the punctures, a double pole at another of the punctures, and holomorphic and nonzero everywhere else. It follows from the fact that the sphere is simply connected that we can take a square root of this function and obtain an analytic homeomorphism (induced by λ) between $\overline{\mathbb{H}^2/\Gamma(2)}$ and $\hat{\mathbb{C}}$. This homomorphism restricts to one between $\mathbb{H}^2/\Gamma(2)$ and the sphere punctured at 0, 1 and ∞. It should be noted that working with eighth rather than fourth powers is a technical device to avoid some computations of κ.

We describe the key step in the above argument. A suitable local coordinate on $\mathbb{H}^2/\Gamma(2)$ at P_∞ is given by $x = \exp(\pi\imath\tau)$, $\tau \in \mathbb{H}^2$ ($\tau \mapsto x$ maps the upper half plane onto a deleted neighborhood of the origin and is invariant under $\tau \mapsto \tau + 2$, a generator of the stabilizer of ∞ in $\Gamma(2)$). The Laurent series expansion of Λ in terms of x has leading term $\frac{c}{x^2}$ with $c \neq 0$. Using the transformation formula with $\gamma(\tau) = -\frac{1}{\tau}$ and $\gamma(\tau) = \frac{\tau}{\tau+1}$ gives us in the first case a power series in x which begins with a nonzero constant term and in the second case, a power series with leading term cx^2 with $c \neq 0$.

For the sake of completeness we describe Picard's theorem and how it now follows. If f is an entire nonconstant function which omits two values, we can assume without loss of generality that the omitted values are the same as those for λ. Since the plane is simply connected we can define $\lambda^{-1}(f(z))$. This defines an entire function in the plane of the complex variable z. The range of this function is a subset of $\mathbb{H}^2$. Following it by a Möbius transformation, we get an entire function whose range is contained in the unit disc. This contradicts Louiville's theorem. We conclude that f must be constant.

7. Spaces of N-th order θ-functions

We begin with a generalization of Theorem 1.8 and its corollary.

Theorem 7.1. *Fix $N \in \mathbb{Z}^+$, ϵ and ϵ' in $\mathbb{R}$. Let $\tau \in \mathbb{H}^2$. Let f be an entire function which satisfies the two functional equations:*

$$f(z+1) = (\exp \pi\imath\epsilon)\ f(z) \tag{2.40}$$

and

$$f(z+\tau) = \left(\exp -\pi\imath\left\{\epsilon' + 2Nz + N\tau\right\}\right)\ f(z),\ z \in \mathbb{C}. \tag{2.41}$$

Then

$$\sum_{l=0}^{N-1}\left(\exp\frac{-\pi\imath l\epsilon}{N}\right)\ f\left(\frac{1-\epsilon'}{2N} + \frac{N-\epsilon}{2N}\tau + \frac{l}{N}\right) = 0 \tag{2.42}$$

and

(2.43)

$$\sum_{l=0}^{N-1}(-1)^l\left(\exp \pi\imath\left\{(l(1-\epsilon)+l^2)\frac{\tau}{N}\right\}\right)\ f\left(\frac{N-\epsilon'}{2N} + \left(\frac{1-\epsilon}{2}+l\right)\frac{\tau}{N}\right) = 0.$$

Proof. The main observations are the following two lemmas.

Lemma 7.2. *For each $\tau \in \mathbb{H}^2$, the entire function on $\mathbb{C}$*

$$z \mapsto \theta\begin{bmatrix}\frac{\epsilon}{N}\\ \epsilon'\end{bmatrix}(Nz, N\tau) \tag{2.44}$$

satisfies the above two functional equations and vanishes to order one at all the lattice points $\left\{\frac{1-\epsilon'}{2N} + \frac{N-\epsilon}{2N}\tau + \frac{n}{N} + m\tau;\ n,\ m \in \mathbb{Z}\right\}$ (and only at these points), hence in particular at the maximal set of (N) inequivalent points

$$\left\{\frac{1-\epsilon'}{2N} + \frac{N-\epsilon}{2N}\tau + \frac{l}{N},\ l = 0,\ \ldots,\ N-1\right\}. \tag{2.45}$$

Lemma 7.3. *For each $\tau \in \mathbb{H}^2$, the entire function*

$$z \mapsto \theta\begin{bmatrix}\epsilon\\ \frac{\epsilon'}{N}\end{bmatrix}\left(z, \frac{\tau}{N}\right) \tag{2.46}$$

satisfies the above two functional equations and vanishes to order one at all the lattice points $\left\{\frac{N-\epsilon'}{2N} + \frac{1-\epsilon}{2N}\tau + n + \frac{m}{N}\tau;\ n,\ m \in \mathbb{Z}\right\}$ (and only at these points), hence in particular at the maximal set of (N) inequivalent points

$$\left\{\frac{N-\epsilon'}{2N} + \frac{1-\epsilon}{2N}\tau + \frac{l}{N}\tau,\ l = 0,\ \ldots,\ N-1\right\}. \tag{2.47}$$

We return to the proof of the theorem. We consider the function

$$z \mapsto g(z) = \frac{f(z)}{\theta\begin{bmatrix}\frac{\epsilon}{N}\\ \epsilon'\end{bmatrix}(Nz, N\tau)},\ z \in \mathbb{C}.$$

It is an elliptic function with periods $1, \tau$ with (at most) simple poles at the points in the list (2.45). The sum of the residues of an elliptic function in a period parallelogram must vanish, so we have

$$\sum_{l=0}^{N-1} \operatorname{Res}_{\frac{1-\epsilon'}{2N}+\frac{N-\epsilon}{2N}\tau+\frac{l}{N}} g = 0.$$

For each l,

$$\operatorname{Res}_{\frac{1-\epsilon'}{2N}+\frac{N-\epsilon}{2N}\tau+\frac{l}{N}} g = \frac{f\left(\frac{1-\epsilon'}{2N}+\frac{N-\epsilon}{2N}\tau+\frac{l}{N}\right)}{N\,\theta'\begin{bmatrix}\frac{\epsilon}{N}\\ \epsilon'\end{bmatrix}\left(\frac{1-\epsilon'}{2}+\frac{N-\epsilon}{2}\tau+l, N\tau\right)},$$

where as usual $' = \frac{\partial}{\partial z}$ (provided the denominator in the above expression does not vanish). Using (2.4), we see that

$$\theta'\begin{bmatrix}\frac{\epsilon}{N}\\ \epsilon'\end{bmatrix}\left(\frac{1-\epsilon'}{2}+\frac{N-\epsilon}{2}\tau+l, N\tau\right)$$

$$= \left(\exp\frac{\pi \imath l\epsilon}{N}\right)\theta'\begin{bmatrix}\frac{\epsilon}{N}\\ \epsilon'\end{bmatrix}\left(\frac{1-\epsilon'}{2}+\frac{N-\epsilon}{2}\tau, N\tau\right).$$

The observation that

$$\theta'\begin{bmatrix}\frac{\epsilon}{N}\\ \epsilon'\end{bmatrix}\left(\frac{1-\epsilon'}{2}+\frac{N-\epsilon}{2}\tau, N\tau\right) \neq 0$$

(since $\theta'\begin{bmatrix}1\\ 1\end{bmatrix} \neq 0$) completes the proof of the first identity of the theorem. The second identity is established by the trick of replacing the function from Lemma 7.2 by the one from Lemma 7.3. The relevant derivative to be computed is

$$\theta'\begin{bmatrix}\epsilon\\ \frac{\epsilon'}{N}\end{bmatrix}\left(\frac{N-\epsilon'}{2N}+\frac{1-\epsilon}{2}\frac{\tau}{N}+l\frac{\tau}{N}, \frac{\tau}{N}\right) = e^{\left(-\pi\imath\{l+(l(1-\epsilon)+l^2)\frac{\tau}{N}\}\right)}$$

$$\times\theta'\begin{bmatrix}\epsilon\\ \frac{\epsilon'}{N}\end{bmatrix}\left(\frac{N-\epsilon'}{2N}+\frac{1-\epsilon}{2}\frac{\tau}{N}, \frac{\tau}{N}\right).$$

Definition 7.4. Let $N \in \mathbb{Z}^+$ and $\begin{bmatrix}\epsilon\\ \epsilon'\end{bmatrix} \in \mathbb{R}^2$. Define $\mathcal{F}_N\begin{bmatrix}\epsilon\\ \epsilon'\end{bmatrix}$ to be the set of entire functions f that satisfy the functional equations (2.40) and (2.41). This set of functions will be called the space of *N-th order θ-functions* with characteristic $\begin{bmatrix}\epsilon\\ \epsilon'\end{bmatrix}$.

It is immediate that $\mathcal{F}_N \begin{bmatrix} \epsilon \\ \epsilon' \end{bmatrix}$ is a nontrivial vector space since it contains the function

$$z \mapsto \theta^N \begin{bmatrix} \frac{\epsilon}{N} \\ \frac{\epsilon'}{N} \end{bmatrix} (z, \tau),$$

as well as the functions (2.44) and (2.46).

Remark 7.5. We have suppressed in the above notation the dependence of the space $\mathcal{F}_N \begin{bmatrix} \epsilon \\ \epsilon' \end{bmatrix}$ on $\tau \in \mathbb{H}^2$. Usually our discussion assumes that we are dealing with a single torus. When it is important to keep track of the dependence of this space on τ, it will be denoted by $\mathcal{F}_N \begin{bmatrix} \epsilon \\ \epsilon' \end{bmatrix} (\tau)$. It is also convenient to think of $\mathcal{F}_N \begin{bmatrix} \epsilon \\ \epsilon' \end{bmatrix}$ as the space of holomorphic functions f on $\mathbb{C} \times \mathbb{H}^2$ with the property that for fixed τ, $f(\cdot, \tau)$ satisfies (2.40) and (2.41).

Proposition 7.6. *Let* $f \in \mathcal{F}_N \begin{bmatrix} \epsilon \\ \epsilon' \end{bmatrix}$. *Then*

$$f(z + n + m\tau) = \left(\exp \pi \imath \left\{n\epsilon - m\epsilon' - 2Nmz - Nm^2\tau\right\}\right) \; f(z),$$

for all $z \in \mathbb{C}$, *and all* n *and* $m \in \mathbb{Z}$.

Proof. The proof is by induction on n and m. The details are left to the reader. The proof of the next proposition is also left to the reader.

Proposition 7.7. (a) *If* $f_i \in \mathcal{F}_{N_i} \begin{bmatrix} \epsilon_i \\ \epsilon_i' \end{bmatrix}$ *for* $i = 1, 2$, *then*

$$f_1 f_2 \in \mathcal{F}_{N_1+N_2} \begin{bmatrix} \epsilon_1 + \epsilon_2 \\ \epsilon_1' + \epsilon_2' \end{bmatrix}.$$

(b) $\mathcal{F}_N \begin{bmatrix} \epsilon + 2n \\ \epsilon + 2m \end{bmatrix} = \mathcal{F}_N \begin{bmatrix} \epsilon \\ \epsilon, \end{bmatrix}$ *for all* n *and* $m \in \mathbb{Z}$.

Proposition 7.8. (a) *Every nontrivial* $f \in \mathcal{F}_N \begin{bmatrix} \epsilon \\ \epsilon' \end{bmatrix}$ *has* N *zeros whose sum is*

$$\frac{N - \epsilon'}{2} + \frac{N - \epsilon}{2}\tau$$

(modulo periods).

(b) *Conversely, given* N *points* x_1, ..., $x_N \in \mathbb{C}$ *whose sum is* $\frac{N-\epsilon'}{2} + \frac{N-\epsilon}{2}\tau$, *there exists an* $f \in \mathcal{F}_N \begin{bmatrix} \epsilon \\ \epsilon' \end{bmatrix}$ *that vanishes precisely at these points (modulo periods).*

Proof. We leave the details of the proof of part (a) to the reader. It follows, for example, by integrating $\frac{zf'}{f}$ over the boundary of the period parallelogram. Alternatively, since the quotient of two linearly independent functions in this space is an elliptic function and we have an example of a function in this space whose set of zeros is known, Theorem 5.1 finishes the proof. To prove part (b), write for $i = 1, \ldots, N$,

$$x_i = \tau \frac{\epsilon_i}{2} + \frac{\epsilon_i'}{2},$$

with ϵ_i and $\epsilon_i' \in \mathbb{R}$. The function $\theta \begin{bmatrix} 1-\epsilon_i \\ 1-\epsilon_i' \end{bmatrix} (\cdot, \tau)$ vanishes only (modulo periods) at x_i. It is easily seen from the previous proposition that $\prod_{i=1}^N \theta \begin{bmatrix} 1-\epsilon_i \\ 1-\epsilon_i' \end{bmatrix} (\cdot, \tau)$ belongs to $\mathcal{F}_N \begin{bmatrix} \epsilon \\ \epsilon' \end{bmatrix} (\tau)$.

Proposition 7.9. *We have*

$$\dim \mathcal{F}_N \begin{bmatrix} \epsilon \\ \epsilon' \end{bmatrix} = N.$$

Proof. Assume that $\dim\, \mathcal{F}_N \begin{bmatrix} \epsilon \\ \epsilon' \end{bmatrix} \geq N+1$. Choose $N-1$ $\tilde{G}_\tau$-inequivalent points $x_1, \ldots, x_{N-1}$ in $\mathbb{C}$. Consider the evaluation map

$$\mathcal{F}_N \begin{bmatrix} \epsilon \\ \epsilon' \end{bmatrix} \to \mathbb{C}^{N-1}$$

that sends the function $f \in \mathcal{F}_N \begin{bmatrix} \epsilon \\ \epsilon' \end{bmatrix}$ to the vector $(f(x_1), \ldots, f(x_{N-1})) \in \mathbb{C}^{N-1}$. Every nonzero function in the kernel K of this map vanishes at $x_1, \ldots, x_{N-1}$ and hence also at

$$x_N = \frac{N-\epsilon'}{2} + \frac{N-\epsilon}{2}\tau - \sum_{i=1}^{N-1} x_i.$$

Since K has dimension at least 2, we can find a nontrivial $f \in K$ that also vanishes at an arbitrary point $y \in \mathbb{C}$. We have reached a contradiction, and hence conclude that $\dim\, \mathcal{F}_N \begin{bmatrix} \epsilon \\ \epsilon' \end{bmatrix} \leq N$. It remains to produce N linearly independent functions in $\mathcal{F}_N \begin{bmatrix} \epsilon \\ \epsilon' \end{bmatrix}$. For $j = 1, \ldots, N$, the function

$$z \mapsto \theta^j \begin{bmatrix} \frac{\epsilon}{N} \\ \frac{\epsilon'}{N} \end{bmatrix} (z, \tau)\, \theta \begin{bmatrix} \frac{\epsilon}{N} \\ \epsilon'\left(1 - \frac{j}{N}\right) \end{bmatrix} ((N-j)z, (N-j)\tau),\ z \in \mathbb{C},$$

belongs to $\mathcal{F}_N\begin{bmatrix}\epsilon\\ \epsilon'\end{bmatrix}$ and has a zero of order j at $\frac{N-\epsilon'}{2N}+\frac{N-\epsilon}{2N}\tau$. Hence these N functions are linearly independent.

We shall see in the sequel how to use these spaces to rederive some of the identities already derived using the theory of elliptic functions and derive in addition some new identities.

Corollary 7.10. *The entire functions*

$$\theta_l : z \mapsto \theta\begin{bmatrix}\frac{\epsilon+2l}{N}\\ \epsilon'\end{bmatrix}(Nz, N\tau),\ l = 0\ ,\ldots,\ N-1,$$

form a basis for $\mathcal{F}_N\begin{bmatrix}\epsilon\\ \epsilon'\end{bmatrix}(\tau)$.

Proof. Obviously $\theta_l \in \mathcal{F}_N\begin{bmatrix}\epsilon+2l\\ \epsilon'\end{bmatrix} = \mathcal{F}_N\begin{bmatrix}\epsilon\\ \epsilon'\end{bmatrix}$. The functions θ_l are linearly independent if and only if the entire functions

$$z \mapsto \exp\{-\pi\imath\epsilon z\}\ \theta_l(z)$$

are. These functions can be expanded in Fourier series

$$\exp\{2\pi\imath l z\}\sum_{n\in\mathbb{Z}} a_{ln}\exp\{2\pi\imath nNz\},$$

with $a_{ln} \neq 0$ for all $n \in \mathbb{Z}$. In terms of the variable $x = \exp\{2\pi\imath z\}$, they define holomorphic functions on the punctured plane $\mathbb{C}^*$ with Laurent series expansions

$$x^l\sum_{n=-\infty}^{\infty} a_{ln}x^{nN}.$$

These N functions are surely linearly independent whenever their holomorphic parts

$$x^l\sum_{n=0}^{\infty} a_{ln}x^{nN}$$

are. An examination of the leading coefficients of the functions

$$a_{00}+a_{01}x^N+a_{02}x^{2N}+\ldots,$$
$$a_{10}x+a_{11}x^{1+N}+a_{12}x^{1+2N}+\ldots,$$
$$\ldots$$
$$a_{N-1,0}x^{N-1}+a_{N-1,1}x^{2N-1}+a_{N-1,2}x^{3N-1}+\ldots$$

finishes the argument.

The positive integer N and characteristic χ determine the linear space $\mathcal{F}_N[\chi]$. The location of zeros determines up to constant multiple a function in this space. It is hence easy to establish the following

Proposition 7.11. *We have for all $z \in \mathbb{C}$,*

$$\theta \begin{bmatrix} \frac{\epsilon}{N} \\ \epsilon' \end{bmatrix} (Nz, N\tau) = c(\tau) \prod_{l=0}^{N-1} \theta \begin{bmatrix} \frac{\epsilon}{N} \\ 1 + \frac{\epsilon' - (1+2l)}{N} \end{bmatrix} (z, \tau),$$

where the constant $c(\tau)$ depends on $\tau \in \mathbb{H}^2$.

Proposition 7.12. *Let $\gamma = \begin{bmatrix} a & b \\ c & d \end{bmatrix} \in \mathrm{SL}(2, \mathbb{Z})$. The map which sends the entire function f to the function g defined by*

$$g(z) = \exp \pi\imath \left\{ \frac{-cNz^2}{c\tau + d} \right\} f\left(\frac{z}{c\tau + d} \right), \quad z \in \mathbb{C},$$

sends the vector space $\mathcal{F}_N \begin{bmatrix} \epsilon \\ \epsilon' \end{bmatrix} \left(\frac{a\tau+b}{c\tau+d} \right)$ to $\mathcal{F}_N \begin{bmatrix} a\epsilon + c\epsilon' - Nac \\ b\epsilon + d\epsilon' + Nbd \end{bmatrix} (\tau)$.

Proof. Write $\hat{\tau} = \frac{a\tau+b}{c\tau+d}$, $\hat{z} = \frac{z}{c\tau+d}$, and let $f \in \mathcal{F}_N \begin{bmatrix} \epsilon \\ \epsilon' \end{bmatrix} (\hat{\tau})$. We rewrite equations (2.19) as

$$\frac{1}{c\tau + d} = a - c\hat{\tau} \text{ and } \frac{\tau}{c\tau + d} = -b + d\hat{\tau}.$$

We compute

$$g(z+1) = e^{\pi\imath \left\{ \frac{-cN(z^2+2z+1)}{c\tau+d} \right\}} f(\hat{z} + a - c\hat{\tau}) = e^{\pi\imath(a\epsilon + c\epsilon' - Nac)}\, g(z)$$

and

$$\begin{aligned} g(z+\tau) &= \exp \pi\imath \left\{ \frac{-cN(z^2+2\tau z+\tau^2)}{c\tau+d} \right\} f(\hat{z} - b + d\hat{\tau}) \\ &= \exp \pi\imath(-b\epsilon - d\epsilon' - Nbd - 2Nz - N\tau)\, g(z). \end{aligned}$$

Remark 7.13. As a result of the Proposition, (2.43) is an immediate consequence of (2.42). Start with $f \in \mathcal{F}_N \begin{bmatrix} \epsilon \\ \epsilon' \end{bmatrix} \left(-\frac{1}{\tau}\right)$ and choose $\gamma = A = \begin{bmatrix} 0 & 1 \\ -1 & 0 \end{bmatrix}$. Now apply (2.42) to the function $g \in \mathcal{F}_N \begin{bmatrix} -\epsilon' \\ \epsilon \end{bmatrix} (\tau)$, and obtain after some simplification that the sum

$$\sum_{l=0}^{N-1} (-1)^l \left(\exp \pi\imath \left\{ (l(1-\epsilon) + l^2) \frac{-1}{N\tau} \right\} \right) f\left(\frac{-N - \epsilon'}{2N} + \left(\frac{1-\epsilon}{2} + l \right) \frac{-1}{N\tau} \right)$$

equals 0, which is readily seen to be equivalent to (2.43) (with τ replaced by $\frac{-1}{\tau}$, which involves no loss of generality).

Remark 7.14. For a specific function in $\mathcal{F}_N\begin{bmatrix}\epsilon\\ \epsilon'\end{bmatrix}$ we can obtain much stronger transformation formulae. For example, for any characteristic $\chi = \begin{bmatrix}\epsilon\\ \epsilon'\end{bmatrix} \in \mathbb{R}^2$, and any element $\gamma = \begin{bmatrix}a & b\\ c & d\end{bmatrix}$ of $\mathrm{SL}(2,\mathbb{Z})$, we have

$$\frac{\exp \pi\imath \left\{\frac{-cNz^2}{c\tau+d}\right\} \theta^N \left[\frac{\chi}{N}\right] \left(\frac{z}{c\tau+d}, \frac{a\tau+b}{c\tau+d}\right)}{\theta^N \left[\frac{\chi_N \gamma}{N}\right](z,\tau)} = \kappa^N \left(\frac{\chi}{N}, \gamma\right)(c\tau+d)^{\frac{N}{2}},$$

for all $z \in \mathbb{C}$, $\tau \in \mathbb{H}^2$, and

$$\frac{\exp \pi\imath \left\{\frac{-cN^2z^2}{cN\tau+d}\right\} \theta\begin{bmatrix}\frac{\epsilon}{N}\\ \epsilon'\end{bmatrix}\left(N\left(\frac{z}{cN\tau+d}\right), N\left(\frac{a\tau+\frac{b}{N}}{cN\tau+d}\right)\right)}{\theta\begin{bmatrix} a\frac{\epsilon}{N} + c\epsilon' - ac\\ b\frac{\epsilon}{N} + d\epsilon' + bd\end{bmatrix}(Nz, N\tau)}$$

$$= \kappa\left(\begin{bmatrix}\frac{\epsilon}{N}\\ \epsilon'\end{bmatrix}, \gamma\right)(c\tau+d)^{\frac{1}{2}}.$$

For the remainder of this section we will specialize to the cases ϵ and ϵ' restricted to the values 1 and 0.

Lemma 7.15. *For each $m \in \mathbb{Z}$,*

$$\theta\begin{bmatrix}\frac{2N-m}{N}\\ 1\end{bmatrix}(Nz, N\tau) = \exp \pi\imath\left(-\frac{m}{N}\right)\theta\begin{bmatrix}\frac{m}{N}\\ 1\end{bmatrix}(-Nz, N\tau)$$

and

$$\theta\begin{bmatrix}\frac{2N-m}{N}\\ 0\end{bmatrix}(Nz, N\tau) = \theta\begin{bmatrix}\frac{m}{N}\\ 0\end{bmatrix}(-Nz, N\tau).$$

Proof. We calculate

$$\theta\begin{bmatrix}\frac{2N-m}{N}\\ 1\end{bmatrix}(Nz, N\tau) = \theta\begin{bmatrix}-\frac{m}{N}\\ 1\end{bmatrix}(Nz, N\tau)$$

$$= \theta\begin{bmatrix}\frac{m}{N}\\ -1\end{bmatrix}(-Nz, N\tau) = \exp \pi\imath\left(-\frac{m}{N}\right)\theta\begin{bmatrix}\frac{m}{N}\\ 1\end{bmatrix}(-Nz, N\tau).$$

This establishes the first equality; the proof of the second is similar. It follows that for N odd, the N entire functions whose values at z are respectively

$$\theta\begin{bmatrix}\frac{1}{N}\\ 1\end{bmatrix}(Nz, N\tau),\ \theta\begin{bmatrix}\frac{3}{N}\\ 1\end{bmatrix}(Nz, N\tau),\ \ldots,$$

$$\theta\begin{bmatrix}\frac{N-2}{N}\\ 1\end{bmatrix}(Nz, N\tau),\ \theta\begin{bmatrix}1\\ 1\end{bmatrix}(Nz, N\tau),$$

$$\theta\begin{bmatrix}\frac{1}{N}\\ 1\end{bmatrix}(-Nz, N\tau),\ \theta\begin{bmatrix}\frac{3}{N}\\ 1\end{bmatrix}(-Nz, N\tau),\ \ldots,\ \theta\begin{bmatrix}\frac{N-2}{N}\\ 1\end{bmatrix}(-Nz, N\tau)$$

(the first two of the above two rows contain theta functions indexed by the characteristics $\begin{bmatrix} \frac{l}{N} \\ 1 \end{bmatrix}$ with $l = 1, 3, \ldots, N$; for the third row, the indices run over $l = 1, 3, \ldots, N-2$) form a basis for the linear space $\mathcal{F}_N \begin{bmatrix} 1 \\ 1 \end{bmatrix}$. For N even, we use the functions

$$\theta \begin{bmatrix} \frac{1}{N} \\ 1 \end{bmatrix} (Nz, N\tau),\ \theta \begin{bmatrix} \frac{3}{N} \\ 1 \end{bmatrix} (Nz, N\tau),\ \ldots,\ \theta \begin{bmatrix} \frac{N-1}{N} \\ 1 \end{bmatrix} (Nz, N\tau),$$

$$\theta \begin{bmatrix} \frac{1}{N} \\ 1 \end{bmatrix} (-Nz, N\tau),\ \theta \begin{bmatrix} \frac{3}{N} \\ 1 \end{bmatrix} (-Nz, N\tau), \ldots,\ \theta \begin{bmatrix} \frac{N-1}{N} \\ 1 \end{bmatrix} (-Nz, N\tau)$$

(in each of the above two rows the appropriate index l takes on the values $1, 3, \ldots, N-1$).

Let us restrict to the case N an odd integer greater than or equal to 3, and $\epsilon = 1 = \epsilon'$. The reason for choosing the above basis is that it leads immediately to a decomposition of our space $\mathcal{F}_N \begin{bmatrix} 1 \\ 1 \end{bmatrix}$ into subspaces of even and odd functions. It is clear that

$$\theta \begin{bmatrix} \frac{1}{N} \\ 1 \end{bmatrix} (Nz, N\tau) + \theta \begin{bmatrix} \frac{1}{N} \\ 1 \end{bmatrix} (-Nz, N\tau),\ \ldots,$$

$$\theta \begin{bmatrix} \frac{N-2}{N} \\ 1 \end{bmatrix} (Nz, N\tau) + \theta \begin{bmatrix} \frac{N-2}{N} \\ 1 \end{bmatrix} (-Nz, N\tau)$$

is a basis for the even subspace and that

$$\theta \begin{bmatrix} 1 \\ 1 \end{bmatrix} (Nz, N\tau),\ \theta \begin{bmatrix} \frac{1}{N} \\ 1 \end{bmatrix} (Nz, N\tau) - \theta \begin{bmatrix} \frac{1}{N} \\ 1 \end{bmatrix} (-Nz, N\tau),\ \ldots,$$

$$\theta \begin{bmatrix} \frac{N-2}{N} \\ 1 \end{bmatrix} (Nz, N\tau) - \theta \begin{bmatrix} \frac{N-2}{N} \\ 1 \end{bmatrix} (-Nz, N\tau)$$

is a basis for the subspace of odd functions. In particular, the even subspace has dimension $\frac{N-1}{2}$ and the odd subspace has dimension $\frac{N+1}{2}$.

We shall denote the subspace of even functions in $\mathcal{F}_N \begin{bmatrix} \epsilon \\ \epsilon' \end{bmatrix}$ by $\mathcal{E}_N \begin{bmatrix} \epsilon \\ \epsilon' \end{bmatrix}$. Each nontrivial function in $\mathcal{E}_N \begin{bmatrix} 1 \\ 1 \end{bmatrix}$ has precisely N (an odd number of) zeros. If x is a zero of such a function, then so is $-x$; so the zeros come in pairs. It thus follows that at least one zero must be a half period, that is, of the form $\tau\frac{n}{2} + \frac{m}{2}$, with n and $m \in \mathbb{Z}$. The perhaps surprising fact is

Theorem 7.16. *For all odd $N \in \mathbb{Z}$, $N \geq 3$, each function in $\mathcal{E}_N \begin{bmatrix} 1 \\ 1 \end{bmatrix}$ vanishes at the three half periods $\frac{1}{2}, \frac{\tau}{2}, \frac{1+\tau}{2}$.*

Proof. The proof is by computation (here l, n, and m are integers):

$$\theta\begin{bmatrix} \frac{2l+1}{N} \\ 1 \end{bmatrix}\left(N\left(\tau\frac{n}{2}+\frac{m}{2}\right), N\tau\right)+\theta\begin{bmatrix} \frac{2l+1}{N} \\ 1 \end{bmatrix}\left(-N\left(\tau\frac{n}{2}+\frac{m}{2}\right), N\tau\right)$$
$$=\left(\exp -\pi\imath\left\{\frac{1}{4}n^2N\tau+\frac{1}{2}n(1+Nm)\right\}\right)\theta\begin{bmatrix} \frac{2l+1}{N}+n \\ 1+Nm \end{bmatrix}(0,N\tau)$$
$$+\left(\exp -\pi\imath\left\{\frac{1}{4}n^2N\tau-\frac{1}{2}n(1-Nm)\right\}\right)\theta\begin{bmatrix} \frac{2l+1}{N}-n \\ 1-Nm \end{bmatrix}(0,N\tau)$$
$$=\left(e^{-\pi\imath\left\{\frac{1}{2}n(1+Nm)\right\}}+e^{\pi\imath\left\{-m-nNm+\frac{1}{2}n-\frac{1}{2}nNm\right\}}\right)$$
$$\times\left(\exp -\pi\imath\left\{\frac{1}{4}n^2N\tau\right\}\right)\theta\begin{bmatrix} \frac{2l+1}{N}+n \\ 1+Nm \end{bmatrix}(0,N\tau).$$

We now observe that the expression enclosed by the first set of parenthesis after the last equal sign vanishes if and only if m and n are not both even.

8. The Jacobi triple product identity

Almost all of the material on theta functions which has appeared until now has its generalization to the several dimensional case where the complex variable z is replaced by a vector $z \in \mathbb{C}^n$, $n \in \mathbb{Z}^+$, and the point $\tau \in \mathbb{H}^2$ is replaced by a matrix τ in the Siegel upper half plane of degree n. In fact, partial motivation for the results discussed so far is to better understand the multivariable case. This is why some of the results discussed till now may not have any application in the sequel. Their application will have to be deferred.

In this section we treat a very one dimensional aspect of the theory of theta functions: their representation as infinite products.

8.1. The triple product identity.

We derive an important formula of Jacobi which expresses the theta function $\theta\begin{bmatrix} 0 \\ 0 \end{bmatrix}(\zeta,\tau)$ as an absolutely convergent infinite product.

Theorem 8.1. *For all z and x in $\mathbb{C}$ with $z \neq 0$ and $|x| < 1$, we have*

$$\prod_{n=1}^{n=\infty}(1-x^{2n})(1+x^{2n-1}z)\left(1+\frac{x^{2n-1}}{z}\right)=\sum_{n=-\infty}^{n=\infty}x^{n^2}z^n.$$

Proof. It suffices to establish the identity under the additional assumption that $x \neq 0$. Define

$$f(z,x)=\sum_{n=-\infty}^{n=\infty}x^{n^2}z^n.$$

We let $\zeta \in \mathbb{C}$ and $\tau \in \mathbb{H}^2$. Our first observation is that the right hand side of the equation in the theorem (that is, the function f) after the change of variables

$$z = \exp(2\pi i\zeta), \ x = \exp(\pi i\tau)$$

becomes $\theta \begin{bmatrix} 0 \\ 0 \end{bmatrix} (\zeta, \tau)$, and therefore we already know that the zeros of this function in terms of the variables (ζ, τ) are at the points $(\frac{m+n\tau}{2}, \tau)$ with m and n odd integers. Thus in terms of the (z, x) variables the function f vanishes at the points $(-x^n, x)$ with $n \in \mathbb{Z}$ odd. It thus follows that

$$\sum_{n=-\infty}^{\infty} x^{n^2} z^n = c(x) \prod_{n=1}^{\infty} (1 + x^{2n-1}z) \left(1 + \frac{x^{2n-1}}{z}\right) = \theta \begin{bmatrix} 0 \\ 0 \end{bmatrix} (\zeta, \tau);$$

our problem is to determine $c(x)$.

If we set for fixed x, $z = 1$, $z = -1$ and $z = x$ which corresponds for fixed τ to $\zeta = 0$, $\zeta = \frac{1}{2}$ and $\zeta = \frac{\tau}{2}$, we get

$$c(x) \prod_{n=1}^{\infty} (1 + x^{2n-1})^2 = \theta \begin{bmatrix} 0 \\ 0 \end{bmatrix} (0, \tau), \ c(x) \prod_{n=1}^{\infty} (1 - x^{2n-1})^2 = \theta \begin{bmatrix} 0 \\ 1 \end{bmatrix} (0, \tau),$$

and

$$2\, c(x) \prod_{n=1}^{\infty} (1 + x^{2n})^2 = x^{-\frac{1}{4}} \theta \begin{bmatrix} 1 \\ 0 \end{bmatrix} (0, \tau).$$

So, we obtain

$$2\, c^3(x) \prod_{n=1}^{\infty} (1 + x^{2n-1})^2 (1 - x^{2n-1})^2 (1 + x^{2n})^2$$

$$= \frac{\theta \begin{bmatrix} 0 \\ 0 \end{bmatrix} (0, \tau) \theta \begin{bmatrix} 0 \\ 1 \end{bmatrix} (0, \tau) \theta \begin{bmatrix} 1 \\ 0 \end{bmatrix} (0, \tau)}{x^{\frac{1}{4}}}.$$

We replace $\prod_{n=1}^{\infty} (1 + x^{2n-1})(1 + x^{2n})$ by $\prod_{n=1}^{\infty} (1 + x^n)$ to obtain

(2.48)

$$2\, c^3(x) \prod_{n=1}^{\infty} (1 + x^n)^2 (1 - x^{2n-1})^2 = \frac{\theta \begin{bmatrix} 0 \\ 0 \end{bmatrix} (0, \tau) \theta \begin{bmatrix} 0 \\ 1 \end{bmatrix} (0, \tau) \theta \begin{bmatrix} 1 \\ 0 \end{bmatrix} (0, \tau)}{x^{\frac{1}{4}}}.$$

The next observation is the obvious identity:

$$\prod_{n=1}^{\infty} (1 + x^n) = \prod_{n=1}^{\infty} \frac{(1 - x^{2n})}{(1 - x^n)} = \prod_{n=1}^{\infty} \frac{1}{(1 - x^{2n-1})},$$

which allows us to replace (2.48) by

$$2\, x^{\frac{1}{4}}\, c^3(x) = \theta\begin{bmatrix}0\\0\end{bmatrix}(0,\tau)\theta\begin{bmatrix}0\\1\end{bmatrix}(0,\tau)\theta\begin{bmatrix}1\\0\end{bmatrix}(0,\tau). \tag{2.49}$$

On the other hand, our original equation for $c(x)$ can be written as

$$c(x)\prod_{n=1}^{\infty}(1+x^{2n-1}z)\prod_{n=2}^{\infty}\left(1+\frac{x^{2n-1}}{z}\right) = \frac{1}{1+\frac{x}{z}}\sum_{n=-\infty}^{\infty} x^{n^2} z^n.$$

We now take the limit as z tends to $-x$ and obtain

$$c(x)\prod_{n=1}^{\infty}(1-x^{2n})^2 = \sum_{n=0}^{\infty}(-1)^n(2n+1)x^{n^2+n} = \frac{\theta'\begin{bmatrix}1\\1\end{bmatrix}(0,\tau)}{-2\pi x^{\frac{1}{4}}}. \tag{2.50}$$

(An easy way to see that the last equation holds is to study for fixed x the holomorphic function

$$z \mapsto g(z) = z\sum_{n=-\infty}^{\infty} x^{n^2} z^n.$$

We need to evaluate

$$\lim_{z\to -x}\frac{g(z)}{z+x}.$$

We observe that

$$f(-x) = -x\sum_{n=-\infty}^{\infty} x^{n^2}(-x)^n = -x\left(\sum_{n=0}^{\infty}(-1)^n x^{n^2+n} + \sum_{n=1}^{\infty}(-1)^n x^{n^2-n}\right)$$

$$= -x\left(-\sum_{n=0}^{\infty}(-1)^{n+1}x^{(n+1)^2-(n+1)} + \sum_{n=1}^{\infty}(-1)^n x^{n^2-n}\right) = 0.$$

Hence by L'Hopital's rule

$$\lim_{z\to -x}\frac{g(z)}{z+x} = (-x)f'(-x) = \sum_{n=0}^{\infty}(-1)^n(2n+1)x^{n^2+n},$$

which concludes the verification of the claim.) Using (2.49), (2.50) and Jacobi's derivative formula, we conclude that $c(x) = \prod_{n=1}^{\infty}(1-x^{2n})$. As a consequence of (2.50), we have also derived the formula of Jacobi (2.1).

The Jacobi triple product formula is a source of many identities between power series on the one hand and infinite products on the other. As an example of the utility of this identity we observe that replacing x by x^k and z by $-x^l$ gives us the formula

(2.51)

$$\prod_{n=0}^{\infty}(1-x^{2kn+k-l})(1-x^{2kn+k+l})(1-x^{2kn+2k}) = \sum_{n=-\infty}^{\infty}(-1)^n x^{kn^2+ln}.$$

Hence, we see immediately that by taking $k = 3,\ l = 1$,

$$\prod_{n=0}^{\infty}(1-x^{6n+2})(1-x^{6n+4})(1-x^{6n+6}) = \sum_{n=-\infty}^{\infty}(-1)^n x^{3n^2+n},$$

or replacing x by $x^{\frac{1}{2}}$, we get the more familiar Euler formula (2.2), the basis of his theorem on partitions of the integers allowing no repetitions. It is also the basis of Ramanujan's work on the partition function; both topics are investigated in Chapter 5.

If we return to our original definition of theta functions and theta constants, we see that

$$\theta\begin{bmatrix}\frac{l}{k}\\1\end{bmatrix}(0,k\tau) = \sum_{n\in\mathbb{Z}}\exp 2\pi\imath\left(\frac{1}{2}\left(n+\frac{l}{2k}\right)^2 k\tau - \left(n+\frac{l}{2k}\right)\frac{1}{2}\right)$$

$$(2.52) \qquad = \exp\left(\frac{\pi\imath l}{2k}\right)x^{\frac{l^2}{4k}}\sum_{n\in\mathbb{Z}}(-1)^n x^{kn^2+ln}$$

$$= \exp\left(\frac{\pi\imath l}{2k}\right)x^{\frac{l^2}{4k}}\prod_{n=0}^{\infty}(1-x^{2kn+k-l})(1-x^{2kn+k+l})(1-x^{2kn+2k}),$$

where $x = \exp(\pi\imath\tau)$, and for the last equality we have used (2.51).

In fact, from (2.3), which relates the theta function with arbitrary characteristic to the one with the zero characteristic, it follows that in terms of the variables $x = \exp(\pi\imath\tau)$, $z = \exp(2\pi\imath\zeta)$, we have

$$(2.53) \qquad \theta\begin{bmatrix}\epsilon\\\epsilon'\end{bmatrix}(\zeta,\tau) = \exp\left(\frac{\pi\imath\epsilon\epsilon'}{2}\right)x^{\frac{\epsilon^2}{4}}z^{\frac{\epsilon}{2}}$$

$$\times\prod_{n=0}^{\infty}(1-x^{2n+2})(1+\exp(\pi\imath\epsilon')x^{2n+1+\epsilon}z)\left(1+\frac{\exp(-\pi\imath\epsilon')x^{2n+1-\epsilon}}{z}\right),$$

so that in particular we have once again (2.52).

We shall return to applications of these ideas in the sequel (particularly in Chapters 5 and 6). At this point we wish to show that the theory of higher order theta functions allows us to obtain rather naturally a generalization of Jacobi's triple product identity. Before we do this, however, we record the following particular cases of (2.53):

$$\theta\begin{bmatrix}0\\0\end{bmatrix}(\zeta,\tau) = \sum_{n\in\mathbb{Z}}x^{n^2}z^n = \prod_{n=1}^{\infty}(1-x^{2n})(1+x^{2n-1}z)\left(1+\frac{x^{2n-1}}{z}\right),$$

$$\theta\begin{bmatrix}0\\1\end{bmatrix}(\zeta,\tau) = \sum_{n\in\mathbb{Z}}(-1)^n x^{n^2}z^n = \prod_{n=1}^{\infty}(1-x^{2n})(1-x^{2n-1}z)\left(1-\frac{x^{2n-1}}{z}\right)$$

and

$$\theta\begin{bmatrix}1\\0\end{bmatrix}(\zeta,\tau)$$

$$= x^{\frac{1}{4}}z^{\frac{1}{2}}\sum_{n\in\mathbb{Z}} x^{n(n+1)}z^n = x^{\frac{1}{4}}z^{\frac{1}{2}}\prod_{n=1}^{\infty}(1-x^{2n})(1+x^{2n}z)\left(1+\frac{x^{2n-2}}{z}\right).$$

We shall need a number of expansions of theta constants. In particular the formula

$$\theta\begin{bmatrix}\frac{1}{3}\\1\end{bmatrix}(0,3\tau) = \exp\left\{\frac{\pi\imath}{6}\right\}x^{\frac{1}{12}}\sum_{n\in\mathbb{Z}}(-1)^n x^{3n^2+n}$$

$$= \exp\left\{\frac{\pi\imath}{6}\right\}x^{\frac{1}{12}}\left(\sum_{n=0}^{\infty}(-1)^n x^{n(3n-1)}(1+x^{2n})\right) = \left(\exp\frac{\pi\imath}{6}\right)x^{\frac{1}{12}}\prod_{n=1}^{\infty}(1-x^{2n}).$$

In the above $x = \exp(\pi\imath\tau)$. If we use the variable $x = \exp\left(\frac{2\pi\imath\tau}{3}\right)$, then

(2.54)

$$\theta\begin{bmatrix}\frac{1}{3}\\1\end{bmatrix}(0,\tau) = \exp\left\{\frac{\pi\imath}{6}\right\}x^{\frac{1}{24}}\sum_{n\in\mathbb{Z}}(-1)^n x^{\frac{3n^2+n}{2}} = \exp\left\{\frac{\pi\imath}{6}\right\}x^{\frac{1}{24}}\prod_{n=1}^{\infty}(1-x^n);$$

its companions are (with $\omega = \exp\left(\frac{2\pi\imath}{3}\right)$)

$$\theta\begin{bmatrix}\frac{1}{3}\\ \\ \frac{1}{3}\end{bmatrix}(0,\tau)$$

$$= \exp\left\{\frac{\pi\imath}{18}\right\}x^{\frac{1}{24}}\sum_{n\in\mathbb{Z}}\exp\left(\frac{\pi\imath n}{3}\right)x^{\frac{3n^2+n}{2}} = \exp\left\{\frac{\pi\imath}{18}\right\}x^{\frac{1}{24}}\prod_{n=1}^{\infty}(1-(\omega x)^n),$$

$$\theta\begin{bmatrix}\frac{1}{3}\\ \\ \frac{5}{3}\end{bmatrix}(0,\tau)$$

$$= \exp\left\{\frac{5\pi\imath}{18}\right\}x^{\frac{1}{24}}\sum_{n\in\mathbb{Z}}\exp\left(\frac{5\pi\imath n}{3}\right)x^{\frac{3n^2+n}{2}} = \exp\left\{\frac{5\pi\imath}{18}\right\}x^{\frac{1}{24}}\prod_{n=1}^{\infty}(1-(\omega^2 x)^n),$$

and

$$\theta\begin{bmatrix}1\\\frac{1}{3}\end{bmatrix}(0,\tau) = \exp\left\{\frac{\pi\imath}{6}\right\}x^{\frac{3}{8}}\sum_{n\in\mathbb{Z}}\exp\left\{\frac{\pi\imath n}{3}\right\}x^{\frac{3n(n+1)}{2}}$$

$$= \omega^{\frac{1}{4}}\ x^{\frac{3}{8}}\ \prod_{n=1}^{\infty}(1-x^{3n})(1+\omega^{\frac{1}{2}}x^{3n})(1+\omega^{\frac{5}{2}}x^{3(n-1)}) = \sqrt{3}\ x^{\frac{3}{8}}\ \prod_{n=1}^{\infty}(1-x^{9n}).$$

Remark 8.2. The last four displayed relations between infinite products and infinite sums were obtained from the Jacobi triple product identity. The uniqueness of Taylor series expansions of analytic functions tells us that they are equivalent in the sense that any three can be obtained from the fourth by algebraic manipulations. We emphasize the dependence of the formulae on theta characteristics to illustrate the "bookkeeping abilities" of characteristics.

8.2. The quintuple product identity. The formula we develop here is not new and has been derived by several different authors in different ways. The derivation we present here seems most natural and incorporates this identity into a larger theory. Further generalizations of this sort are offered in Chapter 4.

In our discussion of N-th order theta functions we encountered the spaces $\mathcal{F}_N\begin{bmatrix}\epsilon\\ \epsilon'\end{bmatrix}$ and in particular the space $\mathcal{F}_N\begin{bmatrix}1\\ 1\end{bmatrix}$ and its decomposition into even and odd subspaces. We saw in Proposition 7.8 that every nontrivial function in this space has N zeros. Moreover, by Theorem 7.16 every function in the even subspace necessarily vanishes at the points $\frac{1}{2}$, $\frac{\tau}{2}$ and $\frac{1+\tau}{2}$.

These remarks (using $N = 3$) are the basis for

Theorem 8.3. *For all $\tau \in \mathbb{H}^2$ and all $\zeta \in \mathbb{C}$,*

$$\theta\begin{bmatrix}\frac{1}{3}\\ 1\end{bmatrix}(3\zeta, 3\tau) + \theta\begin{bmatrix}\frac{1}{3}\\ 1\end{bmatrix}(-3\zeta, 3\tau)$$

$$= c(\tau)\theta\begin{bmatrix}0\\ 0\end{bmatrix}(\zeta,\tau)\theta\begin{bmatrix}0\\ 1\end{bmatrix}(\zeta,\tau)\theta\begin{bmatrix}1\\ 0\end{bmatrix}(\zeta,\tau),$$

where

$$c(\tau) = \frac{2\theta\begin{bmatrix}\frac{1}{3}\\ 1\end{bmatrix}(0,3\tau)}{\theta\begin{bmatrix}0\\ 0\end{bmatrix}(0,\tau)\theta\begin{bmatrix}0\\ 1\end{bmatrix}(0,\tau)\theta\begin{bmatrix}1\\ 0\end{bmatrix}(0,\tau)}.$$

Proof. The function on the left hand side of the equality has three zeros in the fundamental parallelogram, and as also pointed out their locations in the fundamental parallelogram are the three points $\frac{1}{2}, \frac{\tau}{2}, \frac{1+\tau}{2}$. The function on the right has the same three zeros, and the reader can easily check that the quotient of the left hand side by the right hand side is an elliptic function with periods $1, \tau$ (since both numerator and denominator are third order theta functions with the same characteristic) which is holomorphic in the fundamental parallelogram and therefore is a constant. The constant is easily computed by setting $\zeta = 0$, and we find $c(\tau)$ given by the expression in the statement of the theorem.

An important interpretation of the above theorem is obtained by setting $x = \exp(\pi\imath\tau)$ and $z = \exp(2\pi\imath\zeta)$.

Theorem 8.4. *(The quintuple product identity) For all $x \in \mathbb{C}$, $|x| < 1$, and all $z \in \mathbb{C}^*$,*

$$z \sum_{n=-\infty}^{\infty} (-1)^n x^{3n^2+n} z^{3n} + \sum_{n=-\infty}^{\infty} (-1)^n x^{3n^2-n} z^{3n}$$

$$= (1+z) \prod_{n=1}^{\infty} (1-x^{2n})(1-x^{4n-2}z^2)\left(1-\frac{x^{4n-2}}{z^2}\right)(1+x^{2n}z)\left(1+\frac{x^{2n}}{z}\right).$$

Proof. We have

$$\theta\begin{bmatrix}\frac{1}{3}\\1\end{bmatrix}(3\zeta,3\tau) + \theta\begin{bmatrix}\frac{1}{3}\\1\end{bmatrix}(-3\zeta,3\tau)$$

$$= \exp\left(\frac{\pi\imath}{6}\right) x^{\frac{1}{12}} \left(z^{\frac{1}{2}} \sum_{n=-\infty}^{\infty} (-1)^n x^{3n^2+n} z^{3n} + \frac{1}{z^{\frac{1}{2}}} \sum_{n=-\infty}^{\infty} (-1)^n x^{3n^2-n} z^{3n}\right)$$

$$= \frac{\exp(\frac{\pi\imath}{6}) x^{\frac{1}{12}}}{z^{\frac{1}{2}}} \left(z \sum_{n=-\infty}^{\infty} (-1)^n x^{3n^2+n} z^{3n} + \sum_{n=-\infty}^{\infty} (-1)^n x^{3n^2-n} z^{3n}\right).$$

On the other hand, the Jacobi triple product gives

$$c(\tau)\theta\begin{bmatrix}0\\0\end{bmatrix}(\zeta,\tau)\theta\begin{bmatrix}0\\1\end{bmatrix}(\zeta,\tau)\theta\begin{bmatrix}1\\0\end{bmatrix}(\zeta,\tau)$$

$$= \frac{\exp(\frac{\pi\imath}{6}) x^{\frac{1}{12}} \prod_{n=1}^{\infty}(1-x^{2n})}{x^{\frac{1}{4}} \prod_{n=1}^{\infty}(1-x^{2n})^3(1-x^{4n-2})^2(1+x^{2n})^2}$$

$$\times x^{\frac{1}{4}} z^{\frac{1}{2}} (1+\frac{1}{z}) \prod_{n=1}^{\infty} (1-x^{2n})^3(1-x^{4n-2}z^2)\left(1-\frac{x^{4n-2}}{z^2}\right)(1+x^{2n}z)\left(1+\frac{x^{2n}}{z}\right).$$

The equality of the theorem therefore yields

$$z \sum_{n=-\infty}^{\infty} (-1)^n x^{3n^2+n} z^{3n} + \sum_{n=-\infty}^{\infty} (-1)^n x^{3n^2-n} z^{3n}$$

$$= \frac{(1+z)\prod_{n=1}^{\infty}(1-x^{2n})(1-x^{4n-2}z^2)\left(1-\frac{x^{4n-2}}{z^2}\right)(1+x^{2n}z)\left(1+\frac{x^{2n}}{z}\right)}{\prod_{n=1}^{\infty}(1-x^{4n-2})^2(1+x^{2n})^2},$$

and since

$$\prod_{n=1}^{\infty} (1-x^{4n-2})^2(1+x^{2n})^2 = 1$$

we are done.

Before we leave this subject, we point out a consequence of the quintuple product identity. We divide the equality of Theorem 8.3 by $\theta\begin{bmatrix}1\\0\end{bmatrix}(\zeta,\tau)$ and obtain the identity

$$\frac{\theta\begin{bmatrix}\frac{1}{3}\\1\end{bmatrix}(3\zeta,3\tau)+\theta\begin{bmatrix}\frac{1}{3}\\1\end{bmatrix}(-3\zeta,3\tau)}{\theta\begin{bmatrix}1\\0\end{bmatrix}(\zeta,\tau)}$$

$$=\frac{2\theta\begin{bmatrix}\frac{1}{3}\\1\end{bmatrix}(0,3\tau)}{\theta\begin{bmatrix}0\\0\end{bmatrix}(0,\tau)\theta\begin{bmatrix}0\\1\end{bmatrix}(0,\tau)\theta\begin{bmatrix}1\\0\end{bmatrix}(0,\tau)}\theta\begin{bmatrix}0\\0\end{bmatrix}(\zeta,\tau)\theta\begin{bmatrix}0\\1\end{bmatrix}(\zeta,\tau);$$

now we can compute the limit as ζ tends to $\frac{1}{2}$.

The left hand side becomes (the first step is to use L'Hopital's rule)

$$\frac{3\theta'\begin{bmatrix}\frac{1}{3}\\1\end{bmatrix}(\frac{3}{2},3\tau)-3\theta'\begin{bmatrix}\frac{1}{3}\\1\end{bmatrix}(-\frac{3}{2},3\tau)}{\theta'\begin{bmatrix}1\\0\end{bmatrix}(\frac{1}{2},\tau)}=-6\frac{\exp\left(-\frac{\pi\imath}{3}\right)\theta'\begin{bmatrix}\frac{1}{3}\\0\end{bmatrix}(0,3\tau)}{\theta'\begin{bmatrix}1\\1\end{bmatrix}(0,\tau)}$$

$$=-(2\pi\imath)\frac{\exp(-\frac{\pi\imath}{3})x^{\frac{1}{12}}\sum_{n=-\infty}^{\infty}(6n+1)x^{3n^2+n}}{\theta'\begin{bmatrix}1\\1\end{bmatrix}(0,\tau)},\quad x=\exp(\pi\imath\tau).$$

The right hand side becomes

$$\frac{2\theta\begin{bmatrix}\frac{1}{3}\\1\end{bmatrix}(0,3\tau)}{\theta\begin{bmatrix}1\\0\end{bmatrix}(0,\tau)}.$$

Using the Jacobi derivative formula, the identity becomes

$$-(2\pi\imath)\exp\left(-\frac{\pi\imath}{3}\right)x^{\frac{1}{12}}\sum_{n=-\infty}^{\infty}(6n+1)x^{3n^2+n}$$

$$=-2\pi\theta\begin{bmatrix}\frac{1}{3}\\1\end{bmatrix}(0,3\tau)\theta\begin{bmatrix}0\\0\end{bmatrix}(0,\tau)\theta\begin{bmatrix}0\\1\end{bmatrix}(0,\tau),$$

which after using the Jacobi triple product formula and a change of variables (replacing x by $x^{\frac{1}{2}}$) becomes

$$\sum_{n=-\infty}^{\infty}(6n+1)x^{\frac{3n^2+n}{2}}=\prod_{n=1}^{\infty}(1-x^n)^3(1-x^{2n-1})^2.$$

Function theory for the modular group Γ and its subgroups

The Fourier series expansions of $\theta\begin{bmatrix}\epsilon\\ \epsilon'\end{bmatrix} = \theta\begin{bmatrix}\epsilon\\ \epsilon'\end{bmatrix}(0,\tau)$ follow directly from the definitions of these functions. The transformation formula for theta constants and these expansions allow us to use theta functions to construct modular and cusp forms for subgroups of $\mathrm{PSL}(2,\mathbb{Z})$ and to study their divisors. This leads us to the uniformization of surfaces $\mathbb{H}^2/\Gamma(k)$ with small k and generalizations of the Jacobi quartic, for example, to the elegant formula

$$\theta^3\begin{bmatrix}\frac{1}{3}\\ \\ 1\end{bmatrix} = \theta^3\begin{bmatrix}\frac{1}{3}\\ \\ \frac{1}{3}\end{bmatrix} + \theta^3\begin{bmatrix}\frac{1}{3}\\ \\ \frac{5}{3}\end{bmatrix}.$$

For odd k, powers of $\theta\begin{bmatrix}\frac{m}{k}\\ \frac{m'}{k}\end{bmatrix}(0,\tau)$ produce cusp forms[51] for $\Gamma(k)$. Their divisors are supported on the punctures of $\mathbb{H}^2/\Gamma(k)$. For odd primes[52] k, there are $\frac{k^2-1}{2}$ punctures on this surface. This is too many punctures for an analysis of the divisor of the cusp forms encountered, an analysis further complicated by the fact that one needs to use high powers ($8k$, in many

[51] We need to assume that m and m' are integers of the same parity as k.

[52] Assume this for the remainder of these introductory remarks.

cases) of the theta constants to produce cusp forms. Matters simplify considerably in the study of $\theta\begin{bmatrix}\frac{m}{k}\\1\end{bmatrix}(0,k\tau)$ and $\theta'\begin{bmatrix}\frac{m}{k}\\1\end{bmatrix}(0,k\tau)$. In the first of these cases, we obtain automorphic functions for a factor of automorphy for $\Gamma(k)$ that is independent of m. Although we do not have a simple interpretation of these functions on the compact surface $\mathbf{X}_k = \overline{\mathbb{H}^2/\Gamma(k)}$, nontrivial ratios of such functions yield nonconstant meromorphic functions on the compactified surfaces whose divisors are supported on a set of $\frac{k-1}{2}$ distinguished punctures. A further analysis of these forms yields holomorphic mappings Φ of the surfaces into projective spaces and appearances of new Platonic solids based on these surfaces. We conjecture that these mappings are embeddings. We establish the conjecture for small k – in these cases, we also have a good description of the equations satisfied by the curve $\mathbf{X}_k$, and prove, by introducing and analyzing *weights* of finite dimensional spaces of holomorphic functions, that the map Φ is of maximal rank. The analysis of the surfaces $\mathbf{X}_k$ shows the need for theta identities, and a method for obtaining these is described. The case $k = 7$ leads us to the Klein surface and the identity

$$\left(\omega^{-3}\theta\begin{bmatrix}\frac{1}{7}\\1\end{bmatrix}\right)^3\left(\omega^9\theta\begin{bmatrix}\frac{5}{7}\\1\end{bmatrix}\right)+\left(\omega^{-3}\theta\begin{bmatrix}\frac{1}{7}\\1\end{bmatrix}\right)\left(\omega^9\theta\begin{bmatrix}\frac{3}{7}\\1\end{bmatrix}\right)^3$$
$$+\left(\omega^9\theta\begin{bmatrix}\frac{3}{7}\\1\end{bmatrix}\right)\left(\omega^9\theta\begin{bmatrix}\frac{5}{7}\\1\end{bmatrix}\right)^3=0,\ \omega=\exp\left(\frac{\pi\imath}{28}\right).$$

1. Automorphic forms and functions

The material in this section, especially the contents of the first four subsections, can be found, for example, in [**18**]. Proofs are supplied for any material not found in the first three chapters of the last reference.

1.1. Two Banach spaces. Fix a Fuchsian group G and a disc[53] Δ that it leaves invariant. Let r and s be a pair of half integers with the property that their sum is an integer. Let f be a meromorphic function on Δ. We define an *action* of f on functions φ on $f(\Delta)$ as follows:

$$(f^*_{r,s}\varphi)(z)=\varphi(f(z))f'(z)^r\overline{f'(z)}^s,\ z\in\Delta.$$

We have introduced a contravariant action since the identity map acts as the identity operator, and if g is meromorphic on $f(\Delta)$, then

$$(g\circ f)^*_{r,s}=f^*_{r,s}\circ g^*_{r,s}.$$

We abbreviate $f^*_{r,0}$ by f^*_r and f^*_0 by $f^*=\circ f$.

[53]The use of the same symbol to denote a disc on which groups act and a basis for $\mathbb{A}_6(\mathbb{H}^2,\Gamma)$ (up to a constant multiple, the discriminant from elliptic function theory) should not cause any confusion.

Remark 1.1. The *Poincaré density* $\lambda = \lambda_\Delta$ for the metric of constant curvature -1 on the domain Δ may be defined by setting

$$\lambda_{\mathbb{H}^2}(z) = \frac{1}{\Im z},\ z \in \mathbb{H}^2,$$

and requiring that $\lambda_\Delta(z)|dz|$ be a conformal invariant; that is, for every meromorphic univalent function f on Δ,

$$f^*_{\frac{1}{2},\frac{1}{2}}\lambda_{f(\Delta)} = \lambda_\Delta.$$

In particular, we then conclude that

$$g^*_{\frac{1}{2},\frac{1}{2}}\lambda = \lambda, \text{ for all } g \in G.$$

We fix $q \in \mathbb{Z}$. A meromorphic function φ on Δ is called an *automorphic q-form* (also an *automorphic form of weight*[54] q) for the group G if

$$g^*_q\varphi = \varphi, \text{ for all } g \in G.$$

(Next is a technical condition that is automatically satisfied if infinity is not fixed by every element of G.) If $\infty \in \Delta$, then we say that φ is *holomorphic* at ∞ provided

$$\varphi(z) = O(|z|^{-2q}),\ z \to \infty.$$

(This last condition can be rewritten as $\text{ord}_\infty\varphi \geq 2q$.) Assume for the rest of this and the next subsection that $q \geq 2$. The holomorphic automorphic form φ is called *integrable* if its *norm*

$$||\varphi||_{\Delta/G} = \int\int_{\Delta/G} \lambda(z)^{2-q}\ \varphi(z)\frac{dzd\bar{z}}{2}\Big| < \infty,$$

and it is called *bounded* if

$$||\varphi||_{\mathbb{B},\Delta} = \sup\{\lambda(z)^{-q}|\varphi(z)|;\ z \in \Delta\} < \infty.$$

It should be noted that, in the last definition, the same norm would be obtained if we took the supremum over Δ/G, that is, over a fundamental domain ω for the action of G on Δ. Here we must require that the boundary of ω have 2-dimensional Lebesgue measure zero. The Banach space of integrable (bounded) holomorphic q-forms for G on Δ will be denoted by $\mathbb{A}_q(\Delta, G)$ ($\mathbb{B}_q(\Delta, G)$). If G is the trivial group, it is dropped from the notation; thus we will abbreviate for trivial G, $\mathbb{A}_q(\Delta, G)$ to $\mathbb{A}_q(\Delta)$, $||\varphi||_{\Delta/G}$ to $||\varphi||_\Delta$, and $\mathbb{B}_q(\Delta, G)$ to $\mathbb{B}_q(\Delta)$. We define in an analogous manner the Banach spaces of integrable and bounded *measurable* q-forms for G on Δ: $L^1_q(\Delta, G)$ and $L^\infty_q(\Delta, G)$. It is obvious that $\mathbb{A}_q(\Delta, G)$ ($\mathbb{B}_q(\Delta, G)$) is a closed linear subspace of $L^1_q(\Delta, G)$ ($L^\infty_q(\Delta, G)$) because convergence of holomorphic functions in each of our two norms implies uniform convergence on

[54] $2q$ or $-2q$ in some literature.

compact subsets. Similarly, $\mathbb{B}_q(\Delta, G)$ $(L_q^\infty(\Delta, G))$ is a closed linear subspace of $\mathbb{B}_q(\Delta)$ $(L_q^\infty(\Delta))$. The corresponding statements for the spaces of integrable forms are false for infinite groups G, although for finite nontrivial groups G there are nonisometric inclusions of $\mathbb{A}_q(\Delta, G)$ and $L_q^1(\Delta, G)$ into $\mathbb{A}_q(\Delta)$ and $L_q^1(\Delta)$, respectively.

For $\mu \in L_q^1(\Delta, G)$ and $\nu \in L_q^\infty(\Delta, G)$, the *Petersson scalar product* is defined[55] by

$$< \mu, \nu >_{\Delta/G} = \int\int_{\Delta/G} \lambda(z)^{2-2q}\mu(z)\overline{\nu(z)}\left|\frac{dzd\bar{z}}{2}\right|.$$

By the theorem of F. Riesz, the Petersson scalar product is nonsingular; that is, the map that sends $\nu \in L_q^\infty(\Delta, G)$ to the linear functional that maps $\mu \in L_q^1(\Delta, G)$ to $< \mu, \nu >_{\Delta/G} \in \mathbb{C}$ establishes a conjugate $\mathbb{C}$-linear isometric isomorphism between $L_q^\infty(\Delta, G)$ and the dual space $L_q^1(\Delta, G)^*$ of (continuous linear functionals on) $L_q^1(\Delta, G)$.[56]

It is well known, and fairly easy to establish, that the Petersson scalar product also establishes a conjugate $\mathbb{C}$-linear topological (no longer isometric) isomorphism between $\mathbb{B}_q(\Delta, G)$ and $\mathbb{A}_q(\Delta, G)^*$ (a more general theorem is due to L. Bers). The principal tools for establishing this result are integral operators that define adjoint projections

$$\beta : L_q^1(\Delta, G) \to \mathbb{A}_q(\Delta, G)$$
$$\beta : L_q^\infty(\Delta, G) \to \mathbb{B}_q(\Delta, G)$$

that satisfy

$$< \beta\mu, \nu >_{\Delta/G} = < \mu, \beta\nu >_{\Delta/G} \text{ for all } \mu \in L_q^1(\Delta, G) \text{ and all } \nu \in L_q^\infty(\Delta, G).$$

The norms of the projection operators are at most

$$\frac{2q-1}{q-1}.$$

These projection operators may be viewed as machines that manufacture holomorphic automorphic forms from measurable automorphic forms (which are easy enough to manufacture; define them arbitrarily on a fundamental domain and let the group transport them to the rest of Δ). We shall shortly describe more efficient tools (Poincaré series) for constructing holomorphic automorphic forms.

Remark 1.2. The space of Beltrami coefficients for G with support on Δ, $\mathcal{M}(\Delta, G)$ is the open unit ball in the Banach space

$$\lambda^{-2}\overline{L_2^\infty(\Delta, G)} = \{\lambda_\Delta^{-2}\bar{\mu}; \mu \in L_2^\infty(\Delta, G)\}.$$

[55] The definition makes sense for $q \in \mathbb{R}$. We will later use $q \in \mathbb{Q}^+$.

[56] We use the standard notation X^* for the dual (in the appropriate category) of the space X.

It plays a dominant role in the theory of quasiconformal mappings and in work on Teichmüller spaces.

Remark 1.3. Assume that f is a Möbius transformation. It is important to realize that the operator f_q^* behaves well under most of the concepts we have defined. For example,

$$f_q^* : \mathbb{A}_q(f(\Delta), fGf^{-1}) \to \mathbb{A}_q(\Delta, G)$$

and

$$f_q^* : \mathbb{B}_q(f(\Delta), fGf^{-1}) \to \mathbb{B}_q(\Delta, G)$$

are surjective isometries, and

$$< \varphi, \psi >_{f(\Delta)/fGf^{-1}} = < f_q^*(\varphi), f_q^*(\psi) >_{\Delta/G},$$

for all $\varphi \in \mathbb{A}_q(f(\Delta), fGf^{-1})$ and all $\psi \in \mathbb{B}_q(f(\Delta), fGf^{-1})$.

1.2. Poincaré series. We now come to a very important way of producing holomorphic and meromorphic automorphic forms.

The *Poincaré series operator* $\Theta = \Theta_q$ is defined on meromorphic functions φ by

$$\Theta\varphi = \sum_{g \in G} g_q^* \varphi$$

whenever the series converges uniformly and absolutely on compact subsets of Δ, in which case $\Theta\varphi$ is a meromorphic automorphic q-form for G on Δ. It is easy to see[57] that the series converges for meromorphic φ in $L_q^1(\Delta)$ and hence

$$\Theta : \mathbb{A}_q(\Delta) \to \mathbb{A}_q(\Delta, G)$$

is a linear, norm nonincreasing, surjective operator. It should be noted that the inclusion i from $\mathbb{B}_q(\Delta, G)$ into $\mathbb{B}_q(\Delta)$ is the adjoint of the operator Θ as defined above; that is,

$$< \varphi, \psi >_\Delta = < \Theta\varphi, \psi >_{\Delta/G}, \text{ for all } \varphi \in \mathbb{A}_q(\Delta), \text{ and all } \psi \in \mathbb{B}_q(\Delta, G).$$

The surjectivity of Θ is a direct consequence of its adjointness to i.

Remark 1.4. The norm of the Θ operator for the case $q = 2$ is connected to the study of extremal quasiconformal maps. It was shown by C. McMullen that for the most interesting cases the norm is (strictly) less than one.

We introduce a collection of rational functions that play a key role in much of the theory of moduli of Riemann surfaces. Let $a_1, \ldots, a_{2q-1}$ be $2q - 1$ distinct points in $\hat{\mathbb{C}}$. Define for $z \in \mathbb{C}$

$$p(z; a_1, \ldots, a_{2q-1}) = p(z) = \prod_{j=1}^{2q-1} (z - a_j)$$

[57]The main tools are Fubini's theorem and the fact that L^1-convergence of meromorphic functions implies local uniform convergence.

and

$$f(z,\zeta;a_1,\ \ldots,\ a_{2q-1}) = f(z,\zeta) = \frac{-p(z)}{\pi(\zeta - z)p(\zeta)},\ \zeta \in \hat{\mathbb{C}}$$

(we shall, in general, ignore the necessary adjustments when a variable or parameter assumes the value infinity, and leave it to the reader to make the necessary modifications to the formulae – normally interpreting expressions as limits produces the correct modification). We view for the moment z as a parameter and observe that for fixed $z \in \hat{\mathbb{C}} - \{a_1, \ldots, a_{2q-1}\}$ the function $f(z,\cdot)$ is meromorphic and in the Banach space $L_q^1(\Omega)$. It is in $\mathbb{A}_q(\Delta)$ whenever both z and the points a_j are in the limit set Λ of G. Thus we can define a meromorphic automorphic q-form (in the ζ variable) by

$$\begin{aligned}\varphi(z,\zeta) = &\ \varphi(z,\zeta;a_1,\ \ldots,\ a_{2q-1}) \\ = &\ \textstyle\sum_{g\in G} f(z,g(\zeta))g'(\zeta)^q,\ z \in \hat{\mathbb{C}} - \{a_1,\ \ldots, a_{2q-1}\},\end{aligned}$$

for all $\zeta \in \Omega$; that is,

$$\varphi(z,\cdot) = \Theta(f(z,\cdot)).$$

The meromorphic q-form $\varphi(z,\cdot)$ is always an element of $L_q^1(\Omega, G)$ and belongs to $\mathbb{A}_q(\Delta, G)$ whenever both z and the points a_j are in Λ. Further, under the assumption that $a_j \in \Lambda$ and $z \in \Omega$, the form $\varphi(z,\cdot)$ is either regular or has a simple pole at $\zeta = z$ (and points equivalent under G to z). In this case, we let μ be the order of the stabilizer G_z. Then $\varphi(z,\cdot)$ is regular at z if and only if $q \not\equiv 1 \mod \mu$.

1.3. Relative Poincaré series. Let Γ be a subgroup of G. If we start with a Γ invariant form and average appropriately over the Γ-cosets in G, we obtain a G invariant form. Three cases of this construction are of particular interest. Since only the third will be featured prominently in our work, we omit the details and proofs in the first two cases.

1. As before, $q \geq 2 \in \mathbb{Z}$. For loxodromic $A \in G$ with fixed points α and β, define the rational function, a q-differential for the cyclic group $\Gamma = \langle A \rangle$,

$$g(\zeta) = \frac{(\alpha-\beta)^q}{(\zeta-\alpha)^q(\zeta-\beta)^q},\ \zeta \in \hat{\mathbb{C}},$$

and its *relative Poincaré series*

$$\varphi_A(\zeta) = \sum_{\langle A\rangle\backslash G} g(\gamma(\zeta))(\gamma'(\zeta))^q,\ \zeta \in \Omega.$$

2. We now assume that $q \geq 3$. For parabolic $A \in G$ with fixed point α, define

$$g(\zeta) = \frac{1}{(\zeta-\alpha)^{2q}},\ \zeta \in \hat{\mathbb{C}},$$

and as above

$$\psi_\alpha(\zeta) = \sum_{P_\alpha \backslash G} g(\gamma(\zeta))(\gamma'(\zeta))^q, \ \zeta \in \Omega,$$

where $\Gamma = P_\alpha$ is the parabolic stabilizer of α in G. Under appropriate hypothesis on α, the above series also converges for $q = 2$. It yields a holomorphic (modular) form – NOT a cusp form; its projection to Δ/G (see the next subsection), if not identically zero, has a pole of order q at the puncture determined by α.

3. We now consider the case where Γ is of finite index in G and both groups act discontinuously on a set Δ. We may in this case use an arbitrary $q \in \mathbb{Z}$. Let f be a form for Γ and define

$$F(z) = \sum_{\Gamma \backslash G} f(\gamma(z))(\gamma'(z))^q, \ z \in \Delta.$$

Our first observation is that F is well defined. Let

$$g_1, \ \ldots, \ g_n$$

be a set of representatives of the left cosets of Γ in G; that is,

$$G = \bigcup_{i=1}^n \Gamma g_i,$$

and for $1 \le i, j \le n$,

$$g_i \circ g_j^{-1} \in \Gamma \text{ iff } i = j.$$

Our definition reads

$$F(z) = \sum_{i=1}^n f(g_i(z))g_i'(z)^q.$$

Replace g_i by $\gamma_i \circ g_i$ with $\gamma_i \in \Gamma$ and calculate

$$\begin{aligned} \textstyle\sum_{i=1}^n f((\gamma_i \circ g_i)(z))(\gamma_i \circ g_i)'(z)^q &= \textstyle\sum_{i=1}^n f(\gamma_i(g_i(z)))\gamma_i'(g_i(z))g_i'(z)^q \\ &= \textstyle\sum_{i=1}^n f(g_i(z))g_i'(z)^q \end{aligned}.$$

Thus F is independent of the choice of coset representatives. To prove that F is a form for G, we note that for $g \in G$,

$$\begin{aligned} F(g(z))g'(z)^q &= \left(\textstyle\sum_{i=1}^n f(g_i(g(z)))g_i'(g(z))^q\right) g'(z)^q \\ &= \textstyle\sum_{i=1}^n f(g_i(g(z)))(g_i' \circ g)(z)^q = F(z) \end{aligned},$$

since

$$g_1 \circ g, \ \ldots, \ g_n \circ g$$

is another set of representatives of the left cosets of Γ in G because

$$G = \left(\bigcup_{i=1}^n \Gamma g_i\right) g = \bigcup_{i=1}^n \Gamma g_i \circ g,$$

and for $1 \leq i, j \leq n$,

$$(g_i \circ g) \circ (g_j \circ g)^{-1} \in \Gamma \text{ iff } i = j.$$

In the sequel, we will also need to obtain information regarding zeros, poles and bounds for F from the corresponding information for f.

1.4. Projections to the surface. Let G be a finitely generated Fuchsian group of the first kind operating on a disc Δ. Let φ be a meromorphic q-form for G on Δ with $q \in \mathbb{Z}$ arbitrary. The automorphic form φ induces a meromorphic q-form Φ on Δ/G by invariance. If Z is a local coordinate at $x \in \Delta/G$, then Z is viewed as a holomorphic function on some domain $D \subset \Delta$ via the natural projection $\pi : \Delta \to \Delta/G$. We define Φ in terms of Z by

$$\varphi(z) = \Phi(Z)\left(\frac{dZ}{dz}\right)^q, \quad z \in D. \tag{3.1}$$

Let $z_o \in D$ be arbitrary and let $x_o = \pi(z_o)$. We want to relate

$$r = \text{ ord }_{z_o}\varphi \text{ to } R = \text{ ord }_{x_o}\Phi.$$

Let $\mu = |G_{z_o}|$. Without loss of generality, $D = \mathbb{D}$ (the unit disc) and $z_o = 0$. Hence G_0 is generated by $z \mapsto \exp\left\{\frac{2\pi\imath}{\mu}\right\} z$. It follows that we may choose Z so that $Z = z^\mu$ and hence

$$r = R\mu + q(\mu - 1).$$

Since R must be an integer, we conclude that for holomorphic φ,

$$R \geq -\left\lfloor q\left(1 - \frac{1}{\mu}\right)\right\rfloor,$$

where, as before, $\lfloor y \rfloor$ represents the greatest integer less than or equal to the real number y.

We consider next the case where x is a puncture on Δ/G. It is convenient to assume now that (by normalization) $\Delta = \mathbb{H}^2$, and that the parabolic subgroup that determines the puncture is generated by the translation $B = \begin{bmatrix} 1 & 1 \\ 0 & 1 \end{bmatrix}$. A good local coordinate vanishing at $x \in \overline{\mathbb{H}^2/G}$ is $Z = \exp\{2\pi\imath z\}$. If Φ is holomorphic in a punctured neighborhood of x, then from the relation

$$\varphi(z+1) = \varphi(z), \ z \in \mathbb{H}^2,$$

we conclude that φ has a Fourier series expansion

$$\varphi(z) = \sum_{n=-\infty}^{\infty} a_n \exp\{2\pi\imath\, nz\}, \ z \in \mathbb{H}^2. \tag{3.2}$$

If $\varphi \in \mathbb{B}_q(\mathbb{H}^2, G)$, $q \geq 2$, then

$$a_n = 0, \text{ for all } n \leq 0.$$

Definition 1.5. An automorphic form φ on Δ for the group G is *meromorphic* at a cusp z if it is holomorphic in a horodisc determined by z and its projection Φ to Δ/G is regular or has a pole at the puncture determined by the cusp. For such a form, there exists a largest $r \in \mathbb{Z}$ such that for the Fourier series (3.2),

$$a_n = 0, \text{ for all } n \leq r - 1.$$

We hence define

$$\mathrm{ord}_z \varphi = r.$$

It is easily seen that for cusps (and related punctures)

$$R = r - q.$$

Definition 1.6. The *q-canonical ramification divisor* for $\overline{\Delta/G}$ is

$$\mathcal{R}_q = \prod_{x \in \overline{\Delta/G}} x^{-\left\lfloor q\left(1 - \frac{1}{\mu(x)}\right)\right\rfloor},$$

with the understanding (here and in similar contexts subsequently) that

$$-\left\lfloor q\left(1 - \frac{1}{\infty}\right)\right\rfloor = \lim_{\mu \to \infty} - \left\lfloor q\left(1 - \frac{1}{\mu}\right)\right\rfloor = \begin{cases} 1 - q & \text{if } q > 0 \\ -q & \text{if } q \leq 0 \end{cases}.$$

We have seen that for $q \geq 2$, every $\varphi \in \mathbb{B}_q(\Delta, G)$ projects to a meromorphic q-differential Φ on $\overline{\Delta/G}$ whose divisor (Φ) satisfies

$$(\Phi) \geq \mathcal{R}_q. \tag{3.3}$$

Conversely, every meromorphic q-differential Φ on $\overline{\Delta/G}$ whose divisor satisfies the above *integrability* condition lifts via (3.1) to a $\varphi \in \mathbb{B}_q(\Delta, G)$.

Proposition 1.7. *For $q \in \mathbb{Z}$, $q \geq 2$, G finitely generated of the first kind,*

$$\mathbb{A}_q(\Delta, G) = \mathbb{B}_q(\Delta, G),$$

and

$$\dim \mathbb{B}_q(\Delta, G) = (2q-1)(p-1) + \sum_{j=1}^{n} \left\lfloor q - \frac{q}{\mu_j} \right\rfloor,$$

where $(p, n;\ \mu_1, \ldots, \mu_n)$ is the signature of G.

Proof. The statement about dimensions is a consequence of the Riemann-Roch theorem.

Corollary 1.8. *For G of type (p, n)*

$$\dim \mathbb{B}_2(\Delta, G) = 3p - 3 + n.$$

We will also need a version of the last proposition for $q = 1$. It is convenient to establish a more general result.

Definition 1.9. Let $\mathcal{P}$ be the divisor of punctures on Δ/G. We define $\mathbb{A}_q(\Delta, G)$ to be the lift to Δ of the space of meromorphic q-forms Φ whose divisors (Φ) satisfy the integrability condition (3.3) (for $q \geq 2$, the new definition agrees with the old one) and define $\mathbb{A}_q^+(\Delta, G)$ to be the corresponding space where

$$(\Phi) \geq \frac{\mathcal{R}_q}{\mathcal{P}}.$$

The elements of $\mathbb{A}_q(\Delta, G)$ are called *cusp* forms; those of $\mathbb{A}_q^+(\Delta, G)$, *modular* forms.

Proposition 1.10. *Let $q \in \mathbb{Z}$ and let G be of type (p, n). Then*
(a) $\dim\, \mathbb{A}_q^+(\Delta, G) = \dim\, \mathbb{A}_q(\Delta, G) + n$ *if* $q \geq 2$,
(b) $\dim\, \mathbb{A}_1(\Delta, G) = p$,

$$\dim\, \mathbb{A}_1^+(\Delta, G) = \begin{cases} p \text{ if } n = 0 \\ p + n - 1 \text{ if } 2p - 2 + n > 0 \text{ and } n > 0 \\ 0 \text{ if } p = 0 \text{ and } n = 1 \end{cases}$$

(c) $\dim\, \mathbb{A}_0(\Delta, G) = 1$, $1 \leq \dim \mathbb{A}_0^+(\Delta, G) \leq 1 + n$,
and
(d) $\dim\, \mathbb{A}_q^+(\Delta, G) = 0$ *if* $q < 0$.

Remark 1.11. We examine the behavior of automorphic forms at parabolic fixed points in more detail. Let $q \in \mathbb{Z}$, $q \geq 2$. Let G be a finitely generated Fuchsian group of the first kind operating on $\mathbb{H}^2$. If $\mathbb{H}^2/G$ is compact, then $\mathbb{A}_q^+(\mathbb{H}^2, G) = \mathbb{A}_q(\mathbb{H}^2, G)$ and a function φ belongs to this space as soon as it satisfies the invariance equations

$$\varphi(\gamma(z))(\gamma'(z))^q = \varphi(z), \text{ for all } \gamma \in G \text{ and all } z \in \mathbb{H}^2.$$

If G contains a parabolic element with fixed point $x \in \mathbb{R} \cup \{\infty\}$, then $\mathbb{H}^2/G$ is not compact and a modular form needs to satisfy growth conditions at the cusps in addition to the invariance equations.[58] If $x = \infty$, then we may choose $z \mapsto z + b$, $b \in \mathbb{R}^+$, as a generator for G_∞. For $\varphi \in \mathbb{A}_q^+(\mathbb{H}^2, G)$,

$$\varphi(z) = \sum_{n=0}^{\infty} a_n \exp\left(\frac{2\pi \imath n z}{b}\right), \tag{3.4}$$

and we define (the resulting extension is continuous in an appropriate sense)

$$\varphi(\infty) = a_0$$

[58] Because of the invariance equations, there are only finitely many growth conditions, one for each puncture on $\mathbb{H}^2/G$.

and note that if $\varphi \in \mathbb{A}_q(\mathbb{H}^2, G)$, then $\varphi(\infty) = 0$. For finite x, we choose $C \in \mathrm{PSL}(2,\mathbb{R})$ with $C(\infty) = x$ of the form

$$C = \begin{bmatrix} 0 & -1 \\ 1 & 0 \end{bmatrix} \text{ if } x = 0 \text{ and } \begin{bmatrix} x & 0 \\ 1 & \frac{1}{x} \end{bmatrix} \text{ if } x \neq 0.$$

If $\psi \in \mathbb{A}_q^+(\mathbb{H}^2, G)$, then $\varphi = C_q^*(\psi) \in \mathbb{A}_q^+(\mathbb{H}^2, C^{-1}GC)$ and this form has a Fourier expansion (3.4) for some $b \in \mathbb{R}^+$. Since

$$\varphi(z) = \begin{cases} O(1) \text{ for } \psi \in \mathbb{A}_q^+(\mathbb{H}^2, G), \\ O(\exp(-\Im z)) \text{ for } \psi \in \mathbb{A}_q(\mathbb{H}^2, G), \end{cases} \quad z \to \imath\infty,$$

we conclude that as $z \to x$,

$$\psi(z) = \begin{cases} O\left(|z-x|^{-2q}\right) \text{ for } \psi \in \mathbb{A}_q^+(\mathbb{H}^2, G) \\ O\left(|z-x|^{-2q} \exp\left(-\frac{\Im z}{|z-x|^2}\right)|z-x|^{-2q}\right) \text{ for } \psi \in \mathbb{A}_q(\mathbb{H}^2, G) \end{cases}$$

(in the last equation the approach to x must be through an appropriately restricted region in $\mathbb{H}^2$). We hence define

$$\psi(x) = \varphi(\infty) = \lim_{z \to x} (z-x)^{2q} \psi(z).$$

Assume that S generates the stabilizer of x in G:

$$\frac{1}{S(z)-x} = \frac{1}{z-x} - \xi, \ \xi \in \mathbb{R}^+.$$

In this case,

$$C^{-1} \circ S \circ C(z) = x + \xi,$$

and φ has (3.4) with $b = \xi$ as Fourier series expansion. Thus in terms of the local coordinate $Z = \frac{1}{z-x}$,

$$\psi(z) = (z-x)^{-2q} \sum_{n=0}^{\infty} b_n \exp\left[\frac{2\pi \imath n}{\xi}\left(\frac{1}{z-x}\right)\right],$$

with $b_n = \exp\left(-\frac{1}{x}\right) a_n$.

1.5. Factors of automorphy. Let G be a subgroup of $\mathrm{SL}(2,\mathbb{R})$ whose projection (denoted by same symbol) to $\mathrm{PSL}(2,\mathbb{R})$ is a finitely generated Fuchsian group of the first kind.[59] We restrict our attention to the action of G on $\mathbb{H}^2$. A *factor of automorphy* for G is a function

$$e : G \times \mathbb{H}^2 \to \mathbb{C}^*$$

with

$$e(g, \cdot) : \mathbb{H}^2 \to \mathbb{C}^*$$

[59]We also use the same symbol for an element of $\mathrm{SL}(2,\mathbb{R})$ and the Möbius transformation it induces.

holomorphic for all $g \in G$ and

$$e(g_1 g_2, \tau) = e(g_1, g_2(\tau))\; e(g_2, \tau)$$

for all g_1, $g_2 \in G$ and all $\tau \in \mathbb{H}^2$. The most important example of a factor of automorphy is the *canonical* factor $\mathcal{K}$ defined by

$$\mathcal{K}(g, \tau) = g'(\tau),\ g \in G,\ \tau \in \mathbb{H}^2.$$

We will be interested only in a restricted class of factors of automorphy. A factor of automorphy e is called *parabolic* if there exists a real constant q, which we will call the *weight* of e, and for each $g \in G$ there exists a complex number of absolute value 1, $c(g)$, such that

$$e(g, \tau) = c(g)\; g'(\tau)^q, \text{ for all } \tau \in \mathbb{H}^2.$$

This requirement involves only finitely many conditions (one for each generator of G). If $q \in \mathbb{Z}$, then c is a *(normalized) character* for G, that is, is a homomorphism of G into the unit circle, the complex numbers of absolute value 1. It is convenient to write

$$c(g) = \exp\{2\pi\imath\alpha(g)\},\ \alpha = \alpha(g) \in \mathbb{R},\ 0 \le \alpha < 1.$$

A holomorphic function φ on $\mathbb{H}^2$ is e-*automorphic* if it satisfies

$$\varphi(\gamma(\tau))\; e(\gamma, \tau) = \varphi(\tau), \text{ for all } \tau \in \mathbb{H}^2 \text{ and all } \gamma \in G,$$

and has a finite limit as $\tau \in \mathbb{H}^2$ approaches a parabolic fixed point $x \in \mathbb{R} \cup \{\infty\}$ of G through a cusped region determined by x. The vector space of holomorphic e-automorphic functions on $\mathbb{H}^2$ for the group G is denoted by $\mathbb{A}(\mathbb{H}^2, G, e)$.

Let $C \in \mathrm{SL}(2, \mathbb{R})$ and assume that e is a parabolic factor of automorphy for G of weight q.[60] Then

$$\tilde{e}(C^{-1}gC, \tau) = e(g, C(\tau))\frac{C'(\tau)^q}{C'(C^{-1}gC(\tau))^q},\ g \in G, \tau \in \mathbb{H}^2,$$

defines a parabolic factor of automorphy $\tilde{e}$ for $\tilde{G} = C^{-1}GC$ of weight q. Further, the map that sends $\varphi \in \mathbb{A}(\mathbb{H}^2, G, e)$ to $(\varphi \circ C)(C')^q$ establishes a $\mathbb{C}$-linear isomorphism

$$C_*^q : \mathbb{A}(\mathbb{H}^2, G, e) \to \mathbb{A}(\mathbb{H}^2, \tilde{G}, \tilde{e}).$$

If $q \notin \mathbb{Z}$, then we use the same branch of $(C')^q$ in all the above formulae. The resulting factor of automorphy (and its space of automorphic functions) depends in a mild way on this choice.

[60] In this case an element of $\mathbb{A}(\mathbb{H}^2, G, e)$ will be called a q-form.

Remark 1.12. For $g = \begin{bmatrix} a & b \\ c & d \end{bmatrix}$, the map $e(g,\tau) = (c\tau + d)$ defines a factor of automorphy for $\mathrm{SL}(2,\mathbb{R})$. However $(c\tau + d)^{\frac{1}{2}}$ is not a factor of automorphy for (even the smaller) group $\mathrm{SL}(2,\mathbb{Z})$ since it assigns either $\pm\sqrt{\imath}$, an 8-th root of unity, to the pair $\left(\begin{bmatrix} 0 & -1 \\ 1 & 0 \end{bmatrix}, \imath\right)$. This is impossible for a factor of automorphy since $\imath$ is a fixed point of the motion determined by the matrix $\begin{bmatrix} 0 & -1 \\ 1 & 0 \end{bmatrix}$ which has order 4.

In general, for fixed characteristic χ, powers of $\kappa(\chi,\cdot)$ define factors of automorphy for subgroups of $\mathrm{SL}(2,\mathbb{Z})$ that are independent of τ.

1.6. Multiplicative q-forms. Let G be a finitely generated Fuchsian group of the first kind operating on $\mathbb{H}^2$. Let c be a character on G and $q \in \mathbb{Z}$. For the parabolic factor of automorphy $e = c\,\mathcal{K}^q$ for G defined by

$$e(g,\tau) = c(g)g'(\tau)^q,\ \ g \in G,\ \ \tau \in \mathbb{H}^2,$$

we write

$$\mathbb{A}(\mathbb{H}^2, G, e) = \mathbb{A}_q(\mathbb{H}^2, G, c).$$

The trivial character $c = 1$ is, as usual, dropped from the notation.

Let e be a factor of automorphy for G of weight q. A nonzero $\varphi \in \mathbb{A}(\mathbb{H}^2, G, e)$ has a well defined order, $\mathrm{ord}_x\varphi \in \mathbb{R}^+ \cup \{0\}$, at each point $x \in \mathbb{H}^2$ and at each parabolic fixed point $x \in \Lambda_{\mathrm{par}}(G) \subset \mathbb{R} \cup \{\infty\}$ of G. For $x \in \mathbb{H}^2$, $\mathrm{ord}_x\varphi$ is defined as the order of vanishing[61] of φ at x <u>divided</u> by the order of G_x, the stabilizer of x in G. To define the order of φ at a parabolic fixed point x of G requires some preliminaries. We may assume by conjugation that $x = \infty$ and that the stabilizer of x in G is the cyclic group generated by $B = \begin{bmatrix} 1 & 1 \\ 0 & 1 \end{bmatrix}$ (we will call this the *normalized* case). This reduction is achieved by choosing a $C \in \mathrm{PSL}(2,\mathbb{R})$ with $C(\infty) = x$ and sending φ to $C^q_*\varphi$. The resulting function is essentially independent of the choice of C, for if C_1 and C_2 are two real Möbius transformations that send ∞ to x and conjugate a generator of the stabilizer of x in G to B, then $C_2 = C_1 \circ D$ with D a real Möbius transformation that fixes ∞ and commutes with B, hence of the form

$$\tau \mapsto \tau + b,$$

where $b \in \mathbb{R}$. It is easy to see that the resulting change in functions is

$$(C_2)^q_*\varphi(\tau) = c(C_1)^q_*\varphi(\tau + b),$$

where c is a q-th root of unity. This change does not affect the definition of the order of vanishing of the automorphic function under study. In the

[61] A rational, not necessarily an integer.

normalized case

$$\varphi(\tau+1)\exp(2\pi\imath\alpha) = \varphi(\tau), \text{ for all } \tau \in \mathbb{H}^2,$$

where $c(B) = \exp(2\pi\imath\alpha)$ with $\alpha \in \mathbb{R}$, $0 \le \alpha < 1$. Then

$$f(\tau) = \exp(2\pi\imath\alpha\tau)\varphi(\tau),\ \tau \in \mathbb{H}^2,$$

has a Fourier series expansion

$$f(\tau) = \sum_{n\in\mathbb{Z}} a_n \exp(2\pi\imath n\tau),\ \tau \in \mathbb{H}^2.$$

Hence the boundedness assumption on φ translates to a Fourier series expansion for this function of the form (a generalization of (3.4))

$$\varphi(\tau) = \exp(-2\pi\imath\alpha\tau) \sum_{n\in\mathbb{Z},\ n\ge\alpha} a_n \exp(2\pi\imath n\tau),\ \tau \in \mathbb{H}^2.$$

One sets

$$\operatorname{ord}_\infty\varphi = N - \alpha,$$

where N is the smallest (nonnegative) integer n with $a_n \ne 0$. If $\alpha = 0$, then N could be zero. However if $\alpha > 0$, then $N > 0$. For arbitrary $x \in \Lambda_{\text{par}}$,

$$\operatorname{ord}_x\varphi = \operatorname{ord}_\infty C_q^*(\varphi),\ C \in \mathrm{PSL}(2,\mathbb{C}),\ C(\infty) = x.$$

While the order of φ at a parabolic fixed point x is only a nonnegative real number,

$$\operatorname{ord}_x\varphi_1 - \operatorname{ord}_x\varphi_2 \in \mathbb{Z},$$

for all φ_1 and $\varphi_2 \in \mathbb{A}(\mathbb{H}^2, G, e)$. It is routine to prove that if G has signature $(p, n;\ \mu_1,\ \dots,\ \mu_n)$, then (see, for example, [**18**, §III.8] for a similar argument)

$$\sum_{x\in\overline{\mathbb{H}^2/G}} \operatorname{ord}_x\varphi = q\left(2p - 2 + \sum_{j=1}^{n}\left(1 - \frac{1}{\mu_j}\right)\right) = D(q). \tag{3.5}$$

It is useful to observe that for all $\varphi \in \mathbb{A}(\mathbb{H}^2, G, e)$, all $C \in \mathrm{SL}(2,\mathbb{R})$, and all $x \in \mathbb{H}^2 \cup \Lambda_{\text{par}}(C^{-1}GC)$,

$$\operatorname{ord}_{C(x)}\varphi = \operatorname{ord}_x C_q^*(\varphi), \tag{3.6}$$

where q is the weight of e. The possible orders of vanishing of a $0 \ne \varphi \in \mathbb{A}(\mathbb{H}^2, G, e)$ at an ordinary point $x \in \mathbb{H}^2$ are the positive integers

$$0,\ 1,\ \dots,\ D(q).$$

At an elliptic fixed point x, there are restrictions for the possible orders of vanishing of forms $0 \ne \varphi \in \mathbb{A}(\mathbb{H}^2, G, e)$ that can be readily determined. It suffices to observe that these rational numbers satisfy as a consequence of (3.5),

$$0 \le \operatorname{ord}_x\varphi \le D(q),$$

also at parabolic fixed points.

Let $q \in \mathbb{Z}$ and let c be a character on G. If $\varphi \in \mathbb{A}_1(\mathbb{H}^2, G, c) - \{0\}$, then $\varphi^q \in \mathbb{A}_q(\mathbb{H}^2, G, c^q) - \{0\}$, and for all $x \in \mathbb{H}^2 \cup \Lambda_{\text{par}}(G)$,

$$\text{ord}_x \varphi^q = q \ \text{ord}_x \varphi;$$

hence for *rational* characters (characters c with c^N the trivial character for some $N \in \mathbb{Z}^+$), (3.5) is a consequence of the fact that the degree of a canonical divisor on a compact surface of genus $p \geq 0$ is $2p - 2$.

Proposition 1.13. *Let e be a factor of automorphy for the finitely generated Fuchsian group of the first kind G operating on $\mathbb{H}^2$. Then*

$$\dim \mathbb{A}(\mathbb{H}^2, G, e) < \infty.$$

Proof. If $\dim \mathbb{A}(\mathbb{H}^2, G, e) > 0$, choose a nonzero element $\varphi_o \in \mathbb{A}(\mathbb{H}^2, G, e)$. For any $\varphi \in \mathbb{A}(\mathbb{H}^2, G, e)$, the ratio $\frac{\varphi}{\varphi_o}$ projects to a meromorphic function on $\overline{\mathbb{H}^2/G}$ with poles only at the zeros of φ_o. These poles have orders at most $D(q)$.

Remark 1.14. (a) Let c be a character on G. A form $\varphi \in \mathbb{A}_1(\mathbb{H}^2, G, c) - \{0\}$ defines a Prym differential on the compact Riemann surface $\overline{\mathbb{H}^2/G}$ if and only if $c(\gamma) = 1$ for all elliptic and parabolic $\gamma \in G$.

(b) Under appropriate boundedness assumptions we can define the order at cusps of meromorphic forms for the factor of automorphy e so that (3.5) and (3.6) remain valid.

1.7. Residues. Let G be a finitely generated Fuchsian group of the first kind operating on $\mathbb{H}^2$. Let $q \in \mathbb{Z}^+$. Each element $0 \neq \varphi \in \mathbb{A}_q^+(\mathbb{H}^2, G)$ has a well defined residue, $\text{Res}_x \varphi$, at each cusp $x \in \mathbb{R} \cup \{\infty\}$ of G. It is defined easiest by first considering the normalized case: $x = \infty$ stabilized by the group generated by B. In this case φ has a Fourier series expansion (3.4) with $b = 1$, and we define

$$\text{Res}_\infty \varphi = a_0.$$

For an arbitrary cusp $x \in \mathbb{R} \cup \{\infty\}$, we choose a $C \in \text{PSL}(2, \mathbb{R})$ such that

$$C(\infty) = x \text{ and } < B >= C^{-1} G_x C$$

and define

$$\text{Res}_x \varphi = \text{Res}_\infty C_q^*(\varphi).$$

For $q > 1$, not much can be said about the residue. However,

$$\sum_{x \in \Lambda_{\text{par}}(G)} \text{Res}_x \varphi = 0 \text{ for } \varphi \in \mathbb{A}_1^+(\mathbb{H}^2, G).$$

We will be mostly interested in subgroups G of the modular group Γ. For such groups we have a good alternate description of the residue. The stabilizer of ∞ in Γ is generated by $B_{\infty,1}$, where for $p \in \mathbb{Z}$,

$$B_{\infty,p} = \begin{bmatrix} 1 & p \\ 0 & 1 \end{bmatrix};$$

that of $x = \frac{m}{n} \in \mathbb{Q}$ by $B_{x,1}$, where for $p \in \mathbb{Z}$,

$$B_{x,p} = \begin{bmatrix} 1 + pmn & -pm^2 \\ pn^2 & 1 - pmn \end{bmatrix}$$

(in the above, we may and do choose $m \in \mathbb{Z}$ and $n \in \mathbb{Z}^+$ with $(m, n) = 1$). For an arbitrary finite index subgroup G of Γ and $x \in \mathbb{Q} \cup \{\infty\}$, G_x is generated by $B_{x,p}$ for some $p \in \mathbb{Z}^+$. We call p the *width* of the cusp x, in symbols width(x). The width of a cusp is a PSL$(2, \mathbb{Z})$ but not a PSL$(2, \mathbb{R})$ invariant. It follows that

$$\mathrm{Res}_x \varphi = \ \mathrm{width}(x)\varphi(x), \ \ \varphi \in \mathbb{A}_q^+(\mathbb{H}^2, G).$$

Remark 1.15. **(a)** The residue is well defined on the Riemann surface $\overline{\mathbb{H}^2/G}$. If $\pi : \mathbb{H}^2 \cup \Lambda_{\mathrm{par}}(G) \to \overline{\mathbb{H}^2/G}$ is the canonical projection and Φ is the projection of $\varphi \in \mathbb{A}_q^+(\mathbb{H}^2, G)$ to the surface $\overline{\mathbb{H}^2/G}$, then for all $x \in \Lambda_{\mathrm{par}}(G)$,

$$2\pi\imath \mathrm{Res}_{\pi(x)}\Phi = \ \mathrm{Res}_x \varphi.$$

The residue on the surface is independent of the choice of local coordinates at $\pi(x)$ as long as we insist for $q > 1$ that the coordinates vanish at this point.

(b) The concepts in this section extend readily to meromorphic q-forms.

1.8. Weierstrass points for subspaces of $\mathbb{A}(\mathbb{H}^2, G, e)$. We fix a finitely generated Fuchsian group of the first kind G operating on $\mathbb{H}^2$. Let e be a parabolic factor of automorphy of weight q for G. Let V be a nontrivial d-dimensional subspace of $\mathbb{A}(\mathbb{H}^2, G, e)$. We proceed to define an invariant for this space. For $x \in \mathbb{H}^2$ or x a parabolic fixed point of G. Let

$$0 \le r_0 < r_1 < \ldots < r_{d-1}$$

be the possible orders of vanishing[62] at x for elements of $V-\{0\}$. We define the *weight* of the point x with respect to the vector space V as

$$\tau_V(x)=\sum_{i=0}^{d-1}\left(r_i-\frac{i}{\mu}\right)$$

if $x\in\mathbb{H}^2$ and $|G_x|=\mu=\mu_x$, and

$$\tau_V(x)=\sum_{i=0}^{d-1}r_i$$

if x is a parabolic fixed point for G. Note that $r_i\in\mathbb{R}$ for $x\in\Lambda_{\text{par}}(G)$, $r_i\in\mathbb{Q}$ whenever $x\in\mathbb{H}^2$, and $r_i\in\mathbb{Z}$ if in addition $\mu_x=1$. We call x a *Weierstrass* point with respect to V if its weight is above the minimum that it can be, that is, if and only if

$$\tau_V(x)>\begin{cases} 0 \text{ if } x\in\mathbb{H}^2\\ \frac{d(d-1)}{2} \text{ if } x\in\Lambda_{\text{par}}(G)\end{cases}.$$

The *weight* of the vector space V is

$$\tau(V)=\sum_{x\in\overline{\mathbb{H}^2/G}}\tau_V(x).$$

Choose a basis $\varphi_0,\ \ldots,\ \varphi_{d-1}$ for V, and as in [**6**, §III.5] form its *Wronskian*

$$W=\det\begin{bmatrix}\varphi_0 & \cdots & \varphi_{d-1}\\ \varphi_0' & \cdots & \varphi_{d-1}'\\ & \cdots & \\ \varphi_0^{(d-1)} & \cdots & \varphi_{d-1}^{(d-1)}\end{bmatrix}=\det\,[\varphi_0,\ \ldots,\ \varphi_{d-1}].$$

It is easily seen that a change of basis for V results in a nonzero constant multiple of W and that

$$W\in\mathbb{A}(\mathbb{H}^2,G,e^d\kappa^{\frac{d(d-1)}{2}}),$$

where κ is the canonical factor of automorphy for G. Hence the form W has weight $d\left(q+\frac{d-1}{2}\right)$ and

$$\deg\,(W)=d\,\chi(G)\left(q+\frac{d-1}{2}\right),$$

where $\chi(G)$ is the negative Euler characteristic of G.

[62] If $x\in\mathbb{H}^2$ is stabilized by an elliptic subgroup of order μ, then the ordinary order of vanishing of holomorphic functions is divided by μ; at a parabolic fixed point, we use the order of vanishing of the Fourier series in terms of an appropriate local coordinate vanishing at the corresponding puncture. In this context, the weight of a parabolic fixed point can be interpreted as the limiting case of the weight of an elliptic fixed point as $\mu\to\infty$. Thus the second formula for $\tau_V(x)$ should be interpreted as a special case of the first one with $\mu=\infty$. We will use this convention, usually without further remark, throughout the book.

Proposition 1.16. *For all* $x \in \mathbb{H}^2 \cup \Lambda_{par}(G)$,

$$\mathrm{ord}_x W = \tau_V(x).$$

Hence

$$\tau(V) = \deg\,(W) = d\ \chi(G)\left(q + \frac{d-1}{2}\right).$$

Proof. The arguments of [**6**, §III.5], whose notation we employ in this section, easily establish the equality for $x \in \mathbb{H}^2$. For $x \in \mathbb{R} \cup \{\infty\}$ a parabolic fixed point, we use the fact that

$$\det\,[\varphi_0,\ \dots,\ \varphi_{d-1}] = \varphi_0^d \det\left[1,\ \frac{\varphi_1}{\varphi_0},\ \dots,\ \frac{\varphi_{d-1}}{\varphi_0}\right] = \varphi_0^d \det\left[\frac{\varphi_1}{\varphi_0},\ \dots,\ \frac{\varphi_{d-1}}{\varphi_0}\right],$$

and hence by induction on d that

$$\mathrm{ord}_x \det\,[\varphi_0,\ \dots,\ \varphi_{d-1}] = d\ r_0 + \sum_{i=1}^{d-1} r_i - (d-1)r_0.$$

Remark 1.17. The above concepts extend readily to spaces of meromorphic automorphic forms. In this case, care must be exercised in interpreting the order of the Wronskian at points where the forms may have poles.

Proposition 1.18. *Let G be a finitely generated Fuchsian group of the first kind operating on $\mathbb{H}^2$, e be a parabolic factor of automorphy of weight q for G, and V be a nontrivial d-dimensional subspace of $\mathbb{A}(\mathbb{H}^2, G, e)$. Assume V is invariant under γ_q^* for all γ in some Fuchsian group $G' \supset G$. Then for all such γ, $\gamma_Q^* W$, $Q = d\left(q + \frac{d-1}{2}\right)$ is a nonzero constant multiple of W.*

Proof. Let $\gamma \in G'$ and $\{\varphi_i\}$ be a basis for V. Then

$$(\gamma_Q^* W)(\tau) = W(\gamma(\tau))\gamma'(\tau)^Q$$

$$= \det\,[\varphi_0(\gamma(\tau)),\ \varphi_1(\gamma(\tau)),\ \dots, \varphi_{d-1}(\gamma(\tau))]\gamma'(\tau)^q\ \gamma'(\tau)^{q+1} \dots \gamma'(\tau)^{q+d-1}$$

$$= \det \begin{bmatrix} \varphi_0(\gamma(\tau))\gamma'(\tau)^q & \dots & \varphi_{d-1}(\gamma(\tau))\gamma'(\tau)^q \\ \varphi_0'(\gamma(\tau))\gamma'(\tau)^{q+1} & \dots & \varphi_{d-1}'\gamma'(\tau)^{q+1} \\ & \dots & \\ \varphi_0^{(d-1)}(\gamma(\tau))\gamma'(\tau)^{q+d-1} & \dots & \varphi_{d-1}^{(d-1)}(\gamma(\tau))\gamma'(\tau)^{q+d-1} \end{bmatrix},$$

$$= \det \begin{bmatrix} \psi_0(\tau) & \dots & \psi_{d-1}(\tau) \\ \psi_0'(\tau) & \dots & \psi_{d-1}'(\tau) \\ & \dots & \\ \psi_0^{(d-1)}(\tau) & \dots & \psi_{d-1}^{(d-1)}(\tau) \end{bmatrix} = cW(\tau)$$

where $\psi_i(\tau) = \varphi_i(\gamma(\tau))\gamma'(\tau)^q$ and $c \in \mathbb{C}^*$. The next to last equality holds because

$$\psi_i'(\tau) = \varphi_i'(\gamma(\tau))\gamma'(\tau)^{q+1} + q\varphi_i(\gamma(\tau))\gamma'(\tau)^{q-1}\gamma''(\tau)$$

and similar formulae hold for the higher derivatives of the functions ψ_i, the last because $\{\psi_i\}$ is another basis for V.

2. Automorphic forms constructed from theta constants

Let $k \in \mathbb{Z}^+$. Fix as before a $k' \in \mathbb{Z}^+$ that divides k. We embark on a program to study the functions on $\mathbb{H}^2$

$$\tau \mapsto \theta[\chi](0, k'\tau)$$

for various characteristics $\chi \in \mathbb{R}^2$ as forms for $\Gamma(k)$.

2.1. The order of automorphic forms at cusps – Fourier series expansions at $\imath\infty$. We study (the case $k' = 1$) the class of functions on $\mathbb{H}^2$

$$\tau \mapsto \theta^r \begin{bmatrix} \epsilon \\ \epsilon' \end{bmatrix} (0, \tau) = \theta^r \begin{bmatrix} \epsilon \\ \epsilon' \end{bmatrix} = \theta^r[\chi]$$

for various values of r and various characteristics $\chi = \begin{bmatrix} \epsilon \\ \epsilon' \end{bmatrix}$. For this purpose, we need to evaluate the Fourier expansion (at $\imath\infty$) of $\theta \begin{bmatrix} \frac{m}{k} \\ \frac{m'}{k} \end{bmatrix}$ for various integers m and m'. We do most calculations for the surfaces represented by the principal congruence subgroup. The correct local coordinate on $\overline{\mathbb{H}^2/\Gamma(k)}$ at P_∞ is $\zeta = \exp\{\frac{2\pi\imath\tau}{k}\}$. Calculations yield

$$\begin{aligned} \theta \begin{bmatrix} \frac{m}{k} \\ \frac{m'}{k} \end{bmatrix} (0, \tau) &= \sum_{n\in\mathbb{Z}} \exp 2\pi\imath \left\{ \tfrac{1}{2}\left(n + \tfrac{m}{2k}\right)^2 \tau + \left(n + \tfrac{m}{2k}\right)\tfrac{m'}{2k} \right\} \\ &= \sum_{n\in\mathbb{Z}} \zeta^{\frac{1}{2}\left(n+\frac{m}{2k}\right)^2 k} \exp\left\{\pi\imath\left(n + \tfrac{m}{2k}\right)\tfrac{m'}{k}\right\} \\ &= \exp\left\{\pi\imath\tfrac{mm'}{2k^2}\right\} \zeta^{\frac{m^2}{8k}} \sum_{n\in\mathbb{Z}} \exp\left\{\pi\imath\tfrac{m'n}{k}\right\} \zeta^{\frac{k}{2}n(n+\frac{m}{k})} \end{aligned}.$$

We observe that only nonnegative powers of ζ appear in the above sum. We are interested in computing $\mathrm{ord}_\infty\theta \begin{bmatrix} \frac{m}{k} \\ \frac{m'}{k} \end{bmatrix}$. For this calculation, we may replace characteristics by equivalent ones. Hence we may restrict our attention to only those values of m for which $0 \le \frac{m}{k} \le 1$. It also involves no loss of generality to assume that

$$0 \le \frac{m'}{k} < 2 \text{ if } m \ne 0 \text{ nor } k \text{ and } 0 \le \frac{m'}{k} \le 1 \text{ if } m = 0 \text{ or } k.$$

The lowest power of ζ appears when the index of summation n equals either -1 or 0. We consider three cases.

First, we assume that $0 \leq m < k$. In this case, while only nonnegative powers of ζ appear in the sum, the powers that appear may NOT all be integral. Only integral powers appear if k and m have the same parity, and we may, in this case, rewrite our sum for $m = 0$ as

$$\theta\left[\begin{array}{c} 0 \\ \frac{m'}{k} \end{array}\right](0,\tau) = 1 + 2\cos\left\{\frac{\pi m'}{k}\right\}\zeta^{\frac{k}{2}} + \sum_{n\geq 2k}^{\infty} c_n\zeta^n.$$

(Note that k is assumed to be even.) For $0 < m < k$, we have

$$\theta\left[\begin{array}{c} \frac{m}{k} \\ \frac{m'}{k} \end{array}\right](0,\tau)$$

$$= \zeta^{\frac{m^2}{8k}}\left(\exp\left\{\frac{\pi \imath m m'}{2k^2}\right\} + \exp\left\{\frac{\pi\imath(m-2k)m'}{2k^2}\right\}\zeta^{\frac{k}{2}-\frac{m}{2}} + \sum_{n>\frac{k}{2}-\frac{m}{2}}^{\infty} c_n\zeta^n\right),$$

for some (easily computable) constants c_n. If k and m have different parity, then

$$\begin{aligned}\zeta^{-\frac{m^2}{8k}}\theta\left[\begin{array}{c} \frac{m}{k} \\ \frac{m'}{k} \end{array}\right](0,\tau) &= \textstyle\sum_{n\in\mathbb{Z},\ \text{even}}\zeta^{\frac{kn^2+mn}{2}}\exp\left\{\pi\imath\left(n+\frac{m}{2k}\right)\frac{m'}{k}\right\} \\ &\quad +\zeta^{\frac{1}{2}}\textstyle\sum_{n\in\mathbb{Z},\ \text{odd}}\zeta^{\frac{kn^2+mn-1}{2}}\exp\left\{\pi\imath\left(n+\frac{m}{2k}\right)\frac{m'}{k}\right\}\end{aligned},$$

and only integral powers of ζ appear in each of the last two sums. It follows from these formulae that (for the case where k and m have different parity) we also have that

$$\theta\left[\begin{array}{c} \frac{m}{k} \\ \frac{m'}{k} \end{array}\right](0,\tau) = \zeta^{\frac{m^2}{8k}}\left(\exp\left\{\frac{\pi\imath m m'}{2k^2}\right\} + \mathrm{O}(|\zeta|^{\frac{1}{2}})\right), \zeta \to 0.$$

Second, we take up the case $k = m$ (all the above formulae are valid in this case; for convenience we rewrite some of them, providing additional details that will be needed subsequently). We have

$$\theta\left[\begin{array}{c} 1 \\ \frac{m'}{k} \end{array}\right](0,\tau) = \zeta^{\frac{k}{8}}\left(2\cos\left\{\frac{\pi m'}{2k}\right\} + 2\cos\left\{\frac{3\pi m'}{2k}\right\}\zeta^k + \sum_{n\geq 3k}^{\infty} c_n\zeta^n\right).$$

Finally, for $k < m < 2k$ (technically, we could dispense with this case; it is presented for the convenience of the reader), the relevant formulae (for the case where k and m have the same parity) is

$$\theta\left[\begin{array}{c} \frac{m}{k} \\ \frac{m'}{k} \end{array}\right](0,\tau)$$

$$= \zeta^{\frac{m^2}{8k}} \left(\exp\left\{ \frac{\pi\imath(m-2k)m'}{2k^2} \right\} \zeta^{\frac{k}{2}-\frac{m}{2}} + \exp\left\{ \frac{\pi\imath m m'}{2k^2} \right\} + \sum_{n \geq \frac{k}{2}+\frac{m}{2}}^{\infty} c_n \zeta^n \right).$$

We also need to develop the above machinery for the z derivative of the theta function. We start with

$$\theta'\left[\begin{array}{c} \frac{m}{k} \\ \frac{m'}{k} \end{array}\right](0,\tau) = 2\pi\imath \sum\nolimits_{n\in\mathbb{Z}} \left(n+\tfrac{m}{2k}\right) e^{2\pi\imath\left\{\frac{1}{2}\left(n+\frac{m}{2k}\right)^2\tau + \left(n+\frac{m}{2k}\right)\frac{m'}{2k}\right\}}$$

$$= 2\pi\imath \sum\nolimits_{n\in\mathbb{Z}} \left(n+\tfrac{m}{2k}\right) \zeta^{\frac{1}{2}\left(n+\frac{m}{2k}\right)^2 k} \exp\left\{\pi\imath\left(n+\tfrac{m}{2k}\right)\tfrac{m'}{k}\right\} \cdot$$

$$= 2\pi\imath \exp\left\{\pi\imath\tfrac{mm'}{2k^2}\right\}\zeta^{\frac{m^2}{8k}} \sum\nolimits_{n\in\mathbb{Z}} \left(n+\tfrac{m}{2k}\right) \exp\left\{\pi\imath\tfrac{m'n}{k}\right\}\zeta^{\frac{k}{2}n\left(n+\frac{m}{k}\right)}$$

The leading terms for the expansions (where m, m' and k have the same parity) are:

$$\theta'\left[\begin{array}{c} 0 \\ \frac{m'}{k} \end{array}\right](0,\tau) = -4\pi \sin\left\{\frac{\pi m'}{k}\right\}\zeta^{\frac{k}{2}} + \sum_{n\geq 2k}^{\infty} c_n\zeta^n,$$

(next) for $0 < m < k$,

$$\theta'\left[\begin{array}{c} \frac{m}{k} \\ \frac{m'}{k} \end{array}\right](0,\tau) = 2\pi\imath\zeta^{\frac{m^2}{8k}}$$

$$\times \left(\frac{m}{2k}\, e^{\frac{\pi\imath m m'}{2k^2}} + \frac{m-2k}{2k}\, e^{\frac{\pi\imath(m-2k)m'}{2k^2}} \zeta^{\frac{k}{2}-\frac{m}{2}} + \sum_{n > \frac{k}{2}-\frac{m}{2}}^{\infty} c_n\zeta^n \right),$$

and

$$\theta'\left[\begin{array}{c} 1 \\ \frac{m'}{k} \end{array}\right](0,\tau) = -2\pi\zeta^{\frac{k}{8}} \left(\sin\left\{\frac{\pi m'}{2k}\right\} + \sin\left\{\frac{3\pi m'}{2k}\right\}\zeta^k + \sum_{n\geq 3k}^{\infty} c_n\zeta^n \right).$$

We need certain ratios (as before, m, m' and k have the same parity):

$$\frac{\theta'\left[\begin{array}{c} 0 \\ \frac{m'}{k} \end{array}\right](0,\tau)}{\theta\left[\begin{array}{c} 0 \\ \frac{m'}{k} \end{array}\right](0,\tau)} = -4\pi\left(\sin\left\{\frac{\pi m'}{k}\right\}\zeta^{\frac{k}{2}} - \sin\left\{\frac{2\pi m'}{k}\right\}\zeta^k\right) + o(|\zeta|^{2k}),$$

for $0 < m < k$,

$$\frac{\theta'\left[\begin{array}{c} \frac{m}{k} \\ \frac{m'}{k} \end{array}\right](0,\tau)}{\theta\left[\begin{array}{c} \frac{m}{k} \\ \frac{m'}{k} \end{array}\right](0,\tau)} = \pi\imath\left(\frac{m}{k} - 2\exp\left\{-\frac{\pi\imath m'}{k}\right\}\zeta^{\frac{k}{2}-\frac{m}{2}}\right) + o(|\zeta|^{\frac{k}{2}-\frac{m}{2}+1}),$$

and

$$-\frac{1}{\pi}\frac{\theta'\begin{bmatrix}1\\ \frac{m'}{k}\end{bmatrix}(0,\tau)}{\theta\begin{bmatrix}1\\ \frac{m'}{k}\end{bmatrix}(0,\tau)}$$

$$=\tan\left\{\frac{\pi m'}{2k}\right\}+\left(\frac{\sin\left\{\frac{3\pi m'}{2k}\right\}}{\cos\left\{\frac{\pi m'}{2k}\right\}}-\tan\left\{\frac{\pi m'}{2k}\right\}\frac{\cos\left\{\frac{3\pi m'}{2k}\right\}}{\cos\left\{\frac{\pi m'}{2k}\right\}}\right)\zeta^k+o(|\zeta|^{k+1}).$$

We have shown that for $\chi\in X(k)-\{v_1+v_2\}$,

$$\operatorname{ord}_\infty\theta[\chi]=\frac{m(\chi)^2}{8k},\tag{3.7}$$

for $\chi\in X(k)-\{0,\ v_1,\ v_2\}$,

$$\operatorname{ord}_\infty\theta'[\chi]=\begin{cases}\frac{m(\chi)^2}{8k} \text{ if } m(\chi)\neq 0\\ \frac{k}{2} \text{ if } m(\chi)=0\end{cases}\tag{3.8}$$

and for $\chi\in X(k)-X(2)$,

$$\operatorname{ord}_\infty\frac{\theta'[\chi]}{\theta[\chi]}=\begin{cases}0 \text{ if } m(\chi)=k \text{ and } m'(\chi)<k\\ 0 \text{ if } 0<m(\chi)<k\\ \frac{k}{2} \text{ if } m(\chi)=0 \text{ and } m'(\chi)<k\end{cases}.$$

To compute $\operatorname{ord}_x\theta[\chi]$ for an arbitrary cusp $x\in\hat{\mathbb{Q}}$, we use the transformation formula (3.6) to obtain for arbitrary $\gamma\in\Gamma$,

$$\operatorname{ord}_{\gamma(x)}\theta[\chi]=\operatorname{ord}_x\gamma^*_{\frac{1}{4}}\theta[\chi]=\operatorname{ord}_x\theta[\chi\gamma].$$

For fixed x, we can always choose a γ that maps ∞ to x and hence obtain the formula

$$\operatorname{ord}_x\theta[\chi]=\operatorname{ord}_\infty\theta[\chi\gamma]\quad(\gamma\in\Gamma,\ \gamma(\infty)=x)\tag{3.9}$$

most useful for computations. Similar formulae hold for the computation of $\operatorname{ord}_x\theta'[\chi]$.

The basic fact is

Proposition 2.1. *For each $\chi\in X(k)-\{v_1+v_2\}$, $k>1$, $\theta[\chi]$ is a nontrivial e-automorphic modular form for $\Gamma(k)$, with the factor of automorphy $e=e(\chi)$ satisfying*

$$e^4=c\,\mathcal{K}$$

for some character c on $\Gamma(k)$ with $c^{2k} = 1$ (for even k, we have the stronger conclusion $c^{\frac{k}{2}} = 1$). The divisor[63] *of $\theta[\chi]$ is*

$$(\theta[\chi]) = \prod_{\gamma \in \Gamma/G(k)} P_{\gamma(\infty)}^{\frac{m(\chi\gamma)^2}{8k}}.$$

Proof. The fundamental identity (transformation formula) for the motion $\gamma = \begin{bmatrix} a & b \\ c & d \end{bmatrix} \in \mathrm{SL}(2, \mathbb{Z})$ and the point $\tau \in \mathbb{H}^2$ reads

$$\theta[\chi](0, \gamma(\tau)) = \kappa(\chi, \gamma)\ (c\tau + d)^{\frac{1}{2}} \theta[\chi\gamma](0, \tau),$$

where all relevant terms are interpreted as characteristics rather than classes of characteristics. In the above formula $(c\tau + d)^{\frac{1}{2}}$ is a branch of $\gamma'(\tau)^{-\frac{1}{4}}$. In general $\chi\gamma \neq \gamma$ even for $\chi \in X(k)$ and $\gamma \in \Gamma(k)$. However, in this case

$$\chi\gamma = \pm\chi + 2nv_1 + 2n'v_2$$

for some integers n and n'. If $\chi = \begin{bmatrix} \frac{m}{k}, \\ \frac{m'}{k} \end{bmatrix}$, then

$$\theta[\chi\gamma] = \exp \pi\imath \left\{ \frac{m}{k} n \right\} \theta[\chi].$$

It follows that for $\gamma \in \Gamma(k)$, $\tau \in \mathbb{H}^2$, we have

$$\theta[\chi](0, \gamma(\tau)) e(\gamma, \tau) = \theta[\chi](0, \tau),$$

where

$$e(\gamma, \tau) = \kappa(\chi, \gamma)^{-1} \exp \left\{ -\pi\imath \frac{m}{k} n \right\} (c\tau + d)^{-\frac{1}{2}}.$$

The constants and the function appearing in the definition of the factor of automorphy e depend on the representation of $\gamma \in \Gamma(k)$ by a matrix in $\mathrm{SL}(2, \mathbb{Z})$; however, the function $e(\gamma, \cdot)$ is independent of the choice of matrices. The constant $\kappa(\chi, \gamma)$ is always an $8k$ root of unity and a $4k$ root of unity for even k. We rewrite the last displayed equation as

$$e^4(\gamma, \tau) = c(\gamma)\gamma'(\tau),$$

and conclude that c is a character on $\Gamma(k)$. The divisor $(\Theta[\chi])$ is computed using (3.9) and the fact that all punctures on $\mathbb{H}^2/\Gamma(k)$ are $\Gamma/\Gamma(k)$-equivalent to P_∞.

[63] We shall here and in the sequel (until notified to the contrary) follow the convention implied by this notation. We write divisors multiplicatively on the surface corresponding to the group under study. If f (lower case symbol) is an automorphic form (or function) on $\mathbb{H}^2 \cup \mathbb{Q} \cup \{\infty\}$ for $\Gamma(k)$ (or another subgroup of Γ), then F (corresponding upper case symbol) will denote its projection to the compact Riemann surface $S = (\mathbb{H}^2 \cup \mathbb{Q} \cup \{\infty\})/\Gamma(k)$ and (F) will denote the divisor of F. We will use this notation even when F is a multivalued object on S. The divisor of a function φ on $\mathbb{H}^2$ that is an automorphic form for a group G will be written using the points on $\overline{\mathbb{H}^2/G}$ since the latter is G-invariant.

Remark 2.2. Let χ_1 and $\chi_2 \in X(k) - \{v_1 + v_2\}$. It is a nontrivial problem to determine when $e(\chi_1) = e(\chi_2)$.

Definition 2.3. We will call the θ-constant $\theta[\chi]$ for $\chi \in X(k) - \{v_1 + v_2\}$ a *basic* automorphic function for $\Gamma(k)$.

The analogue of the last proposition for the derivative theta constants is contained in

Proposition 2.4. (a) *The theta constant derivative* $\theta' \begin{bmatrix} 1 \\ 1 \end{bmatrix}$ *is a holomorphic automorphic form of weight* $\frac{3}{4}$ *for the group* Γ *with its only zero at the one* Γ*-equivalence class of cusps.*
(b) *For* χ *an even*[64] *integral characteristic,* $\theta'[\chi]$ *is identically zero.*
(c) *Let* $k > 2$. *For* $\chi \in X_o(k)$, $\theta'[\chi]$ *is a holomorphic automorphic form of weight* $\frac{3}{4}$ *for the group* $\Gamma(k)$. *If* k *is odd,* $\frac{\chi(\Gamma(k))}{4}$ *of its zeros are at cusps and* $\frac{\chi(\Gamma(k))}{2}$ *at ordinary points (that is, in* $\mathbb{H}^2$*). If* k *is even,* $\frac{\chi(\Gamma(k))}{4} + \frac{k}{2}\varphi\left(\frac{k}{2}\right)$ *of its zeros are at cusps and the remaining* $\frac{\chi(\Gamma(k))}{2} - \frac{k}{2}\varphi\left(\frac{k}{2}\right)$ *at ordinary points.*
(d) *Let* $k > 2$. *For* $\chi \in X_o(k)$, $\frac{\theta'[\chi]}{\theta[\chi]}$ *is a holomorphic automorphic form of weight* $\frac{1}{2}$ *for the group* $\Gamma(k)$ *with all of its* $\frac{\chi(\Gamma(k))}{2}$ *zeros at ordinary points. Furthermore,* $\left(\frac{\theta'[\chi]}{\theta[\chi]}\right)^2$ *is a modular 1-form*[65] *(with trivial character).*
(e) *For all integers,* $k \geq 2$,

$$\sum_{\chi \in X_o(k)} \left(\frac{\theta'[\chi]}{\theta[\chi]}\right)^2 = 0 = \sum_{\chi \in X(k) - \{v_1 + v_2\}} \left(\frac{\theta'[\chi]}{\theta[\chi]}\right)^2.$$

In the above, $\chi(\Gamma(k))$ is the orbifold Euler characteristic of the group $\Gamma(k)$ and φ, Euler's number theoretic function.

Proof. The basic fact in the proof of the proposition is formula (3.8). We prove parts (c) and (e), leaving the proofs of the other parts of the proposition to the reader. The function $\theta'[\chi]$ is certainly a holomorphic automorphic form of weight $\frac{3}{4}$ for the group $\Gamma(k)$ and hence has $\frac{3\chi(\Gamma(k))}{4}$ zeros. For odd k, the form $\theta'[\chi]$ has the same order at each cusp as the $\frac{1}{4}$-form $\theta[\chi]$. Hence it has $\frac{\chi(\Gamma(k))}{4}$ of its zeros at cusps and it must have the remaining $\frac{\chi(\Gamma(k))}{2}$ of its zeros at ordinary points. For even k, the story is slightly more complicated. At $\varphi\left(\frac{k}{2}\right)$ punctures (those corresponding to characteristics in the tower over 0), the $\frac{3}{4}$-form $\theta'[\chi]$ has order $\frac{k}{2}$, whereas the $\frac{1}{4}$-form $\theta[\chi]$ has order 0; at the other punctures $\theta'[\chi]$ and $\theta[\chi]$ have the same orders. Hence $\theta'[\chi]$ has $\frac{\chi(\Gamma(k))}{4} + \frac{k}{2}\varphi\left(\frac{k}{2}\right)$ zeros at cusps. This finishes the proof of part (c). To prove

[64] In the classical sense.

[65] Its projection to $\mathbb{H}^2$ has simple poles at each of the punctures of $\mathbb{H}^2/\Gamma(k)$.

(e), we note that by part (d), for all $\chi \in X(k)$, $\chi \neq v_1 + v_2$, $\left(\frac{\theta'[\chi]}{\theta[\chi]}\right)^2$ is a 1-form for $\Gamma(k)$. Hence the sum of its residues must be zero. For each cusp x for $\Gamma(k)$,

$$\operatorname{Res}_x \left(\frac{\theta'[\chi]}{\theta[\chi]}\right)^2 = \operatorname{width}(x) \left(\frac{\theta'[\chi\gamma]}{\theta[\chi\gamma]}(0, \imath\infty)\right)^2, \gamma \in \Gamma, \gamma(\infty) = x.$$

Now all the cusps of $\Gamma(k)$ have the same width, and the classes of characteristics in $X_o(k)$ are in one-to-one correspondence with the punctures of $\mathbb{H}^2/\Gamma(k)$. Hence for all integers $k > 2$, we have the first equality in

$$(3.10) \qquad \sum_{\chi \in X_o(k)} \left(\frac{\theta'[\chi]}{\theta[\chi]}(0, \imath\infty)\right)^2 = 0 = \sum_{\chi \in X(k) - \{v_1 + v_2\}} \left(\frac{\theta'[\chi]}{\theta[\chi]}(0, \imath\infty)\right)^2.$$

Since

$$X(k) - \{v_1 + v_2\} = \bigcup_{k' | k, k' \neq 1} X_o(k'),$$

the second equality also holds. Finally the modular 1-forms in part (e) are Γ-invariant. We have shown that the residue (at ∞) vanishes. Hence we have produced cusp forms for the modular group, that is, the zero function.

As a consequence of the last proposition we conclude

Corollary 2.5. *Let $k > 2$ and $\chi \in X_o(k)$.*
(a) *For odd k, the theta constant $\theta'[\chi]$ has precisely $\frac{\chi(\Gamma(k))}{2}$ $\Gamma(k)$-inequivalent zeros in $\mathbb{H}^2$.*
(b) *For even k, the theta constant $\theta'[\chi]$ has $\frac{\chi(\Gamma(k))}{2} - \frac{k}{2}\varphi\left(\frac{k}{2}\right)$ $\Gamma(k)$-inequivalent zeros in $\mathbb{H}^2$.*

It is of interest but quite difficult to determine the location of these zeros. We embark on a slightly different direction and study the residues of the modular 1-forms for $\Gamma(k)$, produced above.

Our computations show that for $k \geq 3$,

$$\frac{\theta'[\chi]}{\theta[\chi]}(0, \imath\infty) = \begin{cases} -\pi \tan\left(\frac{\pi m'(\chi)}{2k}\right) \text{ if } m(\chi) = k \text{ and } m'(\chi) < k \\ \frac{\pi \imath m(\chi)}{k} \text{ if } 0 < m(\chi) < k \\ 0 \text{ if } m(\chi) = 0 \text{ and } m'(\chi) \leq k \end{cases}.$$

Recalling our work on towers, we obtain

Corollary 2.6. (a) *For each odd integer $k \geq 3$,*

$$\sum_{l=0}^{\frac{k-3}{2}} \left(\frac{1+2l}{k}\right)^2 N(k, 1+2l) = \sum_{l'=0,1,\ldots,\frac{k-3}{2},\ (1+2l',k)=1} \tan^2 \frac{\pi(1+2l')}{2k}$$

and[66]

$$\frac{(k-2)(k-1)}{6} = \sum_{l'=0}^{\frac{k-3}{2}} \tan^2 \frac{\pi(1+2l')}{2k}. \tag{3.11}$$

(b) *For all integers* $k \geq 2$,

$$\frac{(k-1)(2k-1)}{3} = \sum_{m'=1}^{k-1} \tan^2 \frac{\pi m'}{2k}.$$

(c) *For even integers* $k \geq 2$,

$$\sum_{l=1}^{k-1} \left(\frac{l}{k}\right)^2 N(2k, 2l) = \sum_{m'=1,3,\ldots,k-1,\ (m',k)=1} \tan^2 \frac{\pi m'}{2k}.$$

(d) *For odd integers* $k \geq 3$,

$$\sum_{l=1}^{k-1} \left(\frac{l}{k}\right)^2 N(2k, 2l) = \sum_{m'=1,2,\ldots,\frac{k-1}{2},\ (m',k)=1} \tan^2 \frac{\pi m'}{k}.$$

Proof. We are evaluating the sums in (3.10). For the first equality in (a), we use the first sum; for the second equality, we use the second sum and recall that the tower over $\frac{m}{k}$, m odd $1 \leq m < k$, contains k elements. For the rest of the claims, we use (3.10) with k replaced by $2k$; for part (b), the second sum; and for parts (c) and (d), the first.

The integers $N(k, m)$ are described by Lemma 4.5 and Remark 4.10, both of Chapter 2.

Remark 2.7. It was not our aim to derive trigonometric identities; they are a byproduct of our work and are presented to show the usefulness of theta function theory. Our work does not explain why the trigonometric sums are rational, in fact, many times, integers.

2.2. Automorphic forms for $\Gamma(k)$.

Theorem 2.8. *Fix an integer* $k > 1$. *Let* $\chi = \begin{bmatrix} \frac{m}{k} \\ \frac{m'}{k} \end{bmatrix} \in Z(k) - \{v_1 + v_2\}$, *with the usual restrictions on the integers* m *and* m'.[67]
(a) *For odd* k, $\theta^{8k}[\chi]$ *is a cusp* $2k$- *form for* $\Gamma(k)$ *(that is, belongs to the Banach space* $\mathbb{A}_{2k}(\mathbb{H}^2, \Gamma(k))$*) provided* $\chi \in X(k)$.

[66] Doron Zeilberger, after a lecture given by one of the authors where the next equality was demonstrated, posted on his web journal on October 5, 1999, a note entitled *The revenge of the plain mathematician* that contains a proof of the equality using properties of Chebychev polynomials. L. Ehrenpreis has also informed us that a more elementary argument than Zeilberger's is also apparently in circulation.

[67] Thus $0 \leq m \leq k$, $0 \leq m' \leq 2k-1$ if $m \neq 0$ and k, and $0 \leq m' \leq k$ for $m = 0$ or k. Further, in the most useful cases ($\chi \in X(k)$), m and m' have the same parity as k.

(b) *For even k, $\theta^{4k}[\chi]$ is a modular k-form for $\Gamma(k)$ (that is, belongs to the vector space $\mathbb{A}_k^+(\mathbb{H}^2,\Gamma(k))$) provided $\chi \in X(k)$. This modular form is a cusp form if and only if $l(\chi)$ is odd. (The level $l(\chi)$ of the characteristic χ has been defined in §2.6 of Chapter 2.)*
(c) *If $k \equiv 0 \bmod 4$ or $k=2$, then $\theta^{2k}[\chi]$ is a modular $\frac{k}{2}$-form for $\Gamma(k)$ (that is, belongs to $\mathbb{A}_{\frac{k}{2}}^+(\mathbb{H}^2,\Gamma(k))$) provided $\chi \in X(k)$. This modular form is a cusp form if and only if $l(\chi)$ is odd.*
(d) *Let $\varphi = \theta^{2k}[\chi]\theta^{2k}[\chi+v_1]\theta^{2k}[\chi+v_2]\theta^{2k}[\chi+v_1+v_2]$. Then for $\gamma = \begin{bmatrix} a & b \\ c & d \end{bmatrix} \in \Gamma(k)$,*

$$\varphi(\gamma(\tau))(\gamma'(\tau))^{2k} = \exp\{\pi \imath k(ab+cd)\}\varphi(\tau),\ \ \tau \in \mathbb{H}^2.$$

Further, for even k, $\varphi \in \mathbb{A}_{2k}(\mathbb{H}^2,\Gamma(k))$, and for odd k, $\varphi^2 \in \mathbb{A}_{4k}(\mathbb{H}^2,\Gamma(k))$.

Proof. Let $r = 8k$ if k is odd and let $r = 4k$ otherwise. Let m and m' be nonnegative integers with the same parity as k. The function $\theta^r \begin{bmatrix} \frac{m}{k}, \\ \frac{m'}{k} \end{bmatrix}$ is holomorphic and nonzero on $\mathbb{H}^2$. Since the theta constant is raised to a power which is a multiple of $2k$, the resulting function depends only on the class of the theta characteristic χ. We also know that $\kappa^r(\chi,\gamma)=1$ for all $\gamma \in \Gamma(k)$ and all characteristics χ whose classes are fixed by $\Gamma(k)$, so $\theta^r \begin{bmatrix} \frac{m}{k}, \\ \frac{m'}{k} \end{bmatrix}$ is an automorphic $\frac{r}{4}$-form for the group $\Gamma(k)$. It remains to show that this function is a holomorphic form. We will do so by showing that $\theta^r \begin{bmatrix} \frac{m}{k}, \\ \frac{m'}{k} \end{bmatrix}$ extends to $\mathbb{H}^2 \cup \mathbb{Q} \cup \{\infty\}$ by examining the divisor $\left(\Theta^r \begin{bmatrix} \frac{m}{k}, \\ \frac{m'}{k} \end{bmatrix}\right)$ of the projection $\Theta^r \begin{bmatrix} \frac{m}{k}, \\ \frac{m'}{k} \end{bmatrix}$ to $(\mathbb{H}^2 \cup \mathbb{Q} \cup \{\infty\})/\Gamma(k)$ of this extension. It is easy to see (from the Fourier series expansion) that $\theta^r \begin{bmatrix} \frac{m}{k}, \\ \frac{m'}{k} \end{bmatrix}$ is holomorphic at ∞. For k odd (so $r = 8k$),

$$\operatorname{ord}_{P_\infty} \Theta^{8k} \begin{bmatrix} \frac{m}{k}, \\ \frac{m'}{k} \end{bmatrix} = \begin{cases} m^2 - 2k \text{ if } 1 \le m \le k \\ (2k-m)^2 - 2k \text{ if } k \le m \le 2k-1 \end{cases}. \tag{3.12}$$

For even k (so $r = 4k$; recall that m is also even in this case),

$$\operatorname{ord}_{P_\infty} \Theta^{4k} \begin{bmatrix} \frac{m}{k}, \\ \frac{m'}{k} \end{bmatrix} = \begin{cases} \frac{m^2}{2} - k \text{ if } 0 \le m \le k \\ \frac{(2k-m)^2}{2} - k \text{ if } k \le m \le 2k-2 \end{cases}. \tag{3.13}$$

In all four of the above formulae, for $m = k$, we must assume that $m' \neq k$. Let $C = \begin{bmatrix} a & b \\ c & d \end{bmatrix} \in \mathrm{SL}(2,\mathbb{Z})$ be arbitrary. The transformation formula for theta functions implies immediately that

$$\theta^r \begin{bmatrix} \frac{m}{k}, \\ \frac{m'}{k} \end{bmatrix} = (C^{-1})^*_{\frac{r}{4}} \left(\theta^r \left(\begin{bmatrix} \frac{m}{k}, \\ \frac{m'}{k} \end{bmatrix} C \right) \right) \kappa^r \left(\begin{bmatrix} \frac{m}{k}, \\ \frac{m'}{k} \end{bmatrix}, C \right)$$

and hence also

$$\mathrm{ord}_z \theta^r \begin{bmatrix} \frac{m}{k}, \\ \frac{m'}{k} \end{bmatrix} = \mathrm{ord}_{C^{-1}(z)} \theta^r \begin{bmatrix} a\frac{m}{k} + c\frac{m'}{k} - ac \\ b\frac{m}{k} + d\frac{m'}{k} + bd \end{bmatrix},$$

for all $z \in \mathbb{H}^2 \cup \mathbb{Q} \cup \{\infty\}$. Similar statements hold for the values and Fourier series expansions for the forms under consideration.

We are now ready to prove that $\theta^r \begin{bmatrix} \frac{m}{k}, \\ \frac{m'}{k} \end{bmatrix} \in \mathbb{A}^+_{\frac{r}{4}}(\mathbb{H}^2, \Gamma(k))$. Since the automorphic form $\theta^r \begin{bmatrix} \frac{m}{k}, \\ \frac{m'}{k} \end{bmatrix}$ is regular (holomorphic and nonzero) on $\mathbb{H}^2$, we need only examine its behavior at the cusps. If $x \in \mathbb{Q} \cup \{\infty\}$ is a parabolic fixed point of $\Gamma(k)$, there exists a $C \in \mathrm{SL}(2,\mathbb{Z})$ with $C^{-1}(x) = \infty$. It follows that for all $\chi \in Z(k) - \{v_1 + v_2\}$, $\mathrm{ord}_x \theta^r[\chi] = \mathrm{ord}_\infty \theta^r[\chi C]$. In particular,

$$\mathrm{ord}_x \theta^r \begin{bmatrix} \frac{m}{k}, \\ \frac{m'}{k} \end{bmatrix} = \mathrm{ord}_\infty \theta^r \begin{bmatrix} \frac{m_1}{k}, \\ \frac{m'_1}{k} \end{bmatrix}$$

for some pair of integers m_1 and m'_1 of the same parity as k. It then also follows that

$$\mathrm{ord}_{P_x} \Theta^r \begin{bmatrix} \frac{m}{k}, \\ \frac{m'}{k} \end{bmatrix} \geq -\frac{r}{4},$$

which shows that $\theta^r \begin{bmatrix} \frac{m}{k}, \\ \frac{m'}{k} \end{bmatrix}$ is a holomorphic $\frac{r}{4}$-form. For odd k we have the stronger statement

$$\mathrm{ord}_{P_x} \Theta^{8k} \begin{bmatrix} \frac{m}{k}, \\ \frac{m'}{k} \end{bmatrix} \geq 1 - 2k,$$

which shows that $\theta^{8k} \begin{bmatrix} \frac{m}{k}, \\ \frac{m'}{k} \end{bmatrix}$ is a $2k$-cusp form. The proof of the first statement in part (c) is similar. In this case $r = 2k$.

We verify next the last statement of parts (b) and (c). If χ is of odd level, then for each $\gamma \in \Gamma$, the characteristic $\chi\gamma$ is also of odd level. In this case $\theta^r[\chi]$ vanishes at all the parabolic fixed points of $\Gamma(k)$ and is hence a cusp form. If χ is of even level l, then there exists a $\gamma \in \Gamma$ with $\chi\gamma = \begin{bmatrix} 0 \\ \frac{\epsilon}{l} \end{bmatrix}$, where $\epsilon = 4$ if $\frac{l}{2}$ is odd and $\epsilon = 2$ if $\frac{l}{2}$ is even. Now θ^r is not zero at the parabolic fixed point $\gamma(\infty)$, and hence cannot be a cusp form. This completes the proof of parts (a), (b) and (c). The proof of part (d) is left to the reader.

Remark 2.9. The automorphic form φ is identically zero when χ is in the adherent quadruple determined by 0 (thus when $\chi = 0,\ v_1$ or v_2).

Remark 2.10. We have used in the preceding proof that for $\chi \in X(k)$ and $\gamma \in \Gamma$,

$$\text{ord}_{\gamma(\infty)}\theta^r[\chi] = \text{ord}_\infty\theta^r[\chi\gamma]. \tag{3.14}$$

The last formula is valid for all $r \in \mathbb{Z}$ once we regard $\theta^r[\chi]$ as a multiplicative $\frac{r}{4}$-form for $\Gamma(k)$. We will use the above formula for the equivalence class of the characteristic χ_o. As γ varies over Γ (in reality only over the set of cosets of $\Gamma/\Gamma(k)$), $\gamma(\infty)$ varies over the punctures of $\mathbb{H}^2/\Gamma(k)$. As γ varies over Γ (in reality only over the set of equivalence classes of $G(k)\backslash\Gamma$), $\chi_o\gamma$ varies over the classes in $X_o(k)$.

We apply the above remark to compute the degree of the divisor of the projection of $\theta^r[\chi_o]$ to $\overline{\mathbb{H}^2/\Gamma(k)}$ and conclude that for all $r \in \mathbb{Z}$,

$$\sum_{\chi\in X_o(k)} \text{ord}_{P_\infty}\Theta^{4r}[\chi] = 2r(p(k) - 1). \tag{3.15}$$

In particular, if we choose $r = 8k$ so that the order at ∞ of $\theta^{8k}[\chi]$ is m^2 for $\chi = \begin{bmatrix} \frac{m}{k} \\ \frac{m'}{k} \end{bmatrix}$, we obtain

Corollary 2.11. *For $k > 1$, we have*

$$\sum_{\chi\in X_o(k)} m(\chi)^2 = \frac{k^2 n(k)}{3}.$$

Proof. We use (3.15), (3.12) or (3.13), and the formula

$$(k-6)n(k) = 12(p(k) - 1).$$

It is the above corollary and the equation which preceded it that shows that the set of characteristics $X_o(k)$ is in fact naturally related to the function theory of $\Gamma(k)$ and not just an accident of numerology. If we consider the case of an odd prime k, the corollary says that

$$k\sum_{l=0}^{\frac{k-3}{2}}(2l+1)^2 + k^2\frac{k-1}{2} = \frac{k^2(k^2-1)}{6},$$

which is, of course, an identity for every odd integer k.

2.3. Meromorphic automorphic functions for $\Gamma(k)$. For fixed $k \in \mathbb{Z}^+$, fixed r, a positive integral multiple of

$$\begin{array}{ll} 8k \text{ for } k \equiv 1 & \text{mod } 2 \\ 4k \text{ for } k \equiv 2 & \text{mod } 2, k \neq 2 \\ 2k \text{ for } k \equiv 0 & \text{mod } 2 \text{ or } k = 2 \end{array},$$

and variable integers m, m', m_1, and m'_1 of the same parity as k, the ratio

$$\frac{\theta^r \begin{bmatrix} \frac{m}{k}, \\ \frac{m'}{k} \end{bmatrix}}{\theta^r \begin{bmatrix} \frac{m_1}{k}, \\ \frac{m'_1}{k} \end{bmatrix}}$$

defines a meromorphic function on $\overline{\mathbb{H}^2/\Gamma(k)}$ whose divisor (of zeros and poles) is supported at the punctures of $\mathbb{H}^2/\Gamma(k)$.

Let k be odd. Fix an equivalence class of characteristics $\chi = \begin{bmatrix} \frac{m}{k}, \\ \frac{m'}{k} \end{bmatrix} \in X(k)$. Then

$$f = \frac{\theta^{6k} \begin{bmatrix} \frac{m}{k}, \\ \frac{m'}{k} \end{bmatrix}}{\theta^{2k} \begin{bmatrix} \frac{m}{k}+1 \\ \frac{m'}{k} \end{bmatrix} \theta^{2k} \begin{bmatrix} \frac{m}{k} \\ \frac{m'}{k}+1 \end{bmatrix} \theta^{2k} \begin{bmatrix} \frac{m}{k},+1 \\ \frac{m'}{k}+1 \end{bmatrix}}$$

defines a meromorphic function on $\mathbb{H}^2$ with the property that for $\gamma = \begin{bmatrix} a & b \\ c & d \end{bmatrix} \in \Gamma(k)$, we have

$$f(\gamma(\tau)) = \exp\{\pi \imath k(ab+bc+cd)\} f(\tau), \text{ for all } \tau \in \mathbb{H}^2.$$

Hence f^2 is a meromorphic automorphic function[68] on $\mathbb{H}^2$ for $\Gamma(k)$. For all $\gamma \in \Gamma$, there exist odd integers m_1 and m'_1 such that

$$f \circ \gamma = \text{constant} \frac{\theta^{6k} \begin{bmatrix} \frac{m_1}{k}, \\ \frac{m'_1}{k} \end{bmatrix}}{\theta^{2k} \begin{bmatrix} \frac{m_1}{k}+1 \\ \frac{m'_1}{k} \end{bmatrix} \theta^{2k} \begin{bmatrix} \frac{m_1}{k} \\ \frac{m'_1}{k}+1 \end{bmatrix} \theta^{2k} \begin{bmatrix} \frac{m_1}{k},+1 \\ \frac{m'_1}{k}+1 \end{bmatrix}}.$$

Using the formula

$$\text{ord}_{P_\infty} F^2 = k(2m-k)$$

[68]The function f itself is not automorphic, because $B^k \in \Gamma(k)$ and

$$f \circ B^k = -f.$$

and the observation that for all $\gamma \in \Gamma$,

$$\operatorname{ord}_{\gamma(\infty)} f = \operatorname{ord}_{\infty} f \circ \gamma,$$

it is routine to compute (F^2). Applying this computation to $\chi = \chi_o \in X(k)$ and using the fact that a principal divisor has degree zero, we obtain for odd k the formula in

Corollary 2.12. *For all integers $k > 2$,*

$$\sum_{\chi \in X_o(k)} m(\chi) = \frac{k}{2} n(k).$$

Proof. The function f could have, alternately, been defined as

$$f = \frac{\theta^{8k}[\chi]}{\theta^{2k}[\chi]\theta^{2k}[\chi + v_1]\theta^{2k}[\chi + v_2]\theta^{2k}[\chi + v_1 + v_2]}.$$

This last definition clearly is valid also for even k, in which case f is an automorphic function for $\Gamma(k)$. Thus we can compute the divisor (F). In the particular case of an odd prime k, the above corollary states

$$k \sum_{l=1}^{\frac{k-1}{2}} (2l+1) + k\frac{k-1}{2} = \frac{k(k^2-1)}{4}.$$

Remark 2.13. The last corollary does *not* hold for $k = 2$. The function f produced is constant in this case, either identically 0 or identically ∞.

2.4. Evaluation of automorphic functions at cusps. In general we are working with analytic functions f on $\mathbb{H}^2$ of the form

$$f : \tau \mapsto \frac{\theta^l \begin{bmatrix} \epsilon \\ \epsilon' \end{bmatrix}}{\theta^l \begin{bmatrix} \delta \\ \delta' \end{bmatrix}}(0, \tau),$$

where $l \in \mathbb{Z}^+$, ϵ, ϵ', δ and δ' are rational numbers. This function has a continuous (with values in $\hat{\mathbb{C}}$) extension to $\mathbb{H}^2 \cup \mathbb{Q} \cup \{\infty\}$. Its value at ∞ ($\imath\infty$ in some notation) is easily computed. We also must compute its value at other rationals. Let $x \in \mathbb{Q}$. Choose a $C \in \mathrm{SL}(2, \mathbb{Z})$ with $C(x) = \infty$. Then for all $\tau \in \mathbb{H}^2 \cup \mathbb{Q} \cup \{\infty\}$, we have

$$f(C(\tau)) = \frac{\theta^l \begin{bmatrix} \epsilon \\ \epsilon' \end{bmatrix}}{\theta^l \begin{bmatrix} \delta \\ \delta' \end{bmatrix}}(0, C(\tau)) = \frac{\theta^l \left[\begin{bmatrix} \epsilon \\ \epsilon' \end{bmatrix} C\right]}{\theta^l \left[\begin{bmatrix} \delta \\ \delta' \end{bmatrix} C\right]}(0, \tau) \frac{\kappa^l\left(\begin{bmatrix} \epsilon \\ \epsilon' \end{bmatrix}, C\right)}{\kappa^l\left(\begin{bmatrix} \delta \\ \delta' \end{bmatrix}, C\right)}.$$

Thus in particular,

$$f(x) = \frac{\theta^l \begin{bmatrix} \epsilon \\ \epsilon' \end{bmatrix}}{\theta^l \begin{bmatrix} \delta \\ \delta' \end{bmatrix}}(0, x) = \frac{\theta^l \left[\begin{bmatrix} \epsilon \\ \epsilon' \end{bmatrix} C^{-1} \right]}{\theta^l \left[\begin{bmatrix} \delta \\ \delta' \end{bmatrix} C^{-1} \right]}(0, \infty) \frac{\kappa^l \left(\begin{bmatrix} \epsilon \\ \epsilon' \end{bmatrix}, C^{-1} \right)}{\kappa^l \left(\begin{bmatrix} \delta \\ \delta' \end{bmatrix}, C^{-1} \right)}.$$

Remark 2.14. There are many complications and some simplifications in using the above formula.
1. The formula is in terms of characteristics rather than classes. This is particularly troublesome for $l = 2$.
2. It is not necessary to evaluate $\kappa^l \left(\begin{bmatrix} 0 \\ 0 \end{bmatrix}, C^{-1} \right)$, normally needed for calculating the constant $\kappa^l \left(\begin{bmatrix} \delta \\ \delta' \end{bmatrix}, C^{-1} \right)$, since the same factor appears in both the numerator and denominator for the expression for $f(x)$.

2.5. Automorphic forms and functions for $G(k)$. Our theory of characteristics and towers over a point $\frac{m}{k}$ allows us to construct in a straightforward manner automorphic forms and functions for $G(k)$. We have seen that $G(k)$ operates on the tower over $\frac{m}{k}$ albeit not necessarily transitively, so that for fixed k and appropriate r, $\prod_{\chi \in \mathcal{T}_{\frac{m}{k}}} \theta^r[\chi] = \omega$ is an automorphic form for $G(k)$. In addition (for appropriate r again), for any $\chi \in \mathcal{T}_1$, $\theta^r[\chi]$ is an automorphic form for $G(k)$. Suitable quotients give rise to automorphic functions.

2.6. Automorphic forms and functions for $\Gamma_o(k)$. The group $\Gamma_o(k)$ has few invariant characteristics. To obtain automorphic functions for these groups, we must study products of basic automorphic functions for $\Gamma(k)$.

Proposition 2.15. *The finite product*

$$\varphi = \prod_{\chi \in \mathcal{T}_{\frac{k}{k}}} \theta[\chi]$$

is a nontrivial e-automorphic modular form for the group $\Gamma_o(k)$ with

$$e = \prod_{\chi \in \mathcal{T}_{\frac{k}{k}}} e(\chi).$$

In particular,

$$e^4 = c\kappa^{N(k,k)}$$

(thus the weight of e is $\frac{N(k,k)}{4}$), and for prime k,

$$(\varphi) = P_0^{\frac{(k-1)(k-2)}{48}} P_\infty^{\frac{k-1}{16}}.$$

Remark 2.16. The computation of (Φ) the divisor of the projection Φ of φ to $\overline{\mathbb{H}^2/\Gamma_o(k)}$, must take into account the torsion in $\Gamma_o(k)$. For primes this can be easily carried out. The fact that $\frac{k-1}{2} \notin \mathbb{Z}$ does not cause any problems. For $k = 3$, for example,

$$(\Phi) = \frac{1}{P_\infty^{\frac{1}{8}} P_0^{\frac{5}{24}} Q^{\frac{1}{6}}},$$

where Q is the projection to $\mathbb{H}^2/\Gamma_o(3)$ of the fixed point of the automorphism of $\mathbb{H}^2/\Gamma(3)$ induced by the motion $\tau \mapsto \tau + 1$. In the general case of an odd prime k,

$$(\Phi) = P_0^{\frac{(k-1)(k-8)}{48}} P_\infty^{\frac{1-k}{16}} \mathcal{Q}_2^{\frac{1-k}{16}} \mathcal{Q}_3^{\frac{1-k}{12}},$$

where $\mathcal{Q}_2$ and $\mathcal{Q}_3$ are the 2 and 3 torsion divisors on $\overline{\mathbb{H}^2/\Gamma_o(k)}$: $\mathcal{Q}_2$ consists of 0 or 2 points, and $\mathcal{Q}_3$ of 0, 1 or 2 points (the branch values of the covering map of $\mathbb{H}^2/\Gamma_o(k)$ by $\mathbb{H}^2/\Gamma(k)$).

If we view φ as a $G(k)$-form,

$$(\varphi) = \prod_{l=1}^{\frac{k-1}{2}} P_{\frac{l}{k}}^{\frac{k-1}{16}} P_{\frac{k}{l}}^{\frac{(k-1)(k-2)}{48}} \text{ and } (\Phi) = \prod_{l=1}^{\frac{k-1}{2}} P_{\frac{l}{k}}^{\frac{1-k}{16}} P_{\frac{k}{l}}^{\frac{(k-1)(k-8)}{48}}.$$

2.7. The structure of $\oplus_{q=0}^{\infty}\mathbb{A}_q(\mathbb{H}^2,\Gamma)$ and $\oplus_{q=0}^{\infty}\mathbb{A}_q^+(\mathbb{H}^2,\Gamma)$. As an illustration, we describe the structure of the graded algebras $\oplus_{q=0}^{\infty}\mathbb{A}_q(\mathbb{H}^2,\Gamma)$ of cusp forms, and $\oplus_{q=0}^{\infty}\mathbb{A}_q^+(\mathbb{H}^2,\Gamma)$ of modular forms for $\Gamma = \mathrm{PSL}(2,\mathbb{Z})$.

Throughout this subsection k and q are integers, with k positive and q nonnegative. It is quite obvious that

$$\mathbb{A}_0(\mathbb{H}^2,\Gamma(k) = \{0\} \text{ and } \mathbb{A}_0^+(\mathbb{H}^2,\Gamma(k) = \mathbb{C}.$$

We hence assume for the rest of the subsection that $q \in \mathbb{Z}^+$. It is a consequence of the Riemann-Roch theorem that for the modular group we have

$$\begin{aligned} \dim\ \mathbb{A}_q(\mathbb{H}^2,\Gamma) &= \max\left\{0, -q + \left\lfloor \tfrac{q}{2} \right\rfloor + \left\lfloor \tfrac{2q}{3} \right\rfloor\right\} \\ &= \begin{cases} 0 & \text{if } q = 1 \\ \left\lfloor \frac{q}{6} \right\rfloor & \text{if } q \not\equiv 1 \mod 6 \\ \left\lfloor \frac{q}{6} \right\rfloor - 1 & \text{if } q \equiv 1 \mod 6 \end{cases} \end{aligned}$$

and

$$\dim\ \mathbb{A}_q^+(\mathbb{H}^2,\Gamma) = 1 - q + \left\lfloor \frac{q}{2} \right\rfloor + \left\lfloor \frac{2q}{3} \right\rfloor = \begin{cases} \left\lfloor \frac{q}{6} \right\rfloor + 1 \text{ if } q \not\equiv 1 \mod 6 \\ \left\lfloor \frac{q}{6} \right\rfloor \text{ if } q \equiv 1 \mod 6 \end{cases}.$$

While for $k > 1$,

$$\dim\ \mathbb{A}_1(\mathbb{H}^2,\Gamma(k)) = p(k),\ \dim\ \mathbb{A}_1^+(\mathbb{H}^2,\Gamma(k)) = p(k) + n(k) - 1,$$

and for $q \geq 2$,

$$\dim\ \mathbb{A}_q(\mathbb{H}^2, \Gamma(k)) = \ (2q-1)(p(k)-1) + (q-1)n(k) = \tfrac{k(2q-1)-6}{12} n(k),$$

$$\dim\ \mathbb{A}_q^+(\mathbb{H}^2, \Gamma(k)) = \ (2q-1)(p(k)-1) + qn(k) = \tfrac{k(2q-1)+6}{12} n(k)$$

There is a hint to the number theoretic significance of the modular group in the next proposition.

Definition 2.17. For $q \in \mathbb{Z}^+$, let $LM(q)$ denote the number of distinct pairs of nonnegative integers (l, m) that solve the equation

$$2l + 3m = q,$$

and let $r(q) = |LM(q)|$ denote the cardinality of this (finite) set.

Proposition 2.18. *For each $q \in \mathbb{Z}^+$, we have*

$$\dim\ \mathbb{A}_q^+(\mathbb{H}^2, \Gamma) = r(q).$$

Proof. It is easily seen that

$$\dim\ \mathbb{A}_q^+(\mathbb{H}^2, \Gamma) = r(q) \text{ for } q = 1, 2, 3, 4, 5, 6.$$

It is also clear that

$$\dim\ \mathbb{A}_{q+6}^+(\mathbb{H}^2, \Gamma) = \dim\ \mathbb{A}_q^+(\mathbb{H}^2, \Gamma) + 1 \text{ for all } q \in \mathbb{Z}^+.$$

We define an injective map $LM(q) \to LM(q+6)$ by sending the ordered pair (l, m) to $(l, m+2)$. The range of this map contains all of $LM(q+6)$ with the exception of pairs of the form $(l, 0)$ and $(l, 1)$. Precisely one such pair belongs to $LM(q+6)$; thus we see that

$$r(q+6) = r(q) + 1,$$

which allows us to conclude the proof of the proposition by induction.

Lemma 2.19. *The values of $\wp$ at the three half periods are*

$$e_1 = \frac{1}{3}\frac{\theta'''\begin{bmatrix}1\\1\end{bmatrix}}{\theta'\begin{bmatrix}1\\1\end{bmatrix}} - \frac{\theta''\begin{bmatrix}1\\0\end{bmatrix}}{\theta\begin{bmatrix}1\\0\end{bmatrix}},\quad e_2 = \frac{1}{3}\frac{\theta'''\begin{bmatrix}1\\1\end{bmatrix}}{\theta'\begin{bmatrix}1\\1\end{bmatrix}} - \frac{\theta''\begin{bmatrix}0\\1\end{bmatrix}}{\theta\begin{bmatrix}0\\1\end{bmatrix}}$$

and

$$e_3 = \frac{1}{3}\frac{\theta'''\begin{bmatrix}1\\1\end{bmatrix}}{\theta'\begin{bmatrix}1\\1\end{bmatrix}} - \frac{\theta''\begin{bmatrix}0\\0\end{bmatrix}}{\theta\begin{bmatrix}0\\0\end{bmatrix}}.$$

Proof. Use (2.8) and the expression of the Weierstrass $\wp$-function in terms of theta functions given in §5 of Chapter 2.

Corollary 2.20. *We have the identities*

$$\varphi_1 = e_1 + e_2 + e_3 = 0$$

and

$$\frac{e_3 - e_2}{\theta^4\begin{bmatrix}1\\0\end{bmatrix}} = \frac{e_1 - e_3}{\theta^4\begin{bmatrix}0\\1\end{bmatrix}} = \frac{e_1 - e_2}{\theta^4\begin{bmatrix}0\\0\end{bmatrix}} = \pi^2.$$

Proof. The first identity follows from Theorem 5.7 of Chapter 2. The last lemma tells us that

$$e_3 - e_2 = \frac{\theta''\begin{bmatrix}0\\1\end{bmatrix}}{\theta\begin{bmatrix}0\\1\end{bmatrix}} - \frac{\theta''\begin{bmatrix}0\\0\end{bmatrix}}{\theta\begin{bmatrix}0\\0\end{bmatrix}}, \quad e_1 - e_3 = \frac{\theta''\begin{bmatrix}0\\0\end{bmatrix}}{\theta\begin{bmatrix}0\\0\end{bmatrix}} - \frac{\theta''\begin{bmatrix}1\\0\end{bmatrix}}{\theta\begin{bmatrix}1\\0\end{bmatrix}}$$

and

$$e_1 - e_2 = \frac{\theta''\begin{bmatrix}0\\1\end{bmatrix}}{\theta\begin{bmatrix}0\\1\end{bmatrix}} - \frac{\theta''\begin{bmatrix}1\\0\end{bmatrix}}{\theta\begin{bmatrix}1\\0\end{bmatrix}}.$$

Theorem 5.3 and Corollary 5.8 of Chapter 2 complete the proof.

Lemma 2.21. *We have*

$$\frac{e_1}{\theta^4\begin{bmatrix}0\\1\end{bmatrix} + \theta^4\begin{bmatrix}0\\0\end{bmatrix}} = \frac{-e_2}{\theta^4\begin{bmatrix}1\\0\end{bmatrix} + \theta^4\begin{bmatrix}0\\0\end{bmatrix}} = \frac{e_3}{\theta^4\begin{bmatrix}1\\0\end{bmatrix} - \theta^4\begin{bmatrix}0\\1\end{bmatrix}} = \frac{\pi^2}{3}.$$

Proof. The last corollary tells us that

$$(e_1 - e_3) + (e_1 - e_2) = \pi^2\left(\theta^4\begin{bmatrix}0\\0\end{bmatrix} + \theta^4\begin{bmatrix}0\\1\end{bmatrix}\right).$$

Since $e_1 + e_2 + e_3 = 0$, we see that $(e_1 - e_3) + (e_1 - e_2) = 3e_1$. This yields the formula for e_1; the other two cases are established analogously.

Corollary 2.22. *Define*

$$\varphi_2 = e_1e_2 + e_1e_3 + e_2e_3 \text{ and } \varphi_3 = e_1e_2e_3.$$

Then

$$\varphi_2 = -\frac{1}{6}\pi^4\left(\theta^8\begin{bmatrix}0\\0\end{bmatrix} + \theta^8\begin{bmatrix}0\\1\end{bmatrix} + \theta^8\begin{bmatrix}1\\0\end{bmatrix}\right)$$

$$= -\frac{1}{3}\pi^4\left(\theta^4\begin{bmatrix}0\\0\end{bmatrix}\theta^4\begin{bmatrix}0\\1\end{bmatrix} + \theta^4\begin{bmatrix}1\\0\end{bmatrix}\theta^4\begin{bmatrix}0\\0\end{bmatrix} - \theta^4\begin{bmatrix}0\\1\end{bmatrix}\theta^4\begin{bmatrix}1\\0\end{bmatrix}\right)$$

and

$$\varphi_3 = -\frac{1}{27}\pi^6$$

$$\times\left(\theta^4\begin{bmatrix}0\\1\end{bmatrix}+\theta^4\begin{bmatrix}0\\0\end{bmatrix}\right)\left(\theta^4\begin{bmatrix}1\\0\end{bmatrix}+\theta^4\begin{bmatrix}0\\0\end{bmatrix}\right)\left(\theta^4\begin{bmatrix}1\\0\end{bmatrix}-\theta^4\begin{bmatrix}0\\1\end{bmatrix}\right).$$

Proof. From the definition

$$\varphi_2=\frac{\pi^2}{9}\left(-\theta^8\begin{bmatrix}0\\0\end{bmatrix}-\theta^8\begin{bmatrix}0\\1\end{bmatrix}-\theta^8\begin{bmatrix}1\\0\end{bmatrix}\right)$$

$$+\frac{\pi^2}{9}\left(\theta^4\begin{bmatrix}0\\1\end{bmatrix}\theta^4\begin{bmatrix}1\\0\end{bmatrix}-\theta^4\begin{bmatrix}0\\0\end{bmatrix}\theta^4\begin{bmatrix}1\\0\end{bmatrix}-\theta^4\begin{bmatrix}0\\0\end{bmatrix}\theta^4\begin{bmatrix}0\\1\end{bmatrix}\right).$$

Corollary 5.5 of Chapter 2 (the quartic identity) yields

$$\theta^8\begin{bmatrix}0\\0\end{bmatrix}+\theta^8\begin{bmatrix}0\\1\end{bmatrix}+\theta^8\begin{bmatrix}1\\0\end{bmatrix}$$

$$=2\left(\theta^4\begin{bmatrix}0\\0\end{bmatrix}\theta^4\begin{bmatrix}0\\1\end{bmatrix}+\theta^4\begin{bmatrix}1\\0\end{bmatrix}\theta^4\begin{bmatrix}0\\0\end{bmatrix}-\theta^4\begin{bmatrix}0\\1\end{bmatrix}\theta^4\begin{bmatrix}1\\0\end{bmatrix}\right),$$

which immediately implies the formula for φ_2. The previous lemma contains the new formula for φ_3.

Lemma 2.23. *For each $k\in\mathbb{Z}^+$, every homogeneous symmetric polynomial in e_1, e_2 and e_3 of degree k is a modular k-form for Γ, that is, belongs to $\mathbb{A}_k^+(\mathbb{H}^2,\Gamma)$.*

Proof. It is a consequence of our work in §4.3 of Chapter 1 (see also Theorem 1.11 of Chapter 2) that for each $\gamma\in\Gamma$, there is a permutation σ_γ of the integers 1, 2 and 3 such that

$$(e_i\circ\gamma)\gamma'=e_{\sigma_\gamma(i)},\ i=1,2,3.$$

If f is a homogeneous symmetric polynomial of degree k in three variables, then

$$f(e_1(\gamma),e_2(\gamma),e_3(\gamma))(\gamma')^k=f(e_{\sigma_\gamma(1)},e_{\sigma_\gamma(2)},e_{\sigma_\gamma(3)})=f(e_1,e_2,e_3).$$

It remains to be shown that $f(e_1,e_2,e_3)$ is holomorphic at $\imath\infty$. Since $f(e_1,e_2,e_3)$ is also a symmetric function of $\varphi_1=0$, φ_2 and φ_3, it suffices to show that these functions are holomorphic at $\imath\infty$. Expansions in terms of $\zeta=\exp\{2\pi\imath\tau\}$, $\tau\in\mathbb{H}^2$, yields

$$\varphi_2(\zeta)=-\frac{\pi^4}{3}\left[1+O(\zeta)\right]\text{ and }\varphi_3(\zeta)=\frac{2\pi^6}{27}\left[1+O(\zeta)\right],\ \zeta\to0.$$

Theorem 2.24. *Let $q\in\mathbb{Z}^+$ and $\varphi\in\mathbb{A}_q^+(\mathbb{H}^2,\Gamma)$. Then there exist constants c_{lm}, $(l,m)\in LM(q)$, such that*

$$\varphi=\sum_{(l,m)\in LM(q)}c_{lm}\varphi_2^l\varphi_3^m.$$

Proof. It suffices to show that the automorphic forms

$$\varphi_2^l \varphi_3^m, \ (l, m) \in LM(q),$$

are linearly independent. Since

$$\deg(\varphi_2) = \frac{1}{3} \text{ and } \deg(\varphi_3) = \frac{1}{2},$$

we conclude that φ_2 has a simple zero at $\frac{1}{2} + \imath\frac{\sqrt{3}}{2}$ and that φ_3 has a simple zero at $\imath$. It follows that

$$\operatorname{ord}_{\frac{1}{2}+\imath\frac{\sqrt{3}}{2}} \varphi_2^l \varphi_3^m = l.$$

Hence the $r(q)$ functions under consideration are linearly independent.

Remark 2.25. We have shown that the graded algebra $\oplus_{q=0}^{\infty} \mathbb{A}_q^+(\mathbb{H}^2, \Gamma)$ is generated over $\mathbb{C}$ by three functions: 1 of degree 0, φ_2 of degree 2 and φ_3 of degree 3. This graded algebra does not contain any nontrivial elements of degree 1.

3. Some special cases ($k' = 1$)

We let

$$P : \mathbb{H}^2 \cup \mathbb{Q} \cup \{\infty\} \to (\mathbb{H}^2 \cup \mathbb{Q} \cup \{\infty\})/\Gamma(k) = \overline{\mathbb{H}^2/\Gamma(k)}$$

denote the canonical projection, and for $x \in \mathbb{H}^2 \cup \mathbb{Q} \cup \{\infty\}$, we abbreviate $P_x = P(x)$. We proceed to derive some function theory for $\overline{\mathbb{H}^2/\Gamma(k)}$ for $k = 1, 2, 3, 4, 5$ and 6. The first five of these surfaces are spheres; the last is a torus.

3.1. $k = 1$. Most of the theory developed here avoided the case where $k = 1$. It can easily be shown (we treat this case in much more detail in the next chapter) that

$$\theta^8 \begin{bmatrix} 0 \\ 0 \end{bmatrix} \theta^8 \begin{bmatrix} 0 \\ 1 \end{bmatrix} \theta^8 \begin{bmatrix} 1 \\ 0 \end{bmatrix}$$

can be taken as a basis for the one dimensional space $\mathbb{A}_6(\mathbb{H}^2, \Gamma)$. In fact, any symmetric function of the three variables

$$\theta^8 \begin{bmatrix} 0 \\ 0 \end{bmatrix}, \ \theta^8 \begin{bmatrix} 0 \\ 1 \end{bmatrix}, \ \theta^8 \begin{bmatrix} 1 \\ 0 \end{bmatrix}$$

will give rise to a modular form. Suitable quotients then give rise to modular functions. Most important in this case are two functions connected with elliptic function theory:

$$\Delta(\tau) = 2^{-13} 3^2 \pi^{-12} \left(g_2^3(\tau) - 27 g_3^2(\tau) \right) \text{ and } \jmath(\tau) = 2^{-6} 3^3 \pi^{-12} \frac{g_2^3(\tau)}{\Delta(\tau)} - 2^7 3.$$

It is known [**27**, §2.2] that $\jmath$ uniformizes $\mathbb{H}^2/\Gamma$; that Δ is a basis for $\mathbb{A}_6(\mathbb{H}^2,\Gamma)$; and, because of our normalizations, that if we expand in terms of the local coordinate $x = \exp(2\pi\imath\tau)$,

$$\jmath(\tau) = \frac{1}{x} + \sum_{n=1}^{\infty} c_n x^n, \ \Delta(\tau) = x + \sum_{n=2}^{\infty} b_n x^n, \tag{3.16}$$

then

$$c_n \in \mathbb{Z}^+ \cup \{0\} \text{ and } b_n \in \mathbb{Z}. \tag{3.17}$$

Remark 3.1. The function $\jmath$ differs by postcomposition by a Möbius transformation from the one introduced in Chapter 1.

The proofs of the properties of the functions Δ and $\jmath$ are based on

Lemma 3.2. *We have the identities*

$$\frac{g_2}{2^2\cdot 3\cdot 5\zeta(4)} = \theta^8\begin{bmatrix}0\\0\end{bmatrix} + \theta^8\begin{bmatrix}0\\1\end{bmatrix} + \theta^8\begin{bmatrix}1\\0\end{bmatrix}$$

and

$$\frac{g_3}{2^2\cdot 5\cdot 7\zeta(6)}$$

$$= \left(\theta^4\begin{bmatrix}0\\0\end{bmatrix} + \theta^4\begin{bmatrix}0\\1\end{bmatrix}\right)\left(\theta^4\begin{bmatrix}0\\0\end{bmatrix} + \theta^4\begin{bmatrix}1\\0\end{bmatrix}\right)\left(\theta^4\begin{bmatrix}0\\1\end{bmatrix} - \theta^4\begin{bmatrix}1\\0\end{bmatrix}\right).$$

Proof. We know that g_2 is a nontrivial element of the one dimensional vector space $\mathbb{A}_2^+(\mathbb{H}^2,\Gamma)$ and that $g_2(\imath\infty) = 2^3 3\cdot 5\zeta(4)$. The group Γ permutes the even integer characteristics. Hence an element $\gamma\in\Gamma$ sends a theta constant for an even integral characteristic to a constant multiple of another such theta constant via the $\gamma^*_{\frac{1}{4}}$-action. Since the constant multiple is an 8-th root of unity, the sum of the 8-th powers of the theta constants with even integral characteristics is a 2-form for the group Γ. Since the value of this 2-form at $\imath\infty$ is 2, the formula for g_2 follows. The proof of the formula for g_3 is similar. The argument is simplified considerably by studying the action only of suitable generators for (rather than an arbitrary element of) Γ, for example, B and A, on the right hand side of the equation defining g_3.

Corollary 3.3. *The function Δ is a basis for the one dimensional vector space $\mathbb{A}_6(\mathbb{H}^2,\Gamma)$ and its Fourier series at $\imath\infty$ is given by* (3.16).

Proof. It is obvious that $g_2^3 - \frac{2^3 3^3 5\zeta^3(4)}{7^2\zeta^2(6)} g_3^2 \in \mathbb{A}_6(\mathbb{H}^2,\Gamma)$, because we have adjusted the constant so that this function vanishes at $\imath\infty$. The Riemann-Roch theorem told us that this vector space has dimension one. It is well known that $\zeta(4) = \frac{\pi^4}{2\cdot 3^2 5}$ and $\zeta(6) = \frac{\pi^6}{3^3 5\cdot 7}$ (see, for example,[69] [**26**, pg. 91]).

[69] The reference [**26**, pg. 91] contains a misprint in the formula for $\zeta(4)$.

Hence $\frac{2^3 3^3 5 \zeta^3(4)}{7^2 \zeta^2(6)} = 3^3$. To conclude that a linear combination of g_2^3 and g_3^2 is a cusp form for Γ, we can avoid reliance on evaluations of the Riemann-zeta function (at positive integers) by expanding the right hand side of the equations defining g_2 and g_3:

$$\begin{aligned} g_2(\tau) &= 2^2 3 \cdot 5\zeta(4)\left(2 + 2^8 x + O(x^2)\right), \\ g_3(\tau) &= 2^2 5 \cdot 7\zeta(6)\left(1 - 2^6 3^2 x + O(x^2)\right) \end{aligned}.$$

Hence

$$g_2^3 - \frac{2^3 3^3 5 \zeta^3(4)}{7^2 \zeta^2(6)} g_3^2 = 2^{16} 3^4 5^3 \zeta^3(4)\left(x + O(x^2)\right).$$

Using the known values of the Riemann zeta function, we see that

$$2^{16} 3^4 5^3 \zeta^3(4) = 2^{13} 3^{-2} \pi^{12}.$$

Corollary 3.4. *The function $\jmath$ is a holomorphic universal covering of the orbifold $\mathbb{H}^2/\Gamma$, and its Fourier series at $\imath\infty$ is given by* (3.16).

Proof. The 6-form Δ for Γ has order 1 at $\imath\infty$. Hence it cannot vanish on $\mathbb{H}^2$. Hence $\jmath$ is holomorphic on $\mathbb{H}^2$. Since the 6-form g_2^3 for Γ has order 0 at $\imath\infty$, $\jmath$ is a nonconstant Γ-invariant function on $\mathbb{H}^2$. Since its projection to $\mathbb{H}^2/\Gamma$ has a simple pole at P_∞ and no other poles, it is a holomorphic universal covering map of $\mathbb{H}^2/\Gamma$. From its Fourier series expansion

$$\frac{g_2^3}{\Delta}(\tau) = 2^9 3^3 5^3 \zeta^3(4)\left(\frac{1}{x} + 2^7 3 + O(x)\right),$$

we conclude that (3.16) holds.

Remark 3.5. (a) We study additional properties of the functions Δ and $\jmath$ (for example, that (3.17) holds) and will supply alternate proofs of many of the facts established above in subsequent chapters.
(b) Using some algebraic preliminaries, we show in §3.5 of Chapter 4, on the basis of our work on theta constants in §2.7, how to eliminate completely the Riemann theta function (as a function of the variable z) from our development. We are however not suggesting that this is a good thing to do. In fact, it is not.

3.2. $k = 2$. The case $k = 2$ differs from the other cases because 2 is the (only) even prime.

We begin with the observation that $\mathbb{H}^2/\Gamma(2)$ is a sphere with 3 punctures: P_∞, P_0, P_1, and that

$$f = \frac{\theta^4\begin{bmatrix}0\\0\end{bmatrix}}{\theta^4\begin{bmatrix}0\\1\end{bmatrix}} \text{ and } g = \frac{\theta^4\begin{bmatrix}1\\0\end{bmatrix}}{\theta^4\begin{bmatrix}0\\1\end{bmatrix}}$$

are two $\Gamma(2)$-invariant functions whose projections to $\overline{\mathbb{H}^2/\Gamma(2)}$ have divisors

$$\frac{P_1}{P_0} \text{ and } \frac{P_\infty}{P_0},$$

respectively. Since

$$f(\infty) = 1 \text{ and } g(1) = -1,$$

it follows that f and g are holomorphic covering maps of

$$\hat{\mathbb{C}} - \{1,\ \infty,\ 0\} \text{ and } \hat{\mathbb{C}} - \{0,\ \infty,\ -1\},$$

respectively. Covering maps of conformally equivalent surfaces are related by conformal maps. It follows that there exists a motion $C \in \mathrm{PSL}(2,\mathbb{C})$ such that $g = C \circ f$. Evaluating this equality at $\tau = 0$ tells us that $C(\infty) = \infty$. Thus we see that $g = \alpha\ f + \beta$ for some $\alpha \in \mathbb{C}^*$, $\beta \in \mathbb{C}$. Evaluating this last equality at $\tau = \infty$ and 1 tells us that $\alpha = 1$ and $\beta = -1$. The identity $g = f - 1$ is the Jacobi quartic (3.18).

The (classical) λ-function (see Chapter 1) is a holomorphic universal covering map of $\mathbb{H}^2/\Gamma(2)$ with divisor $\frac{P_\infty}{P_1}$; hence it is a (nonzero) constant multiple of $\frac{g}{f}$. It is easy to evaluate this constant and obtain that $\lambda = \frac{\theta^4\begin{bmatrix}1\\0\end{bmatrix}}{\theta^4\begin{bmatrix}0\\0\end{bmatrix}}$. Thus we can also express the $\jmath$-invariant in terms of θ-constants.

We now consider the equivalence classes of characteristics represented by vectors of the form $\begin{bmatrix}\frac{m}{2}\\\frac{m'}{2}\end{bmatrix}$ with m and m' even integers. This set consists of the four classical integer characteristics. If we delete from this list the characteristic $\begin{bmatrix}1\\1\end{bmatrix}$, we have left the set of three even characteristics. If we now consider as well all vectors of the form $\begin{bmatrix}\frac{m}{2}\\\frac{m'}{2}\end{bmatrix}$ with m and m' arbitrary integers, we add six additional classes to our list.

The group $\Gamma(2)$ is generated by the motions

$$B^2 : z \mapsto z+2,\ A \circ B^{-2} \circ A : z \mapsto \frac{z}{2z+1},$$

which we denote by S and W, respectively. Each of these (and hence the group $\Gamma(2)$) fixes the four classes of characteristics with integer entries and permutes the other six. Each element of Γ fixes $\begin{bmatrix} 1 \\ 1 \end{bmatrix}$ and permutes the other three integral classes. It therefore follows that we have a homomorphism η of Γ into the permutation group on three elements. In order to use the usual cycle notation for permutations, we identify classes of characteristics with integers as follows:

$$\begin{bmatrix} 0 \\ 0 \end{bmatrix} \sim 1, \ \begin{bmatrix} 0 \\ 1 \end{bmatrix} \sim 2, \text{ and } \begin{bmatrix} 1 \\ 0 \end{bmatrix} \sim 3.$$

Using the generators of Γ introduced above, we find that B induces the permutation, written in cycle notation $(1\ 2)$, and A induces the permutation $(2\ 3)$. These two permutations generate the full permutation group $\mathcal{S}_3$. Since, as we have already mentioned, the subgroup $\Gamma(2)$ is the kernel of the homomorphism η, we get the well known result that $\Gamma/\Gamma(2)$ is isomorphic to $\mathcal{S}_3$.

Let us consider next the action of $\Gamma(2)$ on the set Y consisting of the six additional classes we have described above. Let us denote them (in analogy to what we did with integral classes) by:

$$\begin{bmatrix} 0 \\ \frac{1}{2} \end{bmatrix} \sim 1, \ \begin{bmatrix} \frac{1}{2} \\ 0 \end{bmatrix} \sim 2, \ \begin{bmatrix} \frac{1}{2} \\ \frac{1}{2} \end{bmatrix} \sim 3, \ \begin{bmatrix} \frac{1}{2} \\ \frac{3}{2} \end{bmatrix} \sim 4, \ \begin{bmatrix} 1 \\ \frac{1}{2} \end{bmatrix} \sim 5, \ \begin{bmatrix} \frac{1}{2} \\ 1 \end{bmatrix} \sim 6.$$

We find that S induces the permutation $(2\ 6)(3\ 4)$ and that W induces the permutation $(1\ 5)(2\ 6)$. These generate the Klein 4-group, $\mathbb{Z}_2 \oplus \mathbb{Z}_2$. We have shown that the group $\Gamma(2)$ acts nontrivially on the six classes described above and gives a homomorphism of $\Gamma(2)$ onto the Klein 4-group. It is easy to check that the kernel of this homomorphism is $\Gamma(4)$, so that $\Gamma(2)/\Gamma(4)$ is just the Klein 4-group. More importantl for us though is the fact that the action of $\Gamma(2)$ on the six fractional classes has precisely three distinct orbits and each one can be associated in a natural way with one of the three even integral characteristics. The "natural way" is via the action of Γ, which also acts on the orbits. To be precise, the quotient Y modulo the Klein 4-group consists of the three orbits

$$[1, 1, 5, 5], \ [2, 2, 6, 6], \text{ and } [3, 3, 4, 4]$$

(we have written each element according to its multiplicity). The group Γ acts on these orbits: B fixes the first orbit and permutes the second and third, while A fixes the third and permutes the first and second. If we now compare the actions of Γ on these orbits and the action of Γ on the integral classes of characteristics, it is clear that we should associate

$$\text{the class of } \begin{bmatrix} 0 \\ 0 \end{bmatrix} \text{ with the orbit } [3,3,4,4],$$

$$\text{the class of } \begin{bmatrix} 0 \\ 1 \end{bmatrix} \text{ with the orbit } [2,2,6,6]$$

and

$$\text{the class of } \begin{bmatrix} 1 \\ 0 \end{bmatrix} \text{ with the orbit } [1,1,5,5].$$

Theorem 3.6. *We have the following proportionalities among theta constants:*

$$\frac{\theta^8\begin{bmatrix} 0 \\ \frac{1}{2} \end{bmatrix}\theta^8\begin{bmatrix} 1 \\ \frac{1}{2} \end{bmatrix}}{\theta^4\begin{bmatrix} 1 \\ 0 \end{bmatrix}} = \frac{\theta^8\begin{bmatrix} \frac{1}{2} \\ 0 \end{bmatrix}\theta^8\begin{bmatrix} \frac{1}{2} \\ 1 \end{bmatrix}}{\theta^4\begin{bmatrix} 0 \\ 1 \end{bmatrix}} = \frac{\theta^8\begin{bmatrix} \frac{1}{2} \\ \frac{1}{2} \end{bmatrix}\theta^8\begin{bmatrix} \frac{1}{2} \\ \frac{3}{2} \end{bmatrix}}{\theta^4\begin{bmatrix} 0 \\ 0 \end{bmatrix}}.$$

The first proof that we present of this theorem requires some preliminary results concerning theta functions. Before we actually do this however, let us observe that what we have here is a statement about modular forms for the group $\Gamma(2)$. The statement is that each of the quotients is a modular 3-form for $\Gamma(2)$ and that the three quotients define the same form. The second proof will take advantage of this trivial observation.

As a consequence of the addition formula in Lemma 1.6 of Chapter 2, which should also be a consequence of the Jacobi triple product formula (the reader is invited to supply a proof along these lines) and the transformation formulae for theta functions under equivalent characteristics, one easily obtains the following set of equations:

$$\theta^2\begin{bmatrix} 0 \\ \frac{1}{2} \end{bmatrix}(0,\tau) = \theta\begin{bmatrix} 0 \\ 1 \end{bmatrix}(0,2\tau)\theta\begin{bmatrix} 0 \\ 0 \end{bmatrix}(0,2\tau),$$

$$\theta^2\begin{bmatrix} 1 \\ \frac{1}{2} \end{bmatrix}(0,\tau) = \theta\begin{bmatrix} 0 \\ 1 \end{bmatrix}(0,2\tau)\theta\begin{bmatrix} 1 \\ 0 \end{bmatrix}(0,2\tau),$$

$$\theta^2\begin{bmatrix} \frac{1}{2} \\ 0 \end{bmatrix}(0,\tau) = \theta\begin{bmatrix} \frac{1}{2} \\ 0 \end{bmatrix}(0,2\tau)\left(\theta\begin{bmatrix} 0 \\ 0 \end{bmatrix}(0,2\tau) + \theta\begin{bmatrix} 1 \\ 0 \end{bmatrix}(0,2\tau)\right).$$

$$\theta^2\begin{bmatrix} \frac{1}{2} \\ 1 \end{bmatrix}(0,\tau) = \theta\begin{bmatrix} \frac{1}{2} \\ 2 \end{bmatrix}(0,2\tau)\left(\theta\begin{bmatrix} 0 \\ 0 \end{bmatrix}(0,2\tau) - \theta\begin{bmatrix} 1 \\ 0 \end{bmatrix}(0,2\tau)\right),$$

$$\theta^2\begin{bmatrix} \frac{1}{2} \\ \frac{1}{2} \end{bmatrix}(0,\tau) = \theta\begin{bmatrix} \frac{1}{2} \\ 1 \end{bmatrix}(0,2\tau)\left(\theta\begin{bmatrix} 0 \\ 0 \end{bmatrix}(0,2\tau) - \imath\theta\begin{bmatrix} 1 \\ 0 \end{bmatrix}(0,2\tau)\right)$$

and

$$\theta^2\begin{bmatrix}\frac{1}{2}\\ \\ \frac{3}{2}\end{bmatrix}(0,\tau)=\theta\begin{bmatrix}\frac{1}{2}\\ \\ 3\end{bmatrix}(0,2\tau)\left(\theta\begin{bmatrix}0\\ \\ 0\end{bmatrix}(0,2\tau)+\imath\theta\begin{bmatrix}1\\ \\ 0\end{bmatrix}(0,2\tau)\right).$$

Another application of the previously used transformation formula and some algebraic manipulations yield

$$\theta^4\begin{bmatrix}0\\ \frac{1}{2}\end{bmatrix}(0,\tau)\theta^4\begin{bmatrix}1\\ \frac{1}{2}\end{bmatrix}(0,\tau)=\theta^4\begin{bmatrix}0\\ 1\end{bmatrix}(0,2\tau)\theta^2\begin{bmatrix}0\\ 0\end{bmatrix}(0,2\tau)\theta^2\begin{bmatrix}1\\ 0\end{bmatrix}(0,2\tau),$$

$$\theta^4\begin{bmatrix}\frac{1}{2}\\ 0\end{bmatrix}(0,\tau)\theta^4\begin{bmatrix}\frac{1}{2}\\ 1\end{bmatrix}(0,\tau)$$

$$=-\theta^4\begin{bmatrix}\frac{1}{2}\\ 0\end{bmatrix}(0,2\tau)\left(\theta^2\begin{bmatrix}0\\ 0\end{bmatrix}(0,2\tau)-\theta^2\begin{bmatrix}1\\ 0\end{bmatrix}(0,2\tau)\right)^2$$

and

$$\theta^4\begin{bmatrix}\frac{1}{2}\\ \\ \frac{1}{2}\end{bmatrix}(0,\tau)\theta^4\begin{bmatrix}\frac{1}{2}\\ \\ \frac{3}{2}\end{bmatrix}(0,\tau)$$

$$=-\theta^4\begin{bmatrix}\frac{1}{2}\\ 1\end{bmatrix}(0,2\tau)\left(\theta^2\begin{bmatrix}0\\ 0\end{bmatrix}(0,2\tau)+\theta^2\begin{bmatrix}1\\ 0\end{bmatrix}(0,2\tau)\right)^2.$$

It is important to observe that the right hand sides of the last three equations are equal respectively to

$$\frac{1}{4}\theta^4\begin{bmatrix}0\\ 1\end{bmatrix}(0,2\tau)\theta^4\begin{bmatrix}1\\ 0\end{bmatrix}(0,\tau),\ -\theta^4\begin{bmatrix}\frac{1}{2}\\ 0\end{bmatrix}(0,2\tau)\theta^4\begin{bmatrix}0\\ 1\end{bmatrix}(0,\tau)$$

and

$$-\theta^4\begin{bmatrix}\frac{1}{2}\\ 1\end{bmatrix}(0,2\tau)\theta^4\begin{bmatrix}0\\ 0\end{bmatrix}(0,\tau).$$

Our final formulae are consequences of the general addition formula noted above. They are:

$$\theta\begin{bmatrix}0\\ 0\end{bmatrix}(0,\tau)\theta\begin{bmatrix}0\\ 1\end{bmatrix}(0,\tau)=\theta^2\begin{bmatrix}0\\ 1\end{bmatrix}(0,2\tau),$$

$$\theta\begin{bmatrix}0\\ 0\end{bmatrix}(0,\tau)\theta\begin{bmatrix}1\\ 0\end{bmatrix}(0,\tau)=2\theta^2\begin{bmatrix}\frac{1}{2}\\ 0\end{bmatrix}(0,2\tau)$$

and

$$\theta\begin{bmatrix}0\\ 1\end{bmatrix}(0,\tau)\theta\begin{bmatrix}1\\ 0\end{bmatrix}(0,\tau)=-2\imath\theta^2\begin{bmatrix}\frac{1}{2}\\ 1\end{bmatrix}(0,2\tau).$$

Putting all this together yields the proportionality between the fourth powers of the classical theta constants (integer characteristic) and the products of theta constants with half integer characteristics. We have arrived at some identities among theta constants.

Corollary 3.7. *We have*

$$\theta^8\begin{bmatrix}\frac{1}{2}\\ \frac{1}{2}\end{bmatrix}\theta^8\begin{bmatrix}\frac{1}{2}\\ \frac{3}{2}\end{bmatrix}=\theta^8\begin{bmatrix}\frac{1}{2}\\ 0\end{bmatrix}\theta^8\begin{bmatrix}\frac{1}{2}\\ 1\end{bmatrix}+\theta^8\begin{bmatrix}0\\ \frac{1}{2}\end{bmatrix}\theta^8\begin{bmatrix}1\\ \frac{1}{2}\end{bmatrix}.$$

Proof. The proportionality and the classical identity among the fourth powers of the theta constants with even integral characteristics yields the above formula.

The same arguments establish

$$\frac{\theta^2\begin{bmatrix}0\\ \frac{1}{2}\end{bmatrix}\theta^2\begin{bmatrix}1\\ \frac{1}{2}\end{bmatrix}}{\theta^2\begin{bmatrix}1\\ 0\end{bmatrix}\theta\begin{bmatrix}0\\ 1\end{bmatrix}\theta\begin{bmatrix}0\\ 0\end{bmatrix}}=\frac{1}{2},\qquad \frac{\theta^2\begin{bmatrix}\frac{1}{2}\\ 0\end{bmatrix}\theta^2\begin{bmatrix}\frac{1}{2}\\ 1\end{bmatrix}}{\theta\begin{bmatrix}1\\ 0\end{bmatrix}\theta^2\begin{bmatrix}0\\ 1\end{bmatrix}\theta\begin{bmatrix}0\\ 0\end{bmatrix}}=\frac{\imath}{2}$$

and

$$\frac{\theta^2\begin{bmatrix}\frac{1}{2}\\ \frac{1}{2}\end{bmatrix}\theta^2\begin{bmatrix}\frac{1}{2}\\ \frac{3}{2}\end{bmatrix}}{\theta\begin{bmatrix}1\\ 0\end{bmatrix}\theta\begin{bmatrix}0\\ 1\end{bmatrix}\theta^2\begin{bmatrix}0\\ 0\end{bmatrix}}=-\frac{1}{2}.$$

(The fourth power of the above ratios is constant, $\frac{1}{16}$.) The last three identities are also direct consequences of the Jacobi triple product formula.

Remark 3.8. The methods of the next section yield alternate proofs of the proportionalities established in this section. In Chapter 4 we will describe an alternate method for obtaining identities of the above type. For example, an alternate proof of Theorem 3.6 is obtained by observing that each term in the equalities is a member of the one dimensional space $\mathbb{A}_3(\mathbb{H}^2,\Gamma(2))$.

3.3. $k=3$. The surface $\mathbb{H}^2/\Gamma(3)$ is a four times punctured sphere; the punctures are the images of the parabolic fixed points -1, 0, 1, ∞ under the extension to the parabolic fixed points of the group $\Gamma(3)$ of the canonical projection $\mathbb{H}^2\to\mathbb{H}^2/\Gamma(3)$. We denote these punctures by P_{-1}, P_0, P_1 and P_∞, respectively.

Recall that $\mathrm{PSL}(2,\mathbb{Z})$ is the normalizer of $\Gamma(3)$ as well as the action $\mathrm{SL}(2,\mathbb{R})$ on automorphic forms introduced in the first section of this chapter. For $C\in\mathrm{SL}(2,\mathbb{Z})$, we have that $\psi=C_q^*\varphi$ is a q-form for $\Gamma(3)$, whenever φ is. In the following formulae, we denote the form (for $\Gamma(3)$) θ^{24} by φ. Taking $C=\begin{bmatrix}0&-1\\1&0\end{bmatrix}$, we see that

$$C_6^*\left(\varphi\begin{bmatrix}\frac{1}{3}\\ \frac{1}{3}\end{bmatrix}\right)=\epsilon\varphi\begin{bmatrix}\frac{1}{3}\\ \frac{5}{3}\end{bmatrix},$$

with ϵ a root of unity, and hence

$$\operatorname{ord}_{P_0}\Phi\begin{bmatrix}\frac{1}{3}\\ \\ \frac{1}{3}\end{bmatrix} = \operatorname{ord}_{P_\infty}\Phi\begin{bmatrix}\frac{1}{3}\\ \\ \frac{5}{3}\end{bmatrix} = -5.$$

Similarly, using $C = \begin{bmatrix}-1 & 1\\ -1 & 0\end{bmatrix}$ and $C = \begin{bmatrix}1 & 1\\ -1 & 0\end{bmatrix}$, we see that

$$\operatorname{ord}_{P_1}\Phi\begin{bmatrix}\frac{1}{3}\\ \\ \frac{1}{3}\end{bmatrix} = \operatorname{ord}_{P_\infty}\Phi\begin{bmatrix}\frac{1}{3}\\ \\ \frac{1}{3}\end{bmatrix} = -5$$

and

$$\operatorname{ord}_{P_{-1}}\Phi\begin{bmatrix}\frac{1}{3}\\ \\ \frac{1}{3}\end{bmatrix} = \operatorname{ord}_{P_\infty}\Phi\begin{bmatrix}1\\ \\ \frac{1}{3}\end{bmatrix} = 3,$$

from which it follows that the divisor $\left(\Phi\begin{bmatrix}\frac{1}{3}\\ \\ \frac{1}{3}\end{bmatrix}\right)$ of the projection $\Phi\begin{bmatrix}\frac{1}{3}\\ \\ \frac{1}{3}\end{bmatrix}$ of $\varphi\begin{bmatrix}\frac{1}{3}\\ \\ \frac{1}{3}\end{bmatrix}$ (to the compactification of $\mathbb{H}^2/\Gamma(3)$) is given by

$$\left(\Phi\begin{bmatrix}\frac{1}{3}\\ \\ \frac{1}{3}\end{bmatrix}\right) = \frac{P_{-1}^3}{P_0^5 P_1^5 P_\infty^5}.$$

Similarly

$$\left(\Phi\begin{bmatrix}\frac{1}{3}\\ \\ 1\end{bmatrix}\right) = \frac{P_0^3}{P_{-1}^5 P_1^5 P_\infty^5}, \quad \left(\Phi\begin{bmatrix}\frac{1}{3}\\ \\ \frac{5}{3}\end{bmatrix}\right) = \frac{P_1^3}{P_{-1}^5 P_0^5 P_\infty^5}$$

and

$$\left(\Phi\begin{bmatrix}1\\ \frac{1}{3}\end{bmatrix}\right) = \frac{P_\infty^3}{P_{-1}^5 P_0^5 P_1^5}.$$

Exercise 3.9. Verify the formulae for the divisors of the automorphic forms encountered in this subsection.

It now makes sense to form the ratios

$$\frac{\varphi\begin{bmatrix}\frac{1}{3}\\ \\ \frac{1}{3}\end{bmatrix}}{\varphi\begin{bmatrix}\frac{1}{3}\\ 1\end{bmatrix}} \text{ and } \frac{\varphi\begin{bmatrix}\frac{1}{3}\\ \\ \frac{5}{3}\end{bmatrix}}{\varphi\begin{bmatrix}\frac{1}{3}\\ 1\end{bmatrix}}$$

and observe that these are automorphic functions for the group $\Gamma(3)$ whose projections to the compactification of $\mathbb{H}^2/\Gamma(3)$ have divisors

$$\frac{P_{-1}^8}{P_0^8} \text{ and } \frac{P_1^8}{P_0^8},$$

respectively. It follows that eighth roots (λ and g) of these functions provide covering maps of $\mathbb{H}^2/\Gamma(3)$ (for us there are obvious roots to choose, namely, ratios of cubes of theta constants). We have therefore proven the following

Theorem 3.10. *The maps*

$$\lambda : \tau \mapsto \frac{\theta^3 \begin{bmatrix} \frac{1}{3} \\ \frac{1}{3} \end{bmatrix}}{\theta^3 \begin{bmatrix} \frac{1}{3} \\ 1 \end{bmatrix}} \text{ and } g : \tau \mapsto \frac{\theta^3 \begin{bmatrix} \frac{1}{3} \\ \frac{5}{3} \end{bmatrix}}{\theta^3 \begin{bmatrix} \frac{1}{3} \\ 1 \end{bmatrix}}$$

are holomorphic universal covering maps from $\mathbb{H}^2$ to the four times punctured sphere $\mathbb{H}^2/\Gamma(3)$.

Remark 3.11. Up to choice of signs, the ratios of the third powers of the theta constants that appear in the above theorem depend only on the equivalence classes of the characteristics (not the theta characteristics themselves). It is also easy to conclude directly from the transformation formula that up to sign, the above ratios are automorphic functions on $\mathbb{H}^2$ for $\Gamma(3)$. The elimination of the ambiguities in signs (the fact that we have single rather than multivalued functions) follows from the nature of the singularities that our functions have and the simple connectivity of the compactification of $\mathbb{H}^2/\Gamma(3)$. The choice of the two characteristics is quite arbitrary. The ratio of any two out of the four classes that belong to $X_o(3)$ will give a covering map.

Now, any two covering maps that are automorphic with respect to the same group are post-related by a conformal map (Möbius transformation in our case); thus

$$g = C \circ \lambda,$$

with $C \in \text{PSL}(2, \mathbb{C})$. It remains to evaluate C.

Evaluating the last equation at P_0, we see that C fixes ∞. Thus there are constants $\alpha \neq 0$ and β such that $g = \alpha\lambda + \beta$. It is easiest to proceed by translating this equation to one involving theta constants:

$$\theta^3 \begin{bmatrix} \frac{1}{3} \\ \frac{5}{3} \end{bmatrix} (0, \cdot) = \alpha\theta^3 \begin{bmatrix} \frac{1}{3} \\ \frac{1}{3} \end{bmatrix} (0, \cdot) + \beta\theta^3 \begin{bmatrix} \frac{1}{3} \\ 1 \end{bmatrix} (0, \cdot).$$

By equating leading terms of the Fourier series expansions of these functions we see that $\alpha = -1$ and $\beta = 1$. Thus we conclude that

$$1 = \lambda + g$$

and

$$\theta^3 \begin{bmatrix} \frac{1}{3} \\ \\ \frac{5}{3} \end{bmatrix} (0, \cdot) + \theta^3 \begin{bmatrix} \frac{1}{3} \\ \\ \frac{1}{3} \end{bmatrix} (0, \cdot) = \theta^3 \begin{bmatrix} \frac{1}{3} \\ \\ 1 \end{bmatrix} (0, \cdot).$$

We have therefore proven one half of the following

Theorem 3.12. *For all points τ in $\mathbb{H}^2$ the following two identities hold:*

$$\theta^3 \begin{bmatrix} \frac{1}{3} \\ \\ \frac{5}{3} \end{bmatrix} (0, \tau) + \theta^3 \begin{bmatrix} \frac{1}{3} \\ \\ \frac{1}{3} \end{bmatrix} (0, \tau) = \theta^3 \begin{bmatrix} \frac{1}{3} \\ \\ 1 \end{bmatrix} (0, \tau)$$

and

$$\exp\left(\frac{\pi \imath}{3}\right) \theta^3 \begin{bmatrix} \frac{1}{3} \\ \\ \frac{1}{3} \end{bmatrix} (0, \tau) + \exp\left(\frac{2\pi \imath}{3}\right) \theta^3 \begin{bmatrix} \frac{1}{3} \\ \\ \frac{5}{3} \end{bmatrix} (0, \tau) = \theta^3 \begin{bmatrix} 1 \\ \\ \frac{1}{3} \end{bmatrix} (0, \tau).$$

Proof. The first identity has been derived above; the proof of the second one is similar and left to the reader.

Remark 3.13. It is of course very easy to establish the above cubic identity. The challenge here is not the proof, but the discovery. As we saw in the previous subsection, a similar argument establishes the classical Jacobi quartic identity among the three theta constants with even (integer) characteristics:

$$\theta^4 \begin{bmatrix} 0 \\ 0 \end{bmatrix} = \theta^4 \begin{bmatrix} 0 \\ 1 \end{bmatrix} + \theta^4 \begin{bmatrix} 1 \\ 0 \end{bmatrix}. \tag{3.18}$$

We present several proofs of this identity in this volume. See the end of [**6**] for a proof that is more standard than the ones here.

Till now we have been concentrating on the four classes of characteristics which are fixed pointwise by $\Gamma(3)$. As we saw, representatives for these can be taken to be

$$\begin{bmatrix} 1 \\ \\ \frac{1}{3} \end{bmatrix}, \begin{bmatrix} \frac{1}{3} \\ \\ 1 \end{bmatrix}, \begin{bmatrix} \frac{1}{3} \\ \\ \frac{1}{3} \end{bmatrix}, \begin{bmatrix} \frac{1}{3} \\ \\ \frac{5}{3} \end{bmatrix}.$$

As for the case $k = 2$, we shall label these characteristics by the integers

$$\begin{bmatrix} \frac{1}{3} \\ \\ \frac{1}{3} \end{bmatrix} \sim 1, \begin{bmatrix} \frac{1}{3} \\ \\ 1 \end{bmatrix} \sim 2, \begin{bmatrix} \frac{1}{3} \\ \\ \frac{5}{3} \end{bmatrix} \sim 3, \begin{bmatrix} 1 \\ \\ \frac{1}{3} \end{bmatrix} \sim 4.$$

We recall the generators B and A of the modular group. In the usual notation of permutations, $\eta(B)$, the permutation induced by B is $(1\ 3)(1\ 2)$ and $\eta(A) = (1\ 3)(2\ 4)$. These two even permutations generate the alternating group on four elements, $\mathcal{A}_4$. We have thus shown that $\Gamma/\Gamma(3)$ is isomorphic to $\mathcal{A}_4$. As a matter of fact,

$$\eta(B) = (1\ 3)(1\ 2),\ \eta(A) = (1\ 3)(2\ 4),\ \eta(B^2) = (1\ 2)(1\ 3),$$
$$\eta(B \circ A) = (2\ 3)(2\ 4),\ \eta(A \circ B) = (1\ 2)(1\ 4),\ \eta(B^2 \circ A) = (1\ 4)(1\ 2),$$
$$\eta(A\circ B\circ A) = (1\,4)(1\,3),\ \eta(B\circ A\circ B^2) = (1\,2)(3\,4),\ \eta(B^2\circ A\circ B) = (1\,4)(2\,3),$$
$$\eta(A \circ B^2) = (2\ 4)(2\ 3) \text{ and } \eta(A \circ B^2 \circ A) = (1\ 3)(1\ 4).$$

We have already seen that we have identities among the third powers of the four theta constants with characteristics in the above list. We now show how to use these identities to obtain a representation of $\mathcal{A}_4$ as a group of fractional linear transformations, and, more importantly, to obtain a modular invariant which generalizes the classical λ function. To simplify notation, we let $\theta\begin{bmatrix} \epsilon \\ \epsilon' \end{bmatrix} = \theta_i$, provided we have identified the characteristic $\begin{bmatrix} \epsilon \\ \epsilon' \end{bmatrix}$ with the integer (index) i. Here i ranges over the integers 1, 2, 3 and 4. In this notation (after some trivial substitutions), the identities we have derived are:

$$\theta_3^3 = \theta_2^3 - \theta_1^3 \text{ and } \theta_4^3 = \epsilon^2\theta_2^3 + \theta_1^3, \tag{3.19}$$

where $\epsilon = \exp\frac{\pi\imath}{3}$. Recall the definition $\lambda(\tau) = \frac{\theta_1^3}{\theta_2^3}$, $\tau \in \mathbb{H}^2$. We have already seen that this function provides us with a conformal homeomorphism of $\mathbb{H}^2/\Gamma(3)$ on a four times punctured sphere. One easily computes the punctures to be 0, 1, ϵ^5, and ∞. In general, $\Gamma/\Gamma(k)$ is isomorphic to $\mathrm{Aut}(\mathbb{H}^2/\Gamma(k))$, the group of conformal automorphisms of the Riemann surface $\mathbb{H}^2/\Gamma(k)$. If we choose a holomorphic universal cover

$$f : \mathbb{H}^2 \to \mathbb{H}^2/\Gamma(k),$$

then we obtain a surjective homomorphism

$$\Pi : \Gamma \to \ \mathrm{Aut}(\mathbb{H}^2/\Gamma(k))$$

with kernel $\Gamma(k)$ defined by

$$f \circ \gamma = \Pi(\gamma) \circ f.$$

For $k = 2,\ 3$ and 5, $\mathbb{H}^2/\Gamma(k)$ is a punctured sphere, and hence $\mathrm{Aut}(\mathbb{H}^2/\Gamma(k))$ is a group of Möbius transformations. We proceed to describe the homomorphism Π for the case $k = 3$ and $f = \lambda$.

As already remarked, the group Γ operates (on the right) on the function λ by substitution (composition). Since the action of $\Gamma(3)$ is trivial, we actually have an action of $\Gamma(3)\backslash\Gamma$ on λ. We choose (in the table below) a set of representatives for the nontrivial left cosets of $\Gamma(3)\backslash\Gamma$ and use the transformation formulae to obtain

$$\lambda \circ B = \pm\frac{\theta_3^3}{\theta_1^3},\ \lambda \circ A = \pm\epsilon^5\frac{\theta_3^3}{\theta_4^3},\ \lambda \circ (B^2) = \pm\frac{\theta_2^3}{\theta_3^3},\ \lambda \circ (B \circ A) = \pm\epsilon^5\frac{\theta_1^3}{\theta_3^3},$$

$$\lambda \circ (A \circ B) = \pm\epsilon^4\frac{\theta_2^3}{\theta_4^3},\ \lambda \circ (B^2 \circ A) = \pm\epsilon^5\frac{\theta_4^3}{\theta_1^3},$$

$$\lambda \circ (A \circ B \circ A) = \pm\epsilon^4\frac{\theta_4^3}{\theta_2^3},\ \lambda \circ (B \circ A \circ B^2) = \pm\epsilon^5\frac{\theta_2^3}{\theta_1^3},\ \lambda \circ (B^2 \circ A \circ B) = \pm\frac{\theta_4^3}{\theta_3^3},$$

$$\lambda \circ (A \circ B^2) = \pm\frac{\theta_1^3}{\theta_4^3} \text{ and } \lambda \circ (A \circ B^2 \circ A) = \pm\epsilon^2\frac{\theta_3^3}{\theta_2^3}.$$

We need to eliminate the ambiguous signs and to express the right hand sides of the above equations in terms of λ. We use the identities among the cubes of the theta constants (3.19), the fact that $\Pi(\gamma)$ has finite order for all $\gamma \in \Gamma$ (hence the trace of each of these fractional linear transformations lies in the open real interval $(-2, 2)$), and (when needed) the fact that the Möbius transformation $\Pi(\gamma)$ permutes the four points in $\{\infty,\ 0,\ 1,\ \epsilon^5\}$. We compute

$$\lambda \circ B = \frac{\lambda - 1}{\lambda},\ \lambda \circ A = \frac{\lambda - 1}{\epsilon\lambda - 1},\ \lambda \circ (B^2) = \frac{1}{1 - \lambda},\ \lambda \circ (B \circ A) = \frac{\epsilon^5\lambda}{\lambda - 1},$$

$$\lambda \circ (A \circ B) = \frac{\epsilon}{\lambda + \epsilon^2},\ \lambda \circ (B^2 \circ A) = \frac{\epsilon\lambda - 1}{\epsilon^2\lambda},$$

$$\lambda \circ (A \circ B \circ A) = 1 - \epsilon\lambda,\ \lambda \circ (B \circ A \circ B^2) = \frac{1}{\epsilon\lambda},\ \lambda \circ (B^2 \circ A \circ B) = \frac{\lambda + \epsilon^2}{\lambda - 1},$$

$$\lambda \circ (A \circ B^2) = \frac{\epsilon\lambda}{\epsilon\lambda - 1} \text{ and } \lambda \circ (A \circ B^2 \circ A) = \epsilon^5(1 - \lambda).$$

The above list produces 11 holomorphic functions on $\mathbb{H}^2$. We label them $\{f_1,\ \ldots,\ f_{11}\}$. We label the "identity function" $\tau \mapsto \lambda(\tau)$ by f_{12}. We form for $n = 1,\ 2,\ 3,\ \ldots$,

$$\varphi_n = \sum_{i=1}^{12} f_i^n.$$

Theorem 3.14. *The elementary symmetric functions (more generally any symmetric polynomial) of the* 12 *functions defined above are invariant under the modular group. In particular,*

$$\varphi_1 = 4(1 - \epsilon^2),$$

$$\varphi_2 = -4\epsilon^2,$$

and

$$\varphi_3 = 3\left(\frac{\lambda(\lambda-1)(\epsilon\lambda-1)}{\epsilon} - \frac{(\lambda-1)(\epsilon\lambda-1)}{\lambda^3} + \frac{\epsilon\lambda(\epsilon\lambda-1)}{(\lambda-1)^3} - \frac{\lambda(\lambda-1)}{(\epsilon\lambda-1)^3}\right)$$

is a nontrivial modular function with a simple pole at $\imath\infty$ (in fact φ_3 is a branched holomorphic universal cover of the orbifold $\mathbb{H}^2/\Gamma$ and defines a 12 sheeted cover of $\mathbb{H}^2/\Gamma$ by $\mathbb{H}^2/\Gamma(3)$).

Proof. Note that λ is invariant under $\Gamma(3)$ and $\varphi_n = \sum_{\gamma\in\Gamma(3)\backslash\Gamma} \lambda^n \circ \gamma$ is invariant under Γ. Calculations yield the formulae for the sums of the first three powers of λ. The proofs of the other assertions are routine. We are averaging a function of degree three on $\mathbb{H}^2/\Gamma(3)$ with respect to the group $\Gamma/\Gamma(3)$ of order 12. One expects a function of degree 36. Clearly, we have some cancellation.

Remark 3.15. The function which sends $\tau \in \mathbb{H}^2$ to

$$\lambda + \lambda\circ B + \lambda\circ B^2 = \frac{\lambda^3 - 3\lambda + 1}{\lambda(\lambda-1)}$$

is automorphic for the group $G(3) =< \Gamma(3),\ B >$. This defines a three sheeted cover of $\mathbb{H}^2/G(3)$ by $\mathbb{H}^2/\Gamma(3)$. This fact opens up a series of problems that will be pursued further.

We finally turn our attention to the "complementary set" (consisting of classes represented by nonintegral characteristics of the form $\begin{bmatrix} \frac{m}{3} \\ \frac{m'}{3} \end{bmatrix}$ with m and m' integers, not both odd) of twelve classes of characteristics represented by

$$\begin{bmatrix} 0 \\ \frac{1}{3} \end{bmatrix} \sim 1,\ \begin{bmatrix} 0 \\ \frac{2}{3} \end{bmatrix} \sim 2,\ \begin{bmatrix} \frac{1}{3} \\ 0 \end{bmatrix} \sim 3,\ \begin{bmatrix} \frac{1}{3} \\ \frac{2}{3} \end{bmatrix} \sim 4,\ \begin{bmatrix} \frac{1}{3} \\ \frac{4}{3} \end{bmatrix} \sim 5,\ \begin{bmatrix} \frac{2}{3} \\ \frac{1}{3} \end{bmatrix} \sim 6,$$

$$\begin{bmatrix} \frac{2}{3} \\ 1 \end{bmatrix} \sim 7,\ \begin{bmatrix} \frac{4}{3} \\ \frac{1}{3} \end{bmatrix} \sim 8,\ \begin{bmatrix} \frac{2}{3} \\ \frac{2}{3} \end{bmatrix} \sim 9,\ \begin{bmatrix} \frac{2}{3} \\ \frac{4}{3} \end{bmatrix} \sim 10,\ \begin{bmatrix} \frac{2}{3} \\ 0 \end{bmatrix} \sim 11,\ \begin{bmatrix} 1 \\ \frac{2}{3} \end{bmatrix} \sim 12.$$

The group $\Gamma(3)$ is generated by the motions

$$T_2 : z \mapsto \frac{z}{3z+1},\quad T_3 : z \mapsto \frac{-2z-3}{3z+4},\quad T_4 : z \mapsto \frac{4z-3}{3z-2}$$

and

$$T_4 \circ T_2 \circ T_3 = T_1 : z \mapsto z + 3.$$

It acts on the above equivalence classes of characteristics (by permutation). This defines a homomorphism from $\Gamma(3)$ to $\mathcal{S}_{12}$. It can be shown that $\Gamma(6)$ is the kernel of this homomorphism.

The permutation induced by T_2 is $(2\ 12)(3\ 11)(4\ 10)(5\ 9)$ while the permutation induced by T_3 is $(1\ 12)(3\ 7)(4\ 6)(5\ 8)$. It is not necessary to calculate the permutation induced by T_4 since $T_3 \circ T_4^{-1} : z \mapsto \frac{13z-18}{-18z+25} \in \Gamma(6)$; T_4 will induce the same permutation as T_3. The above permutations induced by T_2 and T_3 generate a subgroup G of order 6 of $\mathcal{S}_{12}$ (the image of $\Gamma(3)$ under the homomorphism).

If we denote by Y the space of equivalence classes of the 12 characteristics in our list, then

$$\begin{aligned} Y/G = \{a = & \ [1,1,12,12,2,2],\ b = [3,11,7,11,7,3], \\ c = & \ [4,10,6,10,4,6],\ d = [5,9,8. 9,8,5]\} \end{aligned}$$

has four elements and Γ acts on Y/G. We check the action of Γ on these four elements. We easily find that B induces the permutation $(d\ c)(d\ b)$ and that A induces the permutation $(a\ b)(c\ d)$. If we now compare this with the action of Γ on the four (fractional) equivalence classes of characteristics fixed pointwise by $\Gamma(3)$, we find that we should associate

the class of the characteristic $\begin{bmatrix} 1 \\ \frac{1}{3} \end{bmatrix}$ with the orbit $[1,1,2,2,12,12]$,

the class of the characteristic $\begin{bmatrix} \frac{1}{3} \\ \frac{1}{3} \end{bmatrix}$ with the orbit [5,5,8,8,9,9],

the class of the characteristic $\begin{bmatrix} \frac{1}{3} \\ \frac{5}{3} \end{bmatrix}$ with the orbit [4,10,6,10,4,6],

and

the class of the characteristic $\begin{bmatrix} \frac{1}{3} \\ 1 \end{bmatrix}$ with the orbit [3,3,7,7,11,11].

The above arguments suggest the following theorem which gives a set of proportionalities with a different flavor than the ones obtained for $k = 2$.

Theorem 3.16. *The quotient of any two of the following products is a constant on the upper half plane:*

$$\theta\begin{bmatrix} 0 \\ \frac{1}{3} \end{bmatrix}\theta\begin{bmatrix} 0 \\ \frac{2}{3} \end{bmatrix}\theta\begin{bmatrix} 1 \\ \frac{2}{3} \end{bmatrix},\quad \theta\begin{bmatrix} \frac{1}{3} \\ 0 \end{bmatrix}\theta\begin{bmatrix} \frac{2}{3} \\ 1 \end{bmatrix}\theta\begin{bmatrix} \frac{2}{3} \\ 0 \end{bmatrix},$$

$$\theta\begin{bmatrix} \frac{1}{3} \\ \frac{2}{3} \end{bmatrix}\theta\begin{bmatrix} \frac{2}{3} \\ \frac{4}{3} \end{bmatrix}\theta\begin{bmatrix} \frac{2}{3} \\ \frac{1}{3} \end{bmatrix},\quad \theta\begin{bmatrix} \frac{1}{3} \\ \frac{4}{3} \end{bmatrix}\theta\begin{bmatrix} \frac{4}{3} \\ \frac{1}{3} \end{bmatrix}\theta\begin{bmatrix} \frac{2}{3} \\ \frac{2}{3} \end{bmatrix}.$$

Proof. We consider the 24-th powers of the above products. Each is a multiplicative (associated to some character (homomorphism of $\Gamma(3)$ into the unit circle)) automorphic 18-form, say φ. The ratio of two of these is hence a multiplicative automorphic function f for the group $\Gamma(3)$. The singularities and zeros of the projection F of f to $\mathbb{H}^2/\Gamma(3)$ are at the punctures. The form Φ has order -9 at each of the punctures. The (multivalued) multiplicative function F is hence regular at the punctures (thus the character associated with F must be trivial). We conclude that F is a (single valued) holomorphic function on the compactification of $\mathbb{H}^2/\Gamma(3)$ and hence constant.

We compute the value of the function λ at the cusps of $\Gamma(3)$. From the divisor of the projection of λ to $\overline{\mathbb{H}^2/\Gamma(3)}$ we obtain,

$$\lambda(-1) = 0 \text{ and } \lambda(0) = \infty.$$

The definition of λ tells us directly that

$$\lambda(\infty) = \exp\left\{\frac{-\pi\imath}{3}\right\}.$$

To compute the fourth puncture on $\mathbb{H}^2/\Gamma(3)$, we need to evaluate $\lambda(1)$. Direct application of the formulae of §2.4 will result in a sign ambiguity. But the first identity of Theorem 3.12 yields

$$\lambda(1) = 1.$$

3.4. $k = 4$. The surface $\mathbb{H}^2/\Gamma(4)$ is of type $(0, 6)$. The six punctures on it are the images under P of the parabolic fixed points

$$\infty,\ 0,\ 1,\ 2,\ 3,\ \frac{1}{2}.$$

As in the case of $k = 3$ we abbreviate $\theta^8 = \varphi$. Using Mathematica computations of the divisors of automorphic forms yield:

$$\left(\Phi\begin{bmatrix}0\\0\end{bmatrix}\right) = (P_\infty^{-1}P_0^{-1}P_1P_2^{-1}P_3P_{\frac{1}{2}}^{-1})^2,$$

$$\left(\Phi\begin{bmatrix}1\\0\end{bmatrix}\right) = (P_\infty P_0^{-1}P_1^{-1}P_2^{-1}P_3^{-1}P_{\frac{1}{2}})^2,$$

$$\left(\Phi\begin{bmatrix}0\\1\end{bmatrix}\right) = (P_\infty^{-1}P_0P_1^{-1}P_2P_3^{-1}P_{\frac{1}{2}}^{-1})^2,$$

$$\left(\Phi\begin{bmatrix}0\\\frac{1}{2}\end{bmatrix}\right) = P_\infty^{-2}P_0^{-1}P_1^{-1}P_2^{-1}P_3^{-1}P_{\frac{1}{2}}^2,$$

$$\left(\Phi\begin{bmatrix}\frac{1}{2}\\0\end{bmatrix}\right) = P_\infty^{-1}P_0^{-2}P_1^{-1}P_2^2P_3^{-1}P_{\frac{1}{2}}^{-1},$$

$$\left(\Phi\begin{bmatrix}\frac{1}{2}\\\\\frac{1}{2}\end{bmatrix}\right) = P_\infty^{-1}P_0^{-1}P_1^{-2}P_2^{-1}P_3^2P_{\frac{1}{2}}^{-1},$$

Value at	f	g	h
∞	$-\imath$	∞	$-\imath$
0	∞	1	-1
1	1	$-\imath$	∞
2	0	-1	1
3	-1	$\imath$	0
$\frac{1}{2}$	$\imath$	0	$\imath$

Table 6. VALUES OF f, g AND h AT THE SIX PUNCTURES.

$$\left(\Phi\begin{bmatrix}\frac{1}{2}\\1\end{bmatrix}\right)=P_\infty^{-1}P_0^{2}P_1^{-1}P_2^{-2}P_3^{-1}P_{\frac{1}{2}}^{-1},$$

$$\left(\Phi\begin{bmatrix}\frac{1}{2}\\ \\ \frac{3}{2}\end{bmatrix}\right)=P_\infty^{-1}P_0^{-1}P_1^{2}P_2^{-1}P_3^{-2}P_{\frac{1}{2}}^{-1},$$

$$\left(\Phi\begin{bmatrix}1\\ \frac{1}{2}\end{bmatrix}\right)=P_\infty^{2}P_0^{-1}P_1^{-1}P_2^{-1}P_3^{-1}P_{\frac{1}{2}}^{-2}.$$

It follows from these computations (or directly) that

$$f=\frac{\theta^2\begin{bmatrix}\frac{1}{2}\\ \\ 0\end{bmatrix}}{\theta^2\begin{bmatrix}\frac{1}{2}\\ \\ 1\end{bmatrix}},\ g=\frac{\theta^2\begin{bmatrix}0\\ \\ \frac{1}{2}\end{bmatrix}}{\theta^2\begin{bmatrix}1\\ \\ \frac{1}{2}\end{bmatrix}}\text{ and }h=\frac{\theta^2\begin{bmatrix}\frac{1}{2}\\ \\ \frac{1}{2}\end{bmatrix}}{\theta^2\begin{bmatrix}\frac{1}{2}\\ \\ \frac{3}{2}\end{bmatrix}}$$

are automorphic functions for $\Gamma(4)$. It is by now routine to conclude that

$$(F)=\frac{P_2}{P_0},\ (G)=\frac{P_{\frac{1}{2}}}{P_\infty},\ (H)=\frac{P_3}{P_1}.$$

Hence f, g, and h are universal covering maps of $\mathbb{H}^2/\Gamma(4)$. Thus there exist Möbius transformations C and D such that

$$g=C\circ f\text{ and }h=D\circ f.$$

To evaluate these Möbius transformations C and D, we compute the values of f, g, and h at representative inequivalent parabolic fixed points (only three such points are needed). These computations are summarized in a short table.

From our table we conclude that $C = \begin{bmatrix} 1 & -\imath \\ 1 & \imath \end{bmatrix}$ and $D = \begin{bmatrix} 1 & 1 \\ -1 & 1 \end{bmatrix}$, and hence we have obtained the theta identities

$$\begin{aligned}&\theta^2\begin{bmatrix}0\\ \frac{1}{2}\end{bmatrix}\theta^2\begin{bmatrix}\frac{1}{2}\\ 0\end{bmatrix}+\imath\theta^2\begin{bmatrix}0\\ \frac{1}{2}\end{bmatrix}\theta^2\begin{bmatrix}\frac{1}{2}\\ 1\end{bmatrix}\\ &=\theta^2\begin{bmatrix}\frac{1}{2}\\ 0\end{bmatrix}\theta^2\begin{bmatrix}1\\ \frac{1}{2}\end{bmatrix}-\imath\theta^2\begin{bmatrix}\frac{1}{2}\\ 1\end{bmatrix}\theta^2\begin{bmatrix}1\\ \frac{1}{2}\end{bmatrix}\end{aligned}$$

and

$$\begin{aligned}&\theta^2\begin{bmatrix}\frac{1}{2}\\ \\ \frac{1}{2}\end{bmatrix}\theta^2\begin{bmatrix}\frac{1}{2}\\ \\ 1\end{bmatrix}-\theta^2\begin{bmatrix}\frac{1}{2}\\ \\ 0\end{bmatrix}\theta^2\begin{bmatrix}\frac{1}{2}\\ \\ \frac{1}{2}\end{bmatrix}\\ &=\theta^2\begin{bmatrix}\frac{1}{2}\\ \\ 0\end{bmatrix}\theta^2\begin{bmatrix}\frac{1}{2}\\ \\ \frac{3}{2}\end{bmatrix}+\theta^2\begin{bmatrix}\frac{1}{2}\\ \\ 1\end{bmatrix}\theta^2\begin{bmatrix}\frac{1}{2}\\ \\ \frac{3}{2}\end{bmatrix}.\end{aligned}$$

Simple calculations show that

$$f(\tau+1)=\frac{\imath}{h(\tau)},\quad g(\tau+1)=-\imath g(\tau),\ \text{and}\ h(\tau+1)=f(\tau),\ \text{for all}\ \tau\in\mathbb{H}^2.$$

We conclude that

$$f^4(\tau+2)=\frac{1}{f^4(\tau)},\quad h^4(\tau+2)=\frac{1}{h^4(\tau)},\quad \text{all}\ \tau\in\mathbb{H}^2,$$

and g^4 is a modular function for $G(4)=\Gamma_o(4)$. Computing on $\mathbb{H}^2/G(4)$, we see that

$$(G^4)=\frac{P_{\frac{1}{2}}}{P_\infty}.$$[70]

This last property of G should not be much of a surprise if the reader recalls the work done on towers of characteristics. We have hence obtained part of

Theorem 3.17. *The two functions on* $\mathbb{H}^2$,

$$\tau\mapsto\frac{\theta^8\begin{bmatrix}0\\ \frac{1}{2}\end{bmatrix}(0,\tau)}{\theta^8\begin{bmatrix}1\\ \frac{1}{2}\end{bmatrix}(0,\tau)}\quad \text{and}\ \tau\mapsto\frac{\theta^8\begin{bmatrix}1\\ 0\end{bmatrix}(0,\tau)}{\theta^8\begin{bmatrix}1\\ \frac{1}{2}\end{bmatrix}(0,\tau)},$$

are holomorphic universal covers of $\mathbb{H}^2/G(4)\cong\mathbb{C}-\{0,\ 1\}$, *and hence we have the theta identity*

$$\theta^8\begin{bmatrix}1\\ \frac{1}{2}\end{bmatrix}\theta^8\begin{bmatrix}1\\ 0\end{bmatrix}+16\theta^8\begin{bmatrix}1\\ \frac{1}{2}\end{bmatrix}\theta^8\begin{bmatrix}0\\ \frac{1}{2}\end{bmatrix}=\theta^8\begin{bmatrix}0\\ \frac{1}{2}\end{bmatrix}\theta^8\begin{bmatrix}1\\ 0\end{bmatrix}.$$

[70] A local coordinate at P_∞ on $\mathbb{H}^2/\Gamma(4)$ is $\exp(\pi\imath\zeta/2)$, $\zeta\in\mathbb{H}^2$ with $\Im\zeta>4$, while on $\mathbb{H}^2/G(4)$, it is $\exp(2\pi\imath\zeta)$, $\zeta\in\mathbb{H}^2$ with $\Im\zeta>1$; that is, the natural projection

$$\overline{\mathbb{H}^2/\Gamma(4)}\to\overline{\mathbb{H}^2/G(4)}$$

is four to one (globally and) in a neighborhood of $P_\infty\in\overline{\mathbb{H}^2/\Gamma(4)}$.

Proof. The second function has divisor $\frac{P_{\frac{1}{2}}}{P_0}$. The value of the first function at 0, $\frac{1}{2}$ and ∞ is 1, 0 and ∞, respectively; of the second, ∞, 0 and 1. The theorem easily follows from these observations.

Theorem 3.18. *For all* $\tau \in \mathbb{H}^2$,

$$\frac{\theta^2\begin{bmatrix} 0 \\ \frac{1}{2} \end{bmatrix}(0,\tau)}{\theta^2\begin{bmatrix} 1 \\ \frac{1}{2} \end{bmatrix}(0,\tau)} = \frac{\theta\begin{bmatrix} 0 \\ 0 \end{bmatrix}(0,2\tau)}{\theta\begin{bmatrix} 1 \\ 0 \end{bmatrix}(0,2\tau)} \quad \textit{and} \quad \frac{\theta^2\begin{bmatrix} 1 \\ 0 \end{bmatrix}(0,\tau)}{\theta^2\begin{bmatrix} 1 \\ \frac{1}{2} \end{bmatrix}(0,\tau)} = 2\frac{\theta\begin{bmatrix} 0 \\ 0 \end{bmatrix}(0,2\tau)}{\theta\begin{bmatrix} 0 \\ 1 \end{bmatrix}(0,2\tau)}.$$

Proof. Both identities are direct consequences of the addition formula for theta constants discussed for the case $k = 2$.

Many times we can reverse the above process and derive relations between functions as a consequence of θ-identities. For example, our work on N-th order theta functions in Chapter 2 (see the derivation of (4.5)) leads to the identity

$$1 = \frac{\theta^4\begin{bmatrix} 1 \\ \frac{1}{2} \end{bmatrix}}{\theta^4\begin{bmatrix} 0 \\ \frac{1}{2} \end{bmatrix}} + \frac{\theta^3\begin{bmatrix} 0 \\ 1 \end{bmatrix}\theta\begin{bmatrix} 0 \\ 0 \end{bmatrix}}{\theta^4\begin{bmatrix} 0 \\ \frac{1}{2} \end{bmatrix}},$$

a relation ($G^2(1 - W) = 1$) between functions G (previously defined) and W on $\overline{\mathbb{H}^2/\Gamma(4)}$ $\left(w = \frac{\theta^3\begin{bmatrix} 0 \\ 1 \end{bmatrix}\theta\begin{bmatrix} 0 \\ 0 \end{bmatrix}}{\theta^4\begin{bmatrix} 0 \\ \frac{1}{2} \end{bmatrix}}, \text{ with divisor } \frac{P_0\, P_2}{P_{\frac{1}{2}}^2} \right)$.

Remark 3.19. The above table giving the values of the functions f, g, h at the cusps shows that under stereographic projection the punctures on $\mathbb{H}^2/\Gamma(4)$ correspond to the vertices on the sphere of a regular octahedron and in particular, the automorphism group $\Gamma/\Gamma(4)$ of $\mathbb{H}^2/\Gamma(4)$ is isomorphic to the octahedral group. A similar claim, involving the tetrahedron, holds for $k = 3$, but the normalization we used for the function λ was not the proper one to see this.

Exercise 3.20. Verify the formulae for the divisors of the automorphic functions and forms encountered in this subsection.

3.5. $k = 5$. The surface $\mathbb{H}^2/\Gamma(5)$ is of type (0,12). The twelve punctures on it are the images of the parabolic fixed points

$$\infty, 0, 1, 2, -1, -2, \frac{1}{2}, \frac{3}{2}, \frac{5}{2}, -\frac{1}{2}, -\frac{3}{2} \text{ and } \frac{2}{5}.$$

If we abbreviate $\theta^{40} = \varphi$, then as in the previous cases, we find that

$$\left(\Phi\begin{bmatrix}\frac{1}{5}\\[4pt]\frac{1}{5}\end{bmatrix}\right) = \frac{P_{-1}^{15}P_{\frac{3}{2}}^{15}}{P_{\infty}^{9}P_{0}^{9}P_{1}P_{2}P_{-2}^{9}P_{\frac{1}{2}}P_{\frac{5}{2}}P_{-\frac{1}{2}}^{9}P_{-\frac{3}{2}}^{9}P_{\frac{2}{5}}},$$

$$\left(\Phi\begin{bmatrix}\frac{1}{5}\\[4pt]\frac{3}{5}\end{bmatrix}\right) = \frac{P_{2}^{15}P_{-\frac{1}{2}}^{15}}{P_{\infty}^{9}P_{0}P_{1}^{9}P_{-1}P_{-2}^{9}P_{\frac{1}{2}}P_{\frac{3}{2}}^{9}P_{\frac{5}{2}}^{9}P_{-\frac{3}{2}}P_{\frac{2}{5}}},$$

$$\left(\Phi\begin{bmatrix}\frac{1}{5}\\[4pt]\frac{7}{5}\end{bmatrix}\right) = \frac{P_{-2}^{15}P_{\frac{1}{2}}^{15}}{P_{\infty}^{9}P_{0}P_{1}P_{2}^{9}P_{-1}^{9}P_{\frac{3}{2}}P_{\frac{5}{2}}^{9}P_{-\frac{1}{2}}P_{-\frac{3}{2}}^{9}P_{\frac{2}{5}}},$$

$$\left(\Phi\begin{bmatrix}\frac{1}{5}\\[4pt]\frac{9}{5}\end{bmatrix}\right) = \frac{P_{1}^{15}P_{-\frac{3}{2}}^{15}}{P_{\infty}^{9}P_{0}^{9}P_{2}^{9}P_{-1}P_{-2}P_{\frac{1}{2}}^{9}P_{\frac{3}{2}}^{9}P_{\frac{5}{2}}P_{-\frac{1}{2}}P_{\frac{2}{5}}},$$

$$\left(\Phi\begin{bmatrix}\frac{3}{5}\\[4pt]\frac{3}{5}\end{bmatrix}\right) = \frac{P_{-1}^{15}P_{\frac{3}{2}}^{15}}{P_{\infty}P_{0}P_{1}^{9}P_{2}^{9}P_{-2}P_{\frac{1}{2}}^{9}P_{\frac{5}{2}}^{9}P_{-\frac{1}{2}}P_{-\frac{3}{2}}P_{\frac{2}{5}}^{9}},$$

$$\left(\Phi\begin{bmatrix}\frac{3}{5}\\[4pt]\frac{7}{5}\end{bmatrix}\right) = \frac{P_{1}^{15}P_{-\frac{3}{2}}^{15}}{P_{\infty}P_{0}P_{2}P_{-1}^{9}P_{-2}^{9}P_{\frac{1}{2}}P_{\frac{3}{2}}P_{\frac{5}{2}}^{9}P_{-\frac{1}{2}}^{9}P_{\frac{2}{5}}^{9}}.$$

We have described above the divisors of cusp forms for only six of the twelve elements of $X_o(5)$ since they are the ones we shall need in what follows. Fifth powers of ratios of theta constants give us meromorphic functions of either degree 7 or degree 5 on the 12-punctured sphere $\mathbb{H}^2/\Gamma(5)$. The divisors of these degree 5 functions are supported at 10 of the 12 punctures. Each of these has 5 simple zeros and 5 simple poles. We illustrate with

$$\left(\frac{\theta^5\begin{bmatrix}\frac{1}{5}\\[4pt]\frac{1}{5}\end{bmatrix}}{\theta^5\begin{bmatrix}\frac{3}{5}\\[4pt]\frac{3}{5}\end{bmatrix}}\right) = \frac{P_1P_2P_{\frac{1}{2}}P_{\frac{5}{2}}P_{\frac{2}{5}}}{P_{\infty}P_0P_{-2}P_{-\frac{1}{2}}P_{-\frac{3}{2}}}.$$

Let us call the function produced above[71] F. Observe that F is regular and nonzero at the two punctures P_{-1} and $P_{\frac{3}{2}}$. Riemann-Hurwitz tells us that

[71]Its lift to $\mathbb{H}^2$ is f.

F has total branch number 8. Let us denote $F(P_{-1})$ by ω_1 and $F(P_{\frac{3}{2}})$ by ω_2. Simple calculations show that

$$\omega_1 = \frac{\theta^5\begin{bmatrix} 1 \\ \frac{1}{5} \end{bmatrix}}{\theta^5\begin{bmatrix} 1 \\ \frac{3}{5} \end{bmatrix}}(0, \imath\infty) = \frac{\cos^5 \frac{\pi}{10}}{\cos^5 \frac{3\pi}{10}} \text{ and } \omega_2 = \frac{\theta^5\begin{bmatrix} 1 \\ \frac{7}{5} \end{bmatrix}}{\theta^5\begin{bmatrix} 1 \\ \frac{1}{5} \end{bmatrix}}(0, \imath\infty) = \frac{\cos^5 \frac{7\pi}{10}}{\cos^5 \frac{\pi}{10}},$$

and that the branch numbers of F at P_{-1} and $P_{\frac{3}{2}}$ must be ≥ 4. It follows that the branch number of F at each of these punctures is precisely four and that F is unbranched elsewhere. Hence a fifth root g of

$$\frac{f - \omega_1}{f - \omega_2}$$

is a holomorphic universal covering map of the 12-punctured sphere $\mathbb{H}^2/\Gamma(5)$ with a simple zero at -1 and a simple pole at $\frac{3}{2}$. By considering several degree 5 maps, one can once again produce relations among theta constants. We have therefore proven most of

Theorem 3.21. *The five single valued functions $(\frac{f-\omega_1}{f-\omega_2})^{\frac{1}{5}}$ are holomorphic universal coverings of the Riemann surface $S = \mathbb{H}^2/\Gamma(5)$. Moreover, there exist complex numbers $\omega_3 \neq \omega_4$ such that the functions*

$$g : \tau \mapsto \left(\frac{\theta^5\begin{bmatrix} 1 \\ \frac{1}{5} \end{bmatrix}(0, \tau) - \omega_3}{\theta^5\begin{bmatrix} 1 \\ \frac{3}{5} \end{bmatrix}(0, \tau) - \omega_4} \right)^{\frac{1}{5}}$$

are also holomorphic universal coverings of S by $\mathbb{H}^2$ and the fifth powers are holomorphic universal coverings of $\mathbb{H}^2/G(5)$.

Proof. The first statement has been proved before the statement of the theorem. The proof of the second statement follows, as does the first after one checks the respective divisors of the functions. The last statement is also an immediate consequence of the structure of the divisors.

Remark 3.22. Instead of the fifth powers of the last function, we could use the ratio $\dfrac{\theta^5\begin{bmatrix} 1 \\ \frac{1}{5} \end{bmatrix}}{\theta^5\begin{bmatrix} 1 \\ \frac{3}{5} \end{bmatrix}}$.

The last theorem tells us that

$$\begin{array}{ccc} \mathbb{H}^2 & \xrightarrow{h} & \mathbb{H}^2 \\ g \big\downarrow & & \big\downarrow g^5 \\ \mathbb{H}^2/\Gamma(5) \cong \mathbb{C} - \{z_1, \ldots, z_{11}\} & \xrightarrow{H} & \mathbb{H}^2/G(5) \cong \mathbb{C} - \{\zeta_1, \zeta_2, \zeta_3\} \end{array}$$

is a commutative diagram, where the eleven z_i and the three ζ_j are distinct complex numbers; each ζ_j is the fifth power of some z_i, in the specified coordinizations of the surfaces $\mathbb{H}^2/\Gamma(5)$ and $\mathbb{H}^2/G(5)$; the map H is given as raising to the fifth power; and h is defined to make the diagram commute.

Exercise 3.23. Compute the values z_i and ζ_j, and the Fourier series expansion of h in terms of the local coordinate $z = \exp\left(\frac{2\pi\imath\tau}{5}\right)$.

There is an alternate (simpler, by virtue of the fact that we have already computed above the divisors of the relevant forms) procedure to get the uniformizing function. It is clear that the quotient

$$\frac{\Phi\begin{bmatrix}\frac{1}{5}\\ \frac{1}{5}\end{bmatrix}\Phi\begin{bmatrix}\frac{1}{5}\\ \frac{3}{5}\end{bmatrix}\Phi\begin{bmatrix}\frac{3}{5}\\ \frac{7}{5}\end{bmatrix}}{\Phi\begin{bmatrix}\frac{1}{5}\\ \frac{7}{5}\end{bmatrix}\Phi\begin{bmatrix}\frac{1}{5}\\ \frac{9}{5}\end{bmatrix}\Phi\begin{bmatrix}\frac{3}{5}\\ \frac{3}{5}\end{bmatrix}}$$

is a meromorphic function on $\mathbb{H}^2/\Gamma(5)$ with a zero and pole of order 40 at P_2 and P_{-2}, respectively. Since the surface is of genus 0 we can extract a fortieth root and obtain

Theorem 3.24. *The function*

$$\frac{\theta\begin{bmatrix}\frac{1}{5}\\ \frac{1}{5}\end{bmatrix}\theta\begin{bmatrix}\frac{1}{5}\\ \frac{3}{5}\end{bmatrix}\theta\begin{bmatrix}\frac{3}{5}\\ \frac{7}{5}\end{bmatrix}}{\theta\begin{bmatrix}\frac{1}{5}\\ \frac{7}{5}\end{bmatrix}\theta\begin{bmatrix}\frac{1}{5}\\ \frac{9}{5}\end{bmatrix}\theta\begin{bmatrix}\frac{3}{5}\\ \frac{3}{5}\end{bmatrix}}$$

is a holomorphic universal covering map from $\mathbb{H}^2$ onto the twelve punctured sphere $\mathbb{H}^2/\Gamma(5)$ with a simple zero at 2 and a simple pole at -2.

3.6. $k = 6$. In this subsection we give a description of the (function theory on the) surface $\mathbb{H}^2/\Gamma(6)$ which we know to be a torus with twelve punctures. We shall compute the period (the τ parameter, a point in the upper half plane) of the compactification of this punctured torus (it turns out to be ω, defined below, – a cube root of -1) and identify the punctures. We will show that, under proper normalizations, the punctures correspond to the following set of 12 points: the origin, the 3 points of order 2, the points of order 3

$$\frac{1+\omega}{3} \text{ and } \frac{2+2\omega}{3},$$

and the 6 points of order 6

$$\frac{1+\omega}{6},\ \frac{1}{6}+\frac{2\omega}{3},\ \frac{1}{3}+\frac{5\omega}{6},\ \frac{2}{3}+\frac{\omega}{6},\ \frac{5}{6}+\frac{\omega}{3} \text{ and } \frac{5+5\omega}{6}.$$

We begin by establishing some known facts (that is, properties derived from the classical theory). Let $\tau \in \mathbb{C}$ and let $B_\tau : z \mapsto z+\tau$. Thus $B_1 = B$. Given a torus T and a point $x \in T$, there exists a $\tau \in \mathbb{H}^2$ and a conformal (surjective) map

$$f : T \to \mathbb{C}/ < B,\ B_\tau >$$

with $T(x) = 0$. The point τ is unique except that it may be replaced by $\gamma(\tau)$ with $\gamma \in \Gamma$ arbitrary. Having chosen τ, the map f is unique except for post composition by an automorphism of $\mathbb{C}/ < B,\ B_\tau >$ that fixes the origin. The stabilizer of a point on a torus is a cyclic group of order 2 except when τ is equivalent to $\imath$ or

$$\omega = \exp\left(\frac{\pi\imath}{3}\right) = \frac{1+\imath\sqrt{3}}{2}$$

modulo Γ (the stabilizer has order 4 or 6, respectively). We may thus identify[72] P_∞ with the origin on $T = \overline{\mathbb{H}^2/\Gamma(6)}$. (Here $\overline{X}$ denotes the compactification of the surface of finite type X obtained by adding the punctures.) The map $B \in \Gamma$ induces an automorphism $\tilde{B}$ of order 6 of the surface T that fixes the origin. This immediately implies that ω is "the" period ($\tau = \omega$) of the torus T, and that the map $\tilde{B}$ is realized in the canonical model of T as

$$\tilde{B} : z \mapsto \omega z.$$

Further (in the canonical model),
1) $\tilde{B}$ fixes only the origin on T,
2) $\tilde{B}^3$ fixes the origin and the 3 half periods[73] on T, and
3) $\tilde{B}^2$ fixes the origin and the 2 points of order three $\frac{1+\omega}{3}$ and $\frac{2(1+\omega)}{3}$.

The 12 punctures on T are (in the Fuchsian model) the images under

$$P : \mathbb{H}^2 \cup \mathbb{Q} \cup \{\infty\} \to T$$

of the parabolic fixed points

$$\infty,\ 0,\ 1,\ 2,\ 3,\ 4,\ 5, \frac{1}{2},\ \frac{3}{2},\ \frac{5}{2},\ \frac{1}{3} \text{ and } \frac{4}{3}.$$

[72]We will hence forth call $\mathbb{C}/ < B,\ B_\tau >$ the *canonical* model of T and by abuse of language describe points of T by their preimages in $\mathbb{C}$. Similarly, we will refer to $\overline{\mathbb{H}^2/\Gamma(6)}$ as the Fuchsian model of T.

[73]The half periods are the points of order 2. A point $x \in T$ has *order* e if e is the smallest positive integer such that $ex = 0$.

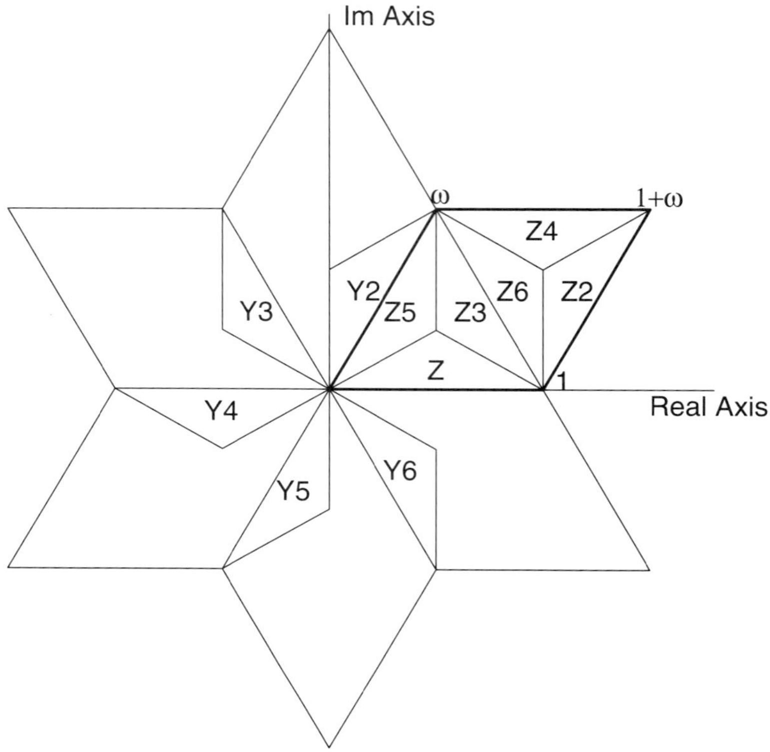

Figure 11. A fundamental domain for T_ω.

For the convenience of the reader, we include a figure that shows the period parallelogram $\mathcal{P}_0$ for $\tilde{G}_\omega =< B, B_\omega >$. It is the parrallelogram enclosed in bold lines in Figure 11. We have also shown the 5 rotations $\tilde{B}^j(\mathcal{P}_0)$, $j = 1, 2, 3, 4, 5$ of $\mathcal{P}_0$. The triangle $\mathrm{Z} \subset \mathcal{P}_0$ is a fundamental domain for the elementary group of signature $(1, 2;\ 2, 3)$ generated by B, B_ω and $\tilde{B}$. We have shown in Figure 11 $\mathrm{Yj} = \tilde{B}^{j-1}\mathrm{Z}$ and Zj, the translate of Yj to a subset of $\mathcal{P}_0$. The reader should also identify the 12 punctures in the figure.

We have already identified $P_\infty \in \overline{\mathbb{H}^2/\Gamma(6)}$ with $0 \in \mathbb{C}/ < B,\ B_\omega >$. Using the fixed points of $\tilde{B}^3$ we conclude that (in the Fuchsian model) the half periods on T are $P_{\frac{1}{2}}$, $P_{\frac{3}{2}}$ and $P_{\frac{5}{2}}$. The three fixed points of $\tilde{B}^2$ are P_∞, $P_{\frac{1}{3}}$ and $P_{\frac{4}{3}}$. Thus $P_{\frac{1}{3}}$ and $P_{\frac{4}{3}}$ are points of order 3. The group $< \tilde{B} >$ acts transitively on the half periods ($\tilde{B}^3$ fixes each of the half periods). It is convenient to follow f by either $\tilde{B}$ or $\tilde{B}^2$ if needed and hence identify $P_{\frac{1}{2}}$ with $\frac{1}{2}$. Hence $P_{\frac{3}{2}}$ and $P_{\frac{5}{2}}$ are identified with $\frac{\omega}{2}$ and $\frac{1+\omega}{2}$, respectively. We have now uniquely fixed the isomorphism between $\overline{\mathbb{H}^2/\Gamma(6)}$ and $\mathbb{C}/ < B,\ B_\omega >$. It should be possible to describe the punctures and automorphisms in the canonical model as well, and we do this now. These are, of course, presented to us in the Fuchsian model.

We have identified $\tilde{B}$ in the canonical model. To describe the full automorphism group $\Gamma/\Gamma(6)$ of $\mathbb{H}^2/\Gamma(6)$ we need only identify $\tilde{A}$ (since B and A generate Γ). Since $\tilde{A}$ is an involution, on T it is of the form

$$z \mapsto \pm z + a$$

for some $a \in \mathbb{C}$. An involution on T has 0 or 4 fixed points. Since A fixes $\imath \in \mathbb{H}^2$ and since the translation $z \mapsto z + a$ acts without fixed points on T, we conclude that

$$\tilde{A} : z \mapsto -z + a.$$

It is now routine to verify that

$$\tilde{A} \circ \tilde{B}^3 : z \mapsto z + a$$

is an element (of order a factor of 12) of $\Gamma/\Gamma(6)$. We need to compute the order of $a \in T$. The order of $a \in T$ is the same as the order of $A \circ B^3 \in \Gamma/\Gamma(6)$, which is easily seen to be 6. In the canonical model $\tilde{A}(0) = a$ and in the Fuchsian model $\tilde{A}(P_\infty) = P_0$. We conclude that P_0 is a point of order 6 on T. We write

$$a = \frac{m + m'\omega}{6},$$

where m and m' are integers to be determined. We may and do assume that these integers satisfy $0 \leq m \leq 5$ and $0 \leq m' \leq 5$. We note that $A \circ B^2 \circ A(\infty) = \frac{1}{2}$ and $A \circ \tilde{B}^2 \circ A(0) = \frac{1}{2}$. Direct computations show that

$$A \circ \tilde{B}^2 \circ A : z \mapsto \omega^2 z + a(1 - \omega^2),$$

from which it follows immediately that $m = 1 = m'$. It is now easy to identify all the punctures. We summarize the calculations in an included table.

Our theory has already shown us how to associate the punctures with the classes of the 12 characteristics (in $X_o(6)$) given in the following table:

$$\begin{matrix} \begin{bmatrix} 0 \\ \frac{1}{3} \end{bmatrix} & \begin{bmatrix} \frac{1}{3} \\ 0 \end{bmatrix} & \begin{bmatrix} \frac{1}{3} \\ \frac{2}{3} \end{bmatrix} & \begin{bmatrix} \frac{1}{3} \\ \frac{4}{3} \end{bmatrix} \\ \begin{bmatrix} 0 \\ \frac{2}{3} \end{bmatrix} & \begin{bmatrix} \frac{2}{3} \\ 1 \end{bmatrix} & \begin{bmatrix} \frac{2}{3} \\ \frac{4}{3} \end{bmatrix} & \begin{bmatrix} \frac{2}{3} \\ \frac{5}{3} \end{bmatrix} \\ \begin{bmatrix} 1 \\ \frac{2}{3} \end{bmatrix} & \begin{bmatrix} \frac{2}{3} \\ 0 \end{bmatrix} & \begin{bmatrix} \frac{2}{3} \\ \frac{1}{3} \end{bmatrix} & \begin{bmatrix} \frac{2}{3} \\ \frac{2}{3} \end{bmatrix} \end{matrix}.$$

Fuchsian model	Canonical model
P_∞	0
P_0	$\frac{1+\omega}{6}$
P_1	$\frac{5}{6}+\frac{\omega}{3}$
P_2	$\frac{2}{3}+\frac{\omega}{6}$
P_3	$\frac{5(1+\omega)}{6}$
P_4	$\frac{1}{6}+\frac{2\omega}{3}$
P_5	$\frac{1}{3}+\frac{5\omega}{6}$
$P_{\frac{1}{2}}$	$\frac{1}{2}$
$P_{\frac{3}{2}}$	$\frac{\omega}{2}$
$P_{\frac{5}{2}}$	$\frac{1+\omega}{2}$
$P_{\frac{1}{3}}$	$\frac{1+\omega}{3}$
$P_{\frac{4}{3}}$	$\frac{2(1+\omega)}{3}$

Table 7. THE PUNCTURES IN THE TWO MODELS.

Recall that our convention is to associate the puncture determined by the parabolic fixed point ∞ with the characteristic $\begin{bmatrix} 1 \\ \frac{2}{3} \end{bmatrix}$. The automorphism group $\Gamma/\Gamma(6)$ (of order 72) of the surface $\mathbb{H}^2/\Gamma(6)$ acts faithfully on this set of 12 classes of characteristics. As we saw above, the group $\Gamma/\Gamma(6)$ contains a cyclic subgroup of order 6 generated by $\tilde{B}$. The subgroup has a unique fixed point, the characteristic $\begin{bmatrix} 1 \\ \frac{2}{3} \end{bmatrix}$. The motion $\tilde{B}^3$ of period 2 has in addition to the fixed point of $\tilde{B}$, 3 additional fixed points. The map $\tilde{B}^2$ is an automorphism of period 3, and has in addition to the fixed point of $\tilde{B}$, 2 additional fixed points. Each of these 3 sets of characteristics and the remaining set of 6 characteristics is invariant under the cyclic group $<\tilde{B}>$. These facts were derived above from the Fuchsian representation of the torus T. They can also be established by examining the action of B on the 12 characteristics (as is done below). Algebraic information regarding the action of B on characteristics can be interpreted geometrically. The automorphism $\tilde{B}$ of $\mathbb{H}^2/\Gamma(6)$ gives a degree 6 map g from $\mathbb{H}^2/\Gamma(6)$ to $\mathbb{H}^2/\Gamma(6)/<\tilde{B}>$. As we have seen, this map (or rather its extension to T) has branching of order 5 at one puncture, branching of order 2 at two punctures, and simple branching at three other punctures. The quotient surface $\mathbb{H}^2/\Gamma(6)/<\tilde{B}>$ must be a punctured sphere (because g is branched) with punctures (coming from the punctures on $\mathbb{H}^2/\Gamma(6)$). Riemann-Hurwitz tells us that we have accounted for all the branching. Hence

$$g : \mathbb{H}^2/\Gamma(6) \to \mathbb{H}^2/\Gamma(6)/<\tilde{B}>$$

is a smooth 6 sheeted cover of a 4-punctured sphere by a 12-punctured torus. We will identify the 4 punctures on the sphere (under proper normalization) as 9, ∞, 1 and 0 after we determine some $< \Gamma(6), B >$-invariant functions on $\mathbb{H}^2$.

There is a second 6 sheeted cover of a 4 times punctured sphere ($\mathbb{H}^2/\Gamma(3)$) by $\mathbb{H}^2/\Gamma(6)$ as a consequence of the fact that $\Gamma(6)$ is a subgroup of index six in $\Gamma(3)$. This cover is not cyclic (its covering group is $\Gamma(3)/\Gamma(6) \cong \Gamma/\Gamma(2) \cong \mathcal{S}_3$, the permutation group on 3 letters). The quotient $\mathbb{H}^2/\Gamma(3)$ is conformally equivalent to the sphere punctured at 4 points. We have already seen in §3.3 that the punctures are $\lambda(-1) = 0$, $\lambda(0) = \infty$, $\lambda(1) = 1$ and $\lambda(\infty) = \frac{1}{2} - \imath\frac{\sqrt{3}}{2}$. By comparing cross ratios we see that the two 4-punctured spheres

$$\mathbb{H}^2/G(6) \cong \mathbb{H}^2/\Gamma(6)/ < \tilde{B} > \text{ and } \mathbb{H}^2/\Gamma(3)$$

are not conformally equivalent. It is also of interest to study the twelve sheeted cover $\mathbb{H}^2/\Gamma(6) \to \mathbb{H}^2/\Gamma(2)$.

We proceed to compute the action of the transformation $B : z \mapsto z + 1$ on the 12 characteristics. Under the action of B we have:

$$\begin{bmatrix} 0 \\ \frac{1}{3} \end{bmatrix} \mapsto \begin{bmatrix} 0 \\ \frac{2}{3} \end{bmatrix} \mapsto \begin{bmatrix} 0 \\ \frac{1}{3} \end{bmatrix}, \quad \begin{bmatrix} 1 \\ \frac{2}{3} \end{bmatrix} \mapsto \begin{bmatrix} 1 \\ \frac{2}{3} \end{bmatrix},$$

$$\begin{bmatrix} \frac{1}{3} \\ \\ 0 \end{bmatrix} \mapsto \begin{bmatrix} \frac{1}{3} \\ \\ \frac{4}{3} \end{bmatrix} \mapsto \begin{bmatrix} \frac{1}{3} \\ \\ \frac{2}{3} \end{bmatrix} \mapsto \begin{bmatrix} \frac{1}{3} \\ \\ 0 \end{bmatrix},$$

$$\begin{bmatrix} \frac{2}{3} \\ \\ 0 \end{bmatrix} \mapsto \begin{bmatrix} \frac{2}{3} \\ \\ \frac{5}{3} \end{bmatrix} \mapsto \begin{bmatrix} \frac{2}{3} \\ \\ \frac{4}{3} \end{bmatrix} \mapsto \begin{bmatrix} \frac{2}{3} \\ \\ 1 \end{bmatrix} \mapsto \begin{bmatrix} \frac{2}{3} \\ \\ \frac{2}{3} \end{bmatrix} \mapsto \begin{bmatrix} \frac{2}{3} \\ \\ \frac{1}{3} \end{bmatrix} \mapsto \begin{bmatrix} \frac{2}{3} \\ \\ 0 \end{bmatrix}.$$

(We finish the picture by showing that the 12 classes of characteristics shown above form a complete $\Gamma/\Gamma(6)$ orbit. Toward this end we note that under $A : z \mapsto -1/z$ we have:

$$\begin{bmatrix} 0 \\ \frac{1}{3} \end{bmatrix} \mapsto \begin{bmatrix} \frac{1}{3} \\ 0 \end{bmatrix} \mapsto \begin{bmatrix} 0 \\ \frac{1}{3} \end{bmatrix}, \quad \begin{bmatrix} 0 \\ \frac{2}{3} \end{bmatrix} \mapsto \begin{bmatrix} \frac{2}{3} \\ 0 \end{bmatrix} \mapsto \begin{bmatrix} 0 \\ \frac{2}{3} \end{bmatrix},$$

$$\begin{bmatrix} 1 \\ \frac{2}{3} \end{bmatrix} \mapsto \begin{bmatrix} \frac{2}{3} \\ 1 \end{bmatrix} \mapsto \begin{bmatrix} 1 \\ \frac{2}{3} \end{bmatrix},$$

which is all that needed to be shown.) The above tables provide us with an alternate method for computing the branching structure of the map $\mathbb{H}^2/\Gamma(6) \to \mathbb{H}^2/\Gamma(6)/ < \tilde{B} >$ at the punctures of the surface $\mathbb{H}^2/\Gamma(6)$.

We abbreviate $\theta^{24} = \varphi$. Computations, as before, of the divisors of automorphic forms yield:

$$\left(\Phi \begin{bmatrix} 0 \\ 0 \end{bmatrix} \right) = (P_\infty^{-1} P_0^{-1} P_1^2 P_2^{-1} P_3^2 P_4^{-1} P_5^2 P_{\frac{1}{2}}^{-1} P_{\frac{3}{2}}^{-1} P_{\frac{5}{2}}^{-1} P_{\frac{1}{3}}^2 P_{\frac{4}{3}}^{-1})^6,$$

$$\left(\Phi\begin{bmatrix} 0 \\ \frac{1}{3} \end{bmatrix}\right) = (P_\infty^{-3}P_0^{-2}P_1P_2^{-2}P_3P_4^{-2}P_5P_{\frac{1}{2}}P_{\frac{3}{2}}P_{\frac{5}{2}}P_{\frac{1}{3}}^{-3}P_{\frac{4}{3}}^{6})^2,$$

$$\left(\Phi\begin{bmatrix} 0 \\ \frac{2}{3} \end{bmatrix}\right) = (P_\infty^{-3}P_0P_1^{-2}P_2P_3^{-2}P_4P_5^{-2}P_{\frac{1}{2}}P_{\frac{3}{2}}P_{\frac{5}{2}}P_{\frac{1}{3}}^{6}P_{\frac{4}{3}}^{-3})^2,$$

$$\left(\Phi\begin{bmatrix} 0 \\ 1 \end{bmatrix}\right) = (P_\infty^{-1}P_0^{2}P_1^{-1}P_2^{2}P_3^{-1}P_4^{2}P_5^{-1}P_{\frac{1}{2}}^{-1}P_{\frac{3}{2}}^{-1}P_{\frac{5}{2}}^{-1}P_{\frac{1}{3}}^{-1}P_{\frac{4}{3}}^{2})^6,$$

$$\left(\Phi\begin{bmatrix} \frac{1}{3} \\ 0 \end{bmatrix}\right) = (P_\infty^{-2}P_0^{-3}P_1P_2P_3^{-3}P_4P_5P_{\frac{1}{2}}^{-2}P_{\frac{3}{2}}^{6}P_{\frac{5}{2}}^{-2}P_{\frac{1}{3}}P_{4/3})^2,$$

$$\left(\Phi\begin{bmatrix} \frac{1}{3} \\ \\ \frac{1}{3} \end{bmatrix}\right) = (P_\infty^{-1}P_0^{-1}P_1^{-1}P_2^{3}P_3^{-1}P_4^{-1}P_5^{3}P_{\frac{1}{2}}^{3}P_{\frac{3}{2}}^{-1}P_{\frac{5}{2}}^{-1}P_{\frac{1}{3}}^{-1}P_{\frac{4}{3}}^{-1})^4,$$

$$\left(\Phi\begin{bmatrix} \frac{1}{3} \\ \\ \frac{2}{3} \end{bmatrix}\right) = (P_\infty^{-2}P_0P_1^{-3}P_2P_3P_4^{-3}P_5P_{\frac{1}{2}}^{-2}P_{\frac{3}{2}}^{-2}P_{\frac{5}{2}}^{6}P_{\frac{1}{3}}P_{\frac{4}{3}})^2,$$

$$\left(\Phi\begin{bmatrix} \frac{1}{3} \\ 1 \end{bmatrix}\right) = (P_\infty^{-1}P_0^{3}P_1^{-1}P_2^{-1}P_3^{3}P_4^{-1}P_5^{-1}P_{\frac{1}{2}}^{-1}P_{\frac{3}{2}}^{3}P_{\frac{5}{2}}^{-1}P_{\frac{1}{3}}^{-1}P_{\frac{4}{3}}^{-1})^4,$$

$$\left(\Phi\begin{bmatrix} \frac{2}{3} \\ 0 \end{bmatrix}\right) = (P_\infty P_0^{-3}P_1^{-2}P_2P_3^{6}P_4P_5^{-2}P_{\frac{1}{2}}P_{\frac{3}{2}}^{-3}P_{\frac{5}{2}}P_{\frac{1}{3}}^{-2}P_{\frac{4}{3}})^2,$$

$$\left(\Phi\begin{bmatrix} \frac{2}{3} \\ \\ \frac{1}{3} \end{bmatrix}\right) = (P_\infty P_0^{-2}P_1^{-3}P_2^{-2}P_3P_4^{6}P_5P_{\frac{1}{2}}P_{\frac{3}{2}}P_{\frac{5}{2}}^{-3}P_{\frac{1}{3}}P_{\frac{4}{3}}^{-2})^2,$$

$$\left(\Phi\begin{bmatrix} \frac{2}{3} \\ \\ \frac{2}{3} \end{bmatrix}\right) = (P_\infty P_0P_1^{-2}P_2^{-3}P_3^{-2}P_4P_5^{6}P_{\frac{1}{2}}^{-3}P_{\frac{3}{2}}P_{\frac{5}{2}}P_{\frac{1}{3}}^{-2}P_{\frac{4}{3}})^2,$$

$$\left(\Phi\begin{bmatrix} \frac{2}{3} \\ 1 \end{bmatrix}\right) = (P_\infty P_0^{6}P_1P_2^{-2}P_3^{-3}P_4^{-2}P_5P_{\frac{1}{2}}P_{\frac{3}{2}}^{-3}P_{\frac{5}{2}}P_{\frac{1}{3}}P_{\frac{4}{3}}^{-2})^2,$$

$$\left(\Phi\begin{bmatrix} 1 \\ 0 \end{bmatrix}\right) = (P_\infty^{2}P_0^{-1}P_1^{-1}P_2^{-1}P_3^{-1}P_4^{-1}P_5^{-1}P_{\frac{1}{2}}^{2}P_{\frac{3}{2}}^{2}P_{\frac{5}{2}}^{2}P_{\frac{1}{3}}^{-1}P_{\frac{4}{3}}^{-1})^6,$$

$$\left(\Phi\begin{bmatrix} 1 \\ \frac{1}{3} \end{bmatrix}\right) = (P_\infty^{3}P_0^{-1}P_1^{-1}P_2^{-1}P_3^{-1}P_4^{-1}P_5^{-1}P_{\frac{1}{2}}^{-1}P_{\frac{3}{2}}^{-1}P_{\frac{5}{2}}^{-1}P_{\frac{1}{3}}^{3}P_{\frac{4}{3}}^{3})^4,$$

$$\left(\Phi\begin{bmatrix} 1 \\ \frac{2}{3} \end{bmatrix}\right) = (P_\infty^{6}P_0P_1P_2P_3P_4P_5P_{\frac{1}{2}}^{-2}P_{\frac{3}{2}}^{-2}P_{\frac{5}{2}}^{-2}P_{\frac{1}{3}}^{-3}P_{\frac{4}{3}}^{-3})^2,$$

$$\left(\Phi\begin{bmatrix} \frac{4}{3} \\ \\ \frac{1}{3} \end{bmatrix}\right) = (P_\infty P_0^{-2}P_1P_2^{6}P_3P_4^{-2}P_5^{-3}P_{\frac{1}{2}}^{-3}P_{\frac{3}{2}}P_{\frac{5}{2}}P_{\frac{1}{3}}P_{\frac{4}{3}}^{-2})^2,$$

$$\left(\Phi\begin{bmatrix} \frac{4}{3} \\ \\ \frac{2}{3} \end{bmatrix}\right) = (P_\infty P_0P_1^{6}P_2P_3^{-2}P_4^{-3}P_5^{-2}P_{\frac{1}{2}}P_{\frac{3}{2}}P_{\frac{5}{2}}^{-3}P_{\frac{1}{3}}^{-2}P_{\frac{4}{3}})^2,$$

$$\left(\Phi\begin{bmatrix} \frac{5}{3} \\ \frac{1}{3} \end{bmatrix}\right) = (P_\infty^{-1} P_0^{-1} P_1^3 P_2^{-1} P_3^{-1} P_4^3 P_5^{-1} P_{\frac{1}{2}}^{-1} P_{\frac{3}{2}}^{-1} P_{\frac{5}{2}}^3 P_{\frac{1}{3}}^{-1} P_{\frac{4}{3}}^{-1})^4,$$

$$\left(\Phi\begin{bmatrix} \frac{5}{3} \\ \frac{2}{3} \end{bmatrix}\right) = (P_\infty^{-2} P_0 P_1 P_2^{-3} P_3 P_4 P_5^{-3} P_{\frac{1}{2}}^6 P_{\frac{3}{2}}^{-2} P_{\frac{5}{2}}^{-2} P_{\frac{1}{3}} P_{\frac{4}{3}})^2.$$

An examination of the above table makes it clear that we should look at ratios of cubes of theta constants with characteristics of the form $\chi = \begin{bmatrix} \frac{\alpha}{3} \\ \frac{\beta}{3} \end{bmatrix}$ with α and β even integers. It is routine to establish (by examining $\kappa^3(\chi, \gamma)$ with $\gamma \in \Gamma(6)$) that these are multiplicative meromorphic functions belonging to a character c whose square is trivial (the homomorphism $c^2 = 1$). We have a number of such functions whose divisors are easily computed:

$$\left(\frac{\Theta^3\begin{bmatrix} 0 \\ \frac{2}{3} \end{bmatrix}}{\Theta^3\begin{bmatrix} \frac{2}{3} \\ 0 \end{bmatrix}}\right) = \frac{P_0 P_{\frac{3}{2}} P_{\frac{1}{3}}^2}{P_\infty P_3^2 P_{\frac{4}{3}}} = \frac{1}{2}, \quad \left(\frac{\Theta^3\begin{bmatrix} 0 \\ \frac{2}{3} \end{bmatrix}}{\Theta^3\begin{bmatrix} \frac{2}{3} \\ \frac{2}{3} \end{bmatrix}}\right) = \frac{P_2 P_{\frac{1}{2}} P_{\frac{1}{3}}^2}{P_\infty P_5^2 P_{\frac{4}{3}}} = \frac{1}{2}(1+\omega),$$

$$\left(\frac{\Theta^3\begin{bmatrix} 0 \\ \frac{2}{3} \end{bmatrix}}{\Theta^3\begin{bmatrix} \frac{4}{3} \\ \frac{2}{3} \end{bmatrix}}\right) = \frac{P_4 P_{\frac{5}{2}} P_{\frac{1}{3}}^2}{P_\infty P_1^2 P_{\frac{4}{3}}} = \frac{\omega}{2}, \quad \left(\frac{\Theta^3\begin{bmatrix} \frac{2}{3} \\ 0 \end{bmatrix}}{\Theta^3\begin{bmatrix} \frac{2}{3} \\ \frac{2}{3} \end{bmatrix}}\right) = \frac{P_2 P_3^2 P_{\frac{1}{2}}}{P_0 P_5^2 P_{\frac{3}{2}}} = \frac{\omega}{2},$$

$$\left(\frac{\Theta^3\begin{bmatrix} \frac{2}{3} \\ 0 \end{bmatrix}}{\Theta^3\begin{bmatrix} \frac{4}{3} \\ \frac{2}{3} \end{bmatrix}}\right) = \frac{P_3^2 P_4 P_{\frac{5}{2}}}{P_0 P_1^2 P_{\frac{3}{2}}} = \frac{1+\omega}{2}, \quad \left(\frac{\Theta^3\begin{bmatrix} \frac{2}{3} \\ \frac{2}{3} \end{bmatrix}}{\Theta^3\begin{bmatrix} \frac{4}{3} \\ \frac{2}{3} \end{bmatrix}}\right) = \frac{P_4 P_5^2 P_{\frac{5}{2}}}{P_1^2 P_2 P_{\frac{1}{2}}} = \frac{1}{2}.$$

In the above we have expressed each of the divisors in both the Fuchsian and canonical model for the torus T. We conclude, by Abel's theorem, that each of the above 6 multiplicative functions are associated with a nontrivial character. To get (single valued) functions on T we may square each of these functions or take appropriate products. Thus, for example, the divisors of two such (degree 8) functions are:

$$\left(\frac{\Theta^6\begin{bmatrix}0\\ \frac{2}{3}\end{bmatrix}}{\Theta^6\begin{bmatrix}\frac{2}{3}\\ 0\end{bmatrix}}\right)=\frac{P_0^2P_{\frac{3}{2}}^2P_{\frac{1}{3}}^4}{P_\infty^2P_3^4P_{\frac{4}{3}}^2}\text{ and }\left(\frac{\Theta^3\begin{bmatrix}0\\ \frac{2}{3}\end{bmatrix}}{\Theta^3\begin{bmatrix}\frac{2}{3}\\ 0\end{bmatrix}}\frac{\Theta^3\begin{bmatrix}\frac{2}{3}\\ \frac{2}{3}\end{bmatrix}}{\Theta^3\begin{bmatrix}\frac{4}{3}\\ \frac{2}{3}\end{bmatrix}}\right)=\frac{P_0P_4P_5^2P_{\frac{3}{2}}P_{\frac{5}{2}}P_{\frac{1}{3}}^2}{P_\infty P_1^2P_2P_3^2P_{\frac{1}{2}}P_{\frac{4}{3}}}.$$

Using the same method, it is easy to show that we can produce two degree 6 functions on T described by

$$\left(\frac{\Theta^4\begin{bmatrix}1\\ \frac{1}{3}\end{bmatrix}}{\Theta^4\begin{bmatrix}1\\ \frac{2}{3}\end{bmatrix}}\right)=\frac{P_{\frac{1}{3}}^3P_{\frac{4}{3}}^3}{P_0P_1P_2P_3P_4P_5}\text{ and }\left(\frac{\Theta^3\begin{bmatrix}1\\ 0\end{bmatrix}}{\Theta^3\begin{bmatrix}1\\ \frac{2}{3}\end{bmatrix}}\right)=\frac{P_{\frac{1}{2}}^2P_{\frac{3}{2}}^2P_{\frac{5}{2}}^2}{P_0P_1P_2P_3P_4P_5}.$$

One easily checks that the first of these two functions on T is invariant under $\tilde{B}$ (its lift to $\mathbb{H}^2$ is invariant under B). The second is invariant under $\tilde{B}$ up to signs. This is sufficient to guarantee that these functions are well defined on $\mathbb{H}^2/<\Gamma(6),\ B>$, with divisors (recall the cover g)

$$\frac{P_{\frac{1}{3}}}{P_0}\text{ and }\frac{P_{\frac{1}{2}}}{P_0},$$

respectively. Thus

$$\lambda=\frac{\theta^4\begin{bmatrix}1\\ \frac{1}{3}\end{bmatrix}}{\theta^4\begin{bmatrix}1\\ \frac{2}{3}\end{bmatrix}}\text{ and }h=\frac{\theta^3\begin{bmatrix}1\\ 0\end{bmatrix}}{\theta^3\begin{bmatrix}1\\ \frac{2}{3}\end{bmatrix}}$$

are holomorphic universal covers of $\mathbb{H}^2/<\Gamma(6),\ B>$, and the punctures are (may be taken as) the values of the function λ at $\infty,\ 0,\ \frac{1}{2}$ and $\frac{1}{3}$. These are easily computed to be $9,\ \infty,\ 1$ and 0. As before, the fact that we have produced two coverings leads to an identity:

$$h+1=\lambda,$$

or its equivalent form

$$\theta^3\begin{bmatrix}1\\ 0\end{bmatrix}\theta\begin{bmatrix}1\\ \frac{2}{3}\end{bmatrix}=\theta^4\begin{bmatrix}1\\ \frac{1}{3}\end{bmatrix}-\theta^4\begin{bmatrix}1\\ \frac{2}{3}\end{bmatrix}.$$

Remark 3.25. The above theta constant identity produced from considerations of uniformization also follows immediately from the addition formula of Chapter 2. It is not clear how to use the methods described above to get functions of degree 2 and 3 on T.

We have attempted to illustrate the power of the construction of cusp forms as a tool to study uniformizations of the surfaces $\mathbb{H}^2/\Gamma(k)$, in particular how theta constants allow one to do this in a relatively simple way. However, the results obtained were ad-hoc and required analysis of special cases (k was restricted to positive integers at most 6). To obtain general results (applicable to all $k \in \mathbb{Z}^+$ or at least to all (odd) primes) we turn to a modification of the previous constructions.

4. Primitive invariant automorphic forms

We begin a study of $\theta[\chi](0, k\tau)$ as a form for $\Gamma(k)$.

4.1. An index 4 subgroup of $\Gamma(k)$ for even k. Assume that k is an even positive integer. We define surjective homomorphisms

$$b : \Gamma(k) \to \mathbb{Z}_2,\ c : \Gamma(k) \to \mathbb{Z}_2 \text{ and } b \oplus c : \Gamma(k) \to \mathbb{Z}_2 \oplus \mathbb{Z}_2$$

as follows. For $\gamma = \begin{bmatrix} a & b \\ c & d \end{bmatrix} \in \Gamma(k)$, $b(\gamma) = 0$ (1) if $\frac{b}{k}$ is even (odd), $c(\gamma) = 0$ (1) if $\frac{c}{k}$ is even (odd), and $(b \oplus c)(\gamma) = (b(\gamma), c(\gamma))$. We let $H(k)$ be the kernel of the homomorphism $b \oplus c$. Thus $H(k) = \Gamma(2k, 2k) \cap \Gamma(k)$ is an index 4 subgroup of $\Gamma(k)$ and $\overline{\mathbb{H}^2/H(k)}$ is a 4-sheeted cover of $\overline{\mathbb{H}^2/\Gamma(k)}$ branched over (some or all of the) punctures. It is also easily seen that $H(k)$ is normal in Γ. We have the obvious normal group inclusions

$$\Gamma(2k) \subset H(k) \subset \Gamma(k) \subset \Gamma(2).$$

Further since $[\Gamma(k) : \Gamma(2k)] = 4$ for $k = 2$ and $= 8$ for $k \geq 4$, we conclude that $H(2) = \Gamma(4)$ and $\Gamma(2k)$ is a subgroup of index 2 of $H(k)$. We define

$$h = \begin{bmatrix} 1 + k + k^2 & 2k \\ -\frac{k^2}{2} & 1 - k \end{bmatrix}, \text{ if } k \equiv 0 \bmod 4,$$

and

$$h = \begin{bmatrix} 1 + k + 3k^2 & 2k \\ \frac{k(2-3k)}{2} & 1 - k \end{bmatrix}, \text{ if } k \equiv 2 \bmod 4,\ k \neq 2.$$

It is easily seen that $h \in H(k) - \Gamma(2k)$. For technical reasons, we define $h = I$ for $k = 2$.

We proceed to describe the 4-sheeted cover $\overline{\mathbb{H}^2/H(k)} \to \overline{\mathbb{H}^2/\Gamma(k)}$. The factor group $\Gamma(k)/H(k)$ is isomorphic to the Klein four group $\mathbb{Z}_2 \oplus \mathbb{Z}_2$. Hence we can pick as representatives for generators for this factor group

$$B^k = \begin{bmatrix} 1 & k \\ 0 & 1 \end{bmatrix} \text{ and } AB^{-k}A^{-1} = \begin{bmatrix} 1 & 0 \\ k & 1 \end{bmatrix}.$$

Let (p, n) be the type of torsion free group $H(k)$. Euler characteristic arguments show that

$$2p - 2 + n = 4(2p(k) - 2 + n(k)).$$

k	p	n
2	0	6
4	3	12
6	13	24
8	41	48
10	85	72
12	145	96

Table 8. TYPE OF $H(k)$.

Each puncture on $\mathbb{H}^2/H(k)$ is fixed by a cyclic subgroup of $\Gamma(k)/H(k)$, hence by either the trivial group or by a group of order 2. If we let m be the number of punctures stabilized by a $\mathbb{Z}_2$, then Riemann-Hurwitz tells us that

$$2p - 2 = 4(2p(k) - 2) + m.$$

Since the punctures on $\mathbb{H}^2/\Gamma(k)$ all come from punctures on $\mathbb{H}^2/H(k)$, we see that

$$n(k) = \frac{m}{2} + \frac{n-m}{4}.$$

The last three displayed equations are not independent (the first is a consequence of the next two); hence they are not sufficient to determine the three unknown integers p, n and m. Since $H(k)$ is normal in Γ, the local picture is the same at each puncture. Since P_∞ is stabilized by B^k in $\Gamma(k)$, we conclude that $m = n$. Hence $H(k)$ has type

$$(4p(k) - 3 + n(k), 2n(k)).$$

We also conclude that each puncture on $\mathbb{H}^2/\Gamma(k)$ comes from ("splits" to) two punctures on $\mathbb{H}^2/H(k)$. The number of punctures on and the genus of $\mathbb{H}^2/H(k)$ grow rapidly with k as a short table suggests.

From x_1, ..., $x_{n(k)}$, a list of cusps that project to the punctures on $\mathbb{H}^2/\Gamma(k)$, we can produce in a systematic way a list of cusps that project to the punctures on $\mathbb{H}^2/H(k)$. Let B_i, $i = 1$, 2, generate the stabilizer of x_i in $\Gamma(k)$. Then I, B_1, B_2, B_2B_1 are representatives for $\Gamma(k)/H(k)$. It is now straightforward to see that the puncture P_{x_1} on $\mathbb{H}^2/\Gamma(k)$ splits to the punctures P_{x_1} and $P_{B_2(x_1)}$ on $\mathbb{H}^2/H(k)$. For computational purposes, we observe that ∞ and 0 are $H(k)$-inequivalent cusps. If P_{x_1} is a fixed point of $\hat{B}^k$, then we may take $B_2 = AB^{-k}A$ and hence $P_{B_2(x_1)} = P_{\frac{x_1}{kx_1+1}}$. If P_{x_1} is not a fixed point of $\hat{B}^k$, then we may take $B_2 = B^k$ and hence $P_{B_2(x_1)} = P_{x_1+k}$. We need to decide when P_{x_1} is a fixed point of $\hat{B}^k$. We digress to obtain an alternate algorithm for computing the punctures of $\mathbb{H}^2/\Gamma(k)$. This procedure will also answer the last question.

Assume that $k \geq 4$. It follows from

$$\frac{p(2k)-1}{p-1} = 2$$

that the 2-sheeted cover $\mathbb{H}^2/\Gamma(2k) \to \mathbb{H}^2/H(k)$ is unramified.[74] Since $h \in H(k) - \Gamma(2k)$, the cusps x and $h(x)$ project to the same puncture on $\mathbb{H}^2/H(k)$. Thus if we start with a list $y_1, \ldots, y_{n(2k)}$ of $n(2k) = 2n = 4n(k)$ cusps that project to the punctures of $\mathbb{H}^2/\Gamma(2k)$, we can renumber the cusps in the list so that

$$y_{n+i} \text{ is } \Gamma(2k)\text{-equivalent to } h(y_i) \text{ for } i = 1, \ldots, n.$$

Then $y_1, \ldots, y_n$ project to the punctures on $\mathbb{H}^2/H(k)$. The punctures on $\mathbb{H}^2/H(2)$ are computable by known methods since $H(2) = \Gamma(4)$.

Now, P_{x_1} is a fixed point of $\hat{B}^k$ if and only if $x_1 + k$ is equivalent modulo $\Gamma(2k)$ to either x_1 or $h(x_1)$.

4.2. A Hilbert space of modified theta constants. Let $V(k)$ be the finite dimensional vector space of holomorphic functions on $\mathbb{H}^2$ spanned by the *modified* theta constants

$$\tau \mapsto \varphi_l(\tau) = \theta[\chi_l](0, k\tau),$$

where the characteristic[75] $\chi_l = \begin{bmatrix} \frac{2l+1}{k} \\ 1 \end{bmatrix}$, $l = 0, \ldots, \frac{k-3}{2}$, for odd k and $\chi_l = \begin{bmatrix} \frac{2l}{k} \\ 0 \end{bmatrix}$, $l = 0, \ldots, \frac{k}{2}$, for even k. Define $V_o(k)$ to be the linear span of those modified theta constants φ_l for which the characteristic $\chi_l \in X_o(k)$.

Lemma 4.1. *For each* $k \in \mathbb{Z}$, $k > 1$,

$$\dim V(k) = \frac{k-1}{2} \textit{ for odd } k \textit{ and } \dim V(k) = \frac{k}{2} + 1 \textit{ for even } k.$$

Proof. It is convenient to work with the local coordinate $\zeta = \exp\left\{\frac{2\pi\imath\tau}{k}\right\}$, $\tau \in \mathbb{H}^2$, at P_∞. Note that with respect to this local coordinate,

$$\text{(3.20)} \qquad \operatorname{ord}_\infty \varphi_l = \frac{(2l+1)^2}{8} \text{ for odd } k \text{ and } \operatorname{ord}_\infty \varphi_l = \frac{l^2}{2} \text{ for even } k.$$

We shall see momentarily that the modular group Γ acts on the vector space $V(k)$ (Lemma 4.2). It does not act on $V_o(k)$ (Remark 4.5). We record

[74] Also for $k = 2$.

[75] The characteristic χ_0 defined above and the characteristic χ_o of the previous chapter are related. For odd k, $\chi_0 A$ is equivalent to χ_o.

for future use the useful observations that for odd k, the modified theta constants have Fourier series expansions that in terms of ζ are given by

$$\Phi_l(\zeta) = \zeta^{\frac{(2l+1)^2}{8}} \left(e^{\left\{\frac{\pi\imath(2l+1)}{2k}\right\}} + e^{\left\{\frac{\pi\imath(2l+1-2k)}{2k}\right\}} \zeta^{k\left(\frac{k-2l-1}{2}\right)} + \ldots \right)$$

$$= \exp\left\{\frac{\pi\imath(2l+1)}{2k}\right\} \zeta^{\frac{(2l+1)^2}{8}} \left(1 - \zeta^{k\left(\frac{k-2l-1}{2}\right)} + \ldots\right),$$

and hence (for l' also in $\mathbb{Z}$ with $0 \le l' \le \frac{k-3}{2}$)

$$\frac{\Phi_l}{\Phi_{l'}}(\zeta) = \zeta^{\frac{(l+l'+1)(l-l')}{2}} e^{\frac{\pi\imath(l-l')}{k}} \left(1 - \zeta^{k\left(\frac{k-2l-1}{2}\right)} + \zeta^{k\left(\frac{k-2l'-1}{2}\right)} + \ldots\right).$$

Lemma 4.2. *For each* $\gamma \in \Gamma$*, the linear operator on functions*

$$\gamma^* = \gamma^*_{\frac{1}{4}} : \; f \mapsto (f \circ \gamma)(\gamma')^{\frac{1}{4}}$$

maps $V(k)$ *onto itself.*

Proof. In defining the operator γ^*, we are making a specific choice for $(\gamma')^{\frac{1}{4}}$. If we view γ as an element of $\mathrm{SL}(2,\mathbb{C})$ (respectively, $\mathrm{PSL}(2,\mathbb{C})$), the operator involves a choice of square (fourth) root of unity. Since for each γ_1 and γ_2 in Γ,

$$(\gamma_1 \circ \gamma_2)^* = c\,(\gamma_2)^* \circ (\gamma_1)^*, \tag{3.21}$$

where c is a fourth root of unity, it suffices to show that the operators defined by the generators B and A of Γ preserve $V(k)$.

Assume k is odd. We start with the generator B. For $\tau \in \mathbb{H}^2$, we find from the transformation formulae for theta constants that

$$\theta\begin{bmatrix} \frac{2l+1}{k} \\ 1 \end{bmatrix}(0, k(\tau+1)) \tag{3.22}$$

$$\begin{aligned} &= \kappa\left(\begin{bmatrix} \frac{2l+1}{k} \\ 1 \end{bmatrix}, B^k\right) \theta\begin{bmatrix} \frac{2l+1}{k} \\ 1 + 2l + 2 + (k-1) \end{bmatrix}(0, k\tau) \\ &= c(B,k) \exp\left\{\tfrac{\pi\imath}{k}(l^2+l)\right\} \theta\begin{bmatrix} \frac{2l+1}{k} \\ 1 \end{bmatrix}(0, k\tau) \end{aligned};$$

that is,

$$B^*(\varphi_l) = c(B,k)\exp\left\{\frac{\pi\imath}{k}(l^2+l)\right\}\varphi_l, \; c(B,k) = \exp\left\{\frac{\pi\imath}{4k}\right\}.$$

In the above and subsequent formulae $c(\gamma, k)$ is a constant of absolute value 1 that depends only on the transformation $\gamma \in \Gamma$ and the integer k. At times we specify the constant for future applications.

The situation for the second generator A of Γ is both more complicated and more interesting. Again for $\tau \in \mathbb{H}^2$,

$$\text{(3.23)} \qquad \tau^{-\frac{1}{2}}\,\theta\left[\begin{array}{c}\frac{2l+1}{k}\\1\end{array}\right]\left(0,\frac{-1}{\frac{\tau}{k}}\right)=c(A,k)\exp\left\{\frac{\pi\imath l}{k}\right\}\theta\left[\begin{array}{c}1\\\frac{2l+1}{k}\end{array}\right]\left(0,\frac{\tau}{k}\right)$$

$$=c(A,k)\exp\left\{\frac{\pi\imath l}{k}\right\}\sum_{j=0}^{k-1}\theta\left[\begin{array}{c}\frac{2j+1}{k}\\2l+1\end{array}\right](0,k\tau),\ c(A,k)=\frac{1}{\sqrt{\imath k}}\exp\left\{\pi\imath\frac{1}{2k}\right\}.$$

Now it is obvious that the last expression can be rewritten as

$$\sum_{j=0}^{\frac{k-3}{2}}c_j\theta\left[\begin{array}{c}\frac{2j+1}{k}\\1\end{array}\right](0,k\tau).$$

In the last formula the numbers c_j (that depend on A, k, and j) are easily computed. We can rewrite our last equation as

$$A^*(\varphi_l)=\frac{1}{\sqrt{\imath k}}\exp\left\{\pi\imath\frac{1}{2k}\right\}\sum_{j=0}^{\frac{k-3}{2}}A_{lj}\varphi_j.$$

Lengthy, but routine, calculations show that

$$\text{(3.24)} \qquad A_{lj}=\exp\left\{2\pi\imath\frac{l(j+1)}{k}\right\}+\exp\left\{-\pi\imath\frac{1}{k}\right\}\exp\left\{-2\pi\imath\frac{(l+1)j}{k}\right\}.$$

For even k, the formulae corresponding to (3.22) and (3.23) are

$$\theta\left[\begin{array}{c}\frac{2l}{k}\\0\end{array}\right](0,k(\tau+1))=c(B,k)\exp\left\{\pi\imath\left(l+\frac{l^2}{k}\right)\right\}\theta\left[\begin{array}{c}\frac{2l}{k}\\0\end{array}\right](0,k\tau)$$

and

$$\tau^{-\frac{1}{2}}\,\theta\left[\begin{array}{c}\frac{2l}{k}\\0\end{array}\right]\left(0,\frac{-1}{\frac{\tau}{k}}\right)=\sum_{j=0}^{\frac{k}{2}}c_j\theta\left[\begin{array}{c}\frac{2j}{k}\\0\end{array}\right](0,k\tau),$$

respectively.

We want to give a Hilbert space structure to $V(k)$; for odd k, we introduce the *Petersson pairing*

$$\text{(3.25)} \qquad <\varphi,\psi>=\int\int_{\mathbb{H}^2/\Gamma(k)}(\Im z)^{-\frac{3}{2}}\varphi(z)\overline{\psi(z)}\left|\frac{dz\overline{dz}}{2}\right|,\ \varphi\text{ and }\psi\in V(k).$$

For even k, we replace $\Gamma(k)$ by $H(k)$. Our first task is to show that this pairing is well defined (independent of the choice of a fundamental domain for the action of $\Gamma(k)$ or $H(k)$ (as appropriate) on $\mathbb{H}^2$). To do so we begin a study of

4.3. Projective representation of Aut $\mathbb{H}^2/\Gamma(k)$. It is a consequence of (3.21) that we can define a homomorphism of Γ into the group of linear transformations of $V(k)$ modulo nonzero multiples of the identity by sending $\gamma \in \Gamma$ to the linear operator $(\gamma^*)^{-1}$. We have a well defined homomorphism of Γ into the projective linear transformations of $\mathbf{P}V(k)$, projectivized $V(k)$, which can be identified, for odd k, with $\mathbf{P}\mathbb{C}^{\frac{k-3}{2}}$. Let

$$\Theta: \Gamma \to \text{Aut } \mathbf{P}V(k)$$

denote this homomorphism. We have already observed that each φ_l is an eigenvector for the operator B^*. Except for $k = 3$, the operator B^* has two or more distinct eigenvalues. Hence B is not in the kernel of Θ for $k \neq 3$. We claim that $\Gamma(k)$ is in the kernel of Θ for odd k. We start with a slightly more general situation and use a classical trick.[76] Let $\gamma \in \Gamma_o(k)$ be represented by the matrix $\begin{bmatrix} a = a(\gamma) & b = b(\gamma) \\ c = c(\gamma) & d = d(\gamma) \end{bmatrix}$ in $\mathrm{SL}(2, \mathbb{Z})$ (hence c is a multiple of k). Therefore for $\tau \in \mathbb{H}^2$, $k\gamma(\tau) = k\frac{a\tau+b}{c\tau+d} = \frac{ak\tau+bk}{\frac{c}{k}k\tau+d}$. It hence makes sense to associate[77] to the motion $\gamma \in \Gamma_o(k)$ another motion $\hat{\gamma} = \begin{bmatrix} a & bk \\ \frac{c}{k} & d \end{bmatrix} \in \Gamma$. Assume that k is odd. Observe that for the characteristic $\chi = \begin{bmatrix} \frac{2l+1}{k} \\ 1 \end{bmatrix}$ and $\tau \in \mathbb{H}^2$, we have

$$\theta[\chi](0, k\gamma(\tau))\gamma'(\tau)^{\frac{1}{4}} = \theta[\chi](0, \hat{\gamma}(k\tau))\hat{\gamma}'(k\tau)^{\frac{1}{4}} = \kappa(\chi, \hat{\gamma})\theta[\chi\hat{\gamma}](0, k\tau).$$

One easily sees that the characteristic $\chi\hat{\gamma}$ is equivalent to $\begin{bmatrix} \frac{2l'+1}{k} \\ 1 \end{bmatrix}$ for some integer l' with $0 \leq l' \leq \frac{k-3}{2}$. Hence there is a permutation $\sigma = \sigma_\gamma$ of the integers in $\left[0, \frac{k-3}{2}\right]$ such that

$$\gamma^*\varphi_l = \tilde{\kappa}(\chi_l, \hat{\gamma})\varphi_{\sigma(l)}, \; l = 0, \; 1, \; ..., \; \frac{k-3}{2}. \tag{3.26}$$

(The constant $\tilde{\kappa}$ is a $2k$-th root of unity, and differs from κ whenever $\chi_{\sigma(l)} - \chi_l\hat{\gamma} \neq 0$ as characteristics.) It is a lengthy, but routine, calculation to show that each φ_l is an eigenvector of γ^* provided that $\gamma \in \Gamma(k)$ and that the eigenvalue is of absolute value 1 and independent of l. For even k, one

[76] See also Lemma 2.1 of Chapter 5.

[77] The map $\gamma \mapsto \hat{\gamma}$ is a group isomorphism of $\Gamma_o(k)$ onto $\Gamma^o(k)$. The definition of this last symbol can surely be left to the reader. It is clear that ^ defines an automorphism of $\mathrm{PSL}(2, \mathbb{C})$ with the property that

$$k\gamma(z) = \hat{\gamma}(kz)$$

and

$$\gamma'(z) = \hat{\gamma}'(kz),$$

for all $\gamma \in \mathrm{PSL}(2, \mathbb{C})$ and all $z \in \mathbb{C} \cup \{\infty\}$. The automorphism ^ is conjugation by $\begin{bmatrix} k^{\frac{1}{2}} & 0 \\ 0 & k^{-\frac{1}{2}} \end{bmatrix}$.

easily sees that the characteristic $\begin{bmatrix} \frac{2l}{k} \\ 0 \end{bmatrix} \hat{\gamma}$ with $l \in \mathbb{Z}$ and $\gamma \in \Gamma_o(k)$ is equivalent to $\begin{bmatrix} \frac{2l'}{k} \\ 0 \end{bmatrix}$ for some $l' \in \mathbb{Z}$. Further, for $l = 0, 1, \ldots, \frac{k}{2}$, each φ_l is an eigenvector of γ^* provided that $\gamma \in H(k)$ and that up to a sign ambivalence, the eigenvalue is independent of l. Hence we have obtained

Lemma 4.3. (a) *For odd k, the vector space $V(k)$ consists of e-automorphic functions for a factor of automorphy e for $\Gamma(k)$ of weight $\frac{1}{4}$.*
(b) *For even k, each φ_l^2, $l = 0, 1, \ldots, \frac{k}{2}$, is an e-automorphic function for a factor of automorphy e for $H(k)$ of weight $\frac{1}{2}$.*

Remark 4.4. If k is an odd prime, and γ_1 and γ_2 belong to $\Gamma_o(k)$, then $\sigma_{\gamma_1} = \sigma_{\gamma_2}$ if and only if $a(\tilde{\gamma}_1 \circ \tilde{\gamma}_2{}^{-1}) \equiv \pm 1 \bmod k$.

Remark 4.5. The above calculations have shown that the Hilbert space $V_o(k)$ is $\Gamma_o(k)$-invariant. The space $V_o(k)$ is not in general Γ-invariant (for example, from (3.24) we see that $A_{01} \neq 0$ for all $k > 3$).

Lemmas 4.2 and 4.3 tell us that for odd k we have a holomorphic map (see §7 for more details)

$$\Phi : \overline{\mathbb{H}^2/\Gamma(k)} \to \mathbf{P}V(k) \cong \mathbf{P}\mathbb{C}^{\frac{k-3}{2}},$$

and, as observed before, we have an induced homomorphism

$$\Theta : \text{ Aut } \mathbb{H}^2/\Gamma(k) \cong \Gamma/\Gamma(k) \to \text{ Aut } \mathbf{P}V(k).$$

The above two maps are related as we shall later see (equation (3.39)). It is important to determine conditions under which Φ (hence also Θ) is injective. We shall return to these questions in §7.

Assume that k is odd. We have produced a homomorphism

$$\sigma : \Gamma_o(k) \to \mathcal{S}_{\frac{k-1}{2}},$$

where $\mathcal{S}_N$ is the permutation group acting on the first N nonnegative integers $\{0, 1, \ldots, N-1\}$, whose kernel contains $G(k)$. For $\gamma \in \Gamma_o(k)$, the $\frac{k-1}{2} \times \frac{k-1}{2}$ matrix γ^* has precisely one nonzero entry in each column. For the column $l = 0, \ldots, \frac{k-3}{2}$, the unique nonzero entry $\tilde{\kappa}(\chi_l, \tilde{\gamma})$ is in the $\sigma_\gamma(l)$ row (rows are numbered 0 to $\frac{k-3}{2}$). Whereas σ_γ depends only on $\gamma \in \Gamma_o(k)/G(k)$, $\tilde{\kappa}$ depends on $\gamma \in \Gamma_o(k)/\Gamma(k)$. Assume for the remainder of this section that k is an odd prime.

Lemma 4.6. *For each odd prime k, $\Gamma_o(k)/G(k)$ is a cyclic group of order $\frac{k-1}{2}$.*

Proof. We have the following inclusions of groups (with each group normal in the succeeding one):

$$\Gamma(k) \lhd G(k) \lhd \Gamma_o(k).$$

By the first isomorphism theorem of group theory

$$\Gamma_o(k)/G(k) \cong (\Gamma_o(k)/\Gamma(k))/(G(k)/\Gamma(k)).$$

The groups $\Gamma_o(k)/\Gamma(k)$ and $G(k)/\Gamma(k)$ can be naturally identified with their images $\tilde{\Gamma}_o(k)$ and $\tilde{G}(k)$ in $\mathrm{SL}(2,\mathbb{Z}_k)$ under reduction mod k. The group $\tilde{G}(k)$ is cyclic of order k with generator B viewed as an element of $\mathrm{SL}(2,\mathbb{Z}_k)$. The elements of the group $\tilde{\Gamma}_o(k)$ are upper triangular. Therefore the problem is reduced to showing that the factor group of upper triangular elements in $\mathrm{SL}(2,\mathbb{Z}_k)$ modulo the upper triangular elements with diagonal elements 1 is a cyclic group. Let $\begin{bmatrix} a & b \\ 0 & d \end{bmatrix}$ be an arbitrary upper triangular matrix in $\mathrm{SL}(2,\mathbb{Z}_k)$. Then since

$$\begin{bmatrix} a & b \\ 0 & d \end{bmatrix} = \begin{bmatrix} a & 0 \\ 0 & d \end{bmatrix} \begin{bmatrix} 1 & ba^{-1} \\ 0 & 1 \end{bmatrix},$$

it is clear that the coset decomposition of $\tilde{\Gamma}_o(k)/\tilde{G}(k)$ is of the form

$$D_1\tilde{G}(k), \ \ldots, \ D_{\frac{k-1}{2}}\tilde{G}(k),$$

with D_j a diagonal matrix in $\mathrm{SL}(2,\mathbb{Z}_k)$. All that remains for us to do is show that the group of diagonal matrices in $\mathrm{SL}(2,\mathbb{Z}_k)$ is cyclic. For the diagonal terms of these matrices $ad \equiv 1$ modulo k. Hence a is a unit and $d = a^{-1}$ in the field $\mathbb{Z}_k$. The group of units in $\mathbb{Z}_k$ is cyclic of order $k-1$, so let a be a generator. Hence $a^{k-1} = 1$ in $\mathbb{Z}_k$, and no smaller integral power of a will be the identity. It thus follows that $a^{\frac{k-1}{2}} = -1$. It follows that

$$\begin{bmatrix} a & 0 \\ 0 & a^{-1} \end{bmatrix}$$

has order $\frac{k-1}{2}$ in $\mathrm{PSL}(2,\mathbb{Z}_k)$.

Let $\gamma \in \Gamma_o(k)$ be a representative of the generator of the cyclic group $\Gamma_o(k)/G(k)$. This group of order $\frac{k-1}{2}$ acts transitively on the set of $\frac{k-1}{2}$ characteristics $\{\chi_0, \ldots, \chi_{\frac{k-3}{2}}\}$. It follows that $\sigma = \sigma_\gamma$ is a cyclic permutation of the first $\frac{k-3}{2}$ nonnegative integers (in some order), and hence that this permutation has no fixed points.

We summarize many of our calculations and conclusions in the next

Lemma 4.7. (a) *The Petersson inner product endows $V(k)$ with a Hilbert space structure; this space is invariant under the $\frac{1}{4}$-action of Γ.*
(b) *For odd k, $V(k)$ consists of automorphic functions for a factor of automorphy of weight $\frac{1}{4}$ for the group $\Gamma(k)$.*

(c) *For even k, $V(k)$ consists of automorphic functions for a factor of automorphy of weight $\frac{1}{4}$ for the group $H(k)$. For each element $\gamma \in \Gamma(k)$, $\gamma^*_{\frac{1}{4}}$ sends φ_l, $l = 0, \ldots, \frac{k}{2}$, to a nonzero constant multiple of itself if γ is in the kernel of c and to a nonzero constant multiple of $\varphi_{\frac{k}{2}-l}$ otherwise.*

Proof. Parts (b) and (c) have already been proven. We have also already established that $\gamma^*_{\frac{1}{4}}$ preserves $V(k)$ for all $\gamma \in \Gamma$. We have also seen that $\gamma^*_{\frac{1}{4}}(\varphi) = c_\gamma \varphi$, where $c_\gamma \in \mathbb{C}$ with $|c_\gamma| = 1$, provided $\gamma \in \Gamma(k)$ ($\in H(k)$) for odd (even) k. Let ω be a fundamental domain for the action of $\Gamma(k)$ on $\mathbb{H}^2$. For odd k, $\gamma \in \Gamma(k)$, φ and $\psi \in V(k)$,

$$\int\int_{\gamma(\omega)} (\Im z)^{-\frac{3}{2}} \varphi(z)\overline{\psi(z)} \left|\frac{dz\overline{dz}}{2}\right|$$

$$= \int\int_{\gamma(\omega)} (\Im\gamma(z))^{-\frac{3}{2}} |\gamma'(z)|^{\frac{3}{2}} c_\gamma^{-1}\varphi(\gamma(z))|\gamma'(z)|^{\frac{1}{4}} \overline{c_\gamma^{-1}\psi(\gamma(z))|\gamma'(z)|^{\frac{1}{4}}} \left|\frac{dz\overline{dz}}{2}\right|$$

$$= \int\int_{\omega} (\Im z)^{-\frac{3}{2}} \varphi(z)\overline{\psi(z)} \left|\frac{dz\overline{dz}}{2}\right|.$$

The proof that the Petersson inner product is well defined for even k is similar.

We record for future use the action of $\gamma \in \Gamma(k)$ on $V(k)$ for even k (ignoring multiplicative constants that depend only on k and γ (that is, that are independent of l)):

(3.27)

$$\gamma^*_{\frac{1}{4}}(\varphi_l) = (-1)^{lb(\gamma)}\varphi_l \text{ if } c(\gamma) = 0 \text{ and } \gamma^*_{\frac{1}{4}}(\varphi_l) = (-1)^{lb(\gamma)}\varphi_{\frac{k}{2}-l} \text{ if } c(\gamma) = 1.$$

It should be noted that the above constants depend not on l but only on its parity. Similarly,

$$B^*_{\frac{1}{4}}(\varphi_l) = \left(\exp \pi\imath \frac{l^2}{k}\right) \varphi_l \text{ and } A^*_{\frac{1}{4}}(\varphi_l) = \varphi_0 + 2\sum_{j=1}^{\frac{k}{2}-1} \left(\cos \frac{2\pi \jmath l}{k}\right) \varphi_l + (-1)^l \varphi_{\frac{k}{2}}.$$

Proposition 4.8. *For each odd prime k, the $\frac{k-1}{2}$ functions $\{\varphi_0, \ldots, \varphi_{\frac{k-3}{2}}\}$ form an orthogonal basis for the Hilbert space $V(k)$. Further, these $\frac{k-1}{2}$ functions have the same norm.*

Proof. We compute for integers l and l' with $0 \le l' < l \le \frac{k-3}{2}$,

$$< \varphi_l, \varphi_{l'} > = < B_*(\varphi_l), B_*(\varphi_{l'}) > = \exp\left\{\frac{\pi\imath}{k}(l^2 + l - l'^2 - l')\right\} < \varphi_l, \varphi_{l'} > .$$

Since k does not divide $(l-l')(l+l'+1)$ we conclude that

$$\exp\left\{\frac{\pi\imath}{k}(l^2+l-l'^2-l')\right\}\neq 1,$$

and hence

$$<\varphi_l,\varphi_{l'}>=0.$$

We have re-proven, in this setting, the well known fact that eigenvectors belonging to distinct eigenvalues are orthogonal. It remains to show that the norm of φ_l is independent of l. A lift to $\Gamma_o(k)$ of a generator for the cyclic group $\Gamma_o(k)/G(k)$ induces an isometry of $V(k)$. For given l, a power of this isometry sends φ_0 to φ_l.

Corollary 4.9. *For all $\gamma\in\Gamma$, $\gamma^*_{\frac{1}{4}}$ is a unitary operator on $V(k)$, for every odd prime k.*

Assume that k is an odd prime. We return to the matrix representation of the operator A^* with respect to the virtually orthonormal basis $\{\varphi_0,\ \ldots,\ \varphi_{\frac{k-3}{2}}\}$ for $V(k)$. Define $\mathcal{A}$ as the $\frac{k-1}{2}\times\frac{k-1}{2}$ matrix whose lj entry, $l,j=0,\ \ldots,\ \frac{k-3}{2}$, is $\left(\exp\left\{\pi\imath\frac{1}{2k}\right\}\right)A_{lj}$. If for $z\in\mathbb{H}^2$ we choose the branch of $z^{\frac{1}{2}}$ to have positive imaginary part, then we conclude that $(A^*)^2=-\imath I$. In particular, we see that $(A^*)^{-1}=\imath A^*$. If we identify the operator A^* with the matrix that represents it with respect to the above basis, then we see that

$$A^*=\frac{1}{\sqrt{\imath k}}\mathcal{A}\text{ and }\mathcal{A}^2=kI.$$

The matrix $\mathcal{A}$ is obviously hermitian (that is, $\mathcal{A}={}^t\overline{\mathcal{A}}$). Since the functions φ_l essentially form an orthonormal basis for $V(k)$, we have shown again that A^* is unitary (that is, $A^{*-1}={}^t\overline{A^*}$). The matrix representing the operator B^* is obviously unitary with respect to any basis. Hence we have obtained an alternate proof that for all $\gamma\in\Gamma$, the matrix representing γ^* is unitary.

Remark 4.10. The transformation formula for B^* shows that φ_l is an eigenvector for this operator; this formula and the one for A^* will facilitate computations of the divisors of certain $\Gamma(k)$-automorphic functions, as well as the images of the punctures on $\mathbb{H}^2/\Gamma(k)$ under the holomorphic map $\Phi:\overline{\mathbb{H}^2/\Gamma(k)}\to\mathbf{P}\mathbb{C}^{\frac{k-3}{2}}$ defined in §4.3.

Remark 4.11. Since the matrix $\frac{1}{\sqrt{k}}\mathcal{A}$ is of order 2, all its eigenvalues are ±1. It follows that the trace of this matrix is an integer. The trace can be computed to be

$$\frac{2}{\sqrt{k}}\sum_{l=0}^{\frac{k-3}{2}}\cos\left(\frac{\pi\imath(2l+1)^2}{2k}\right).$$

It can be shown that if the prime k is congruent to 1 mod 4, then the trace vanishes. If k is congruent to 3 mod 4, the trace equals 1. This is the reflection of the fact that in the former case there are the same number of positive and negative eigenvalues (which is well known for the case $k = 5$) and in the latter case there is exactly one more positive eigenvalue. Sums of the type which appear here are called Gauss sums.

Exercise 4.12. Verify all the assertions in the last remark.

4.4. More Hilbert spaces of modified theta constants. For $k \in \mathbb{Z}^+$, let for the moment W be the finite dimensional vector space of holomorphic functions on $\mathbb{H}^2$ spanned by the modified theta constants

$$\tau \mapsto \psi_l(\tau) = \exp\left\{\pi\imath\frac{1+2l}{2k}\right\}\ \theta[\chi_l]\left(0, \frac{\tau}{k}\right), \tag{3.28}$$

where the characteristic $\chi_l = \begin{bmatrix} 1 \\ \frac{2l+1}{k} \end{bmatrix}$, $l = 0, \ldots, \frac{k-3}{2}$, for odd k and $\chi_l = \begin{bmatrix} 0 \\ \frac{2l}{k} \end{bmatrix}$, $l = 0, \ldots, \frac{k}{2}$, for even k. Let us assume for the remainder of this subsection that k is an odd integer. The appropriate Fourier series expansion for these functions ψ is

$$\Psi_l(\zeta) = 2e^{\pi\imath\frac{1+2l}{2k}}\ \zeta^{\frac{1}{8}}\left(\cos\left\{\frac{\pi(2l+1)}{2k}\right\} + \cos\left\{\frac{3\pi(2l+1)}{2k}\right\}\zeta + \ldots\right).$$

Since $A^*(V(k)) = W$ (because

$$A^*(\varphi_l) = \frac{1}{\sqrt{\imath k}}\ \psi_l \tag{3.29}$$

for $l = 0, \ldots, \frac{k-3}{2}$), and A^* preserves $V(k)$, we conclude that $V(k) = W$. In analogy to (3.27) we have

$$\gamma^*_{\frac{1}{4}}(\psi_l) = (-1)^{lc(\gamma)}\psi_l \text{ if } b(\gamma) = 0 \text{ and } \gamma^*_{\frac{1}{4}}(\psi_l) = (-1)^{lc(\gamma)}\psi_{\frac{k}{2}-l} \text{ if } b(\gamma) = 1.$$

For odd k, let $V'(k)$ be the finite dimensional vector space of holomorphic functions on $\mathbb{H}^2$ spanned by the modified theta constant derivatives

$$\tau \mapsto \varphi'_l(\tau) = \theta'\begin{bmatrix} \frac{2l+1}{k} \\ 1 \end{bmatrix}(0, k\tau),\ l = 0,\ \ldots,\ \frac{k-1}{2}.$$

For even k, let $V'(k)$ be the finite dimensional vector space of holomorphic functions on $\mathbb{H}^2$ spanned by the modified theta constant derivatives

$$\tau \mapsto \varphi'_l(\tau) = \theta'\begin{bmatrix} \frac{2l}{k} \\ 0 \end{bmatrix}(0, k\tau),\ l = 1,\ \ldots,\ \frac{k-2}{2}.$$

Lemma 4.13. *For each $k \in \mathbb{Z},\ k > 1$,*

$$\dim V'(k) = \frac{k+1}{2} \text{ for odd } k \text{ and } \dim V'(k) = \frac{k}{2} - 1 \text{ for even } k.$$

Proof. Note that[78] as automorphic functions for $\Gamma(k)$ (that is, in terms of the local coordinate $\exp\left(\frac{2\pi\imath\tau}{k}\right)$)

$$\mathrm{ord}_\infty \varphi'_l = \frac{(2l+1)^2}{8} \text{ for odd } k \text{ and } \mathrm{ord}_\infty \varphi'_l = \frac{l^2}{2} \text{ for even } k.$$

Lemma 4.14. (a) *The Petersson inner product*

$$< \varphi, \psi >= \int\int_{\mathbb{H}^2/\Gamma(k)} (\Im z)^{-\frac{1}{2}} \varphi(z)\overline{\psi(z)} \left|\frac{dz\overline{dz}}{2}\right|$$

for k odd and

$$< \varphi, \psi >= \int\int_{\mathbb{H}^2/H(k)} (\Im z)^{-\frac{1}{2}} \varphi(z)\overline{\psi(z)} \left|\frac{dz\overline{dz}}{2}\right|$$

for k even endows $V'(k)$ with a Hilbert space structure; this space is invariant under the $\frac{3}{4}$-action of Γ.
(b) *For odd k, $V'(k)$ consists of automorphic functions for a factor of automorphy of weight $\frac{3}{4}$ for the group $\Gamma(k)$.*
(c) *For even k, $V'(k)$ consists of automorphic functions for a factor of automorphy of weight $\frac{3}{4}$ for the group $H(k)$. For each element $\gamma \in \Gamma(k)$, $\gamma^*_{\frac{3}{4}}$ sends φ'_l, $l = 1, \ldots, \frac{k}{2} - 1$, to a nonzero constant multiple of itself if γ is in the kernel of c and to a nonzero constant multiple of $\varphi'_{\frac{k}{2}-l}$ otherwise.*
(d) *For all $\gamma \in \Gamma$, $\gamma^*_{\frac{3}{4}}$ is a unitary operator on $V'(k)$.*

Proof. The proof is similar to that of Lemma 4.7 and left to the reader.

As in the case with $V(k)$, the motions $\gamma \in \Gamma_o(k)/\Gamma(k)$ act by permutations on the basis φ'_l for $V'(k)$:

$$\gamma^*_{\frac{3}{4}} \varphi'_l = \tilde{\kappa}(\chi_l, \hat{\gamma})\varphi'_{\sigma_\gamma(l)},\ l = 0,\ 1,\ \ldots,\ \frac{k-3}{2},\ \gamma^*_{\frac{3}{4}} \varphi'_{\frac{k-1}{2}} = \tilde{\kappa}\varphi'_{\frac{k-1}{2}}.$$

Since the constants appearing in the above formulae have absolute value 1, it is routine to conclude that

Proposition 4.15. *For each odd prime k, the $\frac{k+1}{2}$ functions $\{\varphi'_0, \ldots, \varphi'_{\frac{k-1}{2}}\}$ form an orthogonal basis for the Hilbert space $V'(k)$. Further, the first $\frac{k-1}{2}$ of these functions have the same norm.*

[78]The orders of these modified theta constant derivatives are the same as the corresponding modified theta constants.

Remark 4.16. Although it is difficult to compute $||\varphi_0'||$ and $||\varphi'_{\frac{k-1}{2}}||$, the ratio of these two norms is easily determined.

5. Orders of automorphic forms at cusps

We now assume that k is an odd prime. Our calculations of orders of automorphic forms are simplified by a number of simple observations. For each odd k (which need not be prime) and $l = 0, 1, \ldots,$ or $\frac{k-3}{2}$,

$$\text{ord}_{y+1}\varphi_l = \text{ord}_y B^*(\varphi_l) = \text{ord}_y\varphi_l, \tag{3.30}$$

and

$$\text{ord}_{-\frac{1}{y}}\varphi_l = \text{ord}_y A^*(\varphi_l) = \text{ord}_y\psi_l, \tag{3.31}$$

for all $y \in \mathbb{Q} \cup \{\infty\}$. Similar statements hold for the functions φ_l'.

5.1. Calculations via $\Gamma_o(k)$. As we saw in Chapter 1, it is a consequence of [**27**, Proposition 1.43] that the cusps ∞ and 0 project to the two punctures on $\mathbb{H}^2/\Gamma_o(k)$; thus every cusp in $\mathbb{Q} \cup \{\infty\}$ is $\Gamma_o(k)$-equivalent to either ∞ or 0.

Lemma 5.1. *Let $l = 0, 1, \ldots$ or $\frac{k-3}{2}$. Then (as forms for $\Gamma(k)$)*
(a) $\text{ord}_x\varphi_l = \frac{1}{8}$, *if the cusp $x \in \mathbb{Q} \cup \{\infty\}$ is $\Gamma_o(k)$-equivalent to zero.*
(b) *Let the cusp $x \in \mathbb{Q} \cup \{\infty\}$ be $\Gamma(k)$-equivalent to $\frac{j}{k}$, $j = 1, 2, \ldots, \frac{k-1}{2}$ (this is a complete list of representatives for the $\Gamma(k)$-equivalence classes of cusps that are $\Gamma_o(k)$-equivalent to infinity). Choose a $\gamma \in \Gamma_o(k)$ such that $\gamma(\infty) = x$. Then*

$$\text{ord}_x\varphi_l = \frac{(2\sigma_\gamma(l)+1)^2}{8}.$$

Proof. For $x = 0$, part (a) is a consequence of (3.23). For any $\gamma \in \Gamma$ and every $y \in \mathbb{H}^2 \cup \mathbb{Q} \cup \{\infty\}$, we have

$$\text{ord}_{\gamma(y)}\varphi_l = \text{ord}_y(\gamma_*\varphi_l).$$

If we take $y = 0$ and $\gamma \in \Gamma_o(k)$, then using (3.26) we obtain

$$\text{ord}_{\gamma(0)}\varphi_l = \text{ord}_0\varphi_{\sigma_\gamma(l)},$$

which completes the proof of part (a). We next take $y = \infty$ and $\gamma \in \Gamma_o(k)$. Using (3.26), once again, and (3.20), we obtain

$$\text{ord}_{\gamma(\infty)}\varphi_l = \text{ord}_\infty\varphi_{\sigma_\gamma(l)} = \frac{(2\sigma_\gamma(l)+1)^2}{8}. \tag{3.32}$$

Similar formulae hold for the functions φ_l'. We leave it to the reader to derive them.

Definition 5.2. We call a puncture on $\mathbb{H}^2/\Gamma(k)$ *distinguished* if it is the image under P of a cusp that is $\Gamma_o(k)$-equivalent to ∞. This definition makes sense for arbitrary k. For k an odd prime, the distinguished punctures are $\left\{P_{\frac{j}{k}},\ j = 1,\ ...,\ \frac{k-1}{2}\right\}$.

Remark 5.3. The $\frac{k^2-1}{2}$ punctures on $\mathbb{H}^2/\Gamma(k)$ split up into $\frac{k+1}{2}$ disjoint sets; one of these sets consists of the $\frac{k-1}{2}$ distinguished punctures, each fixed by $\tilde{B}$, and each of the other $\frac{k-1}{2}$ sets consists of k punctures, cyclically permuted by $< \tilde{B} >$. The $k\frac{k-1}{2}$ nondistinguished punctures are $\Gamma_o(k)/\Gamma(k)$-equivalent to P_0.

Remark 5.4. The above lemma (except for the second parenthetical remark) also holds for odd composite k. However in the composite case there are cusps that are $\Gamma_o(k)$-equivalent to neither 0 nor ∞. To compute the order of a modified theta constant at one of these cusps, we must use the methods discussed in next subsection.

Remark 5.5. We saw in the previous chapter that if we identify the puncture P_∞ with the class of the characteristic χ_o, then for each $\gamma \in \Gamma$, the puncture $P_{\gamma^{-1}(\infty)}$ is naturally identified with the class of the characteristic $\chi_o\gamma$. Similarly, if we identify the puncture P_∞ with the class of the characteristic $\chi_o A = \chi_0$, the puncture $P_{\gamma^{-1}(\infty)}$ is naturally identified with the class of the characteristic $\chi_0\gamma$. Further, the distinguished punctures can be listed as $P_{\gamma_j^{-1}(\infty)}$, with $\{\gamma_1,\ ...,\ \gamma_{\frac{k-1}{2}}\}$ a complete set of representatives for the factor group $G(k)\backslash\ \Gamma_c(k)$. The corresponding list of characteristics is then $\mathcal{T}_1 A$ (the characteristics $\{\chi_0,\ ...,\ \chi_{\frac{k-3}{2}}\}$).

Lemma 5.6. *Let l be an integer with $0 \leq l \leq \frac{k-3}{2}$. As x in the last lemma runs over representatives of the distinct $\Gamma(k)$-equivalence classes of cusps $\Gamma_o(k)$-equivalent to ∞, the corresponding integers $\sigma_\gamma(l)$ run over the non-negative integers $\{0,\ 1,\ ...,\ \frac{k-3}{2}\}$.*

Proof. We use the fact that $\Gamma_o(k)/G(k)$ is cyclic of order $\frac{k-1}{2}$ and in one-to-one correspondence with the tower $\mathcal{T}_1$ in $X_o(k)$, the basis for $V(k)$ given by the functions $\{\varphi_0,\ ...,\ \varphi_{\frac{k-3}{2}}\}$, and the distinguished punctures $\{P_\infty = x_1,\ ...,\ x_{\frac{k-1}{2}}\}$ on $\mathbb{H}^2/\Gamma(k)$ determined by the cusps $\Gamma_o(k)$-equivalent to ∞. Choose a generator γ of $\Gamma_o(k)/G(k)$. We reorder the punctures $\{x_2,\ ...,\ x_{\frac{k-1}{2}}\}$, if needed, so that $\gamma(x_i) = x_{i+1}$ (we are using two identifications in this last equation: punctures are identified with cusps that project onto them, and addition is interpreted modulo $\frac{k-1}{2}$). Then we can rewrite (3.32) as

$$\mathrm{ord}_{x_j}\varphi_l = \frac{(2\sigma^{j-1}(l)+1)^2}{8},$$

where $\sigma = \sigma_\gamma$. Since $< \tilde{\gamma} >$ acts cyclically on the distinguished punctures $\{x_1, \ldots, x_{\frac{k-1}{2}}\}$ (or equivalently on the classes of the characteristics $\{\chi_0, \ldots, \chi_{\frac{k-3}{2}}\}$ in some order), for each l, the powers of σ evaluated at l define an element of $\mathcal{S}_{\frac{k-1}{2}}$.

We record the following useful

Corollary 5.7. *Let $\gamma \in \Gamma_o(k)$. If $\sigma_\gamma(l) = l$ for some integer l with $0 \le l \le \frac{k-3}{2}$, then $\gamma \in G(k)$.*

Corollary 5.8. *For each integer l with $0 \le l \le \frac{k-3}{2}$, the orders of φ_l at the $\frac{k-1}{2}$ distinguished punctures are distinct.*

Remark 5.9. The last two lemmas imply that

$$\deg(\varphi_l) = \frac{1}{48}k(k^2-1) = \frac{1}{4}(2p(k) - 2 + n(k)),$$

as expected. It is also of interest to note that as a consequence of the above formula for $\deg(\varphi_l)$, the generic element of $V(k)$ has $\frac{(k^2-1)(k-3)}{48}$ simple $\Gamma(k)$-inequivalent zeros on $\mathbb{H}^2$.

5.2. The general case. In order to develop a general alternate method for computing $\mathrm{ord}_x\varphi_l$ at an arbitrary cusp x for $\Gamma(k)$, we must study in some detail the invariance of modified theta constants under Γ. In order to have elegant formulae, we would need an analogue of the transformation formula for elements $\tilde{\gamma}$ with $\gamma \in \mathrm{PSL}(2, \mathbb{Z})$. Since we do not have such an analogue, we must use other (cruder) methods.

If $C \in \Gamma$, then we know that for every integer l with $0 \le l \le \frac{k-3}{2}$, we have

$$\mathrm{ord}_{C(\infty)}\varphi_l = \mathrm{ord}_\infty(C^*\varphi_l).$$

Now for $\tau \in \mathbb{H}^2$,

$$(C^*\varphi_l)\ (\tau) = \theta\begin{bmatrix} \frac{2l+1}{k} \\ 1 \end{bmatrix}(0, kC(\tau))C'(\tau)^{\frac{1}{4}} = \theta\begin{bmatrix} \frac{2l+1}{k} \\ 1 \end{bmatrix}(0, \hat{C}(k\tau))\hat{C}'(k\tau)^{\frac{1}{4}}.$$

If $C \in \Gamma_o(k)$, we can continue the last equation and obtain

$$(C^*\varphi_l)\ (\tau) = \kappa\left(\begin{bmatrix} \frac{2l+1}{k} \\ 1 \end{bmatrix}, \hat{C}\right)\theta\left[\begin{bmatrix} \frac{2l+1}{k} \\ 1 \end{bmatrix}\hat{C}\right](0, k\tau). \tag{3.33}$$

From which it follows, as seen before, that

$$\mathrm{ord}_{C(\infty)}\varphi_l = \mathrm{ord}_\infty\varphi_{\sigma_C(l)}.$$

However, if $C \in \Gamma - \Gamma_o(k)$, we need an alternate approach. We have seen that[79]

$$B^*\varphi_l = c\ \varphi_l \text{ and } A^*\varphi_l = c\ \psi_l.$$

[79] In each of the next two displayed equations c is a nonzero constant that can easily be computed.

An arbitrary element $C \in \Gamma$ is a word in the generators B and A, hence of the form

$$B^{n_1} \circ A \circ B^{n_2} \circ A \circ B^{n_3} \dots \circ A \circ B^{n_r},$$

for some integer $r > 1$, where the indices

$$n_1,\ n_2,\ \dots\ , n_r$$

are integers with all but the first and last nonzero. This formula, can be used to determine the action of an arbitrary unimodular matrix on the modified theta constants.

6. The field of meromorphic functions on $\overline{\mathbb{H}^2/\Gamma(k)}$

6.1. Functions of small degree. As a consequence of Lemmas 4.3 and 5.1, for odd k, a ratio of any two linearly independent functions in $V(k)$ defines a meromorphic function on $\overline{\mathbb{H}^2/\Gamma(k)}$. In particular, if l and l' are unequal nonnegative integers $\leq \frac{k-3}{2}$, then $\frac{\varphi_l}{\varphi_{l'}}$ projects to such a function. Similar assertions hold for functions in $V'(k)$. We are interested in producing functions of low degree.

Let us assume, as before, that k is an odd prime. The divisor of the projection of $\frac{\varphi_l}{\varphi_{l'}}$ to $\overline{\mathbb{H}^2/\Gamma(k)}$ is supported at the $\frac{k-1}{2}$ punctures that are $\Gamma_o(k)$-equivalent to ∞. Without loss of generality we assume that the number of distinct zeros of the function is less than or equal to the number of distinct poles, thus at most $\frac{k-1}{4}$ distinct zeros. We estimate the degree of such a function by counting the number of zeros (with multiplicities) that it has. An upper bound for the order of a zero of $\frac{\varphi_l}{\varphi_{l'}}$ at a point is obtained by maximizing the order of the zero of φ_l and minimizing the one of $\varphi_{l'}$. We shall use the fact that the orders of the zeros of one of these automorphic forms at the $\frac{k-1}{2}$ distinguished punctures are permutations of the rational numbers

$$\frac{1^2}{8},\ \frac{3^2}{8},\ \dots,\ \frac{(k-2)^2}{8},$$

and we conclude that we can produce such a function of degree at most

$$\frac{1}{64}(k-1)^3, \text{ if } k \equiv 1 \bmod 4 \text{ and } \frac{1}{64}(k-3)(k^2-1), \text{ if } k > 3 \text{ and } k \equiv 3 \bmod 4.$$

The argument proceeds as follows: We consider first the case $k \equiv 1 \bmod 4$. In this case we assume that at most half, $\frac{k-1}{4}$, of the distinguished cusps are zeros of the function. In order to maximize the orders of zeros assume that the form φ_l has orders $\frac{(k-2j)^2}{8}$, $j = 1, .., \frac{k-1}{4}$ and that the form $\varphi_{l'}$ has zeros of orders $\frac{(2j-1)^2}{8}$ at the same set of points. The degree of the quotient $\frac{\varphi_l}{\varphi_{l'}}$ is therefore at most

$$\sum_{j=1}^{\frac{k-1}{4}} \frac{(k-2j)^2}{8} - \frac{(2j-1)^2}{8} = \left(\frac{k-1}{4}\right)^3.$$

In the case of k congruent to 3 mod 4, $k > 3$ we can assume that at most $\frac{(k-3)}{4}$ of equivalence classes of distinguished cusps are zeros of the quotient $\frac{\varphi_l}{\varphi_{l'}}$. An argument similar to the one used in the previous case produces an upper bound for the degree of the function $\frac{\varphi_l}{\varphi_{l'}}$. We have shown that for odd primes k, the compact surface $\overline{\mathbb{H}^2/\Gamma(k)}$ carries a function of degree $O(\frac{k^3}{64})$, $k \to \infty$, which is smaller than its genus: $O(\frac{k^3}{24})$, $k \to \infty$, and an improvement on the result for generic surfaces ($O(\frac{k^3}{48})$, $k \to \infty$; see, for example, [**11**, pg. 261]). The method we have used can surely be modified to guarantee the existence of functions of lower degree.

The Hilbert space $V(k)$ allows us to study a number of interesting divisors[80] on $\overline{\mathbb{H}^2/\Gamma(k)}$.

Proposition 6.1. *For each integer l, $0 \le l \le \frac{k-3}{2}$, let*

$$D_l = \prod_{j=1}^{\frac{k-1}{2}} P_{\frac{j}{k}}^{\operatorname{ord}_{\frac{j}{k}} \varphi_l - \frac{1}{8}}.$$

Then

$$i(D_l) \ge \frac{(k-1)(k-3)(k-5)}{48}.$$

Proof. We first observe, for future use, that

$$\deg D_l = \frac{1}{48}(k^2-1)(k-3) < p,$$

for $k \ge 11$. Since for every $\varphi \in V(k)$, the projection of $\frac{\varphi}{\varphi_l}$ to $\overline{\mathbb{H}^2/\Gamma(k)}$ belongs to $L\left(\frac{1}{D_l}\right)$, we conclude that

$$r\left(\frac{1}{D_l}\right) \ge \frac{k-1}{2}.$$

The inequality on the index of specialty follows from the Riemann-Roch theorem.

Corollary 6.2. *The class of the divisor D_l is independent of l.*

Proof. For l and l' as above $\frac{\Phi_l}{\Phi_{l'}}$ is a meromorphic function on $\overline{\mathbb{H}^2/\Gamma(k)}$ with divisor $\frac{D_l}{D_{l'}}$.

[80] We use, throughout this book, standard (multiplicative) notation ([**6**, Chapter III]) for divisors and the indices appearing in the Riemann-Roch theorem.

Remark 6.3. For certain low primes k, we will re-prove the above proposition in the sequel. Note that for $k = 7$, the divisors D_l have degree 4, and are hence canonical as a consequence of the proposition. The proposition implies trivially the existence of a meromorphic function of degree at most $\frac{1}{48}(k^2 - 1)(k - 3)$ on $\overline{\mathbb{H}^2/\Gamma(k)}$.

6.2. $G(k)$**-invariant functions.** The material of this subsection is valid for odd integers $k \geq 3$. Let $\alpha_i \in \mathbb{Z}$ for $i = 0, 1, \ldots, \frac{k-3}{2}$ be chosen so that

$$\sum_{i=0}^{\frac{k-3}{2}} \alpha_i = 0 \text{ and } k \mid \sum_{i=0}^{\frac{k-3}{2}} (i + i^2)\alpha_i.$$

Then

$$\Pi_{i=0}^{\frac{k-3}{2}} \varphi_i^{\alpha_i}$$

is a $G(k)$-invariant meromorphic function on $\mathbb{H}^2$. It hence makes sense to call $(\alpha_0, \ldots, \alpha_{\frac{k-3}{2}})$ an *admissible* $\frac{k-1}{2}$*-tuple* if it satisfies the above conditions.

For example, for $k = 7$, we obtain admissible triples by choosing the 3 integers

$$(3, -1, -2), (1, 2, -3) \text{ or } (-1, -2, 3).$$

For $k = 11$, we can choose the 5 integers as

$$(2, -1, 0, 0, -1) \text{ or } (-1, 0, 2, -1, 0),$$

and for $k = 13$, the 6 integers as

$$(-1, 1, -1, 0, 0, 1).$$

Assume that $k \geq 7$. Given any collection of $\frac{k-1}{2} - 2$ integers,

$$\alpha_2, \alpha_3, \ldots, \alpha_{\frac{k-3}{2}},$$

it can always be completed to an admissible $\frac{k-1}{2}$-tuple by setting

$$\alpha_1 = kr - \sum_{i=2}^{\frac{k-3}{2}} \frac{i + i^2}{2}\alpha_i \text{ and } \alpha_0 = -\sum_{i=1}^{\frac{k-3}{2}} \alpha_i,$$

where $r \in \mathbb{Z}$ is arbitrary. In particular,

$$\left(\frac{k-3}{2}, -\frac{k-3}{2}, 0, \ldots, 0, -1, 1\right), (-3, 3, 0, \ldots, 0, -1, 1),$$

$$(2, -3, 1, 0, \ldots, 0) \text{ and } (-k + 2, k - 3, 1, 0, \ldots, 0)$$

are always nontrivial admissible $\frac{k-1}{2}$-tuples. For odd $k \geq 5$ one also obtains an admissible $\frac{k-1}{2}$-tuple with entries (in any order) k, $-k$ and $\frac{k-5}{2}$ zeros.

6.3. Generators for the function field $\mathcal{K}(\Gamma(k))$. We need to digress to review some standard material on the algebraic nature of compact Riemann surfaces. Let M be a compact Riemann surface. Let $\mathfrak{M}^*$ be the set of analytic configurations (equivalence classes of convergent Puiseaux series). We are using the language and results from [**6**, §IV.11]. Let z and w be two nonconstant meromorphic functions on M (thus elements of $\mathcal{K}(M)$). These two functions define a nonconstant holomorphic map

$$\varphi : M \to \mathfrak{M}^*$$

such that for all $x \in M$,

$$w(x) = \text{eval}(\varphi(x)) \text{ and } z(x) = \text{proj}(\varphi(x)). \tag{3.34}$$

We are interested in simple conditions that guarantee that the map φ be injective (equivalently, that z and w form a primitive pair on M or that they generate $\mathcal{K}(M)$).

Proposition 6.4. *Let z and w be nonconstant meromorphic functions on M. If* deg z *is prime and* proj *is not injective on $\varphi(M)$, then φ is injective (on M).*

Proof. From the second equation in (3.34) we conclude that

$$\deg z = (\deg \text{proj})(\deg \varphi). \tag{3.35}$$

If proj is one-to-one, $\varphi(M)$ is (conformally equivalent to) the Riemann sphere; otherwise, deg proj $=$ deg z.

Let z and w be nonconstant meromorphic functions on the compact Riemann surface M. The map φ assigns to each point $P \in M$ the Puiseaux series of w at P in terms of z. Assume we know the divisor (z) and the leading terms of the Laurent series expansion of w at the support of (z). Equation (3.34) or equivalently (3.35) is then a useful tool for determining when φ is injective. From the divisor (z) we easily compute deg z. We compute deg proj as the number of zeros (counting multiplicities) of the function proj on $\varphi(M)$. Let P_1, ..., P_r be a complete list of the distinct zeros of z. Assume that w *weakly separates* these points in the sense that for $1 \le i < j \le r$ either

$$w(P_i) \neq w(P_j) \text{ or } b_w(P_i) \neq b_w(P_j), \tag{3.36}$$

and for each i,

$$(b_z(P_i) + 1,\ b_w(P_i) + 1) = 1. \tag{3.37}$$

The first condition (3.36) tells us that the r Puiseaux series $\{\varphi(P_i);\ i = 1, ..., r\}$ are distinct; the second (3.37) allows us to compute the ramification number of proj at each P_i (from purely local data). We conclude that under these hypotheses,

$$\deg \text{proj} = \sum_{i=1}^{r} (b_z(P_i) + 1) = \deg z,$$

which shows, of course, that φ is injective whenever (3.36) and (3.37) are satisfied.

Let t be a local coordinate vanishing at P_1 such that in a neighborhood of P_1,

$$z(t) = t^\alpha$$

for the positive integer $\alpha = \text{ord}_{P_1} z$. Then $\varphi(P_1)$ is a Puiseaux series of the form

$$w(z) = \sum_{j=\beta}^{\infty} c_j z^{\frac{j}{\alpha}}, \quad \beta = \text{ord}_{P_1} w.$$

It now easily follows that if $\varphi(P) = \varphi(P_1)$ for some $P \in M$, then z vanishes at P and hence $P = P_i$ for some i, $i = 1, \ldots, r$. If, for example,

$$\text{ord}_{P_i} w \neq \text{ord}_{P_1} w, \quad i = 2, \ldots, r,$$

and

$$(\text{ord}_{P_i} w, \ \text{ord}_{P_i} z) = 1, \quad i = 1, \ldots, r,$$

then $\deg \varphi = 1$.

With these preliminaries out of the way, we return to the subject under study. Let us restrict our attention to primes $k \geq 7$. It is convenient to work with the functions

$$f_1 = \frac{\varphi_1}{\varphi_0} \text{ and } f_2 = \frac{\varphi_2}{\varphi_1}.$$

The divisors[81] of the functions F_1 and F_2 are supported at the $\frac{k-1}{2}$ distinguished punctures $\{P_{\frac{j}{k}};\ j \in \mathbb{Z},\ 1 \leq j \leq \frac{k-1}{2}\}$. For each such j we can choose integers $b = b(j)$ and $d = d(j)$ such that $jd - bk = 1$. Then the Möbius transformation $\gamma_j = \begin{bmatrix} j & b \\ k & d \end{bmatrix} \in \Gamma_o(k)$, and $\gamma_j(\infty) = \frac{j}{k}$. We have already seen that (the definition of $m(\cdot)$ is given in the last chapter)

$$(F_1) = \prod_{j=1}^{\frac{k-1}{2}} P_{\frac{j}{k}}^{\frac{m(\chi_1\tilde{\gamma}_j)^2 - m(\chi_0\tilde{\gamma}_j)^2}{8}} \text{ and } (F_2) = \prod_{j=1}^{\frac{k-1}{2}} P_{\frac{j}{k}}^{\frac{m(\chi_2\tilde{\gamma}_j)^2 - m(\chi_1\tilde{\gamma}_j)^2}{8}}.$$

[81] In accordance with the conventions followed in much of this book, F_j is the projection to $\mathbb{H}^2/\Gamma(k)$ of the $\Gamma(k)$-invariant function f_j.

An examination of cases shows that there are only two possibilities for the jump in the function m on characteristics; for example:

$$m(\chi_1\tilde{\gamma}_j) = m(\chi_0\tilde{\gamma}_j) + 2j \text{ if } 1 \le j \le \frac{k - m(\chi_0\tilde{\gamma}_j)}{2}$$

and

$$m(\chi_1\tilde{\gamma}_j) = 2(k-j) - m(\chi_0\tilde{\gamma}_j) \text{ if } \frac{k - m(\chi_0\tilde{\gamma}_j)}{2} < j \le \frac{k-1}{2}.$$

We are dealing with characteristics χ of the form $\begin{bmatrix} \frac{m}{k} \\ 1 \end{bmatrix}$, with $m \in \mathbb{Z}$, m odd, and $1 \le m \le k-2$. Thus the equivalence class of χ is completely determined by $m(\chi)$. We conclude that each F_i has either a pole or a zero at each puncture $P_{\frac{j}{k}}$. Let Q_1, ..., Q_r be the complete list of distinct zeros of F_1. Then in some local coordinate t vanishing at Q_i, we can write

$$z = F_1(t) = t^\alpha$$

for some $\alpha \in \mathbb{Z}$, $\alpha > 0$, and

$$w = F_2(z) = \sum_{j=\beta}^{\infty} c_j z^{\frac{j}{\alpha}},$$

where $\beta \in \mathbb{Z}$, $\beta \neq 0$, and the c's are the appropriate Puiseaux series coefficients. Of course,

$$\alpha = \operatorname{ord}_{P_\infty} F_1 \text{ and } \beta = \operatorname{ord}_{P_\infty} F_2.$$

We study the map $\varphi : \overline{\mathbb{H}^2/\Gamma(k)} \to \mathfrak{M}^*$ defined by the functions $z = F_1$ and $w = F_2$. Since

$$\operatorname{ord}_{P_\infty} F_1 = 1 \text{ and } \operatorname{ord}_{P_\infty} F_2 = 2,$$

$\varphi(P_\infty)$ is the Puiseaux series

$$w(z) = cz^2 + \dots .$$

If $\varphi(P) = \varphi(P_\infty)$ for some $P \in \overline{\mathbb{H}^2/\Gamma(k)}$, then we must have

$$\operatorname{ord}_P F_1 = \alpha \text{ and } \operatorname{ord}_P F_2 = 2\alpha,$$

for some positive integer α. Thus $P = P_{\frac{j}{k}}$ for some integer j with $1 \le j \le \frac{k-1}{2}$. If $j \neq 1$, then we must have that

$$\alpha = \frac{m_2^2 - m_1^2}{8} \text{ and } 2\alpha = \frac{m_3^2 - m_2^2}{8},$$

where m_1, m_2, and m_3 are odd integers with

$$1 < m_1 < m_2 < m_3 < k,$$

which leads to the Diophantine equation

$$m_3^2 - 3m_2^2 + 2m_1^2 = 0.$$

Under special circumstances it might be possible to determine solutions of these equations and decide when φ is injective. It is easier to use more terms of the respective Taylor series expansions. In terms of the natural coordinate $\zeta = \exp\left(\frac{2\pi\imath\tau}{k}\right)$ in a neighborhood of P_∞, we have

$$z = \frac{\Phi_1(\zeta)}{\Phi_0(\zeta)} = F_1(\zeta) = \exp\left(\frac{\pi\imath}{k}\right)\,\zeta(1 - \zeta^{k\frac{k-3}{2}} + \ldots)$$

and

$$w = \frac{\Phi_2(\zeta)}{\Phi_1(\zeta)} = F_2(\zeta) = \exp\left(\frac{\pi\imath}{k}\right)\,\zeta^2(1 - \zeta^{k\frac{k-5}{2}} + \ldots).$$

Hence for $\varphi(P_{\frac{1}{k}})$ we have

$$w = w(z) = c_0 z^2\left(1 + c_1 z + \ldots\right).$$

In the above and subsequent equations in this subsection, c_i, with $i \in \mathbb{Z}$, is a nonzero constant (that can easily be computed). The last three displayed equations show that $b_\varphi(P_\infty) = 0$.[82] Thus to show that deg $\varphi = 1$, it suffices[83] to prove that $\varphi(P) \neq \varphi(P_{\frac{1}{k}})$ for all $P \in \overline{\mathbb{H}^2/\Gamma(k)}$, $P \neq P_{\frac{1}{k}}$. If $\varphi(P) = \varphi(P_{\frac{1}{k}})$ for some $P \in \overline{\mathbb{H}^2/\Gamma(k)}$, then it must be the case as above that $P = P_{\frac{j}{k}}$ for some integer j with $1 \leq j \leq \frac{k-1}{2}$. We need to obtain the Taylor series for z and w in terms of a good local coordinate vanishing at $P_{\frac{j}{k}}$. The functions z and w are the projections to $\overline{\mathbb{H}^2/\Gamma(k)}$ of the $\Gamma(k)$-invariant functions $\frac{\varphi_1}{\varphi_0}$ and $\frac{\varphi_2}{\varphi_1}$ on $\mathbb{H}^2 \cup \mathbb{Q} \cup \{\infty\}$. Choose $\gamma \in \Gamma_o(k)$ such that

$$\gamma(\infty) = \frac{j}{k}.$$

Thus we want to study

$$\frac{\varphi_1 \circ \gamma}{\varphi_0 \circ \gamma} \text{ and } \frac{\varphi_2 \circ \gamma}{\varphi_1 \circ \gamma}$$

in a neighborhood of $\imath\infty$. From (3.26) it follows that

$$\frac{\varphi_1 \circ \gamma}{\varphi_0 \circ \gamma} = \frac{\tilde{\kappa}(\chi_1, \tilde{\gamma})}{\tilde{\kappa}(\chi_0, \tilde{\gamma})}\frac{\varphi_{\sigma(1)}}{\varphi_{\sigma(0)}} \text{ and } \frac{\varphi_2 \circ \gamma}{\varphi_1 \circ \gamma} = \frac{\tilde{\kappa}(\chi_2, \tilde{\gamma})}{\tilde{\kappa}(\chi_1, \tilde{\gamma})}\frac{\varphi_{\sigma(2)}}{\varphi_{\sigma(1)}},$$

with $\sigma = \sigma_\gamma$. Thus, for these purposes,

$$z = \alpha\frac{\Phi_l}{\Phi_{l'}} \text{ and } w = \beta\frac{\Phi_m}{\Phi_l},$$

[82] The function z is a perfectly good local coordinate at both $P_\infty \in \overline{\mathbb{H}^2/\Gamma(k)}$ and at $\varphi(P_\infty) \in \mathfrak{M}^*$. In terms of these coordinates φ is the identity map.

[83] We use the symbol P to denote at times a point on $\overline{\mathbb{H}^2/\Gamma(k)}$ and the natural projection $\mathbb{H}^2 \cup \mathbb{Q} \cup \{\infty\} \to \overline{\mathbb{H}^2/\Gamma(k)}$. This identification should cause no confusion to the reader.

for distinct nonnegative integers l, l' and m each at most $\frac{k-3}{2}$ and constants α and β of absolute value 1. Further, if $j \neq 1$, then $\sigma(0) > 0$ (because σ acts without fixed points) and

$$m > l > l' > 0$$

(the first two inequalities are consequences of the fact that both z and w vanish at $P_{\frac{j}{k}}$). It follows that for $\varphi(P)$ we have

$$w = w(z) = c_2 z^{\frac{(m+l+1)(m-l)}{(l+l'+1)(l-l')}} \left(1 + c_3 z^{\frac{2}{(l+l'+1)(l-l')}} + \ldots\right).$$

Hence

$$2 = \frac{(m+l+1)(m-l)}{(l+l'+1)(l-l')} \text{ and } 1 = \frac{2}{(l+l'+1)(l-l')}.$$

We have arrived at a contradiction since

$$(l+l'+1)(l-l') \geq 4,$$

and hence obtain

Theorem 6.5. *For each prime $k \geq 7$, the functions F_1 and F_2 generate the function field $\mathcal{K}(\overline{\mathbb{H}^2/\Gamma(k)})$.*

The $\jmath$-function is Γ-invariant, hence certainly $\Gamma(k)$-invariant. The methods used above also establish

Theorem 6.6. *For each prime $k \geq 7$, the functions J and F_1 generate the function field $\mathcal{K}(\overline{\mathbb{H}^2/\Gamma(k)})$.*

Problem 6.7. Find the algebraic relation satisfied by F_1 and F_2. For $k = 7$ we solve this problem in §8.3. The form of the algebraic equation satisfied by J and F_1 is given in §9.

7. Projective representations

A *coordinate vector* in the finite dimensional projective space $\mathbf{P}\mathbb{C}^{\frac{k-3}{2}}$ is the equivalence class of a vector in $\mathbb{C}^{\frac{k-1}{2}}$ with precisely one nonzero coordinate. For $l = 0, 1, \ldots, \frac{k-5}{2}$,

$$f_l = \frac{\varphi_l}{\varphi_{\frac{k-3}{2}}}$$

is a meromorphic function on the compact surface $\overline{\mathbb{H}^2/\Gamma(k)}$ whose divisor is supported at the distinguished punctures of $\mathbb{H}^2/\Gamma(k)$.

Lemma 7.1. *Let $x \in \mathbb{Q} \cup \{\infty\}$. If*

$$\text{ord}_x \varphi_l > \text{ord}_x \varphi_0 \textit{ for } l = 1, \ldots, \frac{k-3}{2},$$

then $P_x = P_\infty$.

Proof. For each fixed l as above, the function $\frac{\Phi_l}{\Phi_0}$ vanishes at P_x. Hence P_x is a distinguished puncture. Thus $x = \gamma(\infty)$ for some $\gamma \in \Gamma_o(k)$. Now for arbitrary l as above, we have

$$\mathrm{ord}_{\gamma(\infty)}\varphi_l = \mathrm{ord}_\infty \gamma^*(\varphi_l) = \mathrm{ord}_\infty \varphi_{\sigma_\gamma(l)}.$$

We conclude that $\sigma_\gamma(l) > 0$ for all $l > 0$. Thus $\sigma_\gamma(0) = 0$, and hence $\gamma \in G(k)$. It now follows that x and ∞ project to the same puncture on $\mathbb{H}^2/\Gamma(k)$.

Proposition 7.2. *For each odd integer $k \geq 5$, the map*

$$F : \tau \mapsto (f_0(\tau),\ f_1(\tau),\ \ldots,\ f_{\frac{k-5}{2}}(\tau),\ 1)$$

from $\mathbb{H}^2 \cup \mathbb{Q} \cup \{\infty\}$ to $\hat{\mathbb{C}}^{\frac{k-1}{2}}$, or equivalently the map

$$\Phi : \tau \mapsto (\varphi_0(\tau),\ \varphi_1(\tau),\ \ldots,\ \varphi_{\frac{k-5}{2}}(\tau),\ \varphi_{\frac{k-3}{2}}(\tau)) \tag{3.38}$$

from $\mathbb{H}^2 \cup \mathbb{Q} \cup \{\infty\}$ to $\mathbb{C}^{\frac{k-1}{2}}$, defines a holomorphic mapping (also to be called F or Φ, as appropriate) from $\overline{\mathbb{H}^2/\Gamma(k)}$ into $\mathbf{P}\mathbb{C}^{\frac{k-3}{2}}$. If k is prime, the distinguished punctures on the surface $\mathbb{H}^2/\Gamma(k)$ are sent (injectively) onto the coordinate vectors in projective space $\mathbf{P}\mathbb{C}^{\frac{k-3}{2}}$. Further, in this case, the map Φ restricted to the punctures is injective and of maximal rank at each distinguished puncture.

Proof. Only the last three statements need verification.

We need to study the map Φ in some detail. It is convenient to view Φ as a holomorphic map from $\overline{\mathbb{H}^2/\Gamma(k)}$ into $\mathbf{P}V(k)^*$, the space of continuous linear functionals on (lines in) $\mathbf{P}V(k)$. In this setting, $\Phi(P)$ for $P \in \overline{\mathbb{H}^2/\Gamma(k)}$ is the projective equivalence class of the linear functional[84]

$$L_P : \varphi \mapsto \varphi(P),\ \varphi \in V(k).$$

Furthermore, (3.38) identifies L_{P_τ} with the matrix representation of this linear functional with respect to the basis $\{\varphi_0,\ \ldots,\ \varphi_{\frac{k-3}{2}}\}$. If, as can happen (for example, at $P = P_\infty$), $\varphi(P) = 0$ for all $\varphi \in V(k)$, then we let l be the minimum of the orders of vanishing at P of the nonzero elements in $V(k)$ and interpret L_P as evaluation at 0 of the functions in $\frac{V(k)}{\zeta^l}$, where ζ is a local coordinate vanishing at P. A similar interpretation applies to the map F in neighborhoods of points where the meromorphic functions have poles (or vanish simultaneously).

An element $\gamma \in \Gamma/\Gamma(k)$ acts as an automorphism γ of the Riemann surface $\overline{\mathbb{H}^2/\Gamma(k)}$, a projective linear automorphism γ^* (induced by $\gamma^*_{\frac{1}{4}}$) of

[84]In the next and many subsequent formulae we identify points on the surface $\overline{\mathbb{H}^2/\Gamma(k)}$ with their preimages under P in $\mathbb{H}^2 \cup \mathbb{Q} \cup \{\infty\}$, and $\Gamma(k)$-invariant functions on $\mathbb{H}^2 \cup \mathbb{Q} \cup \{\infty\}$ with their projections to $(\mathbb{H}^2 \cup \mathbb{Q} \cup \{\infty\})/\Gamma(k)$.

$\mathbf{P}V(k)$, and a projective linear automorphism γ^* (where $\gamma^*(L) = L(\gamma^*)$ for $L \in \mathbf{P}V(k)^*$) of the projective space $\mathbf{P}V(k)^*$ such that

$$\Phi \circ \gamma = \gamma^* \circ \Phi. \tag{3.39}$$

Assume now that k is a prime. We use representatives for the elements of $\Gamma_o(k)/G(k)$ to compute the images in $\mathbf{P}V(k)^*$ of the distinguished punctures on $\mathbb{H}^2/\Gamma(k)$ under the map Φ. We observe that

$$\Phi(\infty) = (1,\ 0,\ \ldots,\ 0).$$

If the cusp $x \in \mathbb{Q}$ is $\Gamma_o(k)$-equivalent to ∞, then we choose $\gamma \in \Gamma_o(k)$ with $\gamma(\infty) = x$. Then

$$\begin{aligned}\Phi(x) &= L_x = L_{\gamma(\infty)} = \gamma^*(L_\infty)\\ &= (\tilde{\kappa}(\chi_0, \tilde{\gamma})\varphi_{\sigma(0)}(\infty),\ \ldots,\ \tilde{\kappa}(\chi_{\frac{k-3}{2}}, \tilde{\gamma})\varphi_{\sigma(\frac{k-3}{2})}(\infty)),\end{aligned}$$

where $\sigma = \sigma_\gamma$. It follows that $\Phi(x)$ is a coordinate vector with the unique nonzero component in the $\sigma^{-1}(0) + 1$ slot Taking γ to be a generator for the cyclic group $\Gamma_o(k)/G(k)$, we obtain a cyclic permutation σ which implies immediately the injectivity of Φ restricted to the distinguished punctures. For the case of an arbitrary puncture, assume that $\Phi(x) = \Phi(y)$ with x and y in $\mathbb{Q} \cup \{\infty\}$. Write $x = \gamma(\infty)$ with $\gamma \in \Gamma$. Then with $z = \gamma^{-1}(y)$, we have

$$\gamma^*(\Phi(\infty)) = \Phi(\gamma(\infty)) = \Phi(\gamma(z)) = \gamma^*(\Phi(z)).$$

Since γ^* is injective, $\Phi(\infty) = \Phi(z)$. We conclude from the previous lemma that $P_\infty = P_z$. Hence ∞ is $\Gamma(k)$-equivalent to $\gamma^{-1}(y)$; that is, $x = \gamma(\infty)$ is $\Gamma(k)$-equivalent to y. Finally in a neighborhood of P_∞ the map Φ is given in inhomogeneous coordinates by

$$\Phi :\ \zeta \mapsto \left(\frac{\Phi_1(\zeta)}{\Phi_0(\zeta)},\ \ldots,\ \frac{\Phi_{\frac{k-3}{2}}(\zeta)}{\Phi_0(\zeta)}\right).$$

Since $\frac{\Phi_1}{\Phi_0}$ has a simple zero at P_∞, Φ is of maximal rank at this point. It is, by now, routine to reduce the study of the map Φ near an arbitrary distinguished puncture to the case where the puncture is P_∞.

Corollary 7.3. *For every prime $k \geq 7$, the map Φ is also defined by $\frac{k-1}{2}$ linearly independent abelian differentials of the first kind on $\overline{\mathbb{H}^2/\Gamma(k)}$.*

Proof. The corollary is an immediate consequence of the previous proposition and Proposition 6.1.

Remark 7.4. $\Phi(\overline{\mathbb{H}^2/\Gamma(k)})$ is a curve of degree $\frac{(k^2-1)(k-3)}{48}$ in $\mathbf{P}\mathbb{C}^{\frac{k-3}{2}}$. For $k = 7$ it is a curve of genus 3 (see Chapter 1) and degree 4 in 2-dimensional complex projective space, as a consequence of the last corollary (or as can be seen directly (§8.3)), the canonical curve.

Problem 7.5. Is the map Φ injective?

Proposition 7.6. *For each prime $k > 3$, the homomorphism*

$$\Theta : \Gamma/\Gamma(k) \to \mathrm{Aut}\mathbf{P}V(k)$$

is injective.

Proof. If $\gamma^{-1} \in \Gamma$ belongs to the kernel of Θ, then there exists a nonzero complex number λ so that for each integer l with $0 \le l \le \frac{k-3}{2}$, $\gamma^*(\varphi_l) = \lambda\varphi_l$. By Corollary 5.8, γ must map infinity to a $\Gamma(k)$-equivalent cusp. Thus $\gamma \in G(k)$ and hence $\gamma = B^m \circ \gamma_1$, with $m \in \mathbb{Z}$, $0 \le m \le k-1$, and $\gamma_1 \in \Gamma(k)$. Since $\gamma^* = (\gamma_1)^* \circ (B^*)^m$ and γ_1 is in the kernel of Θ, we conclude that γ is in the kernel if and only if B^m is. Using (3.22) for $l = 0$ and 1, we see that $\lambda = c(B,k) = c(B,k)\exp(\frac{2\pi i m}{k})$. Hence $m = 0$.

Remark 7.7. (1) It is probably true that Φ is injective. The injectivity of Θ would follow trivially from this conjecture.
(2) Using the space $V'(k)$ instead of $V(k)$ we can produce maps

$$\Phi' : \overline{\mathbb{H}^2/\Gamma(k)} \to \mathbf{P}\mathbb{C}^{\frac{k-1}{2}} \text{ and } \Theta' : \Gamma/\Gamma(k) \to \ \mathrm{PGL}\left(\frac{k-1}{2}, \mathbb{C}\right)$$

that are analogous to the maps Φ and Θ.

Although we have few tools for the investigation of the maps Φ and Φ' at points $x \in \mathbb{H}^2/\Gamma(k)$, at times enough information at the punctures can be translated to results about interior points. Among the results so obtained are the following two propositions.

Proposition 7.8. *For odd $k > 1$ (even k), each puncture on $\mathbb{H}^2/\Gamma(k)$ ($\mathbb{H}^2/H(k)$) is a Weierstrass point for $V(k)$ and $V'(k)$, and only the punctures are Weierstrass points.*

Proof. Assume that k is odd (> 1). It is obvious that ∞ is a Weierstrass point for $V(k)$ since

$$r_l = \mathrm{ord}_\infty \varphi_l = \frac{(2l+1)^2}{8} > l \text{ for } l = 0, \ ..., \ \frac{k-3}{2}.$$

The same argument shows that ∞ is a Weierstrass point for $V'(k)$. See Proposition 1.18. Since Γ preserves $V(k)$ and $V'(k)$ and acts transitively on the punctures of $\mathbb{H}^2/\Gamma(k)$, we conclude that all the cusps are Weierstrass points of the same weight. Now straightforward calculations (using formulae for the sums of squares of odd integers and the fact that the negative Euler characteristic of $\Gamma(k)$ is $\frac{k^3}{12}\prod_{\mathfrak{P} \text{ prime}, \mathfrak{P}|k}\left(1 - \frac{1}{\mathfrak{P}^2}\right)$) show that

$$\tau_{V(k)}(\infty) = \frac{k(k-1)(k-2)}{48}, \ \tau_{V'(k)}(\infty) = \frac{k(k+1)(k+2)}{48},$$

$$\tau(V(k)) = \frac{k^3(k-1)(k-2)}{96} \prod_{\mathfrak{P} \text{ prime}, \ \mathfrak{P}|k} \left(1 - \frac{1}{\mathfrak{P}^2}\right)$$

and

$$\tau(V'(k)) = \frac{k^3(k+1)(k+2)}{96} \prod_{\mathfrak{P} \text{ prime}, \ \mathfrak{P}|k} \left(1 - \frac{1}{\mathfrak{P}^2}\right).$$

Since the $\frac{k^2}{2} \prod_{\mathfrak{P} \text{ prime}, \ \mathfrak{P}|k} \left(1 - \frac{1}{\mathfrak{P}^2}\right)$ punctures account for the total weight of $V(k)$ and $V'(k)$, the proof is complete in this case. For the odd prime k, we have the simplification

$$\tau(V(k)) = \frac{(k-2)(k-1)^2 k(k+1)}{96} \text{ and } \tau(V'(k)) = \frac{(k-1)k(k+1)^2(k+2)}{96}.$$

The proof for even k is similar.

Proposition 7.9. *For all $k \in \mathbb{Z}^+$, the maps Φ and Φ' have maximal rank everywhere.*

Proof. We need to show that $d\Phi$ and $d\Phi'$ are nonsingular everywhere, equivalently that for all $x \in \mathbb{H}^2 \cup \mathbb{Q} \cup \{\infty\}$ we can find functions f and g in $V(k)$ ($V'(k)$) that are regular at x with $\frac{g}{f}$ having a simple zero at x. Let us assume that k is odd. We have seen that we can choose f and g in both $V(k)$ and $V'(K)$ so that

$$\operatorname{ord}_\infty f = \frac{1}{8} \text{ and } \operatorname{ord}_\infty g = \frac{9}{8}.$$

Hence our maps have maximal rank at the cusps. The previous proposition showed that the weight of each ordinary point $x \in \mathbb{H}^2$ with respect to either $V(k)$ or $V'(k)$ is zero. Hence there certainly are functions f and g in these spaces with

$$\operatorname{ord}_x f = 0 \text{ and } \operatorname{ord}_x g = 1.$$

Thus for odd k, the two maps have maximal rank globally. We leave it to the reader to give the argument for even k.

8. Some special cases ($k' = k$)

8.1. $k = 3$. Combining the results developed here for $V(3)$ with those of §3.3, we get the following identity for $\tau \in \mathbb{H}^2$:

$$\frac{\theta^4 \begin{bmatrix} \frac{1}{3} \\ 1 \end{bmatrix} (0, 3\tau)}{\theta \begin{bmatrix} 1 \\ \frac{1}{3} \end{bmatrix} (0,\tau)\, \theta \begin{bmatrix} \frac{1}{3} \\ \frac{1}{3} \end{bmatrix} (0,\tau)\, \theta \begin{bmatrix} \frac{1}{3} \\ 1 \end{bmatrix} (0,\tau)\, \theta \begin{bmatrix} \frac{1}{3} \\ \frac{5}{3} \end{bmatrix} (0,\tau)} = \frac{1}{2} + \frac{\sqrt{3}}{6}\imath.$$

The above formula for θ-constants is a special case[85] of a more general relation for θ-functions. For all $z \in \mathbb{C}$

$$\theta^3 \begin{bmatrix} \frac{1}{3} \\ 1 \end{bmatrix} (0, 3\tau)\, \theta \begin{bmatrix} \frac{1}{3} \\ 1 \end{bmatrix} (3z, 3\tau)$$

$$= \left(\frac{1}{2} + \frac{\sqrt{3}}{6}\imath\right) \theta \begin{bmatrix} 1 \\ \frac{1}{3} \end{bmatrix} (0, \tau) \theta \begin{bmatrix} \frac{1}{3} \\ \frac{1}{3} \end{bmatrix} (z, \tau)\, \theta \begin{bmatrix} \frac{1}{3} \\ 1 \end{bmatrix} (z, \tau)\, \theta \begin{bmatrix} \frac{1}{3} \\ \frac{5}{3} \end{bmatrix} (z, \tau).$$

Theorem 8.1. *The three functions*

$$\frac{\theta' \begin{bmatrix} \frac{1}{3} \\ 1 \end{bmatrix} (0, 3\tau)}{\theta' \begin{bmatrix} 1 \\ 1 \end{bmatrix} (0, 3\tau)}, \quad \frac{\theta' \begin{bmatrix} 1 \\ 1 \end{bmatrix} (0, \frac{\tau}{3})}{\theta' \begin{bmatrix} 1 \\ 1 \end{bmatrix} (0, 3\tau)} \quad \text{and} \quad \frac{\theta^3 \begin{bmatrix} \frac{1}{3} \\ \frac{5}{3} \end{bmatrix} (0, \tau)}{\theta' \begin{bmatrix} 1 \\ 1 \end{bmatrix} (0, 3\tau)}$$

are holomorphic universal covering maps of $\mathbb{H}^2/\Gamma(3)$ *with respective divisors*

$$\frac{P_x}{P_\infty}, \; \frac{P_0}{P_\infty} \; \text{and} \; \frac{P_1}{P_\infty},$$

with $x \in \mathbb{H}^2$. *These functions and the constant function are linearly dependent yielding the identities*

$$(3.40) \qquad 6 \exp\left(\frac{\pi\imath}{3}\right) \theta' \begin{bmatrix} \frac{1}{3} \\ 1 \end{bmatrix} (0, 3\tau) = \theta' \begin{bmatrix} 1 \\ 1 \end{bmatrix} \left(0, \frac{\tau}{3}\right) + 3\theta' \begin{bmatrix} 1 \\ 1 \end{bmatrix} (0, 3\tau)$$

and

$$6\theta' \begin{bmatrix} \frac{1}{3} \\ 1 \end{bmatrix} (0, 3\tau) = 2\pi \exp\left(\frac{\pi\imath}{3}\right) \theta^3 \begin{bmatrix} \frac{1}{3} \\ \frac{5}{3} \end{bmatrix} (0, \tau) + \frac{1}{2} \exp\left(\frac{5\pi\imath}{6}\right) \theta' \begin{bmatrix} 1 \\ 1 \end{bmatrix} (0, 3\tau).$$

Proof. Call the three functions f, g and h. That f is a $\Gamma(3)$-invariant and the calculation of its divisor are based on the $V'(3)$ theory. The corresponding facts for g are based on the fact that $\theta' \begin{bmatrix} 1 \\ 1 \end{bmatrix} (0, \frac{\tau}{3})$ is a constant multiple of $\theta' \begin{bmatrix} \frac{1}{3} \\ 1 \end{bmatrix} (0, 3\tau)$. For the last function h, we base these conclusions on the development of §3.3. In addition we mention here the identity

$$(3.41) \qquad \theta' \begin{bmatrix} 1 \\ 1 \end{bmatrix} (0, \tau) = 2\pi\imath\theta^3 \begin{bmatrix} \frac{1}{3} \\ 1 \end{bmatrix} (0, 3\tau) = \frac{-2\pi\sqrt{3}}{9} \theta^3 \begin{bmatrix} 1 \\ \frac{1}{3} \end{bmatrix} \left(0, \frac{\tau}{3}\right).$$

The first of these equalities is proved in Chapter 4. The second is a consequence of Lemma 1.4 in Chapter 2. The fact that the functions f, g and h are covering maps follows from the forms of their divisors. Since $\overline{\mathbb{H}^2/\Gamma(3)}$

[85] The relevant theory is developed in §7 of Chapter 2.

is a sphere the identities follow easily from the Fourier series expansions of the functions. The translation of the above identities to power series yields

Corollary 8.2. *We have*

(3.42)
$$\sum_{n=-\infty}^{\infty} (-1)^n (6n+1)\, x^{\frac{3n(3n+1)}{2}}$$
$$= \sum_{n=0}^{\infty}(-1)^n (2n+1)\, x^{\frac{n(n+1)}{2}} - 3x\sum_{n=0}^{\infty}(-1)^n (2n+1)\, x^{\frac{9n(n+1)}{2}}$$

and

$$\sum_{n=0}^{\infty}(-1)^n (6n+1)\, x^{\frac{3n(3n+1)}{2}}$$
$$= \left(\sum_{n=-\infty}^{\infty} e^{\frac{5\pi\imath n}{3}}\, x^{\frac{n(3n+1)}{2}}\right)^3 - 3e^{\frac{4\pi\imath}{3}}\, x\sum_{n=0}^{\infty}(-1)^n (2n+1)\, x^{\frac{9n(n+1)}{2}}.$$

We turn now to an application in number theory. Dedekind's *eta-function* (see, for example, [**22**, pg. 129])

$$\eta(z) = z^{\frac{1}{24}} \prod_{n=1}^{n=\infty} (1-z^n) = z^{\frac{1}{24}} \sum_{n=-\infty}^{\infty} (-1)^n z^{\frac{3n^2+n}{2}}, \ z \in \mathbb{C},\ |z|<1,$$

arises most naturally in the theory of the partition function for the positive integers. If n is a positive integer, we define[86] $P(n)$ to be the number of ways n can be expressed as a sum of nonincreasing positive integers. It is well known [**13**, Ch. XIX] that if we define $P(0) = 1$, then

$$\sum_{n=0}^{\infty} P(n)x^n = \frac{x^{\frac{1}{24}}}{\eta(x)},\ x \in \mathbb{C},\ |x|<1.$$

For $\tau \in \mathbb{H}^2$, we set $x = \exp(2\pi\imath\tau)$. We can hence view η as a function on the upper half plane. It is an interesting observation, nontrivial to verify [**26**, pgs. 95 and 109], that η satisfies a functional equation. It is routine to verify (by comparing Fourier series expansions) that

$$\eta(\tau) = \exp\left(-\frac{\pi\imath}{6}\right)\, \theta\begin{bmatrix} \frac{1}{3} \\ 1 \end{bmatrix}(0, 3\tau),\ \tau \in \mathbb{H}^2.$$

We conclude from (3.22) and (3.23) that

(3.43)
$$\eta(\tau+1) = \exp\left(\frac{\pi\imath}{12}\right)\, \eta(\tau) \text{ and } \eta\left(-\frac{1}{\tau}\right) = \sqrt{-\imath\tau}\ \eta(\tau), \text{ for all } \tau \in \mathbb{H}^2.$$

[86] The functions η and P will be studied extensively in Chapter 5.

8.2. $k = 5$. We begin with a result that, as above, combines ideas from two different approaches. It is the identity for $\tau \in \mathbb{H}^2$:

$$\frac{\theta\begin{bmatrix}\frac{1}{5}\\1\end{bmatrix}(0,5\tau)}{\theta\begin{bmatrix}\frac{3}{5}\\1\end{bmatrix}(0,5\tau)} =$$

$$e^{\frac{4\pi\imath}{5}}\frac{\theta\begin{bmatrix}\frac{1}{5}\\\frac{1}{5}\end{bmatrix}(0,\tau)\,\theta\begin{bmatrix}\frac{1}{5}\\\frac{3}{5}\end{bmatrix}(0,\tau)\,\theta\begin{bmatrix}\frac{1}{5}\\1\end{bmatrix}(0,\tau)\,\theta\begin{bmatrix}\frac{1}{5}\\\frac{7}{5}\end{bmatrix}(0,\tau)\,\theta\begin{bmatrix}\frac{1}{5}\\\frac{9}{5}\end{bmatrix}(0,\tau)}{\theta\begin{bmatrix}\frac{3}{5}\\\frac{1}{5}\end{bmatrix}(0,\tau)\,\theta\begin{bmatrix}\frac{3}{5}\\\frac{3}{5}\end{bmatrix}(0,\tau)\,\theta\begin{bmatrix}\frac{3}{5}\\1\end{bmatrix}(0,\tau)\,\theta\begin{bmatrix}\frac{3}{5}\\\frac{7}{5}\end{bmatrix}(0,\tau)\,\theta\begin{bmatrix}\frac{3}{5}\\\frac{9}{5}\end{bmatrix}(0,\tau)}.$$

We now apply the theory developed for $V(5)$ to construct a special 12 punctured sphere and realize its group of conformal automorphisms as a group of fractional linear transformations isomorphic to $\Gamma/\Gamma(5)$, which in turn is isomorphic to $\mathrm{PSL}(2,\mathbb{Z}_5)$. We shall also recover the classically known fact that this group is isomorphic to the symmetry group of the regular icosahedron [**5**, pg. 2].

We have remarked at least once that the Riemann surface $\mathbb{H}^2/\Gamma(5)$ is a 12 times punctured sphere. The mapping

$$\Phi : \tau \mapsto (\varphi_0(\tau), \varphi_1(\tau)), \tau \in \mathbb{H}^2 \cup \mathbb{Q} \cup \{\infty\},$$

defines an embedding of $\overline{\mathbb{H}^2/\Gamma(5)}$ into $\mathbf{P}\mathbb{C}$. This follows from the observation that for $f = f_0 = \frac{\varphi_0}{\varphi_1}$, we have $(F) = \frac{P_{\frac{2}{5}}}{P_{\frac{1}{5}}}$.

Our formulae for the matrices A^* and B^* give

$$A^* = \frac{1}{\sqrt{5\imath}}\begin{bmatrix}\exp(\frac{\pi\imath}{10}) + \exp(-\frac{\pi\imath}{10}) & \exp(\frac{\pi\imath}{10}) + \exp(-\frac{\pi\imath}{2})\\ \exp(\frac{\pi\imath}{2}) + \exp(-\frac{\pi\imath}{10}) & \exp(\frac{9\pi\imath}{10}) + \exp(-\frac{9\pi\imath}{10})\end{bmatrix}$$

and

$$B^* = c(B,k)\begin{bmatrix}1 & 0\\0 & \exp(\frac{2\pi\imath}{5})\end{bmatrix}.$$

We may identify Φ with f, in this case. Hence for $\gamma \in \Gamma$, we may identify γ^* with $\tilde{\gamma}$, the induced action of γ on $\mathbb{H}^2/\Gamma(5)$.

We shall digress slightly and discuss a **metric on $\mathbf{P}\mathbb{C}^N$**, $N \in \mathbb{Z}^+$. Let $(\cdot,\cdot)$ and $||\cdot||$ represent the usual (Euclidean) complex inner product and norm in $\mathbb{C}^{N+1}$. Then

$$||u - v||^2 = ||u||^2 + ||v||^2 - 2\Re(u,v).$$

If the equivalence classes of u, v are now thought of as points in $\mathbf{P}\mathbb{C}^N$, we can always choose representatives which lie on the unit ball in $\mathbb{C}^{N+1}$. We then define

$$d(u,v) = \inf ||\tilde{u} - \tilde{v}||,$$

where the infimum is taken over all elements $\tilde{u}, \tilde{v}$ in the unit ball of $\mathbb{C}^{N+1}$ equivalent to u, v respectively. Clearly,

$$\inf_{\{\theta,\phi\in\mathbb{R}\}} \left\{2 - 2\Re\left(e^{\imath\theta}u, e^{\imath\phi}v\right)\right\} = \inf \left\{2 - 2\Re(e^{\imath(\theta-\phi)}(u,v))\right\}$$
$$= 2 - 2|(u,v)|.$$

It is interesting to note that when $n = 1$ the distance we have defined is related but not equal to the chordal metric on the sphere. If we identify the sphere with the one point compactification of the plane by stereographic projection

$$z = x + \imath y \mapsto \left(\frac{2x}{x^2+y^2+1}, \frac{2y}{x^2+y^2+1}, \frac{x^2+y^2-1}{x^2+y^2+1}\right), \quad z \in \hat{\mathbb{C}} = \mathbb{C} \cup \{\infty\},$$

then the chordal metric is given by

$$d(z_1, z_2) = \frac{|z_1 - z_2|}{\sqrt{1+|z_1|^2}\sqrt{1+|z_2|^2}}.$$

Under the usual identification of $\mathbf{P}\mathbb{C}$ with the Riemann sphere $(z_1, z_2) \in \mathbf{P}\mathbb{C}$ corresponds to $\frac{z_1}{z_2} \in \hat{\mathbb{C}}$. The chordal metric between (z_1, z_2) and (w_1, w_2) on $\mathbf{P}\mathbb{C}$ is therefore

$$\frac{|\frac{z_1}{z_2} - \frac{w_1}{w_2}|}{\sqrt{1+|\frac{z_1}{z_2}|^2}\sqrt{1+|\frac{w_1}{w_2}|^2}}.$$

If we assume that $|z_1|^2 + |z_2|^2 = |w_1|^2 + |w_2|^2 = 1$, then the above reduces to $|z_1 w_2 - w_1 z_2|$, which can be written as

$$\sqrt{1 - |z_1\bar{w}_1 + z_2\bar{w}_2|^2} = \sqrt{1 - |((z_1, z_2), (w_1, w_2))|^2}.$$

Hence we see that our distance differs from the chordal distance. The common feature which is important however is that both of these distances are invariant under the unitary group.

We can now return to the special case under consideration. Write the matrix $\sqrt{5}\imath A^*$ as $\begin{bmatrix} z_1 & z_2 \\ w_1 & w_2 \end{bmatrix}$ where, of course, $z_2 = \overline{w_1}$, $w_2 = -\overline{z_1} = -z_1$. The 12 points

$$\infty,\ 0,\ 1,\ 2,\ 3,\ 4,\ \frac{1}{2},\ \frac{7}{2},\ \frac{3}{2},\ \frac{9}{2},\ \frac{5}{2},\ \frac{2}{5}$$

on $\mathbb{R} \cup \{\infty\}$ project to the 12 punctures on $\mathbb{H}^2/\Gamma(5)$; they correspond (under the map Φ) to the equivalence classes in $\mathbf{P}\mathbb{C}$ of the following 12 points in $\mathbb{C}^2$:

$$(1,0),\ (0,1),\ (z_1, w_1),\ (z_2, w_2),$$

$$\left(z_1, \exp\left(\frac{2\pi\imath}{5}\right) w_1\right), \left(z_2, \exp\left(\frac{2\pi\imath}{5}\right) w_2\right), \left(z_1, \exp\left(\frac{4\pi\imath}{5}\right) w_1\right),$$

$$\left(z_2, \exp\left(\frac{4\pi\imath}{5}\right) w_2\right), \left(z_1, \exp\left(\frac{6\pi\imath}{5}\right) w_1\right), \left(z_2, \exp\left(\frac{6\pi\imath}{5}\right) w_2\right),$$

$$\left(z_1, \exp\left(\frac{8\pi\imath}{5}\right) w_1\right), \left(z_2, \exp\left(\frac{8\pi\imath}{5}\right) w_2\right).$$

Under the usual identification of projective space $\mathbf{P}\mathbb{C}$ with the extended complex plane $\mathbb{C} \cup \{\infty\}$, the 12 punctures on $f(\mathbb{H}^2 \cup \mathbb{Q} \cup \{\infty\})$ are:

$$\infty, \ 0, \ \frac{z_1}{w_1}, \ \frac{z_2}{w_2},$$

$$\exp\left(\frac{-2\pi\imath}{5}\right)\frac{z_1}{w_1}, \ \exp\left(\frac{-2\pi\imath}{5}\right)\frac{z_2}{w_2}, \ \exp\left(\frac{-4\pi\imath}{5}\right)\frac{z_1}{w_1}, \ \exp\left(\frac{-4\pi\imath}{5}\right)\frac{z_2}{w_2}.$$

$$\exp\left(\frac{-6\pi\imath}{5}\right)\frac{z_1}{w_1}, \ \exp\left(\frac{-6\pi\imath}{5}\right)\frac{z_2}{w_2}, \ \exp\left(\frac{-8\pi\imath}{5}\right)\frac{z_1}{w_1}, \ \exp\left(\frac{-8\pi\imath}{5}\right)\frac{z_2}{w_2}.$$

Pairs of antipodal points on the sphere are given as z and $\frac{-1}{\bar{z}}$. Thus the 12 punctures come in pairs of antipodal points. Furthermore, the five points of absolute value $|\frac{z_i}{w_i}|$, $i = 1, 2$, form a pentagon $\mathcal{P}_i$. The pentagon $\mathcal{P}_2$ ($\mathcal{P}_1$) is inside (outside) the unit circle. It is obvious that $\tilde{B}$ fixes 0 and ∞ and permutes the vertices of each of the pentagons. We view $\hat{\mathbb{C}}$ as the unit sphere in $\mathbb{R}^3$. If the vertices of each of the pentagons are joined by a straight line in $\mathbb{R}^3$ and then the vertices of $\mathcal{P}_2$ ($\mathcal{P}_1$) are connected by a straight line with the origin (∞), we obtain 10 triangles: 10 of the 20 faces of the icosahedron. The remaining 10 faces are obtained by suitably joining the the vertices of the pentagon around ∞ with those of the pentagon around the origin. In fact the entire construction can be summarized by joining the points of minimal distance either in the chordal metric or in the metric we have defined above. The triangles we obtain in this way are equilateral triangles because the transformations $\tilde{\gamma}$, $\gamma \in \Gamma$, are isometries in the metric under consideration.

Exercise 8.3. **(a)** Show that the automorphism groups of the Riemann surfaces $\mathbb{H}^2/\Gamma(4)$ and $\mathbb{H}^2/\Gamma(5)$ are the symmetry groups of the regular octahedron and regular icosahedron respectively.
(b) What is the relation between $\mathbb{H}^2/\Gamma(3)$ and the regular tetrahedron?

Problem 8.4. Explore generalizations of the relationships exhibited in the previous exercise for larger values of k.

8.3. The function field for $\overline{\mathbb{H}^2/\Gamma(7)}$. Using the 3 modified theta constants φ_0, φ_1 and φ_2, we construct three $\Gamma(7)$-automorphic functions

$$f_0 = \frac{\varphi_0}{\varphi_2},\ f_1 = \frac{\varphi_1}{\varphi_2},\ f_2 = \frac{\varphi_0}{\varphi_1}.$$

The divisors of these functions are easily computed.

Proposition 8.5. *We have*

$$(F_0) = \frac{P^2_{\frac{2}{7}} P_{\frac{3}{7}}}{P^3_{\infty}},\ (F_1) = \frac{P^3_{\frac{3}{7}}}{P^2_{\infty} P_{\frac{2}{7}}}\ \textit{and}\ (F_2) = \frac{P^3_{\frac{2}{7}}}{P_{\infty} P^2_{\frac{3}{7}}}.$$

We use $z = F_0$ and $w = F_1$ and study the corresponding holomorphic map

$$\varphi : \overline{\mathbb{H}^2/\Gamma(7)} \to \mathfrak{M}^*.$$

We know that in this case (see Proposition 6.4)

$$3 = \deg z = (\deg \varphi)(\deg \text{proj}).$$

Now $\varphi(P_{\frac{2}{7}})$ is a Puiseaux series of the form

$$\sum_{i=-1}^{\infty} c_i z^{\frac{i}{2}},\ c_{-1} \neq 0,$$

while $\varphi(P_{\frac{3}{7}})$ is of the form

$$\sum_{i=3}^{\infty} d_i z^i,\ d_3 \neq 0,$$

which implies that proj is nonconstant. Hence $\deg \text{proj} = 3$ and $\deg \varphi = 1$. We hence conclude that the meromorphic functions F_0 and F_1 generate $\mathcal{K}(\overline{\mathbb{H}^2/\Gamma(7)})$. (We could also conclude this from Theorem 6.5.) It also follows immediately that P_∞ is a Weierstrass point on the compactified surface $\overline{\mathbb{H}^2/\Gamma(7)}$. Since the automorphism group of $\mathbb{H}^2/\Gamma(7)$ acts transitively on the punctures on this surface, we conclude that each puncture is a simple Weierstrass point.

We can now take appropriate products; the projections to $\overline{\mathbb{H}^2/\Gamma(7)}$ of functions $f_0^2 f_2$, $f_0 f_1^2$ have divisors $\frac{P^7_{\frac{2}{7}}}{P^7_{\infty}}$, $\frac{P^7_{\frac{3}{7}}}{P^7_{\infty}}$.

From the definitions of the functions we know that

$$f_0^2 f_2 = \frac{\varphi_0^3}{\varphi_1 \varphi_2^2},\ f_0 f_1^2 = \frac{\varphi_0 \varphi_1^2}{\varphi_2^3}.$$

Both of the functions in question are lifts of univalent functions on $\mathbb{H}^2/G(7)$ with the pole at the same point (their divisors are $\frac{P_{\frac{2}{7}}}{P_\infty}$ and $\frac{P_{\frac{3}{7}}}{P_\infty}$, respectively), which implies that one of these two functions is obtained from the other by postcomposition by an affine map. This affine map is easily computed to yield the identity

$$\frac{\varphi_0^3}{\varphi_1\varphi_2^2} = \epsilon^{-1}\frac{\varphi_0\varphi_1^2}{\varphi_2^3} + \epsilon^2, \ \epsilon = \exp\left(\frac{\pi\imath}{7}\right),$$

which we can rewrite as

$$(3.44) \quad \left(\omega^{-3}\theta\begin{bmatrix}\frac{1}{7}\\1\end{bmatrix}\right)^3\left(\omega^9\theta\begin{bmatrix}\frac{5}{7}\\1\end{bmatrix}\right) + \left(\omega^{-3}\theta\begin{bmatrix}\frac{1}{7}\\1\end{bmatrix}\right)\left(\omega^9\theta\begin{bmatrix}\frac{3}{7}\\1\end{bmatrix}\right)^3$$
$$+\left(\omega^9\theta\begin{bmatrix}\frac{3}{7}\\1\end{bmatrix}\right)\left(\omega^9\theta\begin{bmatrix}\frac{5}{7}\\1\end{bmatrix}\right)^3 = 0, \ \omega = \exp\left(\frac{\pi\imath}{28}\right).$$

The above identity is also derivable abstractly using methods described in Chapter 4 (see also Theorem 10.3), and it tells us that the Riemann surface $\overline{\mathbb{H}^2/\Gamma(7)}$ is realizable as the locus $X^3Z + Y^3X + Z^3Y = 0$ in $\mathbf{P}\mathbb{C}^2$. Alternatively, starting with the functions $F_0^2F_2$ and F_0, we can describe $\overline{\mathbb{H}^2/\Gamma(7)}$ as the Riemann surface of the algebraic equation $w^7 = z^2(z-1)$. Furthermore, we have realized its automorphism group as a subgroup of the projective linear maps of $\mathbf{P}\mathbb{C}^2$.

8.4. The projective embedding of $\overline{\mathbb{H}^2/\Gamma(7)}$. We use the two degree 3 functions F_0 and F_1 to embed $\overline{\mathbb{H}^2/\Gamma(7)}$ into $\mathbf{P}\mathbb{C}^2$ by the map

$$\varphi : Q \mapsto (F_0(Q),\ F_1(Q),\ 1).$$

We claim that this map φ is injective. We first observe that for the 3 distinguished punctures, we have in homogeneous coordinates

$$\varphi(P_\infty) = (1,0,0),\ \varphi(P_{\frac{2}{7}}) = (0,1,0) \text{ and } \varphi(P_{\frac{3}{7}}) = (0,0,1),$$

while for $Q \in \mathbb{H}^2/\Gamma(7)$ or Q a nondistinguished puncture on this surface, $\varphi(Q)$ never lands in a coordinate hyperplane (where one of the three homogeneous coordinates vanishes). Thus if

$$(F_0(Q_1),\ F_1(Q_1),\ 1) = (F_0(Q_2),\ F_1(Q_2),\ 1)$$

for distinct Q_1 and Q_2 in $\overline{\mathbb{H}^2/\Gamma(7)}$, then for $F_i(Q_1) = z_i \in \mathbb{C}^*$, $i = 0,\ 1$, we have (as divisors) $F_0^{-1}(z_0) = Q_1Q_2R_1$ and $F_1^{-1}(z_1) = Q_1Q_2R_2$ for some points R_1 and R_2 in $\mathbb{H}^2/\Gamma(7)$. The surface $\overline{\mathbb{H}^2/\Gamma(7)}$ has genus 3 and is not hyperelliptic (see, for example, [**6**, Proposition III.7.10]). Hence the Riemann-Roch theorem shows that

$$i(Q_1Q_2) = i(P_\infty^2) = i(P_\infty^3) = i(P_\infty^2P_{\frac{2}{7}}) = 1.$$

We have seen (in the last proposition) that

$$r\left(\frac{1}{P_\infty^3P_{\frac{2}{7}}}\right) \geq 3.$$

Riemann-Roch then tells us that

$$i(P_\infty^3 P_{\frac{2}{7}}) = 1.$$

The divisor $F_0^{-1}(z_1) = Q_1Q_2R_1$ is equivalent ($\sim$) to $F_0^{-1}(\infty) = P_\infty^3$. This implies that $Q_1Q_2R_1P_{\frac{2}{7}} \sim P_\infty^3 P_{\frac{2}{7}}$, which is canonical. In the same way we see that $Q_1Q_2R_2P_\infty \sim P_{\frac{3}{7}}^3 P_\infty$ is canonical. This tells us that $R_1P_{\frac{2}{7}} = R_2P_\infty$, an obvious contradiction.

We already know that Φ is of maximal rank at all the distinguished punctures. The above argument with $Q_1 = Q_2$ gives an alternate proof, in this special case, of the fact that Φ is globally of maximal rank. Hence $\Phi(\overline{\mathbb{H}^2/\Gamma(7)})$ is nonsingular. Further, every hyperplane in $\mathbf{P}\mathbb{C}^2$ intersects $\Phi(\overline{\mathbb{H}^2/\Gamma(7)})$ in 4 points counting multiplicities. By [**6**, Theorem III.10.5], $\Phi(\overline{\mathbb{H}^2/\Gamma(7)}) \subset \mathbf{P}\mathbb{C}^2$ is canonical. Computing the intersection of the image curve with the hyperplane where the first coordinate equals zero, we conclude that

$$i(P_{\frac{2}{7}}^3 P_{\frac{3}{7}}) = 1.$$

Similarly $P_{\frac{3}{7}}^3 P_{\frac{1}{7}}$ and (as we already know) $P_{\frac{1}{7}}^3 P_{\frac{2}{7}}$ are canonical divisors. We have established a seemingly odd duality between $V(7)$ and the vector space $\mathcal{H}$ of abelian differentials of the first kind on the compact Riemann surface $\overline{\mathbb{H}^2/\Gamma(7)}$ leading to a construction of all the integral canonical divisors on this surface. The construction goes as follows. Choose a nonzero $\varphi \in V(7)$. Divide the divisor (φ) by the $\frac{1}{8}$ power of the divisor D of the set of $\Gamma(7)$-inequivalent cusps in $\mathbb{Q} \cup \{\infty\}$. The projection of this divisor to the surface $\overline{\mathbb{H}^2/\Gamma(7)}$ is an integral canonical divisor on this surface, and conversely every canonical integral divisor on $\overline{\mathbb{H}^2/\Gamma(7)}$ is so obtained. It is natural to seek an explanation of this phenomenon. We offer two possible explanations.

Choose a generic element $f_o \in V(7)$. The function f_o will have 4 inequivalent simple zeros at points in $\mathbb{H}^2$ and a zero of order $\frac{1}{8}$ at each cusp. A ratio $\frac{\varphi}{f_o}$ with $\varphi \in V(7)$ will project to F, a constant function or a function of degree 3 or 4 on $\overline{\mathbb{H}^2/\Gamma(7)}$. In the case of a degree 4 function, (F) will be canonical (see Proposition 6.1 and Remark 6.3). Choosing a holomorphic 1-differential ω on $\overline{\mathbb{H}^2/\Gamma(7)}$, we see that an arbitrary holomorphic 1-differential is given as ωF. For the second explanation, let W be the lift to $\mathbb{H}^2$ of a Wronskian of a basis for $\mathcal{H}$. A little thought will show that W is in fact nothing more than $\theta' \begin{bmatrix} 1 \\ 1 \end{bmatrix}$. The divisor of the 6-form W consists of the Weierstrass points of $\overline{\mathbb{H}^2/\Gamma(7)}$ and hence equals D. The map Φ of $\overline{\mathbb{H}^2/\Gamma(7)}$ into $\mathbf{P}\mathbb{C}^2$ is defined by the two triples $(\varphi_0, \varphi_1, \varphi_2)$ and $W^{\frac{1}{8}}(\varphi_0, \varphi_1, \varphi_2)$. One easily sees that $W^{\frac{1}{8}}\ \varphi_l$, $l = 0, 1, 2$, projects to a holomorphic 1-form on $\overline{\mathbb{H}^2/\Gamma(7)}$. Alternately, we have seen that Φ is defined by the 3 functions $(F_0,\ F_1,\ 1)$. We know that the divisor $P_\infty^3 P_{\frac{2}{7}}$ is canonical. We can hence choose a holomorphic 1-form ω on $\overline{\mathbb{H}^2/\Gamma(7)}$ with $(\omega) = P_\infty^3 P_{\frac{2}{7}}$ and conclude that Φ is defined by the 3 linearly independent holomorphic 1-forms $\omega(F_0, F_1,\ 1)$, as expected from Corollary 7.3.

Remark 8.6. The injectivity of the map φ implies at once that $\mathcal{K}(\overline{\mathbb{H}^2/\Gamma(7)})$ is generated by F_1 and F_2. That a primitive pair does not necessarily lead to a projective embedding is illustrated in the next section.

8.5. $k = 11$. In this case, $k = 11$, we are dealing with the compact surface $\overline{\mathbb{H}^2/\Gamma(11)}$ of genus 26 punctured at 60 points. The bookkeeping can already be a bit troublesome. Using the five modified theta constants φ_i, $i = 0,\ 1,\ \ldots,\ 4$, we construct four $\Gamma(11)$-automorphic functions $f_i = \frac{\varphi_i}{\varphi_4}$, $i = 0,\ 1,\ 2,\ 3$. Their divisors are

$$(F_0) = \frac{P_{\frac{2}{11}}^4 P_{\frac{4}{11}}^5 P_{\frac{5}{11}}^3}{P_\infty^{10} P_{\frac{3}{11}}^2},\quad (F_1) = \frac{P_{\frac{3}{11}}^7 P_{\frac{5}{11}}^6}{P_\infty^9 P_{\frac{2}{11}}^3 P_{\frac{4}{11}}},$$

$$(F_2) = \frac{P_{\frac{3}{11}}^3 P_{\frac{4}{11}}^9 P_{\frac{5}{11}}}{P_\infty^7 P_{\frac{2}{11}}^6},\quad (F_3) = \frac{P_{\frac{4}{11}}^2 P_{\frac{5}{11}}^{10}}{P_\infty^4 P_{\frac{2}{11}}^5 P_{\frac{3}{11}}^3}.$$

The material in §6.3 now shows that the functions $z = F_0$ and $w = F_3$ generate $\mathcal{K}(\overline{\mathbb{H}^2/\Gamma(11)})$, and the formulae in §6.2 show that the two functions on $\mathbb{H}^2$

$$\frac{\varphi_0^2}{\varphi_1\varphi_4} = \frac{f_0^2}{f_1} \text{ and } \frac{\varphi_0\varphi_3}{\varphi_2^2} = \frac{f_0 f_3}{f_2^2}$$

project to functions of degree 2 on the torus $\mathbb{H}^2/G(11)$ whose divisors are

$$\frac{P_{\frac{2}{11}} P_{\frac{4}{11}}}{P_\infty P_{\frac{3}{11}}} \text{ and } \frac{P_{\frac{2}{11}} P_{\frac{5}{11}}}{P_{\frac{3}{11}} P_{\frac{4}{11}}},$$

respectively.[87] Even though the meromorphic functions F_0 and F_3 generate the function field on the surface $\overline{\mathbb{H}^2/\Gamma(11)}$, they do not embed this surface into $\mathbf{P}\mathbb{C}^2$ because the two punctures $P_{\frac{2}{11}}$ and $P_{\frac{3}{11}}$ are sent by the map $(1,\ F_0,\ F_3)$ to $(0, 0, 1)$. Note that $F_0F_1 \in L(P_\infty^{-19}) - L(P_\infty^{-18})$, so that all the punctures are Weierstrass points and the lowest "non-gap" is at most 19. The fact that the punctures are all Weierstrass points comes as no surprise since there is an automorphism of the surface which has more than four fixed points (all the distinguished punctures). We do not understand why the Weierstrass "non-gap" sequence contains 19, nor do we know if we

[87] The reader should determine whether these two meromorphic functions generate the function field for the torus $\overline{\mathbb{H}^2/G(11)}$.

can produce a function with a single pole of lower degree. The map Φ is defined by linearly independent holomorphic 1-forms.[88] To see this, choose a holomorphic 1-form ω whose divisor is a multiple of $P_\infty^{10} P_{\frac{2}{11}}^6 P_{\frac{3}{11}}^3 P_{\frac{4}{11}}$, and observe that $\omega(F_0,\ F_1,\ F_2,\ F_3,\ 1)$ also defines Φ. We have once again used Corollary 7.3.

8.6. $k = 13$. The material in Chapter 1 shows that $\overline{\mathbb{H}^2/G(13)}$ is a closed surface of genus 2, hence hyperelliptic (see also [**27**, Proposition 1.43]). The function

$$z = \frac{\varphi_5\ \varphi_1}{\varphi_2\ \varphi_0}$$

is $G(13)$-invariant; a straightforward calculation shows that z is also invariant under the elliptic transformation (of order 2)

$$\gamma = \begin{bmatrix} 5 & -2 \\ 13 & -5 \end{bmatrix} \in \Gamma_o(13).$$

On $\overline{\mathbb{H}^2/G(13)}$,

$$(Z) = \frac{P_{\frac{1}{13}} P_{\frac{5}{13}}}{P_{\frac{4}{13}} P_{\frac{6}{13}}}.$$

We conclude that on $\mathbb{S} = \overline{\mathbb{H}^2/ < G(13),\ \gamma >}$ we have a function of degree one (with a simple zero at $P_{\frac{1}{13}}$ and a simple pole at $P_{\frac{4}{13}}$). Hence $\mathbb{S}$ has genus zero, and γ induces the hyperelliptic involution on $\overline{\mathbb{H}^2/G(13)}$ (Z is a function of degree two on this surface and a function of degree one on $\mathbb{S}$). Moreover, the six fixed points of $\tilde{\gamma}$ are the Weierstrass points on $\overline{\mathbb{H}^2/G(13)}$. We can easily find one fixed point (in $\mathbb{H}^2$) $z_o = \frac{5+\imath}{13}$ of γ; its projection to $\overline{\mathbb{H}^2/G(13)}$ is fixed by the hyperelliptic involution. To find the other fixed points, we introduce the motion

$$\delta = \begin{bmatrix} 2 & 1 \\ 13 & 7 \end{bmatrix} \in \Gamma_o(13)$$

that defines an automorphism of order six of $\overline{\mathbb{H}^2/G(13)}$. Since $\delta^3 \circ \gamma^{-1} \in G(13)$, δ^3 also induces the hyperelliptic involution on $\overline{\mathbb{H}^2/G(13)}$. The projections of the three points $\delta^{j-1}(z_o)$, $j = 1,\ 2,\ 3$, are three of the six Weierstrass points on $\overline{\mathbb{H}^2/G(13)}$. It is not clear how to locate the other three.

We have produced a meromorphic function (Z) of degree 26 on the closed surface $\overline{\mathbb{H}^2/\Gamma(13)}$ of genus 50. It is not clear whether $\overline{\mathbb{H}^2/\Gamma(13)}$ carries a function of lower degree, nor is it obvious how to generalize the material of this subsection to primes > 13. The material of this subsection hints at the complexities involved in studying the general case.

[88] In contrast to the $k = 7$ case, not a maximal set of such forms.

8.7. $k = 9$. The purpose of this subsection is to illustrate what goes wrong when k is not prime and to hint at the possibilities in the study of the composite case. The case $k = 9$ corresponds to a compact surface X of genus 10 punctured at 36 points. The punctures correspond to the characteristics in $X_o(9)$ and are given concretely as the images under $P : \mathbb{H}^2 \cup \mathbb{Q} \cup \{\infty\} \to X = (\mathbb{H}^2 \cup \mathbb{Q} \cup \{\infty\})/\Gamma(9)$ of the rational numbers

$$0,\ 1,\ 2,\ 3,\ 4,\ 5,\ 6,\ 7,\ 8,\quad \frac{1}{2},\ \frac{3}{2},\ \frac{5}{2},\ \ldots,\ \frac{17}{2},$$

$$\frac{1}{3},\ \frac{2}{3},\ \frac{4}{3},\ \frac{5}{3},\ \frac{7}{3},\ \frac{8}{3},\quad \frac{1}{4},\ \frac{3}{4},\ \frac{5}{4},\ \ldots,\ \frac{17}{4},\quad \frac{1}{9},\ \frac{2}{9},\ \frac{4}{9}.$$

The techniques we have developed for the case of primes (§5.1) will allow us to determine the orders of divisors of forms and functions at 30 of the punctures of $\mathbb{H}^2/\Gamma(9)$; the general theory (§5.2) is needed for the remaining six punctures (the projections of the first 6 rationals in the second row above). The punctures on $\mathbb{H}^2/\Gamma(9)$ fall into four disjoint sets [**27**, Proposition 1.43]:
(1) those $\Gamma_o(9)/\Gamma(9)$-equivalent to P_∞ ($P_{\frac{1}{9}}$, $P_{\frac{2}{9}}$, $P_{\frac{4}{9}}$),
(2) those $\Gamma_o(9)/\Gamma(9)$-equivalent to $P_{\frac{1}{3}}$ ($P_{\frac{1}{3}}$, $P_{\frac{4}{3}}$, $P_{\frac{7}{3}}$),
(3) those $\Gamma_o(9)/\Gamma(9)$-equivalent to $P_{\frac{2}{3}}$ ($P_{\frac{2}{3}}$, $P_{\frac{5}{3}}$, $P_{\frac{8}{3}}$), and
(4) those $\Gamma_o(9)/\Gamma(9)$-equivalent to P_0 (the remaining 27 punctures).

Because of (3.30) and the fact that the divisor of each φ_l has degree $\frac{27}{2}$, we need only compute, in addition to the answers provided by Lemma 5.1, $\mathrm{ord}_x\varphi_l$ for $x = \frac{1}{3}$, $\frac{2}{3}$, $l = 0, \ldots, 3$. The results are summarized in the table in this subsection with entries $8\mathrm{ord}_x\varphi_l$. The calculations at $x = \frac{1}{3}$ and $\frac{2}{3}$ can be based on the fact that

$$C_1 = A \circ B^{-3} \circ A = \begin{bmatrix} 1 & 0 \\ 3 & 1 \end{bmatrix} \text{ and } C_2 = B \circ A \circ B^3 \circ A = \begin{bmatrix} 2 & -1 \\ 3 & -1 \end{bmatrix}$$

map ∞ to these points, respectively. It follows that the matrices $(C_1)^*_{\frac{1}{4}}$ and $(C_2)^*_{\frac{1}{4}}$ are both of the form (the blank spots represent nonzero entries)

$$\begin{bmatrix} & 0 & & \\ 0 & & 0 & 0 \\ & 0 & & \\ & 0 & & \end{bmatrix}.$$

We are dealing with the map

$$\Phi : \tau \mapsto (\varphi_0(\tau),\ \varphi_1(\tau),\ \varphi_2(\tau),\ \varphi_3(\tau)),\ \tau \in \mathbb{H}^2 \cup \mathbb{Q} \cup \{\infty\}.$$

As with primes, we can compute the images under Φ of the cusps $\Gamma_o(k)$-equivalent to ∞. We see that

$$P_{\frac{1}{9}} \mapsto (1,0,0,0),\ P_{\frac{2}{9}} \mapsto (0,0,1,0) \text{ and } P_{\frac{4}{9}} \mapsto (0,0,0,1).$$

cusp x	$l=0$	$l=1$	$l=2$	$l=3$
$\frac{1}{9}$	1	9	25	49
$\frac{2}{9}$	49	9	1	25
$\frac{4}{9}$	25	9	49	1
$\frac{1}{3},\ \frac{2}{3},\ \frac{4}{3},\ \frac{5}{3},\ \frac{7}{3},\ \frac{8}{3}$	1	9	1	1
$0,\ \ldots,\ 8,\ \frac{1}{2},\ \ldots,\ \frac{17}{2},\ \frac{1}{4},\ \ldots,\ \frac{17}{4}$	1	1	1	1

Table 9. VALUES OF $8\ \mathrm{ord}_x\varphi_l$.

Table 9 also shows that $(0,1,0,0) \notin \Phi(\overline{\mathbb{H}^2/\Gamma(9)})$. The map Φ is also defined by the meromorphic functions $(F_0,\ F_1,\ F_2,\ 1)$ and the holomorphic 1-forms $\omega(F_0,\ F_1,\ F_2,\ 1)$, where ω is a holomorphic differential on $\overline{\mathbb{H}^2/\Gamma(9)}$ whose divisor is a multiple of $P_\infty^6 P_{\frac{2}{9}}^3$. Since the surface $\overline{\mathbb{H}^2/\Gamma(9)}$ has genus ten, there is always such a ω. More information on the map Φ is given in the proof of Theorem 10.5.

We have observed that for primes k, the support of the divisor of any quotient $F = \frac{\Phi_l}{\Phi_{l'}}$ is precisely the set of distinguished punctures,

$$\left\{P_{\frac{1}{k}},\ P_{\frac{2}{k}},\ \ldots,\ P_{\frac{\frac{k-1}{2}}{k}}\right\}.$$

This fails[89] for $k=9$ as illustrated by $l=1,\ l'=3$. We find that in this case, the function F has a pole of order 5 at $P_{\frac{1}{9}} = P_\infty$ and a pole of order 2 at $P_{\frac{2}{9}}$. If the support of the divisor (F) were at the 3 distinguished punctures, then F would be forced to have a zero of order 7 at the point $P_{\frac{4}{9}}$. We find however that F indeed has a zero at $P_{\frac{4}{9}}$ but only a simple zero. It follows that the function F has at least 6 additional zeros. Table 9 allows us to conclude that F has simple zeros at the 6 points $\left\{P_{\frac{1}{3}},\ P_{\frac{2}{3}},\ P_{\frac{4}{3}},\ P_{\frac{5}{3}},\ P_{\frac{7}{3}},\ P_{\frac{8}{3}}\right\}$, and is regular elsewhere. We summarize the above and related calculations (for $l=0,\ l'=3$ and for $l=2,\ l'=3$) in

$$\left(\frac{\Phi_1}{\Phi_3}\right) = \frac{P_{\frac{4}{9}} P_{\frac{1}{3}} P_{\frac{2}{3}} P_{\frac{4}{3}} P_{\frac{5}{3}} P_{\frac{7}{3}} P_{\frac{8}{3}}}{P_{\frac{1}{9}}^5 P_{\frac{2}{9}}^2},\quad \left(\frac{\Phi_0}{\Phi_3}\right) = \frac{P_{\frac{2}{9}}^3 P_{\frac{4}{9}}^3}{P_{\frac{1}{9}}^6} \text{ and } \left(\frac{\Phi_2}{\Phi_3}\right) = \frac{P_{\frac{4}{9}}^6}{P_{\frac{1}{9}}^3 P_{\frac{2}{9}}^3}.$$

Since we have produced a meromorphic function of degree 6 on X which is holomorphic on $X - \{P_\infty\}$, we conclude that P_∞ is a Weierstrass point and thus all 36 punctures on $\mathbb{H}^2/\Gamma(9)$ are Weierstrass points. In fact we can

[89] Recall that for $k=9$, we have only 3 distinguished punctures: $\left\{P_{\frac{1}{9}},\ P_{\frac{2}{9}},\ P_{\frac{4}{9}}\right\}$.

determine eight of the first ten "non-gaps" at P_∞; they are: 6, 9, 11, 12, 15, 17, 18, 20.

There is an additional observation to be made. We consider in place of Φ, the map

$$\Phi_1 : \tau \mapsto (\varphi_0(\tau),\ \varphi_2(\tau),\ \varphi_3(\tau)),\ \tau \in \mathbb{H}^2 \cup \mathbb{Q} \cup \{\infty\}.$$

The reader can check that the mapping in this case is certainly not injective, and in fact that here we have a map from $Y = \overline{\mathbb{H}^2/\Gamma(9)/G}$ to $\mathbf{P}\mathbb{C}^2$, where G is the group generated by $\tilde{B}^3$. Now $\tilde{B}^3$ is an automorphism of $\mathbb{H}^2/\Gamma(9)$ of period three with exactly nine fixed points, all at the punctures of $\mathbb{H}^2/\Gamma(9)$. The fixed points are the images under P of the points

$$\frac{1}{3}, \frac{2}{3}, \frac{4}{3}, \frac{5}{3}, \frac{7}{3}, \frac{8}{3}, \frac{1}{9}, \frac{2}{9}, \frac{4}{9},$$

and the quotient $\mathbb{H}^2/\Gamma(9)/G$ is a torus with 18 punctures.

The equations of X and Y are given in §10.4 as an example of a general theory. We now use the functions $z = \frac{\Phi_0}{\Phi_3}$ and $w = \frac{\Phi_3}{\Phi_2}$ on Y. The divisors of these functions are

$$(z) = \frac{P_{\frac{2}{9}} P_{\frac{4}{9}}}{P^2_{\frac{1}{9}}} \text{ and } (w) = \frac{P_{\frac{1}{9}} P_{\frac{2}{9}}}{P^2_{\frac{4}{9}}}.$$

For the resulting map φ, we have $(\mathrm{proj} \circ \varphi)(P_{\frac{2}{9}}) = 0 = (\mathrm{proj} \circ \varphi)(P_{\frac{4}{9}})$. Hence by Proposition 6.4, φ is injective. Since B induces an automorphism of period 3 of Y with a fixed point, Y is conformally equivalent to $T_{\frac{1}{2}+\imath\frac{\sqrt{3}}{2}}$. It might be interesting to determine the involution with the same fixed point as $\tilde{B}$.

We return to a study of the two maps into projective space. The map $\Phi_1 : Y \to \mathbf{P}\mathbb{C}^2$ is injective[90] and of maximal rank. For if not, we can derive a linear relation

$$c_1\varphi_0\varphi_2 + c_2\varphi_0\varphi_3 + c_3\varphi_2\varphi_3 + c_4\varphi_3^2 = 0,\ c_i \in \mathbb{C},\ i = 1, 2, 3, 4,$$

which is seen to be a contradiction by examining the orders of the zeros at ∞ of the functions appearing in the above equation. Since the map Φ_1 may be viewed as the map Φ followed by a projection,[91] we conclude once again (without using Proposition 7.9) that Φ is of maximal rank. As a map from $\overline{\mathbb{H}^2/\Gamma(9)}$, it is injective. For if $\Phi(x) = \Phi(y)$ with x and y in $\overline{\mathbb{H}^2/\Gamma(9)}$, then $\Phi_1(x_1) = \Phi_1(y_1)$, where x_1 and x_2 are the projections of x and y under the

[90] This assertion implies at once that the map φ is injective.

[91] This assertion follows from the fact that for each cusp $x \in \mathbb{Q} \cup \{\infty\}$, there is at least one $j = 0, 2$ or 4, such that

$$\mathrm{ord}_x\varphi_1 \geq \mathrm{ord}_x\varphi_j.$$

canonical map $\overline{\mathbb{H}^2/\Gamma(9)} = X \to Y$. It follows that $y = \tilde{B}^{3j}$ with $j = 0, 1$ or 2. As a consequence of (3.22), $j = 0$.

9. The function field of $\overline{\mathbb{H}^2/\Gamma(k)}$ over $\overline{\mathbb{H}^2/\Gamma}$

The following theorem is an immediate corollary of a result of E. Artin. It is independent of Theorem 6.5.

Theorem 9.1. *For each prime $k \geq 5$, the meromorphic functions*

$$f_l = \frac{\varphi_l}{\varphi_{\frac{k-3}{2}}}, \ l = 0, \ 1, \ \ldots, \ \frac{k-5}{2},$$

generate the function field of $\overline{\mathbb{H}^2/\Gamma(k)}$ over $\overline{\mathbb{H}^2/\Gamma}$. Furthermore, $\mathcal{K}(\overline{\mathbb{H}^2/\Gamma(k)}$ is a Galois extension of $\mathcal{K}(\overline{\mathbb{H}^2/\Gamma})$ with Galois group $\Gamma/\Gamma(k)$.

Proof. Recall that the $\jmath$-invariant generates the function field $\mathcal{K}(\overline{\mathbb{H}^2/\Gamma})$ of $\overline{\mathbb{H}^2/\Gamma}$; that is, every meromorphic function on the Riemann sphere $\overline{\mathbb{H}^2/\Gamma}$ is a rational function of $\jmath$. Let $\mathbb{K}$ denote the field over $\mathbb{C}$ generated by $\jmath$ and the $\frac{k-3}{2}$ functions listed above. The group $\Gamma/\Gamma(k) \cong \mathrm{PSL}(2, \mathbb{Z}_k)$ acts as a group of automorphisms of $\mathbb{K}$ as a consequence of Proposition 7.6; its fixed field is precisely $\mathbb{C}(\jmath)$. By a theorem of E. Artin in [**19**, pg. 194], $\mathbb{K}$ is a Galois extension of $\mathcal{K}(\overline{\mathbb{H}^2/\Gamma})$ of degree $|\Gamma/\Gamma(k)|$. So is $\mathcal{K}(\overline{\mathbb{H}^2/\Gamma(k)})$. Since $\mathbb{K} \subset \mathcal{K}(\overline{\mathbb{H}^2/\Gamma(k)})$, we conclude that $\mathbb{K} = \mathcal{K}(\overline{\mathbb{H}^2/\Gamma(k)})$.

Remark 9.2. As a consequence of Theorem 6.6, the single function $w = \frac{\varphi_1}{\varphi_0}$ generates the function field of $\overline{\mathbb{H}^2/\Gamma(k)}$ over $\overline{\mathbb{H}^2/\Gamma}$. Furthermore, w satisfies an equation of the form

$$\prod_{j=1}^{kn(k)} (w - a_j) = 0, \ a_j \in \mathbb{C}(\jmath).$$

10. Equations that are satisfied by the embedding

Let k be an odd integer. Having mapped $\mathbb{H}^2/\Gamma(k)$ into $\mathbf{P}\mathbb{C}^{\frac{k-3}{2}}$, we would like to find the equations that define the image. In this section we will obtain a method for deriving such equations and then use the method to obtain some quartic relations between the functions defining some of the mappings.

10.1. The residue theorem. The key tool for obtaining theta identities to be used in the current situation is a special case ($N = k$, $\epsilon = 1, \epsilon' = k$) of Theorem 7.1 of Chapter 2.

10.2. The algorithm. We describe a general method (consisting of three steps) for obtaining theta identities among sums of products of theta constants of the form $\theta\left[\begin{array}{c}\frac{i}{k}\\1\end{array}\right]$ (with variable $i \in \mathbb{Z}$ and k fixed and odd).

Step 0. Start with the following set of characteristics $\left[\begin{array}{c}\frac{i}{k}\\1\end{array}\right]$, where $i \in \mathbb{Z}$, $-k+2 \leq i \leq k$, and i is odd. Denote this set by $\mathbb{X}$. Note that $|\mathbb{X}| = k$. To simplify notation, we shall abuse language and identify the characteristic $\left[\begin{array}{c}\frac{i}{k}\\1\end{array}\right]$ with the integer i.

Step 1. Choose $m \in \mathbb{Z}$, $1 < m < k$. Pick $l_r \in \mathbb{Z}$ such that $1 \leq r \leq m$ and $\sum_{r=1}^{m} l_r \equiv 0 \bmod k$. Without loss of generality (see next displayed equation), we may and hence do assume that $0 \leq l_r < k$ for each r.

Step 2. For each r, $1 \leq r \leq m$, delete from $\mathbb{X}$ the unique characteristic $\left[\begin{array}{c}\frac{k_r}{k}\\1\end{array}\right]$ for which

$$\theta\left[\begin{array}{c}\frac{k_r}{k}\\1\end{array}\right]\left(\frac{l_r}{k}\tau, \tau\right) = 0.$$

This means that

$$k_r = k - 2l_r.$$

Denote the deleted set by $\mathbb{A}$. From the characteristics in $\mathbb{X} - \mathbb{A}$, we choose[92] m characteristics α_r with the property that $\sum_{r=1}^{m} \alpha_r \equiv 0 \bmod k$. The α_r's are not necessarily distinct. We define two functions on $\mathbb{C}$:

$$g(z) = \prod_{r=1}^{m} \theta\left[\begin{array}{c}\frac{\alpha_r}{k}\\1\end{array}\right](z, \tau)$$

and

$$f(z) = \prod_{j \in \mathbb{X}-\mathbb{A}} \theta\left[\begin{array}{c}\frac{j}{k}\\1\end{array}\right](z, \tau) g(z).$$

The transformation formulae tell us that the function f satisfies the conditions of the last theorem (because $\sum_{r=1}^{m}(k_r - \alpha_r) \equiv 0 \bmod 2k$), and vanishes at $k - m$ points $\{\frac{j}{k}\tau;\ j \in \mathbb{Z},\ 0 \leq j < k,\ j \neq l_r\}$. Hence the theorem yields an identity involving a sum of m products of k theta functions evaluated at proper points. (There may, of course, be the possibility of lots of cancellation, and we can by changing the variable and the characteristic replace theta functions by theta constants, as we do in the next subsection.)

---

[92]We do not know if such a choice is always possible. However, if $l_r \neq \frac{k-1}{2}$, $l_r \neq \frac{m-1}{2}$ for odd m, and $l_r \neq \frac{k+m-1}{2}$ for even m (this means that $k_r \neq 1$, $k_r \neq k - (m-1)$ for odd m, and $k_r \neq 1 - m$ for even m), then we can choose $\alpha_i = 1$ for $i = 1, \ldots, m-1$, $\alpha_m = k - (m-1)$ for odd m, and $\alpha_m = -(m-1)$ for even m. This prescription could lead to trivial identities, for example if $m = 2$.

10.3. Three term identities. We use the ($m =$) 3 step algorithm outlined above to produce three term identities. Choose $l_1 = 0,\ l_2 = j,\ l_3 = k - j$, with $j \in \mathbb{Z},\ 0 < j \le \frac{k-1}{2}$. The characteristics to be deleted by step 2 are: $\begin{bmatrix} 1 \\ 1 \end{bmatrix}$, $\begin{bmatrix} \frac{-k+2j}{k} \\ 1 \end{bmatrix}$ and $\begin{bmatrix} \frac{k-2j}{k} \\ 1 \end{bmatrix}$. From the remaining set of $k - 3$ characteristics, we choose 3 characteristics $\alpha_r,\ r = 1,\ 2,\ 3$, such that $\sum_{i=1}^{3} \alpha_r \equiv 0 \bmod k$. The function f is defined as

$$f(z) = \prod_{i \in \mathbb{X}} \theta^{n_i} \begin{bmatrix} \frac{i}{k} \\ 1 \end{bmatrix} (z, \tau),$$

where $n_i \in \mathbb{Z}$ for $i \in \mathbb{X}$ is computed as $n_i = 1+$ the number of r for which $\alpha_r = i$ if $i \neq -k + 2j,\ k - 2j,\ k$ and $n_i = 0$ if $i = -k + 2j,\ k - 2j$, or k. We need to calculate the values of the function f at the three points $0,\ \frac{j}{k}\tau$ and $\frac{k-j}{k}\tau$. The last theorem therefore yields

$$\begin{aligned} &\prod_{i \in \mathbb{X},\ i \neq k, -k+2j, k-2j} \theta^{n_i} \begin{bmatrix} \frac{i}{k} \\ 1 \end{bmatrix} (0, \tau) \\ &+ (-1)^j \exp \pi \imath \left\{ \frac{j^2}{k} \tau \right\} \prod_{i \in \mathbb{X},\ i \neq k, -k+2j, k-2j} \theta^{n_i} \begin{bmatrix} \frac{i}{k} \\ 1 \end{bmatrix} \left(\frac{j}{k}\tau, \tau \right) \\ &+ (-1)^{k-j} \exp \pi \imath \left\{ \frac{(k-j)^2}{k} \tau \right\} \prod_{i \in \mathbb{X},\ i \neq k, -k+2j, k-2j} \theta^{n_i} \begin{bmatrix} \frac{i}{k} \\ 1 \end{bmatrix} \left(\frac{k-j}{k}\tau, \tau \right) = 0. \end{aligned}$$

As usual, we change the above formula to a relation among only theta constants:

$$\begin{aligned} &\prod_{i \in \mathbb{X},\ i \neq k, k-2j, 2j-k} \theta^{n_i} \begin{bmatrix} \frac{i}{k} \\ 1 \end{bmatrix} (0, \tau) + \prod_{i \in \mathbb{X},\ i \neq k, k-2j, 2j-k} \theta^{n_i} \begin{bmatrix} \frac{i+2j}{k} \\ 1 \end{bmatrix} (0, \tau) \\ &+ \prod_{i \in \mathbb{X},\ i \neq k, k-2j, 2j-k} \theta^{n_i} \begin{bmatrix} \frac{i-2j}{k} \\ 1 \end{bmatrix} (0, \tau) = 0. \end{aligned}$$

The terms involving the exponential function miraculously canceled out because $\sum_{i \in \mathbb{X}} n_i = k$. The last formula for the identity can be reduced by use of the periodicity of the theta constants to one involving only the characteristics $\{1,\ 3,\ \ldots,\ k-2\}$. This involves introducing some $2k$-th roots of unity and leads to cancellation of most terms. We start with the observation that for any integer l translation of the characteristics in $\mathbb{X}$ by the characteristic $\begin{bmatrix} \frac{2l}{k} \\ 1 \end{bmatrix}$ produces an automorphism of $\mathbb{X}$, provided addition is interpreted modulo $2k$, that does not otherwise alter any of the above formulae. This reduces the set of characteristics in the formula for the identity to $\{-k+2,\ \ldots, -1,\ 1,\ 3,\ \ldots,\ k\}$. Next, using periodicity once more, we replace the term $\theta\begin{bmatrix}\frac{l}{k}\\1\end{bmatrix}$ (where l is an odd integer with $-k+2\le l\le -1$) by the term $\exp\left(\pi\imath\frac{l}{k}\right)\theta\begin{bmatrix}\frac{-l}{k}\\1\end{bmatrix}$. We have reduced the set of characteristics appearing in our formulae to

$$\{1,\ 3,\ \ldots,\ k\}.$$

Our final observation is that the characteristic k never appears. We then proceed to cancel common terms. We have obtained

Theorem 10.1. *Let $j\in\mathbb{Z}$, $0<j\le\frac{k-1}{2}$. For $i=1,\ 2,\ 3$, choose odd integers α_i with $|\alpha_i|\le k-2$, $|\alpha_i|\ne k-2j$, and $\sum_{i=1}^{3}\alpha_i\equiv 0\mod k$. The following quartic relation holds among theta constants:*

$$\theta\begin{bmatrix}\frac{k-4j}{k}\\1\end{bmatrix}\prod_{i=1,\ 2,\ 3}\theta\begin{bmatrix}\frac{\alpha_i}{k}\\1\end{bmatrix}+e^{\pi\imath\frac{k-4j}{k}}\theta\begin{bmatrix}\frac{k-2j}{k}\\1\end{bmatrix}\prod_{i=1,\ 2,\ 3}\theta\begin{bmatrix}\frac{\alpha_i+2j}{k}\\1\end{bmatrix}$$
$$+e^{\pi\imath\frac{-k+2j}{k}}\theta\begin{bmatrix}\frac{k-2j}{k}\\1\end{bmatrix}\prod_{i=1,\ 2,\ 3}\theta\begin{bmatrix}\frac{\alpha_i-2j}{k}\\1\end{bmatrix}=0.$$

Problem 10.2. For a fixed odd integer k, the above identity is determined by the quadruple $(j;\alpha_1,\alpha_2,\alpha_3)$. It is obvious that both $(j;\alpha_1,\alpha_2,\alpha_3)$ and $(j;-\alpha_1,-\alpha_2,-\alpha_3)$ always determine the same identity. There are other, less clear, relations among the identities. For $k=9$, for example, the quadruples $(1;1,3,5)$ and $(2;-7,-3,1)$ lead to the same identity. It is of interest to determine minimal generators for the ideal of identities and whether the identities determine the curve $\Phi(\overline{\mathbb{H}^2/\Gamma(k)})\subset\mathbf{P}\mathbb{C}^{\frac{k-3}{2}}$.

10.4. Examples of equations. We conclude this section by constructing explicit examples for the cases $k=7$, 9 and 13. We define[93] $\theta_l=\theta\begin{bmatrix}\frac{2l+1}{k}\\1\end{bmatrix}$ for $l=0,\ 1,\ \ldots,\ \frac{k-3}{2}$. For $k=7$, we choose $j=1$, $\alpha_1=\alpha_2=3$, and $\alpha_3=1$. From the last theorem we obtain

Theorem 10.3. *We have for $k=7$,*

$$\theta_0^3\theta_2+\epsilon^9\theta_2^3\theta_1+\epsilon^6\theta_1^3\theta_0=0,\ \epsilon=\exp\left(\frac{\pi\imath}{7}\right).$$

Remark 10.4. We have re-proven (3.44).

We consider next the case $k=9$. We have seen that $\mathbb{H}^2/\Gamma(9)$ is a closed surface of genus ten punctured at 36 points. The compact surface $\overline{\mathbb{H}^2/\Gamma(9)}$ is mapped into $\mathbf{P}\mathbb{C}^3$ by $\Phi=(\varphi_0,\ \varphi_1,\ \varphi_2,\ \varphi_3)$. Omitting the function φ_1 we

[93] Obviously the functions θ_l and φ_l are closely related. We add this definition to emphasize that an identity among theta constants in the variable $k\tau$ (or $\frac{\tau}{k}$) is equivalent to an identity for the variable τ.

obtain a mapping of the surface $\mathbb{H}^2/<\Gamma(9), B^3>$ into $\mathbf{P}\mathbb{C}^2$. We seek the equation of the image, that is, equations among the φ_l (equivalently among the θ_l), $l \neq 1$. Choosing $j = 3$, $\alpha_1 = \alpha_2 = 1$, $\alpha_3 = 7$, we get the equation

$$\theta_0^2\theta_3 + \epsilon^{13}\theta_3^2\theta_2 + \epsilon^8\theta_2^2\theta_0 = 0, \ \epsilon = \exp\left(\frac{\pi\imath}{9}\right).$$

This is a nonsingular equation of a torus, the torus $\overline{\mathbb{H}^2/<\Gamma(9), B^3>}$ sitting in $\mathbf{P}\mathbb{C}^2$. The zero set of this equation is irreducible (hence connected); see, for example, [**15**, pg. 27].

We obtain equations involving the rest of the characteristics by choosing $j = 1, 2, 4$ respectively and $\alpha_1 = \alpha_2 = \alpha_3 = 3$:

$$\theta_2\theta_1^3 + \epsilon^5\theta_3(\theta_2^3 + \epsilon^6\theta_0^3) = 0, \ \theta_0\theta_1^3 + \epsilon\theta_2(\theta_3^3 - \theta_0^3) = 0, \ \theta_3\theta_1^3 - \theta_0(\epsilon^6\theta_3^3 + \theta_2^3) = 0.$$

We can however use ψ_l instead of φ_l to map our surface into projective space. As a consequence of (3.29) we are led to the same identities in terms of φ_l and in terms of ψ_l. The introduction of the change of variable

$$x_l = \exp\left\{-\frac{\pi\imath l}{9}\right\} \psi_l, \ l = 0, \ ..., \ 3,$$

leads to more elegant formulae (with ± 1 as coefficients) and the

Theorem 10.5. *The intersection of the following five hypersurfaces in* $\mathbf{P}\mathbb{C}^3$ *define the surface* $\overline{\mathbb{H}^2/\Gamma(9)}$*:*

$$\begin{gathered} x_0^2x_3 + x_3^2x_2 - x_2^2x_0 = 0, \ -x_0x_1^3 + x_2(x_3^3 + x_0^3) = 0, \\ -x_2x_1^3 + x_3(x_2^3 + x_0^3) = 0, \ -x_3x_1^3 + x_0(x_2^3 - x_3^3) = 0, \\ x_1^9 - x_2^3x_0^6 + x_3^3x_0^6 - x_3^3x_2^6 + x_2^3x_3^6 + x_0^3x_3^6 - x_0^3x_2^6 = 0. \end{gathered}$$

Proof. The first 4 relations are the translates to the x_l variables of our last 4 three term identities (given in the θ_l variables). The last relation is, of course, not a three term identity. It is obtained through the use of the residue theorem. More precisely, we use the fact that the sum of the residues of the elliptic function (for fixed $\tau \in \mathbb{H}^2$)

$$z \mapsto \frac{\theta^3\begin{bmatrix} 1 \\ \frac{3}{9} \end{bmatrix}(z,\tau)\ \theta^6\begin{bmatrix} 1 \\ \frac{15}{9} \end{bmatrix}(z,\tau)}{\theta\begin{bmatrix} 1 \\ 1 \end{bmatrix}(9z, 9\tau)}, z \in \mathbb{C},$$

is zero. The point $(0, 1, 0, 0) \in \mathbf{P}\mathbb{C}^3$ which does not belong to $\Phi(\overline{\mathbb{H}^2/\Gamma(9)})$ will satisfy every three term identity derived from Theorem 10.1. Hence another type of identity is needed to describe the curve $\Phi(\overline{\mathbb{H}^2/\Gamma(9)})$ in $\mathbf{P}\mathbb{C}^3$. The fifth equation serves this purpose.

Denote by H^0 the set defined by the first equation in $\mathbf{P}\mathbb{C}^2$, and by H^1 the set in $\mathbf{P}\mathbb{C}^3$ defined by the entire set of five equations. We have already observed that H^0 defines the torus $\overline{\mathbb{H}^2/<\Gamma(9), B^3>}$. To show that H^1 defines the closed surface $\overline{\mathbb{H}^2/\Gamma(9)}$, we study the commutative diagram

$$\begin{array}{ccc} \overline{\mathbb{H}^2/\Gamma(9)} & \xrightarrow{X} & \mathbf{P}\mathbb{C}^3 - \{(0,1,0,0)\} \\ \pi \downarrow & & \downarrow \omega, \\ \overline{\mathbb{H}^2/<\Gamma(9), B^3>} & \xrightarrow{X_1} & \mathbf{P}\mathbb{C}^2 \end{array}$$

where π is the canonical projection of $\overline{\mathbb{H}^2/\Gamma(9)}$ onto $\overline{\mathbb{H}^2/\Gamma(9)}/<\tilde{B}^3>$,

$$X(\tau) = (x_0(\tau), x_1(\tau), x_2(\tau), x_3(\tau)),\ \tau \in \mathbb{H}^2,$$

$$X_1(\tau) = (x_0(\tau), x_2(\tau), x_3(\tau)),\ \tau \in \mathbb{H}^2,$$

and

$$\omega(x_0, x_1, x_2, x_3) = (x_0, x_2, x_3),\ (x_0, x_1, x_2, x_3) \in \mathbf{P}\mathbb{C}^3 - \{(0,1,0,0)\}.$$

The map ω is well defined since its domain excludes the point $(0,1,0,0)$. We already pointed out that the image of the map X avoids this point. We know that $\deg \pi = 3$, X is injective, and that

$$X(\overline{\mathbb{H}^2/\Gamma(9)}) \subset H^1.$$

The proof of the theorem can now be completed by showing that every point $x \in H^0$ has (at most) three preimages (counting multiplicities) in H^1. If the i-th component of x is nonzero, then equation $i+1$ shows that $\omega^{-1}(x)$ has 3 preimages in H^1 (counting multiplicity). It is easier to deal with set theoretic preimages. The preimage in $\mathbf{P}\mathbb{C}^3 - \{(0,1,0,0)\}$ under π of a point $x = (x_0, x_2, x_3) \in \mathbf{P}\mathbb{C}^2$ consists of either three distinct points or a single point. If at least one of the three pairs

$$(x_0, x_2(x_3^3 + x_0^3)),\ (x_2, x_3(x_2^3 + x_0^3)),\ (x_3, x_0(x_2^3 - x_3^3))$$

has two nonzero entries, then $\pi^{-1}(x_0, x_2, x_3)$ contains three distinct points. The points

$$(1,0,0),\ (0,1,0) \text{ and } (0,0,1)$$

belong to H^0; their (unique) preimages in H^1 are the points

$$(1,0,0,0),\ (0,0,1,0) \text{ and } (0,0,0,1),$$

respectively. The remaining points $x = (x_0, x_2, x_3)$ in H^0 have 3 nonzero components. Thus it involves no loss of generality to assume that $x_0 = -1$. For the preimage of such an x to consist of a single point, we must have that both x_2 and x_3 are cube roots of unity. We conclude there are at most 9 points (in addition to the coordinate vectors) in H^1 which are preimages of branch values of the map π that is, points of the form

$$(-1, \eta^j, \eta^l),\ \eta = \exp \frac{2\pi\imath}{3},\ j, l = 0, 1, 2.$$

However, not all 9 points belong to H^0. Only the 6 pairs $(j, l) = (0,1)$, $(0,2)$, $(1,0)$, $(1,1)$, $(2,0)$ and $(2,2)$ correspond to points in H^0. We have shown that with 9 exceptions (corresponding to the punctures on the Riemann surface $(\mathbb{H}^2/\Gamma(9))/ < \tilde{B}^3 >$) the preimage of a point in H^0 contains precisely 3 points of H^1. The only remaining possibility is for H^1 to contain the point $(0, 1, 0, 0)$. This possibility is eliminated by the fifth equation.

Corollary 10.6. *The map X is injective and of maximal rank. Hence its image is a nonsingular irreducible subvariety of* $\mathbf{P}\mathbb{C}^3$.

Proof. In the proof of the theorem we already used that X is injective; it is of maximal rank because X_1 is. The image of X is nonsingular by Chow's theorem (see, for example, [**11**, pg. 167]); it is irreducible by the results of the theorem or as a consequence of the fact that it is the image under X of a compact Riemann surface.

Remark 10.7. If we are willing to use algebraic sets AND their complements, then we can get a much simpler description of the surface $\overline{\mathbb{H}^2/\Gamma(9)} \subset \mathbf{P}\mathbb{C}^3$ as the set of those $X = (x_0, x_1, x_2, x_3) \in \mathbf{P}\mathbb{C}^3$ that satisfy the first 4 equations of the theorem and $X \neq (0, 1, 0, 0)$.

We have already observed that $\overline{\mathbb{H}^2/\Gamma(13)}$ is a closed surface of genus 50. It is convenient to work with the map

$$X = (x_0, \ ..., \ x_5) : \overline{\mathbb{H}^2/\Gamma(13)} \to \mathbf{P}\mathbb{C}^5,$$

where

$$x_l = \exp\left\{-\frac{\pi i l}{13}\right\} \psi_l, \ l = 0, \ ..., \ 5.$$

We define a map from

$$\mathbb{Y} = \{(x_0, \ ..., \ x_5) \in \mathbf{P}\mathbb{C}^5; \ (x_0x_2, x_1x_5, x_3x_4) \neq 0\}$$

to $\mathbf{P}\mathbb{C}^2$ by

$$\Omega(x_0, x_1, x_2, x_3, x_4, x_5) = (x_0x_2, x_1x_5, x_3x_4) = X_1.$$

It is easy to see that (x_0x_2, x_1x_5, x_3x_4) is invariant under B and $\gamma = \begin{bmatrix} 5 & -2 \\ 13 & -5 \end{bmatrix}$; moreover from Theorem 10.1, we obtain (by setting $j = 1$, $\alpha_1 = 1$, $\alpha_2 = 5$, $\alpha_3 = 7$) the identity $(x_0x_2)(x_1x_5)+(x_1x_5)(x_3x_4)-(x_3x_4)(x_0x_2) = 0$. The equation gives a quadratic curve in $\mathbf{P}\mathbb{C}^2$. It is well known that a nonsingular quadratic curve is a Riemann surface of genus 0 (the result also follows from §8.6). Even though $X(\overline{\mathbb{H}^2/\Gamma(13)}) \not\subset \mathbb{Y}$, we have obtained a projective embedding $X_1 : \overline{\mathbb{H}^2/(< G(13), \gamma >)} \to \mathbf{P}\mathbb{C}^2$. We have seen in §8.6 that $\overline{\mathbb{H}^2/ < G(13), \gamma >}$ is conformally equivalent to $\mathbf{P}\mathbb{C}$.

11. Some special cases (restricted characteristics)

Our main objective is to study the invariance of the functions

$$f_l = \frac{\theta^k \begin{bmatrix} \frac{2l+1}{k} \\ 1 \end{bmatrix}}{\theta^k \begin{bmatrix} 1 \\ \frac{2l+1}{k} \end{bmatrix}}, \ l = 0, \ 1, \ \ldots, \ \frac{k-3}{2},$$

and related functions under the group Γ and its normal subgroup $\Gamma(k)$. We limit the discussion to the case of odd primes k.

11.1. Characteristics with $m' = k$. The functions

$$\theta^k \begin{bmatrix} \frac{2l-1}{k} \\ 1 \end{bmatrix}, \ l = 0, \ 1, \ \ldots, \ \frac{k-3}{2},$$

span a $\frac{k-1}{2}$-dimensional Hilbert space (with Petersson scalar product) of forms belonging to a fixed factor of automorphy e of weight $\frac{k}{4}$ for the group generated by $\Gamma(k)$ and the motion $C = \begin{bmatrix} 1 & 0 \\ 1 & 1 \end{bmatrix}$. This space is invariant under $\Gamma^o(k)$; it is not invariant under A.

11.2. Characteristics with $m = k$. The previous subsection suggests that we study the functions

$$\theta_l = \theta \begin{bmatrix} 1 \\ \frac{2l+1}{k} \end{bmatrix}, \ l = 0, \ 1, \ \ldots, \ \frac{k-3}{2}.$$

There is a homomorphism[94]

$$\sigma : \Gamma_o(k) \to \mathcal{S}_{\frac{k-1}{2}}$$

whose kernel contains $\Gamma(k)$ such that for all $\gamma \in \Gamma_o(k)$ (we write σ_γ for $\sigma(\gamma)$)

$$\gamma^*_{\frac{1}{4}} \theta_l = c(\gamma, l) \theta_{\sigma_\gamma(l)},$$

for some constant $c(\gamma, l)$ of absolute value 1.

For $\gamma \in \Gamma(k)$, $c(\gamma, l)^k$ is independent of l. It follows that for $k \geq 5$, $\left(\frac{\theta_0}{\theta_1}\right)^k \in \mathcal{K}(\overline{\mathbb{H}^2/\Gamma(k)})$. We have not taken advantage of the fact that we are working with characteristics in a single tower. It is easily seen that the kernel of σ contains $G(k)$ and since $c(B, l)$ is independent of l, $\left(\frac{\theta_0}{\theta_1}\right)^k \in \mathcal{K}(\overline{\mathbb{H}^2/G(k)})$. For $k = 5$ (a four punctured sphere) and 7 (a six punctured

[94]Even though we use the same letters, θ and σ, to designate functions and homomorphisms previously encountered, the new entities do not coincide, although they are related, with those previously defined.

sphere), the divisors of the projections of these functions to the respective surfaces are

$$\frac{P_{\frac{1}{3}}}{P_0} \text{ and } \frac{P_{\frac{1}{5}}^3}{P_0\, P_{\frac{1}{4}}^2}.$$

We have already established most of

Theorem 11.1. *The functions*

$$\left\{\theta_l^k;\ l = 0,\ 1,\ \ldots,\ \frac{k-3}{2}\right\}$$

span a $\frac{k-1}{2}$-dimensional Hilbert space $W(k)$ (with Petersson scalar product) of forms belonging to a fixed factor of automorphy e of weight $\frac{k}{4}$ for the group $G(k)$. This space is invariant under $\Gamma_o(k)$; it is not invariant under A.

Proof. It is clear that

$$A^*_{\frac{k}{4}}\theta_l^k = c_l\theta^k \begin{bmatrix} \frac{2l+1}{k} \\ 1 \end{bmatrix}, \tag{3.45}$$

for some $c_l \in \mathbb{C}$, $|c_l| = 1$, and hence

$$\text{ord}_0\theta_l^k = \frac{(2l+1)^2}{8}.$$

It is thus obvious that the $\frac{k-1}{2}$ spanning functions are linearly independent. Note that $\text{ord}_\infty\theta_l^k = \frac{k^2}{8}$, and hence

$$\text{ord}_\infty\theta \geq \frac{k^2}{8}, \text{ for all } \theta \in W(k). \tag{3.46}$$

Now if $A^*_{\frac{k}{4}}(W(k)) = W(k)$, then the linearly independent functions

$$A^*_{\frac{k}{4}}\theta_l^k;\ l = 0,\ 1,\ \ldots,\ \frac{k-3}{2},$$

would form a basis for $W(k)$. From (3.45) we conclude that

$$\text{ord}_\infty A^*_{\frac{k}{4}}\theta_l^k = \frac{(2l+1)^2}{k}.$$

This contradicts (3.46).

We know that $G(3) = \Gamma_o(3)$; this group is of signature $(0,3;\ 3,\infty,\infty)$. For primes $k > 3$, the group $G(k)$ is torsion free of type $\left(\frac{(k-5)(k-7)}{24}, k-1\right)$. The $k-1$ punctures on the surface $\mathbb{H}^2/G(k)$ are

$$P_{\frac{\alpha}{k}},\ P_{\frac{k}{\alpha}};\ \alpha = 1,\ 2,\ \ldots,\frac{k-1}{2}.$$

An easy way to see this is to recall that $\Gamma_o(k)/G(k)$ is a cyclic group of order $\frac{k-1}{2}$, and its generator may be taken to be $\begin{bmatrix} 2 & k \\ k & \frac{k^2+1}{2} \end{bmatrix}$. Each of the two punctures P_∞ and P_0 on $\mathbb{H}^2/\Gamma_o(k)$ lifts to $\frac{k-1}{2}$ punctures on $\mathbb{H}^2/G(k)$. The lifts of P_∞ are $P_{\frac{\alpha}{k}}$; the lifts of P_0 are $P_{\frac{k}{\alpha}}$. For computations, we need to know what the stabilizers of the various cusps look like. If γ is the generator of the stabilizer of $\frac{\alpha}{k}$ in $\Gamma(k)$, $\gamma^{\frac{1}{k}}$ is its stabilizer in $G(k)$, whereas the cusp $\frac{k}{\alpha}$ has the same stabilizer in both $\Gamma(k)$ and $G(k)$.

Lemma 11.2. *Let $l = 0, 1, \ldots,$ or $\frac{k-3}{2}$. Then*
(a) $\text{ord}_x\theta_l = \frac{1}{8}$, *if the cusp $x \in \mathbb{Q} \cup \{\infty\}$ is $\Gamma_o(k)$-equivalent to ∞ (in particular for the cusps $\frac{\alpha}{k}$, $\alpha = 1, 2, \ldots, \frac{k-1}{2}$).*
(b) *Let the cusp $x \in \mathbb{Q} \cup \{\infty\}$ be $\Gamma_o(k)$-equivalent to $\frac{k}{\alpha}$, $\alpha = 1, 2, \ldots, \frac{k-1}{2}$ (this is a complete list of representatives for the $G(k)$-equivalence classes of cusps that are $\Gamma_o(k)$-equivalent to 0). Choose a $\gamma \in \Gamma_o(k)$ such that $\gamma(0) = x$. Then as $G(k)$- forms,*

$$\text{ord}_x\theta_l = \frac{(2\sigma_\gamma(l)+1)^2}{8k}.$$

Corollary 11.3. *The divisor of the projection to $\mathbb{H}^2/G(k)$ of the $G(k)$-automorphic function on $\mathbb{H}^2$, $\frac{\theta_l^k}{\theta_{l'}^k}$, $0 \le l \ne l' \le \frac{k-3}{2}$ is supported at the $\frac{k-1}{2}$ punctures*

$$P_{\frac{k}{1}},\ P_{\frac{k}{2}},\ \ldots,\ P_{\frac{2k}{k-1}}.$$

As an application of the previous lemma, we establish

Theorem 11.4. *Let*

$$f(\tau) = \left(\frac{\theta\begin{bmatrix} \frac{1}{5} \\ 1 \end{bmatrix}(0, 5\tau)}{\theta\begin{bmatrix} \frac{3}{5} \\ 1 \end{bmatrix}(0, 5\tau)} \right)^5 \quad \text{and } g(\tau) = \left(\frac{\theta\begin{bmatrix} 1 \\ \frac{1}{5} \end{bmatrix}(0, \tau)}{\theta\begin{bmatrix} 1 \\ \frac{3}{5} \end{bmatrix}(0, \tau)} \right)^5.$$

There exists a Möbius transformation E such that $f = E \circ g$.

Evaluating the Möbius transformation E and rewriting the resulting identity in terms of the local coordinate $x = \exp(2\pi\imath\tau)$ leads to the interesting power series identity contained in

Corollary 11.5. *For all $x \in \mathbb{C}$, $|x| < 1$,*

$$\frac{1}{x}\prod_{n=0}^{\infty} \frac{(1-x^{5n+2})^5(1-x^{5n+3})^5}{(1-x^{5n+1})^5(1-x^{5n+4})^5}$$

k	possible l
3,5,7	none
9	0
11	1
13	2
15	3
17	0,4
19	1,5
21	2,6

Table 10. VALUES OF l THAT PRODUCE $\Gamma(k)$-AUTOMORPHIC FUNCTIONS IN THEOREM 11.7.

$$= \frac{-\frac{1}{2}\left(11+5\sqrt{5}\right)\prod_{n=1}^{\infty}\frac{\left(1+2\cos\left(\frac{\pi}{5}\right)x^n+x^{2n}\right)^5}{\left(1+2\cos\left(\frac{3\pi}{5}\right)x^n+x^{2n}\right)^5}+\frac{1}{2}\left(11-5\sqrt{5}\right)}{1-\prod_{n=1}^{\infty}\frac{\left(1+2\cos\left(\frac{\pi}{5}\right)x^n+x^{2n}\right)^5}{\left(1+2\cos\left(\frac{3\pi}{5}\right)x^n+x^{2n}\right)^5}}.$$

Proof. The product expansions of the functions f and g are obtained from the Jacobi triple product. Note that

$$\cos\left(\frac{\pi}{5}\right) = \frac{1+\sqrt{5}}{4} \text{ and } \cos\left(\frac{3\pi}{5}\right) = \frac{1-\sqrt{5}}{4}.$$

Problem 11.6. We do not know whether the k functions on $\mathbb{C} \times \mathbb{H}^2$

$$\theta^k\begin{bmatrix}\frac{2l+1}{k}\\ 1\end{bmatrix},\ \theta^k\begin{bmatrix}1\\ \frac{2l+1}{k}\end{bmatrix},\ l=0,\ 1,\ \ldots,\ \frac{k-3}{2},$$

which belong to $\mathcal{F}_k\begin{bmatrix}1\\1\end{bmatrix}$ are linearly independent.

11.3. Ratios. Returning to the functions introduced at the beginning of this section, we have

Theorem 11.7. *For every odd integer k, and $l \in \mathbb{Z}$, $0 \le l \le \frac{k-3}{2}$, f_l is a $\Gamma(k)$-automorphic function provided*

$$l \equiv \frac{k-1}{2} \mod 4.$$

For small values of k, the theorem produces few functions as summarized in Table 10.

Chapter 4

Theta constant identities

In this chapter, we use the theory of theta functions and theta constants developed in Chapter 2 to derive identities among the theta constants. The identities we present are not exhaustive but are meant to be representative of the types of identities that are obtainable. Some are more surprising than others, and some have more number theoretic content than others. Our motivation here is deriving identities which are either aesthetically pleasing to look at or have significant number theoretic content. A striking example is the following. For each positive integer $N \geq 3$, we have

$$\sum_{l=1}^{N}(-1)^l\theta^N\left[\begin{array}{c}\frac{1}{N}\\ \frac{2N-2l+1}{N}\end{array}\right]=0. \tag{4.1}$$

We postpone the discussion of the number theoretic content of some of the identities obtained to Chapter 7.

As indicated above, this chapter is concerned with deriving identities among the theta functions and theta constants we have met until now. We have already encountered some of these identities in Chapter 2. There we already came across the quartic identity (Corollary 5.5), the Jacobi derivative formula (Corollary 5.8), the Jacobi triple product identity and the quintuple product identity (Theorem 8.4). Those identities were derived there either because we wanted to use the result or we just could not contain ourselves and presented them as a taste of what is to come. Similarly, in Chapter 3 we have already come across some of the identities which arose in connection with uniformizations of the compactifications of the Riemann surfaces

$\mathbb{H}^2/\Gamma(k)$. Here we present in a systematic way a number of interesting identities.

There are three basic techniques which we use in order to derive identities. The first technique, which is also the most classical, is based on considerations of linear algebra. One is given a finite dimensional vector space and a finite set of vectors in the space with cardinality greater than the dimension of the space. There necessarily is a linear relation between the elements of the set. Most of the examples of theta constant or theta function identities we have encountered until now were all derived in this way. There are two subcases to this technique. The first uses N-th order theta functions; the second, finite dimensional spaces of automorphic forms for subgroups of the modular group.

A second technique is via uniformization theory. Suppose we have two analytic homeomorphisms of a Riemann surface S onto the sphere:

$$f : S \to \mathbb{C} \cup \{\infty\} \text{ and } g : S \to \mathbb{C} \cup \{\infty\}.$$

It is then clear that $g \circ f^{-1}$ is an analytic automorphism of the sphere and there is a Möbius transformation $\begin{bmatrix} a & b \\ c & d \end{bmatrix}$ so that $g = \frac{af+b}{cf+d}$. More generally, two universal covering maps automorphic with respect to the same group are post related by a Möbius transformation.

This technique is limited but does give easy derivations of some identities which have interesting consequences. We have already encountered some of the identities which arise in this way in Chapter 3.

The most powerful and newest technique we use is related to elliptic function theory. It comes from the well known property of elliptic functions which asserts that the sum of the residues of an elliptic function in a period parallelogram vanishes. In order to demonstrate the power of this technique we show how it gives easy proofs of both the Jacobi quartic identity and the Jacobi derivative formula.

There are also ad-hoc techniques which one can use in order to derive theta constant identities. We use one of these to begin our discussion and rederive once again the Jacobi quartic identity. Our point of departure is Lemma 1.6 of Chapter 2. The equality given by the lemma yields:

$$\theta^2 \begin{bmatrix} 0 \\ 0 \end{bmatrix} (0, \tau) = \theta^2 \begin{bmatrix} 0 \\ 0 \end{bmatrix} (0, 2\tau) + \theta^2 \begin{bmatrix} 1 \\ 0 \end{bmatrix} (0, 2\tau),$$

$$\theta^2 \begin{bmatrix} 0 \\ 1 \end{bmatrix} (0, \tau) = \theta^2 \begin{bmatrix} 0 \\ 0 \end{bmatrix} (0, 2\tau) - \theta^2 \begin{bmatrix} 1 \\ 0 \end{bmatrix} (0, 2\tau),$$

$$\theta^2 \begin{bmatrix} 1 \\ 0 \end{bmatrix} (0, \tau) = 2\theta \begin{bmatrix} 0 \\ 0 \end{bmatrix} (0, 2\tau) \theta \begin{bmatrix} 1 \\ 0 \end{bmatrix} (0, 2\tau).$$

It thus follows immediately that

$$\theta^4 \begin{bmatrix} 0 \\ 0 \end{bmatrix} (0,\tau) = \theta^4 \begin{bmatrix} 0 \\ 1 \end{bmatrix} (0,\tau) + \theta^4 \begin{bmatrix} 1 \\ 0 \end{bmatrix} (0,\tau),$$

which is of course the Jacobi quartic identity.

While we have termed the above an ad-hoc proof, it does suggest that the identity has a combinatorial spirit. We shall return to this theme later in the chapter. We also point out that the lemma used above can also be used to construct new identities from old ones.

Equation (4.18), which we shall soon meet, defines an important number theoretic function known as Ramanujan's τ-function; we call it here T. It is connected with identities of the type considered here since the left hand side of (4.18) is η^{24}, a nontrivial element of $\mathbb{A}_6(\mathbb{H}^2,\Gamma)$. The space of cusp 6-forms for the modular group Γ is one dimensional; a basis for this space is hence given by η^{24}. The fact that a form in this space can be constructed by properly averaging 6-forms for the principal congruence subgroup $\Gamma(k)$ yields the theta identity

$$k^{12}\eta^{24}(k\tau) + \sum_{l=0}^{k-1} \eta^{24}\left(\frac{\tau+l}{k}\right) = k\ T(k)\ \eta^{24}(\tau);$$

that translates to the power series identity (here $\eta^{24}(\tau) = \sum_{n=1}^{\infty} T(n)x^n$, $x = \exp(2\pi\imath\tau)$)

$$k^{11}\sum_{n=1}^{\infty} T(n)x^{kn} = \sum_{n=1}^{\infty}(T(k)T(n) - T(kn))x^n.$$

The multiplicative property (a result of Mordell) of the T-function follows immediately: that is, if $n \in \mathbb{Z}^+$ is not a multiple of the positive prime k, then $T(kn) = T(k)T(n)$; and if n is a multiple of k, say $n = k^r l$ with $r \geq 1$ and l not a multiple of k, then

$$T(k^{r+1}l) = T(k)T(k^r l) - k^{11}T(k^{r-1}l).$$

In particular, setting $l = 1$ gives a recursion formula

$$T(k^{r+1}) = T(k)T(k^r) - k^{11}T(k^{r-1}).$$

Despite the fact that we have a closed expression for the constant $T(k)$ (in terms of theta constants), it is not at all obvious that our methods can answer Lehmer's question whether $T(k) = 0$ for some k. Congruence statements cannot easily be used to answer this question, since it is clear from the above formulae that $T(5n) \equiv 0 \mod 5$ and $T(7n) \equiv 0 \mod 7$. We obtain several proofs of these two congruences as well as many others in the next chapter. We will also generalize the T-function and study a number of multiplicative number theoretic functions.

1. Dimension considerations

We use theta functions and theta constants to construct finite dimensional vector spaces of meromorphic functions and differentials on compact Riemann surfaces. We then look for relations among the objects constructed. We have already used this idea in Chapter 2 in the following way.

Recall Theorem 5.3 of Chapter 2 in whose proof four identities for the theta functions with integer characteristics are derived. The reader will recall that the identities follow from the simple fact that the dimension of a certain space of meromorphic functions is 2. These identities, by specialization, then yielded the Jacobi quartic identity.

An even simpler argument tells us that if we have a meromorphic function on a compact Riemann surface of genus $g > 0$ with at most a simple pole at one point, the function is necessarily a constant; that is, the vector space of such functions is one dimensional. This sufficed to give the Jacobi derivative formula, which was proven in Theorem 5.7 of Chapter 2.

For each $N \in \mathbb{Z}^+$, we defined in Chapter 2 the vector spaces $\mathcal{F}_N \begin{bmatrix} \epsilon \\ \epsilon' \end{bmatrix}$. These spaces were shown to be N-dimensional, and bases for these spaces were constructed. Furthermore, we saw how to decompose these spaces into the subspaces of even and odd functions. It turned out that for N odd and at least 3, the dimension of the subspace of even functions is $\frac{N-1}{2}$ and the dimension of the subspace of odd functions is $\frac{N+1}{2}$. In particular, for $N = 3$ the subspace of even functions is one dimensional, and this was enough to give us the quintuple product identity.

1.1. The septuple product identity.

1.1. The septuple product identity. We have already seen that we have a decomposition of the space of N-th order theta functions $\mathcal{F}_N \begin{bmatrix} 1 \\ 1 \end{bmatrix}$ into the spaces $\mathcal{E}_N \begin{bmatrix} 1 \\ 1 \end{bmatrix}$ and $\mathcal{O}_N \begin{bmatrix} 1 \\ 1 \end{bmatrix}$, the spaces of even and odd functions in $\mathcal{F}_N \begin{bmatrix} 1 \\ 1 \end{bmatrix}$. We have also seen that when $N \geq 3$ is an odd integer, which we assume from now on, the dimensions of these spaces are N, $\frac{N-1}{2}$, $\frac{N+1}{2}$ and that we can take

$$\theta \begin{bmatrix} \frac{2l+1}{N} \\ 1 \end{bmatrix} (Nz, N\tau), \quad l = 0, \ ..., \ \frac{N-3}{2},$$

$$\theta \begin{bmatrix} \frac{2l+1}{N} \\ 1 \end{bmatrix} (-Nz, N\tau), \quad l = 0, ..., \frac{N-3}{2}, \ \theta \begin{bmatrix} 1 \\ 1 \end{bmatrix} (Nz, N\tau)$$

as a basis for the space $\mathcal{F}_N\begin{bmatrix}1\\1\end{bmatrix}$. Furthermore

$$\frac{1}{2}\left(\theta\begin{bmatrix}\frac{2l+1}{N}\\1\end{bmatrix}(Nz,N\tau)+\theta\begin{bmatrix}\frac{2l+1}{N}\\1\end{bmatrix}(-Nz,N\tau)\right),\quad l=0,\ \ldots,\ \frac{N-3}{2},$$

can be taken as a basis for $\mathcal{E}_N\begin{bmatrix}1\\1\end{bmatrix}$ and

$$\frac{1}{2}\left(\theta\begin{bmatrix}\frac{2l+1}{N}\\1\end{bmatrix}(Nz,N\tau)-\theta\begin{bmatrix}\frac{2l+1}{N}\\1\end{bmatrix}(-Nz,N\tau)\right),\quad l=0,\ldots,\frac{N-3}{2},$$

$$\theta\begin{bmatrix}1\\1\end{bmatrix}(Nz,N\tau)$$

can be taken as a basis for $\mathcal{O}_N\begin{bmatrix}1\\1\end{bmatrix}$.

It is easy to construct a function in $\mathcal{E}_N\begin{bmatrix}1\\1\end{bmatrix}$ with a zero of order $N-3$ at the origin.

$$z\mapsto\theta\begin{bmatrix}0\\0\end{bmatrix}(z,\tau)\theta\begin{bmatrix}0\\1\end{bmatrix}(z,\tau)\theta\begin{bmatrix}1\\0\end{bmatrix}(z,\tau)\theta^{N-3}\begin{bmatrix}1\\1\end{bmatrix}(z,\tau)$$

is such a function. In fact, this function is unique up to multiplication by a constant. It follows that we can find constants

$$c_1,\ c_2,\ \ldots,\ c_{\frac{N-1}{2}}$$

with the property that

$$\frac{1}{2}\sum_{l=1}^{\frac{N-1}{2}}c_l\left(\theta\begin{bmatrix}\frac{2l-1}{N}\\1\end{bmatrix}(Nz,N\tau)+\theta\begin{bmatrix}\frac{2l-1}{N}\\1\end{bmatrix}(-Nz,N\tau)\right)$$
$$=\theta\begin{bmatrix}0\\0\end{bmatrix}(z,\tau)\theta\begin{bmatrix}0\\1\end{bmatrix}(z,\tau)\theta\begin{bmatrix}1\\0\end{bmatrix}(z,\tau)\theta^{N-3}\begin{bmatrix}1\\1\end{bmatrix}(z,\tau).$$

Our problem reduces to linear algebra, finding the constants c_l.

We recall that we are dealing with even functions of z, so that the odd order derivatives vanish automatically at the origin. It thus follows that the equations satisfied by the constants c_l are the following ones. For $m=0,\ \ldots,\ \frac{N-3}{2}$, consider the equations

$$\sum_{l=0}^{\frac{N-3}{2}}c_l\theta^{(2m)}\begin{bmatrix}\frac{2l+1}{N}\\1\end{bmatrix}(0,N\tau)$$

$$=\delta_{l,\frac{N-3}{2}}c\theta\begin{bmatrix}0\\0\end{bmatrix}(0,\tau)\theta\begin{bmatrix}0\\1\end{bmatrix}(0,\tau)\theta\begin{bmatrix}1\\0\end{bmatrix}(0,\tau)(\theta')^{N-3}\begin{bmatrix}1\\1\end{bmatrix}(0,\tau),$$

where δ is the Kronecker function[95], c is a universal constant and $\theta^{(2m)}$ denotes the $2m$-th derivative with respect to the variable z.

The above system of equations has a unique solution given by Cramer's rule. If we denote by W the matrix (with $\frac{N-1}{2}$ rows and columns) of the system, we have the solution (provided W is nonsingular – as we show below)

$$c_l = \frac{c\theta\begin{bmatrix}0\\0\end{bmatrix}\theta\begin{bmatrix}0\\1\end{bmatrix}\theta\begin{bmatrix}1\\0\end{bmatrix}\left(\theta'\begin{bmatrix}1\\1\end{bmatrix}\right)^{N-3} W_{\frac{N-1}{2},l}}{\det(W)},$$

where $W_{\frac{N-1}{2},l}$ denotes the cofactor of the element in row $\frac{N-1}{2}$, column l.

To show that W is nonsingular and evaluate its determinant, we use the heat equation to convert the derivatives with respect to z to derivatives with respect to the variable τ. This will introduce another constant. The crucial observation is that the determinant now is up to this nonzero constant multiple precisely the Wronskian

$$W\left(\theta\begin{bmatrix}\frac{1}{N}\\1\end{bmatrix}(0,N\tau),\ \ldots,\ \theta\begin{bmatrix}\frac{N-2}{N}\\1\end{bmatrix}(0,N\tau)\right).$$

We have already observed in Proposition 7.8 of Chapter 3 that this Wronskian is an $\frac{(N-1)(N-2)}{8}$-form for $\Gamma(k)$. Since the zeros of this form are at the cusps and have the same order at each cusp, we conclude that

$$\det(W) = M\left(\theta'\begin{bmatrix}1\\1\end{bmatrix}(0,\tau)\right)^{\frac{(N-1)(N-2)}{6}},$$

where M is a constant (in particular the Wronskian is a form for the full modular group Γ). Thus for some universal constant K,

$$c_l = K\frac{\theta\begin{bmatrix}0\\0\end{bmatrix}\theta\begin{bmatrix}0\\1\end{bmatrix}\theta\begin{bmatrix}1\\0\end{bmatrix}W_{\frac{N-1}{2},l}}{\theta'\begin{bmatrix}1\\1\end{bmatrix}(0,\tau)^{\frac{(N-4)(N-5)}{6}}}.$$

The above leads to a sequence of identities which generalize the quintuple product identity which we have already seen in Chapter 2. For that case all that we have said here is unnecessary. The first interesting case from the current point of view is $N = 5$, which we now work out in detail. In this case our formulae become

$$c_1 = -K\theta\begin{bmatrix}0\\0\end{bmatrix}\theta\begin{bmatrix}0\\1\end{bmatrix}\theta\begin{bmatrix}1\\0\end{bmatrix}\theta\begin{bmatrix}\frac{3}{5}\\1\end{bmatrix}(0,5\tau),$$

95

$$\delta_{i,j} = \begin{cases} 1 \text{ if } i = j \\ 0 \text{ if } i \neq j \end{cases}.$$

$$c_2 = K\theta\begin{bmatrix}0\\0\end{bmatrix}\theta\begin{bmatrix}0\\1\end{bmatrix}\theta\begin{bmatrix}1\\0\end{bmatrix}\theta\begin{bmatrix}\frac{1}{5}\\1\end{bmatrix}(0,5\tau).$$

We have therefore derived the identity

$$\begin{aligned}
&c_1\left(\theta\begin{bmatrix}\frac{1}{5}\\1\end{bmatrix}(5\zeta,5\tau)+\theta\begin{bmatrix}\frac{1}{5}\\1\end{bmatrix}(-5\zeta,5\tau)\right)\\
&+c_2\left(\theta\begin{bmatrix}\frac{3}{5}\\1\end{bmatrix}(5\zeta,5\tau)+\theta\begin{bmatrix}\frac{3}{5}\\1\end{bmatrix}(-5\zeta,5\tau)\right)\\
&=\theta\begin{bmatrix}0\\0\end{bmatrix}(\zeta,\tau)\theta\begin{bmatrix}0\\1\end{bmatrix}(\zeta,\tau)\theta\begin{bmatrix}1\\0\end{bmatrix}(\zeta,\tau)\theta^2\begin{bmatrix}1\\1\end{bmatrix}(\zeta,\tau).
\end{aligned}$$

We rewrite the previous identity as

$$\begin{aligned}
&-\theta\begin{bmatrix}\frac{3}{5}\\1\end{bmatrix}(0,5\tau)\left(\frac{\theta\begin{bmatrix}\frac{1}{5}\\1\end{bmatrix}(5\zeta,5\tau)+\theta\begin{bmatrix}\frac{1}{5}\\1\end{bmatrix}(-5\zeta,5\tau)}{2}\right)\\
&+\theta\begin{bmatrix}\frac{1}{5}\\1\end{bmatrix}(0,5\tau)\left(\frac{\theta\begin{bmatrix}\frac{3}{5}\\1\end{bmatrix}(5\zeta,5\tau)+\theta\begin{bmatrix}\frac{3}{5}\\1\end{bmatrix}(-5\zeta,5\tau)}{2}\right)\\
&=2K'\frac{\theta\begin{bmatrix}0\\0\end{bmatrix}(\zeta,\tau)\theta\begin{bmatrix}0\\1\end{bmatrix}(\zeta,\tau)\theta\begin{bmatrix}1\\0\end{bmatrix}(\zeta,\tau)}{\theta\begin{bmatrix}0\\0\end{bmatrix}\theta\begin{bmatrix}0\\1\end{bmatrix}\theta\begin{bmatrix}1\\0\end{bmatrix}}\theta^2\begin{bmatrix}1\\1\end{bmatrix}(\zeta,\tau),
\end{aligned}$$

where K' is a constant which needs to be calculated. It is easily seen that $K' = \exp\left(\frac{2\pi\imath}{5}\right)$.

We now make the change of variables $x = \exp(\pi\imath\tau)$, $z = \exp(2\pi\imath\zeta)$, use the property that $\theta\begin{bmatrix}\epsilon\\\epsilon'\end{bmatrix}(-\zeta,\tau) = \theta\begin{bmatrix}-\epsilon\\-\epsilon'\end{bmatrix}(\zeta,\tau)$ and the Jacobi triple product formula to conclude that our identity can be written as

$$\begin{aligned}
&\sum_{n=-\infty}^{\infty}(-1)^n x^{5n^2+n}\left(\sum_{n=-\infty}^{\infty}(-1)^n x^{5n^2+3n}z^{5n+3}+\sum_{n=-\infty}^{\infty}(-1)^n x^{5n^2-3n}z^{5n}\right)\\
&-\sum_{n=-\infty}^{\infty}(-1)^n x^{5n^2+3n}\left(\sum_{n=-\infty}^{\infty}(-1)^n x^{5n^2+n}z^{5n+2}+\sum_{n=-\infty}^{\infty}(-1)^n x^{5n^2-n}z^{5n+1}\right)\\
&=(z+1)(z-1)^2\prod_{n=1}^{\infty}(1-x^{2n})^2(1-x^{2n}z)\left(1-\frac{x^{2n}}{z}\right)(1-x^{2n}z^2)\left(1-\frac{x^{2n}}{z^2}\right).
\end{aligned}$$

We call the last equation the *septuple product identity*. It is a generalization of the quintuple product identity.

1.2. Further generalizations. We have reduced the derivation of the septuple product identity to a problem in linear algebra in the two dimensional vector space $\mathcal{E}_5\begin{bmatrix}1\\1\end{bmatrix}$. As we have already mentioned, a similar calculation can be made for any $\mathcal{E}_N\begin{bmatrix}1\\1\end{bmatrix}$, for every odd $N \geq 5$. For that matter, there is no reason to restrict ourselves to $\mathcal{E}_N\begin{bmatrix}1\\1\end{bmatrix}$. We could just as well work in $\mathcal{O}_N\begin{bmatrix}1\\1\end{bmatrix}$. We proceed to demonstrate this for the case of $\mathcal{O}_3\begin{bmatrix}1\\1\end{bmatrix}$, a two dimensional space with basis

$$\frac{\theta\begin{bmatrix}\frac{1}{3}\\1\end{bmatrix}(3\zeta,3\tau)-\theta\begin{bmatrix}\frac{1}{3}\\1\end{bmatrix}(-3\zeta,3\tau)}{2},\ \theta\begin{bmatrix}1\\1\end{bmatrix}(3\zeta,3\tau).$$

The function $\theta^3\begin{bmatrix}1\\1\end{bmatrix}(\zeta,\tau)$ is obviously in $\mathcal{O}_3\begin{bmatrix}1\\1\end{bmatrix}$, so there are constants c_1, c_2 so that

$$\frac{c_1}{2}\theta\begin{bmatrix}\frac{1}{3}\\1\end{bmatrix}(3\zeta,3\tau)-\theta\begin{bmatrix}\frac{1}{3}\\1\end{bmatrix}(-3\zeta,3\tau)+c_2\theta\begin{bmatrix}1\\1\end{bmatrix}(3\zeta,3\tau)=\theta^3\begin{bmatrix}1\\1\end{bmatrix}(\zeta,\tau).$$

The equations for c_1, c_2 now are

$$c_1\theta'\begin{bmatrix}\frac{1}{3}\\1\end{bmatrix}(0,3\tau)+c_2\theta'\begin{bmatrix}1\\1\end{bmatrix}(0,3\tau)=0,$$

$$c_1\theta'''\begin{bmatrix}\frac{1}{3}\\1\end{bmatrix}(0,3\tau)+c_2\theta'''\begin{bmatrix}1\\1\end{bmatrix}(0,3\tau)=c\left(\theta'\begin{bmatrix}1\\1\end{bmatrix}(0,3\tau)\right)^3,$$

where c is a constant. The determinant of the matrix of our system is again the Wronskian

$$W\left(\theta'\begin{bmatrix}\frac{1}{3}\\1\end{bmatrix}(0,3\tau),\ \theta'\begin{bmatrix}1\\1\end{bmatrix}(0,3\tau)\right)=m\left(\theta'\begin{bmatrix}1\\1\end{bmatrix}(0,\tau)\right)^{\frac{10}{3}},$$

for some $m \in \mathbb{C}^*$.

Remark 1.1. From Proposition 7.8 of Chapter 3 we see that for odd positive N,

$$W\left(\theta'\begin{bmatrix}\frac{1}{N}\\1\end{bmatrix}(0,N\tau),\ \theta'\begin{bmatrix}\frac{3}{N}\\1\end{bmatrix}(0,N\tau),\ \ldots,\ \theta'\begin{bmatrix}1\\1\end{bmatrix}(0,N\tau)\right)$$

is a form of weight $\frac{(N+1)(N+2)}{8}$ for $\Gamma(N)$ and, as the reader can check is a multiple of $(\eta(\tau))^{\frac{(N+1)(N+2)}{2}}$.

We continue with our calculation obtaining

$$c_1 = -K\frac{\theta'\begin{bmatrix} 1 \\ 1 \end{bmatrix}(0,3\tau)}{\left(\theta'\begin{bmatrix} 1 \\ 1 \end{bmatrix}(0,\tau)\right)^{\frac{1}{3}}},\quad c_2 = K\frac{\theta'\begin{bmatrix} \frac{1}{3} \\ 1 \end{bmatrix}(0,3\tau)}{\left(\theta'\begin{bmatrix} 1 \\ 1 \end{bmatrix}(0,\tau)\right)^{\frac{1}{3}}}.$$

We now use the fact that $\left(\theta'\begin{bmatrix} 1 \\ 1 \end{bmatrix}(0,\tau)\right)^{\frac{1}{3}}$ is up to a constant multiple the same as $\theta\begin{bmatrix} \frac{1}{3} \\ 1 \end{bmatrix}(0,3\tau)$ and can therefore write our identity as

$$\begin{aligned}\frac{1}{2}\theta'\begin{bmatrix} 1 \\ 1 \end{bmatrix}(0,3\tau)&\left(\theta\begin{bmatrix} \frac{1}{3} \\ 1 \end{bmatrix}(3\zeta,3\tau) - \theta\begin{bmatrix} \frac{1}{3} \\ 1 \end{bmatrix}(-3\zeta,3\tau)\right)\\ &-\theta'\begin{bmatrix} \frac{1}{3} \\ 1 \end{bmatrix}(0,3\tau)\theta\begin{bmatrix} 1 \\ 1 \end{bmatrix}(3\zeta,3\tau)\\ &= M\theta\begin{bmatrix} \frac{1}{3} \\ 1 \end{bmatrix}(0,3\tau)\theta^3\begin{bmatrix} 1 \\ 1 \end{bmatrix}(\zeta,\tau),\end{aligned}$$

where once again M is a constant which we shall compute.

We set $\zeta = \frac{1}{2}$ in the last identity and find that $M = \frac{\pi\imath}{3}$ so that our fundamental identity in this case is

$$\begin{aligned}\frac{1}{2}\theta'\begin{bmatrix} 1 \\ 1 \end{bmatrix}(0,3\tau)&\left(\theta\begin{bmatrix} \frac{1}{3} \\ 1 \end{bmatrix}(3\zeta,3\tau) - \theta\begin{bmatrix} \frac{1}{3} \\ 1 \end{bmatrix}(-3\zeta,3\tau)\right)\\ &-\theta'\begin{bmatrix} \frac{1}{3} \\ 1 \end{bmatrix}(0,3\tau)\theta\begin{bmatrix} 1 \\ 1 \end{bmatrix}(3\zeta,3\tau)\\ &= \frac{\pi\imath}{3}\theta\begin{bmatrix} \frac{1}{3} \\ 1 \end{bmatrix}(0,3\tau)\theta^3\begin{bmatrix} 1 \\ 1 \end{bmatrix}(\zeta,\tau).\end{aligned}$$

We rewrite this identity in the coordinates $x = \exp(2\pi\imath\tau)$, $z = \exp(2\pi\imath\zeta)$ and use in addition the Jacobi triple product formula to obtain

$$\begin{aligned}3\sum_{n=0}^{\infty}(-1)^n(2n+1)x^{\frac{3(n^2+n)}{2}}&\left[\sum_{n=-\infty}^{\infty}(-1)^n x^{\frac{3n^2-n}{2}}z^{3n} - \sum_{n=-\infty}^{\infty}(-1)^n x^{\frac{3n^2+n}{2}}z^{3n+1}\right]\\ &+\sum_{n=-\infty}^{\infty}(-1)^n(6n+1)x^{\frac{3n^2+n}{2}}\sum_{n=-\infty}^{\infty}(-1)^n x^{\frac{3(n^2+n)}{2}}z^{3n+2}\end{aligned}$$

$$= \frac{(z-1)^3}{z}\prod_{n=1}^{\infty}(1-x^n)^4\left(1-\frac{x^n}{z}\right)^3(1-x^n z)^3. \tag{4.2}$$

We view the above identity as one whose depth is of the same order of magnitude as of the septuple product identity derived earlier.

2. Uniformization considerations

In Chapter 3 we derived two uniformizations of the Riemann surface $\mathbb{H}^2/\Gamma(3)$. These uniformizations gave rise to cubic theta identities given by Theorem 3.12. Similar results were obtained for $k = 4$ and 5 since uniformizations in these cases were easy to obtain and, as explained above, uniformizations give rise to identities. We do not repeat these calculations but ask the reader to recall this material.

3. Elliptic functions as quotients of N-th order theta functions

3.1. The Jacobi quartic and derivative formula revisited. In our discussion of N-th order theta functions we saw that quotients of functions in the same linear space gave rise to elliptic functions. We recall $\mathcal{F}_3\begin{bmatrix}1\\1\end{bmatrix}$, the space of third order functions with characteristic $\begin{bmatrix}1\\1\end{bmatrix}$. It is clear that the two functions

$$\theta^3\begin{bmatrix}1\\1\end{bmatrix}(z,\tau) \text{ and } \theta\begin{bmatrix}0\\0\end{bmatrix}(z,\tau)\theta\begin{bmatrix}0\\1\end{bmatrix}(z,\tau)\theta\begin{bmatrix}1\\0\end{bmatrix}(z,\tau)$$

are both in the space $\mathcal{F}_3\begin{bmatrix}1\\1\end{bmatrix}$, so that their quotient

$$f(z) = \frac{\theta^3\begin{bmatrix}1\\1\end{bmatrix}(z,\tau)}{\theta\begin{bmatrix}0\\0\end{bmatrix}(z,\tau)\theta\begin{bmatrix}0\\1\end{bmatrix}(z,\tau)\theta\begin{bmatrix}1\\0\end{bmatrix}(z,\tau)}$$

is an elliptic function with poles at the points $\frac{1}{2}, \frac{\tau}{2}$ and $\frac{1+\tau}{2}$ and a triple zero at the origin. The residue theorem gives

$$\operatorname{Res}_{\frac{1}{2}} f + \operatorname{Res}_{\frac{\tau}{2}} f + \operatorname{Res}_{\frac{1+\tau}{2}} f = 0.$$

Computing the residues in question yields

$$\operatorname{Res}_{\frac{1}{2}} f = \frac{\theta^3\begin{bmatrix}1\\1\end{bmatrix}(\frac{1}{2},\tau)}{\theta\begin{bmatrix}0\\0\end{bmatrix}(\frac{1}{2},\tau)\theta\begin{bmatrix}0\\1\end{bmatrix}(\frac{1}{2},\tau)\theta'\begin{bmatrix}1\\0\end{bmatrix}(\frac{1}{2},\tau)}$$

$$= \frac{\theta^3\begin{bmatrix}1\\2\end{bmatrix}(0,\tau)}{\theta\begin{bmatrix}0\\1\end{bmatrix}(0,\tau)\theta\begin{bmatrix}0\\2\end{bmatrix}(0,\tau)\theta'\begin{bmatrix}1\\1\end{bmatrix}(0,\tau)}.$$

The simple form of the denominator in the first quotient is due to the fact that $\theta\begin{bmatrix}1\\0\end{bmatrix}(\cdot,\tau)$ vanishes at $\frac{1}{2}$; the second equality is obtained from the first by replacing values of theta functions by the corresponding theta constants.

In the same fashion we obtain

$$\operatorname{Res}_{\frac{\tau}{2}} f = \frac{\theta^3\begin{bmatrix}1\\1\end{bmatrix}(\frac{\tau}{2},\tau)}{\theta\begin{bmatrix}0\\0\end{bmatrix}(\frac{\tau}{2},\tau)\theta'\begin{bmatrix}0\\1\end{bmatrix}(\frac{\tau}{2},\tau)\theta\begin{bmatrix}1\\0\end{bmatrix}(\frac{\tau}{2},\tau)}$$

$$= \frac{-\theta^3\begin{bmatrix}2\\1\end{bmatrix}(0,\tau)}{\theta\begin{bmatrix}1\\0\end{bmatrix}(0,\tau)\theta\begin{bmatrix}2\\0\end{bmatrix}(0,\tau)\theta'\begin{bmatrix}1\\1\end{bmatrix}(0,\tau)}$$

and

$$\operatorname{Res}_{\frac{1+\tau}{2}} f = \frac{\theta^3\begin{bmatrix}1\\1\end{bmatrix}(\frac{1+\tau}{2},\tau)}{\theta'\begin{bmatrix}0\\0\end{bmatrix}(\frac{1+\tau}{2},\tau)\theta\begin{bmatrix}0\\1\end{bmatrix}(\frac{1+\tau}{2},\tau)\theta\begin{bmatrix}1\\0\end{bmatrix}(\frac{1+\tau}{2},\tau)}$$

$$= \frac{-\theta^3\begin{bmatrix}2\\2\end{bmatrix}(0,\tau)}{\theta\begin{bmatrix}1\\2\end{bmatrix}(0,\tau)\theta\begin{bmatrix}2\\1\end{bmatrix}(0,\tau)\theta'\begin{bmatrix}1\\1\end{bmatrix}(0,\tau)}.$$

The sum of the last three residues equals zero, and (2.9) of Chapter 2 gives the Jacobi quartic identity.

It is a rather amusing fact that the Jacobi derivative formula is also a consequence of the same idea using almost the same function. Rather than the function f employed above, we consider its reciprocal, $g = \frac{1}{f}$; again, an elliptic function. It has a third order pole at the origin and is finite elsewhere (on the torus). The fact that the residue of g at the origin is zero gives the equation

$$\frac{\theta'''\begin{bmatrix}1\\1\end{bmatrix}(0,\tau)}{\theta'\begin{bmatrix}1\\1\end{bmatrix}(0,\tau)} = \frac{\theta''\begin{bmatrix}0\\0\end{bmatrix}(0,\tau)}{\theta\begin{bmatrix}0\\0\end{bmatrix}(0,\tau)} + \frac{\theta''\begin{bmatrix}0\\1\end{bmatrix}(0,\tau)}{\theta\begin{bmatrix}0\\1\end{bmatrix}(0,\tau)} + \frac{\theta''\begin{bmatrix}1\\0\end{bmatrix}(0,\tau)}{\theta\begin{bmatrix}1\\0\end{bmatrix}(0,\tau)}.$$

The above equation is simply equation (2.34) of Chapter 2 previously derived, and from it we obtain as before Corollary 5.8 of that chapter.

3.2. More identities - revisited. The material in the last subsection was a rederivation of the two most classical of the theta identities to illustrate the method. Using N-th order theta functions and linear algebra we can recover many previously established results and more importantly prove new ones. We present them as

Exercise 3.1. This exercise provides an alternate way to obtain some of our previous results and leads to new θ-identities. The basic tool is the finite dimensionality of spaces of N-th order theta functions.

1. The 4 entire functions

$$\theta^2\begin{bmatrix}0\\0\end{bmatrix}(\cdot,\tau),\ \theta^2\begin{bmatrix}0\\1\end{bmatrix}(\cdot,\tau),\ \theta^2\begin{bmatrix}1\\0\end{bmatrix}(\cdot,\tau),\ \theta^2\begin{bmatrix}1\\1\end{bmatrix}(\cdot,\tau)$$

belong to the 2-dimensional vector space $\mathcal{F}_2\begin{bmatrix}0\\0\end{bmatrix}$. Any 2 distinct functions in this set are linearly independent because they have distinct zero sets. Any 3 are linearly dependent. Derive these linear dependencies. For example, we conclude that there are nonzero constants c_1, c_2 and c_3, such that for all $z \in \mathbb{C}$ and all $\tau \in \mathbb{H}^2$,

$$c_1(\tau)\,\theta^2\begin{bmatrix}0\\0\end{bmatrix}(z,\tau)+c_2(\tau)\,\theta^2\begin{bmatrix}0\\1\end{bmatrix}(z,\tau)+c_3(\tau)\,\theta^2\begin{bmatrix}1\\0\end{bmatrix}(z,\tau)=0.$$

Evaluating the last expression at $z=\frac{1}{2}$, $z=\frac{\tau}{2}$ and $z=\frac{1+\tau}{2}$ computes the constants. We obtain the identity

(4.3)

$$\theta^2\begin{bmatrix}0\\0\end{bmatrix}\theta^2\begin{bmatrix}0\\0\end{bmatrix}(z,\tau)=\theta^2\begin{bmatrix}0\\1\end{bmatrix}\theta^2\begin{bmatrix}0\\1\end{bmatrix}(z,\tau)+\theta^2\begin{bmatrix}1\\0\end{bmatrix}\theta^2\begin{bmatrix}1\\0\end{bmatrix}(z,\tau),$$

$$\text{for all } z\in\mathbb{C}, \text{ and all } \tau\in\mathbb{H}^2.$$

Setting $z=0$, we obtain once again the Jacobi quartic identity. Similarly we can establish by this method

(4.4)

$$\theta^2\begin{bmatrix}0\\0\end{bmatrix}\theta^2\begin{bmatrix}1\\1\end{bmatrix}(z,\tau)=\theta^2\begin{bmatrix}1\\0\end{bmatrix}\theta^2\begin{bmatrix}0\\1\end{bmatrix}(z,\tau)-\theta^2\begin{bmatrix}0\\1\end{bmatrix}\theta^2\begin{bmatrix}1\\0\end{bmatrix}(z,\tau).$$

2. The 4 entire functions

$$\theta^2\begin{bmatrix}0\\\frac{\epsilon'}{2}\end{bmatrix}(\cdot,\tau),\ \theta^2\begin{bmatrix}1\\\frac{\epsilon'}{2}\end{bmatrix}(\cdot,\tau),$$

$$\theta\begin{bmatrix}0\\0\end{bmatrix}(\cdot,\tau)\,\theta\begin{bmatrix}0\\\epsilon'\end{bmatrix}(\cdot,\tau),\ \theta\begin{bmatrix}1\\0\end{bmatrix}(\cdot,\tau)\,\theta\begin{bmatrix}1\\\epsilon'\end{bmatrix}(\cdot,\tau)$$

belong to the 2-dimensional vector space $\mathcal{F}_2\begin{bmatrix}0\\\epsilon'\end{bmatrix}$. Once again, any 2 distinct functions in this set are linearly independent because they have distinct zero sets. Any 3 are linearly dependent. Derive these linear dependencies. Working with the first 3 functions, we obtain

$$\theta^2\begin{bmatrix}0\\1+\frac{\epsilon'}{2}\end{bmatrix}\theta^2\begin{bmatrix}0\\\frac{\epsilon'}{2}\end{bmatrix}(z,\tau)=\theta^2\begin{bmatrix}1\\1+\frac{\epsilon'}{2}\end{bmatrix}\theta^2\begin{bmatrix}1\\\frac{\epsilon'}{2}\end{bmatrix}(z,\tau)$$

$$+\theta^2\begin{bmatrix}0\\1\end{bmatrix}\theta\begin{bmatrix}0\\0\end{bmatrix}(z,\tau)\,\theta\begin{bmatrix}0\\\epsilon'\end{bmatrix}(z,\tau),\text{ for all } z\in\mathbb{C},\text{ and for all }\tau\in\mathbb{H}^2,$$

leading to the identity

$$\theta^2\begin{bmatrix}0\\1+\frac{\epsilon'}{2}\end{bmatrix}\theta^2\begin{bmatrix}0\\\frac{\epsilon'}{2}\end{bmatrix}=\theta^2\begin{bmatrix}1\\1+\frac{\epsilon'}{2}\end{bmatrix}\theta^2\begin{bmatrix}1\\\frac{\epsilon'}{2}\end{bmatrix}+\theta^2\begin{bmatrix}0\\1\end{bmatrix}\theta\begin{bmatrix}0\\0\end{bmatrix}\theta\begin{bmatrix}0\\\epsilon'\end{bmatrix}$$

among constants. For $\epsilon'=1$, this identity reduces to

$$\theta^4\begin{bmatrix}0\\\frac{1}{2}\end{bmatrix}=\theta^4\begin{bmatrix}1\\\frac{1}{2}\end{bmatrix}+\theta^3\begin{bmatrix}0\\1\end{bmatrix}\theta\begin{bmatrix}0\\0\end{bmatrix}. \tag{4.5}$$

3. The 4 entire functions

$$\theta^3\begin{bmatrix}\frac{\epsilon}{3}\\\frac{\epsilon'}{3}\end{bmatrix}(\cdot,\tau),\ \theta^3\begin{bmatrix}\frac{\epsilon}{3}+\frac{2}{3}\\\frac{\epsilon'}{3}\end{bmatrix}(\cdot,\tau),\ \theta^3\begin{bmatrix}\frac{\epsilon}{3}+\frac{4}{3}\\\frac{\epsilon'}{3}\end{bmatrix}(\cdot,\tau),$$

$$\theta\begin{bmatrix}\frac{\epsilon}{3}\\\frac{\epsilon'}{3}\end{bmatrix}(\cdot,\tau)\,\theta\begin{bmatrix}\frac{\epsilon}{3}+\frac{2}{3}\\\frac{\epsilon'}{3}\end{bmatrix}(\cdot,\tau)\,\theta\begin{bmatrix}\frac{\epsilon}{3}+\frac{4}{3}\\\frac{\epsilon'}{3}\end{bmatrix}(\cdot,\tau)$$

belong to the 3-dimensional vector space $\mathcal{F}_3\begin{bmatrix}\epsilon\\\epsilon'\end{bmatrix}$. Derive the linear dependence between them.

4. Obtain identities similar to the ones derived above for arbitrary N.

3.3. More identities - new results. We proceed to describe how the residue theorem leads to some new results as well as some additional classical theorems.

In Chapter 3, §3.3, we proved, using uniformization theory, an identity among the cubes of theta constants related to the Riemann surface $\mathbb{H}^2/\Gamma(3)$. Specifically we proved Theorem 3.12. We re-prove this theorem and generalize it considerably.

Theorem 3.2. *For each $N\in\mathbb{Z}^+$, equation (4.1) holds.*

Proof. The case $N=1$ is trivial and the case $N=3$ was already established. We consider the function

$$\frac{\theta^N\begin{bmatrix}1\\\frac{1}{N}\end{bmatrix}(z,\tau)}{\theta\begin{bmatrix}\frac{1}{N}\\1\end{bmatrix}(Nz,N\tau)}.$$

Recalling Lemma 7.2 and Proposition 7.11 of Chapter 2, we see that the above function is elliptic with periods 1 and τ whose inequivalent poles are at the N points

$$\frac{N-1}{N}\frac{\tau}{2},\ \frac{N-1}{N}\frac{\tau}{2}+\frac{1}{N},\ \ldots,\ \frac{N-1}{N}\frac{\tau}{2}+\frac{N-1}{N}.$$

The residue theorem gives

$$\sum_{l=0}^{N-1}\frac{\theta^N\begin{bmatrix}1\\ \frac{1}{N}\end{bmatrix}\left(\frac{N-1}{N}\frac{\tau}{2}+\frac{l}{N},\tau\right)}{\theta'\begin{bmatrix}\frac{1}{N}\\ 1\end{bmatrix}\left(\frac{N-1}{N}\frac{N\tau}{2}+l,N\tau\right)}=0.$$

An appeal to equations (2.8), (2.9) and the fact that $\theta'\begin{bmatrix}1\\ 1\end{bmatrix}$ never vanishes complete the proof.

The above theorem, as already mentioned, is a rather far reaching generalization of the cubic identity previously obtained as a consequence of a uniformization result. It is fairly clear that the method used to prove the last theorem is a more powerful tool than uniformization for obtaining theta identities.

We consider another application. Let $\begin{bmatrix}a\\ b\end{bmatrix}$ and $\begin{bmatrix}c\\ d\end{bmatrix}$ be arbitrary vectors in $\mathbb{R}^2$. Define

$$f(z)=\frac{\theta\begin{bmatrix}a\\ b\end{bmatrix}(z,\tau)\theta\begin{bmatrix}-a\\ -b\end{bmatrix}(z,\tau)\theta\begin{bmatrix}c\\ d\end{bmatrix}(z,\tau)\theta\begin{bmatrix}-c\\ -d\end{bmatrix}(z,\tau)}{\theta\begin{bmatrix}0\\ 0\end{bmatrix}(z,\tau)\theta\begin{bmatrix}0\\ 1\end{bmatrix}(z,\tau)\theta\begin{bmatrix}1\\ 0\end{bmatrix}(z,\tau)\theta\begin{bmatrix}1\\ 1\end{bmatrix}(z,\tau)}.$$

The fundamental properties of theta functions imply that f is an elliptic function with periods 1 and τ whose poles are at the origin and at the three points of order 2 of the torus, namely, the points 0, $\frac{1}{2}$, $\frac{\tau}{2}$ and $\frac{1+\tau}{2}$. The residue theorem yields

$$\begin{aligned}\theta^2\begin{bmatrix}a\\ b\end{bmatrix}\theta^2\begin{bmatrix}c\\ d\end{bmatrix}&-\theta\begin{bmatrix}a+1\\ b\end{bmatrix}\theta\begin{bmatrix}-a+1\\ -b\end{bmatrix}\theta\begin{bmatrix}c+1\\ d\end{bmatrix}\theta\begin{bmatrix}-c+1\\ -d\end{bmatrix}\\ &-\theta\begin{bmatrix}a\\ b+1\end{bmatrix}\theta\begin{bmatrix}-a\\ -b+1\end{bmatrix}\theta\begin{bmatrix}c\\ d+1\end{bmatrix}\theta\begin{bmatrix}-c\\ -d+1\end{bmatrix}\\ &+\theta\begin{bmatrix}a+1\\ b+1\end{bmatrix}\theta\begin{bmatrix}-a+1\\ -b+1\end{bmatrix}\theta\begin{bmatrix}c+1\\ d+1\end{bmatrix}\theta\begin{bmatrix}-c+1\\ -d+1\end{bmatrix}=0.\end{aligned}$$

If we set $(a,b)=(c,d)=(0,0)$, we obtain once again the classical Jacobi quartic identity. Arbitrary z and $w\in\mathbb{C}$ can be written as $z=a\frac{\tau}{2}+\frac{b}{2}$ and $w=c\frac{\tau}{2}+\frac{d}{2}$ with a, b, c and $d\in\mathbb{R}$. Thus the previous identity can be recast as a theta function identity.

Theorem 3.3. *For all z and $w \in \mathbb{C}$,*

$$\theta^2\begin{bmatrix}0\\0\end{bmatrix}(z,\tau)\theta^2\begin{bmatrix}0\\0\end{bmatrix}(w,\tau)-\theta^2\begin{bmatrix}1\\0\end{bmatrix}(z,\tau)\theta^2\begin{bmatrix}1\\0\end{bmatrix}(w,\tau)$$

$$-\theta^2\begin{bmatrix}0\\1\end{bmatrix}(z,\tau)\theta^2\begin{bmatrix}0\\1\end{bmatrix}(w,\tau)+\theta^2\begin{bmatrix}1\\1\end{bmatrix}(z,\tau)\theta^2\begin{bmatrix}1\\1\end{bmatrix}(w,\tau)=0.$$

Remark 3.4. Recall equation (2.33) that we derived in the process of proving Theorem 5.3 of Chapter 2. We recover that formula as the special case $w = 0$ of the last theorem. Many other identities can be derived by choosing special values for z and w. We leave this task to the reader.

3.4. More first order applications. In analogy to Theorem 7.1 of Chapter 2 we establish the following result.

Theorem 3.5. *For $i = 1, \ldots, N$, let ϵ_i and $\epsilon_i' \in \mathbb{R}$ satisfy*

$$\epsilon_i \neq \epsilon_j \text{ and } \epsilon_i' \neq \epsilon_j' \text{ for } i \neq j,\ \sum_{i=1}^N \epsilon_i = \epsilon,\ \sum_{i=1}^N \epsilon_i' = \epsilon'.$$

Then for all $f \in \mathcal{F}_N\begin{bmatrix}\epsilon\\\epsilon'\end{bmatrix}$,

$$\sum_{i=1}^N\left(\exp \pi\imath\left\{\frac{1}{4}(1-\epsilon_i)^2N\tau+\frac{1}{2}(1-\epsilon_i)(\epsilon'+N(1-\epsilon_i))\right\}\right)\tag{4.6}$$

$$\times\frac{f\left(\frac{1-\epsilon_i}{2}\tau+\frac{1-\epsilon_i'}{2}\right)}{\prod_{j=1,\ \ldots,\ N,\ j\neq i}\theta\begin{bmatrix}1+\epsilon_j-\epsilon_i\\1+\epsilon_j'-\epsilon_i'\end{bmatrix}}=0.$$

Proof. The theorem is obtained by applying the residue theorem to the elliptic function

$$\frac{f}{\prod_{j=1}^N\theta\begin{bmatrix}\epsilon_j\\\epsilon_j'\end{bmatrix}(\cdot,\tau)}.$$

Corollary 3.6. *Under the hypothesis of the theorem,*

$$\sum_{i=1}^N\frac{\theta^N\begin{bmatrix}1+\frac{\epsilon}{N}-\epsilon_i\\1+\frac{\epsilon'}{N}-\epsilon_i'\end{bmatrix}}{\prod_{j=1,\ \ldots,\ N,\ j\neq i}\theta\begin{bmatrix}1+\epsilon_j-\epsilon_i\\1+\epsilon_j'-\epsilon_i'\end{bmatrix}}=0.$$

Proof. We use

$$f = \theta^N \begin{bmatrix} \frac{\epsilon}{N} \\ \frac{\epsilon'}{N} \end{bmatrix} (\cdot, \tau).$$

As an application of the last corollary, we take the special case $N = 3$,

$$\epsilon_1 = \frac{\epsilon}{6},\ \epsilon_2 = \frac{\epsilon}{3},\ \epsilon_3 = \frac{\epsilon}{2},\ \epsilon_1' = \frac{\epsilon'}{6},\ \epsilon_2' = \frac{\epsilon'}{3},\ \epsilon_3' = \frac{\epsilon'}{2},$$

and obtain the identity

$$\frac{\theta^2 \begin{bmatrix} 1+\frac{\epsilon}{6} \\ 1+\frac{\epsilon'}{6} \end{bmatrix}}{\theta \begin{bmatrix} 1+\frac{\epsilon}{3} \\ 1+\frac{\epsilon'}{3} \end{bmatrix}} + \frac{\theta^2 \begin{bmatrix} 1-\frac{\epsilon}{6} \\ 1-\frac{\epsilon'}{6} \end{bmatrix}}{\theta \begin{bmatrix} 1-\frac{\epsilon}{3} \\ 1-\frac{\epsilon'}{3} \end{bmatrix}} = 0,$$

and the case $N = 4$,

$$\epsilon_1 = \frac{\epsilon}{16},\ \epsilon_2 = \frac{\epsilon}{8},\ \epsilon_3 = \frac{\epsilon}{4},\ \epsilon_4 = \frac{9\epsilon}{16},\ \epsilon_1' = \frac{\epsilon'}{16},\ \epsilon_2' = \frac{\epsilon'}{8},\ \epsilon_3' = \frac{\epsilon'}{4},\ \epsilon_4' = \frac{9\epsilon'}{16},$$

which leads to

$$\frac{\theta^3 \begin{bmatrix} 1+\frac{3\epsilon}{16} \\ 1+\frac{3\epsilon'}{16} \end{bmatrix}}{\theta \begin{bmatrix} 1+\frac{\epsilon}{16} \\ 1+\frac{\epsilon'}{16} \end{bmatrix} \theta \begin{bmatrix} 1+\frac{\epsilon}{2} \\ 1+\frac{\epsilon'}{2} \end{bmatrix}} + \frac{\theta^3 \begin{bmatrix} 1+\frac{\epsilon}{8} \\ 1+\frac{\epsilon'}{8} \end{bmatrix}}{\theta \begin{bmatrix} 1-\frac{\epsilon}{16} \\ 1-\frac{\epsilon'}{16} \end{bmatrix} \theta \begin{bmatrix} 1+\frac{7\epsilon}{16} \\ 1+\frac{7\epsilon'}{16} \end{bmatrix}}$$

$$+ \frac{\theta^3 \begin{bmatrix} 1-\frac{5\epsilon}{16} \\ 1-\frac{5\epsilon'}{16} \end{bmatrix}}{\theta \begin{bmatrix} 1-\frac{\epsilon}{2} \\ 1-\frac{\epsilon'}{2} \end{bmatrix} \theta \begin{bmatrix} 1-\frac{7\epsilon}{16} \\ 1-\frac{7\epsilon'}{16} \end{bmatrix}} = 0.$$

As a further indication of how the results of Chapter 2 can lead to identities, we derive an immediate consequence of Proposition 7.11 of that chapter.

Theorem 3.7. *For all n and $m \in \mathbb{R}$ and all $N \in \mathbb{Z}^+$,*

$$\frac{\theta \begin{bmatrix} \epsilon+m \\ \epsilon'+Nn \end{bmatrix} (0, N\tau)}{\prod_{l=0}^{N-1} \theta \begin{bmatrix} \epsilon+m \\ 1+n+\frac{\epsilon'-(1+2l)}{N} \end{bmatrix} (0,\tau)} = \frac{\theta \begin{bmatrix} \epsilon \\ \epsilon' \end{bmatrix} (0, N\tau)}{\prod_{l=0}^{N-1} \theta \begin{bmatrix} \epsilon \\ 1+\frac{\epsilon'-(1+2l)}{N} \end{bmatrix} (0,\tau)}.$$

Proof. Proposition 7.11 of Chapter 2 tells us that the entire function

$$z \mapsto \frac{\theta \begin{bmatrix} \frac{\epsilon}{N} \\ \epsilon' \end{bmatrix} (Nz, N\tau)}{\prod_{l=0}^{N-1} \theta \begin{bmatrix} \frac{\epsilon}{N} \\ 1+\frac{\epsilon'-(1+2l)}{N} \end{bmatrix} (z,\tau)}$$

is constant. Hence its value at $z = \frac{n}{2} + \frac{m}{2}\tau$ coincides with its value at 0. We thus obtain an identity involving theta functions. Using (2.14), we reduce it to an identity involving only theta constants. Finally, we replace $\frac{\epsilon}{N}$ by ϵ.

Setting $m = 0$ and $n = \frac{2\alpha}{N}$ with $\alpha \in \mathbb{Z}$ in the theorem, one obtains a formula for $\exp \pi \imath \epsilon \alpha$, which the reader is invited to derive.

Theorem 7.16 of Chapter 2 has many consequences which we now begin to explore. As before $N \geq 3$ is an odd integer. If a function in $\mathcal{E}_N \begin{bmatrix} 1 \\ 1 \end{bmatrix}$ is divided by the product of the three even theta functions with integer characteristics, we obtain an entire function. We start with the case $N = 3$. In this situation $\dim \mathcal{E}_3 \begin{bmatrix} 1 \\ 1 \end{bmatrix} = 1$, and we have identified the location of the three zeros for every nontrivial function in this space.

Theorem 3.8. *We have the identity in $z \in \mathbb{C}$ and $\tau \in \mathbb{H}^2$,*

$$\theta \begin{bmatrix} \frac{1}{3} \\ 1 \end{bmatrix} (3z, 3\tau) + \theta \begin{bmatrix} \frac{1}{3} \\ 1 \end{bmatrix} (-3z, 3\tau)$$

$$= c(\tau)\, \theta \begin{bmatrix} 0 \\ 0 \end{bmatrix} (z, \tau)\, \theta \begin{bmatrix} 0 \\ 1 \end{bmatrix} (z, \tau)\, \theta \begin{bmatrix} 1 \\ 0 \end{bmatrix} (z, \tau),$$

where

$$c(\tau) = 2^{\frac{2}{3}} \exp\left\{\frac{\pi \imath}{6}\right\} \theta^{-\frac{2}{3}} \begin{bmatrix} 0 \\ 0 \end{bmatrix} (0, \tau)\, \theta^{-\frac{2}{3}} \begin{bmatrix} 0 \\ 1 \end{bmatrix} (0, \tau)\, \theta^{-\frac{2}{3}} \begin{bmatrix} 1 \\ 0 \end{bmatrix} (0, \tau).$$

Appropriate choices of cube roots of the theta constants must, of course, be used in the last of the above formulae.

Proof. The first statement (including the fact that $c(\tau)$ is never zero) follows from the remarks preceding the statement of the theorem. It also follows from these remarks that $c(\tau)$ can be evaluated by setting $z = 0$; hence

$$c(\tau) = \frac{2\, \theta \begin{bmatrix} \frac{1}{3} \\ 1 \end{bmatrix} (0, 3\tau)}{\theta \begin{bmatrix} 0 \\ 0 \end{bmatrix} (0, \tau)\, \theta \begin{bmatrix} 0 \\ 1 \end{bmatrix} (0, \tau)\, \theta \begin{bmatrix} 1 \\ 0 \end{bmatrix} (0, \tau)}.$$

The above remark was in fact the content of Theorem 8.2 of Chapter 2. The calculation of alternate formulae for the function $c(\tau)$ is more complicated. We differentiate twice with respect to z ($\frac{\partial}{\partial z} = '$) the equality in the statement of the theorem, set $z = 0$, and obtain

$$\frac{18}{c(\tau)} \theta'' \begin{bmatrix} \frac{1}{3} \\ 1 \end{bmatrix} (0, 3\tau) = \theta'' \begin{bmatrix} 0 \\ 0 \end{bmatrix} (0, \tau)\, \theta \begin{bmatrix} 0 \\ 1 \end{bmatrix} (0, \tau)\, \theta \begin{bmatrix} 1 \\ 0 \end{bmatrix} (0, \tau)$$

$$+ \theta'' \begin{bmatrix} 0 \\ 1 \end{bmatrix} (0, \tau)\, \theta \begin{bmatrix} 0 \\ 0 \end{bmatrix} (0, \tau)\, \theta \begin{bmatrix} 1 \\ 0 \end{bmatrix} (0, \tau)$$

$$+\theta''\left[\begin{array}{c}1\\0\end{array}\right](0,\tau)\,\theta\left[\begin{array}{c}0\\0\end{array}\right](0,\tau)\,\theta\left[\begin{array}{c}0\\1\end{array}\right](0,\tau).$$

The first derivatives do not appear on the right hand side of the above equation because every odd function vanishes at the origin. It immediately follows that

$$9\,\frac{\theta''\left[\begin{array}{c}\frac{1}{3}\\1\end{array}\right](0,3\tau)}{\theta\left[\begin{array}{c}\frac{1}{3}\\1\end{array}\right](0,3\tau)}=\frac{\theta''\left[\begin{array}{c}0\\0\end{array}\right](0,\tau)}{\theta\left[\begin{array}{c}0\\0\end{array}\right](0,\tau)}+\frac{\theta''\left[\begin{array}{c}0\\1\end{array}\right](0,\tau)}{\theta\left[\begin{array}{c}0\\1\end{array}\right](0,\tau)}+\frac{\theta''\left[\begin{array}{c}1\\0\end{array}\right](0,\tau)}{\theta\left[\begin{array}{c}1\\0\end{array}\right](0,\tau)}.$$

We now use the heat equation to transform the last equation to ($\frac{\partial}{\partial\tau}=\dot{}$)

$$3\,\frac{\dot\theta\left[\begin{array}{c}\frac{1}{3}\\1\end{array}\right](0,3\tau)}{\theta\left[\begin{array}{c}\frac{1}{3}\\1\end{array}\right](0,3\tau)}=\frac{\dot\theta\left[\begin{array}{c}0\\0\end{array}\right](0,\tau)}{\theta\left[\begin{array}{c}0\\0\end{array}\right](0,\tau)}+\frac{\dot\theta\left[\begin{array}{c}0\\1\end{array}\right](0,\tau)}{\theta\left[\begin{array}{c}0\\1\end{array}\right](0,\tau)}+\frac{\dot\theta\left[\begin{array}{c}1\\0\end{array}\right](0,\tau)}{\theta\left[\begin{array}{c}1\\0\end{array}\right](0,\tau)}.$$

Integrating (logarithmically) we obtain

$$\theta^3\left[\begin{array}{c}\frac{1}{3}\\1\end{array}\right](0,3\tau)=c\,\theta\left[\begin{array}{c}0\\0\end{array}\right](0,\tau)\,\theta\left[\begin{array}{c}0\\1\end{array}\right](0,\tau)\,\theta\left[\begin{array}{c}1\\0\end{array}\right](0,\tau),\tag{4.7}$$

where c is a constant of integration which is easily computed using (2.35), (2.36), (2.37) and (2.39); its value is $\frac{\imath}{2}$. As a consequence of Jacobi's derivative formula we see that

$$\theta^3\left[\begin{array}{c}\frac{1}{3}\\1\end{array}\right](0,3\tau)=\frac{-\imath}{2\pi}\theta'\left[\begin{array}{c}1\\1\end{array}\right](0,\tau).\tag{4.8}$$

In fact, we shall soon derive this identity again.

It is perhaps interesting to reinterpret the last result in terms of theta constants alone. By setting $z=\frac{m\tau+n}{2}$ we obtain

Corollary 3.9. *For all n and $m\in\mathbb{R}$ and all $\tau\in\mathbb{H}^2$,*

$$\theta\left[\begin{array}{c}\frac{1}{3}+m\\1+3n\end{array}\right](0,3\tau)+(\exp\{\pi\imath m\})\theta\left[\begin{array}{c}\frac{1}{3}-m\\1-3n\end{array}\right](0,3\tau)$$

$$=c(\tau)\,\theta\left[\begin{array}{c}m\\n\end{array}\right](0,\tau)\,\theta\left[\begin{array}{c}m\\1+n\end{array}\right](0,\tau)\,\theta\left[\begin{array}{c}1+m\\n\end{array}\right](0,\tau);$$

in particular, (for $n=0$)

$$\theta\left[\begin{array}{c}\frac{1}{3}+m\\1\end{array}\right](0,3\tau)+(\exp\{\pi\imath m\})\theta\left[\begin{array}{c}\frac{1}{3}-m\\1\end{array}\right](0,3\tau)$$

$$=c(\tau)\,\theta\left[\begin{array}{c}m\\0\end{array}\right](0,\tau)\,\theta\left[\begin{array}{c}m\\1\end{array}\right](0,\tau)\,\theta\left[\begin{array}{c}1+m\\0\end{array}\right](0,\tau),$$

and (for $n = \frac{2}{3}$)

$$\begin{aligned}&\left(\exp\left\{\pi\imath\left(m+\frac{1}{3}\right)\right\}\right)\theta\begin{bmatrix}\frac{1}{3}-m\\1\end{bmatrix}(0,3\tau)\\&+\left(\exp\left\{\pi\imath\left(2m-\frac{1}{3}\right)\right\}\right)\theta\begin{bmatrix}\frac{1}{3}-m\\1\end{bmatrix}(0,3\tau)\\=&\ c(\tau)\,\theta\begin{bmatrix}m\\\frac{2}{3}\end{bmatrix}(0,\tau)\,\theta\begin{bmatrix}m\\\frac{5}{3}\end{bmatrix}(0,\tau)\,\theta\begin{bmatrix}1+m\\\frac{2}{3}\end{bmatrix}(0,\tau).\end{aligned}$$

Hence also

$$\frac{\theta\begin{bmatrix}\frac{1}{3}+m\\1\end{bmatrix}(0,3\tau)+e^{\pi\imath m}\theta\begin{bmatrix}\frac{1}{3}-m\\1\end{bmatrix}(0,3\tau)}{e^{\pi\imath(m+\frac{1}{3})}\theta\begin{bmatrix}\frac{1}{3}+m\\1\end{bmatrix}(0,3\tau)+e^{\pi\imath(2m-\frac{1}{3})}\theta\begin{bmatrix}\frac{1}{3}-m\\1\end{bmatrix}(0,3\tau)}$$

$$=\frac{\theta\begin{bmatrix}m\\0\end{bmatrix}(0,\tau)\,\theta\begin{bmatrix}m\\1\end{bmatrix}(0,\tau)\,\theta\begin{bmatrix}1+m\\0\end{bmatrix}(0,\tau)}{\theta\begin{bmatrix}m\\\frac{2}{3}\end{bmatrix}(0,\tau)\,\theta\begin{bmatrix}m\\\frac{5}{3}\end{bmatrix}(0,\tau)\,\theta\begin{bmatrix}1+m\\\frac{2}{3}\end{bmatrix}(0,\tau)}.$$

Before leaving this aspect of the subject we consider the case $N = 5$. Here the result is a bit more complicated since we have more freedom in the choice of zeros. We can arbitrarily assign the location of one zero of a function in $\mathcal{E}_5\begin{bmatrix}1\\1\end{bmatrix}$. Let us place one zero at the origin. This of course implies that the function will necessarily have a double zero at the origin. An example of such a function is given by

$$\begin{aligned}f(z)=&\frac{\theta\begin{bmatrix}\frac{1}{5}\\1\end{bmatrix}(5z,5\tau)+\theta\begin{bmatrix}\frac{1}{5}\\1\end{bmatrix}(-5z,5\tau)}{\theta\begin{bmatrix}\frac{1}{5}\\1\end{bmatrix}(0,5\tau)}\\&-\frac{\theta\begin{bmatrix}\frac{3}{5}\\1\end{bmatrix}(5z,5\tau)+\theta\begin{bmatrix}\frac{3}{5}\\1\end{bmatrix}(-5z,5\tau)}{\theta\begin{bmatrix}\frac{3}{5}\\1\end{bmatrix}(0,5\tau)}.\end{aligned}$$

It is obvious that

$$f(z)=c(\tau)\theta\begin{bmatrix}0\\0\end{bmatrix}(z,\tau)\theta\begin{bmatrix}0\\1\end{bmatrix}(z,\tau)\theta\begin{bmatrix}1\\0\end{bmatrix}(z,\tau)\theta^2\begin{bmatrix}1\\1\end{bmatrix}(z,\tau),$$

for some nonzero constant $c(\tau)$.

Remark 3.10. These are the functions that gave rise to the septuple product identity proven at the end of the previous section.

Using the first definition of $f(z)$ we conclude that

$$f\left(\frac{1}{5}\right) = 2\left(\cos\frac{\pi}{5} - \cos\frac{3\pi}{5}\right) = \sqrt{5} = f\left(\frac{2}{5}\right),$$

$$f\left(\frac{3}{5}\right) = f\left(\frac{4}{5}\right) = -f\left(\frac{1}{5}\right).$$

Corollary 3.11. *For all $\tau \in \mathbb{H}^2$,*

$$\frac{\theta^2\begin{bmatrix}1\\\frac{3}{5}\end{bmatrix}(0,\tau)}{\theta^2\begin{bmatrix}1\\\frac{1}{5}\end{bmatrix}(0,\tau)} = \frac{\theta\begin{bmatrix}0\\\frac{4}{5}\end{bmatrix}(0,\tau)\theta\begin{bmatrix}0\\\frac{1}{5}\end{bmatrix}(0,\tau)\theta\begin{bmatrix}1\\\frac{4}{5}\end{bmatrix}(0,\tau)}{\theta\begin{bmatrix}0\\\frac{2}{5}\end{bmatrix}(0,\tau)\theta\begin{bmatrix}0\\\frac{3}{5}\end{bmatrix}(0,\tau)\theta\begin{bmatrix}1\\\frac{2}{5}\end{bmatrix}(0,\tau)}.$$

Proof. It follows from the second expression given above for $f(z)$ that

$$f\left(\frac{1}{5}\right) = c(\tau)\theta\begin{bmatrix}0\\\frac{2}{5}\end{bmatrix}(0,\tau)\theta\begin{bmatrix}0\\\frac{3}{5}\end{bmatrix}(0,\tau)\theta\begin{bmatrix}1\\\frac{2}{5}\end{bmatrix}(0,\tau)\theta^2\begin{bmatrix}1\\\frac{3}{5}\end{bmatrix}(0,\tau)$$

and

$$f\left(\frac{2}{5}\right) = c(\tau)\theta\begin{bmatrix}0\\\frac{4}{5}\end{bmatrix}(0,\tau)\theta\begin{bmatrix}0\\\frac{1}{5}\end{bmatrix}(0,\tau)\theta\begin{bmatrix}1\\\frac{4}{5}\end{bmatrix}(0,\tau)\theta^2\begin{bmatrix}1\\\frac{1}{5}\end{bmatrix}(0,\tau).$$

The equality of these two expressions gives the result.

3.5. Some modular equations. In this subsection we use the residue theorem to derive some *modular* equations. For the current purposes a modular equation may be taken to mean a relation involving theta constants evaluated at τ and $k\tau$ with $k \in \mathbb{Z}^+$.

The function

$$f(z) = \frac{\theta\begin{bmatrix}1\\1\end{bmatrix}(3z,3\tau)}{\theta\begin{bmatrix}0\\0\end{bmatrix}(z,\tau)\theta\begin{bmatrix}0\\1\end{bmatrix}(z,\tau)\theta\begin{bmatrix}1\\0\end{bmatrix}(z,\tau)}$$

is elliptic with period τ and poles at the points $\frac{1}{2}$, $\frac{\tau}{2}$ and $\frac{1+\tau}{2}$. As we have seen previously,

$$\operatorname{Res}_{\frac{1}{2}} f = \frac{\theta\begin{bmatrix}1\\1\end{bmatrix}(\frac{3}{2},3\tau)}{\theta\begin{bmatrix}0\\0\end{bmatrix}(\frac{1}{2},\tau)\theta\begin{bmatrix}0\\1\end{bmatrix}(\frac{1}{2},\tau)\theta'\begin{bmatrix}1\\0\end{bmatrix}(\frac{1}{2},\tau)}$$

$$= \frac{\theta\begin{bmatrix}1\\0\end{bmatrix}(0,3\tau)}{\theta\begin{bmatrix}0\\0\end{bmatrix}(0,\tau)\theta\begin{bmatrix}0\\1\end{bmatrix}(0,\tau)\theta'\begin{bmatrix}1\\1\end{bmatrix}(0,\tau)},$$

$$\operatorname{Res}_{\frac{\tau}{2}} f = \frac{\theta\begin{bmatrix}0\\1\end{bmatrix}(0,3\tau)}{\theta\begin{bmatrix}0\\0\end{bmatrix}(0,\tau)\theta\begin{bmatrix}1\\0\end{bmatrix}(0,\tau)\theta'\begin{bmatrix}1\\1\end{bmatrix}(0,\tau)}$$

and

$$\operatorname{Res}_{\frac{1+\tau}{2}} f = \frac{-\theta\begin{bmatrix}0\\0\end{bmatrix}(0,3\tau)}{\theta\begin{bmatrix}0\\1\end{bmatrix}(0,\tau)\theta\begin{bmatrix}1\\0\end{bmatrix}(0,\tau)\theta'\begin{bmatrix}1\\1\end{bmatrix}(0,\tau)}.$$

We have proven

Theorem 3.12. *For all $\tau \in \mathbb{H}^2$,*

$$\theta\begin{bmatrix}0\\0\end{bmatrix}(0,\tau)\theta\begin{bmatrix}0\\0\end{bmatrix}(0,3\tau) \tag{4.9}$$
$$= \theta\begin{bmatrix}0\\1\end{bmatrix}(0,\tau)\theta\begin{bmatrix}0\\1\end{bmatrix}(0,3\tau) + \theta\begin{bmatrix}1\\0\end{bmatrix}(0,\tau)\theta\begin{bmatrix}1\\0\end{bmatrix}(0,3\tau).$$

The above theorem can be called a cubic modular equation. As an exercise the reader is invited to derive the following quintic modular equation

$$\frac{\theta\begin{bmatrix}0\\0\end{bmatrix}(0,5\tau)}{\theta\begin{bmatrix}0\\0\end{bmatrix}(0,\tau)} - \frac{\theta\begin{bmatrix}0\\1\end{bmatrix}(0,5\tau)}{\theta\begin{bmatrix}0\\1\end{bmatrix}(0,\tau)} - \frac{\theta\begin{bmatrix}1\\0\end{bmatrix}(0,5\tau)}{\theta\begin{bmatrix}1\\0\end{bmatrix}(0,\tau)} \tag{4.10}$$
$$= -5\frac{\theta\begin{bmatrix}0\\0\end{bmatrix}(0,5\tau)\theta\begin{bmatrix}0\\1\end{bmatrix}(0,5\tau)\theta\begin{bmatrix}1\\0\end{bmatrix}(0,5\tau)}{\theta\begin{bmatrix}0\\0\end{bmatrix}(0,\tau)\theta\begin{bmatrix}0\\1\end{bmatrix}(0,\tau)\theta\begin{bmatrix}1\\0\end{bmatrix}(0,\tau)}$$

by applying the residue theorem to the elliptic function

$$\frac{\theta\begin{bmatrix}1\\1\end{bmatrix}(5z,5\tau)}{\theta\begin{bmatrix}0\\0\end{bmatrix}(z,\tau)\theta\begin{bmatrix}0\\1\end{bmatrix}(z,\tau)\theta\begin{bmatrix}1\\0\end{bmatrix}(z,\tau)\theta^2\begin{bmatrix}1\\1\end{bmatrix}(z,\tau)}.$$

We continue our excursion by deriving a modular equation for the prime 7 which is different from the ones discussed above in the sense that it involves square root irrationalities.

Theorem 3.13. *For every $\tau \in \mathbb{H}^2$ we have,*

$$\sqrt{\theta\begin{bmatrix}0\\0\end{bmatrix}(0,\tau)\theta\begin{bmatrix}0\\0\end{bmatrix}(0,7\tau)} \tag{4.11}$$

$$= \sqrt{\theta \begin{bmatrix} 0 \\ 1 \end{bmatrix} (0,\tau) \theta \begin{bmatrix} 0 \\ 1 \end{bmatrix} (0,7\tau)} + \sqrt{\theta \begin{bmatrix} 1 \\ 0 \end{bmatrix} (0,\tau) \theta \begin{bmatrix} 1 \\ 0 \end{bmatrix} (0,7\tau)}.$$

Proof. The proof consists of two parts. The first part uses the description of the zero set of a theta function which allows us to conclude that the 7-th order theta function

$$f(z) = \theta \begin{bmatrix} 1 \\ 1 \end{bmatrix} (7z, 7\tau)$$

has simple zeros at the points $\{0, \frac{1}{7}, \frac{2}{7}, ..., \frac{6}{7}\}$ and therefore

$$f(z) = c(\tau) \prod_{l=1}^{l=7} \theta \begin{bmatrix} 1 \\ \frac{2l-1}{7} \end{bmatrix} (z, \tau),$$

where $c(\tau) \in \mathbb{C}^*$ is a constant which depends on τ whose value will not be needed. We compute

$$f\left(\frac{1}{2}\right) = \theta \begin{bmatrix} 1 \\ 1 \end{bmatrix} \left(\frac{7}{2}, 7\tau\right) = \theta \begin{bmatrix} 1 \\ 8 \end{bmatrix} (0, 7\tau) = c(\tau) \prod_{l=1}^{l=7} \theta \begin{bmatrix} 1 \\ \frac{2l+6}{7} \end{bmatrix} (0, \tau),$$

$$f\left(\frac{\tau}{2}\right) = \theta \begin{bmatrix} 1 \\ 1 \end{bmatrix} \left(\frac{7\tau}{2}, 7\tau\right) = \exp\left(2\pi\imath \left[-\frac{7\tau}{8} - \frac{1}{4}\right]\right) \theta \begin{bmatrix} 2 \\ 1 \end{bmatrix} (0, 7\tau)$$

$$= c(\tau) \prod_{l=1}^{l=7} \exp\left(2\pi\imath \left[-\frac{\tau}{8} - \frac{1}{4}\frac{2l-1}{7}\right]\right) \theta \begin{bmatrix} 2 \\ \frac{2l-1}{7} \end{bmatrix} (0, \tau),$$

and

$$f\left(\frac{1+\tau}{2}\right) = \theta \begin{bmatrix} 1 \\ 1 \end{bmatrix} \left(\frac{7\tau+7}{2}, 7\tau\right) = \exp\left(2\pi\imath \left[-\frac{7\tau}{8} - 2\right]\right) \theta \begin{bmatrix} 2 \\ 8 \end{bmatrix} (0, 7\tau)$$

$$= c(\tau) \prod_{l=1}^{l=7} \exp\left(2\pi\imath \left[-\frac{\tau}{8} - \frac{1}{4}\frac{2l+6}{7}\right]\right) \theta \begin{bmatrix} 2 \\ \frac{2l+6}{7} \end{bmatrix} (0, \tau).$$

It follows from the last three equations that

$$(A) \qquad \theta \begin{bmatrix} 1 \\ 0 \end{bmatrix} (0, 7\tau) = c(\tau) \prod_{l=1}^{l=7} \theta \begin{bmatrix} 1 \\ \frac{2l+6}{7} \end{bmatrix} (0, \tau),$$

$$(B) \qquad \theta \begin{bmatrix} 0 \\ 0 \end{bmatrix} (0, 7\tau) = c(\tau) \prod_{l=1}^{l=7} \exp\left(2\pi\imath \left[-\frac{2l+6}{28}\right]\right) \theta \begin{bmatrix} 0 \\ \frac{2l+6}{7} \end{bmatrix} (0, \tau)$$

and

$$(C) - \imath\theta \begin{bmatrix} 0 \\ 1 \end{bmatrix} (0, 7\tau) = c(\tau) \prod_{l=1}^{l=7} \exp\left(2\pi\imath \left[-\frac{2l-1}{28}\right]\right) \theta \begin{bmatrix} 0 \\ \frac{2l-1}{7} \end{bmatrix} (0, \tau).$$

The terms on the right hand side of the equalities simplify considerably. Since

$$\theta\begin{bmatrix} 1 \\ \frac{2(8-l)+6}{7} \end{bmatrix} = \theta\begin{bmatrix} -1 \\ \frac{2l-22}{7} \end{bmatrix} = \theta\begin{bmatrix} 1-2 \\ \frac{2l+6}{7}-4 \end{bmatrix} = \theta\begin{bmatrix} 1 \\ \frac{2l+6}{7} \end{bmatrix}$$

and

$$\theta\begin{bmatrix} 0 \\ \frac{2(8-l)-1}{7} \end{bmatrix} = \theta\begin{bmatrix} 0 \\ \frac{2l-15}{7} \end{bmatrix} = \theta\begin{bmatrix} 0 \\ \frac{2l-1}{7}-2 \end{bmatrix} = \theta\begin{bmatrix} 0 \\ \frac{2l-1}{7} \end{bmatrix},$$

equations (A), (B) and (C) can be rewritten as

$$(A') \qquad \theta\begin{bmatrix} 1 \\ 0 \end{bmatrix}(0,7\tau) = -c(\tau)\theta\begin{bmatrix} 1 \\ 0 \end{bmatrix}(0,\tau)\prod_{l=1}^{l=3}\theta^2\begin{bmatrix} 1 \\ \frac{2l+6}{7} \end{bmatrix}(0,\tau),$$

$$(B') \qquad \theta\begin{bmatrix} 0 \\ 0 \end{bmatrix}(0,7\tau) = -c(\tau)\theta\begin{bmatrix} 0 \\ 0 \end{bmatrix}(0,\tau)\prod_{l=1}^{l=3}\theta^2\begin{bmatrix} 0 \\ \frac{2l+6}{7} \end{bmatrix}(0,\tau)$$

and

$$(C') \qquad \theta\begin{bmatrix} 0 \\ 1 \end{bmatrix}(0,7\tau) = -c(\tau)\theta\begin{bmatrix} 0 \\ 1 \end{bmatrix}(0,\tau)\prod_{l=1}^{l=3}\theta^2\begin{bmatrix} 0 \\ \frac{2l-1}{7} \end{bmatrix}(0,\tau).$$

In particular, multiplication of (A'), (B') and (C') by

$$\theta\begin{bmatrix} 1 \\ 0 \end{bmatrix}(0,\tau),\ \theta\begin{bmatrix} 0 \\ 0 \end{bmatrix}(0,\tau) \text{ and } \theta\begin{bmatrix} 0 \\ 1 \end{bmatrix}(0,\tau),$$

respectively makes the right hand side of each equality a product of squares.

To obtain the identity of the theorem, we begin the second part of the proof. The elliptic function

$$g(z) = \frac{\theta\begin{bmatrix} 1 \\ \frac{5}{7} \end{bmatrix}(z,\tau)\theta\begin{bmatrix} 1 \\ \frac{3}{7} \end{bmatrix}(z,\tau)\theta\begin{bmatrix} 1 \\ \frac{13}{7} \end{bmatrix}(z,\tau)}{\theta\begin{bmatrix} 0 \\ 0 \end{bmatrix}(z,\tau)\theta\begin{bmatrix} 0 \\ 1 \end{bmatrix}(z,\tau)\theta\begin{bmatrix} 1 \\ 0 \end{bmatrix}(z,\tau)}$$

with periods 1 and τ has poles at $\frac{1}{2}$, $\frac{\tau}{2}$ and $\frac{1+\tau}{2}$. By now standard computations yield

$$\operatorname{Res}_{\frac{1}{2}} g = \frac{\theta\begin{bmatrix} 1 \\ \frac{12}{7} \end{bmatrix}\theta\begin{bmatrix} 1 \\ \frac{10}{7} \end{bmatrix}\theta\begin{bmatrix} 1 \\ \frac{20}{7} \end{bmatrix}}{\theta'\begin{bmatrix} 1 \\ 1 \end{bmatrix}\theta\begin{bmatrix} 0 \\ 0 \end{bmatrix}\theta\begin{bmatrix} 0 \\ 1 \end{bmatrix}},\quad \operatorname{Res}_{\frac{\tau}{2}} g = \frac{\theta\begin{bmatrix} 0 \\ \frac{5}{7} \end{bmatrix}\theta\begin{bmatrix} 0 \\ \frac{3}{7} \end{bmatrix}\theta\begin{bmatrix} 0 \\ \frac{13}{7} \end{bmatrix}}{\theta'\begin{bmatrix} 1 \\ 1 \end{bmatrix}\theta\begin{bmatrix} 1 \\ 0 \end{bmatrix}\theta\begin{bmatrix} 0 \\ 0 \end{bmatrix}}$$

and

$$\operatorname{Res}_{\frac{\tau+1}{2}} f = \frac{-\theta\begin{bmatrix} 0 \\ \frac{12}{7} \end{bmatrix}\theta\begin{bmatrix} 0 \\ \frac{10}{7} \end{bmatrix}\theta\begin{bmatrix} 0 \\ \frac{20}{7} \end{bmatrix}}{\theta'\begin{bmatrix} 1 \\ 1 \end{bmatrix}\theta\begin{bmatrix} 1 \\ 0 \end{bmatrix}\theta\begin{bmatrix} 0 \\ 1 \end{bmatrix}}.$$

The sum of the last three residues is zero. This fact along with equations (A'), (B') and (C') imply (4.11).

Remark 3.14. It is important to observe that the residue theorem sometimes gives trivial results. Consider the elliptic function

$$f(z) = \frac{\theta\begin{bmatrix} \frac{2l+1}{N} \\ 1 \end{bmatrix}(Nz, N\tau)}{\theta\begin{bmatrix} 1 \\ 1 \end{bmatrix}(Nz, N\tau)},$$

where $N \geq 3$ is a prime, $0 \leq l \leq \frac{N-3}{2}$. The fact that the sum of the residues of f vanishes is the trivial observation that the sum of the N-th roots of unity vanishes.

In the course of the proof of the Jacobi triple product identity we proved the formula of Jacobi

$$\prod_{n=1}^{\infty}(1-x^n)^3 = \sum_{n=0}^{\infty}(-1)^n(2n+1)x^{\frac{n^2+n}{2}}. \tag{4.12}$$

At the beginning of this chapter we derived Jacobi's derivative formula from the residue theorem. We shall now show that the above formula of Jacobi also follows from these considerations. We begin as always with an elliptic function. The only singularity of

$$f(z) = \frac{\theta\begin{bmatrix} \frac{1}{3} \\ 1 \end{bmatrix}(3z, 3\tau)}{\theta^3\begin{bmatrix} 1 \\ 1 \end{bmatrix}(z, \tau)}$$

is at the origin. Setting the residue of f at 0 equal to zero yields

$$9\frac{\theta''\begin{bmatrix} \frac{1}{3} \\ 1 \end{bmatrix}(0, 3\tau)}{\theta\begin{bmatrix} \frac{1}{3} \\ 1 \end{bmatrix}(0, 3\tau)} = \frac{\theta'''\begin{bmatrix} 1 \\ 1 \end{bmatrix}(0, \tau)}{\theta'\begin{bmatrix} 1 \\ 1 \end{bmatrix}(0, \tau)}.$$

The heat equation and integration give

$$\theta^3\begin{bmatrix} \frac{1}{3} \\ 1 \end{bmatrix}(0, 3\tau) = c\theta'\begin{bmatrix} 1 \\ 1 \end{bmatrix}(0, \tau).$$

The constant c is easily computed to be $-\frac{\imath}{2\pi}$. We have already seen that in the variable $x = \exp(2\pi\imath\tau)$,

$$\theta\begin{bmatrix}\frac{1}{3}\\1\end{bmatrix}(0,3\tau) = \exp\left(\frac{\pi\imath}{6}\right)x^{\frac{1}{24}}\sum_{n\in\mathbb{Z}}(-1)^n x^{\frac{3n^2+n}{2}} = \exp\left(\frac{\pi\imath}{6}\right)x^{\frac{1}{24}}\prod_{n=1}^{\infty}(1-x^n)$$

(the last equality is a consequence of the Jacobi triple product identity and is due to Euler). There is a combinatorial proof of Euler's identity which we will present in Chapter 7.

In the variable x, $\theta'\begin{bmatrix}1\\1\end{bmatrix}(0,\tau) = -2\pi x^{\frac{1}{8}}\sum_{n=0}^{\infty}(-1)^n(2n+1)x^{\frac{n^2+n}{2}}$. Hence we have re-proven (4.8) which also gives the formula of Jacobi (4.12).

This is a good place to reintroduce the reader to the important number theoretic function $\eta(\tau)$. In §8.1 of Chapter 3 we defined the *eta*-function. Its value at $\tau \in \mathbb{H}^2$ is given by the formula

$$(4.13)\quad \eta(\tau) = x^{\frac{1}{24}}\sum_{n=-\infty}^{\infty}(-1)^n x^{\frac{n(3n+1)}{2}} = x^{\frac{1}{24}}\prod_{n=1}^{\infty}(1-x^n),\ x = \exp(2\pi\imath\tau).$$

Note that we have identities connecting three important functions:

$$(4.14)\qquad \theta'\begin{bmatrix}1\\1\end{bmatrix}(0,\tau) = 2\pi\imath\theta^3\begin{bmatrix}\frac{1}{3}\\1\end{bmatrix}(0,3\tau) = -2\pi\eta^3(\tau).$$

Another connection of the η-function to θ-functions is given by

Proposition 3.15. [**16**, Th. 12 of Ch. 3] *For all* $\tau \in \mathbb{H}^2$,

$$\theta(\tau) = \frac{\eta^2\left(\frac{\tau+1}{2}\right)}{\eta(\tau+1)}.$$

Proof. We compute after using (4.13)

$$\frac{\eta^2\left(\frac{\tau+1}{2}\right)}{\eta(\tau+1)} = \prod_{n=1}^{\infty}(1-x^{2n})(1+x^{2n-1})^2 = \theta(\tau).$$

In Chapter 1, in our discussion of elliptic function theory, we showed that associated with the field of elliptic functions with periods 1, τ there is a cubic polynomial

$$(4.15)\qquad 4x^3 - g_2x - g_3 = 4(x-e_1)(x-e_2)(x-e_3),$$

with distinct roots e_1, e_2 and e_3. The algebraic curve determined by this polynomial is conformally equivalent to the torus corresponding to τ. The *discriminant* of this cubic is $\tilde{\Delta} = g_2^3 - 27g_3^2$, a nonzero (because the roots are distinct) function of the variable τ on the upper half plane. We now will show that $\tilde{\Delta}$ is a constant multiple of $\theta'^8\begin{bmatrix}1\\1\end{bmatrix}$, thus relating it to the identities (4.14). Since

$$e_1 + e_2 + e_3 = 0,\ g_2 = -4(e_1e_2 + e_1e_3 + e_2e_3) \text{ and } g_3 = 4e_1e_2e_3,$$

we see after some algebraic simplification that

$$\tilde{\Delta} = 16(e_1 - e_2)^2(e_1 - e_3)^2(e_2 - e_3)^2 = 16(e_1 - e_2)^2(2e_1 + e_2)^2(e_1 + 2e_2)^2.$$

Actually, a general cubic is solvable as was established, among others, by the Italian mathematician Girolama Cardano[96] in the sixteenth century. Thus not only can we describe $\tilde{\Delta} = 16(e_1 - e_2)^2(e_1 - e_3)^2(e_2 - e_3)^2$ in terms of g_2 and g_3, but we can obtain a formula for each e_i in terms of the coefficients of the cubic. For the sake of completeness we go through the details of Cardan's classical solution to the cubic[97] $4x^3 - g_2x - g_3$.

The case $g_2 = 0$ is, of course, trivial. So we assume that $g_2 \neq 0$. Let us write a zero x of the cubic as $x = y + z$. This leads to the transformed equation

$$4(y^3 + z^3) + (y + z)(12yz - g_2) - g_3 = 0.$$

We have written our desired root of (4.15) as a sum of two variables. We can clearly, in general, impose another condition on these variables. It is convenient to set $12yz - g_2 = 0$ (and therefore $y^3z^3 = \frac{g_2^3}{4^3 \cdot 3^3}$). The transformed equation becomes $y^3 + z^3 = \frac{g_3}{4}$. We are led to the two equations

$$y^3 + z^3 = \frac{g_3}{4}, \qquad y^3z^3 = \frac{g_2^3}{4^3 \cdot 3^3}$$

that have a unique solution for y^3 and z^3, the roots of the quadratic equation

$$t^2 - \frac{g_3}{4}t + \frac{g_2^3}{4^3 \cdot 3^3} = 0.$$

It thus follows that

$$\tilde{\Delta} = g_2^3 - 27g_3^2 = -16 \cdot 27((y^3 + z^3)^2 - 4y^3z^3) = -16 \cdot 27(y^3 - z^3)^2.$$

Cardan's solution to the cubic described above can be written as[98]

$$e_1 = y + z, e_2 = \omega y + \omega^2 z, e_3 = \omega^2 y + \omega z,$$

where ω is a cube root of unity. Computing from this incomplete information provided by the last line, we see that

$$(e_1 - e_2)^2(e_1 - e_3)^2(e_2 - e_3)^2 = -27(y^3 - z^3)^2,$$

[96] Whose last name is often Anglecized to Cardan.

[97] The general cubic $a_0x^3 + a_1x^2 + a_2x + a_3$, $a_0 \neq 0$, is reduced to (approximately) this form by replacing x by $\frac{x}{a_0^{\frac{1}{3}}} - \frac{a_1}{3a_0^{\frac{1}{3}}}$.

[98] We leave it to the reader to compute y and z in terms of g_2 and g_3.

establishing once again the equivalence of the two expressions (one in terms of the coefficients of the cubic; the second in terms of its roots) for $\tilde{\Delta}$.

Having established a formula for the discriminant in terms of the roots of the cubic, we invoke Corollary 2.20 of Chapter 3 to obtain

$$(e_1 - e_2)^2(e_1 - e_3)^2(e_2 - e_3)^2 = \pi^{12}\theta^8\begin{bmatrix} 0 \\ 0 \end{bmatrix}\theta^8\begin{bmatrix} 0 \\ 1 \end{bmatrix}\theta^8\begin{bmatrix} 1 \\ 0 \end{bmatrix} = \pi^4\theta'^8\begin{bmatrix} 1 \\ 1 \end{bmatrix}.$$

Remark 3.16. The above presentation shows that $\tilde{\Delta}$ is a nonzero constant multiple of Δ. The arguments involved avoided the use of the Riemann ζ-function, fulfilling the promise made in Remark 3.5 of Chapter 3.

4. Identities which arise from modular forms

In Chapter 1, §6.4, we introduced the principal congruence subgroups $\Gamma(k)$ and subsequently the various subgroups $G(k)$, $\Gamma(p,q)$ and $\Gamma_o(k)$. Most important for our considerations will be the case $p = q = k$. In the above situation we have coverings

$$\mathbb{H}^2/\Gamma(k) \to \mathbb{H}^2/\Gamma(k,k),\ \mathbb{H}^2/\Gamma(k,k) \to \mathbb{H}^2/\Gamma_o(k) \text{ and } \mathbb{H}^2/\Gamma(k) \to \mathbb{H}^2/\Gamma_o(k).$$

The work of Chapter 3 has laid the function theoretic basis for constructing identities, and in fact several identities have already been established there.

The transformation theory of theta constants shows that η^{24} is a cusp 6-form for Γ and hence for any finite index subgroup, in particular for $\Gamma(k)$ with $k \in \mathbb{Z}^+$ arbitrary. This fact together with the ability to explicitly compute the divisors of these forms for various examples allows us to derive identities. We shall begin with almost trivial proofs of the two identities already derived, the Jacobi derivative formula and the formula of Jacobi (4.12). We view η^{24} as a cusp 6-form for $\Gamma(2)$ and its projection to $\overline{\mathbb{H}^2/\Gamma(2)}$ as a 6-differential. The divisor of this 6-differential on this compact Riemann surface is precisely $P_\infty^{-4}P_0^{-4}P_1^{-4}$. Furthermore, both $\theta^8\begin{bmatrix} 0 \\ 0 \end{bmatrix}(0,\tau)\theta^8\begin{bmatrix} 0 \\ 1 \end{bmatrix}(0,\tau)\theta^8\begin{bmatrix} 1 \\ 0 \end{bmatrix}(0,\tau)$ and $\theta^{24}\begin{bmatrix} \frac{1}{3} \\ 1 \end{bmatrix}(0,3\tau)$ also project to 6-differentials on the compactification of $\mathbb{H}^2/\Gamma(2)$ with precisely the same divisor. We remind the reader that the correct local coordinate at P_∞ on $\mathbb{H}^2/\Gamma(2)$ is $x = \exp(\pi\imath\tau)$.

It follows that the quotient of any two of the three 6-differentials must be a constant. Thus 8-th roots of these ratios are also constant:

(4.16)

$$\frac{\theta'\begin{bmatrix} 1 \\ 1 \end{bmatrix}(0,\tau)}{\theta\begin{bmatrix} 0 \\ 0 \end{bmatrix}(0,\tau)\theta\begin{bmatrix} 0 \\ 1 \end{bmatrix}(0,\tau)\theta\begin{bmatrix} 1 \\ 0 \end{bmatrix}(0,\tau)} = c_1 \text{ and } \frac{\theta'\begin{bmatrix} 1 \\ 1 \end{bmatrix}(0,\tau)}{\theta^3\begin{bmatrix} \frac{1}{3} \\ 1 \end{bmatrix}(0,3\tau)} = c_2.$$

Evaluating the constants c_1 and c_2 yields the two results of Jacobi in a rather economical manner.

Exercise 4.1. The same methods yield a proof of the Jacobi quartic identity. Prove that

$$B_1^*\left(\theta^4\begin{bmatrix}0\\0\end{bmatrix}\right)=\theta^4\begin{bmatrix}0\\1\end{bmatrix},\ B_1^*\left(\theta^4\begin{bmatrix}0\\1\end{bmatrix}\right)=\theta^4\begin{bmatrix}0\\0\end{bmatrix},$$

$$B_1^*\left(\theta^4\begin{bmatrix}1\\0\end{bmatrix}\right)=-\theta^4\begin{bmatrix}1\\0\end{bmatrix},\ A_1^*\left(\theta^4\begin{bmatrix}0\\0\end{bmatrix}\right)=-\theta^4\begin{bmatrix}0\\0\end{bmatrix},$$

$$A_1^*\left(\theta^4\begin{bmatrix}0\\1\end{bmatrix}\right)=-\theta^4\begin{bmatrix}1\\0\end{bmatrix}\text{ and }A_1^*\left(\theta^4\begin{bmatrix}1\\0\end{bmatrix}\right)=-\theta^4\begin{bmatrix}0\\1\end{bmatrix}.$$

It follows that for the function

$$f=\theta^4\begin{bmatrix}0\\0\end{bmatrix}-\theta^4\begin{bmatrix}0\\1\end{bmatrix}-\theta^4\begin{bmatrix}1\\0\end{bmatrix}$$

we have

$$B_1^*f=-f\text{ and }A_1^*f=-f.$$

Hence $f^2\in\mathbb{A}_2^+(\mathbb{H}^2,\Gamma)$. Since f vanishes at $\imath\infty$, we have the stronger conclusion $f^2\in\mathbb{A}_2(\mathbb{H}^2,\Gamma)$. Since this last vector space is trivial, so is the automorphic form f.

4.1. Multiplicative meromorphic forms. The above discussion required the evaluation of divisors of cusp forms for the group $\Gamma(2)$. These calculations can in fact be avoided because the ratios encountered are multiplicative meromorphic functions for Γ. We need

Lemma 4.2. (a) *If $f\neq 0$ is a meromorphic multiplicative q-form for Γ, then*

$$\deg(f)=\sum_{x\in\overline{\mathbb{H}^2/\Gamma}}\operatorname{ord}_x f=\frac{q}{6}.$$

(b) *Let f be a holomorphic Γ-automorphic function on $\mathbb{H}^2$ for some character χ. Assume that f extends meromorphically to the point at infinity. If f is nonvanishing, then f is constant and (hence) $\chi=1$.*

Proof. Part (a) has been established in Chapter 3. We proceed to the proof of (b). Since f is holomorphic except perhaps at one Γ-orbit, it either has a pole at infinity or is finite there. It cannot have a pole there, for then it must have a zero someplace on $\mathbb{H}^2$. If it in fact vanishes at ∞, it must be identically zero. In any event it must be constant. To see this, observe that $\frac{f'}{f}$ projects to the Riemann surface $\overline{\mathbb{H}^2/\Gamma}$ as an abelian differential of the first kind. Since $\overline{\mathbb{H}^2/\Gamma}$ has genus zero, this differential must be trivial. Thus $f'=0$ and f is constant.

Part (b) of the above lemma tells us at once that the two quotients in (4.16) are constants, since the respective numerators and denominators are automorphic forms of weight $\frac{3}{4}$ for Γ (the factor of automorphy is $\kappa^{\frac{3}{4}}$, with κ canonical).

In Chapter 2 we defined the notion of rational classes of characteristics. For each positive integer k, we decomposed a certain set of classes $Z(k)$ into two sets which we called $X(k)$ and $Y(k)$. We recall that $X(k)$ consisted of classes represented by vectors $\chi \in \mathbb{R}^2$ of the form $\begin{bmatrix} \frac{m}{k}, \\ \frac{m'}{k} \end{bmatrix}$ with m and $m' \in \mathbb{Z}$ of the same parity as k. We also introduced a distinguished subset $X_o(k) \subset X(k)$ that is in a one to one correspondence with the set of $n(k)$ punctures on $\mathbb{H}^2/\Gamma(k)$. The Jacobi derivative formula is generalized by the next

Theorem 4.3. *For each integer $k > 1$, there is a constant $c(k) \in \mathbb{C}^*$ such that*

$$\left(\theta' \begin{bmatrix} 1 \\ 1 \end{bmatrix}\right)^{n(k)} = c(k) \prod_{\chi \in X_0(k)} \theta^3 [\chi] .$$

Proof. The transformation theory tells us that the 8-th power of each side of the equality is a $6n(k)$-form for $\Gamma(k)$. We can explicitly compute the divisors on each side and see that they are identical. To avoid the routine but time consuming calculation of divisors, we offer a second proof of the theorem. The holomorphic function $\frac{(\theta')^{n(k)}\begin{bmatrix} 1 \\ 1 \end{bmatrix}}{\prod_{\chi\in X_o(k)} \theta^3[\chi]}$ is Γ-automorphic with some character, does not vanish on $\mathbb{H}^2$, and has its only possible singularity at points equivalent to ∞. Hence it must be a nonzero constant.

As remarked above, the last theorem should be viewed as a generalization of the Jacobi derivative formula. In this vein one can ask, what is the generalization of its companion, the second equation of (4.16)? The answer is

Theorem 4.4. *For every odd prime k, there exists a nonzero constant $c(k)$ such that*

$$\theta' \begin{bmatrix} 1 \\ 1 \end{bmatrix} (0,\tau) \left(\theta' \begin{bmatrix} 1 \\ 1 \end{bmatrix} (0, k\tau)\right)^{\frac{k-3}{2}} = c(k) \prod_{l=0}^{\frac{k-3}{2}} \theta^3 \begin{bmatrix} \frac{2l+1}{k} \\ 1 \end{bmatrix} (0, k\tau).$$

Proof. It is clear that both sides of the identity are automorphic forms of the same weight for the group $\Gamma_o(k)$; hence it suffices to show that they have the same divisors. This requires checking only the cusp at zero and the cusp at infinity. It is an easy task to compute the divisors of the two functions on

$\mathbb{H}^2/\Gamma(k)$. We see that the left hand side has a zero of order $\frac{k}{8}+\frac{k-3}{2}\frac{k^2}{8}$ at each distinguished cusp (these are the cusps for $\Gamma(k)$ that are $\Gamma_o(k)$-equivalent to ∞) and a zero of order $\frac{k}{8}+\frac{k-3}{16}$ at each nondistinguished cusp. The right hand side has a zero of order $\frac{3}{8}\sum_{l=0}^{\frac{k-3}{2}}(2l+1)^2 = \frac{3}{8}\frac{k(k-1)(k-2)}{6} = \frac{k(k-1)(k-2)}{16}$ at ∞ and thus the same order at each distinguished cusp. The order at a nondistinguished cusp of the right hand side is also clearly $\frac{3}{8}\frac{k-1}{2}$.

Remark 4.5. Theorem 4.4 can also be proven (the reader is encouraged to do so) as a direct consequence of the Jacobi triple product formula since each side has a product expansion that is known.

4.2. Cusp forms for Γ. The identities obtained up until this point used basically the fact that a meromorphic function on a compact Riemann surface which had no poles must be constant. Our next result will use a variant of this idea: the fact that dim $\mathbb{A}_6(\mathbb{H}^2,\Gamma)=1$. That $\left(\theta'\begin{bmatrix}1\\1\end{bmatrix}\right)^8=(2\pi)^8\eta^{24}$ is a cusp 6-form for Γ (thus $\eta^{24}=\Delta$) has already been observed in equation (4.14). Our next theorem produces another formula for the same differential.

Theorem 4.6. *For every prime k and all $\tau\in\mathbb{H}^2$,*

$$k^{12}\eta^{24}(k\tau)+\sum_{l=0}^{k-1}\eta^{24}\left(\frac{\tau+l}{k}\right)=C\ \eta^{24}(\tau), \tag{4.17}$$

where $C=C(k)$ is a constant independent of τ.

Proof. In order to establish the identity it is sufficient to prove that the function f defined by the left hand side is a cusp 6-form for the group Γ. It obviously vanishes at $\imath\infty$. It remains to use the transformation formula to establish invariance. The group Γ is generated by the two Möbius transformations $A:\tau\mapsto\frac{-1}{\tau}$ and $B:\tau\mapsto\tau+1$, so it suffices to study the action induced by these two motions.

It is clear that the two functions on $\mathbb{H}^2$, $\tau\mapsto k^{12}\eta^{24}(k\tau)$ and $\tau\mapsto\sum_{l=0}^{k-1}\eta^{24}\left(\frac{\tau+l}{k}\right)$ are each invariant under B since B operates as a shift operator on the terms appearing in the sum. Hence we have shown that $f(\tau+1)=f(\tau)$, for all $\tau\in\mathbb{H}^2$. Let us index the first modified theta constant defining f by k and the terms appearing in the sum by $0,\ 1,\ \ldots,\ k-1$. We identify a motion with the permutation it induces on these $k+1$ functions; hence the operator B corresponds to the permutation

$$\sigma_B=(0,\ 1,\ 2,\ \ldots,\ k-1).$$

We now consider the operator A. Observe that

$$\eta^{24}\left(k\frac{-1}{\tau}\right)=\eta^{24}\left(\frac{-1}{\frac{\tau}{k}}\right)=\left(\frac{\tau}{k}\right)^{12}\eta^{24}\left(\frac{\tau}{k}\right),\text{ for all }\tau\in\mathbb{H}^2.$$

It is thus clear that the motion A permutes the first two terms (because $A^2=I$) defining f. It remains to see what happens to the terms corresponding to $l=1,\ \ldots,\ k-1$. We claim that each of these terms is either fixed or permuted with another such term. Let us fix such an integer l. Since k is a prime, we can choose an $m=m(l)\in\mathbb{Z}$ such that $lm\equiv -1 \bmod k$ (we may and do choose m to be a positive integer $\leq k-1$), and set

$$a=l,\ b=-\frac{1+lm}{k},\ c=k,\ d=-m.$$

Then $\begin{bmatrix} a & b\\ c & d\end{bmatrix}$ is an element of Γ, and

$$\eta^{24}\left(\frac{\frac{-1}{\tau}+l}{k}\right)=\eta^{24}\left(\frac{-1+l\tau}{k\tau}\right)=\eta^{24}\left(\frac{a\frac{\tau+m}{k}+b}{c\frac{\tau+m}{k}+d}\right).$$

The above identity therefore tells us that $\eta^{24}\left(\frac{\frac{-1}{\tau}+l}{k}\right)=\tau^{12}\eta^{24}\left(\frac{\tau+m}{k}\right)$. We have shown that the permutation σ_A consists of $\frac{k+1}{2}$ transpositions if -1 is not a square in $\mathbb{Z}_k$, is the transposition $(0,2)$ if $k=2$, and consists of $\frac{k-1}{2}$ transpositions if $k\neq 2$ and -1 is a square in $\mathbb{Z}_k$ (σ_A has two fixed points in this case). We conclude that for any motion $\gamma\in\Gamma$ we have $f(\gamma\tau)\gamma'(\tau)^6=f(\tau)$, for all $\tau\in\mathbb{H}^2$, or that f is a cusp 6-form for the modular group. The method of proof we have given here also yields most of

Corollary 4.7. *For each positive prime k, there is a homomorphism of Γ into the permutation group on $k+1$ objects with kernel $\Gamma(k)$.*

Proof. The permutation σ_γ induced by $\gamma\in\Gamma$ has been described in the proof of the above theorem, and we need only show that the kernel of the homomorphism $\gamma\mapsto\sigma_\gamma$ is $\Gamma(k)$. It suffices for the case $k>3$ (because $\Gamma/\Gamma(k)$ is a simple group in these cases) to show that $\Gamma(k)$ is contained in the kernel (the kernel is not all of Γ because B does not belong to it). This essentially follows from the identity

$$\frac{\frac{a\tau+b}{c\tau+d}+l}{k}=\frac{(a+lc)\frac{\tau+l}{k}+\frac{l(d-a)+b-l^2c}{k}}{kc\frac{\tau+l}{k}+d-cl},\text{ for all }\tau\in\mathbb{H}^2,$$

for $\gamma=\begin{bmatrix} a & b\\ c & d\end{bmatrix}\in\Gamma(k)$. Since $\hat{\gamma}=\begin{bmatrix} a+lc & \frac{l(d-a)+b-l^2c}{k}\\ kc & d-cl\end{bmatrix}\in\Gamma$ and

$$\frac{\gamma(\tau)+l}{k}=\hat{\gamma}\left(\frac{\tau+l}{k}\right),\ \gamma'(\tau)=\hat{\gamma}'\left(\frac{\tau+l}{k}\right),$$

we conclude that for all $\tau \in \mathbb{H}^2$,

$$\eta^{24}\left(\frac{\gamma(\tau)+l}{k}\right)\gamma'(\tau)^6 = \eta^{24}\left(\hat{\gamma}\left(\frac{\tau+l}{k}\right)\right)\hat{\gamma}'\left(\frac{\tau+l}{k}\right)^6 = \eta^{24}\left(\frac{\tau+l}{k}\right).$$

Hence the permutation σ_γ fixes the integers 0, 1, ..., $k-1$. Hence also k. In other words, the permutation induced by an element of $\Gamma(k)$ is the identity. For $k=2$ and 3, we must show that the kernel of the map from $\Gamma/\Gamma(k)$ to the permutation group is trivial. We show this by listing representatives for the quotient group and computing their images. For $k=2$, we have

$$\sigma_B = (0,1),\ \sigma_A = (0,2).$$

Hence

$$\sigma_{BA} = (0,2),\ \sigma_{AB} = (0,1,2) \text{ and } \sigma_{ABA} = (1,2).$$

For the case $k=3$,

$$\sigma_B = (0,1,2),\ \sigma_A = (0,3)(1,2).$$

Hence

$$\sigma_{B^2} = (0,2,1),\ \sigma_{BA} = (0,3,1),\ \sigma_{AB} = (0,2,3),$$

$$\sigma_{B^2A} = (0,3,2),\ \sigma_{ABA} = (1,3,2),\ \sigma_{BAB^2} = (1,3),$$

$$\sigma_{B^2AB} = (0,1)(2,3),\ \sigma_{AB^2} = (0,1,3) \text{ and } \sigma_{AB^2A} = (1,2,3).$$

The fact that $\Gamma/\Gamma(k)$ acts as a permutation group on $k+1$ objects is neither new nor surprising. We know that $\Gamma/\Gamma(k)$ is isomorphic to $\mathrm{PSL}(2,\mathbb{Z}_k)$. The group $\mathrm{SL}(2,\mathbb{Z}_k)$ acts in a natural linear way on the finite plane $\mathbb{Z}_k \times \mathbb{Z}_k$. If we projectivize this cartesian product, we obtain a space with $k+1$ elements, namely the projective equivalence classes of

$$\{(0,1),\ (1,0),\ (1,1),\ \ldots,\ (1,k-1)\}.$$

The group $\mathrm{PSL}(2,\mathbb{Z}_k)$ acts on this projective plane and permutes the points in it. It is hence a representation of $\Gamma/\Gamma(k)$ as a permutation group on $k+1$ objects. We have, by our theory, shown that there is an analytic counterpoint to this abstract result in terms of a subspace of the space of cusp 6-forms for the group $\Gamma(k)$ or, alternatively, of a space of meromorphic 6-differentials on the Riemann surface $\overline{\mathbb{H}^2/\Gamma(k)}$. For every finitely generated Fuchsian group of the first kind G acting on $\mathbb{H}^2$, the automorphism group $\mathrm{Aut}(\mathbb{H}^2/G)$ acts on $\mathbb{A}_q(\mathbb{H}^2, G)$, for all $q \in \mathbb{Z}^+$, as well as on various subspaces of this Banach space. Our case is distinguished by the fact that the group $\Gamma/\Gamma(k) \cong \mathrm{Aut}(\mathbb{H}^2/\Gamma(k))$ induces a permutation on the finite set of $k+1$ 6-differentials described above.

4.3. Some special results for the primes 5 and 7. The next chapter will contain many identities obtained by averaging functions. These identities will contain interesting combinatorial information. For the present in order to whet the reader's appetite we list two such results, whose proofs will be found in Chapter 5.

Theorem 4.8. *For all* $\tau \in \mathbb{H}^2$,

$$\sum_{l=0}^{4} \frac{\eta(5(\tau+l))}{\eta(\frac{\tau+l}{5})} = 5^2 \left(\frac{\eta(5\tau)}{\eta(\tau)}\right)^6 .$$

Theorem 4.9. *For all* $\tau \in \mathbb{H}^2$,

$$\sum_{l=0}^{6} \frac{\eta(7(\tau+l))}{\eta(\frac{\tau+l}{7})} = 7^2 \left(\frac{\eta(7\tau)}{\eta(\tau)}\right)^4 + 7^3 \left(\frac{\eta(7\tau)}{\eta(\tau)}\right)^8 .$$

5. Ramanujan's τ-function

The purpose of this section is to explain the combinatorial and number theoretic content of Theorem 4.6. The key is equation (4.8) which allows us to translate Theorem 4.6 to an infinite product identity. We begin with the following

Definition 5.1. Let $T(n)$, $n \in \mathbb{Z}^+$, be defined by the series

$$x \prod_{n=1}^{\infty} (1-x^n)^{24} = \sum_{n=1}^{\infty} T(n) x^n = \eta^{24}(\tau), \ x = \exp(2\pi\imath\tau). \tag{4.18}$$

Remark 5.2. We have defined a function

$$T : \mathbb{Z}^+ \to \mathbb{Z};$$

in the literature, it is sometimes called *Ramanujan's τ-function.*

Since $\sum_{n=1}^{\infty} T(n)x^n$ is the Fourier series expansion of η^{24} in terms of the variable $x = \exp(2\pi\imath\tau)$, the identity of Theorem 4.6 takes the form

Proposition 5.3. *For all* $x \in \mathbb{C}$, $|x| < 1$,

$$k^{12} \sum_{n=1}^{\infty} T(n) x^{kn} + k \sum_{n=1}^{\infty} T(kn) x^n = C \sum_{n=1}^{\infty} T(n) x^n .$$

Setting $n = 1$ we evaluate the constant $C = kT(k)$, and thus the identity of the theorem is simply

Theorem 5.4 (Mordell). *For all primes k and all $x \in \mathbb{C}$, $|x| < 1$,*

$$k^{11} \sum_{n=1}^{\infty} T(n) x^{kn} = \sum_{n=1}^{\infty} (T(k)T(n) - T(kn)) x^n . \tag{4.19}$$

As a simple consequence of Proposition 5.3, we therefore also have

Corollary 5.5. *If $n \in \mathbb{Z}^+$ is not a multiple of the positive prime k, then $T(kn) = T(k)T(n)$. If n is a multiple of k, say $n = k^r l$ with $r \geq 1$ and l not a multiple of k, then*

$$T(k^{r+1}l) = T(k)T(k^r l) - k^{11}T(k^{r-1}l).$$

In particular, setting $l = 1$ gives a recursion formula

$$T(k^{r+1}) = T(k)T(k^r) - k^{11}T(k^{r-1}).$$

Setting $r = 0$ gives the next to last equality as

$$T(kl) = T(k)T(l) - k^{11}T\left(\frac{l}{k}\right)$$

that can be assumed to hold for all positive primes k and all positive integers l if one understands T to be defined on $\mathbb{Q}$ with $T(r) = 0$ for all $r \in \mathbb{Q} - \mathbb{Z}^+$.

Corollary 5.6. *If m and n are relatively prime positive integers, then*

$$T(mn) = T(m)T(n).$$

Proof. We first show that if l is not congruent to zero modulo k, then $T(k^r l) = T(k^r)T(l)$. We show this by induction on r. It is clearly true for $r = 1$ since this is in fact the content of the first part of the previous corollary. Assume that the result is true for $r \leq m$; then by the above corollary, $T(k^{m+1}l) = T(k)T(k^m l) - k^{11}T(k^{m-1}l)$. By the induction hypothesis this is equal to $T(k)T(k^m)T(l) - k^{11}T(k^{m-1})T(l)$. This is the same as $T(l)(T(k)T(k^m) - k^{11}T(k^{m-1}))$, which again by the second part of the previous corollary is equal to $T(l)T(k^{m+1})$.

Now assume that m and n are relatively prime. It follows that mn can be written as $p_1^{n_1}p_2^{n_2} \ldots p_r^{n_r}$ where the p_j are distinct primes and the primary decomposition of m consists of the first $r' < r$ terms and the primary decomposition of n consists of the complementary terms. It follows therefore from the above result that

$$T(p_1^{n_1}p_2^{n_2} \ldots p_r^{n_r}) = T(p_1^{n_1})T(p_2^{n_2} \ldots p_r^{n_r})$$

and therefore by induction that

$$T(p_1^{n_1}p_2^{n_2} \ldots p_r^{n_r}) = T(p_1^{n_1})\ldots T(p_r^{n_r}).$$

Remark 5.7. In Chapter 6, we will study generalizations of Ramanujan's τ-function.

6. Identities among infinite products

In Chapter 2 we derived the Jacobi triple product identity which allows us to express theta functions as infinite product expansions. In particular, we recall equation (2.53) of that chapter where we wrote out the theta constant with characteristic $\begin{bmatrix} \epsilon \\ \epsilon' \end{bmatrix}$ as an infinite product. In this section we translate some of the identities obtained in the previous section to identities among infinite products. The translation of an identity to an infinite product (here, in this section) involves choosing an appropriate local coordinate and then canceling a common factor. This is the plan for the proof of each of the next three propositions. The operations we perform here are all algebraic rather than function theoretic. We begin with the Jacobi quartic identity. Applications are deferred to a later chapter.

Proposition 6.1. *For all x of modulus less than* 1 *we have,*

$$\prod_{n=1}^{\infty}(1+x^{2n-1})^8 = \prod_{n=1}^{\infty}(1-x^{2n-1})^8 + 16x\prod_{n=1}^{\infty}(1+x^{2n})^8. \tag{4.20}$$

Proof. The appropriate local coordinate is $x = \exp(\pi\imath\tau)$ and the common factor to be canceled is $\prod_{n=1}^{\infty}(1-x^{2n})^4$.

The translation of Theorem 3.12 yields.

Proposition 6.2. *For all x of modulus less than* 1 *we have,*

$$\prod_{n=1}^{\infty}(1+x^{2n-1})^2(1+x^{6n-3})^2 \tag{4.21}$$

$$= \prod_{n=1}^{\infty}(1-x^{2n-1})^2(1-x^{6n-3})^2 + 4x\prod_{n=1}^{\infty}(1+x^{2n})^2(1+x^{6n})^2.$$

Proof. The appropriate local coordinate in this case is $x = \exp\left(\frac{2\pi\imath\tau}{3}\right)$ and the common factor to be canceled is $\prod_{n=1}^{\infty}(1-x^{2n})(1-x^{6n})$.

For the last in this series, we turn to a translation of Theorem 3.13.

Proposition 6.3. *For all x of modulus less than* 1 *we have,*

$$\prod_{n=1}^{\infty}(1+x^{2n-1})(1+x^{14n-7}) \tag{4.22}$$

$$= \prod_{n=1}^{\infty}(1-x^{2n-1})(1-x^{14n-7}) + 2x\prod_{n=1}^{\infty}(1+x^{2n})(1+x^{14n}).$$

Proof. The appropriate local coordinate in this case is $x = \exp\left(\frac{2\pi\imath\tau}{7}\right)$ and the common factor to be canceled is $\sqrt{\prod_{n=1}^{\infty}(1-x^{2n})(1-x^{14n})}$.

In order to continue translating identities involving theta constants to identities among infinite products we need an infinite product expansion for the η-function. Fortunately equation (4.8) provides us with the infinite product. Again, we only sketch the major ideas since this topic is treated in great detail in the next chapter. Note that for every prime $k \geq 5$, $\frac{k^2-1}{24} \in \mathbb{Z}^+$. We define

$$H(x) = \frac{x^{\frac{k^2-1}{24}}}{\prod_{n=1}^{\infty}(1-x^n)} = x^{\frac{k^2-1}{24}} \sum_{n=0}^{\infty} P(n)x^n.$$

We refer to P as *Ramanujan's partition function*; for all $n \in \mathbb{Z}^+$, $P(n)$ counts the number of partitions of n. We shall return to the function P (and its generalizations) in the next chapter. Here we simply observe that we have

$$\sum_{l=0}^{k-1} H(\epsilon^l x) = k \sum_{m=1}^{\infty} P\left(km - \frac{k^2-1}{24}\right) x^{km}, \ \epsilon = \exp\left(\frac{2\pi\imath}{k}\right).$$

We can now translate Theorems 4.8 and 4.9 to identities involving infinite products. These identities give the celebrated level one Ramanujan congruences for the primes 5 and 7. We will re-prove these identities in Chapter 5.

Proposition 6.4. *Let* $\epsilon = \exp\left(\frac{2\pi\imath}{5}\right)$. *For all* x *of modulus less than one we have*

$$\sum_{l=0}^{4} \frac{(\epsilon^l x)}{\prod_{n=1}^{\infty}(1-(\epsilon^l x)^{\frac{n}{5}})} = 5\sum_{n=1}^{\infty} P(5n-1)x^n = 5^2 \prod_{n=1}^{\infty} \frac{(1-x^{5n})^5}{(1-x^n)^6}. \tag{4.23}$$

Proof. Use the infinite product expansion of the η-function to translate Theorem 4.8. We will see in the next chapter that it is most convenient to prove Theorem 4.8 up to an undetermined constant before establishing this result. The computation of the constant then follows from the fact that $P(4) = 5$. In a similar fashion we translate Theorem 4.9 to obtain

Proposition 6.5. *Let* $\epsilon = \exp\left(\frac{2\pi\imath}{7}\right)$. *For all* x *of modulus less than one we have*

$$\sum_{l=0}^{6} \frac{(\epsilon^l x)^2}{\prod_{n=1}^{\infty}(1-(\epsilon^l x)^{\frac{n}{7}})} = 7\sum_{n=1}^{\infty} P(7n-2)x^n \tag{4.24}$$

$$= 7^2 \prod_{n=1}^{\infty} \frac{(1-x^{7n})^3}{(1-x^n)^4} + 7^3 \prod_{n=1}^{\infty} \frac{(1-x^{7n})^7}{(1-x^n)^8}.$$

7. Identities via logarithmic differentiation

One of the very useful things about infinite product expansions is that they are amenable to logarithmic differentiation. In this section we derive some identities which follow from this operation. The basic idea is to construct a meromorphic function on a Riemann surface which is expressible as a product, and differentiate logarithmically, obtaining in this way a meromorphic differential on the surface with specified poles and residues. If we are also able to construct the same meromorphic differential in a different way, then we have derived an identity. Among the important number theoretic functions we will encounter is the classical σ-function

$$\sigma(n) = \sum_{d \in \mathbb{Z}^+, d|n} d, \; n \in \mathbb{Z}^+,$$

the sum of the positive divisors of a positive integer. This function will reappear prominently in Chapter 7.

We begin with a very classical example. Consider the Riemann surface $\mathbb{H}^2/\Gamma(2)$ of type $(0,3)$. The function $\dfrac{\theta^4\begin{bmatrix}0\\0\end{bmatrix}}{\theta^4\begin{bmatrix}0\\1\end{bmatrix}}$ is a holomorphic universal covering map of this surface (that we identify with the Riemann sphere punctured at 0, 1 and ∞) that takes the points 1 and 0 on the real axis to the punctures 0 and ∞ on the sphere. It follows that the logarithmic derivative (with respect to τ) of this function projects to a meromorphic differential (abelian differential of the third kind) on $\overline{\mathbb{H}^2/\Gamma(2)}$ with simple poles at the points P_1 and P_0 with residues ± 1. To obtain an alternate expression for this differential, we prove

Lemma 7.1. *The function $f(\tau) = \frac{\eta^8(2\tau)}{\eta^4(\tau)}$ projects to a meromorphic differential on $\overline{\mathbb{H}^2/\Gamma(2)}$ with simple poles at the points P_1 and P_0 (that is holomorphic elsewhere).*

Proof. The transformation theory tells us that f transforms like an automorphic 1-form but may possibly have a twist; that is, $f(\gamma(\tau))\gamma'(\tau)c_\gamma = f(\tau)$ with $c_\gamma \in \mathbb{C}^*$, $|c_\gamma| = 1$. Furthermore f is regular except possibly at the cusps, $\deg(f) = 1$ and $\mathrm{ord}_\infty f = 1$. It follows that f is also regular at the cusps not equivalent to ∞. A high enough power of f is then certainly a modular form for $\Gamma(2)$. Since $\overline{\mathbb{H}^2/\Gamma(2)}$ is simply connected we can extract roots and conclude that f itself is a modular form; that is $c_\gamma = 1$ for all $\gamma \in \Gamma(2)$. It follows that f projects to $\overline{\mathbb{H}^2/\Gamma(2)}$ as an abelian differential of the third kind with simple poles and these only at P_0 and P_1.

We have therefore derived the following

Theorem 7.2. *There exists a nonzero constant c such that for all $\tau \in \mathbb{H}^2$ we have*

$$\frac{d}{d\tau}\log\frac{\theta^4\begin{bmatrix}0\\0\end{bmatrix}(0,\tau)}{\theta^4\begin{bmatrix}0\\1\end{bmatrix}(0,\tau)} = c\frac{\left(\theta'\begin{bmatrix}1\\1\end{bmatrix}(0,2\tau)\right)^{\frac{8}{3}}}{\left(\theta'\begin{bmatrix}1\\1\end{bmatrix}(0,\tau)\right)^{\frac{4}{3}}}. \tag{4.25}$$

The infinite product translation of the identity of the last theorem is

Proposition 7.3. *For all x of modulus less than one we have*

(4.26)

$$x\prod_{n=1}^{\infty}(1-x^{2n})^4(1+x^{2n})^8 = \frac{1}{2}\sum_{n=1}^{\infty}\left(\frac{(2n-1)x^{2n-1}}{1+x^{2n-1}}+\frac{(2n-1)x^{2n-1}}{1-x^{2n-1}}\right).$$

Proof. We write the various theta constants in the equality of the last theorem in terms of their infinite product expansions as we have done previously. The side of the equation involving θ' simplifies, ignoring the constant to the left hand side of equation (4.26). We logarithmically differentiate the right hand side using the change of variables $x = \exp(\pi\imath\tau)$ (hence $\frac{d}{d\tau} = \pi\imath x\frac{d}{dx}$). The constant c is computed by equating the coefficients of x on each side in the power series representations.

Remark 7.4. (**a**) The reader will possibly observe that we can replace the right hand side of (4.26) with the power series $\sum_{n=1}^{\infty}\sigma(2n-1)x^{2n-1}$, where $\sigma(n)$ is the sum of the positive divisors of $n \in \mathbb{Z}^+$, and that the left hand side is essentially $\theta^4\begin{bmatrix}1\\0\end{bmatrix}(0,\tau)$, which leads to a remarkable relation between the σ-function and the representation of a positive integer as a sum of four triangular numbers. We shall return to this in Chapter 7.
(**b**) There are many other identities that can be derived for the $k = 2$ theory by applying the same method to different uniformizations. Given an identity based on the uniformization of a group $G \subset \Gamma$, applying an inner automorphism to it will, of course, reproduce the identity. However an outer automorphism will, in general, produce a new identity.

In order to obtain our next result we use the uniformization of $\mathbb{H}^2/\Gamma(3)$ obtained in Chapter 3. There we showed that the meromorphic function $g(\tau) = \frac{\theta^3\begin{bmatrix}\frac{1}{3}\\\frac{1}{3}\end{bmatrix}(0,\tau)}{\theta^3\begin{bmatrix}\frac{1}{3}\\\frac{5}{3}\end{bmatrix}(0,\tau)}$ is a holomorphic universal covering map of the four times punctured sphere $\overline{\mathbb{H}^2/\Gamma(3)}$ with $g(-1) = 0$ and $g(1) = \infty$. It follows that the logarithmic derivative of g will project to $\overline{\mathbb{H}^2/\Gamma(3)}$ as an abelian differential of the third kind with simple poles at P_{-1} and P_1. We are once again faced with the problem of constructing this meromorphic differential in another way.

Lemma 7.5. *The automorphic form defined by* $f(\tau) = \frac{\eta^3(3\tau)\eta^3(\frac{\tau}{3})}{\eta^2(\tau)}$ *projects to a meromorphic differential on* $\overline{\mathbb{H}^2/\Gamma(3)}$ *with simple poles at the points* P_{-1} *and* P_1.

Proof. The transformation theory and the material in Chapter 3 tells us that f^{24} is a 24-form that is holomorphic at the cusps and does not vanish at -1 and 1. Since the genus of $\overline{\mathbb{H}^2/\Gamma(3)}$ is zero we can extract a 24-th root to obtain a 1-form. The lemma follows readily.

We now use either Euler's identity or the Jacobi triple product formula to write g in the variable $x = \exp(\frac{2\pi\imath\tau}{3})$ as

$$g(\tau) = c\prod_{n=1}^{\infty}\frac{(1-(\omega x)^n)^3}{(1-(\omega^2 x)^n)^3}, \quad \omega = \exp\left(\frac{2\pi\imath}{3}\right). \tag{4.27}$$

We have derived

Theorem 7.6. *For all* $\tau \in \mathbb{H}^2$ *we have*

$$\frac{d}{d\tau}\log\frac{\theta^3\begin{bmatrix}\frac{1}{3}\\ \frac{1}{3}\end{bmatrix}(0,\tau)}{\theta^3\begin{bmatrix}\frac{1}{3}\\ \frac{5}{3}\end{bmatrix}(0,\tau)} = c\frac{\theta'\begin{bmatrix}1\\1\end{bmatrix}(0,3\tau)\theta'\begin{bmatrix}1\\1\end{bmatrix}(0,\frac{\tau}{3})}{\left(\theta'\begin{bmatrix}1\\1\end{bmatrix}(0,\tau)\right)^{\frac{2}{3}}}. \tag{4.28}$$

Proposition 7.7. *For all* x *of modulus less than one we have*

$$x\prod_{n=1}^{\infty}\frac{(1-x^{9n})^3(1-x^n)^3}{(1-x^{3n})^2} = -\sqrt{3}\imath\sum_{n=1}^{\infty}\left(\frac{n(\omega^2 x)^n}{1-(\omega^2 x)^n} - \frac{n(\omega x)^n}{1-(\omega x)^n}\right). \tag{4.29}$$

Remark 7.8. The above theorems and propositions are examples of the type of results which are obtainable from an analysis involving the relation between functions and differentials on surfaces of low genera. We shall now be a bit more general and perhaps make clearer how we are able to arrive at the given expressions for meromorphic differentials.

We begin by considering the following three automorphic forms:

$$\eta^{24}(k\tau),\ \eta^{24}(\tau) \text{ and } \eta^{24}\left(\frac{\tau}{k}\right).$$

The transformation theory and our work in Chapter 3 shows that the middle one is a 6-form for the entire group $\Gamma = \text{PSL}(2,\mathbb{Z})$, the first is a 6-form for the subgroup $\Gamma_o(k)$ and the last for the subgroup $\Gamma^o(k)$. It follows that all three are 6-forms for $\Gamma(k,k) = \Gamma_o(k) \cap \Gamma^o(k)$, hence also for $\Gamma(k)$. The divisors of the projections of these 6-differentials to $\overline{\mathbb{H}^2/\Gamma(k,k)}$ are respectively

$$P_\infty^{k^2-6} P_0^{-5} P_1^{-5} \dots P_{k-1}^{-5},\ P_\infty^{k-6} P_0^{k-6} P_1^{k-6} \dots P_{k-1}^{k-6} \text{ and } P_\infty^{-5} P_0^{k^2-6} P_1^{-5} \dots P_{k-1}^{-5}.$$

If we now take three integers, s, t, u, and consider $\eta^{24s}(k\tau)\eta^{24t}(\tau)\eta^{24u}\left(\frac{\tau}{k}\right)$, we find that it projects to a meromorphic $6(s+t+u)$-differential on $\overline{\mathbb{H}^2/\Gamma(k,k)}$ with divisor

$$P_\infty^{(k^2-6)s+(k-6)t-5u} P_0^{(k^2-6)u+(k-6)t-5s} P_1^{-5s+(k-6)t-5u} \dots P_{k-1}^{-5s+(k-6)t-5u}.$$

The differential constructed in the proof of Theorem 7.2 corresponds to choosing $s = 2$, $t = -1$, $u = 0$ after extracting a sixth root.

A different way to proceed is to construct an abelian differential of the third kind whose only singularities are simple poles at P_∞ and P_0. This is achieved by choosing $s = u = -2$, $t = 5$ and again extracting a sixth root. There are two ways to construct univalent functions on $\overline{\mathbb{H}^2/\Gamma(2)} = \overline{\mathbb{H}^2/\Gamma(2,2)}$ with zeros and poles only at P_∞ and P_0 leading to

Theorem 7.9. *There exist constants c_1 and $c_2 \in \mathbb{C}^*$ such that*

$$\frac{\eta^{20}(\tau)}{\eta^8(2\tau)\eta^8\left(\frac{\tau}{2}\right)} = c_1 \frac{d}{d\tau} \log \frac{\theta^4\begin{bmatrix}1\\0\end{bmatrix}(0,\tau)}{\theta^4\begin{bmatrix}0\\1\end{bmatrix}(0,\tau)} = c_2 \frac{d}{d\tau} \log \frac{\eta^8(2\tau)}{\eta^8\left(\frac{\tau}{2}\right)}$$

for all $\tau \in \mathbb{H}^2$.

Another way to proceed is to construct a meromorphic differential with simple poles at P_1 and P_∞. The result that can now be established is

Theorem 7.10. *There exist constants c such that for all $\tau \in \mathbb{H}^2$,*

$$c\frac{\eta^8\left(\frac{\tau}{2}\right)}{\eta^4(\tau)} = \frac{d}{d\tau} \log \frac{\theta^4\begin{bmatrix}0\\0\end{bmatrix}(0,\tau)}{\theta^4\begin{bmatrix}1\\0\end{bmatrix}(0,\tau)}.$$

The above results are basically consequences of the $\Gamma(2)$ theory which is classical and has been understood for a long time. In order to illustrate this fact we return now to a result of Chapter 2 where we proved Theorem 5.3. Recall that this theorem was the source of one of our proofs of the quartic identity and the Jacobi derivative formula. Now that we already know the Jacobi derivative formula and the infinite product expansions of theta functions we can say more. The above theorem and the Jacobi derivative formula tell us that

$$\pi^2 = \frac{\dfrac{\theta''\begin{bmatrix}0\\1\end{bmatrix}}{\theta\begin{bmatrix}0\\1\end{bmatrix}} - \dfrac{\theta''\begin{bmatrix}1\\0\end{bmatrix}}{\theta\begin{bmatrix}1\\0\end{bmatrix}}}{\theta^4\begin{bmatrix}0\\0\end{bmatrix}} = \frac{\dfrac{\theta''\begin{bmatrix}0\\0\end{bmatrix}}{\theta\begin{bmatrix}0\\0\end{bmatrix}} - \dfrac{\theta''\begin{bmatrix}1\\0\end{bmatrix}}{\theta\begin{bmatrix}1\\0\end{bmatrix}}}{\theta^4\begin{bmatrix}0\\1\end{bmatrix}} = \frac{\dfrac{\theta''\begin{bmatrix}0\\1\end{bmatrix}}{\theta\begin{bmatrix}0\\1\end{bmatrix}} - \dfrac{\theta''\begin{bmatrix}0\\0\end{bmatrix}}{\theta\begin{bmatrix}0\\0\end{bmatrix}}}{\theta^4\begin{bmatrix}1\\0\end{bmatrix}}.$$

In our work on the Jacobi triple product we derived the following identities involving power series and infinite product expansions in terms of the variable $x = \exp(\pi\imath\tau)$ (these are expanded versions of (2.35), (2.36) and (2.37)):

$$\theta\begin{bmatrix}0\\0\end{bmatrix}(0,\tau) = 1 + 2\sum_{n=1}^{\infty} x^{n^2} = \prod_{n=1}^{\infty}(1-x^{2n})(1+x^{2n-1})^2,$$

$$\theta\begin{bmatrix}0\\1\end{bmatrix}(0,\tau) = 1 + 2\sum_{n=1}^{\infty} (-1)^n x^{n^2} = \prod_{n=1}^{\infty}(1-x^{2n})(1-x^{2n-1})^2,$$

$$\theta\begin{bmatrix}1\\0\end{bmatrix}(0,\tau) = 2x^{\frac{1}{4}}\sum_{n=0}^{\infty} x^{n(n+1)} = 2x^{\frac{1}{4}}\prod_{n=1}^{\infty}(1-x^{2n})(1+x^{2n})^2.$$

An immediate consequence of the last quoted theorem and the above infinite product expansion is

Corollary 7.11. *Let $S_4(n)$ be the number of representations of $n \in \mathbb{Z}^+$ as a sum of four squares*[99] *and $T_4(n)$, the number of representations of n as a sum of four triangular numbers (of the form $\frac{m(m+1)}{2}$ with $m \in \mathbb{Z}$). Then,*

$$1 + \sum_{n=1}^{\infty} S_4(n)x^n = \left(\sum_{n=-\infty}^{\infty} x^{n^2}\right)^4 = \prod_{n=1}^{\infty}(1-x^{2n})^4(1+x^{2n-1})^8$$

$$= 1 + 8\left(\sum_{n=1}^{\infty}\frac{(2n-1)x^{2n-1}}{1-x^{2n-1}} + \sum_{n=1}^{\infty}\frac{2nx^{2n}}{1+x^{2n}}\right)$$

and

$$1 + \frac{1}{16}\sum_{n=1}^{\infty} T_4(n)x^n = \frac{1}{16}\left(\sum_{n=-\infty}^{\infty} x^{\frac{n(n+1)}{2}}\right)^4 \tag{4.30}$$

$$= \prod_{n=1}^{\infty}(1-x^n)^4(1+x^n)^8 = \sum_{n=1}^{\infty}\frac{(2n-1)x^{n-1}}{1-x^{2n-1}}.$$

[99] The order of the squares is material; also, for $a \neq 0$, a^2 and $(-a)^2$ count as different squares. Similarly, each triangular number has two representations.

Proof. We work with the identities of Theorem 5.3 of Chapter 2 listed above. We use the heat equation to replace the second derivative with respect to z by the first derivative with respect to τ and use the observation that they are thus transformed to equations involving logarithmic derivatives. We then use the power series and product expansions of the theta functions with integral characteristics in terms of the variable $x = \exp(\pi\imath\tau)$. In particular,

$$\pi^2\theta^4\begin{bmatrix}0\\0\end{bmatrix} = \frac{\theta''\begin{bmatrix}0\\1\end{bmatrix}}{\theta\begin{bmatrix}0\\1\end{bmatrix}} - \frac{\theta''\begin{bmatrix}1\\0\end{bmatrix}}{\theta\begin{bmatrix}1\\0\end{bmatrix}}$$

$$= 4\pi\imath\frac{\dot\theta\begin{bmatrix}0\\1\end{bmatrix}}{\theta\begin{bmatrix}0\\1\end{bmatrix}} - \frac{\dot\theta\begin{bmatrix}1\\0\end{bmatrix}}{\theta\begin{bmatrix}1\\0\end{bmatrix}}$$

$$= 4\pi\imath\frac{d}{d\tau}\log\frac{\theta\begin{bmatrix}0\\1\end{bmatrix}}{\theta\begin{bmatrix}1\\0\end{bmatrix}}.$$

Similarly,

$$\pi\theta^4\begin{bmatrix}0\\1\end{bmatrix} = 4\imath\frac{d}{d\tau}\log\frac{\theta\begin{bmatrix}0\\0\end{bmatrix}}{\theta\begin{bmatrix}1\\0\end{bmatrix}} \text{ and } \pi\theta^4\begin{bmatrix}1\\0\end{bmatrix} = 4\imath\frac{d}{d\tau}\log\frac{\theta\begin{bmatrix}0\\1\end{bmatrix}}{\theta\begin{bmatrix}0\\0\end{bmatrix}}.$$

The first and third of the last three formulae yield the two identities of the corollary. We expand the left hand side of each identity in a power series and the right hand side (before differentiating) as an infinite product.

Remark 7.12. In consonance with the remark following Theorem 7.2 we show that

$$\sum_{n=1}^{\infty}\frac{(2n-1)x^{n-1}}{1-x^{2n-1}} = \sum_{n=0}^{\infty}\sigma(2n+1)x^n.$$

This follows from the fact that for any three numbers N, n and m which satisfy the condition

$$N = n - 1 + (2n-1)m$$

it is also true that

$$2N + 1 = (2n-1)(2m+1).$$

In fact, the second equation in the corollary is equivalent to Theorem 7.2.

The above $\Gamma(2)$ theory is classical; the one for $\Gamma(3)$ is newer. The surface $\mathbb{H}^2/\Gamma(3)$ is a sphere punctured at four points; these are P_∞, P_0, P_1 and P_2. The divisor of the $6(s+t+u)$-form $\eta^{24s}(3\tau)\eta^{24t}(\tau)\eta^{24u}\left(\frac{\tau}{3}\right)$, that we have been studying, on $\overline{\mathbb{H}^2/\Gamma(3,3)}$ is

$$P_\infty^{3s-3t-5u}P_0^{3u-3t-5s}P_1^{-5s-3t-5u}P_2^{-5s-3t-5u}.$$

We construct a meromorphic 6-differential which is regular at the points P_1, P_2 and which has the same order at the points P_0, P_∞. The conditions on the integers s, t, u are $5s+3t+5u=0$ and $s=u$. A simple choice is $s=u=-3$ and $t=10$, giving rise to the 24-form $\frac{\eta^{240}(\tau)}{\eta^{72}(3\tau)\ \eta^{72}\left(\frac{\tau}{3}\right)}$ whose projection to $\overline{\mathbb{H}^2/\Gamma(3,3)}$ has divisor $P_\infty^{-24}P_0^{-24}$. Extracting a 24-th root gives us the 1-form $\frac{\eta^{10}(\tau)}{\eta^{3}(3\tau)\ \eta^{3}\left(\frac{\tau}{3}\right)}$ whose projection to the surface has simple poles at P_∞ and P_0.

The function $f(\tau)=\frac{\eta^3(3\tau)}{\eta^3\left(\frac{\tau}{3}\right)}$ projects to a meromorphic function on the same Riemann surface with divisor $\frac{P_\infty}{P_0}$. Hence $\frac{d}{d\tau}\log\frac{\eta^3(3\tau)}{\eta^3\left(\frac{\tau}{3}\right)}$ projects to a meromorphic 1-form with simple poles at these points. It follows that we have derived the identity in

Theorem 7.13. *There is a nonzero constant c such that for all $\tau\in\mathbb{H}^2$,*

$$c\frac{\eta^{10}(\tau)}{\eta^3(3\tau)\ \eta^3\left(\frac{\tau}{3}\right)}=\frac{d}{d\tau}\log\frac{\eta^3(3\tau)}{\eta^3\left(\frac{\tau}{3}\right)}.$$

The translation to infinite products is

Proposition 7.14. *For all x of modulus less than one we have* [100]

$$(4.31)\qquad \prod_{n=1}^{\infty}\frac{(1-x^{3n})^{10}}{(1-x^n)^3(1-x^{9n})^3}=1+3\sum_{n=1}^{\infty}n\left(\frac{x^n}{1-x^n}-\frac{9x^{9n}}{1-x^{9n}}\right)$$

$$=1+3\sum_{m=1}^{\infty}\sigma(m)(x^m-9x^{9m})=1+3\sum_{m=1}^{\infty}\left(\sigma(m)-9\sigma\left(\frac{m}{9}\right)\right)x^m.$$

In the above expression, as usual, we extend the function σ to be zero on the noninteger positive rationals.

Remark 7.15. (**a**) The form used to prove Theorem 7.6 which is regular at P_0 and P_∞ and has poles at P_1 and P_2 corresponds to the choice of $s=u=3$ and $t=-2$.
(**b**) The above methods need to be modified to produce results for primes ≥ 5 because the groups involved represent surfaces with many punctures and our methods do not handle the complicated bookkeeping involved.

[100] $\sigma(n)$ is the sum of the positive divisors of $n\in\mathbb{Z}^+$.

8. Averaging automorphic forms

In §4 of this chapter we saw examples where the theory of modular forms gives rise to identities among theta constants. In this section we continue this approach and obtain some additional identities. The same three forms which were crucial in the discussion of §7 will be used now.

We begin with the observation that for $k = 2$ and 3, the automorphic form $\left(\frac{\eta^{\frac{k^2+1}{k}}(\tau)}{\eta(k\tau)\ \eta\left(\frac{\tau}{k}\right)}\right)^{\frac{24}{k^2-1}}$ projects to a meromorphic differential on $\overline{\mathbb{H}^2/\Gamma(k,k)}$ with singularities (simple poles) at P_0 and P_∞. This follows from an evaluation of the divisors of the $\Gamma(k,k)$ forms given in §7. We have chosen $s = u = \frac{-1}{k^2-1}$ and $t = \frac{k^2+1}{k(k^2-1)}$. For $k = 2, 3$ these are 1-forms. We can average these forms over $\Gamma(k,k)\backslash\Gamma_o(k)$, as we have done on previous occasions, to obtain the $\Gamma_o(k)$-form $\sum_{l=0}^{k-1}\left(\frac{\eta^{\frac{k^2+1}{k}}(\tau+l)}{\eta(k(\tau+l))\ \eta\left(\frac{\tau+l}{k}\right)}\right)^{\frac{24}{k^2-1}}$.

The interesting sum defined above projects to a meromorphic differential on $\overline{\mathbb{H}^2/\Gamma_o(k)}$ with simple poles at P_0 and P_∞ (viewed, of course, as points on $\mathbb{H}^2/\Gamma_o(k)$ rather than as points on $\mathbb{H}^2/\Gamma(k,k)$).

We can, as we have done many times in this chapter, construct this differential in another way. We start with a meromorphic function on $\overline{\mathbb{H}^2/\Gamma_o(k)}$ with a simple pole at P_0 and a simple zero at P_∞, take its logarithmic derivative and obtain a differential of the type needed. The required function is $\left(\frac{\eta(k\tau)}{\eta(\tau)}\right)^{\frac{24}{k-1}}$. This is, of course, not the same function that was used in the previous section. We have arrived at

Theorem 8.1. *For $k = 2$ and 3 there is a constant $c = c(k)$ such that*

$$c\sum_{l=0}^{k-1}\left(\frac{\eta^{\frac{k^2+1}{k}}(\tau+l)}{\eta(k(\tau+l))\ \eta\left(\frac{\tau+l}{k}\right)}\right)^{\frac{24}{k^2-1}} = \frac{d}{d\tau}\log\left(\frac{\eta(k\tau)}{\eta(\tau)}\right)^{\frac{24}{k-1}} \tag{4.32}$$

for all $\tau \in \mathbb{H}^2$.

Remark 8.2. We have used very strongly in the above discussion the fact that for $k = 2$ and 3, $\overline{\mathbb{H}^2/\Gamma(k,k)}$ and $\overline{\mathbb{H}^2/\Gamma_o(k)}$ are of genus zero (which allowed us to extract appropriate roots). The reader may wonder why we did not include the case $k = 5$ in the above discussion since in this case as well the genera of the relevant surfaces are zero. Of course, the form with which we began the discussion, $\left(\frac{\eta^{\frac{k^2+1}{k}}(\tau)}{\eta(k\tau)\ \eta\left(\frac{\tau}{k}\right)}\right)^{\frac{24}{k^2-1}}$, is not a 1-form for $k = 5$. We could perhaps have changed the formula. The reason we did not even try to do this is that for this case, the problem is also from another source. It is due to the presence of torsion. The orbifolds $\mathbb{H}^2/\Gamma(5,5)$ and $\mathbb{H}^2/\Gamma_o(5)$ have signatures $(0,8;\ 2,2,\infty,\infty,\infty,\infty,\infty,\infty)$ and $(0,4;\ 2,2,\infty,\infty)$, respectively. A holomorphic 1-form for these groups must vanish at the elliptic fixed points of the group; the function $\frac{\eta^{\frac{26}{5}}(\tau)}{\eta(5\tau)\eta(\frac{\tau}{5})}$ does not vanish on the upper half plane. It hence cannot be a power of an automorphic 1-form.

Using the ideas already introduced in this chapter, we see that the left hand side of equation (4.32) aside from multiplication by a constant is

$$\prod_{n=1}^{\infty}\frac{\left(1-x^{kn}\right)^{\frac{24(k^2+1)}{k(k^2-1)}}}{\left(1-x^{k^2n}\right)^{\frac{24}{k^2-1}}}\sum_{l=0}^{k-1}\frac{1}{\prod_{n=1}^{\infty}\left(1-\left(\exp\left(\frac{2\pi\imath l}{k}\right)x\right)^n\right)^{\frac{24}{k^2-1}}},\ x=\exp\left(\frac{2\pi\imath\tau}{k}\right).$$

The same arguments give the right hand side of equation (4.32), again aside from multiplication by a constant, as

$$1+\frac{24}{k-1}\sum_{n=1}^{\infty}n\left(\frac{x^{kn}}{1-x^{kn}}-\frac{kx^{k^2n}}{1-x^{k^2n}}\right).$$

In order to simplify our notation we introduce the function P_N defined by (see (5.1) of Chapter 5, devoted to a study of these partition functions) $\sum_{n=0}^{\infty}P_N(n)x^n=\prod_{n=1}^{\infty}\frac{1}{(1-x^n)^N}$, and observe that our last two displayed formulae define functions of x^k (so we can replace x^k by x). We obtain

Corollary 8.3. *For $k=2$ and 3 we have*

(4.33)
$$\prod_{n=1}^{\infty}\frac{\left(1-x^{n}\right)^{\frac{24(k^2+1)}{k(k^2-1)}}}{\left(1-x^{kn}\right)^{\frac{24}{k^2-1}}}\sum_{n=0}^{\infty}P_{\frac{24}{k^2-1}}(kn)x^n=1+\frac{24}{k-1}\sum_{n=1}^{\infty}n\left(\frac{x^{n}}{1-x^{n}}-\frac{kx^{kn}}{1-x^{kn}}\right)$$

for all $x\in\mathbb{C}$, $|x|<1$.

We can obtain additional results related to the above discussion. For an odd prime k, the sum $\sum_{l=0}^{\frac{k-3}{2}}\left(\frac{\theta'\begin{bmatrix}1\\ \frac{2l+1}{k}\end{bmatrix}}{\theta\begin{bmatrix}1\\ \frac{2l+1}{k}\end{bmatrix}}\right)^2$ defines a holomorphic 1-form for $\Gamma_o(k)$. We hence conclude

Theorem 8.4. *For $k = 3,\ 5,\ 7$ and 13, there exists a constant $c = c(k)$ such that*

$$c\sum_{l=0}^{\frac{k-3}{2}}\left(\frac{\theta'\begin{bmatrix}1\\ \frac{2l+1}{k}\end{bmatrix}(0,\tau)}{\theta\begin{bmatrix}1\\ \frac{2l+1}{k}\end{bmatrix}(0,\tau)}\right)^2 = \frac{d}{d\tau}\log\left(\frac{\eta(k\tau)}{\eta(\tau)}\right)^{\frac{24}{k-1}}$$

for all $\tau \in \mathbb{H}^2$.

Corollary 8.5. *For $k = 3,\ 5,\ 7$ and 13,*

$$\sum_{l=0}^{\frac{k-3}{2}}\left(\frac{\sum_{n=0}^{\infty}\left(n+\frac{1}{2}\right)\sin\left(\frac{\pi(2l+1)(2n+1)}{2k}\right)x^{\frac{n(n+1)}{2}}}{\sum_{n=0}^{\infty}\cos\left(\frac{\pi(2l+1)(2n+1)}{2k}\right)x^{\frac{n(n+1)}{2}}}\right)^2$$

$$= \frac{(k-2)(k-1)}{24}\left(1+\frac{24}{k-1}\sum_{n=1}^{\infty}n\left(\frac{x^n}{1-x^n}-\frac{kx^{kn}}{1-x^{kn}}\right)\right)$$

for all $x \in \mathbb{C}$, $|x| < 1$.

Proof. We have used Corollary 2.6 of Chapter 3, where we showed that for every odd integer $k \geq 3$, $\sum_{l=0}^{\frac{k-3}{2}}\tan^2\left(\frac{\pi(2l+1)}{2k}\right) = \frac{(k-2)(k-1)}{6}$.

Exercise 8.6. Obtain the analogue of the last theorem and its corollary for the even prime 2.

Two questions are immediately evident as a result of the last theorem. Is there another, easier, way to produce the holomorphic 1-forms appearing on the left hand side of the equation of the theorem? Why is the prime 11 left out?

As far as the first question, $f = \dfrac{\theta'\begin{bmatrix}1\\ \frac{1}{5}\end{bmatrix}}{\theta\begin{bmatrix}1\\ \frac{1}{5}\end{bmatrix}}\dfrac{\theta'\begin{bmatrix}1\\ \frac{3}{5}\end{bmatrix}}{\theta\begin{bmatrix}1\\ \frac{3}{5}\end{bmatrix}}$ is a candidate for a holomorphic 1-form for $\Gamma_o(5)$; it certainly satisfies

$$f(\gamma(\tau))\gamma'(\tau)c_\gamma = f(\tau),\ \gamma \in \Gamma_o(5),\ \tau \in \mathbb{H}^2,$$

with $c_\gamma \in \mathbb{C}$, $|c_\gamma| = 1$. It is also holomorphic and nonzero at the cusps. It might however not vanish at the elliptic fixed points (of order 2) of $\Gamma_o(5)$. If f were a modular 1-form for $\Gamma_o(5)$, then for some nonzero constant c, we would have

$$c\frac{\theta'\begin{bmatrix}1\\ \frac{1}{5}\end{bmatrix}}{\theta\begin{bmatrix}1\\ \frac{1}{5}\end{bmatrix}}\frac{\theta'\begin{bmatrix}1\\ \frac{3}{5}\end{bmatrix}}{\theta\begin{bmatrix}1\\ \frac{3}{5}\end{bmatrix}} = \left(\frac{\theta'\begin{bmatrix}1\\ \frac{1}{5}\end{bmatrix}}{\theta\begin{bmatrix}1\\ \frac{1}{5}\end{bmatrix}}\right)^2 + \left(\frac{\theta'\begin{bmatrix}1\\ \frac{3}{5}\end{bmatrix}}{\theta\begin{bmatrix}1\\ \frac{3}{5}\end{bmatrix}}\right)^2.$$

A comparison of the first two terms of the respective Fourier series expansions shows that no such constant can exist. The group $\Gamma_o(5)$ of signature $(0,4;\ 2,2,\infty,\infty)$ has the presentation (See the next section for more details on the presentation of $\Gamma_o(5)$.)

$$< E_1, E_2, P_1, P_2; E_i^2 = I, P_i \text{ is parabolic}, E_1E_2P_1P_2 = I > .$$

It must be the case that $c_\gamma \neq 1$ for at least one generator γ. It is easily seen that $c_{P_i} = 1$. Hence $c_{E_i} = -1$. We also conclude[101] that at any elliptic fixed point of $\Gamma_o(5)$ either $\theta'\begin{bmatrix} 1 \\ \frac{1}{5} \end{bmatrix}$ or $\theta'\begin{bmatrix} 1 \\ \frac{3}{5} \end{bmatrix}$ does not vanish, but the sum

$$\left(\frac{\theta'\begin{bmatrix} 1 \\ \frac{1}{5} \end{bmatrix}}{\theta\begin{bmatrix} 1 \\ \frac{1}{5} \end{bmatrix}}\right)^2 + \left(\frac{\theta'\begin{bmatrix} 1 \\ \frac{3}{5} \end{bmatrix}}{\theta\begin{bmatrix} 1 \\ \frac{3}{5} \end{bmatrix}}\right)^2 \text{ does.}$$

For the case $k = 11$, we present a small preview of the next chapter, where the details of the proofs will be found. The function $\left(\frac{\eta(11\tau)}{\eta(\tau)}\right)^{12} \in \mathcal{K}(\Gamma_o(11))$ with divisor $\frac{P_\infty^5}{P_0^5}$. The sum $\sum_{l=0}^4 \left(\frac{\theta'\begin{bmatrix} 1 \\ \frac{2l+1}{11} \end{bmatrix}}{\theta\begin{bmatrix} 1 \\ \frac{2l+1}{11} \end{bmatrix}}\right)^2$ defines a modular 1-form for $\Gamma_o(11)$ whose projection to $\mathbb{H}^2/\Gamma_o(11)$ has simple poles at the two punctures P_∞ and P_0 of $\mathbb{H}^2/\Gamma_o(11)$. The space of such forms has dimension 2. We must also identify the cusp 1-form for $\Gamma_o(11)$. It is given by $\eta(11\tau)^2\eta(\tau)^2$. We conclude that

Theorem 8.7. *There exist constants c_1 and c_2 such that*

$$c_1 \sum_{l=0}^{4} \left(\frac{\theta'\begin{bmatrix} 1 \\ \frac{2l+1}{11} \end{bmatrix}(0,\tau)}{\theta\begin{bmatrix} 1 \\ \frac{2l+1}{11} \end{bmatrix}(0,\tau)}\right)^2 + c_2(\eta(11\tau)\eta(\tau))^2 = \frac{d}{d\tau}\log\left(\frac{\eta(11\tau)}{\eta(\tau)}\right)^{12}$$

for all $\tau \in \mathbb{H}^2$.

Remark 8.8. We will see below (Remark 9.9 **(b)**) that $c_2 = 0$.

---

[101]The sum φ (because it is a 1-form for $\Gamma_o(5)$) must vanish at each elliptic fixed point (because they are of order two). Each elliptic fixed point contributes a positive integral multiple $\frac{1}{2}$ to the degree of the divisor (φ). Since deg $(\varphi) = 2$, the zeros of φ at the elliptic fixed points must be simple and φ has no other zeros. If both $\theta'\begin{bmatrix} 1 \\ \frac{1}{5} \end{bmatrix}$ and $\theta'\begin{bmatrix} 1 \\ \frac{3}{5} \end{bmatrix}$ were to vanish at an elliptic fixed point, then φ would have at least a double zero there (contributing at least 1 to deg (φ)) and the other (inequivalent) elliptic fixed point would still contribute at least $\frac{1}{2}$ to deg (φ), reaching a contradiction.

9. The groups $G(k)$

Our first aim is to establish function theoretically the beautiful identity of Ramanujan given in Corollary 9.2. We hence concentrate our attention, for the moment, on the prime 5. The infinite product expansion in terms of the local coordinate $x = \exp(2\pi\imath\tau)$, of the function $\omega(\tau) = \frac{\eta(\tau)^5}{\eta(5\tau)}$ is given by $\prod_{n=1}^{\infty} \frac{(1-x^n)^5}{(1-x^{5n})}$. Thus we have an obvious function theoretic interpretation of the left hand side of the equality in Corollary 9.2. The function ω is certainly a holomorphic multiplicative 1-form for $\Gamma_o(5)$. To describe this form more precisely, we consider the inclusion of groups

$$\Gamma(5) \subset G(5) \subset \Gamma_o(5)$$

(the first is of index 5 and the second normal of index 2); the corresponding orbifolds

$$\overline{\mathbb{H}^2/\Gamma(5)} \xrightarrow{\pi_1} \overline{\mathbb{H}^2/G(5)} \xrightarrow{\pi_2} \overline{\mathbb{H}^2/\Gamma_o(5)} \tag{4.34}$$

have signatures

$$(0,12;\ \infty,\ \ldots,\ \infty),\ (0,4;\ \infty,\ \infty,\ \infty,\ \infty) \text{ and } (0,4;\ 2,\ 2,\ \infty,\ \infty),$$

respectively. It is easiest to understand the above tower of covers in terms of the picture of $\mathbb{H}^2/\Gamma(5)$ in the previous chapter. The twelve punctures on $\mathbb{H}^2/\Gamma(5)$ are best organized into four groups:

$$P_\infty = P_{\frac{1}{5}};\ P_{\frac{2}{5}};\ P_{-2},\ P_{-1},\ P_0,\ P_1,\ P_2;\ P_{-\frac{3}{2}},\ P_{-\frac{1}{2}},\ P_{\frac{1}{2}},\ P_{\frac{3}{2}},\ P_{\frac{5}{2}}.$$

The motion $B = \begin{bmatrix} 1 & 1 \\ 0 & 1 \end{bmatrix} \in G(5)$ generates $G(5)$ over $\Gamma(5)$; it produces an automorphism of period five of $\overline{\mathbb{H}^2/\Gamma(5)}$ that fixes pointwise the first two punctures listed above for $\mathbb{H}^2/\Gamma(5)$ and permutes cyclically the punctures in the third and fourth group of the list to produce the punctures

$$P_\infty,\ P_{\frac{2}{5}},\ P_0,\ P_{\frac{1}{2}} \text{ on } \mathbb{H}^2/G(5).$$

On the standard realization (model) of the sphere $\mathbb{H}^2/\Gamma(5)$ this map of period 5 corresponds to a power of the rotation $z \mapsto \exp\left(\frac{2\pi\imath}{5}\right) z$; the induced projection π_1 is thus $z \mapsto z^5$. An involution acts on $\mathbb{H}^2/G(5)$ to produce $\mathbb{H}^2/\Gamma_o(5)$; it identifies the first and second pair of punctures on the last list (it also fixes two points on $\mathbb{H}^2/G(5)$). An analysis of the cover π_2 shows that in the standard model it is of the form $z \mapsto z + \frac{\alpha}{z}$ for some $\alpha \in \mathbb{C}^*$. The fact that $G(5)$ is torsion free allows us to conclude, by analyzing its singularities, that ω is a holomorphic 1-form for $G(5)$. It is regular and nonzero at both $P_{\frac{1}{5}}$ and $P_{\frac{2}{5}}$. It has simple zeros at P_0 and $P_{\frac{1}{2}}$. The divisor (ω) of ω as a $G(5)$ modular 1-form is $P_0\ P_{\frac{1}{2}}$; its projection to $\overline{\mathbb{H}^2/G(5)}$ has divisor $P_{\frac{1}{5}}^{-1}\ P_{\frac{2}{5}}^{-1}$.

Our next task is to construct a meromorphic function on $\overline{\mathbb{H}^2/G(5)}$ with divisor $\frac{P_{\frac{1}{5}}}{P_{\frac{2}{5}}}$. Using the constructions of the previous chapter we see that

$$\left(\frac{\theta\begin{bmatrix}\frac{3}{5}\\1\end{bmatrix}(0,5\tau)}{\theta\begin{bmatrix}\frac{1}{5}\\1\end{bmatrix}(0,5\tau)}\right)^5$$

is such a function. Using logarithmic differentiation we conclude that

Theorem 9.1. *There exists a $c \in \mathbb{C}^*$ such that for all $\tau \in \mathbb{H}^2$,*

$$\frac{\eta(\tau)^5}{\eta(5\tau)} = c\frac{d}{d\tau}\log\left(\frac{\theta\begin{bmatrix}\frac{3}{5}\\1\end{bmatrix}(0,5\tau)}{\theta\begin{bmatrix}\frac{1}{5}\\1\end{bmatrix}(0,5\tau)}\right)$$

for all $\tau \in \mathbb{H}^2$.

The definitions of the theta constants and the Jacobi triple product formula tell us that with $x = \exp(2\pi\imath\tau)$, as above,

$$\frac{\theta\begin{bmatrix}\frac{3}{5}\\1\end{bmatrix}(0,5\tau)}{\theta\begin{bmatrix}\frac{1}{5}\\1\end{bmatrix}(0,5\tau)} = \exp\left(\frac{\pi\imath}{5}\right)x^{\frac{1}{5}}\frac{\sum_{n\in\mathbb{Z}}\left(\exp(\pi\imath)x^{\frac{3}{2}}\right)^n x^{\frac{5n^2}{2}}}{\sum_{n\in\mathbb{Z}}\left(\exp(\pi\imath)x^{\frac{1}{2}}\right)^n x^{\frac{5n^2}{2}}}$$

$$= \exp\left(\frac{\pi\imath}{5}\right)x^{\frac{1}{5}}\frac{\prod_{n=0}^{\infty}(1-x^{5n+1})(1-x^{5n+4})}{\prod_{n=0}^{\infty}(1-x^{5n+2})(1-x^{5n+3})}.$$

Using the fact that $\frac{d}{d\tau} = 2\pi\imath x\frac{d}{dx}$, we see that

$$\frac{d}{d\tau}\log\left(\frac{\theta\begin{bmatrix}\frac{3}{5}\\1\end{bmatrix}(0,5\tau)}{\theta\begin{bmatrix}\frac{1}{5}\\1\end{bmatrix}(0,5\tau)}\right) = 2\pi\imath x\left(\frac{1}{5} - \sum_{n=1}^{\infty}\frac{\chi(n)nx^{n-1}}{1-x^n}\right).$$

The constant c in the theorem is easily evaluated to yield

Corollary 9.2. *For all $x \in \mathbb{C}$ with $|x| < 1$,*

$$\prod_{n=1}^{\infty}\frac{(1-x^n)^5}{(1-x^{5n})} = 1 - 5\sum_{n=1}^{\infty}\frac{\chi(n)nx^n}{1-x^n},$$

where

$$\chi(n) = \begin{cases} +1 \text{ if } n \equiv \pm 1 \mod 5 \\ -1 \text{ if } n \equiv \pm 2 \mod 5 \\ 0 \text{ if } n \equiv 0 \mod 5 \end{cases} = \left(\frac{n}{5}\right). \tag{4.35}$$

Remark 9.3. In the above, we have extended in an obvious manner the definition of the Legendre symbol $\left(\frac{\cdot}{\cdot}\right)$. A combinatorial interpretation of the corollary is obtained when we observe that the power series representation of the right hand side is

$$1 - 5\sum_{n=1}^{\infty} f(n)x^n \text{ with } f(n) = \sum_{d|n} \chi(d)d.$$

It follows that f is multiplicative (in the sense of (4.36)). We conclude that $f(5^m) = 1$ and $f(5^m n) = f(n)$ for all $m \in \mathbb{Z}^+$ and all $n \in \mathbb{Z}^+$ with $(5, n) = 1$.

For all primes k (see the next chapter for the definition of $\alpha(k)$),

$$\omega(\tau) = \left(\frac{\eta(\tau)^k}{\eta(k\tau)}\right)^{\alpha(k)} \quad \text{and } \omega_1(\tau) = \left(\frac{\eta(k\tau)^k}{\eta(\tau)}\right)^{\alpha(k)}$$

are multiplicative $\alpha(k)\frac{k-1}{4}$-modular forms for $\Gamma_o(k)$ with respective divisors

$$P_0^{\alpha(k)\frac{k^2-1}{24}} \text{ and } P_\infty^{\alpha(k)\frac{k^2-1}{24}}.$$

This information is enough to conclude that for those k for which $G(k)$ is torsion free and $\overline{\mathbb{H}^2/G(k)}$ has genus zero, the factor of automorphy for these forms are powers of the canonical factors, that is, for $k = 5$ and 7,

$$\omega(\gamma(\tau))\left(\gamma'(\tau)\right)^{\alpha(k)\frac{k-1}{4}} = \omega(\tau) \text{ and } \omega_1(\gamma(\tau))\left(\gamma'(\tau)\right)^{\alpha(k)\frac{k-1}{4}} = \omega_1(\tau),$$

$$\text{for all } \gamma \in G(k) \text{ and all } \tau \in \mathbb{H}^2.$$

Theorem 9.4. *There exists a $c \in \mathbb{C}^*$ such that*

$$\frac{\eta(5\tau)^5}{\eta(\tau)} = c\frac{d}{d\tau}\log\left(\frac{\theta\begin{bmatrix}1\\\frac{3}{5}\end{bmatrix}(0,\tau)}{\theta\begin{bmatrix}1\\\frac{1}{5}\end{bmatrix}(0,\tau)}\right)$$

for all $\tau \in \mathbb{H}^2$.

Proof. The projection of the $G(5)$-invariant modular form ω_1 to $\overline{\mathbb{H}^2/G(5)}$ has divisor $P_0^{-1}P_{\frac{1}{2}}^{-1}$. Since the projection of the $G(5)$-invariant function $\left(\frac{\theta\begin{bmatrix}1\\\frac{3}{5}\end{bmatrix}}{\theta\begin{bmatrix}1\\\frac{1}{5}\end{bmatrix}}\right)^5$ has divisor $P_0P_{\frac{1}{2}}^{-1}$, the theorem follows readily.

As usual we rewrite the theta identity of the theorem in terms of local coordinates ($x = \exp(2\pi\imath\tau)$). Now,

$$\frac{\eta(5\tau)^5}{\eta(\tau)} = x\prod_{n=0}^{\infty}\frac{(1-x^{5n})^5}{(1-x^n)}$$

and

$$\frac{\theta\begin{bmatrix}1\\ \frac{3}{5}\end{bmatrix}(0,\tau)}{\theta\begin{bmatrix}1\\ \frac{1}{5}\end{bmatrix}(0,\tau)} = \exp\left(\frac{\pi\imath}{5}\right)\frac{\sum_{n\in\mathbb{Z}}\left(\exp\left(\frac{3\pi\imath}{5}\right)x^{\frac{1}{2}}\right)^n x^{\frac{n^2}{2}}}{\sum_{n\in\mathbb{Z}}\left(\exp\left(\frac{\pi\imath}{5}\right)x^{\frac{1}{2}}\right)^n x^{\frac{n^2}{2}}}$$

$$= \exp\left(\frac{\pi\imath}{5}\right)\frac{\prod_{n=0}^{\infty}\left(1+\exp\left(\frac{3\pi\imath}{5}\right)x^{n+1}\right)\left(1+\exp\left(-\frac{3\pi\imath}{5}\right)x^n\right)}{\prod_{n=0}^{\infty}\left(1+\exp\left(\frac{\pi\imath}{5}\right)x^{n+1}\right)\left(1+\exp\left(-\frac{\pi\imath}{5}\right)x^n\right)}.$$

Therefore

$$c\prod_{n=0}^{\infty}\frac{(1-x^{5n})^5}{(1-x^n)} = \sum_{n=0}^{\infty}\frac{(n+1)\exp\left(\frac{3\pi\imath}{5}\right)x^n}{1+\exp\left(\frac{3\pi\imath}{5}\right)x^{n+1}} + \sum_{n=0}^{\infty}\frac{n\exp\left(-\frac{3\pi\imath}{5}\right)x^{n-1}}{1+\exp\left(-\frac{3\pi\imath}{5}\right)x^n}$$

$$-\sum_{n=0}^{\infty}\frac{(n+1)\exp\left(\frac{\pi\imath}{5}\right)x^n}{1+\exp\left(\frac{\pi\imath}{5}\right)x^{n+1}} - \sum_{n=0}^{\infty}\frac{n\exp\left(-\frac{\pi\imath}{5}\right)x^{n-1}}{1+\exp\left(-\frac{\pi\imath}{5}\right)x^n}.$$

After some algebraic simplification and evaluation of the constant c, we are led to

$$\prod_{n=0}^{\infty}\frac{(1-x^{5n})^5}{(1-x^n)} = \frac{1}{\sqrt{5}}\sum_{n=0}^{\infty}(n+1)x^n$$

$$\times\left(\frac{1+\sqrt{5}+4x^{n+1}}{2+(1+\sqrt{5})x^{n+1}+2x^{2n+2}} - \frac{1-\sqrt{5}+4x^{n+1}}{2+(1-\sqrt{5})x^{n+1}+2x^{2n+2}}\right)$$

$$= \sum_{n=1}^{\infty}nx^{n-1}\frac{(1-x^n)(1-x^{2n})}{(1-x^{5n})}.$$

We have hence established

Corollary 9.5. *For all $x \in \mathbb{C}$ with $|x| < 1$,*

$$\prod_{n=1}^{\infty}\frac{(1-x^{5n})^5}{(1-x^n)} = \sum_{n=1}^{\infty}nx^{n-1}\frac{(1-x^n)(1-x^{2n})}{(1-x^{5n})}.$$

An interesting number theoretic interpretation is obtained by modifying the above series.

Corollary 9.6. *Write for $x \in \mathbb{C}$ with $|x| < 1$,*

$$x\prod_{n=1}^{\infty}\frac{(1-x^{5n})^5}{(1-x^n)} = \sum_{n=1}^{\infty}f(n)x^n.$$

Then f is a multiplicative function; that is,

$$f(mn) = f(m)f(n) \text{ whenever } (m,n) = 1. \tag{4.36}$$

Furthermore, $f(5^n) = 5^n$ for all $n \geq 1$. If n is a prime, then

$$f(n) = \begin{cases} n+1 \text{ for } n \equiv \pm 1 \mod 5, \\ n-1 \text{ for } n \equiv \pm 2 \mod 5, \\ 5 \text{ for } n = 5. \end{cases}$$

Proof. From the previous corollary we see that

$$x \prod_{n=1}^{\infty} \frac{(1-x^{5n})^5}{(1-x^n)} = \sum_{n=1}^{\infty} nx^n \frac{(1-x^n)(1-x^{2n})}{(1-x^{5n})};$$

we rewrite the right hand side as

$$\sum_{n=1}^{\infty} \frac{nx^n}{1-x^{5n}} - \sum_{n=1}^{\infty} \frac{nx^{2n}}{1-x^{5n}} - \sum_{n=1}^{\infty} \frac{nx^{3n}}{1-x^{5n}} + \sum_{n=1}^{\infty} \frac{nx^{4n}}{1-x^{5n}}.$$

It is clear that for $m = 1, 2, 3, 4$

$$\sum_{n=1}^{\infty} \frac{nx^{mn}}{1-x^{5n}} = \sum_{n=1}^{\infty} f_m(n)x^n,$$

where

$$f_m(n) = \sum_{d|n,\ \frac{n}{d} \equiv m \mod 5} d.$$

It thus follows that

$$x \prod_{n=1}^{\infty} \frac{(1-x^{5n})^5}{(1-x^n)} = \sum_{n=1}^{\infty} f(n)x^n,$$

with

$$f(n) = \sum_{d|n} \chi\left(\frac{n}{d}\right) d$$

and χ is defined by (4.35). An elementary number theoretic argument establishes the multiplicativity of $f : \mathbb{Z}^+ \to \mathbb{Z}$. Its claimed other properties are also easily established.

Let k be an odd prime and let $N \in \mathbb{Z}$ be such that $(k-1)|12N$. Under these conditions, we show in the next chapter that $\left(\frac{\eta(k\tau)}{\eta(\tau)}\right)^{\frac{24N}{k-1}}$ is a $\Gamma_o(k)$-invariant function with divisor $P_\infty^N P_0^{-N}$. Thus $\frac{d}{d\tau} \log \left(\frac{\eta(k\tau)}{\eta(\tau)}\right)^{\frac{24N}{k-1}} d\tau$ projects to $\overline{\mathbb{H}^2/\Gamma_o(k)}$ as an abelian differential of the third kind whose only singularities are simple poles at P_∞ and P_0 with residues $\pm N$, respectively. The usual calculations (with $x = \exp(2\pi\imath\tau)$) yield

$$\frac{d}{d\tau}\log\left(\frac{\eta(k\tau)}{\eta(\tau)}\right)^{\frac{24N}{k-1}} = 2\pi\imath N\left[1 - \frac{24}{k-1}\sum_{n=1}^{\infty}\left(\frac{knx^{kn}}{1-x^{kn}} - \frac{nx^n}{1-x^n}\right)\right].$$

We need a second expression for this function. For $l \in \mathbb{Z}$, $0 \le l \le \frac{k-3}{2}$,
$\left(\frac{\theta'\begin{bmatrix}1\\ \frac{1+2l}{k}\end{bmatrix}}{\theta\begin{bmatrix}1\\ \frac{1+2l}{k}\end{bmatrix}}\right)^2$ is a modular 1-form for $G(k)$ and hence

$$\sum_{l=0}^{\frac{k-3}{2}}\left(\frac{\theta'\begin{bmatrix}1\\ \frac{1+2l}{k}\end{bmatrix}}{\theta\begin{bmatrix}1\\ \frac{1+2l}{k}\end{bmatrix}}\right)^2$$

is a modular 1-form for $\Gamma_o(k)$.

Remark 9.7. This fact has already been shown in the previous section. Here we extend slightly those results.

One readily computes (again with $x = \exp(2\pi\imath\tau)$ and $z = \exp(2\pi\imath\zeta)$) that

$$\frac{\theta'\begin{bmatrix}1\\ \frac{1+2l}{k}\end{bmatrix}}{\theta\begin{bmatrix}1\\ \frac{1+2l}{k}\end{bmatrix}}(0,\tau) = \left[\frac{d}{d\zeta}\log\theta\begin{bmatrix}1\\ \frac{1+2l}{k}\end{bmatrix}(\zeta,\tau)\right]_{\zeta=0}$$

$$= \pi\imath\left(1 + 2\sum_{n=0}^{\infty}\left(\frac{\exp\left(\frac{\pi\imath(1+2l)}{k}\right)x^{n+1}}{1+\exp\left(\frac{\pi\imath(1+2l)}{k}\right)x^{n+1}} - \frac{\exp\left(-\frac{\pi\imath(1+2l)}{k}\right)x^{n}}{1+\exp\left(-\frac{\pi\imath(1+2l)}{k}\right)x^{n}}\right)\right)$$

$$= -\pi\left(\tan\left(\frac{\pi(1+2l)}{2k}\right) + 4\sum_{n=1}^{\infty}\frac{\sin\left(\frac{\pi(1+2l)}{k}\right)x^n}{1+2\cos\left(\frac{\pi(1+2l)}{k}\right)x^n + x^{2n}}\right).$$

To obtain alternate descriptions of the above automorphic form, we let $\lambda_l = \exp\left(\frac{\pi\imath(1+2l)}{k}\right)$ and write

$$-\frac{1}{\pi}\frac{\theta'\begin{bmatrix}1\\ \frac{1+2l}{k}\end{bmatrix}}{\theta\begin{bmatrix}1\\ \frac{1+2l}{k}\end{bmatrix}} = -\imath\left(1 + 2\sum_{n=0}^{\infty}\left(\frac{\lambda_l x^{n+1}}{1+\lambda_l x^{n+1}} - \frac{\overline{\lambda_l}x^n}{1+\overline{\lambda_l}x^n}\right)\right)$$

$$= -\imath\left(\frac{1-\overline{\lambda_l}}{1+\overline{\lambda_l}} + 2\sum_{n=1}^{\infty}\left(\frac{\lambda_l x^n}{1+\lambda_l x^n} - \frac{\overline{\lambda_l}x^n}{1+\overline{\lambda_l}x^n}\right)\right)$$

$$= -\imath\left(\frac{1-\overline{\lambda_l}}{1+\overline{\lambda_l}} + 2\sum_{n=1}^{\infty} x^n \left(\lambda_l \sum_{m=0}^{\infty}(-\lambda_l x^n)^m - \overline{\lambda_l}\sum_{m=0}^{\infty}\left(-\overline{\lambda_l}x^n\right)^m\right)\right)$$

$$= \tan\left(\frac{\pi(1+2l)}{2k}\right) + 4\sum_{m=1}^{\infty}(-1)^{m-1}\sin\left(\frac{\pi(1+2l)m}{k}\right)\frac{x^m}{1-x^m}$$

$$= \tan\left(\frac{\pi(1+2l)}{2k}\right) + 4\sum_{n=1}^{\infty}\left(\sum_{1\le m\le n;\ m|n}(-1)^{m-1}\sin\left(\frac{\pi(1+2l)m}{k}\right)\right)x^n.$$

Except for evaluations of constants, we have (once again) established much of

Theorem 9.8. *For each odd prime k,*

$$\frac{d}{d\tau}\log\left(\frac{\eta(k\tau)}{\eta(\tau)}\right) + \frac{1}{2\pi\imath(k-2)}\sum_{l=0}^{\frac{k-3}{2}}\left(\frac{\theta'\begin{bmatrix}1\\ \frac{1+2l}{k}\end{bmatrix}}{\theta\begin{bmatrix}1\\ \frac{1+2l}{k}\end{bmatrix}}\right)^2(0,\tau),\quad \tau\in\mathbb{H}^2,$$

is a cusp 1-form for $\Gamma_o(k)$. This form is identically zero provided $k \le 13$, $k \ne 11$.

Proof. The constant, whose calculation is left to the reader, in front of the sum is determined by the condition that the Fourier series expansion (at infinity) have zero constant term. The resulting modular form projects to an abelian differential ω of the third kind on $\overline{\mathbb{H}^2/\Gamma_o(k)}$ with at worse simple poles at P_∞ and P_0. Since we set things up so that ω has zero residue at P_∞, its residue at P_0 must also vanish. For $k \le 13$, $k \ne 11$, $\omega = 0$ since surfaces of genus zero do not carry nontrivial abelian differentials of the first kind.

Remark 9.9. **(a)** Let k be an odd prime. The power series version of the equation in the theorem is

$$\frac{(k-2)(k-1)}{6}\left[1+\frac{24}{k-1}\sum_{n=1}^{\infty}\left(\frac{nx^n}{1-x^n}-\frac{knx^{kn}}{1-x^{kn}}\right)\right]$$

$$= \sum_{l=0}^{\frac{k-3}{2}}\left(\tan\frac{\pi(1+2l)}{2k} + 4\sum_{n=1}^{\infty}\sum_{1\le m\le n;\ m|n}(-1)^{m-1}\sin\frac{\pi(1+2l)m}{k}x^n\right)^2.$$

The fact that the constant terms of the Taylor series expansions of the forms appearing in the above equation are equal is equivalent to the trigonometric identity (3.11).

In general, it is convenient to define

$$f(l,n) = \sum_{1\le m\le n;\ m|n} (-1)^{m-1} \sin\left(\frac{\pi(1+2l)m}{k}\right).$$

The equality of the coefficients of x tells us that

$$k-2 = 2\sum_{l=0}^{\frac{k-3}{2}} \tan\left(\frac{\pi(2l+1)}{2k}\right) \sin\left(\frac{\pi(2l+1)}{k}\right)$$

$$= 2\sum_{l=0}^{\frac{k-3}{2}} 2\sin^2\left(\left(\frac{\pi(2l+1)}{2k}\right) = 2\sum_{l=0}^{\frac{k-3}{2}} \left(1-\cos\left(\frac{\pi(2l+1)}{k}\right)\right).$$

This equality can easily be verified directly for all odd positive integers k. The equality of the coefficients of x^n, $n \ge 2$, tells us that

$$(k-2)\left(\sigma(n) - k\sigma\left(\frac{n}{k}\right)\right)$$

$$= 2\sum_{l=0}^{\frac{k-3}{2}} \tan\left(\frac{\pi(2l+1)}{2k}\right) f(l,n) + 4\sum_{l=0}^{\frac{k-3}{2}}\sum_{p=1}^{n-1} f(l,p)f(l,n-p).$$

We have established the above equalities for primes $k \le 19$ (see below).

(b) The modular forms appearing in the above theorem seem to be basic objects in our study. Let w be the cusp 1-form produced by the theorem. One way to establish that $w = 0$ is to show that its projection to $\overline{\mathbb{H}^2/\Gamma_o(k)}$ vanishes to high enough order at P_∞. The surfaces $\overline{\mathbb{H}^2/\Gamma_o(k)}$, $k \ge 11$, $k \ne 13$ (recall that k is assumed to be prime), have positive genera. In general for the prime k,

$$\text{(4.37)} \qquad \deg\,(w) = \frac{k+1}{6}, \text{ provided } w \ne 0.$$

Calculations using MATHEMATICA show that

$$\text{ord}_\infty w \ge \begin{cases} 3 \text{ for } k = 11 \\ 4 \text{ for } k = 17 \\ 4 \text{ for } k = 19 \end{cases}.$$

If w were nontrivial, its order at ∞ would contradict (4.37). Thus we conclude that w is the trivial differential for $k = 11$, 17 and 19. Because of rounding errors, one rapidly loses confidence in the accuracy of these calculations. Nevertheless, we

Conjecture 9.10. For each odd prime k and all $\tau \in \mathbb{H}^2$,

$$\frac{d}{d\tau}\log\left(\frac{\eta(k\tau)}{\eta(\tau)}\right)+\frac{1}{2\pi\imath(k-2)}\sum_{l=0}^{\frac{k-3}{2}}\left(\frac{\theta'\begin{bmatrix}1\\ \frac{1+2l}{k}\end{bmatrix}}{\theta\begin{bmatrix}1\\ \frac{1+2l}{k}\end{bmatrix}}\right)^2(0,\tau)=0.$$

We examine in detail the first special case: $k=3$. In this case we can exploit the fact that $G(3)=\Gamma_o(3)$. Let $w=\left(\frac{\theta'\begin{bmatrix}1\\ \frac{1}{3}\end{bmatrix}}{\theta\begin{bmatrix}1\\ \frac{1}{3}\end{bmatrix}}\right)^2$. We know that $w\in\mathbb{A}_1^+(\mathbb{H}^2,\Gamma_o(3))$ and that $\deg(w)=\frac{2}{3}$. Thus as a 1-form for $\Gamma_o(3)$, $(w)=P^{\frac{2}{3}}$, where P is an elliptic fixed point of (order three) for the group $\Gamma_o(3)$. Since $\begin{bmatrix}2 & -1\\ 3 & -1\end{bmatrix}$ is an elliptic element of this group, we conclude

Theorem 9.11. *The function $\theta'\begin{bmatrix}1\\ \frac{1}{3}\end{bmatrix}$ has simple zeros at the $\Gamma_o(3)$-orbit of $\frac{3}{2}+\frac{\sqrt{3}}{2}\imath$ and is nonzero elsewhere on $\mathbb{H}^2$.*

In general it is quite difficult to locate the zeros of $\theta'\begin{bmatrix}\epsilon\\ \epsilon'\end{bmatrix}$.

The function $\dfrac{\theta^3\begin{bmatrix}1\\ \frac{1}{3}\end{bmatrix}}{\theta\begin{bmatrix}\frac{1}{3}\\ \frac{1}{3}\end{bmatrix}\theta\begin{bmatrix}\frac{1}{3}\\ 1\end{bmatrix}\theta\begin{bmatrix}\frac{1}{3}\\ \frac{5}{3}\end{bmatrix}}$ is not $\Gamma(3)$-invariant; however, its third power is $G(3)=\Gamma_o(3)$-invariant with divisor $\frac{P_\infty}{P_0}$. Thus

Theorem 9.12. *There exists a nonzero constant c such that*

$$\frac{\theta^3\begin{bmatrix}1\\ \frac{1}{3}\end{bmatrix}}{\theta\begin{bmatrix}\frac{1}{3}\\ \frac{1}{3}\end{bmatrix}\theta\begin{bmatrix}\frac{1}{3}\\ 1\end{bmatrix}\theta\begin{bmatrix}\frac{1}{3}\\ \frac{5}{3}\end{bmatrix}}(0,\tau)=c\left(\frac{\eta(3\tau}{\eta(\tau)}\right)^4.$$

Exercise 9.13. Give an alternate proof of the above theorem based on the triple product formula.

We briefly turn our attention to the case $k=4$ and the torsion free group $\Gamma_o(4)=G(4)$ of type $(0,3)$. The three punctures on the surface $\mathbb{H}^2/G(4)$ are

P_∞, P_0 and $P_{\frac{1}{2}}$. The functions $\left(\dfrac{\theta'\begin{bmatrix}0\\ \frac{1}{2}\end{bmatrix}}{\theta\begin{bmatrix}0\\ \frac{1}{2}\end{bmatrix}}\right)^2$ and $\left(\dfrac{\vartheta'\begin{bmatrix}1\\ \frac{1}{2}\end{bmatrix}}{\vartheta\begin{bmatrix}1\\ \frac{1}{2}\end{bmatrix}}\right)^2$ are modular 1-forms for $\Gamma_o(4)$, with divisors P_∞ and $P_{\frac{1}{2}}$, respectively. The $G(4)$-invariant functions $\left(\frac{\eta(4\tau)}{\eta(\tau)}\right)^8$ and $\left(\dfrac{\theta\begin{bmatrix}0\\ \frac{1}{2}\end{bmatrix}}{\theta\begin{bmatrix}1\\ \frac{1}{2}\end{bmatrix}}\right)^8$ have divisors $\frac{P_\infty}{P_0}$ (Chapter 5, §10.7) and $\frac{P_{\frac{1}{2}}}{P_\infty}$ (Chapter 3, §3.4), respectively. Hence

Theorem 9.14. *There exist constants c_1 and $c_2 \in \mathbb{C}^*$ such that*

$$\left(\frac{\theta'\begin{bmatrix}1\\ \frac{1}{2}\end{bmatrix}(0,\tau)}{\theta\begin{bmatrix}1\\ \frac{1}{2}\end{bmatrix}(0,\tau)}\right)^2 = c_1\frac{d}{d\tau}\log\left(\frac{\eta(4\tau)}{\eta(\tau)}\right)$$

and

$$\left(\frac{\theta'\begin{bmatrix}0\\ \frac{1}{2}\end{bmatrix}(0,\tau)}{\theta\begin{bmatrix}0\\ \frac{1}{2}\end{bmatrix}(0,\tau)}\right)^2 = c_2\frac{d}{d\tau}\log\left(\frac{\eta(4\tau)}{\eta(\tau)}\frac{\theta\begin{bmatrix}0\\ \frac{1}{2}\end{bmatrix}(0,\tau)}{\theta\begin{bmatrix}1\\ \frac{1}{2}\end{bmatrix}(0,\tau)}\right).$$

The above constants are easily evaluated (for example, $c_1 = -4\pi\imath$) and the theorem translated to the equalities in the next corollary once one uses the Jacobi triple product to compute the power series expansion of one more differential:

$$\frac{\theta'\begin{bmatrix}0\\ \frac{1}{2}\end{bmatrix}}{\theta\begin{bmatrix}0\\ \frac{1}{2}\end{bmatrix}}(0,\tau) = \left[\frac{d}{d\zeta}\log\theta\begin{bmatrix}0\\ \frac{1}{2}\end{bmatrix}(\zeta,\tau)\right]_{\zeta=0} = -4\pi x^{\frac{1}{2}}\sum_{n=0}^{\infty}\frac{x^n}{1+x^{2n+1}}.$$

Corollary 9.15. *For all $x \in \mathbb{C}$, $|x| < 1$,*

$$\left(1+4\sum_{n=1}^{\infty}\frac{x^n}{1+x^{2n}}\right)^2 = 1+8\sum_{n=1}^{\infty}n\left(\frac{x^n}{1-x^n}+2\frac{x^{2n}}{1+x^{2n}}\right)$$

and

$$2\left(\sum_{n=0}^{\infty}\frac{x^n}{1+x^{2n+1}}\right)^2 = \sum_{n=0}^{\infty}\left(\frac{(n+1)x^n}{1+x^{n+1}}+\frac{(2n-1)x^{2n}}{1+x^{2n+1}}\right).$$

We should also consider, for the sake of completeness, the case of the even prime $k = 2$. The motion $\begin{bmatrix} 1 & -1 \\ 2 & -1 \end{bmatrix}$ generates $\Gamma_o(2)$ over $\Gamma_o(4)$ and permutes the classes of the characteristics $\begin{bmatrix} 0 \\ \frac{1}{2} \end{bmatrix}$ and $\begin{bmatrix} 1 \\ \frac{1}{2} \end{bmatrix}$. Hence

$$\left(\frac{\theta'\begin{bmatrix} 0 \\ \frac{1}{2} \end{bmatrix}}{\theta\begin{bmatrix} 0 \\ \frac{1}{2} \end{bmatrix}}\right)^2 + \left(\frac{\theta'\begin{bmatrix} 1 \\ \frac{1}{2} \end{bmatrix}}{\theta\begin{bmatrix} 1 \\ \frac{1}{2} \end{bmatrix}}\right)^2$$

is a modular 1-form for $\Gamma_o(2)$. Its value at $\imath\infty$ is easily seen to be nonzero (it equals π^2); hence its divisor is $P^{\frac{1}{2}}$, where P is an elliptic fixed point of $\Gamma_o(2)$. We have proven

Theorem 9.16. *There exists a nonzero constant c such that*

$$\left(\frac{\theta'\begin{bmatrix} 0 \\ \frac{1}{2} \end{bmatrix}}{\theta\begin{bmatrix} 0 \\ \frac{1}{2} \end{bmatrix}}\right)^2 (0,\tau) + \left(\frac{\theta'\begin{bmatrix} 1 \\ \frac{1}{2} \end{bmatrix}}{\theta\begin{bmatrix} 1 \\ \frac{1}{2} \end{bmatrix}}\right)^2 (0,\tau) = c\frac{d}{d\tau}\log\left(\frac{\eta(2\tau)}{\eta(\tau)}\right).$$

The value of c is easily established (it equals $-12\pi\imath$) and the theorem translated to

Corollary 9.17. *For all $x \in \mathbb{C}$, $|x| < 1$,*

$$2x\left(\sum_{n=0}^{\infty}\frac{x^n}{1+x^{2n+1}}\right)^2 + 2\left(\frac{1}{4}+\sum_{n=1}^{\infty}\frac{x^n}{1+x^{2n}}\right)^2 = 3\left(\frac{1}{24}+\sum_{n=1}^{\infty}n\frac{x^n}{1+x^n}\right).$$

We return for one more application to the coverings (4.34). Using the fact that we know π_2 and holomorphic universal covering maps of both $\mathbb{H}^2/G(5)$ and $\mathbb{H}^2/\Gamma_o(5)$, we have produced two coverings of this last orbifold. It yields

Theorem 9.18. *For some constants $\alpha \in \mathbb{C}$, $a \in \mathbb{C}^*$ and $b \in \mathbb{C}$,*

$$\left(\frac{\theta\begin{bmatrix} \frac{3}{5} \\ 1 \end{bmatrix}(0,5\tau)}{\theta\begin{bmatrix} \frac{1}{5} \\ 1 \end{bmatrix}(0,5\tau)}\right)^5 + \alpha\left(\frac{\theta\begin{bmatrix} \frac{1}{5} \\ 1 \end{bmatrix}(0,5\tau)}{\theta\begin{bmatrix} \frac{3}{5} \\ 1 \end{bmatrix}(0,5\tau)}\right)^5 = a\left(\frac{\eta(\tau)}{\eta(5\tau)}\right)^6 + b.$$

Evaluation of constants leads to the identity

Corollary 9.19. *For all $x \in \mathbb{C}$ with $|x| < 1$,*

$$x\prod_{n=0}^{\infty}\frac{(1-x^{5n+1})^5(1-x^{5n+4})^5}{(1-x^{5n+2})^5(1-x^{5n+3})^5} - \frac{1}{x}\prod_{n=0}^{\infty}\frac{(1-x^{5n+2})^5(1-x^{5n+3})^5}{(1-x^{5n+1})^5(1-x^{5n+4})^5}$$

$$= -\frac{1}{x}\prod_{n=1}^{\infty}\frac{(1-x^n)^6}{(1-x^{5n})^6} - 11.$$

Chapter 5

Partition theory: Ramanujan congruences

The purpose of this chapter is to use complex function theory, particularly the theories of theta functions and theta constants as developed in the preceding Chapters 2, 3 and 4, to study the *(Euler-Ramanujan) partition functions*

$$P_N : \mathbb{Z}^+ \cup \{0\} \to \mathbb{Z},\ N \in \mathbb{Z},$$

and the related analytic functions on the unit disc

$$\mathbf{P}^N : \mathbb{D} = \{x \in \mathbb{C};\ |x| < 1\} \to \mathbb{C}^* = \mathbb{C} - \{0\}$$

defined by[102]

$$\mathbf{P}^N(x) = \prod_{n=1}^{\infty} \frac{1}{(1-x^n)^N} = \sum_{n=0}^{\infty} P_N(n)x^n. \tag{5.1}$$

With the usual local coordinate change $x = \exp(2\pi\imath\tau)$,

$$\mathbf{P}^{-1}(x) = \exp\left(-\frac{\pi\imath\tau}{12}\right)\eta(\tau) = \exp\left(-\frac{\pi\imath}{12}(\tau+2)\right)\theta\begin{bmatrix} \frac{1}{3} \\ 1 \end{bmatrix}(0, 3\tau).$$

Although we have encountered this function previously, this chapter presents the more or less full treatment it deserves. We call $P_N(n)$ a *partition coefficient*. Although we use mostly formal properties of these infinite products and series, they do converge absolutely and uniformly on compact

[102]We will also, on occasion, assume that $N \in \mathcal{F}$, for some field $\mathcal{F} \supset \mathbb{Z}$, in which case $P_N(n) \in \mathcal{F}$, the cases $\mathcal{F} = \mathbb{Q}$, $\mathbb{R}$ and $\mathbb{C}$ being most important.

subsets of the unit disc $\mathbb{D}$. Obviously for positive N, $P_N(n) \in \mathbb{Z}^+$ for all $n \in \mathbb{Z}^+ \cup \{0\}$. It is routine to verify that

$$P_N(0) = 1,\ P_N(1) = N,\ P_N(2) = \frac{N(N+3)}{2!},\ P_N(3) = \frac{N(N+1)(N+8)}{3!},$$

$$P_N(4) = \frac{N(N+1)(N+3)(N+14)}{4!},$$

(we will use these and more complicated similar formulae) and that for fixed $n \in \mathbb{Z} - \{0\}$, $P_N(n)$ is a polynomial in N of degree n of the form

$$\sum_{i=1}^{n} a_i N^i;\ n!a_i \in \mathbb{Z}^+.$$

In general, it is quite difficult to compute these polynomials. The first 15 polynomials, $P_N(n)$, $n = 0,\ 1,\ \ldots,\ 14$, are described in §1. We also find their integral roots (extended to include the cases $15 \le n \le 40$) since these are needed for applications. These and most other calculations were performed using the symbolic manipulation programs MATHEMATICA and MAPLE; they would have been almost prohibitively expensive only a few years ago. We extend the definition of the function P_N by setting

$$P_N(n) = 0,\ \text{for all } n \in \mathbb{Q} - \mathbb{Z}^+ - \{0\}.$$

The most interesting cases are

$$P_1 = P,$$

the *classical partition* function studied by Ramanujan, the case $N = -1$ considered (earlier) by Euler (equation (2.2)), $N = -3$ studied (also before Ramanujan) by Jacobi ((2.1)),[103] and $N = -24$ investigated by Lehmer because of its relation to the τ-function (here called T) by the formula

$$P_{-24}(n) = T(n+1),\ n \in \mathbb{Z}^+ \cup \{0\}.$$

Jacobi's original derivation of identities for P_{-3} and much of our work need a basic relation between products and sums; it can and will be based on Jacobi's identity, which gives a closed expression for the function P_{-3}. This theorem of Jacobi can be obtained (as we did in Chapter 2) as a consequence of the Jacobi triple product formula. We also obtained a new proof of Euler's identity, another closed form for partition (here $N = -1$) functions.

There are two main related tools available to us. Constructions of meromorphic functions on compactifications of Riemann surfaces represented by the action of a finite index subgroup of the modular group Γ on the upper half plane $\mathbb{H}^2$ and constructions of automorphic forms for the same groups.

[103] As far as we are aware, the only known closed forms for the functions P_N are for $N = -1$ and -3.

Our presentation is motivated by an attempt to understand, using theta constants, the function theory for subgroups of Γ. Only properties of the theta constant derivative function on $\mathbb{H}^2$, $\theta'[\chi](0,\cdot)$, with χ the unique equivalence class of odd integral characteristics, are used in most of this chapter. However much of the material in previous chapters involved both $\theta[\chi]$ and $\theta'[\chi]$ with rational characteristics χ. Since the ratio of $\theta'\begin{bmatrix}1\\1\end{bmatrix}$ to η^3 is constant, we may and do base this chapter on the properties of the η-function. Although we write most formulae in terms of this function,[104] we point out that much of the structure is missed by an approach that relies solely on this classical function. The subject is enriched by the inclusion of theta functions with rational characteristics; these correspond to points of finite order on certain tori. In the more classical literature only points of order two were studied.

We need to know the multiplier system for the η-function to construct single valued functions on Riemann surfaces. This information can be obtained from the properties of κ using number theoretic arguments. Since we want to avoid these, we rely instead on [**16**, Th. 2, Ch. 4].

A good place to start an introduction to partition theory is with Ramanujan, who proved that

$$P(5m+4) \equiv 0 \bmod 5,\ P(7m+5) \equiv 0 \bmod 7 \text{ and } P(11m+6) \equiv 0 \bmod 11.$$

Bases for the first two of these congruences are the elegant formulae obtained by Ramanujan

$$\sum_{m=1}^{\infty} P(5m-1)x^m = 5x\frac{\prod_{n=1}^{\infty}(1-x^{5n})^5}{\prod_{n=1}^{\infty}(1-x^n)^6} \tag{5.2}$$

and

$$\sum_{m=1}^{\infty} P(7m-2)x^m = 7x\frac{\prod_{n=1}^{\infty}(1-x^{7n})^3}{\prod_{n=1}^{\infty}(1-x^n)^4}\left[7x\frac{\prod_{n=1}^{\infty}(1-x^{7n})^4}{\prod_{n=1}^{\infty}(1-x^n)^4}+1\right]. \tag{5.3}$$

Ramanujan more or less conjectured and partially proved the following more general statements, one for each fixed positive integer $n \in \mathbb{Z}^+$.
(a) If $24m \equiv 1 \bmod 5^n$, then $P(m) \equiv 0 \bmod 5^n$.
(b) If $24m \equiv 1 \bmod 7^n$, then $P(m) \equiv 0 \bmod 7^{1+\lfloor\frac{n}{2}\rfloor}$.[105]
(c) If $24m \equiv 1 \bmod 11^n$, then $P(m) \equiv 0 \bmod 11^n$.

We refer to these three results as the *(Ramanujan) level n congruences for the primes* 5, 7 *and* 11, respectively, for the partition function P_1. Similar congruence statements also hold for the general partition functions P_N as

[104]NOT only for topographical economy.
[105]The floor $\lfloor\cdot\rfloor$ and ceiling $\lceil\cdot\rceil$ functions will be defined in this chapter in §4.2.

well as for other primes, for example, for 2, 3, 5, 7 and 13. To state these results, it is convenient to introduce some more notation. For the positive prime k, let

$$\alpha(k) = \begin{cases} 8 \text{ for } k = 2 \\ 3 \text{ for } k = 3 \\ 1 \text{ for } k \geq 5 \end{cases} \quad \text{and } \beta(k) = \begin{cases} 1 \text{ for } k \leq 5 \\ \frac{k^2-1}{24} \text{ for } k \geq 5 \end{cases}.$$

We note that

$$\alpha(k) = \max\left\{1, \frac{24}{k^2-1}\right\} \text{ and } \beta(k) = \max\left\{1, \frac{k^2-1}{24}\right\}$$

and

$$(5.4) \qquad \frac{\beta(k)}{\alpha(k)} = \frac{k^2-1}{24}.$$

One of the main results on partition congruences may be summarized as follows: Let $k = 2, 3, 5, 7$ or 13. Let $N \in \mathbb{Z}^+$. If $\frac{24}{\alpha(k)} m \equiv N \mod k^n$, then $P_{\alpha(k)N}(m) \equiv 0 \mod k^{\lfloor \frac{\alpha n}{2} + \epsilon \rfloor}$, where $\alpha = \alpha(k, N)$ depends only on $R\left(\frac{\alpha(k)N}{24}\right)$, the residue of $\alpha(k)N$ modulo 24, and $\epsilon = \epsilon(N) = O(\log N)$. For small values of N, these results can be improved. For example,
(d) If $8m \equiv 1 \mod 3^n$, then $P_3(m) \equiv 0 \mod 3^{\lfloor \frac{3n}{2} \rfloor + 1}$.
(e) If $3m \equiv 1 \mod 2^n$, then $P_8(m) \equiv 0 \mod 2^{3\lceil \frac{n+1}{2} \rceil}$.
We prove general congruence results especially for low values of N, and although we try to avoid too much emphasis on special cases, we have found that these add to the general understanding of an incomplete picture. We have hence included summaries of many calculations.

We begin with a proof of the level one and two congruences for the prime 5. This is an application of Corollary 9.6 of Chapter 4, which states that

$$H(x) = x \prod_{n=1}^{\infty} \frac{(1-x^{5n})^5}{1-x^n} = \sum_{n=1}^{\infty} f(n) x^n,$$

with $f(n)$ a multiplicative function with the property $f(5^n) = 5^n$ for all $n \in \mathbb{Z}^+$. It thus follows that

$$\frac{1}{5} \sum_{l=0}^{4} H(e^{\frac{2\pi i l}{5}} x) = \prod_{n=1}^{\infty} (1-x^{5n})^5 \sum_{n=1}^{\infty} P(5n-1) x^{5n} = \sum_{n=1}^{\infty} f(5n) x^{5n}.$$

The above identity can be rewritten as

$$\sum_{n=1}^{\infty} P(5n-1) x^n = \sum_{n=1}^{\infty} f(5n) x^n \sum_{n=0}^{\infty} P_5(n) x^n = 5 \sum_{n=1}^{\infty} f(n) x^n \sum_{n=0}^{\infty} P_5(n) x^n;$$

the last equality follows from the fact that $f(5n) = 5f(n)$. The equality of the extreme terms in the last equation is the level one congruence for the prime 5.

Using the Cauchy product, the right hand side of the last equality equals

$$5\sum_{n=1}^{\infty}\left(\sum_{j=0}^{n-1} P_5(j)f(n-j)\right)x^n.$$

Repeating the above averaging process yields

$$\sum_{n=1}^{\infty} P(25n-1)x^n = 25\sum_{n=1}^{\infty}\left(\sum_{j=0}^{5n-1}\frac{P_5(j)f(5n-j)}{5}\right)x^n.$$

For $j \equiv 0$ modulo 5, we use the fact that $5|f(5n-j)$ (Corollary 9.6 of Chapter 4). If $j \not\equiv 0$ modulo 5, then $5|P_5(j)$ (Proposition 10.3 of this chapter). We have hence obtained the level two congruence for the prime 5. Unfortunately, this extremely simple idea seems not to yield the higher level congruences, nor does it seem to extend to the prime 7.

Throughout these introductory remarks, unless noted to the contrary, k represents a positive prime, n a positive integer and N an arbitrary integer. For given k, n and N, we construct in more than one way a meromorphic function $F_{k,n,N}$ (we consistently abbreviate $F_{k,1,N} = F_{k,N}$) on the compact Riemann surface $\overline{\mathbb{H}^2/\Gamma_o(k)}$. The construction starts with a function invariant under the group $\Gamma(k^n,k)$ that is averaged over the left cosets $\Gamma(k^n,k)\backslash\Gamma_o(k)$. Our work takes advantage of the availability of simple representatives for the cosets $\Gamma(k^n,k)\backslash\Gamma_o(k)$. The construction of the function $F_{k,n,N}$, with $N \in \mathbb{Z}^+$, leads naturally to the study of the power series

$$\sum_{m=l_1(k,n,N)}^{\infty} P_{\alpha(k)N}(k^n m - l_2(k,n,N))x^m,$$

where $l_1(k,n,N)$ and $l_2(k,n,N) \in \mathbb{Z}^+$, and to Laurent series expansions for these functions in terms of partition coefficients. Similar formulae, with stronger consequences, hold for negative N. Expressing the function $F_{k,n,N}$ in terms of simpler functions (possible for many values of k) leads to partition identities. Our approach produces identities that explain the function theoretic reason why partition identities should hold. We need additional number theoretic and combinatorial tools to establish many of them.

We have shown above how to prove the level one and two congruences for the prime 5 by a method with limited applicability. It is easy to obtain the level one congruences for the primes[106] 5 and 7 by methods that have

[106]We concentrate in these introductory remarks mostly on the primes 5, 7 and 11 because of their prevalence in the literature on partitions.

wider applicability. The basic tool is the construction of the meromorphic functions $F_{5,1}$ and $F_{7,1}$. The level two congruences for the primes 5 and 7 are almost direct consequences of the work needed to derive the corresponding level one congruences. The same method produces many other congruences, for example, the three level one and two level two congruences:

$$P_3(3m+2) \equiv 0 \mod 3^2,$$
$$P_2(5m+3) \equiv 0 \mod 5,\ P_2(5^2m+23) \equiv 0 \mod 5^2,$$
$$P_4(7m+6) \equiv 0 \mod 7 \text{ and } P_4(7^2m+41) \equiv 0 \mod 7^2.$$

A general method, involving the reduction to a finite set of calculations, for obtaining such level one congruences modulo some primes is derived. The same method also yields the level two congruences described above. Higher level congruences for primes ≤ 7 and 13 are based on so-called modular equations and detailed combinatorial arguments. We present only some sample cases with the expectation and hope that more transparent function theoretic arguments will be found.

The approaches discussed so far[107] have not produced any of the congruences for the prime 11. A study of the partition functions P_N for negative N leads to the construction of meromorphic functions $F_{k,N}$ on the surfaces $\overline{\mathbb{H}^2/\Gamma_o(k)}$ that in many interesting cases turn out to be constant. The study of these functions leads us to a unified way of establishing the level one congruences for the primes 5, 7 and 11. The method produces many equalities for the Taylor coefficients of the partition functions P_N with negative N, among them

$$P_{-8}(5m+3) = -5^3 P_{-8}\left(\frac{m-1}{5}\right) \text{ and } P_{-26}(11m+9) = 11^2 P_{-26}\left(\frac{m-11}{11}\right).$$

It also produces three term identities, for example,

$$P_{-6}(5m+1) = -5^2 P_{-6}\left(\frac{m-1}{5}\right) - 2\cdot 3P_{-6}(m).$$

Each of the equalities obtained has higher level analogues. Starting with a constant function $F_{k,N}$, for a single fixed pair (k,N), we produce many congruences for partition coefficients $P_{\tilde{N}}(k^n m + l_3(k,\tilde{N}))$ (all $n \in \mathbb{Z}^+$, many $\tilde{N} \in -\mathbb{Z}^+$, $l_3(k,\tilde{N}) \in \mathbb{Z}$ depends only on k, n and $\tilde{N}$) modulo a power the prime k. The method does not, however, produce any information modulo powers of the prime for positive $\tilde{N}$.

For $k = 2, 3, 5$ and 7, the first of the methods described above for proving the level one congruences gives a formula for the relevant function $F_{k,N}$ and its Taylor series at ∞. The second method does not. By studying the

[107]The main difference between the prime 11 and the smaller primes is that $\overline{\mathbb{H}^2/\Gamma_o(k)}$ is a sphere for primes $k \leq 7$ and a torus for $k = 11$.

function $G_{k,N}(\tau) = F_{k,N}\left(-\frac{1}{k\tau}\right)$, obtained from $F_{k,N}$ by an automorphism of $\mathbb{H}^2/\Gamma_o(k)$, we are able to describe a third method for establishing congruence results. This method works for the three primes 5, 7 and 11 and gives formulae for the relevant functions.

The literature on the partition coefficients $P_N(m)$ studies divisibility properties of arithmetic sequences of such coefficients modulo powers of primes. There are two parts to the story. The establishment of such a divisibility property and then showing that it is sharp, that is, no larger power of the prime divides all the terms in the sequence. We, in general, ignore the issue of sharpness because we are interested in determining the gcd (thus sharpness is a by product) of arithmetic sequences of partition coefficients. A general result gives a finite algorithm for computing

$$\gcd\left\{P_N(km + l_4(k,N));\ m \in \mathbb{Z}^+ \cup \{0\}\right\},$$

for many primes k and many (for most primes we consider all) $N \in \mathbb{Z}$, where $l_4(k,N) \in \mathbb{Z}$ is not arbitrary but depends on k and N. It has as a consequence the surprising fact (at least for us) that for the prime 13

$$\gcd\left\{P_{-15}(13m+1);\ m \in \mathbb{Z}^+ \cup \{0\}\right\} = 5.$$

1. Calculations of $P_N(n)$

We start with a combinatorial interpretation of the partition functions. It is, of course, obvious that for all $N \in \mathbb{Z}$, $P_N(0) = 1$ and $P_N(1) = N$. It is easily seen that in general for $n \in \mathbb{Z}^+$, $P_N(n)$ is a polynomial in N of degree n, zero constant term (thus with a root at $N = 0$) and all the rest of the coefficients are positive rationals (hence no roots in $\mathbb{R}^+$) . These polynomials may, however, have roots in $-\mathbb{Z}^+$, the negative integers. It is a nontrivial matter of some interest to compute these polynomials and their real roots. Assume that $n \in \mathbb{Z}^+$. For $N \in \mathbb{Z}^+$, $P_N(n)$ counts the total number of partitions of n using positive integers of N colors. For $-N \in \mathbb{Z}^+$, $P_N(n)$ counts the difference between the total number of even and odd partitions of n WITHOUT REPETITIONS using positive integers of $-N$ colors. A *partition* of $n \in \mathbb{Z}^+$ is a sum of positive integers totaling n. Thus the partitions of 4 are: 4, 3+1, 2+2, 2+1+1, 1+1+1+1. The *parity* of the partition is the parity of the number of summands. One easily checks from the definitions or the last enumeration that

$$P(4) = 5,\ P_{-1}(4) = 0,\ P_{-3}(4) = 0,\ P_2(4) = 20,\ P_{-2}(4) = 1.$$

Nontrivial calculations of the coefficients $P_N(n)$ require the use of computer programs. We have relied on MATHEMATICA and MAPLE. To obtain a recursive formula for $P_N(n)$, we use the second equality of (5.1). Logarithmic differentiation (followed by some algebraic simplification; multiplication of the identity by x) leads to

$$N\left(\sum_{n=0}^{\infty}\frac{nx^n}{1-x^n}\right)\left(\sum_{n=0}^{\infty}P_N(n)x^n\right)=\sum_{n=0}^{\infty}nP_N(n)x^n.$$

Now,

$$\sum_{n=0}^{\infty}\frac{nx^n}{1-x^n}=\sum_{n=1}^{\infty}\sum_{m=1}^{\infty}nx^{nm}=\sum_{n=1}^{\infty}\sigma(n)x^n,$$

where $\sigma(n)$ is the sum of the positive divisors of n. It follows that

$$N\sum_{m=1}^{n}\sigma(m)P_N(n-m)=nP_N(n). \tag{5.5}$$

Lemma 1.1. *For each $n\in\mathbb{Z}^+$, $n!P.(n)$ is a polynomial of degree n with zero constant term and all its other coefficients are positive integers; that is,*

$$n!P_N(n)=\sum_{i=1}^{n}\alpha_iN^i,\ \ \alpha_i\in\mathbb{Z}^+,\alpha_n=1.$$

Proof. We know that $P_N(0)=1$ and $P_N(1)=N$. Hence the lemma is true for $n=1$. Let us write $n!P_N(n)=\sum_{i=1}^{n}\alpha_i^{(n)}N^i$. For induction we assume that $n>1$ and that the lemma holds for all positive integers less than n. From (5.5) and the induction hypothesis

$$\begin{aligned}n!P_N(n)&=N(n-1)!\sum_{m=1}^{n}\sigma(m)P_N(n-m)\\&=N\sum_{m=1}^{n}\frac{(n-1)!}{(n-m)!}\sigma(m)(n-m)!P_N(n-m)\\&=N\sum_{m=1}^{n}\frac{(n-1)!}{(n-m)!}\sigma(m)\sum_{i=1}^{n-m}\alpha_i^{(n-m)}N^i,\end{aligned}$$

with $\alpha_i^{(n-m)}\in\mathbb{Z}^+$.

The first of the two tables in this subsection summarizes the calculations of the polynomials $P.(n)$.

Remark 1.2. For most of our application we do not need the polynomial $P.(n)$, but only its integral zeros. The once formidable task of determining the rational roots of the polynomials encountered here is now handled easily and relatively quickly using modern workstations and symbolic manipulation programs. We summarize some of the calculations of roots in Table 12.

n	$n!\,P_N(n)$
0	1
1	N
2	$N(3+N)$
3	$N(1+N)(8+N)$
4	$N(1+N)(3+N)(14+N)$
5	$N(3+N)(6+N)(8+21N+N^2)$
6	$N(1+N)(10+N)(144+181N+34N^2+N^3)$
7	$N(2+N)(3+N)(8+N)(120+529N+50N^2+N^3)$
8	$N(1+N)(3-N)(6+N)(4200+9994N+1571N^2+74N^3+N^4)$
9	$N(1+N)(3+N)(4+N)(14+N)(26+N)(120+491N+60N^2+N^3)$
10	$N(1+N)(6531840+29758896N+28014804N^2+10035116N^3+1709659N^4$ $+147854N^5+6496N^6+134N^7+N^8)$
11	$N(1+N)(2+N)(3+N)(8+N)$ $(907200+7260120N+1983222N^2+190049N^3+8057N^4+151N^5+N^6)$
12	$N(2+N)(3+N)(186278400+832270800N+1364062148N^2+574762916N^3$ $+105693297N^4+9998529N^5+513162N^6+14154N^7+193N^8+N^9)$
13	$N(1+N)(3+N)(8+N)(10+N)(27941760+304993872N+344183076N^2$ $+104464688N^3+11870389N^4+631628N^5+16774N^6+212N^7+N^8)$
14	$N(1+N)(3+N)(4+N)(6+N)(2075673600+14914586880N+19207997676N^2$ $+5178575464N^3+615548731N^4+38627729N^5+1353814N^6+26246N^7+259N^8+N^9)$

Table 11. RAMANUJAN POLYNOMIALS (Factored to show all integral roots).

It is important for us to observe that the argument given above in the proof of the lemma does not depend on N being an integer. We can replace N by a real or even complex variable t and the conclusion remains that $n!P_t(n)$ is a monic polynomial in t with coefficients in the nonnegative integers. This observation has an important corollary which we state as

Lemma 1.3. *For each fixed $n \in \mathbb{Z}^+$, the real roots of the polynomial $P_t(n)$ are nonpositive. The real rational roots are in $\{0\} \cup -\mathbb{Z}^+$. In particular, there are no rational roots in the interval $(-1, 0)$.*

Proof. The fact that there are no real positive roots is immediate from the fact that the coefficients are positive. Assume that we have a nonintegral rational root $\frac{p}{q}$, $q \geq 2$ (with p and q relatively prime). We can assume that n is at least 2 so that this would imply

$$\sum_{i=1}^{n} \alpha_i \left(\frac{p}{q}\right)^i = 0.$$

Multiplying by q^{n-1} leads to $\frac{p^n}{q}$ being an integer, which contradicts that $\frac{p}{q}$ is not an integer.

2. Some preliminaries

2.1. $\Gamma(p, q)$-invariant functions. Much of our work here is based on the following elementary

n	Negatives of integral roots of $P_\cdot(n)$
1	0
2	0, 3
3	0, 1, 8
4	0, 1, 3, 14
5	0, 3, 6
6	0, 1, 10
7	0, 2, 3, 8
8	0, 1, 3, 6
9	0, 1, 3, 4, 14, 26
10	0, 1
11	0, 1, 2, 3, 8
12	0, 2, 3
13	0, 1, 3, 8, 10
14	0, 1, 3, 4, 6
15	0, 8, 14
16	0, 1, 3
17	0, 1, 2, 3, 6, 10
18	0, 1, 2, 3, 8
19	0, 1, 3, 4, 6, 8, 14
20	0, 1, 3, 26
21	0, 1, 2
22	0, 2, 3
23	0, 1, 3, 6, 8
24	0, 1, 3, 4, 14
25	0, 1, 2, 3
26	0, 3, 6, 14
27	0, 1, 3, 8, 10
28	0, 1, 8, 10
29	0, 1, 3, 8, 14
30	0, 1, 3
31	0, 1, 3, 4, 8, 26
32	0, 1, 2, 3, 6, 14
33	0, 1, 3, 6
34	0, 1, 3, 4, 10, 14
35	0, 3, 6, 8
36	0, 1, 10
37	0, 1, 2, 3, 14
38	0, 1, 3, 8
39	0, 1, 2, 3, 4, 8, 10
40	0, 3, 6

Table 12. INTEGRAL ROOTS OF $P_\cdot(n)$.

Lemma 2.1. *Let p, q and $N \in \mathbb{Z}^+$ and let $r \in \mathbb{Z}$. A function $\varphi(\tau)$ is a $\Gamma(Np,q)$ r-form if and only if $\varphi(N\tau)$ is a $\Gamma(p,Nq)$ r-form. Further, the map $\varphi(\tau) \mapsto \varphi(N\tau)$ preserves cusp forms, modular forms and meromorphic forms as well as orders (of zeros and poles).*

Proof. Observe that

$$\gamma = \begin{bmatrix} a & b \\ c & d \end{bmatrix} \in \Gamma(Np,q) \text{ iff } \hat{\gamma} = \begin{bmatrix} a & \frac{b}{N} \\ Nc & d \end{bmatrix} \in \Gamma(p,Nq).$$

We first present a classical (old fashioned) approach to the proof of part of the theorem. Assume that $\varphi(\tau)$ is an r-form for $\Gamma(Np,q)$. Then

$$\begin{aligned} \varphi(N\hat{\gamma}(\tau))\hat{\gamma}'(\tau)^r &= \varphi\left(\tfrac{aN\tau+b}{cN\tau+d}\right)(cN\tau+d)^{-2r} \\ &= \varphi(\gamma(N\tau))\gamma'(N\tau)^r = \varphi(N\tau) \end{aligned}.$$

Thus $\varphi(N\tau)$ is an r-form for $\Gamma(p,Nq)$. The proof of the converse is similar. Now if $\varphi(\tau)$ is holomorphic (and vanishes) at the cusp x for $\Gamma(Np,q)$, then $\varphi(N\tau)$ is holomorphic (and vanishes) at the cusp $\frac{x}{N}$ for $\Gamma(p,Nq)$. Note that the stabilizers of ∞ in $\Gamma(Np,q)$ and $\Gamma(p,Nq)$ are generated by $\begin{bmatrix} 1 & Np \\ 0 & 1 \end{bmatrix}$ and $\begin{bmatrix} 1 & p \\ 0 & 1 \end{bmatrix}$, respectively, while if $\begin{bmatrix} 1+\alpha x & -\alpha x^2 \\ \alpha & 1-\alpha x \end{bmatrix}$, $\alpha \in \mathbb{C}^*$, generates the stabilizer of the finite cusp x for $\Gamma(Np,q)$, then $\begin{bmatrix} 1+\alpha N\frac{x}{N} & -\alpha N\left(\frac{x}{N}\right)^2 \\ \alpha N & 1-\alpha N\frac{x}{N} \end{bmatrix}$ generates the corresponding stabilizer of the cusp $\frac{x}{N}$ for $\Gamma(p,Nq)$. Good local coordinates at P_∞ are $x = \exp\left(\frac{2\pi\imath\tau}{Np}\right)$ and $x^N = \exp\left(\frac{2\pi\imath\tau}{p}\right)$ on $\Gamma(Np,q)$ and $\Gamma(p,Nq)$, respectively. It follows that if

$$\varphi(\tau) = \sum_{i=n}^{\infty} a_i x^i$$

is the Laurent series expansion of $\varphi(\tau)$, then

$$\varphi(N\tau) = \sum_{i=n}^{\infty} a_i (x^N)^i$$

is the Laurent series expansion of $\varphi(N\tau)$ and hence

$$\text{ord}_\infty \varphi(\tau) = \text{ord}_\infty \varphi(N\tau).$$

The corresponding argument (calculation) for finite cusps is more complicated. To avoid reliance on computations, we turn to a modern (functorial) approach and observe that $C \circ \gamma \circ C^{-1} = \hat{\gamma}$, where $C = \begin{bmatrix} 1 & 0 \\ 0 & N \end{bmatrix}$; that is, conjugation by C induces an isomorphism of $\Gamma(Np,q)$ onto $\Gamma(p,Nq)$. The map C_r^* takes $\Gamma(p, Nq)$-invariant r-forms to $\Gamma(Np, q)$-invariant r-forms. Further for such meromorphic $\Gamma(p, Nq)$-invariant functions φ, and all $x \in \mathbb{H}^2$ or x a cusp (parabolic fixed point) for $\Gamma(p, Nq)$,

$$\text{ord}_x \varphi = \text{ord}_{C^{-1}(x)} C_r^*(\varphi). \tag{5.6}$$

The observation

$$C_r^*(\varphi)(\tau) = N^{-r} \varphi\left(\frac{\tau}{N}\right)$$

completes the proof.

Remark 2.2. In particular, the lemma tells us that for all $k \in \mathbb{Z}^+$, a function $f(\tau)$ is $\Gamma_o(k^2)$-invariant if and only if $f(\frac{\tau}{k})$ is $\Gamma(k,k)$-invariant.

Corollary 2.3. *The function $f(\frac{\tau}{k})$ is $\Gamma(k,k)$-invariant whenever $f(\tau)$ is $\Gamma_o(k)$-invariant.*

Proof. A $\Gamma_o(k)$-invariant function is certainly $\Gamma_o(k^2)$-invariant.

Remark 2.4. The last corollary indicates the usefulness of the lemma. In general we have a map from r-forms $f(\tau)$ for $\Gamma(p,q) \supset \Gamma(p, Nq)$ to r-forms $f\left(\frac{\tau}{N}\right)$ for $\Gamma(Np, q) \subset \Gamma\left(Np, \frac{q}{N}\right)$. The image consists of $\Gamma\left(Np, \frac{q}{N}\right)$-invariant forms. In general this latter group is contained in $\text{PSL}(2, \mathbb{Q})$ (not $\text{PSL}(2,\mathbb{Z})$).

To apply the above ideas to the situation of interest to us, we recall that $\eta^{24}(\tau) \in \mathbb{A}_6(\mathbb{H}^2, \Gamma)$. Thus for all positive integers p and q, $\eta^{24}(q\tau) \in \mathbb{A}_6(\mathbb{H}^2, \Gamma_o(q))$ and $\eta^{24}\left(\frac{\tau}{p}\right) \in \mathbb{A}_6(\mathbb{H}^2, \Gamma^o(p))$. Thus the functions

$$\varphi^{24}(\tau) = \left(\frac{\theta' \begin{bmatrix} 1 \\ 1 \end{bmatrix} (0, q\tau)}{\theta' \begin{bmatrix} 1 \\ 1 \end{bmatrix} (0, \frac{\tau}{p})} \right)^8 = \left(\frac{\eta(q\tau)}{\eta(\frac{\tau}{p})} \right)^{24}, \quad \phi^{24}(\tau) = \left(\frac{\eta(q\tau)}{\eta(\tau)} \right)^{24}$$

and

$$\psi^{24}(\tau) = \left(\frac{\eta(\tau)}{\eta(\frac{\tau}{p})} \right)^{24}$$

are $\Gamma(p,q)$-invariant with

$$\text{ord}_{P_\infty} \varphi^{24} = qp - 1, \ \ \text{ord}_{P_\infty} \phi^{24} = p(q-1) \text{ and } \text{ord}_{P_\infty} \psi^{24} = p - 1.$$

The above orders are of course calculated in the appropriate local coordinate: $x = \exp(\frac{2\pi\imath\tau}{p})$.

We will be using the series and product expansion of the η-function many times. The reader should at this point recall equation (4.13).

The calculations of the orders of the functions φ, ϕ and ψ at P_0 and other points are simplified by the transformation formula equation (5.6) for

r-forms. Our forms $\theta'\begin{bmatrix}1\\1\end{bmatrix}(0, s\cdot)$ have weight $\frac{3}{4}$. We need to know how these forms transform under $A^*_{\frac{3}{4}}$ and $B^*_{\frac{3}{4}}$, and the images of the groups $\Gamma(p,q)$ under conjugation by A and B. Since[108] $A_0^*\varphi^{24} = c\varphi^{-24}$, $A_0^*\phi^{24} = c\left(\frac{\eta(\frac{\cdot}{q})}{\eta(\cdot)}\right)^{24}$ and $A_0^*\psi^{24} = c\left(\frac{\eta(\cdot)}{\eta(p\cdot)}\right)^{24}$ and $A\Gamma(p,q)A = \Gamma(q,p)$, we conclude that as $\Gamma(p,q)$-forms,

$$\mathrm{ord}_{P_0}\varphi^{24} = 1 - pq,\ \mathrm{ord}_{P_0}\phi^{24} = 1 - q \text{ and } \mathrm{ord}_{P_0}\psi^{24} = q(1-p).$$

We illustrate some of the nontrivial consequences of these simple concepts. The divisor (η^{24}) of $\eta^{24} \in \mathbb{A}_6(\mathbb{H}^2, \Gamma)$ on $\overline{\mathbb{H}^2/\Gamma}$ has degree one. Before continuing we need a preliminary

Remark 2.5. We emphasize the conventions we are using. When we say that the divisor (η^{24}) has degree one, we are computing its degree (as a form) on the compactified upper half ($\mathbb{H}^2 \cup \mathbb{Q} \cup \{\infty\}$) plane modulo the action of the group Γ and not as a differential on the associated Riemann surface $\overline{\mathbb{H}^2/\Gamma}$. It is convenient, however, to write the divisor in terms of points on the surface. The projection of η^{24} to the surface $\overline{\mathbb{H}^2/\Gamma}$ has poles of order 5, 3 and 4 at P_∞, P_i and $P_{\exp(\frac{\pi\imath}{6})}$ respectively. When computing divisors of forms (and their degrees) in subsequent sections, we will be using the same conventions. As the example shows, it is more economical to work on the holomorphic universal cover of the orbifold. In addition, the reader is reminded that the correct local coordinate must be used in all cases and this is determined by the stability subgroup of the cusp in question. When we computed the order of the $\Gamma(p,q)$-invariant forms at ∞, we used the local coordinate $x = \exp(\frac{2\pi\imath\tau}{p})$. For the computation of the order at 0, we first conjugated the group to map 0 to ∞, transformed the form appropriately and then used the local coordinate $x = \exp(\frac{2\pi\imath\tau}{q})$.

Since the Γ-form η^{24} has a simple zero at P_∞, this must be its divisor. Let k be a prime. Since η^{24} is a cusp 6-form for Γ, it is certainly one for $\Gamma_o(k)$ and $\Gamma(k,k)$ and

$$(\eta^{24}) = P_\infty P_0^k \text{ on } \overline{\mathbb{H}^2/\Gamma_o(k)},\ (\eta^{24}) = P_\infty^k P_0^k \ ...\ P_{k-1}^k \text{ on } \overline{\mathbb{H}^2/\Gamma(k,k)}.$$

The 6-form $\eta^{24}(k\tau)$ for $\Gamma_o(k)$ can only vanish at P_∞ and P_0. Since

$$\deg\left(\eta^{24}(k\tau)\right) = k+1 \text{ and } \mathrm{ord}_{P_\infty}\eta^{24} = k,$$

---

[108] In the next three equations c is a constant of absolute value 1. We will follow the convention in much of the book that unless indicated to the contrary, the symbols c or *const* appearing in a formula refer to constants (usually nonzero) whose exact value is not material (perhaps only for the moment).

we conclude that

$$(\eta^{24}(k\tau)) = P_\infty^k \; P_0 \text{ on } \overline{\mathbb{H}^2/\Gamma_o(k)}.$$

From the fact that $\pi : \overline{\mathbb{H}^2/\Gamma(k,k)} \to \overline{\mathbb{H}^2/\Gamma_o(k)}$ is completely ramified over P_∞ and completely unramified over P_0, we conclude that

$$(\eta^{24}(k\tau)) = P_\infty^{k^2} \; P_0 \; P_1 \; ... \; P_{k-1} \text{ on } \overline{\mathbb{H}^2/\Gamma(k,k)}.$$

The facts
1. $A \in N(\Gamma(k,k))$,
2. $\tilde{A}$ interchanges[109] the punctures P_∞ and P_0 on $\mathbb{H}^2/\Gamma(k,k)$ and permutes the punctures P_1, ..., P_{k-1}, and
3. $A_6^* \left(\eta^{24}(k\tau)\right) = c\eta^{24}\left(\frac{\tau}{k}\right)$

show that

$$\left(\eta^{24}\left(\frac{\tau}{k}\right)\right) = P_\infty \; P_0^{k^2} \; P_1 \; ... \; P_{k-1} \text{ on } \overline{\mathbb{H}^2/\Gamma(k,k)}.$$

We now take up a more general situation by specializing to the case that $q = k$, a prime, and $p = k^m$ with $m \in \mathbb{Z}^+$ arbitrary. The natural projection map $\pi : \mathbb{H}^2/\Gamma(k^m,k) \to \mathbb{H}^2/\Gamma_o(k)$ is completely ramified over P_∞ and only partially ramified over P_0. Let P_{x_0}, ..., P_{x_t} be a complete preimage of P_0 under π. Assume that π is locally ν_i to one in a neighborhood of P_{x_i}. Without loss of generality, $x_0 = 0$. Then $\nu_0 = 1$ and $\sum_{i=0}^t \nu_i = k^m$. We need the following

Lemma 2.6. *Let $k \in \mathbb{Z}^+$ be prime and $m \in \mathbb{Z}^+$ be arbitrary. The divisor of $\varphi^{24} = \left(\frac{\eta(k\tau)}{\eta\left(\frac{\tau}{k^m}\right)}\right)^{24}$ (respectively, $\psi^{24} = \left(\frac{\eta(\tau)}{\eta\left(\frac{\tau}{k^m}\right)}\right)^{24}$) on $\overline{\mathbb{H}^2/\Gamma(k^m,k)}$ is $P_\infty^{k^{m+1}-1} P_0^{1-k^{m+1}} \prod_{i=1}^t P_{x_i}^{\mu_i}$ $(P_\infty^{k^m-1} P_0^{k(1-k^m)} \prod_{i=1}^t P_{x_i}^{\sigma_i})$, where $|\mu_i| \le k^m - 2$ $(|\sigma_i| \le k^m - 2)$ and $\sum_{i=1}^t \mu_i = 0$ $(\sum_{i=1}^t \sigma_i = (k-1)(k^m-1))$. Further, $\mu_i = 0$ $(\sigma_i = k-1)$ for $i = 1$, ..., $t = k-1$, if $m = 1$.*

Proof. We view η and related functions as forms of weight $\frac{1}{4}$. The degrees of their divisors as forms for $\Gamma_o(k)$ and $\Gamma(k^m,k)$ are $\frac{1}{24}(k+1)$ and $\frac{1}{24}k^m(k+1)$, respectively. Since we know that as $\Gamma_o(k)$-forms,

$$\text{ord}_\infty \eta(k\cdot) = \frac{k}{24} \text{ and } \text{ord}_0 \eta(k\cdot) = \frac{1}{24},$$

we conclude that this form has no other zeros on $\overline{\mathbb{H}^2/\Gamma_o(k)}$ and that its divisor is $P_\infty^{\frac{k}{24}} P_0^{\frac{1}{24}}$. Thus its divisor on $\overline{\mathbb{H}^2/\Gamma(k^m,k)}$ is

$$(\eta(k\cdot)) = P_\infty^{\frac{k^{m+1}}{24}} \; P_0^{\frac{1}{24}} \prod_{i=1}^t P_{x_i}^{\frac{\nu_i}{24}}. \tag{5.7}$$

[109] $\tilde{A}$ is the automorphism induced by A on the surface $\overline{\mathbb{H}^2/\Gamma(k,k)}$.

It is harder to compute the divisor of $\eta\left(\frac{\cdot}{k^m}\right)$. However, its divisor on $\overline{\mathbb{H}^2/\Gamma(k^m,k)}$ is

$$\left(\eta\left(\frac{\cdot}{k^m}\right)\right) = P_\infty^{\frac{1}{24}} P_0^{\frac{k^{m+1}}{24}} \prod_{i=1}^{t} P_{x_i}^{\frac{\alpha_i}{24}} \text{ on } \overline{\mathbb{H}^2/\Gamma(k^m,k)}, \tag{5.8}$$

where $\alpha_i \geq 1$ because η vanishes at each cusp and $\sum_{i=1}^{t} \alpha_i = k^m - 1$. Thus $1 \leq \alpha_i \leq k^m - 1$. In particular,

$$k^m - 2 \geq \mu_i = \nu_i - \alpha_i \geq 2 - k^m.$$

This gives us the form of the divisor of φ^{24}. To finish the proof for the case $m = 1$ (for this function), use the fact that $\eta\left(\frac{\cdot}{k}\right)$ is a $\Gamma^o(k)$-form. The arguments for the function ψ^{24} are similar using

$$(\eta) = P_\infty^{\frac{k^m}{24}} P_0^{\frac{k}{24}} \prod_{i=1}^{t} P_{x_i}^{\frac{\beta_i}{24}} \text{ on } \overline{\mathbb{H}^2/\Gamma(k^m,k)},$$

where $\beta_i \geq 1$ and $\sum_{i=1}^{t} \beta_i = k(k^m - 1)$.

Remark 2.7. Formulae (5.7) and (5.8) are also valid for $m = 0$ provided we replace the products by 1.

It is of interest to determine when one can extract a 24-th root of the function φ^{24} and obtain a (single valued) $\Gamma(p,q)$-invariant function. A necessary condition is that $24|(pq-1)$. To obtain a sufficient condition, we assume this divisibility condition (thus both p and q are necessarily odd) and we study the invariance of the function φ under $\Gamma(p,q)$. We use the notation and results (the basic fact is contained in the next proposition) of [**16**, Ch. 4, Th. 2] for the multiplier system v_η for the η-function.

Proposition 2.8. *For all* $\tau \in \mathbb{H}^2$ *and all* $\gamma = \begin{bmatrix} a & b \\ c & d \end{bmatrix} \in \mathrm{SL}(2,\mathbb{Z})$,

$$\eta(\gamma(\tau)) = v_\eta(\gamma)(c\tau + d)^{\frac{1}{2}}\eta(\tau),$$

where

$$v_\eta(\gamma) = \begin{cases} \left(\frac{d}{c}\right)^* \exp\left\{\frac{\pi\imath}{12}[(a+d)c - bd(c^2-1) - 3c]\right\} & \text{for } c \text{ odd} \\ \left(\frac{c}{d}\right)_* \exp\left\{\frac{\pi\imath}{12}[(a+d)c - bd(c^2-1) + 3d - 3 - 3cd]\right\} & \text{for } c \text{ even} \end{cases}.$$

The symbols $\left(\frac{\cdot}{\cdot}\right)^*$ and $\left(\frac{\cdot}{\cdot}\right)_*$, whose values are ± 1, are defined in [**16**, Ch. 4]. They are closely related to Legendre's and Jacobi's symbols $\left(\frac{\cdot}{\cdot}\right)$.

The multiplier system of a closely related function, $\theta(\tau) = \theta\begin{bmatrix} 0 \\ 0 \end{bmatrix}(0,\tau)$, is also known. It is described in the next proposition, which is [**16**, Ch. 4, Th. 3]. The reader is invited to develop a theory for θ that parallels our work on η.

Proposition 2.9. *For all* $\tau \in \mathbb{H}^2$ *and all* $\gamma = \begin{bmatrix} a & b \\ c & d \end{bmatrix} \in \Gamma_\theta$,

$$\theta(\gamma(\tau)) = v_\theta(\gamma)(c\tau + d)^{\frac{1}{2}}\theta(\tau),$$

where

$$v_\theta(\gamma) = \begin{cases} \left(\frac{d}{c}\right)^* \exp\left\{-\frac{\pi\imath c}{4}\right\} & \text{if } b \equiv 1 \equiv c,\ a \equiv 0 \equiv d \mod 2 \\ \left(\frac{c}{d}\right)_* \exp\left\{\frac{\pi\imath(d-1)}{4}\right\} & \text{if } a \equiv 1 \equiv d,\ b \equiv 0 \equiv c \mod 2 \end{cases}.$$

Proceeding with the study of the functions of interest, for

$$\gamma = \begin{bmatrix} a & bp \\ cq & d \end{bmatrix} \in \Gamma(p,q),$$

$$\varphi(\gamma(\tau)) = \frac{\eta(q\gamma(\tau))\ \gamma'(\tau)^{\frac{1}{4}}}{\eta\left(\frac{\gamma(\tau)}{p}\right)\gamma'(\tau)^{\frac{1}{4}}} = \frac{\eta(\overline{\gamma}(q\tau))\ \overline{\gamma}'(q\tau)^{\frac{1}{4}}}{\eta\left(\underline{\gamma}(\frac{\tau}{p})\right)\underline{\gamma}'\left(\frac{\tau}{q}\right)^{\frac{1}{4}}} = \frac{v_\eta(\overline{\gamma})}{v_\eta(\underline{\gamma})}\ \frac{\eta(q\tau)}{\eta\left(\frac{\tau}{p}\right)},$$

where $\overline{\gamma} = \begin{bmatrix} a & bpq \\ c & d \end{bmatrix}$ and $\underline{\gamma} = \begin{bmatrix} a & b \\ cqp & d \end{bmatrix}$. If c is odd, then

$$\frac{v_\eta(\overline{\gamma})}{v_\eta(\underline{\gamma})} = \frac{\left(\frac{d}{c}\right)^*}{\left(\frac{d}{cqp}\right)^*}\ \exp\left(\frac{\pi\imath}{12}(qp-1)[-(a+d)c + bd(1+c^2qp) + 3c]\right).$$

In the above formula the two symbols[110] are equal provided pq is a square and the exponential yields 1 (since $24|(qp-1)$). For even c, under the same restrictions,

$$\frac{v_\eta(\overline{\gamma})}{v_\eta(\underline{\gamma})} = \frac{\left(\frac{c}{d}\right)_*}{\left(\frac{cqp}{d}\right)_*}\ \exp\left(\frac{\pi\imath}{12}(qp-1)[-(a+d)c + bd(1+c^2qp) + 3cd]\right) = 1.$$

We have established part of

Proposition 2.10. *Let* p *and* $q \in \mathbb{Z}^+$. *Let*

$$\varphi(\tau) = \frac{\eta(q\tau)}{\eta(\frac{\tau}{p})},\ \phi(\tau) = \frac{\eta(q\tau)}{\eta(\tau)}\ \text{and}\ \psi(\tau) = \frac{\eta(\tau)}{\eta(\frac{\tau}{p})},\ \tau \in \mathbb{H}^2.$$

Sufficient conditions for φ *(respectively,* ϕ *and* ψ*) to define a* $\Gamma(p,q)$*-invariant function is that* $24|(pq-1)$ *(*$24|(q-1)$ *and* $24|(p-1)$*) and that* pq *(*q *and* p*) be a square. The divisibility conditions are also necessary. The function* φ^2 *(*ϕ^2 *and* ψ^2*) is* $\Gamma(p,q)$*-invariant as soon as the respective divisibility condition holds.*

[110]The most important case, Legendre's symbol, denoted by $\left(\frac{\cdot}{\cdot}\right)$ is defined for primes k and integers $a \not\equiv 0 \mod k$ by

$$\left(\frac{a}{k}\right) = \begin{cases} +1 \text{ if } a \text{ is a square} \mod k \\ -1 \text{ otherwise} \end{cases}.$$

To simplify notation, we extend the definition of $\left(\frac{a}{k}\right)$ to all integers by setting $\left(\frac{a}{k}\right) = 0$ for $a \equiv 0 \mod k$.

Proof. The arguments before the statement of the proposition establish the result for φ. The proofs for the other two functions are similar.

Corollary 2.11. *Let $N \in \mathbb{Z}^+$ divide 24. Then the N-th powers of the above functions are $\Gamma(p,q)$-invariant provided the divisibility condition holds with 24 replaced by $\frac{24}{N}$ and the square condition holds if N is odd.*

It is important to set up convenient notation that recognizes the dependence of the above functions on the positive integers p and q. We need

Definition 2.12. For the positive integers p and q, we define the (nonvanishing holomorphic) function

$$f_{q,p}(\tau) = \frac{\eta(q\tau)}{\eta(\frac{\tau}{p})} = \varphi(\tau),\ \tau \in \mathbb{H}^2.$$

Note that in terms of the local coordinate $x = \exp(2\pi\imath\tau)$ at P_∞ on $\overline{\mathbb{H}^2/\Gamma_o(k)}$,

$$f_{q,p}(\tau) = x^{\frac{1}{24}\left(q-\frac{1}{p}\right)} \frac{\prod_{n=1}^{\infty}(1-x^{qn})}{\prod_{n=1}^{\infty}(1-x^{\frac{n}{p}})}.$$

Remark 2.13. The last proposition and its corollary give sufficient conditions for the functions $f_{q,p}^N$, $f_{q,1}^N$ and $f_{1,p}^N$ with arbitrary $N \in \mathbb{Z}$ to be $\Gamma(p,q)$-invariant. The following examples will turn out to be important. Let $k > 1$ be a prime and m and n nonnegative integers. If $N \in \mathbb{Z}^+$ is chosen as the smallest positive integer with the property that $24|(k^{m+n}-1)N$, then f_{k^n,k^m}^N (f_{k^n,k^m}^{2N}) is $\Gamma(k^m,k^n)$-invariant, provided $m+n$ is even (odd). Special cases of this remark, most important for our applications, are the next three propositions.

Recall the definitions of $\alpha(k)$ and $\beta(k)$ given in the introduction to this chapter.

Proposition 2.14. *For each prime k, the function $f_{k,k}^{\alpha(k)}(\tau) = \left(\frac{\eta(k\tau)}{\eta(\frac{\tau}{k})}\right)^{\alpha(k)}$ is $\Gamma(k,k)$-invariant with divisor $\left(\frac{P_\infty}{P_0}\right)^{\beta(k)}$.*

Proof. For all primes k, $\frac{24}{\alpha(k)}\,\big|(k^2-1)$. The formula for the divisor of the function follows from the preliminary remarks for this subsection. The reader should supply the necessary argument.

Corollary 2.15. (a) *For k prime and ≥ 13, P_∞ and P_0 are Weierstrass points on the compact Riemann surface $\overline{\mathbb{H}^2/\Gamma(k,k)}$.*
(b) *For k prime and ≥ 11, the surface $\overline{\mathbb{H}^2/\Gamma(k,k)}$ is not hyperelliptic provided $\frac{k^2-1}{24}$ is odd, in particular, for $k = 11$, 13 and 19.*

The following propositions summarize the computations of the divisors of the forms studied at the beginning of this section.

Proposition 2.16. *For each prime k and all $N \in \mathbb{Z}$ such that $(k-1)|12N$, $f_{k,1}^{\frac{24N}{k-1}}(\tau) = f_k^N(\tau) = \left(\frac{\eta(k\tau)}{\eta(\tau)}\right)^{\frac{24N}{k-1}}$ is a $\Gamma_o(k)$-invariant function with divisor $P_\infty^N P_0^{-N}$.*

Remark 2.17. For primes $k < 11$, the proposition applies to all $N \in \mathbb{Z}^+$ (in particular to $N = 1$). For $k = 11$, $N = 5$ is the smallest positive integer for which the proposition produces a function. It is easily seen that the fact that $P_\infty^5 P_0^{-5}$ is a principal divisor on the compact surface of genus 1, $\overline{\mathbb{H}^2/\Gamma_o(11)}$, implies that $P_\infty^N P_0^{-N}$ cannot be principal on this surface for $N = 1$, 2, 3 and 4, and that for every such principal divisor $5|N$.

Proposition 2.18. *For each prime k and all $N \in \mathbb{Z}$ such that $(k-1)|12N$, $f_{1,k}^{\frac{24N}{k-1}}(\tau) = \left(\frac{\eta(\tau)}{\eta(\frac{\tau}{k})}\right)^{\frac{24N}{k-1}}$ is a $\Gamma(k,k)$-invariant function with divisor*

$$P_\infty^N \ P_1^N \ P_2^N \ \ldots \ P_{k-1}^N \ P_0^{-kN}.$$

Exercise 2.19. Verify the formulae for the divisors in the last two propositions.

Table 13 of $\Gamma(k^m, k^n)$-invariant functions of low degree summarizes the calculations for most of the cases that will be needed.

2.2. Calculation of divisor of $\eta(N\cdot)$. We know that η is a multiplicative $\frac{1}{4}$-form for Γ. Hence for $N \in \mathbb{Z}^+$, $\eta(N\cdot)$ is also a multiplicative $\frac{1}{4}$-form for $\Gamma_o(N)$. We need to compute its order at an arbitrary cusp $x \in \mathbb{Q} \cup \{\infty\}$. Let $\varphi = \theta'\begin{bmatrix} 1 \\ 1 \end{bmatrix}$. We know that

$$\mathrm{ord}_x \varphi(N\cdot) = 3 \ \mathrm{ord}_x \eta(N\cdot).$$

The function φ is one of a family of functions that appear in a transformation theory. Let

$$\varphi_l = \theta'\begin{bmatrix} \frac{1+2l}{N} \\ 1 \end{bmatrix}(0, N\cdot) \text{ and } \psi_l = \theta'\begin{bmatrix} \frac{1+2l}{N} \\ 0 \end{bmatrix}(0, N\cdot).$$

In terms of the local coordinate $x = \exp(\frac{2\pi\imath\tau}{N})$,

$$\mathrm{ord}_\infty \varphi_l = \frac{(1+2l)^2}{8} = \mathrm{ord}_\infty \psi_l.$$

The action of generators of Γ on the forms under study shows

$$B_{\frac{3}{4}}^*(\varphi_l) = \text{ const } \varphi_l \text{ and } B_{\frac{3}{4}}^*(\psi_l) = \text{ const } \psi_l$$

function φ	prime k	case	lowest positive N
$f^N_{k^n,k^m}$	≥ 5	$m+n$ even	1
		$m+n$ odd	$*(k-1)$
	3	$m+n$ even	3
		$m+n$ odd	$*(3^{m+n}-1)$
	2	$m+n$ even	8
		$m+n$ odd	$*(2^{m+n}-1)$
$f^N_{k^n,1}$	≥ 5	n even	1
		n odd	$*(k-1)$
	3	n even	3
		n odd	$*(3^n-1)$
	2	n even	8
		n odd	24
f^N_{1,k^m}	≥ 5	m even	1
		m odd	$*(k-1)$
	3	m even	3
		m odd	$*(3^m-1)$
	2	m even	8
		m odd	24

Table 13. $\Gamma(k^m,k^n)$-INVARIANT FUNCTIONS OF LOW DEGREE. (In the above table $*r$ is the smallest positive <u>even</u> integer N such that $24|rN$.)

(here and below each undetermined constant (c or const) is of absolute value 1) and

$$A^*_{\frac{3}{4}}\left(\theta'\begin{bmatrix}\frac{1+2l}{N}\\ 1\end{bmatrix}(0,N\cdot)\right) = cN^{-\frac{3}{2}}\theta'\begin{bmatrix}1\\ \frac{1+2l}{N}\end{bmatrix}\left(0,\frac{\cdot}{N}\right)$$

$$= cN^{-\frac{1}{2}}\sum_{j=0}^{N-1}\theta'\begin{bmatrix}\frac{1+2j}{N}\\ 1+2l\end{bmatrix}(0,N\cdot) = cN^{-\frac{1}{2}}\sum_{j=0}^{[\frac{N-1}{2}]}c_j\theta'\begin{bmatrix}\frac{1+2j}{N}\\ 1\end{bmatrix}(0,N\cdot),$$

where c_j is a nonzero constant. Similarly,

$$A^*_{\frac{3}{4}}\left(\theta'\begin{bmatrix}\frac{1+2l}{N}\\ 0\end{bmatrix}(0,N\cdot)\right) = cN^{-\frac{3}{2}}\theta'\begin{bmatrix}0\\ \frac{1+2l}{N}\end{bmatrix}\left(0,\frac{\cdot}{N}\right)$$

$$= cN^{-\frac{1}{2}}\sum_{j=0}^{N-1}\theta'\begin{bmatrix}\frac{2j}{N}\\ 1+2l\end{bmatrix}(0,N\cdot) = cN^{-\frac{1}{2}}\sum_{j=0}^{[\frac{N-1}{2}]}c_j\theta'\begin{bmatrix}\frac{1+2j}{N}\\ 1\end{bmatrix}(0,N\cdot).$$

Using the transformation formula given by equation (5.6) and writing an arbitrary $C\in\Gamma$ as a word in A and B allow us to compute $\mathrm{ord}_x\eta(N\cdot)$ at an arbitrary cusp x.

2.3. Coset representatives. We will often encounter the following situation of group inclusions: $G_1 \subset G_2 \subset \Gamma$. We will need to average a G_1-form (or function) over the left G_1 cosets in G_2, $G_1\backslash G_2$, to obtain[111] a G_2-form (see §1.3 of Chapter 3). We consider the cases:

(a) $G_1 = \Gamma_o(k)$, $G_2 = \Gamma$, k **prime**: This case was encountered in Theorem 4.6 of Chapter 4. We have $[\Gamma : \Gamma_o(k)] = k+1$. The $k+1$ motions

$$I = \begin{bmatrix} 1 & 0 \\ 0 & 1 \end{bmatrix}, \ AB^l = \begin{bmatrix} 0 & -1 \\ 1 & l \end{bmatrix}, \ l = 0,\ 1,\ \dots,\ k-1, \tag{5.9}$$

certainly belong to Γ, and since for the relevant values of l and l',

$$AB^l(AB^{l'})^{-1} = \begin{bmatrix} * & * \\ l'-l & * \end{bmatrix} \in \Gamma_o(k)$$

if and only if $l = l' \bmod k$, the $k+1$ motions in (5.9) are representatives for $\Gamma_o(k)\backslash\Gamma$.

For every integer $q \geq 2$, the relative Poincaré series operator $\Theta_{\Gamma_o(k)\backslash\Gamma}$ maps $\mathbb{A}_q(\mathbb{H}^2, \Gamma_o(k))$ to $\mathbb{A}_q(\mathbb{H}^2, \Gamma)$; in our case we have

$$\Theta_{\Gamma_o(k)\backslash\Gamma} = I + \sum_{l=0}^{k-1} (A \circ B^l)_q^* = I + \sum_{l=0}^{k-1} (B^l)_q^* \circ A_q^*.$$

In Theorem 4.6 of Chapter 4,

$$\varphi(\tau) = k^{12}\Delta(k\tau) \in \mathbb{A}_6(\mathbb{H}^2, \Gamma_o(k)),$$

$$A_6^*(\varphi)(\tau) = k^{12}\eta^{24}\left(\frac{-1}{\frac{\tau}{k}}\right)\tau^{-12} = \Delta\left(\frac{\tau}{k}\right).$$

Hence

$$\left((B^l)_6^* \circ A_6^*\right)(\varphi)(\tau) = \Delta\left(\frac{\tau+l}{k}\right).$$

(b) $G_1 = \Gamma(k^n, k)$, $G_2 = \Gamma_o(k) = \Gamma(1,k)$, k **prime**, $n \in \mathbb{Z}^+$: From $[\Gamma_o(k) : \Gamma(k^n, k)] = k^n$ we conclude that the motions B^l, $l = 0,\ \dots,\ k^n - 1$, are representatives for $\Gamma(k^n, k)\backslash\Gamma_o(k)$.

[111] The second group inclusion ($G_2 \subset \Gamma$) is irrelevant for most of our work. The most important cases involve subgroups G_1 of finite index in G_2.

3. Generalities on constructions of $\Gamma_o(k)$-invariant functions

For the rest of this chapter, unless otherwise indicated, $k \in \mathbb{Z}^+$ is a prime. Our primary current interest is the construction of $\Gamma_o(k)$-invariant meromorphic functions whose Laurent series in terms of distinguished local coordinates have integral coefficients. Our methods yield the desired results for small primes k. For large primes, we pose more questions than we give answers. Most of this chapter is based on the properties of the classical η-function and its multiplier system. Since $\eta(\tau)$ is a constant multiple of both $\theta' \begin{bmatrix} 1 \\ 1 \end{bmatrix}^{\frac{1}{3}} (0, \tau)$ and $\theta \begin{bmatrix} \frac{1}{3} \\ 1 \end{bmatrix} (0, 3\tau)$, we could alternatively base our development on either of these functions. These latter approaches would be more in line with the point of view in the rest of this book. The use of the η-function underlines the classical foundation of most of this chapter.

3.1. The basic problems. For any finitely generated Fuchsian group G of the first kind operating on $\mathbb{H}^2$, $\overline{\mathbb{H}^2/G}$ represents the compactification (as a Riemann surface) of the orbifold $\mathbb{H}^2/G$ (obtained by filling in the punctures); both $\mathcal{K}(G)$ and $\mathcal{K}(\overline{\mathbb{H}^2/G})$ represent its function field. We let, as before, $P : \mathbb{H}^2 \cup \Lambda_G \to \overline{\mathbb{H}^2/G}$ denote the canonical projection, where Λ_G denotes the set of parabolic fixed points of G; usually $\Lambda_G = \mathbb{Q} \cup \{\infty\}$. We write P_x for $P(x)$, $x \in \mathbb{H}^2 \cup \Lambda_G$. We study two linear subspaces of $\mathcal{K}(\Gamma_o(k))$.

Definition 3.1. We let

$$\mathcal{K}(\Gamma_o(k))_0$$
$$= \{f \in \mathcal{K}(\Gamma_o(k));\ f \text{ is holomorphic except at } P_0 \text{ and } f \text{ vanishes at } P_\infty\}$$

and

$$\mathcal{K}(\Gamma_o(k))_\infty$$
$$= \{f \in \mathcal{K}(\Gamma_o(k));\ f \text{ is holomorphic except at } P_\infty \text{ and } f \text{ vanishes at } P_0\}.$$

Precomposition with $A_k = \begin{bmatrix} 0 & -1 \\ k & 0 \end{bmatrix}$ establishes a $\mathbb{C}$-linear isomorphism between these two spaces. We need bases for them. It is easier, in most cases, to work directly with $\mathcal{K}(\Gamma_o(k))_\infty$ and then translate the results to the case of greater interest, $\mathcal{K}(\Gamma_o(k))_0$. It is convenient to introduce the following definition for primes k and arbitrary $N \in \mathbb{Z}^+$:

$$D(k, N) = \dim\{f \in \mathcal{K}(\Gamma_o(k))_0;\ \deg f \le N\}.$$

Obviously $D(k, N) \le N$ with equality if and only if $p = p(\overline{\mathbb{H}^2/\Gamma_o(k)}) = 0$, and in general, as a consequence of the Weierstrass gap theorem,

$$D(k, N) = N - p \text{ for } N \ge 2p.$$

3.2. Some generalities. As pointed out above, this subsection should be considered as motivational; it is not needed for our subsequent presentation. Assume that $k \in \mathbb{Z},\ k \geq 2$. The fact that $\eta^{24} \in \mathbb{A}_6(\mathbb{H}^2, \Gamma)$ shows at once that the ratio $f_{k,1}^{24}(\tau) = \left(\frac{\eta(k\tau)}{\eta(\tau)}\right)^{24}$ is a $\Gamma_o(k)$-invariant meromorphic function whose divisor on $\overline{\mathbb{H}^2/\Gamma_o(k)}$ (provided that k is a prime) is $\frac{P_\infty^{k-1}}{P_0^{k-1}}$. For a prime k, it is possible to extract a $(k-1)$-root of the above function if and only if $\mathbb{H}^2/\Gamma_o(k)$ has genus zero, thus if and only if $k = 2, 3, 5, 7$ and 13. For these values of k, we will obtain below alternate constructions[112] for the function

(5.10)
$$f_k(\tau) = f_{k,1}^{\frac{24}{k-1}}(\tau) = \left(\frac{\theta'\begin{bmatrix}1\\1\end{bmatrix}(0,k\tau)}{\theta'\begin{bmatrix}1\\1\end{bmatrix}(0,\tau)}\right)^{\frac{8}{k-1}} = \left(\frac{\eta(k\tau)}{\eta(\tau)}\right)^{\frac{24}{k-1}} \text{ with divisor } \frac{P_\infty}{P_0}.$$

In terms of the local coordinate $x = \exp(2\pi\imath\tau)$ on $\overline{\mathbb{H}^2/\Gamma_o(k)}$ vanishing at P_∞, for all $N \in \mathbb{Z}$,

$$f_k(\tau)^N = x^N \left(\prod_{n=1}^{\infty}(1-x^{kn})^{\frac{24N}{k-1}}\right)\left(\prod_{n=1}^{\infty}(1-x^n)^{\frac{24N}{1-k}}\right) \tag{5.11}$$
$$= x^N \left(\sum_{m=0}^{\infty} P_{\frac{24N}{1-k}}(m)x^{km}\right)\left(\sum_{m=0}^{\infty} P_{\frac{24N}{k-1}}(m)x^m\right).$$

In general, for primes $k \geq 5$, $f_{k,1}^{*(k-1)} \in \mathcal{K}(\Gamma_o(k))$. In particular, f_{11}^5 is singlevalued (see Corollary 2.11).

The material in Chapter 3 showed that for odd k, the meromorphic function $f_{k,k}^3(\tau) = \dfrac{\theta'\begin{bmatrix}1\\1\end{bmatrix}(0,k\tau)}{\theta'\begin{bmatrix}1\\1\end{bmatrix}(0,\frac{\tau}{k})}$ is $\Gamma(k)$-invariant and its divisor on $\overline{\mathbb{H}^2/\Gamma(k)}$ is $\dfrac{\left(P_{\frac{1}{k}}P_{\frac{2}{k}}\ \cdots\ P_{\frac{k-1}{2k}}\right)^{\frac{k^2-1}{8}}}{A\left(P_{\frac{1}{k}}P_{\frac{2}{k}}\ \cdots\ P_{\frac{k-1}{2k}}\right)^{\frac{k^2-1}{8}}}$, provided k is also a prime. Recall that A is the fractional linear transformation $\tau \mapsto \frac{-1}{\tau}$ and that a Möbius transformation (for example, A) acts on divisors pointwise. The above formula needs to be modified to produce a $\Gamma(k)$-invariant function for even k.

[112] We will use the same definition of f for all $k \in \mathbb{Z}^+$ (even when it defines a multivalued function).

We can and will proceed in a slightly different manner, for all k, $f^{24}_{k,k}(\tau) = \left(\dfrac{\theta'\begin{bmatrix}1\\1\end{bmatrix}(0,k\tau)}{\theta'\begin{bmatrix}1\\1\end{bmatrix}(0,\frac{\tau}{k})}\right)^8 = \left(\frac{\eta(k\tau)}{\eta(\frac{\tau}{k})}\right)^{24}$ defines a $\Gamma(k,k) = \Gamma_o(k) \cap \Gamma^o(k)$-invariant function with divisor $\frac{P_\infty^{k^2-1}}{P_0^{k^2-1}}$ provided k is a prime. We can in this case extract a k^2-1 root of our last function provided $k = 2$, 3 or 5. For $k = 7$ one can show directly that we are able to extract a $\frac{k^2-1}{2} = 24$-th root. One of our immediate aims is to produce functions of low degree on the compact surface $\overline{\mathbb{H}^2/\Gamma(k,k)}$. In this regard, we will also need to study another function. For primes k, $f^{24}_{1,k}(\tau) = \left(\frac{\eta(\tau)}{\eta(\frac{\tau}{k})}\right)^{24}$ is $\Gamma^o(k)$-invariant; hence certainly $\Gamma(k,k)$-invariant and its divisor on $\overline{\mathbb{H}^2/\Gamma(k,k)}$ is $\frac{P_\infty^{k-1}\ P_1^{k-1}\ P_2^{k-1}\ \ldots\ P_{k-1}^{k-1}}{P_0^{k(k-1)}}$.

4. Constructions of (group) $\Gamma_o(k)$-invariant functions

We describe three methods for constructions of $\Gamma_o(k)$-invariant functions. The first (§4.1) is classical and well known; the second (§4.2) is an extension of a known method; the third (§4.4) appears to be new to the literature, but can easily be derived from existing work. We work almost exclusively with the classical η-function. Our development, as we have already pointed out, could also be based on either $\theta'\begin{bmatrix}1\\1\end{bmatrix}(0,\tau)$ or $\theta\begin{bmatrix}\frac{1}{3}\\1\end{bmatrix}(0,3\tau)$. As pointed out in equation (4.14), these three functions are intimately related.

4.1. The direct construction.

Proposition 4.1. *For each prime k,*
(a) $f_k(\tau) = f_{k,1}^{\frac{24}{k-1}}(\tau) = \left(\frac{\eta(k\tau)}{\eta(\tau)}\right)^{\frac{24}{k-1}}$ *is a $\Gamma_o(k)$-multiplicative function with divisor $\frac{P_\infty}{P_0}$;*
(b) *for all $N \in \mathbb{Z}$, f_k^N is $\Gamma_o(k)$-invariant if $(k-1)|12N$; in particular*
(c) *(for $N = 1$) f_k is $\Gamma_o(k)$-invariant if and only if $k = 2$, 3, 5, 7 or 13 (the cases where the genus of $\overline{\mathbb{H}^2/\Gamma_o(k)}$ is zero); and*
(d) $f_k \circ A_k = k^{\frac{12}{1-k}} f_k^{-1}$.

Recall that $A_k = \begin{bmatrix} 0 & -1 \\ k & 0 \end{bmatrix}$.

Proof. Part (a) follows from our previous observation that f_k^{k-1} is $\Gamma_o(k)$-invariant with divisor $\left(\frac{P_\infty}{P_0}\right)^{k-1}$. Part (b) is the content of the previously established Proposition 2.16. Part (c) is a direct consequence of (a); (d) is a consequence of the transformation formula for the η-function.

The Laurent series expansions for the functions f_k^N are given above in (5.11).

Problem 4.2. Obtain an alternate useful characterization of the subgroups G of $\Gamma_o(k)$ for which f_k is single valued on $\overline{\mathbb{H}^2/G}$.

The last proposition shows that for $k = 2, 3, 5, 7$ and 13, $f_k^N \in \mathcal{K}(\Gamma_o(k))_0$ provided $N \in \mathbb{Z}^+$ and $f_k^N \in \mathcal{K}(\Gamma_o(k))_\infty$ provided $-N \in \mathbb{Z}^+$. Thus

$$f_k,\ f_k^2,\ \dots,\ f_k^N,\ \dots$$

is a basis for $\mathcal{K}(\Gamma_o(k))_0$ and

$$f_k^{-1},\ f_k^{-2},\ \dots,\ f_k^{-N},\ \dots$$

is a basis for $\mathcal{K}(\Gamma_o(k))_\infty$. For $k = 11$ and 17, for example, we must use other sets of functions to produce bases for $\mathcal{K}(\Gamma_o(k))_0$ and $\mathcal{K}(\Gamma_o(k))_\infty$. We now turn to other, more elaborate, procedures for manufacturing such functions.

Remark 4.3. 1. There are two more or less obvious $\Gamma_o(k)$-invariant meromorphic functions. Since $\jmath$ is Γ-invariant, it is certainly $\Gamma_o(k)$-invariant. Further, as a $\Gamma_o(k)$-function,

$$\mathrm{ord}_\infty \jmath = -1 \text{ and } \mathrm{ord}_0 \jmath = -k.$$

The second $\Gamma_o(k)$-invariant function[113] is $J(\tau) = \jmath(k\tau)$. It is easily seen that

$$J\left(-\frac{1}{k\tau}\right) = \jmath(\tau) \text{ and hence } \mathrm{ord}_\infty J = -k \text{ and } \mathrm{ord}_0 J = -1.$$

For primes k, we have computed the polar divisors of $\jmath$ and J. However, since these functions necessarily have zeros on $\mathbb{H}^2/\Gamma_o(k)$, they are not as useful as powers of f_k.

2. We will need to study Laurent series expansions at P_∞ and P_0 of $\Gamma_o(k)$-invariant meromorphic functions on $\mathbb{H}^2$. Good local coordinates at these two points on $\mathbb{H}^2/\Gamma_o(k)$ are $x = \exp(2\pi\imath\tau)$ and $y = \exp\left(-\frac{2\pi\imath}{k\tau}\right)$, respectively. Let φ be $\Gamma_o(k)$-invariant with Laurent series expansions

$$\varphi(\tau) = \sum_{n=N}^{\infty} a_n x^n = \sum_{n=M}^{\infty} b_n y^n.$$

[113] We are abandoning the convention previously used for the same letter written in lower and upper case. Thus J is *not* the projection of $\jmath$ or j to $\mathbb{H}^2/\Gamma_o(k)$.

Then $\psi(\tau) = \varphi\left(-\frac{1}{k\tau}\right)$ is also $\Gamma_o(k)$-invariant with Laurent series expansions

$$\psi(\tau) = \sum_{n=M}^{\infty} b_n x^n = \sum_{n=N}^{\infty} a_n y^n.$$

4.2. Averaging $\Gamma(k^n, k)$-invariant functions. Let k be a prime. The most important groups for our development are the Hecke groups $\Gamma_o(k)$, and we need to construct functions invariant with respect to these groups. We have already described one method. To get identities we need at least another method. To produce $\Gamma_o(k)$-invariant functions, we may (and will) start with a $\Gamma(k^n, k)$-invariant function (for some $n \in \mathbb{Z}^+$) and average it over the left cosets $\Gamma(k^n, k)\backslash\Gamma_o(k)$. The most important case for us is $n = 1$.

Definition 4.4. We define $\beta(\Gamma(k,k))$ as the smallest positive integer β with $\left(\frac{P_\infty}{P_0}\right)^\beta$ a principal divisor on the compact Riemann surface $\overline{\mathbb{H}^2/\Gamma(k,k)}$. Similarly, we define $\beta(\Gamma_o(k))$ as the smallest positive integer β with $\left(\frac{P_\infty}{P_0}\right)^\beta$ a principal divisor on the compact Riemann surface $\overline{\mathbb{H}^2/\Gamma_o(k)}$.

It follows easily that for all $N \in \mathbb{Z}$, a meromorphic function (an element of the field $\mathcal{K}(\overline{\mathbb{H}^2/\Gamma(k,k)})$) with divisor $P_\infty^{\beta(\Gamma(k,k))N} P_0^{-\beta(\Gamma(k,k))N}$ is given by

$$f_{k,k}^{\frac{\beta(\Gamma(k,k))N}{\beta(k)}}(\tau) = \left(\frac{\eta(k\tau)}{\eta(\frac{\tau}{k})}\right)^{\frac{\beta(\Gamma(k,k))N}{\beta(k)}}.$$

Obviously, $\beta(\Gamma(k,k)) = 1$ if and only if $p(\Gamma(k,k)) = 0$ (thus only for $k = 2$, 3 and 5).

Consider the set of positive integers γ such that $\frac{P_\infty^\gamma}{P_0^\gamma}$ is a principal divisor on $\overline{\mathbb{H}^2/\Gamma(k,k)}$. It is obvious from the fact that the divisor class $\frac{P_\infty}{P_0}$ of degree zero is a point of order $\beta(\Gamma(k,k))$ (in the Jacobi variety of $\overline{\mathbb{H}^2/\Gamma(k,k)}$), that such a γ is a multiple of $\beta(\Gamma(k,k))$. Hence we conclude that if we produce a principal divisor $\frac{P_\infty^\gamma}{P_0^\gamma}$ as above, then $\beta(\Gamma(k,k))$ must be a factor of γ. Proposition 2.14 shows that $\beta(\Gamma(k,k))|\beta(k)$. It is also obvious that if $\overline{\mathbb{H}^2/\Gamma(k,k)}$ has positive genus and we have produced a function with divisor $\frac{P_\infty^\gamma}{P_0^\gamma}$, for some prime γ, then $\beta(\Gamma(k,k)) = \gamma$. This remark is helpful for $k = 7$, 11 and 13, but not, for example, for $k = 17$, 19, 23 and 29. The limited applicability of this remark is illustrated by Table 14 illustrating some factorizations.

It seems reasonable to conjecture that for all primes k, $\beta(k) = \beta(\Gamma(k,k))$. We shall prove this for $k \leq 89$ (Theorem 4.9). For $\beta(\Gamma_o(k))$, results similar to those described for $\beta(\Gamma(k,k))$ hold, in particular, $\beta(\Gamma_o(k))|(k-1)$.

prime k	$\frac{k^2-1}{24}$	Nontrivial factorization of $\frac{k^2-1}{24}$
5	1	
7	2	
11	5	
13	7	
17	12	$2^2 3$
19	15	$3 \cdot 5$
23	22	$2 \cdot 11$
29	35	$5 \cdot 7$
31	40	$2^3 5$
37	57	$3 \cdot 19$
41	70	$2 \cdot 5 \cdot 7$
67	187	$11 \cdot 17$

Table 14. FACTORIZATIONS.

Define (for convenience and consistent with earlier usage) for $x \in \mathbb{R}$, the *floor* function $\lfloor x \rfloor$ to be the largest integer $\leq x$ and the *ceiling* function $\lceil x \rceil$ to be the smallest integer $\geq x$. Note that

$$\lceil x \rceil = -\lfloor -x \rfloor$$

and that $\lfloor x \rfloor$ is the integral part of x. For $a \in \mathbb{Z}$ and $b \in \mathbb{Z}^+$, we let $R\left(\frac{a}{b}\right)$ be the remainder of dividing b into a: that is, $R\left(\frac{a}{b}\right) \in \mathbb{Z}$, $0 \leq R\left(\frac{a}{b}\right) \leq b-1$, and $a \equiv R\left(\frac{a}{b}\right) \mod b$. Hence

$$a = \left\lfloor \frac{a}{b} \right\rfloor b + R\left(\frac{a}{b}\right).$$

We now average a $\Gamma(k^n, k)$-invariant function with singularities only at points $\Gamma_o(k)$-equivalent to P_∞ or P_0 over the group (viewed as a subgroup of $\mathrm{PSL}(2, \mathbb{Z}_{k^n})$) of order k^n generated by B to obtain a $\Gamma_o(k)$-invariant function. Assume that k is a prime ≥ 5 and that $n \in \mathbb{Z}^+$ is odd. Then k^{n+1} is a square and $24|(k^{n+1}-1)$. Similarly if $n \in \mathbb{Z}^+$ is even, then k^n is a square and $24|(k^n - 1)$. Thus Lemma 2.6 and Proposition 2.10 apply in these cases. Similar statements hold for the primes 2 and 3.

Theorem 4.5. *Let k be a prime and let $N \in \mathbb{Z}$.*

(a) *Let $n \geq 0$ be an integer. The formula*[114]

$$F_{k,2n+1,N}(\tau) = \frac{1}{k^{2n+1}} \sum_{l=0}^{k^{2n+1}-1} \left(\frac{\eta(k(\tau+l))}{\eta(\frac{\tau+l}{k^{2n+1}})} \right)^{\alpha(k)N}$$

[114]The most important case for us is $n = 0$, and we hence abbreviate $F_{k,1,N} = F_{k,N}$.

defines a $\Gamma_o(k)$-invariant function that is holomorphic, except possibly at the punctures, with

(5.12)

$$\operatorname{ord}_{P_\infty} F_{k,2n+1,N} \geq \left\lceil \frac{\alpha(k)N}{24}\left(k - \frac{1}{k^{2n+1}}\right)\right\rceil = \left\lceil \beta(k)N\frac{k^{2n+2}-1}{k^{2n+1}(k^2-1)}\right\rceil;$$

equality holds provided $N > 0$. Further

$$\operatorname{ord}_{P_0} F_{k,2n+1,N} = \frac{\alpha(k)N}{24}(1-k^{2n+2}) = -\beta(k)N\frac{k^{2n+2}-1}{k^2-1} \text{ for } N > 0.$$

(b) *Let $n \in \mathbb{Z}^+$. The formula*

$$F_{k,2n,N}(\tau) = \frac{1}{k^{2n}}\sum_{l=0}^{k^{2n}-1}\left(\frac{\eta(\tau+l)}{\eta(\frac{\tau+l}{k^{2n}})}\right)^{\alpha(k)N}$$

defines a $\Gamma_o(k)$-invariant function that is holomorphic, except possibly at the punctures, with

$$\operatorname{ord}_{P_\infty} F_{k,2n,N} = \left\lceil \frac{\alpha(k)N}{24}\left(1 - \frac{1}{k^{2n}}\right)\right\rceil = \left\lceil \beta(k)N\frac{k^{2n}-1}{k^{2n}(k^2-1)}\right\rceil,$$

and

$$\operatorname{ord}_{P_0} F_{k,2n,N} = \frac{\alpha(k)Nk}{24}(1-k^{2n}) = -k\beta(k)N\frac{k^{2n}-1}{k^2-1} \text{ for } N > 0.$$

(c) *For all $n \in \mathbb{Z}^+$, $F_{k,n,0} = 1$ (the function is constant).*
(d) *For $N < 0$, $F_{k,N}$ is holomorphic except at P_∞ and*

$$\operatorname{ord}_{P_\infty} F_{k,N} \geq \left\lceil \frac{\beta(k)N}{k}\right\rceil.$$

In particular, this function is constant, with value $P_{\alpha(k)N}(-\beta(k)N)$, if

$$-\beta(k)N < \begin{cases} k \text{ for } k \leq 7 \text{ or } k = 13, \\ 2k \text{ for } k = 11 \text{ or } k \geq 17, \\ 3k \text{ for } k \geq 17 \text{ and } \overline{\mathbb{H}^2/\Gamma_o(k)} \text{ is not hyperelliptic.} \end{cases}$$

Proof. We treat the case of odd n. We start with the observation that $\varphi = f_{k,k^n}^{\alpha(k)}$ is a meromorphic function on $\overline{\mathbb{H}^2/\Gamma(k^n,k)}$ and in terms of the local coordinate $x = \exp\left\{\frac{2\pi\imath\tau}{k^n}\right\}$ at P_∞ on that surface,

$$\varphi(\tau) = \left(\frac{\eta(k\tau)}{\eta(\frac{\tau}{k^n})}\right)^{\alpha(k)} = \left(x^{\frac{k^{n+1}-1}{24}}\frac{\prod_{m=1}^\infty(1-x^{k^{n+1}m})}{\prod_{m=1}^\infty(1-x^m)}\right)^{\alpha(k)}.$$

In terms of this local coordinate the map $\tau \mapsto \tau + 1$ is given by $x \mapsto \exp\left(\frac{2\pi\imath}{k^n}\right) x$. Hence

$$\frac{1}{k^n}\sum_{l=0}^{k^n-1}\left(\frac{\eta(k(\tau+l))}{\eta(\frac{\tau+l}{k^n})}\right)^{\alpha(k)N} = \prod_{m=1}^{\infty}(1-x^{k^{n+1}m})^{\alpha(k)N}$$

$$\times \sum_{m=\left\lceil \frac{\alpha(k)N}{24}\left(k-\frac{1}{k^n}\right)\right\rceil}^{\infty} P_{\alpha(k)N}\left(k^n m - \frac{\alpha(k)N(k^{n+1}-1)}{24}\right) x^{k^n m}.$$

Thus (5.12) (including the statement about equality). It remains to study the behavior of $F_{k,2n+1,N}$ at P_0 for positive N (here and below n is a non-negative integer). Observe that for all $N \in \mathbb{Z}$,

$$F_{k,2n+1,N} = \frac{1}{k^{2n+1}}\sum_{l=0}^{k^{2n+1}-1}\varphi^N \circ B^l \text{ and } \operatorname{ord}_0\varphi^N = \alpha(k)N\frac{1-k^{2n+2}}{24}.$$

Finally, if $N > 0$, then from Lemma 2.6 we conclude

$$\operatorname{ord}_{B^l(0)}\varphi^N \geq \alpha(k)N\frac{2-k^{2n+1}}{24} > \alpha(k)N\frac{1-k^{2n+2}}{24}, \text{ for } l = 1, \ldots, k^{2n+1}-1.$$

For even indices, the appropriate power series is

$$F_{k,2n,N}(\tau) = \prod_{m=1}^{\infty}(1-x^{k^{2n}m})^{\alpha(k)N}$$

$$\times \sum_{m=\left\lceil \frac{\alpha(k)N}{24}\left(1-\frac{1}{k^{2n}}\right)\right\rceil}^{\infty} P_{\alpha(k)N}\left(k^{2n} m - \frac{\alpha(k)N(k^{2n}-1)}{24}\right) x^{k^{2n} m}.$$

The order of this function at P_∞ is easily computed from this formula. We need to compute the order of the function $F_{k,2n,N}$ at P_0 (for positive N only). The argument is similar to the one used for the case of odd integers $2n+1$. This completes the proofs of parts (a) and (b). Part (c) requires no comments.

In our study of the functions $F_{k,n,N}$ we need to distinguish between even and odd n since the definitions of these functions depend on the parity of n.

It is useful to rewrite the equations for $F_{k,n,N}$ in terms of the correct local coordinate $x = \exp(2\pi\imath\tau)$ at P_∞ on $\overline{\mathbb{H}^2/\Gamma_o(k)}$. This is accomplished by replacing x^{k^n} by x. In particular, for $n = 1$,

$$(5.13) \quad F_{k,N}(\tau) = \prod_{m=1}^{\infty}(1-x^{km})^{\alpha(k)N} \sum_{m=\left\lceil \frac{\beta(k)N}{k}\right\rceil}^{\infty} P_{\alpha(k)N}(km - \beta(k)N)x^m$$

$$= \sum_{m=0}^{\infty} P_{-\alpha(k)N}(m)x^{km} \sum_{m=\left\lceil \frac{\beta(k)N}{k} \right\rceil}^{\infty} P_{\alpha(k)N}(km - \beta(k)N)x^m.$$

The leading term of this expression is

$$P_{\alpha(k)N}\left(R\left(\frac{-\beta(k)N}{k}\right)\right) x^{\left\lceil \frac{\beta(k)N}{k} \right\rceil}.$$

It is hence obvious that

$$\text{(5.14)} \qquad \operatorname{ord}_{P_\infty} F_{k,N} = \left\lceil \frac{\beta(k)N}{k} \right\rceil \text{ provided } P_{\alpha(k)N}\left(R\left(\frac{-\beta(k)N}{k}\right)\right) \neq 0.$$

This last inequality holds for any positive N, because $P_N(n) > 0$ for all $N \in \mathbb{Z}^+$ and all $n \in \mathbb{Z}^+ \cup \{0\}$.

For $N < 0$ the order of $F_{k,N}$ at ∞ is sometimes given by (5.14), and in general by

$$\operatorname{ord}_{P_\infty} F_{k,N} = \left\lceil \frac{\beta(k)N}{k} \right\rceil$$
$$+ \min \left\{ a \in \mathbb{Z}^+ \cup \{0\};\ P_{\alpha(k)N}\left(ka + R\left(\frac{-\beta(k)N}{k}\right)\right) \neq 0 \right\}.$$

The proof of (d) is completed by consideration of the genus of the compact surface $\overline{\mathbb{H}^2/\Gamma_o(k)}$.

Remark 4.6. **1**. In the above construction of $F_{k,1,N} = F_{k,N}$, under appropriate hypothesis, N may assume some rational values. This remark is a key step in the proof of Theorem 4.9. The construction of constant functions $F_{k,N}$ (dealt with in a portion of part (d) of the theorem) will be pursued further in §6.

2. For $-N \in \mathbb{Z}^+$, all that one can conclude is $\operatorname{ord}_{P_0} F_{k,N} \geq 0$. See also §6.

3. We used sums (rather than products) of $\Gamma(k^n, k)$-invariant functions to produce $\Gamma_o(k)$-invariant functions. Products seem to lead to identities easily proven using the Jacobi triple product formula.

4. The above theorem is the basis for obtaining partition identities; we also need to investigate the properties of the constants $\{c_i\}$ appearing in the next theorem.

5. We need to explain the reason for distinguishing odd and even positive integers n in our definition of the function $F_{k,n,N}$. The difference is accounted for by the starting point in our construction. We need to work with $\Gamma(k^n, k)$-functions. We are using $\left(\frac{\eta(k\tau)}{\eta(\frac{\tau}{k^n})}\right)^{\alpha(k)N}$ and $\left(\frac{\eta(\tau)}{\eta(\frac{\tau}{k^n})}\right)^{\alpha(k)N}$. For odd n, the first of these is $\Gamma(k^n, k)$-invariant for all N, while for even n, we must impose a condition on N in accordance with Table 13 (for example, for $k = 5$, N must be a multiple of 6). Similarly, for even n, the second of our functions is $\Gamma(k^n, k)$-invariant for all N, while for odd n, the condition imposed on N is that it again be a multiple of 6 (if $k = 5$). For the study of Ramanujan congruences, the case $N = 1$ is most important.

6. The definition of the function $F_{k,N}(\tau)$ can be rewritten as

$$\frac{1}{k}(\eta(k\tau))^{\alpha(k)N}\sum_{l=0}^{k-1}\exp\left(\frac{\pi\imath k\alpha(k)Nl}{6}\right)\left(\eta\left(\frac{\tau+l}{k}\right)\right)^{-\alpha(k)N}.$$

As a consequence of the fact that equation (5.10) defines a $\Gamma_o(k)$-invariant function for small k, we have

Theorem 4.7. *Let k be a positive prime such that $p(\overline{\mathbb{H}^2/\Gamma_o(k)}) = 0$ (thus $k = 2$, 3, 5, 7 or 13).*
(a) *Let $N \in \mathbb{Z}^+$. There exists an appropriate collection of constants $\{c_i\}$ such that*

$$F_{k,n,N} = \sum_{i=\left\lceil\frac{\alpha(k)N}{24}\left(k-\frac{1}{k^n}\right)\right\rceil}^{\frac{\alpha(k)N}{24}(k^{n+1}-1)} c_i f_k^i, \text{ for } n \text{ an odd positive integer,}$$

and

$$F_{k,n,N} = \sum_{i=\left\lceil\frac{\alpha(k)N}{24}\left(1-\frac{1}{k^n}\right)\right\rceil}^{\frac{\alpha(k)N}{24}(k^{n+1}-k)} c_i f_k^i, \text{ for } n \text{ an even positive integer.}$$

Further, the coefficients for the extreme indices

$$c_{\left\lceil\frac{\alpha(k)N}{24}\left(k-\frac{1}{k^n}\right)\right\rceil} \text{ and } c_{\frac{\alpha(k)N}{24}(k^{n+1}-1)},$$

$$c_{\left\lceil\frac{\alpha(k)N}{24}\left(1-\frac{1}{k^n}\right)\right\rceil} \text{ and } c_{\frac{\alpha(k)N}{24}(k^{n+1}-k)}$$

for odd and even n, respectively are nonzero. Thus for odd n (in terms of the local coordinate $x = \exp(2\pi\imath\tau)$ at P_∞ on $\overline{\mathbb{H}^2/\Gamma_o(k)}$),

$$F_{k,n,N}(\tau)$$

$$= \prod_{m=1}^{\infty}(1-x^{km})^{\alpha(k)N} \sum_{m=\left\lceil\frac{\alpha(k)N}{24}\left(k-\frac{1}{k^n}\right)\right\rceil}^{\infty} P_{\alpha(k)N}\left(k^n m - \frac{\alpha(k)N(k^{n+1}-1)}{24}\right)x^m$$

$$(5.15) \qquad = \sum_{i=\left\lceil \frac{\alpha(k)N}{24}\left(k-\frac{1}{k^n}\right)\right\rceil}^{\frac{\alpha(k)N}{24}(k^{n+1}-1)} c_i \left[x\frac{\prod_{m=1}^{\infty}(1-x^{km})^{\frac{24}{k-1}}}{\prod_{n=1}^{\infty}(1-x^m)^{\frac{24}{k-1}}}\right]^i,$$

$$c_{\left\lceil \frac{\alpha(k)N}{24}\left(k-\frac{1}{k^n}\right)\right\rceil} = P_{\alpha(k)N}\left(k^n - \frac{\alpha(k)N(k^{n+1}-1)}{24}\right).$$

For even n,

$$F_{k,n,N}(\tau)$$

$$= \prod_{m=1}^{\infty}(1-x^m)^{\alpha(k)N} \sum_{m=\left\lceil \frac{\alpha(k)N}{24}\left(1-\frac{1}{k^n}\right)\right\rceil}^{\infty} P_{\alpha(k)N}\left(k^n m - \frac{\alpha(k)N(k^n-1)}{24}\right)x^m$$

$$(5.16) \qquad = \sum_{i=\left\lceil \frac{\alpha(k)N}{24}\left(1-\frac{1}{k^n}\right)\right\rceil}^{\frac{\alpha(k)Nk}{24}(k^n-1)} c_i \left[x\frac{\prod_{m=1}^{\infty}(1-x^{km})^{\frac{24}{k-1}}}{\prod_{m=1}^{\infty}(1-x^m)^{\frac{24}{k-1}}}\right]^i,$$

$$c_{\left\lceil \frac{\alpha(k)N}{24}\left(1-\frac{1}{k^n}\right)\right\rceil} = P_{\alpha(k)N}\left(k^n - \frac{\alpha(k)N(k^n-1)}{24}\right).$$

(b) *Let $N \in \mathbb{Z}$. There exist constants c_i, $i = \left\lceil \frac{\beta(k)N}{k}\right\rceil$, ..., $\beta(k)\max\{0,N\}$ (with $c_{\left\lceil \frac{\beta(k)N}{k}\right\rceil} \neq 0$ for $N > 0$), such that*

$$F_{k,N} = \sum_{i=\left\lceil \frac{\beta(k)N}{k}\right\rceil}^{\beta(k)\max\{0,N\}} c_i f_k^i;$$

that is,

$$F_{k,N}(\tau) = \frac{1}{k}\sum_{l=0}^{k-1}\left(\frac{\eta(k(\tau+l))}{\eta\left(\frac{\tau+l}{k}\right)}\right)^{\alpha(k)N} = \sum_{i=\left\lceil \frac{\beta(k)N}{k}\right\rceil}^{\beta(k)\max\{0,N\}} c_i\left(\frac{\eta(k\tau)}{\eta(\tau)}\right)^{\frac{24i}{k-1}}.$$

The special case $n = 1$ is important enough for the related power series identity to be recorded as

Corollary 4.8. *Let k be a positive prime such that $p(\overline{\mathbb{H}^2/\Gamma_o(k)}) = 0$ and let $N \in \mathbb{Z}$. Then in terms of the local coordinate $x = \exp(2\pi\imath\tau)$ at P_∞ on $\overline{\mathbb{H}^2/\Gamma_o(k)}$,*

$$(5.17) \qquad \prod_{m=1}^{\infty}(1-x^{km})^{\alpha(k)N} \sum_{m=\left\lceil \frac{\beta(k)N}{k}\right\rceil}^{\infty} P_{\alpha(k)N}(km-\beta(k)N)x^m$$

$$= \sum_{i=\left\lceil \frac{\beta(k)N}{k} \right\rceil}^{\beta(k)\max\{0,N\}} c_i \left[x \prod_{m=1}^{\infty} \left(\frac{1-x^{km}}{1-x^m} \right)^{\frac{24}{k-1}} \right]^i .$$

Proof. That for $N > 0$, the constants c_i, for $i < \left\lceil \frac{\beta(k)N}{k} \right\rceil$, vanish follows immediately from the power series expansions of the functions in the identities of the corollary.

The above corollary carries important information that will be explored and exploited in later sections of this text. We will be studying the ideal generated by the infinite set of integers $\{P_N(km+R);\ m \in \mathbb{Z}^+ \cup \{0\}\}$, for various positive primes k, integers N and remainders R ($R \in \mathbb{Z};\ 0 \leq R < k$). It is a serious limitation of our methods that R depends on k and N.

The functions $F_{k,N}$ do not, generally, form a basis for the vector spaces of meromorphic functions $\mathcal{K}(\Gamma_o(k))_0$ and $\mathcal{K}(\Gamma_o(k))_\infty$ on the compact surface $\overline{\mathbb{H}^2/\Gamma_o(k)}$, even when this surface is the sphere. We will see later (§4.4) how to obtain such a basis.

We have several immediate consequences of Theorem 4.5. Part (d) yields much information which we describe in §6 that contains a more systematic analysis of conditions that insure that $F_{k,N}$ is constant. We remind the reader that N in the above quoted theorem does not have to be an integer. The quantities P_N are also defined for N, a rational or real number.

As a sample of the possibilities, we show for primes k with $5 \leq k \leq 89$, the number $\frac{k^2-1}{24}$ is the minimal value of $r \in \mathbb{Z}^+$ for which we can find a meromorphic function on the compactification of $\mathbb{H}^2/\Gamma(k,k)$ with divisor $(\frac{P_\infty}{P_0})^r$.

Theorem 4.9. *For every prime $k \leq 89$, $\beta(\Gamma(k,k)) = \beta(k)$.*

Proof. Without loss of generality we can assume that $k \geq 7$ since for $k = 2$, 3 or 5, $\beta(\Gamma(k,k)) = \beta(k) = 1$. For primes k with $7 \leq k < 17$, the number $\frac{k^2-1}{24}$ is prime and the surface $\overline{\mathbb{H}^2/\Gamma(k,k)}$ has positive genus. Hence $\beta(\Gamma(k,k)) = \frac{k^2-1}{24}$. Let $k \geq 7$ be a prime so that $\beta(k) = \frac{k^2-1}{24}$. We know that $1 < \beta(\Gamma(k,k)) | \beta(k)$. Say $\beta(\Gamma(k,k)) = \frac{k^2-1}{24d}$ with $d \in \mathbb{Z}$, $d \geq 2$. The construction in Theorem 4.5 works with $N \in \mathbb{Q}$ for $n = 1$ as long as $\left(\frac{\eta(k\tau)}{\eta(\frac{\tau}{k})}\right)^N$ is a meromorphic function on $\overline{\mathbb{H}^2/\Gamma(k,k)}$. It follows that we can take $-N = \frac{1}{d}$ and obtain a function $F_{k,-\frac{1}{d}}$ whose only singularity is a pole of order at most $\left\lfloor \frac{k^2-1}{24dk} \right\rfloor$ at P_∞. Thus $F_{k,-\frac{1}{d}}$ is constant[115] if $\left\lfloor \frac{k^2-1}{24dk} \right\rfloor < 2$. This condition is

[115] This conclusion and hence the theorem can be extended to higher values of the prime k after establishing the nonhyperellipticity of most of the surfaces $\overline{\mathbb{H}^2/\Gamma_o(k)}$.

easily verified for primes ≤ 89. The fact that $F_{k,-\frac{1}{d}}$ is constant tells us that (this is established in Corollary 6.5)

$$P_{-\frac{1}{d}}\left(km + \frac{k^2-1}{24d}\right) = P_{-\frac{1}{d}}\left(\frac{k^2-1}{24d}\right) P_{-\frac{1}{d}}\left(\frac{m}{k}\right).$$

In particular, for $m = 1$,

$$P_{-\frac{1}{d}}\left(k + \frac{k^2-1}{24d}\right) = 0$$

(since $P_{-\frac{1}{d}}\left(\frac{1}{k}\right) = 0$). This is of course a contradiction to Lemma 1.3 since there we showed that $P_y(n)$ does not vanish for n a positive integer and $y \in (-1, 0)$ rational.

Remark 4.10. Theorem 4.7 and its corollaries have many other consequences. For example, consider the case $k = 5$ and $N = -8$. Here $P_{-8}(3) = 0$ and $P_{-8}(8) = -5^3$. It follows that

$$\frac{\sum_{m=-1}^{\infty} P_{-8}(5m+8)x^m}{\sum_{m=0}^{\infty} P_{-8}(m)x^{5m}} = P_{-8}(8),$$

and hence

$$P_{-8}(5m+3) = -5^3 P_{-8}\left(\frac{m-1}{5}\right).$$

Other consequences are explored in §6. The equalities arising from constant functions can also be obtained by different methods. We illustrate this for $k = 11$ (in §10.5).

Working with the function $f_{1,k}^{\frac{24}{k-1}}$ instead of $f_{k,k}^{\frac{24}{k^2-1}}$ produces analogous results. For the prime k and all $N \in \mathbb{Z}$ such that $(k-1)|12N$,

$$Y_{k,N}(\tau) = \frac{1}{k}\sum_{l=0}^{k-1} f_{1,k}(\tau+l)^{\frac{24N}{k-1}}$$

defines a $\Gamma_o(k)$-invariant function (that is holomorphic except possibly at P_∞ and P_0) whose Laurent series expansion at P_∞ in terms of the local coordinate $x = \exp(2\pi\imath\tau)$ is given by

$$\prod_{n=1}^{\infty}(1-x^n)^{\frac{24N}{k-1}} \sum_{m=\lceil\frac{N}{k}\rceil}^{\infty} P_{\frac{24N}{k-1}}(km-N)x^m.$$

If $N > 0$, then

$$\mathrm{ord}_{P_0} Y_{k,N} = -kN \text{ and } \mathrm{ord}_{P_\infty} Y_{k,N} \geq \left\lceil \frac{N}{k} \right\rceil,$$

with equality if $(k-1)|24N$. The function has additional zeros in most cases. If $N<0$, then

$$\mathrm{ord}_{P_\infty} Y_{k,N} \geq \left\lfloor N - \frac{N}{k} \right\rfloor \text{ and } \mathrm{ord}_{P_0} Y_{k,N} \geq N(k-1).$$

If, in addition, $p(\overline{\mathbb{H}^2/\Gamma_o(k)}) = 0$ and $N \in \mathbb{Z}^+$, then $Y_{k,N} = \sum_{i=\lceil \frac{N}{k} \rceil}^{kN} c_i f_k^i$ and hence

$$\sum_{m=\lceil \frac{N}{k} \rceil}^{\infty} P_{\frac{24N}{k-1}}(km-N)x^m = \sum_{i=\lceil \frac{N}{k} \rceil}^{kN} c_i x^i \frac{\prod_{n=1}^{\infty}(1-x^{kn})^{\frac{24i}{k-1}}}{\prod_{n=1}^{\infty}(1-x^n)^{\frac{24(N+i)}{k-1}}}.$$

The functions $F_{k,N}$ and $Y_{k,N}$ need not be independent. We have

Theorem 4.11. *For every prime k and all $N \in \mathbb{Z}$ such that $(k-1)|12N$,*

$$F_{k,\frac{N(k+1)}{\beta(k)}} = f_k^N Y_{k,N}.$$

Proof. The proof is an easy computation in terms of the local coordinate x. It is left to the reader.

If we introduce the *averaging operator* (here k is a prime and $n \in \mathbb{Z}^+$)

$$U_{k,n}(f(\tau)) = \frac{1}{k^n}\sum_{l=0}^{k^n-1} f(\tau+l) = \frac{1}{k^n}\sum_{l=0}^{k^n-1} f(B^l(\tau)), \tag{5.18}$$

we see that

$$F_{k,n,N} = \begin{cases} U_{k,n}\left(\left(\frac{\eta(k\cdot}{\eta(\frac{\cdot}{k^n})}\right)^{\alpha(k)N}\right), & n \text{ odd} \\ U_{k,n}\left(\left(\frac{\eta(\cdot)}{\eta(\frac{\cdot}{k^n})}\right)^{\alpha(k)N}\right), & n \text{ even} \end{cases},$$

$$U_{k,n} = \frac{1}{k^n}\Theta_{\Gamma(k^n,k)\backslash\Gamma_o(k)},$$

where Θ is the Poincaré series operator, and

$$U_{k,n} : \mathcal{K}(\Gamma(k^n,k)) \to \mathcal{K}(\Gamma_o(k)). \tag{5.19}$$

We abbreviate $U_{k,1}$ by $U(k)$. These operators preserve the constant function 1. Since $\Gamma(k^n,k) \subset \Gamma_o(k)$, $\mathcal{K}(\Gamma(k^n,k)) \supset \mathcal{K}(\Gamma_o(k))$; the operator $U_{k,n}$ restricted to $\mathcal{K}(\Gamma_o(k))$ is the identity. Further,

$$U_{k,n}(gf) = gU_{k,n}(f), \text{ for all } g \in \mathcal{K}(\Gamma_o(k)) \text{ and all } f \in \mathcal{K}(\Gamma(k^n,k));$$

that is, $U_{k,n}$ is a linear operator from the field $\mathcal{K}(\Gamma(k^n,k))$ to the field $\mathcal{K}(\Gamma_o(k))$ viewed as vector spaces over $\mathcal{K}(\Gamma_o(k))$.

k	N	$\text{ord}_\infty F_{k,-N}$
2	$\in \mathbb{Z}^+$	$-\lfloor \frac{N}{2} \rfloor$
3	$\in \mathbb{Z}^+$	$-\lfloor \frac{N}{3} \rfloor$
5	5, 6, 7, 9	-1
7	4, 5	-1
13	2	-1

Table 15. ORDERS OF POLES OF SOME FUNCTIONS $F_{k,-N}$.

It is necessary for us to also study the family of operators $V_{k,M}$, $M \in \mathbb{Z}$ ($V_k = V_{k,0}$), defined on functions f on $\mathbb{H}^2$ by

$$(5.20)\qquad V_{k,M}(f)(\tau) = \frac{1}{k}\sum_{l=0}^{k-1}\left(\frac{\eta(k(\tau+l))}{\eta\left(\frac{\tau+l}{k}\right)}\right)^{\alpha(k)M} f\left(\frac{\tau+l}{k}\right),\ \tau \in \mathbb{H}^2.$$

Observe that for every function f,

$$V_{k,M}(f) = V_k(f_{k^2,1}^{\alpha(k)M} f).$$

Proposition 4.12. *For all primes k and all $M \in \mathbb{Z}$,*

$$V_{k,M} : \mathcal{K}(\Gamma_o(k)) \to \mathcal{K}(\Gamma_o(k)).$$

Proof. If $f(\tau) \in \mathcal{K}(\Gamma_o(k))$, then $f\left(\frac{\tau}{k}\right) \in \mathcal{K}(\Gamma(k,k))$. Since the function $\frac{\eta(k\tau)}{\eta(\frac{\tau}{k})}$ belongs to the field $\mathcal{K}(\Gamma(k,k))$ and we are averaging a $\Gamma(k,k)$-invariant function over $\Gamma(k,k)\backslash\Gamma_o(k)$, the result is a $\Gamma_o(k)$-invariant function.

If $k = 2, 3, 5, 7$ or 13, then $f_k^N \in \mathcal{K}(\Gamma_o(k))_0$ ($f_k^N \in \mathcal{K}(\Gamma_o(k))_\infty$) provided $N > 0$ ($N < 0$). Calculations (left to the reader) tell us that[116] in these cases

$$V_{k,M}(f_k^N) = f_k^{-N} F_{k,M+\frac{k+1}{\beta(k)}N}.$$

We establish a more general result in

Theorem 4.13. *For all primes k, all $N \in \mathbb{Z}$ and all $n \in \mathbb{Z}^+$,*

$$V_k(F_{k,2n-1,N}) = F_{k,2n,N} \text{ and } V_{k,N}(F_{k,2n,N}) = F_{k,2n+1,N}.$$

[116] Note that

$$\frac{k+1}{\beta(k)} = \begin{cases} k+1 \text{ for } k = 2,\ 3,\ 5 \\ \frac{24}{k-1} \text{ for } k = 5,\ 7,\ 13 \end{cases}.$$

Proof. We leave the proof of the first equality to the reader and establish the second one since in this case the proof has some extra twists.

Recalling that $\eta(\tau+l)=\exp\left(\frac{\pi i l}{12}\right)\eta(\tau)$ for all $l\in\mathbb{Z}$, we see that

$$\begin{aligned} f_{k^2,1}^{\alpha(k)N}(\tau)F_{k,2n,N}(\tau) &= \tfrac{1}{k^{2n}}\left(\tfrac{\eta(k^2\tau)}{\eta(\tau)}\right)^{\alpha(k)N}\textstyle\sum_{l=0}^{k^{2n}-1}\left(\tfrac{\eta(\tau+l)}{\eta(\frac{\tau+l}{k^{2n}})}\right)^{\alpha(k)N} \\ &= \tfrac{1}{k^{2n}}\textstyle\sum_{l=0}^{k^{2n}-1}\left(\tfrac{\eta(k^2\tau+l)}{\eta(\frac{\tau+l}{k^{2n}})}\right)^{\alpha(k)N}. \end{aligned}$$

Thus

$$\begin{aligned} V_k(f_{k^2,1}^{\alpha(k)N}F_{k,2n,N})(\tau) &= \tfrac{1}{k^{2n+1}}\textstyle\sum_{L=0}^{k-1}\sum_{l=0}^{k^{2n}-1}\left(\tfrac{\eta(k(\tau+L)+l)}{\eta(\frac{\frac{\tau+L}{k}+l}{k^{2n}})}\right)^{\alpha(k)N} \\ &= \tfrac{1}{k^{2n+1}}\textstyle\sum_{L=0}^{k-1}\sum_{l=0}^{k^{2n}-1}\left(\tfrac{\eta(k(\tau+kl+L)-k^2l+l)}{\eta(\frac{\frac{\tau+L}{k}+l}{k^{2n}})}\right)^{\alpha(k)N} \\ &= \tfrac{1}{k^{2n+1}}\textstyle\sum_{L=0}^{k-1}\sum_{l=0}^{k^{2n}-1}\left(\tfrac{\eta(k(\tau+kl+L))}{\eta(\frac{\frac{\tau+L}{k}+l}{k^{2n}})}\right)^{\alpha(k)N} \\ &= \tfrac{1}{k^{2n+1}}\textstyle\sum_{l=0}^{k^{2n+1}-1}\left(\tfrac{\eta(k\tau+l)}{\eta(\frac{\tau+l}{k^{2n+1}})}\right)^{\alpha(k)N} = F_{k,2n+1,N}. \end{aligned}$$

In the third of the last sequence of equalities we have used the invariance of the η-function and the fact that for the prime k, $24|(k^2-1)\alpha(k)$.

We return to the cases where $p(\Gamma_o(k))=0$. Assume that $N>0$ and also that $\beta(k)M+N>0$. In this case

$$\mathrm{ord}_\infty V_{k,M}(f_k^N)=\left\lceil\frac{\beta(k)M+N}{k}\right\rceil\geq 1$$

and

$$\mathrm{ord}_0 V_{k,M}(f_k^N)=-(\beta(k)M+kN)<0;$$

hence the operator $V_{k,M}$ preserves $\mathcal{K}(\Gamma_o(k))_0$. In general,

$$\mathrm{ord}_\infty V_{k,M}(f_k^N)\geq\left\lceil\frac{\beta(k)M+N}{k}\right\rceil$$

and

$$\mathrm{ord}_0 V_{k,M}(f_k^N)\begin{cases} =-(\beta(k)M+kN) \text{ if } \beta(k)M+(1+k)N\geq 0\\ \geq N \text{ if } \beta(k)M+(1+k)N<0\end{cases}.$$

We are most interested in the operators V_k and $V_{k,N}$. Both preserve the field $\mathcal{K}(\Gamma_o(k))_0$; they need not preserve $\mathcal{K}(\Gamma_o(k))_\infty$.

4.3. Bases for $\mathcal{K}(\Gamma_o(k))_0$ and $\mathcal{K}(\Gamma_o(k))_\infty$. The functions f_k^n with $n \in \mathbb{Z}^+$ (respectively, $-n \in \mathbb{Z}^+$) form a basis for $\mathcal{K}(\Gamma_o(k))_0$ ($\mathcal{K}(\Gamma_o(k))_\infty$) for $k = 2$, 3, 5, 7 and 13 (the cases where $p(\mathbb{H}^2\Gamma_o(k) = 0)$. To obtain bases for these linear spaces for more values of k, we began an exploration of alternate constructions of automorphic functions.

4.3.1. *The case of genus* 0. We consider the cases where $p(\overline{\mathbb{H}^2/\Gamma_o(k)}) = 0$. One easily sees from Theorem 4.7 part (b), that for $k = 2$, 3 and 5,

$$F_{k\,1} = c_k f_k. \tag{5.21}$$

Thus we have produced an alternate basis in these cases for $\mathcal{K}(\Gamma_o(k))_0$. This new method does not give additional information for $k = 7$ and 13.

We can do better for $\mathcal{K}(\Gamma_o(k))_\infty$. For $k = 2, 3, 5, 7$ and 13, the respective functions

$$F_{2,-2} - F_{2,-2}(0),\ F_{3,-3} - F_{3,-3}(0),\ F_{5,-5} - F_{5,-5}(0),$$
$$F_{7,-4} - F_{7,-4}(0),\ F_{13,-2} - F_{13,-2}(0)$$

on $\overline{\mathbb{H}^2/\Gamma_o(k)}$ have simple poles at P_∞ and simple zeros at P_0.

4.3.2. *The case of genus* 1. Next in line are the cases where $p(\overline{\mathbb{H}^2/\Gamma_o(k)}) =$ 1; that is, $k = 11$, 17 and 19.

$k = 11$. The functions

$$F_2 = F_{11,-5} - F_{11,-5}(0)$$

and

$$F_3 = F_{11,-7} - F_{11,-7}(0)$$

belong to $\mathcal{K}(\Gamma_o(11))_\infty$ and have poles of order 2 and 3 at P_∞, respectively. Hence a basis for $\mathcal{K}(\Gamma_o(11))_\infty$ is

$$F_2,\ F_3,\ F_2^2,\ F_2F_3,\ F_2^3,\ F_2^2F_3, ...,\ F_2^n,\ F_2^{n-1}F_3,\ F_2^{n+1},\ ... .$$

$k = 17$. The situation for the prime 17 seems more complicated. As for $k = 11$, $\mathcal{K}(\Gamma_o(17))_\infty$ contains functions with poles at P_∞ of orders 2 and 3. The orders of the poles of $F_{17,-N}$ for $N = 1, 2, 3, ...$ are

$$0,\ 0,\ 0,\ 0,\ 3,\ 4,\ 4,\ 0,\ 6,\ 7,\ 7,\ 8,\ 9,\ 0,\ 10,\ 11,\ 12,\ 12,\ ... ,$$

respectively. The linear span of

$$\{F_{17,-N}; N = 5,\ 6,\ 7\}$$

(functions with poles of orders 3, 4 and 4, respectively) does not contain a function with a pole of order 2. (It hence produces a linear dependence when the constant function is adjoined and thus also a partition identity.) We persevere and consider the linear span of the functions

$$\{F_{17,-N}; N = 5,\ 6,\ 9,\ 10,\ 11\}$$

(functions with poles of orders 3, 4, 6, 7 and 7, respectively). It contains a function with a pole of order 5 in the linear span of

$$\{F_{17,-N}; N = 5,\ 6,\ 9,\ 10,\ 11\}.$$

The function with a pole of order 2 at P_∞ seems illusive.

$k = 19$. The functions $F_{19,-4}$ and $F_{19,-5}$ both have poles of order 3 at P_∞. Simple calculations show that

$$5^2 F_{19,-4} + 2^2 F_{19,-5}$$

has a pole of order 2 at P_∞. Hence

$$F_2,\ F_3,\ F_2^2,\ F_2F_3, \ldots,\ F_2^n,\ F_2^{n-1}F_3,\ F_2^{n+1},\ \ldots,$$

is a basis for $\mathcal{K}(\Gamma_o(19))_\infty$, where

$$F_2 = 5^2 F_{19,-4} + 2^2 F_{19,-5} - 5^2 F_{19,-4}(0) - 2^2 F_{19,-5}(0)$$

and

$$F_3 = F_{19,-4} - F_{19,-4}(0).$$

4.3.3. *Hyperbolic surfaces.* For most primes k, the surface $\overline{\mathbb{H}^2/\Gamma_o(k)}$ is hyperbolic, that is, of genus at least two. At this point we merely hint at the difficulties encountered in trying to construct functions of small degree on these surfaces.

$k = 23$. The surface $\overline{\mathbb{H}^2/\Gamma_o(23)}$ has genus 2 ($\beta(23) = 22$). The orders of the poles of $F_{23,-N}$ for $N = 1, 2, 3, \ldots$, are

$$0,\ 0,\ 0,\ 0,\ 4,\ 0,\ 6,\ 0,\ 8,\ 0,\ 10,\ 11,\ 12,\ 0,\ 14,\ 15,\ 16,\ 17,\ \ldots,$$

respectively. It thus appears that the case of the prime 23 is particularly difficult.

$k = 29$. Here once again we have a surface of genus 2 ($\beta(29) = 35$). The orders of the poles of $F_{29,-N}$ for $N = 1, 2, 3, \ldots$, are

$$0,\ 0,\ 0,\ 0,\ 6,\ 7,\ 8,\ 0,\ 10,\ 12,\ \ldots,$$

respectively.

$k = 31$. Here once again we have a surface of genus 2 ($\beta(31) = 40$). The orders of the poles of $F_{29,-N}$ for $N = 1, 2, 3, \ldots$, are

$$0,\ 0,\ 0,\ 5,\ 6,\ 0,\ 9,\ 10,\ 11,\ 0,\ \ldots,$$

respectively.

Remark 4.14. 1. The Laurent series at P_∞ on $\overline{\mathbb{H}^2/\Gamma_o(k)}$ in terms of the local coordinate $x = \exp(2\pi\imath\tau)$ of each of the functions considered so far is expressible in terms of the partition functions and hence has integer coefficients.

2. In some sense the above attempt to find bases for $\mathcal{K}(\Gamma_o(k))_\infty$ has very limited use for us. It would be more useful for our applications (as will be seen later) to find bases for subspaces of $\mathcal{K}(\Gamma_o(k))_0$ from the functions $\{F_{k,N}; N \in \mathbb{Z}^+\}$ (these functions vanish at P_∞) or equivalently bases for subspaces of $\mathcal{K}(\Gamma_o(k))_\infty$ from the functions $\{G_{k,N} - G_{k,N}(0); -N \in \mathbb{Z}^+\}$ defined in the next subsection.

4.4. Precomposing with A_k. Throughout this subsection $k \in \mathbb{Z}^+$ is a prime. We study the involution of $\mathcal{K}(\Gamma_o(k))$ induced by precomposition of functions with the involution A_k of $\Gamma_o(k)$. We note that

$$A_k(\mathcal{K}(\Gamma_o(k))_0) = \mathcal{K}(\Gamma_o(k))_\infty.$$

It hence makes sense to define[117] for $n \in \mathbb{Z}^+ \cup \{0\}$ and $N \in \mathbb{Z}$,

$$G_{k,n,N}(\tau) = F_{k,n,N}\left(-\frac{1}{k\tau}\right) \text{ and } G_{k,N}(\tau) = F_{k,N}\left(-\frac{1}{k\tau}\right),\ \tau \in \mathbb{H}^2.$$

Since

$$\mathrm{ord}_{P_\infty} F_{k,N} = \mathrm{ord}_{P_0} G_{k,N} \text{ and } \mathrm{ord}_{P_0} F_{k,N} = \mathrm{ord}_{P_\infty} G_{k,N},$$

we see that for positive N, the function $G_{k,N}$ is holomorphic except at P_∞, where it has pole of order $\beta(k)N$ (at P_0 it has a zero of order $\left\lceil \frac{\beta(k)N}{k} \right\rceil$); for negative N, $G_{k,N}$ is holomorphic except at P_0, where it has a pole of order at most $\left\lfloor \frac{-\beta(k)N}{k} \right\rfloor$; for $N = 0$, $G_{k,0}$ is the constant function 1. It follows that for primes $k \leq 7$ and 13,

$$G_{k,N} = \sum_{i=\left\lceil \frac{\beta(k)N}{k} \right\rceil}^{\beta(k)\max\{0,N\}} d_i f_k^{-i}.$$

Our first task is to establish the Laurent series expansions of the function $G_{k,N}$ at P_∞. From the definitions we see that

$$G_{k,N}(\tau) = \frac{1}{k}\sum_{l=0}^{k-1}\left(\frac{\eta(-\frac{1}{k^2\tau} + \frac{l}{k})}{\eta(-\frac{1}{\tau} + kl)}\right)^{-\alpha(k)N}.$$

[117] In line with previous conventions, $G_{k,1,N} = G_{k,N}$.

The invariance properties of the η-function lead to

Proposition 4.15. *For all $\tau \in \mathbb{H}^2$ the following identities hold (as long as $l \in \mathbb{Z}$).*

$$\eta\left(-\frac{1}{\tau}+kl\right) = \left(\exp\frac{\pi\imath}{12}(kl-3)\right)\tau^{\frac{1}{2}}\eta(\tau),$$

$$\eta\left(-\frac{1}{k^2\tau}\right) = \left(\exp\frac{-\pi\imath}{4}\right)k\tau^{\frac{1}{2}}\eta(k^2\tau),$$

$$\eta\left(-\frac{1}{2^2\tau}+\frac{1}{2}\right) = \left(\exp\frac{-\pi\imath}{4}\right)2^{\frac{1}{2}}\tau^{\frac{1}{2}}\eta\left(\tau+\frac{1}{2}\right),$$

and for odd k, $0<l<k$,

$$\eta\left(-\frac{1}{k^2\tau}+\frac{l}{k}\right)$$

$$=\left(\frac{-m}{k}\right)e^{\frac{\pi\imath}{12}\left\{(l-m)k-\left(\frac{1+lm}{k}\right)m(k^2-1)-3k\right\}}k^{\frac{1}{2}}\tau^{\frac{1}{2}}\eta\left(\tau+\frac{m}{k}\right),$$

where $m \in \mathbb{Z}^+$ satisfies $0<m<k$ and $lm \equiv -1 \mod k$.

Proof. The first two equations are easily seen to be correct; the derivation of the third and fourth equations are is similar. So we consider the hypothesis of the fourth equation. Choose the unique $m = m(l) \in \mathbb{Z}$ such that $lm \equiv -1 \mod k$ and $1 \le m \le k-1$. Then $\gamma = \begin{bmatrix} l & -\frac{1+lm}{k} \\ k & -m \end{bmatrix}$ is an element of Γ, and by Proposition 2.8

$$\eta\left(-\frac{1}{k^2\tau}+\frac{l}{k}\right) = \eta\left(\gamma\left(\tau+\frac{m}{k}\right)\right) = v_\eta(\gamma)(k\tau)^{\frac{1}{2}}\eta\left(\tau+\frac{m}{k}\right).$$

The fourth of the equations in the proposition for $k=3$ simplifies to

$$\eta\left(-\frac{1}{3^2\tau}+\frac{1}{3}\right) = \left(\exp\frac{-\pi\imath}{3}\right)3^{\frac{1}{2}}\tau^{\frac{1}{2}}\eta\left(\tau+\frac{2}{3}\right),$$

$$\eta\left(-\frac{1}{3^2\tau}+\frac{2}{3}\right) = \left(\exp\frac{-\pi\imath}{6}\right)3^{\frac{1}{2}}\tau^{\frac{1}{2}}\eta\left(\tau+\frac{1}{3}\right)$$

and for primes $k \ge 5$ to

$$\eta\left(-\frac{1}{k^2\tau}+\frac{l}{k}\right) = \left(\frac{-m}{k}\right)\left(\exp\frac{\pi\imath}{12}\{(l-m)k-3k\}\right)k^{\frac{1}{2}}\tau^{\frac{1}{2}}\eta\left(\tau+\frac{m}{k}\right).$$

Hence

$$G_{2,N}(\tau) = 2^{-1-8N}\left[\left(\frac{\eta(2^2\tau)}{\eta(\tau)}\right)^{-8N} + 2^{4N}\exp\frac{4N\pi\imath}{3}\left(\frac{\eta\left(\tau+\frac{1}{2}\right)}{\eta(\tau)}\right)^{-8N}\right],$$

$$3^{1+3N}G_{3,N}(\tau)$$
$$= \left(\frac{\eta(3^2\tau)}{\eta(\tau)}\right)^{-3N} + 3^{\frac{5N}{2}}(-1)^N \left(e^{\frac{N\pi\imath}{4}}\left(\frac{\eta\left(\tau+\frac{1}{3}\right)}{\eta(\tau)}\right)^{-3N} + \left(\frac{\eta\left(\tau+\frac{2}{3}\right)}{\eta(\tau)}\right)^{-3N}\right)$$

and for $k \geq 5$,

$$k^{1+N}G_{k,N}(\tau)$$
$$= \left(\frac{\eta(k^2\tau)}{\eta(\tau)}\right)^{-N} + k^{\frac{N}{2}} \sum_{m=1}^{k-1} \left(\frac{-m}{k}\right)^{-N} e^{\frac{\pi\imath}{12}N(3(k-1)+mk)} \left(\frac{\eta(\tau+\frac{m}{k})}{\eta(\tau)}\right)^{-N}.$$

We can write in general

$$k^{1+\alpha(k)N}G_{k,N}(\tau)$$
$$= \left[\left(\frac{\eta(k^2\tau)}{\eta(\tau)}\right)^{-\alpha(k)N} + k^{\frac{\alpha(k)N}{2}} \sum_{m=1}^{k-1} c_{k,m} \left(\frac{\eta(\tau+\frac{m}{k})}{\eta(\tau)}\right)^{-\alpha(k)N}\right].$$

Thus

$$k^{2+\alpha(k)N}V_k(G_{k,N})(\tau)$$
$$= \sum_{l=0}^{k-1} \left(\frac{\eta(k(\tau+l))}{\eta\left(\frac{\tau+l}{k}\right)}\right)^{-\alpha(k)N} + k^{\frac{\alpha(k)N}{2}} \sum_{m=1}^{k-1}\sum_{l=0}^{k-1} c_{k,m} \left(\frac{\eta\left(\frac{\tau+l+m}{k}\right)}{\eta\left(\frac{\tau+l}{k}\right)}\right)^{-\alpha(k)N}.$$

In terms of the local coordinate $x = \exp(2\pi\imath\tau)$ on $\overline{\mathbb{H}^2/\Gamma_o(k)}$ at P_∞,

$$G_{2,N}(\tau) = 2^{-1-4N} \sum_{j=0}^{\infty} P_{-8N}(j)x^j$$
$$\times \left[2^{-4N}x^{-N} \sum_{j=0}^{\infty} P_{8N}(j)x^{2^2 j} + (-1)^N \sum_{j=0}^{\infty}(-1)^j P_{8N}(j)x^j\right],$$

$$G_{3,N}(\tau) = 3^{-1-\frac{3N}{2}} \sum_{j=0}^{\infty} P_{-3N}(j)x^j$$
$$\times \left[\begin{array}{c} 3^{-\frac{3N}{2}}x^{-N}\sum_{j=0}^{\infty} P_{3N}(j)x^{3^2 j} \\ +\sum_{j=0}^{\infty}\left(\exp\left\{\frac{(7N+4j)\pi\imath}{6}\right\} + \exp\left\{\frac{(5N+8j)\pi\imath}{6}\right\}\right) P_{3N}(j)x^j \end{array}\right],$$

and for $k \geq 5$,

$$G_{k,N}(\tau) = k^{-1-\frac{N}{2}} \sum_{j=0}^{\infty} P_{-N}(j)x^j$$
$$\times \left[\begin{array}{c} k^{-\frac{N}{2}}x^{-\beta(k)N}\sum_{j=0}^{\infty} P_N(j)x^{k^2 j} \\ +e^{\frac{\pi\imath}{4}N(k-1)}\sum_{m=1}^{k-1} c_m \sum_{j=0}^{\infty} e^{\frac{2\pi\imath mj}{k}} P_N(j)x^j \end{array}\right],$$

where

$$c_m = \left(\frac{-m}{k}\right)^{-N} \exp\frac{2\pi\imath mN\beta(k)}{k}.$$

It is not obvious when the coefficients of these Laurent series expansions are integers. It will turn out that in many cases of interest, we will obtain power series with integral coefficients. Hence we introduce the following

Definition 4.16. A function $f \in \mathcal{K}(\Gamma_o(k))$ is said to have a *good* expansion if its Laurent series at P_∞ in terms of the local coefficient $x = \exp(2\pi\imath\tau)$ and at P_0 in terms of $y = \exp\left(\frac{2\pi\imath}{k\tau}\right)$ have integer coefficients.

The formula for the Laurent series expansion for $G_{k,N}$ can be simplified considerably by interchanging some orders of summation and observing that (for a fixed m) the exponential, $\exp\left\{\frac{2\pi\imath m j}{k}\right\}$, depends only on the residue J of j modulo k. We need

Definition 4.17. Let k be a prime and $N \in \mathbb{Z}$ be fixed. For $J \in \mathbb{Z}$, define for $k = 2$,

$$\kappa_J = (-1)^{N+J};$$

for $k = 3$,

$$\kappa_J = \begin{cases} 3^{\frac{1}{2}}\left\{\exp\left(\frac{(7N+4J)\pi\imath}{6}\right) + \exp\left(\frac{(5N+8J)\pi\imath}{6}\right)\right\} & \text{if } N \text{ is odd} \\ \exp\left(\frac{(7N+4J)\pi\imath}{6}\right) + \exp\left(\frac{(5N+8J)\pi\imath}{6}\right) & \text{if } N \text{ is even} \end{cases};$$

and for $k \geq 5$,

$$\kappa_J = \begin{cases} k^{\frac{1}{2}} e^{\frac{\pi\imath N(k-1)}{4}} \sum_{m=1}^{k-1} \left(\frac{-m}{k}\right)^{-N} e^{\frac{2\pi\imath m}{k}(\beta(k)N+J)} & \text{if } N \text{ is odd} \\ e^{\frac{\pi\imath N(k-1)}{4}} \sum_{m=1}^{k-1} e^{\frac{2\pi\imath m}{k}(\beta(k)N+J)} & \text{if } N \text{ is even} \end{cases}.$$

Thus for $k \geq 5$

$$\left(\sum_{j=0}^{\infty} P_{-N}(j)x^j\right)^{-1} G_{k,N}(\tau)$$

$$= k^{-\lfloor\frac{2+N}{2}\rfloor}\left(k^{-\lfloor\frac{1+N}{2}\rfloor} x^{-\beta(k)N} \sum_{j=0}^{\infty} P_N(j)x^{k^2 j} + \sum_{J=0}^{k-1} \kappa_J \sum_{j=0}^{\infty} P_N(kj+J)x^{kj+J}\right).$$

We compute the constants κ_J for $k = 2$ and 3 and summarize the results in tabular form.

Let us assume that $k \geq 5$ and set $a = \beta(k)N + J$. Then (see, for example, [**20**, Ch. IV, §3]) for odd N and $(a,k) = 1$,

$$\sum_{m=1}^{k-1}\left(\frac{-m}{k}\right)^{-N} e^{\frac{2\pi\imath a m}{k}} = \sum_{m=1}^{k-1}\left(\frac{-m}{k}\right) e^{\frac{2\pi\imath a m}{k}} = \left(\frac{-a}{k}\right)\sum_{m=1}^{k-1}\left(\frac{m}{k}\right) e^{\frac{2\pi\imath m}{k}}$$

$$= \begin{cases} \left(\frac{-a}{k}\right) k^{\frac{1}{2}} \text{ for } k \equiv 1 \mod 4 \\ \left(\frac{-a}{k}\right) k^{\frac{1}{2}}\imath \text{ for } k \equiv 3 \mod 4 \end{cases}.$$

k	N	J	κ_J
2		$\equiv N \mod 2$	1
		$\not\equiv N \mod 2$	-1
3	$\equiv 0 \mod 12$	$\equiv 0 \mod 3$	2
		$\not\equiv 0 \mod 3$	-1
	$\equiv 1 \mod 12$	$\equiv 0 \mod 3$	-3
		$\equiv 1 \mod 3$	3
		$\equiv 2 \mod 3$	0
	$\equiv 2 \mod 12$	$\not\equiv 1 \mod 3$	1
		$\equiv 1 \mod 3$	-2
	$\equiv 3 \mod 12$	$\equiv 0 \mod 3$	0
		$\equiv 1 \mod 3$	3
		$\equiv 2 \mod 3$	-3
	$\equiv 4 \mod 12$	$\not\equiv 2 \mod 3$	-1
		$\equiv 2 \mod 3$	2
	$\equiv 5 \mod 12$	$\equiv 0 \mod 3$	3
		$\equiv 1 \mod 3$	0
		$\equiv 2 \mod 3$	-3
	$\equiv 6 \mod 12$	$\equiv 0 \mod 3$	-2
		$\not\equiv 0 \mod 3$	1
	$\equiv 7 \mod 12$	$\equiv 0 \mod 3$	3
		$\equiv 1 \mod 3$	-3
		$\equiv 2 \mod 3$	0
	$\equiv 8 \mod 12$	$\not\equiv 1 \mod 3$	-1
		$\equiv 1 \mod 3$	2
	$\equiv 9 \mod 12$	$\equiv 0 \mod 3$	0
		$\equiv 1 \mod 3$	-3
		$\equiv 2 \mod 3$	3
	$\equiv 10 \mod 12$	$\not\equiv 2 \mod 3$	1
		$\equiv 2 \mod 3$	-2
	$\equiv 11 \mod 12$	$\equiv 0 \mod 3$	-3
		$\equiv 1 \mod 3$	0
		$\equiv 2 \mod 3$	3

Table 16. CALCULATIONS OF κ_J FOR $k = 2$ AND 3.

For odd N and $(a,k) = k$,

$$\sum_{m=1}^{k-1}\left(\frac{-m}{k}\right)^N e^{\frac{2\pi i a m}{k}} = \sum_{m=1}^{k-1}\left(\frac{-m}{k}\right) e^{\frac{2\pi i a m}{k}} = \sum_{m=1}^{k-1}\left(\frac{-m}{k}\right) = 0.$$

For even N, the evaluation of κ_J is trivial.

Proposition 4.18. *For all primes k and all $N \in \mathbb{Z}$,*
(a) $\kappa_J \in \mathbb{Z}$, *as a matter of fact* $\kappa_J \in \{0,\ \pm 1,\ \pm(k-1),\ \pm k\}$, *and*
(b) $\sum_{J=0}^{k-1} \kappa_J = 0$.

Proof. We have established part (a); we leave the proof of (b) to the reader.

The proof of the proposition established somewhat more: For primes $k \geq 5$,

$$\kappa_J = \begin{cases} \pm k \text{ for } (a,k) = 1 \text{ and } N \text{ odd} \\ 0 \text{ for } (a,k) = k \text{ and } N \text{ odd} \\ \pm 1 \text{ for } (a,k) = 1 \text{ and } N \text{ even} \\ \pm(k-1) \text{ for } (a,k) = k \text{ and } N \text{ even} \end{cases},$$

where, as before, $a = \beta(k)N + J$.

Hence we have almost proved

Theorem 4.19. *For all primes $k \geq 5$ and all integers $N \leq -1$, the functions $G_{k,N}$ have good expansions; furthermore, the Laurent series coefficients in their expansions at P_∞ are divisible by $k^{-1+\lceil -\frac{N}{2}\rceil}$. The same conclusion holds for the functions $G_{k,N} - G_{k,N}(\infty)$.*

Proof. The last displayed equation shows that the Laurent coefficients of $G_{k,N}$ at P_∞ are integers that satisfy the divisibility condition. It remains to show that the Laurent coefficients of $G_{k,N}$ at P_0 are integers. But the Laurent coefficients of $G_{k,N}$ at P_0 in terms of $y = \exp\left(\frac{2\pi\imath}{k\tau}\right)$ are the same as the Laurent coefficients of $F_{k,N}$ at P_∞ in terms of $x = \exp(2\pi\imath\tau)$, known to be integers. To prove these assertions for $G_{k,N} - G_{k,N}(\infty)$, we merely observe that $G_{k,N}(\infty) = k^{-1+\lceil -\frac{N}{2}\rceil}\kappa_0$.

4.4.1. *The case of genus 0, revisited.* In the following we denote the expression $G_{k,N} - G_{k,N}(\infty)$ by $H_{k,N}$. Some equations of interest here are[118]

$$F_{2,1} = cH_{2,-2} = cH_{2,-3},\ F_{3,1} = cH_{3,-3} = cH_{3,-4} = cH_{3,-5}, \tag{5.22}$$

$$F_{5,1} = cH_{5,-5} = cH_{5,-6} = cH_{5,-7} = cH_{5,-9}, \tag{5.23}$$

$$f_7 = cH_{7,-4} = cH_{7,-5},\ f_{13} = cH_{13,-2}.$$

We have produced new bases for $\mathcal{K}(\Gamma_o(k))_0$ for $k = 2,\ 3,\ 5,\ 7$ and 13. These are important for applications since the Taylor series expansions of the functions $H_{k,N}$ have integral coefficients.

[118]Throughout this book c, *const* and *constant* represent generic nonzero constants, usually of absolute value 1, that may change (even) within a single equation.

4.4.2. *The case of genus* 1, *revisited.* We begin with

$k = 11$. Let

$$H_2 = G_{11,-5} - G_{11,-5}(\infty) \text{ and } H_3 = G_{11,-7} - G_{11,-7}(\infty).$$

These functions have good Taylor series expansions, and

$$H_2,\ H_3,\ H_2^2,\ H_2H_3,\ H_2^3,\ H_2^2H_3, ...,\ H_2^n,\ H_2^{n-1}H_3,\ H_2^{n+1},\ \dots ,$$

form a basis for $\mathcal{K}(\Gamma_o(11))_0$.

$k = 19$. The function

$$5^2G_{19,-4} + 2^2G_{19,-5}$$

has a pole of order 2 at P_0. Hence

$$H_2,\ H_3,\ H_2^2,\ H_2H_3, ...,\ H_2^n,\ H_2^{n-1}H_3,\ H_2^{n+1},\ \dots ,$$

is a basis for $\mathcal{K}(\Gamma_o(19))_0$, where

$$H_2 = 19^{-2}(5^2G_{19,-4} + 2^2G_{19,-5} - 5^2G_{19,-4}(\infty) - 2^2G_{19,-5}(\infty))$$

and

$$H_3 = 19^{-3}(G_{19,-4} - G_{19,-4}(\infty)).$$

For applications, one would need to describe the orders (of vanishing) at P_∞ of functions in the finite (134) dimensional vector space

$$\{f \in \mathcal{K}(\Gamma_o(19))_0;\ \deg f \leq 135\}.$$

5. Partition identities

We will find the following definition useful whenever the resulting function has alternate descriptions (as in the current situation).

Definition 5.1. The *generating function* for the sequence $\{a_n\}_{n\in\{0\}\cup\mathbb{Z}^+}$ is the formal power series $\sum_{n=0}^{\infty} a_nx^n$. The power series defines the germ of an analytic function at 0 provided it has a positive radius of convergence.

We have been studying the generating functions for the sequences of partition coefficients $\{P_N(n)\}_{n\in\{0\}\cup\mathbb{Z}^+}$, $N \in \mathbb{Z}$ fixed. We will study partition congruences later in this chapter. For the moment we record some immediate consequences of some of the equations we have derived. Equation (5.21) translates immediately to the power series (partition) identities

(5.24)
$$\sum_{m=1}^{\infty} P_{\alpha(k)}(km-1)x^m = P_{\alpha(k)}(k-1)\ x\prod_{n=1}^{\infty}(1-x^{kn})^{\frac{24}{k-1}-\alpha(k)}\sum_{m=0}^{\infty} P_{\frac{24}{k-1}}(m)x^m$$

$$= P_{\alpha(k)}(k-1)\, x \prod_{n=1}^{\infty}(1-x^{kn})^{\frac{24}{k-1}-\alpha(k)} \prod_{n=1}^{\infty}(1-x^n)^{\frac{24}{1-k}},$$

valid for $k = 2$, 3 and 5. Since $P_8(1) = 8$, $P_3(2) = 9$ and $P(4) = 5$, we have established

Theorem 5.2 (Level one congruences for the primes 2, 3, 5). *For every positive integer m, we have the congruences*

$$P_8(2m-1) \equiv 0 \mod 8, \; P_3(3m-1) \equiv 0 \mod 9, \; P(5m-1) \equiv 0 \mod 5.$$

The power series identity (5.24) tells us how to compute, using Cauchy products, some partition coefficients in terms of other partition coefficients.

Corollary 5.3. *For all $m \in \mathbb{Z}^+$,*

$$P_8(2m-1) = 8 \sum_{j=0}^{\lfloor \frac{m-1}{2} \rfloor} P_{-16}(j) P_{24}(m-1-2j),$$

$$P_3(3m-1) = 9 \sum_{j=0}^{\lfloor \frac{m-1}{3} \rfloor} P_{-9}(j) P_{12}(m-1-3j)$$

and

$$P(5m-1) = 5 \sum_{j=0}^{\lfloor \frac{m-1}{5} \rfloor} P_{-5}(j) P_6(m-1-5j). \tag{5.25}$$

Remark 5.4. The theorem and its corollary can be expressed in more compact form as: For the primes $k = 2$, 3 and 5,

$$P_{\alpha(k)}(km-1) \equiv 0 \mod P_{\alpha(k)}(k-1)$$

and

$$P_{\alpha(k)}(km-1) = P_{\alpha(k)}(k-1) \sum_{j=0}^{\lfloor \frac{m-1}{k} \rfloor} P_{\alpha(k)-\frac{24}{k-1}}(j)\; P_{\frac{24}{k-1}}(m-1-kj).$$

Similarly, (5.22) and (5.23) tell us that

$$\prod_{n=1}^{\infty}(1-x^{2n})^8 \sum_{m=1}^{\infty} P_8(2m-1)x^m$$

$$= \frac{1}{2^2} \prod_{n=1}^{\infty}(1-x^n)^{-16} \left[2^8 x^2 \sum_{j=0}^{\infty} P_{-16}(j) x^{2^2 j} + \sum_{j=0}^{\infty} (-1)^j P_{-16}(j) x^j \right] - \frac{1}{2^2}$$

$$= \frac{-1}{2\cdot 3}\prod_{n=1}^{\infty}(1-x^n)^{-24}\left[2^{12}x^3\sum_{j=0}^{\infty}P_{-24}(j)x^{2^2j}-\sum_{j=0}^{\infty}(-1)^jP_{-24}(j)x^j\right]-\frac{1}{2\cdot 3},$$

$$\prod_{n=1}^{\infty}(1-x^{3n})^3\sum_{m=1}^{\infty}P_3(3m-1)x^m=\prod_{n=1}^{\infty}(1-x^n)^{-9}$$

$$\times\left[3^4x^3\sum_{j=0}^{\infty}P_{-9}(j)x^{3^2j}+\sum_{j=0}^{\infty}\left(P_{-9}(3j+2)x^{3j+2}-P_{-9}(3j+1)x^{3j+1}\right)\right]$$

$$=-\frac{1}{4}-\frac{1}{4}\prod_{n=1}^{\infty}(1-x^n)^{-12}$$

$$\times\left[3^6x^4\sum_{j=0}^{\infty}P_{-12}(j)x^{3^2j}+\sum_{j=0}^{\infty}\left(-P_{-12}(j)x^j+3P_{-12}(3j+1)x^{3j+1}\right)\right]$$

$$=-\frac{3}{2\cdot 5}+\frac{3}{2\cdot 5}\prod_{n=1}^{\infty}(1-x^n)^{-15}$$

$$\times\left[3^7x^5\sum_{j=0}^{\infty}P_{-15}(j)x^{3^2j}+\sum_{j=0}^{\infty}\left(P_{-15}(3j)x^{3j}-P_{-15}(3j+1)x^{3j+1}\right)\right],$$

$$\prod_{n=1}^{\infty}(1-x^{5n})\sum_{m=1}^{\infty}P(5m-1)x^m$$

$$=\prod_{n=1}^{\infty}(1-x^n)^{-5}\times\left[\begin{array}{c}5^2x^5\sum_{j=0}^{\infty}P_{-5}(j)x^{5^2j}\\ +\sum_{j\in\mathbb{Z}^+,j\equiv 2,3 \mod 5}P_{-5}(j)x^j\\ -\sum_{j\in\mathbb{Z}^+,j\equiv 1,4 \mod 5}P_{-5}(j)x^j\end{array}\right]$$

$$=\frac{5}{6}\prod_{n=1}^{\infty}(1-x^n)^{-6}\left[5^2x^6\sum_{j=0}^{\infty}P_{-6}(j)x^{5^2j}+\sum_{j=0}^{\infty}P_{-6}(5j+1)x^{5j+1}\right]$$

$$=\frac{5}{2\cdot 7}\left(\prod_{n=1}^{\infty}(1-x^n)^{-7}\times\left[\begin{array}{c}5^3x^7\sum_{j=0}^{\infty}P_{-7}(j)x^{5^2j}\\ +\sum_{j\in\mathbb{Z}^+\cup\{0\},j\equiv 0,4 \mod 5}P_{-7}(j)x^j\\ -\sum_{j\in\mathbb{Z}^+,j\equiv 1,3 \mod 5}P_{-7}(j)x^j\end{array}\right]-1\right)$$

$$=\frac{-5}{2\cdot 3^2}\left(\prod_{n=1}^{\infty}(1-x^n)^{-9}\times\left[\begin{array}{c}5^4x^9\sum_{j=0}^{\infty}P_{-9}(j)x^{5^2j}\\ +\sum_{j\in\mathbb{Z}^+,j\equiv 1,2 \mod 5}P_{-9}(j)x^j\\ -\sum_{j\in\mathbb{Z}^+\cup\{0\},j\equiv 0,3 \mod 5}P_{-9}(j)x^j\end{array}\right]+1\right),$$

$$x\prod_{n=1}^{\infty}(1-x^{7n})^4 = -\frac{1}{2^2}\left[7x^8\sum_{j=0}^{\infty}P_{-4}(j)x^{7^2j}+\sum_{j=0}^{\infty}P_{-4}(7j+1)x^{7j+1}\right]$$

$$= \frac{1}{2\cdot 5}\prod_{n=1}^{\infty}(1-x^n)^{-1}\times\left[\begin{array}{c} 7^2x^{10}\sum_{j=0}^{\infty}P_{-5}(j)x^{7^2j} \\ +\sum_{j\in\mathbb{Z}^+\cup\{0\},j\equiv 0,4,5 \mod 7}P_{-5}(j)x^j \\ -\sum_{j\in\mathbb{Z}^+,j\equiv 1,2,6 \mod 7}P_{-5}(j)x^j \end{array}\right]$$

$$-\frac{1}{2\cdot 5}\prod_{n=1}^{\infty}(1-x^n)^4$$

and

$$\prod_{n=1}^{\infty}(1-x^n)^2 = -\frac{1}{2}\left[x\sum_{j=0}^{\infty}P_{-2}(j)x^{13j}+\sum_{j=0}^{\infty}P_{-2}(13j+1)x^j\right].$$

The above formulae give us alternate methods for computing $P_{\alpha(k)}(km-1)$ for $k=2$, 3 and 5 and much other interesting information. Several examples follow.

The first equation for $k=2$ tells us that

$$2\prod_{n=1}^{\infty}(1+x^n)^8(1-x^n)^{24}\sum_{m=0}^{\infty}P_8(2m+1)x^{m+1}$$

$$=2^7x^2\sum_{m=0}^{\infty}P_{-16}(m)x^{4m}-\sum_{m=0}^{\infty}P_{-16}(2m+1)x^{2m+1}.$$

In particular, for all integers n with $1\le n\le 6$, $2^n|P_8(2m+1)$, for all $m\in\mathbb{Z}^+\cup\{0\}$, if and only if $2^{n+1}|P_{-16}(2m+1)$, for all $m\in\mathbb{Z}^+\cup\{0\}$. We will show in §10.1 that $2^3|P_8(2m+1)$. Hence also $2^4|P_{-16}(2m+1)$. This last divisibility condition is different from the ones we shall obtain from Theorem 6.7 (the case $k=2=N$). Using equation (5.37), we obtain the Taylor series for an infinite product:

$$2^4\prod_{n=1}^{\infty}(1+x^n)^{24}(1-x^n)^{16}=2^7x\sum_{m=0}^{\infty}P_{-16}(m)x^{4m}-\sum_{m=0}^{\infty}P_{-16}(2m+1)x^{2m}.$$

The second equation for $k=2$ implies

$$3\prod_{n=1}^{\infty}(1+x^n)^8(1-x^n)^{32}\sum_{m=0}^{\infty}P_8(2m+1)x^{m+1}$$

$$=-2^{11}x^3\sum_{m=0}^{\infty}P_{-24}(m)x^{4m}-\sum_{m=0}^{\infty}P_{-24}(2m+1)x^{2m+1}.$$

Thus for each positive odd integer m,

$$3 \left| \left(P_{-24}(m) - P_{-24}\left(\frac{m-3}{4}\right)\right)\right.$$

and, as before, a relation between products and sums

$$24\prod_{n=1}^{\infty}(1+x^n)^{24}(1-x^n)^{24} = -2^{11}x^2\sum_{m=0}^{\infty}P_{-24}(m)x^{4m} - \sum_{m=0}^{\infty}P_{-24}(2m+1)x^{2m}.$$

Similarly for $k=3$ we conclude

$$\prod_{n=1}^{\infty}(1-x^n)^9(1-x^{3n})^3\sum_{m=1}^{\infty}P_3(3m-1)x^m$$

$$= 3^4x^3\sum_{m=0}^{\infty}P_{-9}(m)x^{9m} + \sum_{m=0}^{\infty}(P_{-9}(3m+2)x^{3m+2} - P_{-9}(3m+1)x^{3m+1}),$$

$$4\prod_{n=1}^{\infty}(1-x^n)^{12}(1-x^{3n})^3\sum_{m=1}^{\infty}P_3(3m-1)x^m$$

$$= -3^6x^4\sum_{m=0}^{\infty}P_{-12}(m)x^{9m} - 3\sum_{m=0}^{\infty}P_{-12}(3m+1)x^{3m+1}$$

and

$$10\prod_{n=1}^{\infty}(1-x^n)^{15}(1-x^{3n})^3\sum_{m=1}^{\infty}P_3(3m-1)x^m$$

$$= \begin{bmatrix} 3^8x^5\sum_{m=0}^{\infty}P_{-15}(m)x^{9m} \\ -3\sum_{m=0}^{\infty}(2P_{-15}(3m+1)x^{3m+1} + P_{-15}(3m+2)x^{3m+2}) \end{bmatrix}.$$

Using the second of the above equations and the fact that $3|P_{-12}(3m+1)$ (Theorem 6.8), we conclude that $3^2|P_3(3m+1)$ (a consequence of (5.41)). This is the level one congruence for the prime 3. Conversely, the last equation and the fact that $3^2|P_3(3m+1)$ tell us that
(a) if $m \equiv 1 \mod 3$, then $P_{-15}(m) \equiv 0 \mod 15$,
(b) if $m \equiv 2$ or $8 \mod 9$, then $P_{-15}(m) \equiv 0 \mod 30$ and
(c) if $m \equiv 5 \mod 3$, then $3P_{-15}\left(\frac{m-5}{9}\right) + P_{-15}(m) \equiv 0 \mod 30$.

In view of equation (5.41), the last three displayed equations tell us that

$$3^2x\prod_{n=1}^{\infty}(1-x^n)^{-3}(1-x^{3n})^{12}$$

$$= 3^4x^3\sum_{m=0}^{\infty}P_{-9}(m)x^{9m} + \sum_{m=0}^{\infty}(P_{-9}(3m+2)x^{3m+2} - P_{-9}(3m+1)x^{3m+1}),$$

$$2^23x\prod_{n=1}^{\infty}(1-x^{3n})^{12} = -3^5x^4\sum_{m=0}^{\infty}P_{-12}(m)x^{9m} - \sum_{m=0}^{\infty}P_{-12}(3m+1)x^{3m+1}$$

and

$$2\cdot 3\cdot 5x\prod_{n=1}^{\infty}(1-x^n)^3(1-x^{3n})^{12}$$

$$=\left[\begin{array}{r} 3^7x^5\sum_{m=0}^{\infty}P_{-15}(m)x^{9m} \\ -\sum_{m=0}^{\infty}(2P_{-15}(3m+1)x^{3m+1}+P_{-15}(3m+2)x^{3m+2}) \end{array}\right].$$

The $k=5$ equations imply

$$\prod_{n=1}^{\infty}(1-x^{5n})(1-x^n)^5\sum_{m=1}^{\infty}P(5m-1)x^m=5^2x^5\sum_{m=0}^{\infty}P_{-5}(m)x^{5^2m}$$
$$+\sum_{m\in\mathbb{Z}^+,m\equiv 2,3 \mod 5}P_{-5}(m)x^m-\sum_{m\in\mathbb{Z}^+,m\equiv 1,4 \mod 5}P_{-5}(m)x^m,$$

$$2\cdot 3\prod_{n=1}^{\infty}(1-x^{5n})(1-x^n)^6\sum_{m=1}^{\infty}P(5m-1)x^m$$
$$=5^3x^6\sum_{m=0}^{\infty}P_{-6}(m)x^{5^2m}+5\sum_{m=0}^{\infty}P_{-6}(5m+1)x^{5m+1},$$

$$2\cdot 7\prod_{n=1}^{\infty}(1-x^{5n})(1-x^n)^7\sum_{m=1}^{\infty}P(5m-1)x^m=5^4x^7\sum_{m=0}^{\infty}P_{-7}(m)x^{5^2m}$$
$$-5\sum_{m\in\mathbb{Z}^+,m\equiv 2 \mod 5}P_{-7}(m)x^m-2\cdot 5\sum_{m\in\mathbb{Z}^+,m\equiv 1,3 \mod 5}P_{-7}(m)x^m$$

and

$$2\cdot 3^2\prod_{n=1}^{\infty}(1-x^{5n})(1-x^n)^9\sum_{m=1}^{\infty}P(5m-1)x^m=-5^5x^9\sum_{m=0}^{\infty}P_{-9}(m)x^{5^2m}$$
$$2-\cdot 5\sum_{m\in\mathbb{Z}^+,m\equiv 1,2 \mod 5}P_{-9}(m)x^m-5\sum_{m\in\mathbb{Z}^+,m\equiv 4 \mod 5}P_{-9}(m)x^m.$$

The second of the above equations implies that $5|P(5m-1)$, the level one congruence for the prime 5 as well as the less studied conclusion that for $m\equiv 1 \mod 5$, $P_{-6}\left(\frac{m-1}{5}\right)+P_{-6}(m)\equiv 0 \mod 6$. As a consequence of equation (5.2) (see also §10.3),

$$5x\prod_{n=1}^{\infty}(1-x^{5n})^6(1-x^n)^{-1}=5^2x^5\sum_{m=0}^{\infty}P_{-5}(m)x^{5^2m}$$
$$+\sum_{m\in\mathbb{Z}^+,m\equiv 2,3 \mod 5}P_{-5}(m)x^m-\sum_{m\in\mathbb{Z}^+,m\equiv 1,4 \mod 5}P_{-5}(m)x^m,$$

$$2\cdot 3x\prod_{n=1}^{\infty}(1-x^{5n})^6=5^2x^6\sum_{m=0}^{\infty}P_{-6}(m)x^{5^2m}+\sum_{m=0}^{\infty}P_{-6}(5m+1)x^{5m+1},$$

$$2 \cdot 7x \prod_{n=1}^{\infty} (1-x^{5n})^6 (1-x^n) = 5^3 x^7 \sum_{m=0}^{\infty} P_{-7}(m) x^{5^2 m}$$
$$- \sum_{m \in \mathbb{Z}^+, m \equiv 2 \mod 5} P_{-7}(m) x^m - 2 \sum_{m \in \mathbb{Z}^+, m \equiv 1,3 \mod 5} P_{-7}(m) x^m$$

and

$$2 \cdot 3^2 x \prod_{n=1}^{\infty} (1-x^{5n})^6 (1-x^n)^3 = -5^4 x^9 \sum_{m=0}^{\infty} P_{-9}(m) x^{5^2 m}$$
$$-2 \sum_{m \in \mathbb{Z}^+, m \equiv 1,2 \mod 5} P_{-9}(m) x^m - \sum_{m \in \mathbb{Z}^+, m \equiv 4 \mod 5} P_{-9}(m) x^m.$$

The first of the $k = 7$ equations gives us the Taylor series expansion of an infinite product, the second formula can be translated to

$$2 \cdot 5x \prod_{n=1}^{\infty} (1-x^{7n})^4 (1-x^n)$$
$$= 7^2 x^{10} \sum_{m=0}^{\infty} P_{-5}(m) x^{49m} - \sum_{m \in \mathbb{Z}, m \equiv 3 \mod 7} P_{-5}(m) x^m$$
$$-2 \sum_{m \in \mathbb{Z}^+, m \equiv 1,2,6 \mod 7} P_{-5}(m) x^m - \sum_{m=0}^{\infty} P_{-5}(m) x^m,$$

and the last gives a formula for P_{-2} which is equivalent to equation (5.28) and contains equation (5.29).

6. Production of constant functions

Most of this section is devoted to the study of the production of constant functions $F_{k,N}$. In particular, we determine the partition congruences implied by the fact that such a function is constant. Throughout the section k represents a prime.

6.1. The Frobenius automorphism. We start with the well known algebraic fact that for any prime k,

$$\prod_{n=1}^{\infty} (1-x^n)^k \equiv \prod_{n=1}^{\infty} (1-x^{kn}), \quad \mod k,$$

which allows us to obtain many congruences modulo primes. We need an elementary consequence of this last congruence. Its proof is reproduced for the reader's convenience.

Proposition 6.1. [**16**, Th. 8 of Ch. 7] *Let k be a positive prime and let N, s and $t \in \mathbb{Z}$. Assume that $N \equiv s \mod k$. If for all $n \in \mathbb{Z}^+$ with $n \equiv t \mod k$, $P_{-s}(n) \equiv 0 \mod k$, then for all such n, $P_{-N}(n) \equiv 0 \mod k$.*

Proof.

$$\sum_{n=0}^{\infty} P_{-N}(n)x^n = \prod_{n=1}^{\infty}(1-x^n)^N = \prod_{n=1}^{\infty}(1-x^n)^{Qk+s} = \prod_{n=1}^{\infty}(1-x^n)^{Qk}(1-x^n)^s$$

$$\equiv \prod_{n=1}^{\infty}(1-x^{kn})^Q\,(1-x^n)^s = \sum_{n=0}^{\infty} P_{-Q}(n)x^{kn}\sum_{n=0}^{\infty} P_{-s}(n)x^n \mod k.$$

Hence from the last equality

$$\sum_{n=0}^{\infty} P_{-N}(n)x^n \equiv \sum_{n=0}^{\infty} P_{-Q}\left(\frac{n}{k}\right)x^n \sum_{n=0}^{\infty} P_{-s}(n)x^n \mod k;$$

comparing coefficients in this last equation yields

$$P_{-N}(n) \equiv \sum_{j=0}^{n} P_{-Q}\left(\frac{j}{k}\right)\; P_{-s}(n-j) = \sum_{j=0}^{\lfloor\frac{n}{k}\rfloor} P_{-Q}(j)\; P_{-s}(n-kj) \mod k.$$

The last equality is the consequence of the fact that only those j's that are multiples of k contribute to the sum.

Assume that $N\in\mathbb{Z}^+$. The function $F_{k,-N}$ is constant for many choices of k and N. Since the only possible pole[119] of this function is at P_∞, it will be constant if and only if it is regular there. The condition that the function $F_{k,-N}$ be regular at P_∞ is that the Laurent series coefficients for negative indices in (5.13) (with N replaced by $-N$, of course) vanish. This is our next

Theorem 6.2. *Let k be a positive prime and let $N\in\mathbb{Z}^+$. The following are equivalent:*
(a) *$F_{k,-N}$ is a constant function.*
(b) *$P_{-\alpha(k)N}(mk+\beta(k)N)=0$ for $m=-\left\lfloor\frac{\beta(k)N}{k}\right\rfloor$, ..., -2, -1.*
(c) *the function $F_{k,-N}$ is invariant under the map $A_k:\tau\mapsto-\frac{1}{k\tau}$.*
Under these conditions, the constant value of $F_{k,-N}(\tau)$ is $P_{-\alpha(k)N}(\beta(k)N)$.

Proof. Only that (c) implies (a) requires some comment. If $F_{k,-N}$ is invariant under the automorphism A_k of $\overline{\mathbb{H}^2/\Gamma_o(k)}$, then (since this function is holomorphic at P_0) it is also holomorphic at P_∞, hence constant.

Remark 6.3. 1. The list in (b) is to be regarded as empty if $\left\lfloor\frac{\beta(k)N}{k}\right\rfloor=0$. In this case $F_{k,-N}$ is of course constant.
2. The list of $\left\lfloor\frac{\beta(k)N}{k}\right\rfloor$ arguments for the function $P_{-\alpha(k)N}$,

$$-\left\lfloor\frac{\beta(k)N}{k}\right\rfloor k+\beta(k)N,\; -\left(\left\lfloor\frac{\beta(k)N}{k}\right\rfloor+1\right)k+\beta(k)N,\; ...,\; -k+\beta(k)N,$$

[119] Of order at most $\lfloor\frac{\beta(k)N}{k}\rfloor$.

can also be written as

$$R\left(\frac{\beta(k)N}{k}\right),\ R\left(\frac{\beta(k)N}{k}\right)+k,\ \ldots,\ \beta(k)N-k.$$

Once again this list is to be considered as the empty set for integers N with $1 \leq \beta(k)N < k$.

3. It is a corollary of the last theorem that

$$P_{-\alpha(k)N}\left(R\left(\frac{\beta(k)N}{k}\right)\right)=0 \Longleftrightarrow \operatorname{ord}_\infty F_{k,-N} > -\left\lfloor\frac{\beta(k)N}{k}\right\rfloor.$$

This observation provides a reformulation of Lehmer's question on the vanishing of coefficients of the Fourier series expansion of the discriminant from elliptic function theory; however, it does not seem to bring us any closer to a resolution.

Theorem 6.4. *For k and N satisfying one (hence all three) of the conditions of the previous theorem*

$$\frac{\sum_{m=0}^{\infty} P_{-\alpha(k)N}(km+\beta(k)N)x^m}{\sum_{m=0}^{\infty} P_{-\alpha(k)N}(m)x^{km}} = P_{-\alpha(k)N}(\beta(k)N).$$

Corollary 6.5. *Let $m \in \mathbb{Z}$, $m \geq -\left\lfloor\frac{\beta(k)N}{k}\right\rfloor$. For k and N as in the theorem,*[120]

$$P_{-\alpha(k)N}(km+\beta(k)N) = P_{-\alpha(k)N}(\beta(k)N)P_{-\alpha(k)N}\left(\frac{m}{k}\right),$$

or equivalently

$$P_{-\alpha(k)N}\left(km+R\left(\frac{\beta(k)N}{k}\right)\right) = P_{-\alpha(k)N}(\beta(k)N)P_{-\alpha(k)N}\left(\frac{m-\left\lfloor\frac{\beta(k)N}{k}\right\rfloor}{k}\right). \tag{5.26}$$

Equation (5.26) has as an immediate consequence that if $F_{k,N}$ is constant, then

$$P_{-\alpha(k)N}\left(R\left(\frac{\beta(k)N}{k}\right)\right) = \begin{cases} P_{-\alpha(k)N}(\beta(k)N) \text{ if } \left\lfloor\frac{\beta(k)N}{k}\right\rfloor = 0 \\ 0 \text{ if } \left\lfloor\frac{\beta(k)N}{k}\right\rfloor \neq 0 \end{cases}$$

and generally, for all $m \in \mathbb{Z}^+ \cup \{0\}$,

$$P_{-\alpha(k)N}(k^2m+\beta(k)N) = P_{-\alpha(k)N}(\beta(k)N)P_{-\alpha(k)N}(m),$$

$$P_{-\alpha(k)N}(km+\beta(k)N) = 0 \text{ if } m \not\equiv 0 \mod k.$$

---

[120]We continue to use the convention that $P_N(n)=0$ if $n \notin \mathbb{Z}^+ \cup \{0\}$.

6.2. Constant functions. We are interested in determining for which k and $N \neq 0$ is $F_{k,N}$ a constant function. A necessary condition is that $-N \in \mathbb{Z}^+$ (we know that $F_{k,0}$ is constant); we now assume than $N \in \mathbb{Z}^+$. When $F_{k,-N}$ is constant, we need to know (to get good applications) when $k|P_{-\alpha(k)N}(\beta(k)N)$. Our basic tool is Theorem 6.2. There are two reasons why a particular function $F_{k,-N}$ reduces to be a constant. The first, trivial, reason is because a function of degree $\left\lfloor \frac{\beta(k)N}{k} \right\rfloor$ does not exist on the surface $\overline{\mathbb{H}^2/\Gamma_o(k)}$; the second, computational, reason is that the combinatorial number theory of the partition function $P_{-\alpha(k)N}$ forces enough Laurent series coefficients to vanish.

We have not made an exhaustive study of when we obtain constant functions. We summarize the conclusions of our calculations below. We have included the cases we will need in subsequent sections.

(a) For $k = 2,\ 3,\ 5,\ \beta(k) = 1$. Hence in these cases for $N = 1,\ 2,\ \ldots, k-1$, $F_{k,-N}$ is constant since it is regular at P_∞.

(a1) For $k = 2$, only $F_{2,-1}$ is constant. This follows from the fact that $P_{-8N}(1) = -8N$ (used for odd N) and $P_{-8N}(0) = 1$ (used for even N).

(a2) For $k = 3$, only $F_{3,-1}$ and $F_{3,-2}$ are constant. This follows from the fact that $P_{-3N}(1) = -3N$ (used for $N \equiv 1 \mod 3$), $P_{-3N}(2) = \frac{9}{2}N(N-1)$ (used for $N \equiv 2 \mod 3$) and $P_{-3N}(0) = 1$ (used for $N \equiv 0 \mod 3$).

(a3) For $k = 5$, the functions $F_{5,-N}$ for $N = 1,\ 2,\ 3,\ 4$, are constant for trivial reasons. In addition, only $F_{5,-8}$ and $F_{5,-14}$ are constant. This follows from two facts. The first of these are the general formulae $P_{-N}(1) = -N$ (used for $N \equiv 1 \mod 5$), $P_{-N}(2) = \frac{N(N-3)}{2}$ (used for $N \equiv 2 \mod 5$), $P_{-N}(3) = -\frac{N(N-1)(N-8)}{3!}$ (used for $N \equiv 3 \mod 5$), $P_{-N}(4) = \frac{N(N-1)(N-3)(N-14)}{4!}$ (used for $N \equiv 4 \mod 5$) and $P_{-N}(0) = 1$ (used for $N \equiv 0 \mod 5$). The second fact we need is the calculation $0 = P_{-14}(4) = P_{-14}(9)$. We will also use below that $P_{-8}(8) = -5^3$ and $P_{-14}(14) = -5^6$.

(b) For $k = 7$, we have $\beta(7) = 2$ and[121] $p(\overline{\mathbb{H}^2/\Gamma_o(7)}) = 0$. The functions $F_{7,-N}$, $N = 1,\ 2,\ 3$, are constant for trivial reasons. The functions $F_{7,-6}$ and $F_{7,-10}$ are constant as a result of the calculations $0 = P_{-6}(5) = P_{-10}(6) = P_{-10}(13)$. We also will need to know that $P_{-6}(12) = 7^2$ and $P_{-10}(20) = 7^4$. There are no other constant functions $F_{7,-N}$ as a consequence of the tabulation of $P_N(n)$ in §1.

(c) For $k = 11$, $\beta(11) = 5$ and $p(\overline{\mathbb{H}^2/\Gamma_o(11)}) = 1$. Hence for $N = 1,\ 2,\ 3,\ 4$, the functions $F_{11,-N}$ are constant for trivial reasons. The calculations

$$0 = P_{-6}(8) = P_{-6}(19),\ 0 = P_{-8}(7) = P_{-8}(18) = P_{-8}(29),$$

[121]We continue to denote the genus of the closed surface X by $p(X)$.

$$0 = P_{-10}(6) = P_{-10}(17) = P_{-10}(28) = P_{-10}(39),$$

$$0 = P_{-14}(4) = P_{-14}(15) = ... = P_{-14}(59),$$

and

$$0 = P_{-26}(9) = P_{-26}(20) = ... = P_{-26}(120)$$

show that $F_{11,-N}$ are constant functions for $N = 6$, 8, 10, 14 and 26. We also have

$$P_{-6}(30) = 11^2,\ P_{-8}(40) = -11^3,\ P_{-10}(50) = 11^4,\ P_{-14}(70) = 11^6,$$

and

$$P_{-26}(130) = 11^{12}.$$

We have accounted for all the constant functions $F_{11,N}$ as a result of the tabulations of $P_N(n)$.

(d) For $k = 13$ ($\beta(13) = 7$ and $p(\overline{\mathbb{H}^2/\Gamma_o(13)}) = 0$) there are only two constant functions $F_{13,-1}$ (for trivial reasons) and $F_{13,-3}$ (because $P_{-3}(8) = 0$ but $P_{-3}(21) = 13$). For $F_{13,-N}$ with $N > 1$ to be constant, it is necessary that $P_{-N}\left(R\left(\frac{7N}{13}\right)\right) = 0$. The tabulations of §1 show that this occurs only for $N = 3$.

(e) For $k = 17$ ($\beta(17) = 12$ and $p(\overline{\mathbb{H}^2/\Gamma_o(17)}) = 1$) there are exactly six constant functions: $F_{17,-1}$, $F_{17,-2}$ (for trivial reasons) and $F_{17,-3}$ (because $P_{-3}(2) = 0 = P_{-3}(19)$ but $P_{-3}(36) = 17$). The functions $F_{17,-N}$ for $N = 4$, 8 and 14 are also constant. The calculations of the Ramanujan coefficients $P_N(n)$ show that there are no other constant functions for this prime.

(f) For $k = 19$ ($\beta(19) = 15$ and $p(\overline{\mathbb{H}^2/\Gamma_o(19)}) = 1$) there are three constant functions: $F_{19,-1}$, $F_{19,-2}$ (for trivial reasons) and $F_{19,-3}$ (because $P_{-3}(7) = 0 = P_{-3}(26)$ but $P_{-3}(45) = -19$). So are $F_{19,-6}$ and $F_{19,-10}$ (with $P_{10}(150) = 19^4$), but $F_{19,-18}$ is not constant.

(g) For $k = 23$ ($\beta(23) = 22$ and $p(\overline{\mathbb{H}^2/\Gamma_o(23)}) = 2$) there are three constant functions: $F_{23,-1}$, $F_{23,-2}$ (for trivial reasons) and $F_{23,-3}$ (because $P_{-3}(20) = 0 = P_{-3}(43)$ but $P_{-3}(66) = -23$). The functions $F_{23,-N}$ for $N = 4$, 6, 8, 10 and 14 are also constant.

(h) For $k = 29$ ($\beta(29) = 35$ and $p(\overline{\mathbb{H}^2/\Gamma_o(19)}) = 1$) there are three constant functions $F_{29,-1}$ (for trivial reasons), $F_{29,-2}$ (because $P_{-2}(12) = 0 = P_{-2}(41)$ but $P_{-2}(70) = -1$) and $F_{29,-3}$ (because $P_{-3}(18) = 0 = P_{-3}(47) = P_{-3}(76)$ but $P_{-3}(105) = 29$). The functions $F_{29,-N}$ for $N = 4$, 8 and 14 are also constant.

(i) We have determined all the constant functions for positive primes $k \leq 29$. As a result of our tabulation in Table 12, the reader can extend the calculations to include the primes 31 and 37. For larger primes, we need to expand the calculations for Table 12.

(j) It is possible to determine at this point all the primes k for which $F_{k,N}$

is constant for $N = -1$, -2 and -3; for example, $F_{67,-1}$ is constant. See also Theorem 6.10.

6.3. Congruences. From Theorem 6.4, its corollary and the calculations of the previous paragraph, we conclude many congruences. Some of these results also follow directly from part (d) of Theorem 4.5. It is best to summarize the results in tabular form using the notation of equation (5.26), listing the cases and specifying the parameters for that equation to hold.

Remark 6.6. Some comments about Table 17 follow. The list for primes $k \leq 29$ is complete in the sense that it includes the equalities obtained from all constant functions $F_{k,N}$ for such k. We have listed only the cases we have actually computed. To complete the list and obtain similar lists for bigger primes, we need information about the polynomials $P_{\cdot}(n)$ for all $n \in \mathbb{Z}^+$. We have calculated these polynomials only for $0 \leq n \leq 40$. We have also included an illustrative identity for the prime 67 (the case $N = 1$). There are many more similar identities for P_{-N} with $N = 1$, 2 and 3 involving primes $k \geq 31$.

Theorem 6.7. *Let k be a prime and $N \in \mathbb{Z}^+$ be such that $F_{k,-N}$ is a constant function. Then for all n and $m \in \mathbb{Z}^+ \cup \{0\}$,*

$$P_{-\alpha(k)N}\left(k^{2n+2}m + \beta(k)N\frac{k^{2n+2}-1}{k^2-1}\right) = P^{n+1}_{-\alpha(k)N}(\beta(k)N)P_{-\alpha(k)N}(m).$$

In particular,

$$P_{-\alpha(k)N}\left(\beta(k)N\frac{k^{2n+2}-1}{k^2-1}\right) = P^{n+1}_{-\alpha(k)N}(\beta(k)N).$$

Furthermore,

$$P_{-\alpha(k)N}\left(k^{n+1}m + R\left(\frac{\beta(k)N}{k}\right)\right) = 0$$

for $n = 0$ if $m \not\equiv \left\lfloor \frac{\beta(k)N}{k} \right\rfloor \mod k$ and for $n > 0$ if $0 \neq \left\lfloor \frac{\beta(k)N}{k} \right\rfloor$.

Proof. We rewrite our basic equality, equation (5.26), with $\tilde{N} = -\alpha(k)N$, $R = R\left(\frac{\beta(k)N}{k}\right)$, $M = \left\lfloor \frac{\beta(k)N}{k} \right\rfloor$ and $P = P_{\tilde{N}}(\beta(k)N)$ as

$$P_{\tilde{N}}(km + R) = PP_{\tilde{N}}\left(\frac{m-M}{k}\right),$$

which immediately also (replacing m by $km + M$) tells us that (the case $n = 0$ of the first equation in the theorem)

$$P_{\tilde{N}}(k^2m + \beta(k)N) = PP_{\tilde{N}}(m).$$

k	$\alpha(k)N$	$R\left(\frac{\beta(k)N}{k}\right)$	$P_{-\alpha(k)N}(\beta(k)N)$	$\left\lfloor\frac{\beta(k)N}{k}\right\rfloor$
2	8	1	-2^3	0
3	3	1	-3	0
	6	2	3^2	0
5	1	1	-1	0
	2	2	-1	0
	3	3	5	0
	4	4	-5	0
	8	3	-5^3	1
	14	4	5^6	2
7	1	2	-1	0
	2	4	1	0
	3	6	-7	0
	6	5	7^2	1
	10	6	7^4	2
11	1	5	1	0
	2	10	1	0
	3	4	-11	1
	4	9	-11	1
	6	8	11^2	2
	8	7	-11^3	3
	10	6	11^4	4
	14	4	11^6	6
	26	9	11^{12}	11
13	1	7	1	0
	3	8	13	1
17	1	12	-1	0
	2	7	-1	1
	3	2	17	2
	4	14	-17	2
	8	11	17^3	5
	14	15	17^6	9
19	1	15	-1	0
	2	11	1	1
	3	7	19	2
	6	14	19^2	4
	10	17	19^4	7
23	1	22	1	0
	2	21	1	1
	3	20	-23	2
	4	19	-23	3
	6	17	23^2	5
	8	15	23^3	7
	10	13	23^4	9
	14	9	23^6	13
29	1	6	-1	1
	2	12	1	2
	3	18	29	3
	4	24	29	4
	8	19	-29^3	9
	14	26	-29^6	26
67	1	53	-1	2

Table 17. CONSTANTS APPEARING IN RECURSION FORMULAE FOR PARTITION COEFFICIENTS. The lines in the above table contain one entry for each pair (k, N), k a prime ≤ 29 for which $F_{k,-N}$ is constant.

We define for $n \in \mathbb{Z}^+ \cup \{0\}$,

$$l_n = \beta(k)N(k^{2n} + k^{2n-2} + \ldots + k^2 + 1) = \beta(k)N\frac{k^{2n+2} - 1}{k^2 - 1}.$$

It follows by induction that

$$\begin{aligned} P_{\tilde{N}}(k^{2n+4}m + l_{n+1}) &= P_{\tilde{N}}(k^2(k^{2n+2}m + l_n) + \beta(k)N) \\ &= PP_{\tilde{N}}(k^{2n+2}m + l_n) = P^{n+2}P_{\tilde{N}}(m), \end{aligned}$$

establishing the first equality. The second equality is a special case ($m = 0$) of the first. The case $n = 0$ of the last equation is a consequence of equation (5.26). The recursion formula tells us that for $n \in \mathbb{Z}^+$,

$$P_{\tilde{N}}(k^{n+1}m + R) = P_{\tilde{N}}(k(k^n m) + R) = PP_{\tilde{N}}\left(k^{n-1}m - \frac{M}{k}\right),$$

which completes the proof of the last equality.

The above calculations together with Proposition 6.1 yield immediately the following

Theorem 6.8. *For all $N \in \mathbb{Z}$ and all $m \in \mathbb{Z}^+ \cup \{0\}$,*

$$P_{2N}(2m+1) \equiv 0 \mod 2,$$
$$P_{3N}(3m+1) \equiv 0 \equiv P_{3N}(3m+2) \mod 3,$$
$$P_{5N+2}(5m+3) \equiv 0 \equiv P_{5N+1}(5m+4) \mod 5,$$
$$P_{7N+4}(7m+6) \equiv 0 \equiv P_{7N+1}(7m+5) \mod 7,$$
$$P_{11N+8}(11m+4) \equiv 0 \equiv P_{11N+7}(11m+9) \equiv P_{11N+5}(11m+8) \mod 11,$$
$$P_{11N+3}(11m+7) \equiv 0 \equiv P_{11N+1}(11m+6) \equiv P_{11N+7}(11m+9) \mod 11,$$
$$P_{13N+10}(13m+8) \equiv 0 \mod 13,$$
$$P_{17N+14}(17m+2) \equiv P_{17N+13}(17m+14) \equiv P_{17N+9}(17m+11) \mod 17,$$
$$P_{17N+3}(17m+15) \equiv 0 \mod 17,$$
$$P_{19N+16}(19m+7) \equiv P_{19N+13}(19m+14) \equiv P_{19N+9}(19m+17) \equiv 0 \mod 19,$$
$$P_{23N+20}(23m+20) \equiv P_{23N+19}(23m+19) \equiv P_{23N+17}(23m+17) \equiv 0 \mod 23,$$
$$P_{23N+15}(23m+15) \equiv P_{23N+13}(23m+13) \equiv P_{23N+9}(23m+9) \equiv 0 \mod 23,$$
$$P_{29N+26}(29m+18) \equiv P_{29N+25}(29m+24) \equiv P_{29N+21}(29m+19) \equiv 0 \mod 29,$$

and

$$P_{29N+15}(29m+26) \equiv 0 \mod 29.$$

The mysterious patterns described above are clearly in need of further investigation and explanation. The second mod 5 and 7 and fifth mod 11 congruences listed above are the level one Ramanujan congruences for the primes 5, 7 and 11.

We have produced congruences mod k only for small primes k. To obtain a general result, we first need

Lemma 6.9. *For all odd primes k, $\frac{k^2-1}{8}$ is a triangular number, but $\frac{k^2-1}{8} - \alpha k$ is not for all $\alpha \in \mathbb{Z}$ with $0 < \alpha < \frac{k^2-1}{8k}$.*

Proof. Let $T_n = \frac{n(n+1)}{2} = 1+2+\ldots+n$ denote the n-th triangular number ($n \in \mathbb{Z}^+ \cup \{0\}$). Obviously $\frac{k^2-1}{8} = T_{\frac{k-1}{2}}$. If $\frac{k^2-1}{8} - \alpha k = T_{\frac{k-2-\beta}{2}}$, for some odd positive integer $\beta \leq k-2$, then

$$\frac{k^2-1}{8} - \alpha k = \frac{k^2-1}{8} - \left[\left(\frac{k-1}{2}\right) + \left(\frac{k-3}{2}\right) + \ldots + \left(\frac{k-\beta}{2}\right)\right].$$

Thus

$$2(\beta+1-4\alpha)k = (1+\beta)^2.$$

Thus $k|(1+\beta)$. It is here that we use the hypothesis that k is prime. Since $1+\beta \leq k-1 < k$, we have arrived at a contradiction.

Theorem 6.10. *For all primes $k \geq 5$, $F_{k,-3}$ is a constant function,*

$$F_{k,-3} = P_{-3}\left(\frac{k^2-1}{8}\right) = (-1)^{\frac{k-1}{2}} k.$$

As a consequence of Proposition 6.1 we have

Corollary 6.11. *For all primes $k \geq 5$, all $N \in \mathbb{Z}$ and all $m \in \mathbb{Z}^+ \cup \{0\}$,*

$$P_{kN-3}\left(km + \frac{k^2-1}{8}\right) \equiv 0 \mod k.$$

Remark 6.12. For $N = 0$, the corollary should follow directly from Jacobi's identity.

6.4. Functions $F_{k,n,N}$ for negative N. It is obvious that $F_{k,n,0} = 1$ for all primes k and all $n \in \mathbb{Z}^+$. As a consequence of the averaging process which takes us from $F_{k,n,N}$ to $F_{k,n+1,N}$, we see that if for some odd n, $F_{k,n,N} = c$ (thus $N \leq 0$), then also $F_{k,n+1,N} = c$; a similar conclusion holds for even n (provided $F_{k,N}$ is constant). This observation is the motivation for the next theorem and provides an alternative way to establish Theorem 6.7.

Theorem 6.13. *If $F_{k,-N}$ is constant $(= P_{-\alpha(k)N}(\beta(k)N))$ for the prime k and $N \in \mathbb{Z}^+$, then so is $F_{k,n,-N}$ $(= P_{-\alpha(k)N}(\beta(k)N)^{\lfloor \frac{n+1}{2} \rfloor})$ for all $n \in \mathbb{Z}^+$.*

Proof. Let $c = P_{-\alpha(k)N}(\beta(k)N)$. The theorem certainly holds for $n = 1$. Assume by induction that $F_{k,n,-N}$ is constant and equals $c^{\lfloor \frac{n+1}{2} \rfloor}$ for $n \in \mathbb{Z}^+$. If n is odd, then

$$F_{k,n+1,-N} = V_k(c^{\lfloor \frac{n+1}{2} \rfloor}) = c^{\lfloor \frac{n+2}{2} \rfloor}.$$

For even n,

$$F_{k,n+1,-N} = V_k(c^{\lfloor \frac{n+1}{2} \rfloor} f_{k^2,1}^{-\alpha(k)N}) = c^{\lfloor \frac{n}{2} \rfloor} F_{k,-N} = c^{\lfloor \frac{n+2}{2} \rfloor}.$$

The last theorem implies a strengthened form of Theorem 6.2.

Corollary 6.14. *Let k be a positive prime and let $N \in \mathbb{Z}^+$. The following are equivalent:*
(a) *$F_{k,-N}$ is a constant function.*
(b) *$F_{k,n,-N}$ is a constant function for all $n \in \mathbb{Z}^+$.*
(c) *$P_{-\alpha(k)N}(mk+\beta(k)N) = 0$ for $m = -\left\lfloor \frac{\beta(k)N}{k} \right\rfloor, \ldots, -2, -1$.*
(d) *For each odd positive integer n,*

$$P_{-\alpha(k)N}\left(k^n m + \frac{\alpha(k)N(k^{n+1}-1)}{24}\right) = 0$$

for $m = -\left\lfloor \frac{\alpha(k)N}{24}\left(k - \frac{1}{k^n}\right) \right\rfloor, \ldots, -2, -1,$
and for each even positive integer n,

$$P_{-\alpha(k)N}\left(k^n m + \frac{\alpha(k)N(k^n-1)}{24}\right) = 0$$

for $m = -\left\lfloor \frac{\alpha(k)N}{24}\left(1 - \frac{1}{k^n}\right) \right\rfloor, \ldots, -2, -1.$
(e) *$P_{-\alpha(k)N}(m) = 0$ for $m =$*
(in case of odd n)

$$R\left(\frac{\frac{\alpha(k)N(k^{n+1}-1)}{24}}{k^n}\right),\ R\left(\frac{\frac{\alpha(k)N(k^{n+1}-1)}{24}}{k^n}\right) + k^n,\ \ldots,$$

$$\frac{\alpha(k)N(k^{n+1}-1)}{24} - k^n = \frac{k^n(k\alpha(k)N-24)-\alpha(k)N}{24} = r(k,N,n)$$

(in case of even n)

$$R\left(\frac{\frac{\alpha(k)N(k^{n}-1)}{24}}{k^n}\right),\ R\left(\frac{\frac{\alpha(k)N(k^{n}-1)}{24}}{k^n}\right) + k^n,\ \ldots,$$

$$\frac{\alpha(k)N(k^{n}-1)}{24} - k^n = \frac{k^n(\alpha(k)N-24)-\alpha(k)N}{24} = r(k,N,n).$$

Each of the lists in (e) are to be regarded as empty if the corresponding last entry is negative, that is, if and only if

$$\frac{\alpha(k)N}{24} < \begin{cases} \frac{1}{k-k^{-n}} & \text{for } n \text{ odd} \\ \frac{1}{1-k^{-n}} & \text{for } n \text{ even} \end{cases}.$$

A similar condition is imposed on the lists in (d). Note that

$$\lim_{n\to\infty} r(k,N,n) = \begin{cases} \infty \text{ if } k\alpha(k)N > 24 \\ -\frac{1}{k} \text{ if } k\alpha(k)N = 24 \\ -\infty \text{ if } k\alpha(k)N < 24 \end{cases}.$$

Corollary 6.15. *Let k be a positive prime and let $N \in \mathbb{Z}^+$ be such that $F_{k,-N}$ is a constant function. Then for all $n \in \mathbb{Z}^+$,*

$$P_{-\alpha(k)N}\left(\frac{\alpha(k)N(k^{2\lfloor\frac{n+1}{2}\rfloor}-1)}{24}\right) = \left[P_{-\alpha(k)N}\left(\frac{\alpha(k)N(k^2-1)}{24}\right)\right]^{\lfloor\frac{n+1}{2}\rfloor}.$$

Let[122] $c = P_{-\alpha(k)N}(\beta(k)N)$. The fact that $F_{k,-N}$ is constant yields, for all $n \in \mathbb{Z}^+ \cup \{0\}$, the power series identities

$$\frac{\sum_{m=\left\lceil\frac{\alpha(k)N}{24}\left(\frac{1}{k^{2n+1}}-k\right)\right\rceil}^{\infty} P_{-\alpha(k)N}\left(k^{2n+1}m + \frac{\alpha(k)N(k^{2(n+1)}-1)}{24}\right)x^m}{\prod_{m=1}^{\infty}(1-x^{km})^{\alpha(k)N}} = c^{n+1}$$

and

$$\frac{\sum_{m=\left\lceil\frac{\alpha(k)N}{24}\left(1-\frac{1}{k^{2n}}\right)\right\rceil}^{\infty} P_{-\alpha(k)N}\left(k^{2n}m + \frac{\alpha(k)N(k^{2n}-1)}{24}\right)x^m}{\prod_{m=1}^{\infty}(1-x^{m})^{\alpha(k)N}} = c^{n}.$$

The first of these last two equalities tells us that for each $m \in \mathbb{Z}^+ \cup \{0\}$, "the partition coefficient" $c^{-(n+1)}P_{-\alpha(k)N}\left(k^{2n+1}m + \frac{\alpha(k)N(k^{2(n+1)}-1)}{24}\right)$ is independent of $n \in \mathbb{Z}^+ \cup \{0\}$ and (we use the case $n=0$) that

$$P_{-\alpha(k)N}(km + \beta(k)N) = 0 \text{ for } m \not\equiv 0 \mod k; \tag{5.27}$$

the second equality says $c^{-n}P_{-\alpha(k)N}\left(k^{2n}m + \frac{\alpha(k)N(k^{2n}-1)}{24}\right)$ is independent of $n \in \mathbb{Z}^+ \cup \{0\}$.

These equations have additional consequences. The definitions give us for all $N \in \mathbb{Z}$ the power series expansions

$$\prod_{m=1}^{\infty}(1-x^m)^N = \sum_{m=0}^{\infty} P_{-N}(m)x^m.$$

If we assume that the function $F_{k,-N}$ is constant, then we obtain nontrivial identities among the coefficients P_{-N} for the values N listed in the second column of Table 17. These relations tell us that $P_{-N}(m) = 0$ for many positive integers m and thus leads to

Proposition 6.16. *Each of the negative integers $-M = -1, -2, -3, -4, -6, -8, -10, -14$ and -26 are roots of infinitely many polynomials $P.(n)$.*

Proof. Each M as above is of the form $\alpha(k)N$ with k prime, $N \in \mathbb{Z}^+$ and $\alpha(k)N$ an entry in the second column of Table 17. By (5.27), $-M$ is a root of $P.(km + \beta(k)N)$ for all positive integers m not congruent to 0 mod k. (We know from the Jacobi and Euler identities that -1 and -3 are roots of infinitely many polynomials $P.(n)$.)

[122]Recall (5.4).

Remark 6.17. **(a)** We conclude, for example, for $k = 5$ and $N = 4$ (because $F_{5,-4}$ is constant) that

$$P_{-4}\left(\frac{5^{2\lfloor\frac{n+1}{2}\rfloor}-1}{6}\right) = [P_{-4}(4)]^{\lfloor\frac{n+1}{2}\rfloor} = (-5)^{\lfloor\frac{n+1}{2}\rfloor}$$

and

$$P_{-4}(5m+4) = 0 \text{ for } m \not\equiv 0 \mod 5.$$

But for $k = 5$ and $N = 5$ (where $F_{5,-5}$ is not constant),

$$P_{-5}(5) = -6 \text{ and } P_{-5}(130) = -89,$$

exhibiting no obvious relation.
(b) The next interesting case for the prime 5 is $N = 8$. The fact that $F_{5,-8}$ is constant showed that

$$P_{-8}\left(\frac{5^{2\lfloor\frac{n+1}{2}\rfloor}-1}{3}\right) = [P_{-8}(8)]^{\lfloor\frac{n+1}{2}\rfloor} = (-5^3)^{\lfloor\frac{n+1}{2}\rfloor}.$$

That $F_{5,-8}$ is constant is equivalent to $0 = P_{-8}(3)$. But this single equality has infinitely many consequences. It tells us, for example, that

$$P_{-8}(5m+3) = 0 \text{ for } m \not\equiv 1 \mod 5.$$

(c) We have used the information in Table 17 and equation (5.27) to prove Proposition 6.16. We can establish a weaker proposition with slightly less calculation. Pick a positive integer N so that $F_{k,-N}$ is constant, $k \geq 5$ and $kN > 24$. We know that $P_{-N}(r(k,N,n)) = 0$. Since $\lim_{n\to\infty} r(k,N,n) = \infty$, we conclude that $-N$ is a root of infinitely many polynomials $P.(m)$.

6.5. Functions $F_{k,-N}$ of small degree. As a consequence of the previous material in this section, we obtained information for the $F_{k,-N}$ with $k = 2$, 3, 5 and $1 \leq N < k$ (also some information for other primes). The next interesting case is basically a restatement of Theorem 4.7 .

Theorem 6.18. (a) *For $k = 2,\ 3,\ 5$, and $N = k,\ k+1,\ \ldots,\ 2k-1$,*

$$F_{k,-N}(\tau) = \frac{1}{k}\sum_{l=0}^{k-1}\left(\frac{\eta(\frac{\tau+l}{k})}{\eta(k(\tau+l))}\right)^{\alpha(k)N} = c_0 + c_1\left(\frac{\eta(\tau)}{\eta(k\tau)}\right)^{\frac{24}{k-1}}.$$

(b) *(The case $k = 7$.) For $N = 4,\ 5$ and 6,*

$$F_{7,-N}(\tau) = \frac{1}{7}\sum_{l=0}^{6}\left(\frac{\eta(\frac{\tau+l}{7})}{\eta(7(\tau+l))}\right)^{N} = c_0 + c_1\left(\frac{\eta(\tau)}{\eta(7\tau)}\right)^{4}.$$

(c) *(The case $k = 13$.) For $N = 2$ and 3,*

$$F_{13,-N}(\tau) = \frac{1}{13}\sum_{l=0}^{12}\left(\frac{\eta(\frac{\tau+l}{13})}{\eta(13(\tau+l))}\right)^{N} = c_0 + c_1\left(\frac{\eta(\tau)}{\eta(13\tau)}\right)^{2}.$$

Corollary 6.19. (a) *For* $k = 2,\ 3,\ 5,$ *and* $N = k,\ k+1,\ \ldots,\ 2k-1,$

$$\frac{\sum_{m=-1}^{\infty} P_{-\alpha(k)N}(km+N)x^m}{\prod_{n=1}^{\infty}(1-x^{kn})^{\alpha(k)N}} = c_0 + c_1\left[x^{-1}\prod_{n=1}^{\infty}\left(\frac{1-x^{kn}}{1-x^n}\right)^{\frac{24}{1-k}}\right].$$

(b) *(The case* $k = 7$.*)* *For* $N = 4,\ 5$ *and* $6,$

$$\prod_{n=1}^{\infty}(1-x^{7n})^{-N}\sum_{m=-1}^{\infty} P_{-N}(7m+2N)x^m = c_0 + c_1\left[x^{-1}\prod_{n=1}^{\infty}\left(\frac{1-x^{7n}}{1-x^n}\right)^{-4}\right].$$

(c) *(The cases* $k = 13$.*)* *For* $N = 2$ *and* $3,$

$$\prod_{n=1}^{\infty}(1-x^{13n})^{-N}\sum_{m=-1}^{\infty} P_{-N}(13m+7N)x^m = c_0+c_1\left[x^{-1}\prod_{n=1}^{\infty}\left(\frac{1-x^{13n}}{1-x^n}\right)^{-2}\right].$$

It is easily seen that in case (a)

$$c_0 = P_{-\alpha(k)N}(N) + \frac{24}{k-1}P_{-\alpha(k)N}(-k+N) \text{ and } c_1 = P_{-\alpha(k)N}(-k+N),$$

in case (b)

$$c_0 = P_{-N}(2N) + 4P_{-N}(-7+2N) \text{ and } c_1 = P_{-N}(-7+2N),$$

and in case (c)

$$c_0 = P_{-N}(7N) + 2P_{-N}(-13+7N) \text{ and } c_1 = P_{-N}(-13+7N).$$

In some cases (for example $k = 5$ with $N = 8$, $k = 7$ with $N = 6$ and $k = 13$ with $N = 3$) $c_1 = 0$, producing a constant function. The special cases in the last corollary; $N = k+1$ for $k = 2$, 3 and 5; $N = 4$ for $k = 7$; and $N = 2$ for $k = 13$ yield (functions of degree one and)

Corollary 6.20. (a) *For* $k = 2,\ 3,\ 5,$

$$\sum_{m=-1}^{\infty} P_{\frac{24}{1-k}}(k(m+1)+1)x^m = \left(P_{\frac{24}{1-k}}(k+1) - \left(\frac{24}{k-1}\right)^2\right)\sum_{m=0}^{\infty} P_{\frac{24}{1-k}}(m)x^{km}$$

$$+\frac{24}{1-k}\sum_{m=0}^{\infty} P_{\frac{24}{1-k}}(m)x^{m-1}.$$

In particular,
(a1) *(*$k = 2$*)*

$$P_{-24}(2m+1) = -2^{11}P_{-24}\left(\frac{m-1}{2}\right) - 2^3 3P_{-24}(m),$$

(a2) *(*$k = 3$*)*

$$P_{-12}(3m+1) = -3^5 P_{-12}\left(\frac{m-1}{3}\right) - 2^2 3P_{-12}(m),$$

(a3) *($k = 5$)*

$$P_{-6}(5m+1) = -5^2 P_{-6}\left(\frac{m-1}{5}\right) - 2 \cdot 3P_{-6}(m).$$

(b) *(The case $k = 7$.)*

$$\sum_{m=-1}^{\infty} P_{-4}(7m+8)x^m = -7 \sum_{m=0}^{\infty} P_{-4}(m)x^{7m} - 2^2 \sum_{m=0}^{\infty} P_{-4}(m)x^{m-1};$$

hence

$$P_{-4}(7m+1) = -7P_{-4}\left(\frac{m-1}{7}\right) - 2^2 P_{-4}(m).$$

(c) *(The case $k = 13$.)*

(5.28)

$$\sum_{m=-1}^{\infty} P_{-2}(13m+14)x^m = -\sum_{m=0}^{\infty} P_{-2}(m)x^{13m} - 2\sum_{m=0}^{\infty} P_{-2}(m)x^{m-1};$$

hence

$$P_{-2}(13m+1) = -P_{-2}\left(\frac{m-1}{13}\right) - 2P_{-2}(m). \tag{5.29}$$

Remark 6.21. It is of interest to determine whether the above list of *three term recursions* for partition coefficients P_N is complete. The case $k = 2$ of the previous corollary is a simple application of the multiplicativity of the T-function, whose generalizations will be investigated in Chapter 6.

7. Averaging operators

Let k be a positive prime. In this section we study several linear operators defined on function spaces. For G a finite index subgroup of Γ, the field of meromorphic functions $\mathcal{K}(\overline{\mathbb{H}^2/G})$ on $\overline{\mathbb{H}^2/G}$ may be identified with the field $\mathcal{K}(G)$ of G-invariant meromorphic functions on $\mathbb{H}^2$ that extend continuously to the cusps of G.

7.1. Automorphisms of $\mathcal{K}(\Gamma_o(k))$. We have introduced the linear operators $U_{k,n}$ given by equation (5.18) and $V_{k,M}$ given by equation (5.20). We introduce next some linear operators that can be iterated. For this purpose, we recall that for $c \in \mathbb{C}^*$ we defined the fractional linear transformation $M_c = \begin{bmatrix} c & 0 \\ 0 & 1 \end{bmatrix}$. We identify a fractional linear transformation M with the linear operator $M^* = M_0^*$ it induces on function spaces by precomposition. We also define $\mathcal{M}_f$ as multiplication by the function f. Since

$$M_{\frac{1}{k^n}} : \mathcal{K}(\Gamma_o(k^{n+1})) \to \mathcal{K}(\Gamma(k^n, k))$$

function φ	group G	prime k	condition on $N \in \mathbb{Z}$	formula for $\varphi(\tau)$
$f_k^N = f_{k,1}^{\frac{24N}{k-1}}$	$\Gamma_o(k)$	≤ 7 & 13 ≥ 7	none $(k-1)$ divides $12N$	$\left(\frac{\eta(k\tau)}{\eta(\tau)}\right)^{\frac{24N}{k-1}}$
$f_{k,k}^{\frac{24N}{k^2-1}}$	$\Gamma(k,k)$	≤ 5 ≥ 5	none a multiple of $\frac{k^2-1}{24}$	$\left(\frac{\eta(k\tau)}{\eta(\frac{\tau}{k})}\right)^{\frac{24N}{k^2-1}}$
$F_{k,N}$	$\Gamma_o(k)$	≤ 5 ≥ 5	none a multiple of $\frac{k^2-1}{24}$	$\frac{1}{k}\sum_{l=0}^{k-1} f_{k,k}^{\frac{24N}{k^2-1}}(\tau+l)$
$f_{1,k}^{\frac{24N}{k-1}}$	$\Gamma(k,k)$	≤ 7 & 13 ≥ 7	none $(k-1)$ divides $12N$	$\left(\frac{\eta(\tau)}{\eta(\frac{\tau}{k})}\right)^{\frac{24N}{k-1}}$
$Y_{k,N}$	$\Gamma_o(k)$	≤ 7 & 13 ≥ 7	none $(k-1)$ divides $12N$	$\frac{1}{k}\sum_{l=0}^{k-1} f_{1,k}^{\frac{24N}{k-1}}(\tau+l)$

Table 18. SUFFICIENT CONDITIONS FOR SINGLE VALUEDNESS OF φ ON $\mathbb{H}^2/G$.

is an isomorphism as a consequence of Lemma 2.1, we can study the linear operator

$$\tilde{U}_{k,n} = U_{k,n} \circ M_{\frac{1}{k^n}} : \mathcal{K}(\Gamma_o(k^{n+1})) \to \mathcal{K}(\Gamma_o(k))$$

that satisfies

$$\tilde{U}_{k,n}(gM_{k^n}(f)) = f\tilde{U}_{k,n}(g), \text{ for all } f \in \mathcal{K}(\Gamma_o(k)) \text{ and } g \in \mathcal{K}(\Gamma_o(k^{n+1})).$$

Since $\mathcal{K}(\Gamma_o(k)) \subset \mathcal{K}(\Gamma_o(k^{n+1}))$, we can view $\tilde{U}_{k,n}$ as a $\mathbb{C}$-linear automorphism of $\mathcal{K}(\Gamma_o(k))$.

We note that in this setting (recall equation (5.20))

$$U_k(f_{k,k}^{\alpha(k)N}) = V_k(f_{k^2,1}^{\alpha(k)N}) = F_{k,N}$$

and

$$V_k(f_k^N) = V_k\left(f_{k,1}^{\frac{24N}{k-1}}\right) = Y_{k,N} = f_k^{-N} F_{k,\frac{N(k+1)}{\beta(k)}}.$$

For the second sequence of equalities, which is generalized below, we require that $(k-1)|12N$.

We have encountered several group invariant functions. For the convenience of the reader, we summarize our conclusions in Table 18.

We need to study one more set of functions (mostly for easier bookkeeping). A generalization of Theorems 4.5 and 4.11 is provided by

Proposition 7.1. (a) *For all $M \in \mathbb{Z}$ and all $N \in \mathbb{Z}$ such that $(k-1)|12N$,*

$$V_k\left(f_{k^2,1}^{\alpha(k)M} f_{k,1}^{\frac{24N}{k-1}}\right) = U_k\left(f_{k,k}^{\alpha(k)M} f_{1,k}^{\frac{24N}{k-1}}\right) \in \mathcal{K}(\Gamma_o(k))$$

is holomorphic except possibly at P_0 and P_∞. Its Taylor series expansion at P_∞ in terms of the local coordinate $x = \exp(2\pi\imath\tau)$ is

$$\frac{\prod_{n=1}^{\infty}(1-x^n)^{\frac{24N}{k-1}}\sum_{m=\left\lceil\frac{\beta(k)M+N}{k}\right\rceil}^{\infty}P_{\alpha(k)M+\frac{24N}{k-1}}(km-\beta(k)M-N)x^m}{\prod_{n=1}^{\infty}(1-x^{kn})^{-\alpha(k)M}};$$

hence

$$\operatorname{ord}_{P_\infty}V_k\left(f_{k^2,1}^{\alpha(k)M}f_{k,1}^{\frac{24N}{k-1}}\right)\geq\left\lceil\frac{\beta(k)M+N}{k}\right\rceil,$$

with equality if $\alpha(k)M+\frac{24N}{k-1}>0$. The function has a pole of order $\beta(k)M+kN$ if $\beta(k)M+(k+1)N>0$.
(b) *Assume that $k = 2$, 3, 5, 7 or 13. If $\alpha(k)M+\frac{24N}{k-1}>0$ and $\beta(k)M+(k+1)N>0$, then there exist constants c_i with $c_{\left\lceil\frac{\beta(k)M+N}{k}\right\rceil}\neq 0$ such that*

$$V_k\left(f_{k^2,1}^{\alpha(k)M}f_{k,1}^{\frac{24N}{k-1}}\right)=\sum_{i=\left\lceil\frac{\beta(k)M+N}{k}\right\rceil}^{\beta(k)M+kN}c_if_k^i=\sum_{i=\left\lceil\frac{\beta(k)M+N}{k}\right\rceil}^{\beta(k)M+kN}c_if_{k,1}^{\frac{24i}{k-1}}.$$

Proof. We have already seen the relevant arguments twice; the details are left to the reader.

Corollary 7.2. *Under the general hypothesis of the proposition,*

$$V_k\left(f_{k^2,1}^{\alpha(k)M}f_{k,1}^{\frac{24N}{k-1}}\right)=V_k\left(f_{k^2,1}^{\alpha(k)M}f_k^N\right)=f_k^{-N}F_{k,M+\frac{k+1}{\beta(k)}N}=V_{k,M}(f_k^N).$$

Proof. In the local coordinate x,

$$V_k\left(f_{k^2,1}^{\alpha(k)M}f_{k,1}^{\frac{24N}{k-1}}\right)=\prod_{n=1}^{\infty}(1-x^{kn})^{\alpha(k)M}$$

$$\times\prod_{n=1}^{\infty}(1-x^n)^{\frac{24N}{k-1}}\sum_{m=\left\lceil\frac{\beta(k)M+N}{k}\right\rceil}^{\infty}P_{\alpha(k)M+\frac{24N}{k-1}}(km-\beta(k)M-N)x^m$$

$$=x^{-N}\frac{\prod_{n=1}^{\infty}(1-x^n)^{\frac{24N}{k-1}}}{\prod_{n=1}^{\infty}(1-x^{kn})^{\frac{24N}{k-1}}}\prod_{n=1}^{\infty}(1-x^{kn})^{\alpha(k)\left(M+\frac{k+1}{\beta(k)}N\right)}$$

$$\times\sum_{m=\left\lceil\frac{\beta(k)M+N(k+1)}{k}\right\rceil}^{\infty}P_{\alpha(k)\left(M+\frac{k+1}{\beta(k)}N\right)}\left(km-\beta(k)\left(M+\frac{N(k+1)}{\beta(k)}\right)\right)x^m$$

$$=f_k^{-N}F_{k,M+\frac{k+1}{\beta(k)}N}.$$

It is of interest to determine the extent to which the operators U_k and V_k preserve the linear spaces $\mathcal{K}(\Gamma_o(k))_0$ and $\mathcal{K}(\Gamma_o(k))_\infty$. From $U_k(f_k^N) = F_{k,N}$, we see that

$$\mathrm{ord}_\infty U_k(f_k^N) = \left\lceil \frac{\beta(k)}{k} N \right\rceil \text{ and } \mathrm{ord}_0 U_k(f_k^N) = -\beta(k)N, \; N > 0,$$

and

$$\mathrm{ord}_\infty U_k(f_k^N) \geq \left\lceil \frac{\beta(k)}{k} N \right\rceil \text{ and } \mathrm{ord}_0 U_k(f_k^N) \geq 0, \; N \leq 0.$$

Assume that $(k-1)|12N$. From $V_k(f_k^N) = f_k^{-N} F_{k, \frac{k+1}{\beta(k)}N}$, we conclude

$$\mathrm{ord}_\infty V_k(f_k^N) = \left\lceil \frac{k+1}{k} N \right\rceil - N \text{ and } \mathrm{ord}_0 V_k(f_k^N) = kN, \; N > 0,$$

and

$$\mathrm{ord}_\infty V_k(f_k^N) \geq \left\lceil \frac{k+1}{k} N \right\rceil - N \text{ and } \mathrm{ord}_0 V_k(f_k^N) \geq -N, \; N \leq 0.$$

For $k = 2$, 3, 5, 7 and 13, a basis for $\mathcal{K}(\Gamma_o(k))_0$ ($\mathcal{K}(\Gamma_o(k))_\infty$, respectively) consists of the functions f_k^N, $N = 1, 2, \ldots$ ($N = -1, -2, \ldots$). Hence we see that

$$V_k : \mathcal{K}(\Gamma_o(k))_0 \to \mathcal{K}(\Gamma_o(k))_0 \text{ and } V_k : \mathcal{K}(\Gamma_o(k))_\infty \to \mathcal{K}(\Gamma_o(k))_\infty,$$

for $k = 2$, 3, 5, 7 and 13. Note that these linear cperators are neither surjective nor injective. For the operators U_k, we conclude that

$$U_k : \mathcal{K}(\Gamma_o(k))_0 \to \mathcal{K}(\Gamma_o(k))_0$$

is injective but not surjective and that the linear space $\mathcal{K}(\Gamma_o(k))_\infty$ is not preserved under U_k. Using the results of §8, we can express $U_k(f_k^N)$ and $V_k(f_k^N)$ as polynomials in f_k and f_k^{-1} for all $N \in \mathbb{Z}$ as long as $k = 2$, 3, 5 or 7. If $N \in \mathbb{Z}^+$, only powers of f_k appear. It should be emphasized that the coefficients of these polynomials can be computed recursively.

7.2. Other linear maps. For m and $n \in \mathbb{Z}^+ \cup \{0\}$,

$$[\Gamma(k^m, k) : \Gamma(k^{m+n}, k)] = k^n$$

and the k^n motions $B^{k^m l}$, $l = 0, 1, \ldots, k^n - 1$, generate $\Gamma(k^m, k)$ over $\Gamma(k^{m+n}, k)$. Hence

$$W_{k;m,n}(f) = \frac{1}{k^n} \sum_{l=0}^{k^n-1} f \circ B^{k^m l}$$

defines a linear operator from $\mathcal{K}(\Gamma(k^{m+n}, k))$ to $\mathcal{K}(\Gamma(k^m, k))$. Further,

$$U_{k,n} = W_{k;0,n},$$

and if n' is also a nonnegative integer, then

$$W_{k;m,n+n'} = W_{k;m+n,n'} \circ W_{k;m,n} : \mathcal{K}(\Gamma(k^{m+n+n'}, k)) \to \mathcal{K}(\Gamma(k^m, k)).$$

8. Modular equations

For every inclusion of finite index subgroups of the modular group

$$G_1 \subset G_2 \subset \Gamma,$$

we have a natural finite sheeted holomorphic branched covering map

$$\overline{\mathbb{H}^2/G_1} \to \overline{\mathbb{H}^2/G_2} \to \overline{\mathbb{H}^2/\Gamma}$$

and a natural inclusion of the respective fields of meromorphic functions

$$\mathcal{K}(G_1) \supset \mathcal{K}(G_2) \supset \mathcal{K}(\Gamma).$$

Further, $\mathcal{K}(G_1)$ is a finite algebraic extension of $\mathcal{K}(G_2)$, and $\mathcal{K}(G_2)$ is a finite algebraic extension of $\mathcal{K}(\Gamma)$. Each of the fields $\mathcal{K}(G_1)$ and $\mathcal{K}(G_2)$ is generated over $\mathbb{C}$ by one (if the corresponding surface is of genus zero) or two (if the corresponding surface is of positive genus) elements; the field $\mathcal{K}(\Gamma)$ is generated by one function (usually one uses the $\jmath$-invariant). If both $\overline{\mathbb{H}^2/G_1}$ and $\overline{\mathbb{H}^2/G_2}$ are spheres, then we can let x be a generator of $\mathcal{K}(G_2)$ and y be a generator of $\mathcal{K}(G_1)$ and conclude that $y \in \mathbb{C}[x]$. The remaining task is the computation of the coefficients of this polynomial f, to be called the *modular equation for y over x*, that satisfy $y = f(x)$. The polynomial f is then the (generally, ramified) covering map from $\overline{\mathbb{H}^2/G_1}$ to $\overline{\mathbb{H}^2/G_2}$. If $\overline{\mathbb{H}^2/G_2}$ is the sphere (but $\overline{\mathbb{H}^2/G_1}$ is arbitrary), then we can let x be a generator of $\mathcal{K}(G_2)$. Any $y \in \mathcal{K}(G_1)$ will satisfy a polynomial equation over $\mathbb{C}(x)$; it is much more difficult to determine this polynomial.

Every $\Gamma_o(k)$-invariant function is of course $\Gamma(k,k)$-invariant. In particular, for $k = 2$, 3, 5, 7 and 13, $f_k^{-1} = f_{k,1}^{\frac{24}{1-k}}$ is a $\Gamma_o(k)$-invariant function with divisor $P_0 P_\infty^{-1}$. Its divisor on $\overline{\mathbb{H}^2/\Gamma(k,k)}$ is $P_0 P_1 ... P_{k-1} P_\infty^{-k}$. For $k = 2$, 3 and 5, $h_k^{-1} = f_{k,k}^{\frac{24}{1-k^2}}$ is a $\Gamma(k,k)$-invariant function with divisor $P_0 P_\infty^{-1}$. Hence for these k, f_k^{-1} is a polynomial of degree k in h_k^{-1} with zero constant term. We call these and similar relations for functions on $\Gamma(k,k)$ *modular equations for the prime k*. It turns out that these equations have many important applications. It is of interest to also consider the $\Gamma^o(k)$-invariant function $f_{1,k}^{\frac{24}{k-1}}$ with divisor $P_\infty P_0^{-1}$; its projection to $\overline{\mathbb{H}^2/\Gamma(k,k)}$ has divisor $P_\infty P_1 ... P_{\frac{1}{k-1}} P_0^{-k}$. For $k = 2$, 3 and 5 it is a polynomial of degree k in h_k with zero constant term.

8.1. $k=2$. A modular equation in this case is

$$f_2^{-1} = 2^4 h_2^{-1} + h_2^{-2} \text{ or } h_2^{2+N} = f_2(2^4 h_2^{1+N} + h_2^N);$$

the first of these translates to the power series identity

$$\prod_{n=1}^{\infty}(1-x^{2n})^{24} = 2^4 x \prod_{n=1}^{\infty}(1-x^n)^8 \prod_{n=1}^{\infty}(1-x^{4n})^{16} + \prod_{n=1}^{\infty}(1-x^n)^{16} \prod_{n=1}^{\infty}(1-x^{4n})^8.$$

This power series identity, as can be checked, is equivalent to the Jacobi quartic identity which we have seen many times. The second identity tells us (each of the following two statements implies the other) that

(5.30)

$$F_{2,2+N} = f_2(2^4 F_{2,1+N} + F_{2,N}) \text{ and } F_{2,N} = -2^4 F_{2,1+N} + f_2^{-1} F_{2,2+N},$$

which along with (the previously computed expressions)

$$F_{2,0} = 1 \text{ and } F_{2,1} = 2^3 f_2$$

gives us a recursive way for computing $F_{2,N}$. It is convenient to introduce the following

Definition 8.1. For each prime k and each $n \in \mathbb{Z}^+$, we let $\pi_k(n)$ be the highest power of k that divides n.

Remark 8.2. Each π_k is a discrete valuation of rank one on the integers; that is,

$$\pi_k : \mathbb{Z}^* = \mathbb{Z} - \{0\} \to \mathbb{Z}^+$$

is a multiplicative surjective homomorphism

$$\pi_k(nm) = \pi_k(n)\pi_k(m), \text{ for all } n, m \in \mathbb{Z}^*$$

and

$$\pi_k(n+m) \geq \min\{\pi_k(n), \pi_k(m)\}, \text{ for all } n, m \in \mathbb{Z}^*.$$

Lemma 8.3. *For each* $N \in \mathbb{Z}^+$,

$$F_{2,N} = \sum_{i=\lceil \frac{N}{2} \rceil}^{N} c_{N,i} f_2^i \text{ with } c_{N,i} \in \mathbb{Z}^+,$$

and

$$\pi_2(c_{N,i}) \begin{cases} = 0 \text{ if } N \text{ is even and } i = \frac{N}{2} \\ \geq 7 + 8\left(i - \frac{N}{2} - 1\right) \text{ if } N \text{ is even and } i > \frac{N}{2} \\ = 3 \text{ if } N \text{ is odd and } i = \frac{N+1}{2} \\ \geq 3 + 8\left(i - \frac{N+1}{2}\right) \text{ if } N \text{ is odd and } i > \frac{N+1}{2} \end{cases}.$$

In particular,

$$\pi_2(c_{N,i}) \geq 3 + 8\left(i - \frac{N+1}{2}\right).$$

Proof. Only the divisibility conditions need to be verified. Since (as previously computed)

$$F_{2,2} = f_2 + 2^7 f_2^2,$$

the lemma holds for $N = 1$ and 2. For an induction argument, assume that the divisibility conditions hold for $c_{n,i}$ with $1 \leq n \leq N+1$. For $N > 0$,

$$F_{2,2+N} = f_2\left(2^4 \sum_{i=\lceil\frac{N+1}{2}\rceil}^{N+1} c_{N+1,i} f_2^i + \sum_{i=\lceil\frac{N}{2}\rceil}^{N} c_{N,i} f_2^i\right) = \sum_{i=\lceil\frac{N+2}{2}\rceil}^{N+2} c_{N+2,i} f_2^i.$$

Hence we have the recursion relation

$$c_{N+2,i} = 2^4 c_{N+1,i-1} + c_{N,i-1}.$$

It follows that

$$\pi_2(c_{N+2,i}) \geq \min\{4 + \pi_2(c_{N+1,i-1}),\ \pi_2(c_{N,i-1})\},$$

with equality if the last two terms in the expression for the minimum are unequal. If $N+2$ is odd, then

$$\begin{aligned}\pi_2(c_{N+2,1+\lceil\frac{N}{2}\rceil}) &\geq \min\left\{4 + \pi_2(c_{N+1,\lceil\frac{N}{2}\rceil}),\ \pi_2(c_{N,\lceil\frac{N}{2}\rceil})\right\}\\ &= \min\{4,\ 3\} = 3.\end{aligned}$$

Also for $2 + \lceil\frac{N}{2}\rceil \leq i \leq N+2$,

$$\begin{aligned}\pi_2(c_{N+2,i}) &\geq \min\left\{11 + 8\left(i - 1 - \frac{N+1}{2}\right),\ 3 + 8\left(i - 1 - \frac{N+1}{2}\right)\right\}\\ &= 3 + 8\left(i - \frac{N+3}{2}\right).\end{aligned}$$

If $N+2$ is even, then

$$\pi_2(c_{N+2,1+\frac{N}{2}}) \geq \min\{4 + \infty,\ 0\} = 0.$$

For $2 + \frac{N}{2} \leq i \leq N+2$,

$$\begin{aligned}\pi_2(c_{N+2,i}) &\geq \min\left\{7 + 8\left(i - 1 - \frac{N+2}{2}\right),\ 7 + 8\left(i - 1 - \frac{N}{2} - 1\right)\right\}\\ &= 7 + 8\left(i - \frac{N+2}{2} - 1\right).\end{aligned}$$

We record for subsequent use the following formulae which show that the induction argument (hence the lemma) does not in general produce sharp bounds:

$$F_{2,3} = 2^3(1+2)f_2^2 + 2^{11}f_2^3, \ F_{2,4} = f_2^2 + 2^9 f_2^3 + 2^{15} f_2^4,$$
$$F_{2,5} = 2^3(1+2^2)f_2^3 + 2^{11}(1+2^2)f_2^4 + 2^{19} f_2^5,$$
$$F_{2,6} = f_2^3 + 2^7(1+2^3)f_2^4 + 2^{16}(1+2)f_2^5 + 2^{23} f_2^6$$

and

$$F_{2,7} = 2^3(1+2+2^2)f_2^4 + 2^{12}(1+2+2^2)f_2^5 + 2^{19}(1+2+2^2)f_2^6 + 2^{27} f_2^7.$$

For negative indices N we have

$$F_{2,-1} = -2^3, \ F_{2,-2} = 2^7 + f_2^{-1} \text{ and } F_{2,-3} = -2^{11} - 2^3(1+2)f_2^{-1};$$

only the middle formula does not appear among the direct calculations of §10.

8.2. $k = 3$**.** The basic equation

$$f_3^{-1} = 3^3 h_3^{-1} + 3^2 h_3^{-2} + h_3^{-3}$$

leads to the power series identity

$$\prod_{n=1}^{\infty}(1-x^{3n})^{12} = 3^3 x^2 \prod_{n=1}^{\infty}(1-x^n)^3 \prod_{n=1}^{\infty}(1-x^{9n})^9 \tag{5.31}$$
$$+3^2 x \prod_{n=1}^{\infty}(1-x^n)^6 \prod_{n=1}^{\infty}(1-x^{9n})^6 + \prod_{n=1}^{\infty}(1-x^n)^9 \prod_{n=1}^{\infty}(1-x^{9n})^3.$$

Our basic equation says that for all $N \in \mathbb{Z}$,

$$h_3^{3+N} = f_3(3^3 h_3^{2+N} + 3^2 h_3^{1+N} + h_3^N);$$

thus since we shall see in §10 that

$$F_{3,0} = 1, \ F_{3,1} = 3^2 f_3 \text{ and } F_{3,2} = 2\cdot 3 f_3 + 3^5 f_3^2, \tag{5.32}$$

we obtain the recursion formulae that for each $N \in \mathbb{Z}$,

$$F_{3,3+N} = f_3(3^3 F_{3,2+N} + 3^2 F_{3,1+N} + F_{3,N})$$

and

$$F_{3,N} = -3^2 F_{3,1+N} - 3^3 F_{3,2+N} + f_3^{-1} F_{3,3+N}.$$

As in the case $k = 2$, for each $N \in \mathbb{Z}^+ \cup \{0\}$,

$$F_{3,N} = \sum_{i=\lceil \frac{N}{3} \rceil}^{N} c_{N,i} f_3^i \text{ with } c_{N,i} \in \mathbb{Z}^+.$$

We can study the divisibility, by powers of 3, properties of these coefficients $c_{N,i}$ using (5.32) and (the consequence of the recursion formulae)

$$\pi_3(c_{N+3,i}) \geq \min\{3 + \pi_3(c_{N+2,i-1}),\ 2 + \pi_3(c_{N+1,i-1}),\ \pi_3(c_{N,i-1})\}.$$

Calculations show that

$$F_{3,-1} = -3,\ F_{3,-2} = 3^2 \text{ and } F_{3,-8} = -3^{11} + 2^2 3^2 7 f_3^{-2}.$$

These facts plus the recursion formulae allow us to investigate the coefficients $c_{N,i}$ for negative N (these can be zero or negative).

The above development was based on the group inclusion $\Gamma(3,3) \subset \Gamma_o(3)$. If we use the inclusion $\Gamma(3) \subset G(3)$, we get related results. The constructions and results of Chapter 3 produce the $\Gamma(3)$-invariant functions

$$w = \frac{\theta^3\begin{bmatrix}\frac{1}{3}\\ 1\end{bmatrix}}{\theta^3\begin{bmatrix}1\\ \frac{1}{3}\end{bmatrix}} \text{ and } W = \frac{\theta^3\begin{bmatrix}\frac{1}{3}\\ \frac{1}{3}\end{bmatrix}\theta^3\begin{bmatrix}\frac{1}{3}\\ 1\end{bmatrix}\theta^3\begin{bmatrix}\frac{1}{3}\\ \frac{5}{3}\end{bmatrix}}{\theta^9\begin{bmatrix}1\\ \frac{1}{3}\end{bmatrix}}$$

with respective divisors

$$\frac{P_0}{P_\infty} \text{ and } \frac{P_{-1}P_0P_1}{P_\infty^3}.$$

We compute the relation satisfied by these two functions to be

$$W = w^3 + \sqrt{3}\imath w^2 - w$$

which leads us to the theta identity

$$\theta^3\begin{bmatrix}\frac{1}{3}\\ \frac{1}{3}\end{bmatrix}\theta^3\begin{bmatrix}\frac{1}{3}\\ \frac{5}{3}\end{bmatrix} = \theta^6\begin{bmatrix}\frac{1}{3}\\ 1\end{bmatrix} + \sqrt{3}\imath\theta^3\begin{bmatrix}\frac{1}{3}\\ 1\end{bmatrix}\theta^3\begin{bmatrix}1\\ \frac{1}{3}\end{bmatrix} - \theta^6\begin{bmatrix}1\\ \frac{1}{3}\end{bmatrix}:$$

the translation of this identity to power series leads us once again to (5.31). The above identity is an algebraic consequence of the identities proven in Theorem 3.12 of Chapter 3. This is another example of how the use of theta constants with rational characteristics leads to interesting identities.

8.3. $k = 5$. In this case

$$f_5^{-1} = \sum_{i=1}^{5} c_i h_5^{-i};$$

the evaluation of constants leads to

$$f_5^{-1} = 5^2 h_5^{-1} + 5^2 h_5^{-2} + 3 \cdot 5 h_5^{-3} + 5 h_5^{-4} + h_5^{-5} \tag{5.33}$$

and hence also

$$\prod_{n=1}^{\infty}(1 - x^{5n})^6 = 5^2 x^4 \prod_{n=1}^{\infty}(1 - x^n) \prod_{n=1}^{\infty}(1 - x^{25n})^5$$

$$+5^2x^3\prod_{n=1}^{\infty}(1-x^n)^2\prod_{n=1}^{\infty}(1-x^{25n})^4+3\cdot 5x^2\prod_{n=1}^{\infty}(1-x^n)^3\prod_{n=1}^{\infty}(1-x^{25n})^3$$

$$+5x\prod_{n=1}^{\infty}(1-x^n)^4\prod_{n=1}^{\infty}(1-x^{25n})^2+\prod_{n=1}^{\infty}(1-x^n)^5\prod_{n=1}^{\infty}(1-x^{25n}).$$

Remark 8.4. The above identity should be compared with the identities obtained in the last section of Chapter 4.

We rewrite the modular equations in more general form (for $N \in \mathbb{Z}$),

$$h_5^{5+N} = f_5 h_5^N(5^2h_5^4 + 5^2h_5^3 + 3\cdot 5h_5^2 + 5h_5 + 1).$$

Hence (applying the operator U_5)

(5.34) $F_{5,5+N} = f_5(5^2F_{5,4+N} + 5^2F_{5,3+N} + 3\cdot 5F_{5,2+N} + 5F_{5,1+N} + F_{5,N})$

and also

(5.35) $f_5^{-1}F_{5,5+N} - 5^2F_{5,4+N} - 5^2F_{5,3+N} - 3\cdot 5F_{5,2+N} - 5F_{5,1+N} = F_{5,N}.$

We shall see in §10 that

$$F_{5,0} = 1,\ F_{5,1} = 5f_5,\ F_{5,2} = 5^3f_5^2 + 2\cdot 5f_5,$$

$F_{5,3} = 5^5f_5^3 + 3\cdot 5^3f_5^2 + 3^2f_5$ and $F_{5,4} = 5^7f_5^4 + 2^25^5f_5^3 + 2\cdot 5^211f_5^2 + 2^2f_5$.

We have seen that for all $N \in \mathbb{Z}^+$,

$$F_{5,N} = \sum_{i=\lceil \frac{N}{5}\rceil}^{N} c_{N,i}f_5^i$$

with $c_{N,i} \in \mathbb{C}$. We can now conclude that $c_{N,i} \in \mathbb{Z}^+$, and we have a recursive way for computing these constants. We will need the following stronger version of the last formula.

Lemma 8.5. [**16**, Ch. 8, Lemma 4] *For each* $N \in \mathbb{Z}^+ \cup \{0\}$,

$$F_{5,N} = \sum_{i=\lceil \frac{N}{5}\rceil}^{N} c_{N,i}f_5^i \text{ with } c_{N,i}\in\mathbb{Z}^+ \text{ and } 5^{\lfloor \frac{5i-N-1}{2}\rfloor}|c_{N,i}.$$

In particular, $5|c_{N,i}$ *for all* i *provided* $N \equiv 1$ *or* $2 \mod 5$, *and* $5|c_{N,i}$ *for all* $i > \lceil \frac{N}{5}\rceil$.

The sum is to be interpreted as 1 for $N = 0$, and $5^{\lfloor \frac{5i-N-1}{2}\rfloor}$ is to be replaced by 1 for $5i - N \le 2$.

	$i=0$	$i=1$	$i=2$	$i=3$	$i=4$	$i=5$	$i=6$
$N=0$	1						
$N=1$		5					
$N=2$		$2\cdot 5$	5^3				
$N=3$		3^2	$3\cdot 5^3$	5^5			
$N=4$		2^2	$2\cdot 5^2 11$	$2^2 5^5$	5^7		
$N=5$		1	$2^2 5^3$	$2^3 5^5$	5^8	5^9	
$N=6$			$3^2 5\cdot 7$	$2\cdot 5^4 23$	$3^2 5^6 7$	$2\cdot 3\cdot 5^9$	5^{11}

Table 19. TABULATIONS OF $c_{N,i}$ FOR THE PRIME 5.

Proof. Direct computations established the claim for $0 \le N \le 4$. For an induction (on N) argument, assume that the index i in the sum runs from 1 to N (with $c_{N,i} = 0$ for $i < \lceil \frac{N}{5} \rceil$). From the basic recursion formula (5.34),

$$c_{N+5,i} = 5^2 c_{N+4,i-1} + 5^2 c_{N+3,i-1} + 3\cdot 5 c_{N+2,i-1} + 5 c_{N+1,i-1} + c_{N,i-1}.$$

Thus by induction $5^{\lfloor \frac{5i-(N+5)-1}{2} \rfloor} | c_{N+5,i}$.

Remark 8.6. **(a)** We can of course use (5.34) and backwards induction to get a formula for nonpositive $-N$. It reads (with $N \in \mathbb{Z}^+$)

$$F_{5,-N} = \sum_{i=0}^{\lfloor \frac{N}{5} \rfloor} c_{-N,i} f_5^{-i} \text{ with } c_{-N,i} \in \mathbb{Z}.$$

(b) Applying the operator A to (5.33), we obtain another modular equation

$$f_{1,5}^6 = h_5 + 5h_5^2 + 3\cdot 5h_5^3 + 5^2 h_5^4 + 5^2 h_5^5.$$

We will need some of the values $c_{N,i}$ from the last lemma. We summarize our calculations in tabular form.

8.4. $k = 7$. This is a more complicated case because h_7 is not $\Gamma(7,7)$-invariant ($\beta(7) = 2$), but h_7^2 is.[123] On $\overline{\mathbb{H}^2/\Gamma(7,7)}$,

$$(f_7^{-1}) = P_0\ P_1\ \ldots\ P_6\ P_\infty^{-7} \text{ and } (h_7^{-2}) = P_0^2\ P_\infty^{-2}.$$

We can certainly choose constants c_i and d_j so that

$$H = f_7^{-2} - f_7^{-1} \sum_{i=0}^{3} c_i h_7^{-2i} - \sum_{j=1}^{7} d_j h_7^{-2j}$$

has a pole of order ≤ 5 at P_∞ and its Laurent series expansion at this point in terms of the local coordinate $x = \exp\left(\frac{2\pi\imath\tau}{7}\right)$ has zero coefficients for the exponents -4 and -2. It turns out that with this choice of constants, the

[123]The surface $\overline{\mathbb{H}^2/\Gamma(7,7)}$ has genus one.

coefficients for the exponents -5, -3, -1 and 0 also vanish. So the function H is identically zero for this choice of coefficients. We thus have obtained the identity

$$f_7^{-2} = f_7^{-1}\left(7h_7^{-6} + 5\cdot 7h_7^{-4} + 7^2h_7^{-2}\right)$$
$$+h_7^{-14} + 7h_7^{-12} + 3\cdot 7h_7^{-10} + 7^2h_7^{-8} + 3\cdot 7^2h_7^{-6} + 7^3h_7^{-4} + 7^3h_7^{-2}.$$

The resulting recursion formula for the function $F_{7,N}$ reads

$$F_{7,7+N} = \left\{\begin{array}{r} f_7\left(7F_{7,4+N} + 5\cdot 7F_{7,5+N} + 7^2F_{7,6+N}\right) \\ +f_7^2\left(F_{7,N} + 7F_{7,1+N} + 3\cdot 7F_{7,2+N}\right. \\ \left. +\, 7^2F_{7,3+N} + 3\cdot 7^2F_{7,4+N} + 7^3F_{7,5+N} + 7^3F_{7,6+N}\right) \end{array}\right. .$$

8.5. $k = 13$. The invariant functions on the genus eight surface $\overline{\mathbb{H}^2/\Gamma(13,13)}$,

$$(v) = (f_{13}^{-1}) = P_0\ P_1\ \ldots\ P_{12}\ P_\infty^{-13} \text{ and } (u) = (h_{13}^{-7}) = P_0^7\ P_\infty^{-7},$$

are certainly algebraically related. The form of the relation is

$$\sum_{0\le i\le 13, 0\le j\le 7} a_{ij}u^iv^j,$$

with the (so far) undetermined coefficients $a_{ij} \in \mathbb{C}$ (the indexing set runs over integers only). Since we know the Taylor series expansions of the functions u and v at infinity, using modern computers we should be able to calculate these 112 constants. It is not clear how to efficiently perform these calculations manually (as was presumably done by Weber [**28**]).

9. The ideal of partition identities

There is a unifying theme behind our results. It is contained in the next theorem. Throughout this section we assume that k is a prime such that $p(\overline{\mathbb{H}^2/\Gamma_o(k)}) = 0$; that is, $k = 2$, 3, 5, 7 or 13. We fix an $n \in \mathbb{Z}^+ \cup \{0\}$ and an $N \in \mathbb{Z}$. We study the function $F_{k,n,N}$ for arbitrary N for $n = 1$ and only for nonnegative N for $n > 1$. We note that

$$(5.36)\qquad F_{k,n,N} = \left\{\begin{array}{l} \sum_{i=\left\lceil \frac{\alpha(k)N}{24}\left(k-\frac{1}{k^n}\right)\right\rceil}^{\frac{\alpha(k)N}{24}(k^{n+1}-1)} c_if_k^i \text{ for } n \text{ odd and } N\ge 0, \\ \sum_{i=\left\lceil \frac{\alpha(k)N}{24}\left(1-\frac{1}{k^n}\right)\right\rceil}^{\frac{\alpha(k)N}{24}(k^{n+1}-k)} c_if_k^i \text{ for } n \text{ even and } N\ge 0, \\ \sum_{i=\left\lceil \frac{\beta(k)N}{k}\right\rceil}^{\beta(k)\max\{0,N\}} c_if_k^i \text{ for } n=1 \text{ and all } N, \end{array}\right.$$

where each $c_i \in \mathbb{C}$. In terms of the distinguished local coordinate $x = \exp(2\pi\imath\tau)$ the last identity can be rewritten as equations similar to (5.17). To strengthen our results, on the coefficients c_i, we begin with a sequence of definitions.

We start with an indexing set

$$I = \{I_1,\ I_1+1,\ \ldots,\ I_2\},$$

where

$$I_1 = \begin{cases} \left\lceil \alpha(k)\frac{N}{24}\left(k - \frac{1}{k^n}\right)\right\rceil \text{ for } n \text{ odd and } N \geq 0 \\ \left\lceil \alpha(k)\frac{N}{24}\left(1 - \frac{1}{k^n}\right)\right\rceil \text{ for } n \text{ even and } N \geq 0 \\ \left\lceil \frac{\beta(k)N}{k} \right\rceil \text{ for } n = 1 \text{ and all } N \end{cases}$$

and

$$I_2 = \begin{cases} \alpha(k)\frac{N}{24}(k^{n+1} - 1) \text{ for } n \text{ odd and } N \geq 0 \\ \alpha(k)\frac{Nk}{24}(k^n - 1) \text{ for } n \text{ even and } N \geq 0 \\ 0 \text{ for } n = 1 \text{ and } N \leq 0 \end{cases}.$$

Let c_i be the coefficients defined by equation (5.36). We then have

Theorem 9.1. *Let $k = 2,\ 3,\ 5,\ 7$ or 13. Let $n \in \mathbb{Z}^+ \cup \{0\}$ and $N \in \mathbb{Z}$. If $n > 1$, assume that $N \geq 0$. Then $c_i \in \mathbb{Z}$ and*

$$c = d_1 = d,$$

where

$$c = c(k, n, N) = \gcd\{c_i; i \in I\},$$

$$d_1 = d_1(k, n, N) = \gcd\left\{ P_{\alpha(k)N}\left(k^n m - \alpha(k)N\frac{k^{2\lfloor\frac{n+1}{2}\rfloor} - 1}{24}\right); m \in I \right\}$$

and

$$d = d(k, n, N) = \gcd\left\{ P_{\alpha(k)N}\left(k^n m - \alpha(k)N\frac{k^{2\lfloor\frac{n+1}{2}\rfloor} - 1}{24}\right); m \geq I_1 \right\}.$$

Proof. The case $N = 0$ is of course trivial.[124] Obviously $d|d_1$. Assume that $N > 0$ and that n is odd. Our basic tool is equation (5.15). The left hand side of this equation is a power series in x of the form $\sum_{m=I_1}^{\infty} a_m x^m$ with $a_m \in \mathbb{Z}$; the right hand side is a power series in x of the form $\sum_{m=I_1}^{\infty} b_m x^m$ with $b_m \in \mathbb{C}$ depending on the undermined constants c_i. Obviously $b_{I_1} = c_{I_1} = a_{I_1} \in \mathbb{Z}$. For $I_1 < r \leq I_2$, the coefficient of x^r in the right hand side of (5.15) is of the form

$$c_r + \sum_{i=I_1}^{r-1} \alpha_i c_i,\ \alpha_i \in \mathbb{Z}.$$

[124] It is obvious that there exists a finite indexing set J such that

$$d = \gcd\left\{ P_{\alpha(k)N}\left(k^n m - \alpha(k)N\frac{k^{2\lfloor\frac{n+1}{2}\rfloor} - 1}{24}\right); m \in J \right\}.$$

The theorem identifies J and gives an alternate formula for d.

Since it equals a_r, we conclude by induction that $c_r \in \mathbb{Z}$. Our basic identity (5.15) can be rewritten as

$$\sum_{m=\left\lceil \frac{\alpha(k)N}{24}\left(k-\frac{1}{k^n}\right)\right\rceil}^{\infty} P_{\alpha(k)N}\left(k^n m - \frac{\alpha(k)N(k^{n+1}-1)}{24}\right) x^m$$

$$= \prod_{m=1}^{\infty}(1-x^{km})^{-\alpha(k)N} \sum_{i=\left\lceil \frac{\alpha(k)N}{24}\left(k-\frac{1}{k^n}\right)\right\rceil}^{\frac{\alpha(k)N}{24}(k^{n+1}-1)} c_i \left[x \frac{\prod_{m=1}^{\infty}(1-x^{km})^{\frac{24}{k-1}}}{\prod_{n=1}^{\infty}(1-x^m)^{\frac{24}{k-1}}}\right]^i,$$

which implies that $d|c_r$ for $I_1 \le r \le I_2$ and thus $d|c$. The original form of the above equation tells us that $c|P_{\alpha(k)N}\left(k^n m - \frac{\alpha(k)N(k^{n+1}-1)}{24}\right)$ for all $m \ge I_1$ and thus $c|d$. We conclude that $c = d$. It remains to show that $c = d_1$. We need to be a bit more explicit in our analysis.

The constants c_i in our basic power series identity (5.15) are obtained from

$$\sum_{m=I_1}^{I_2} a_m x^m = \sum_{i=I_1}^{I_2} c_i Q_i(x),$$

where:

1. The constants a_m are integral linear combinations of the partition coefficients

$$\left\{P_{\alpha(k)N}\left(k^n m' - \frac{\alpha(k)N(k^{n+1}-1)}{24}\right);\ m' = I_1,\ I_1+1,\ \ldots,\ I_2\right\}$$

and hence

$$d_1 | a_m.$$

2. (We do not need this equation for the moment; however, it facilitates calculations of examples.)

$$a_m = P_{\alpha(k)N}\left(k^n m - \frac{\alpha(k)N(k^{n+1}-1)}{24}\right) \text{ for } I_1 \le m < k.$$

3. The polynomial Q_i is of the form

$$Q_i(x) = x^i + \sum_{j=i+1}^{I_2} \alpha_{ij} x^j,\ \alpha_{ij} \in \mathbb{Z}.$$

4.

$$Q_i(x) = x^i \sum_{m=0}^{I_2-i} P_{\frac{24i}{k-1}}(m) x^m \text{ if } I_2 < k.$$

The equations to be solved (for the c_i) are

$a_{I_1} = c_{I_1},\ a_{I_1+1} = c_{I_1+1} + \beta_{2,1} c_{I_1},\ \ldots,\ a_{I_2} = c_{I_2} + \beta_{N,I_2-1} c_{I_2-1} + \ldots + \beta_{N,1} c_{I_1},$

with $\beta_{ij} \in \mathbb{Z}$. It follows by induction that $c_i \in \mathbb{Z}$ and that

$$d_1|c_i,\ i = I_1,\ I_1+1,\ \ldots,\ I_2.$$

Thus $d_1|c$. Since $c = d|d_1$, we have established the equality of the three integers c, d_1 and d.

The modifications required for even n are left to the reader; equation (5.16) replaces equation (5.15), of course. The argument for negative N is similar (here we consider only the case $n = 1$) to the one given above. We outline the key steps. For convenience, we work with $-N$ for $N \in \mathbb{Z}^+$. In this case $I_1 = -\left\lfloor \frac{\beta(k)N}{k} \right\rfloor$ and $I_2 = 0$. The basic identity from equation (5.17) is now

$$\sum_{m=I_1}^{0} a_m x^m = \sum_{i=0}^{-I_1} c_i Q_i(x),$$

where:
1. The constants a_m are integral linear combinations of the partition coefficients

$$\left\{P_{-\alpha(k)N}(km' + \beta(k)N);\ m' = -\left\lfloor \frac{\beta(k)N}{k} \right\rfloor,\ -\left\lfloor \frac{\beta(k)N}{k} \right\rfloor + 1,\ \ldots,\ 0\right\}$$

and hence (as before) $d_1|a_m$.
2.

$$a_m = P_{-\alpha(k)N}(km + \beta(k)N),\ \text{if } -m < k.$$

3. The polynomial Q_i (in $\frac{1}{x}$) is of the form

$$Q_i(x) = x^{-i} + \sum_{j=-i+1}^{0} \alpha_{ij} x^j,\ \alpha_{ij} \in \mathbb{Z}.$$

Thus the equations to be solved are

$$a_{I_1} = c_{-I_1},\ a_{I_1+1} = c_{-I_1-1} + \beta_{-I_1-1,-I_1} c_{-I_1},\ \ldots,$$

$$a_0 = c_0 + \beta_{0,1}\, c_1 + \ \ldots\ + \beta_{0,-I_1} c_{-I_1},$$

with $\beta_{ij} \in \mathbb{Z}$. It follows by (backwards) induction that $d|c_i$, $i = 0,\ 1,\ \ldots,\ -I_1$.

Remark 9.2. 1. It is obvious that

$$d(k,1,N) = \ \gcd\left\{P_{\alpha(k)N}\left(km + R\left(\frac{-\beta(k)N}{k}\right)\right); m \in \mathbb{Z}^+ \cup \{0\}\right\}.$$

The last theorem tells us that (in the most interesting cases),

$$d(k,1,N) = \gcd\left\{P_{\alpha(k)N}(R),\ P_{\alpha(k)N}(R+k),\ \ldots,\ P_{\alpha(k)N}((k-1)\beta(k)N)\right\},$$

k	N	$d(k,1,N)$
2	N odd, $-5 \le N \le 95$	2^3
3	$N \equiv 1 \mod 3$, $-8 \le N \le 94$	3^2
3	$N \equiv 2 \mod 3$, $-16 \le N \le 95$	3
5	-14	5^6
5	-8	5^3
5	11, 17, 36, 42, 61, 67, 86, 92	5^2
5	-48, -24, -23, -19, -18, -13, -9, -4, -3, 1, 2, 6, 7, 12, 16, 21, 22, 26, 27, 31, 32, 37, 41, 46, 47, 51, 52, 56, 57, 62, 66, 71, 72, 76, 77, 81, 82, 87, 91, 92	5
7	-10	7^4
7	-6, 39, 43	7^2
7	-48, -24, -20, -17, -13, -3 1, 4, 8, 11, 15, 18, 22, 25, 29, 32, 36, 46, 50, 53, 57, 60, 64, 67, 71, 74, 78, 81, 85	7
11	-4, -3	11
11	-6	11^2
11	-8	11^3
11	-10	11^4
13	-16, -3, 10	13
13	-15	5

Table 20. SUMMARY OF CALCULATIONS: GENERATORS OF THE IDEALS OF PARTITION COEFFICIENTS.

where $R = \left(R\left(\frac{-\beta(k)N}{k}\right)\right)$, if $N > 0$ (this involves the evaluation of $\beta(k)N + 1 + \left\lfloor -\frac{\beta(k)N}{k} \right\rfloor$ partition constants), and

$$d(k,1,N) = \gcd\left\{P_{\alpha(k)N}(R),\ P_{\alpha(k)N}(R+k),\ ...,\ P_{\alpha(k)N}(-\beta(k)N)\right\}$$

if $N < 0$ (this involves the evaluation of $1 + \left\lfloor -\frac{\beta(k)N}{k} \right\rfloor$ partition constants).
2. If $N \in \mathbb{Z} - \{0\}$ is a multiple of k, then $d(k,1,N) = 1$ because $P_{\alpha(k)N}(0) = 1$.

Computations (for $n = 1$ and small $|N|$), using MATHEMATICA, of $d(k,1,N)$ based on the last theorem are summarized in Table 20.

We have excluded from this table the many calculations that result in a common divisor $d = 1$. These include:
$k = 2$, $-6 \le N \le 96$ and N even,
$k = 3$, $-6 \le N \le 96$ and $N \equiv 0 \mod 3$,

$k = 5$, $-22 \leq N \leq 95$ and not in above table,
$k = 7$, $-23 \leq N \leq 87$ and not in above table, and
$k = 11$, $N = -1$ and -2,
$k = 13$, $N = -48$ and $-24 \leq N \leq 15$, N not in above table.
Table 20 contains only entries for which calculations were performed. We did NOT take advantage of Proposition 6.1 in compiling the table. That proposition implies that for $k \geq 5$, if k divides $d(k, N)$, then k also divides $d(k, N + k)$. Thus whenever N appears in our table, so do $N \pm k$ (provided they are within our calculation limits). The results for $k = 11$ are consequences of our work in §6, not the last theorem.

Remark 9.3. The calculations summarized above include the proof of the Ramanujan level one congruences for the primes 2, 3, 5 and 7. One of the surprises of these calculations is that $d(13, 1, -15) = 5$.

The calculations and the modular equations suggest the following theorem. It is a consequence of the modular equations.

Theorem 9.4. *Let $N \in \mathbb{Z}$.*
(a) $d(2, 1, N) = 1$ *for* $N \equiv 0 \mod 2$ *and*
$\pi_2(d(2, 1, N)) = 3$ *for* $N \equiv 1 \mod 2$.
(b) $d(3, 1, N) = 1$ *for* $N \equiv 0 \mod 3$ *and*
$$\pi_3(d(3,1,N)) = \begin{cases} 2 \text{ for } N \equiv 1 \mod 3 \\ 1 \text{ for } N \equiv 2 \mod 3 \end{cases}.$$
(c) $d(5, 1, N) = 1$ *for* $N \equiv 0,\ 3,\ 4 \mod 5$ *and*
$5|d(5, 1, N)$ *for* $N \equiv 1,\ 2 \mod 5$.
(d) $d(7, 1, N) = 1$ *for* $N \equiv 0,\ 2,\ 3,\ 5,\ 6 \mod 7$ *and*
$7|d(7, 1, N)$ *for* $N \equiv 1,\ 4 \mod 7$.

Proof. We consider only the case $k = 5$; the other cases are left as exercises for the reader. The calculations show that $d(5, 1, 0) = 1 = d(5, 1, 3) = d(5, 1, 4)$ and $d(5, 1, 1) = 5 = d(5, 1, 2)$. Equations (5.34) and (5.35) say that for all $N \in \mathbb{Z}$, $c(5, 1, N + 5)$ divides the gcd of

$$\{5^2c(5,1,N+4),\ 5^2c(5,1,N+3),\ 5c(5,1,N+2),\ 5c(5,1,N+1),\ c(5,1,N)\}$$

and $c(5, 1, N - 5)$ divides the gcd of

$$\{5c(5,1,N-4),\ 5c(5,1,N-3),\ 5^2c(5,1,N-2),\ 5^2c(5,1,N-1),\ c(5,1,N)\}.$$

We know that $c(5, 1, N) = 1$ for $N = 0$, 3 and 4. The last two displayed equations tell us that if $c(5, 1, N) = 1$, then also $c(5, 1, N \pm 5) = 1$, and if $5|c(5, 1, N)$, then also $5|c(5, 1, N \pm 5)$. The results of the calculations in Table 20,$c(5, 1, N) = 5$ for $N = 1$ and 2, finish the argument for the case we considered.

Remark 9.5. The modular equations alone are insufficient to compute $d(k,1,N)$. We have seen that they imply that $\pi_5(d(5,1,11)) \geq 1$, but our calculations showed that $d(5,1,11) = 5^2$.

10. Examples: Calculations for small k

We assemble in this section the results of various calculations of $F_{k,N}$ for small values of k and $|N|$. We emphasize theta identities, their power series versions and the resulting partition coefficient congruences. We are mostly exploiting Theorem 4.7 and its corollaries. We make, however, several digressions to discuss related questions (usually general properties suggested by a particular example).

10.1. $k = 2$. The equation $F_{2,1} = 2^3 f_2$ is equivalent to the theta constant identity

$$\left(\frac{\eta(2\tau)}{\eta(\frac{\tau}{2})}\right)^8 + \left(\frac{\eta(2(\tau+1))}{\eta(\frac{\tau+1}{2})}\right)^8 = 2^4 \left(\frac{\eta(2\tau)}{\eta(\tau)}\right)^{24}.$$

Its power series version is

$$\sum_{m=1}^{\infty} P_8(2m-1)x^m = 2^3 x \frac{\prod_{n=1}^{\infty}(1-x^{2n})^{16}}{\prod_{n=1}^{\infty}(1-x^n)^{24}}; \tag{5.37}$$

the case $k = 2$ of Corollary 5.3, and is algebraically equivalent to the Jacobi quartic. It yields the level one partition congruence for the prime 2.

As a consequence of our work in §9, we are able to draw additional consequences from Corollary 5.3 for the first three primes by taking a small detour.

Theorem 10.1 (The level two congruences for the primes 2, 3 and 5). *For each* $m \in \mathbb{Z}^+$,

$$P_8(2^2 m - 1) \equiv 0 \mod 2^6,$$
$$P_3(3^2 m - 1) \equiv 0 \mod 3^4$$

and

$$P(5^2 m - 1) \equiv 0 \mod 5^2,$$

or more compactly

$$P_{\alpha(k)}(k^2 m - 1) \equiv 0 \mod P^2_{\alpha(k)}(k-1), \ k = 2,\ 3,\ 5.$$

Proof. Let $k = 2$, 3 or 5. If $m \equiv 1 \mod 2$, then $P_{24}(m) \equiv 0 \mod 2^3$ because $d(2,1,3) = 2^3$. Similarly, if $m \equiv 2 \mod 3$ ($m \equiv 4 \mod 5$), then $P_{12}(m) \equiv 0 \mod 3^2$ ($P_6(m) \equiv 0 \mod 5$) because $d(3,1,4) = 3^2$ ($d(5,1,6) = 5$). The result now follows directly from the formulae contained in Corollary 5.3.

The results for $F_{2,2}$ are not so elegant. The theta identity obtained,

$$\left(\frac{\eta(2\tau)}{\eta(\frac{\tau}{2})}\right)^{16} + \left(\frac{\eta(2(\tau+1))}{\eta(\frac{\tau+1}{2})}\right)^{16} = 2\left(\frac{\eta(2\tau)}{\eta(\tau)}\right)^{24} + 2^8\left(\frac{\eta(2\tau)}{\eta(\tau)}\right)^{48},$$

translates to power series identity

$$\prod_{n=1}^{\infty}(1-x^n)^{16}\sum_{m=1}^{\infty}P_{16}(2m-2)x^m = x\prod_{n=1}^{\infty}(1+x^n)^8 + 2^7x^2\prod_{n=1}^{\infty}(1+x^n)^{32},$$

an equality that does not seem to have any obvious congruence implications, but is equivalent to the square of the Jacobi quartic.

The general power series formula for $F_{2,-N}$ with $N \in \mathbb{Z}^+$ reads

$$\prod_{n=1}^{\infty}(1-x^{2n})^{-8N}\sum_{m=-\lfloor\frac{N}{2}\rfloor}^{\infty}P_{-8N}(2m+N)x^m \tag{5.38}$$

$$= \sum_{i=0}^{\lfloor\frac{N}{2}\rfloor} c_i\left[x^{-1}\prod_{n=1}^{\infty}\left(\frac{1-x^{2n}}{1-x^n}\right)^{-24}\right]^i.$$

For $N = 3$, it gives us the theta identity

$$\frac{1}{2}\sum_{l=0}^{1}\left(\frac{\eta(\frac{\tau+l}{2})}{\eta(2(\tau+l))}\right)^{24} = -2^3\left[3\left(\frac{\eta(\tau)}{\eta(2\tau)}\right)^{24} + 2^8\right]$$

and power series identity (the case $k = 2$ of Mordell's theorem)

$$\prod_{n=1}^{\infty}(1-x^{2n})^{-24}\sum_{m=-1}^{\infty}P_{-24}(2m+3)x^m = -2^3\left[3x^{-1}\prod_{n=1}^{\infty}(1+x^n)^{-24} + 2^8\right].$$

Remark 10.2. This last power series identity should be compared with the case $k = 2$ of the last corollary prior to Section 7; it is algebraically equivalent to cubing the Jacobi quartic.

The general theory discussed above was based on the fact that $\left(\frac{\eta(2\tau)}{\eta(\frac{\tau}{2})}\right)^8$ is a covering map for $\mathbb{H}^2/\Gamma(2) = \mathbb{H}^2/\Gamma(2,2)$ with divisor $\frac{P_\infty}{P_0}$. Another such covering map is given by $\dfrac{\theta^4\begin{bmatrix}1\\0\end{bmatrix}}{\theta^4\begin{bmatrix}0\\1\end{bmatrix}}$. Thus

$$2^3\left(\frac{\theta'\begin{bmatrix}1\\1\end{bmatrix}(0,2\tau)}{\theta'\begin{bmatrix}1\\1\end{bmatrix}(0,\frac{\tau}{2})}\right)^2 = \left(\frac{\theta\begin{bmatrix}1\\0\end{bmatrix}(0,\tau)}{\theta\begin{bmatrix}0\\1\end{bmatrix}(0,\tau)}\right)^3,$$

which is equivalent to the obvious identity

$$\prod_{n=1}^{\infty}(1+x^n) = \prod_{n=1}^{\infty}\frac{1}{1-x^{2n-1}}.$$

We can draw some more consequences of Theorem 4.5 (for $k = 2$ and negative N). Since for all $N \in \mathbb{Z}$, $P_{-8N}(1) = -8N$ and $P_{-8N}(0) = 1$, we conclude that

$$\text{(5.39)} \qquad \operatorname{ord}_{P_\infty} F_{2,-N} = -\left\lfloor \frac{N}{2} \right\rfloor, \text{ for all } N \in \mathbb{Z},$$

and that the functions $\{F_{2,-2N};\ N \in \mathbb{Z}^+ \cup \{0\}\}$ can be used as a basis for the space of meromorphic functions on $\overline{\mathbb{H}^2/\Gamma_o(2)}$ with poles only at P_∞. In particular, for each $N \in \mathbb{Z}^+ \cup \{0\}$,

$$F_{2,-(2N+1)} = \sum_{i=0}^{N} c_i F_{2,-2i}.$$

For $N = 0$ we have $F_{2,-1} = 1$; the expansion of F using equation (5.13) is a special case of a previous result. Going back to the basic definitions, we are led to the theta identity

$$\eta^8\left(\frac{\tau}{2}\right) + \eta^8\left(\frac{\tau+1}{2}\right) = -2^4\eta^8(2\tau),$$

equivalently (once again the Jacobi quartic) to the infinite product identity (here $x = \exp\{\pi\imath\tau\}$)

$$\text{(5.40)} \qquad \prod_{n=0}^{\infty}(1-x^{2n+1})^8 - \prod_{n=0}^{\infty}(1+x^{2n+1})^8 = -2^4\ x \prod_{n=1}^{\infty}(1+x^{2n})^8.$$

When combined with the formula in Theorem 4.6 of Chapter 4, we obtain from the last identity

$$\eta^8\left(\frac{\tau}{2}\right)\eta^8\left(\frac{\tau+1}{2}\right)\left[\eta^8\left(\frac{\tau}{2}\right) + \eta^8\left(\frac{\tau+1}{2}\right)\right] = 2^4\eta^{24}(\tau),$$

which translates to the infinite product identity equivalent to equation (5.40). The case $N = 1$ ($2N + 1 = 3$ and $F_{2-3} = c_0 + c_1 F_{2,-2}$) tells us that

$$\sum_{m=0}^{\infty} P_{-24}(2m+1)x^{m-1} = 2^{10}\sum_{m=0}^{\infty} P_{-24}(m)x^{2m}$$

$$-2^3 3\sum_{m=0}^{\infty} P_{-8}(m)x^{2m}\sum_{m=0}^{\infty} P_{-16}(2m)x^{m-1}$$

and the recursion formula

$$P_{-24}(2m+3) = 2^{10} P_{-24}\left(\frac{m}{2}\right) - 2^3 3 \sum_{j=0}^{\lfloor \frac{m+1}{2} \rfloor} P_{-8}(j) P_{-16}(2m+2-4j).$$

If we use instead the definitions of the terms involved in this equation, we get the identity

$$\prod_{n=0}^{\infty}(1-x^{2n+1})^{24} - \prod_{n=0}^{\infty}(1+x^{2n+1})^{24} = 2^{11}\ x^3 \prod_{n=1}^{\infty}(1+x^{2n})^{24}$$

$$-2^3 3\ x \prod_{n=1}^{\infty}(1+x^{2n})^8 \left[\prod_{n=0}^{\infty}(1-x^{2n+1})^{16} + \prod_{n=0}^{\infty}(1+x^{2n+1})^{16}\right].$$

When combined with (5.40), it yields

$$\prod_{n=0}^{\infty}(1-x^{2n+1})^{24} - \prod_{n=0}^{\infty}(1-x^{2n+1})^{16}(1+x^{2n+1})^8$$

$$+\prod_{n=0}^{\infty}(1-x^{2n+1})^8(1+x^{2n+1})^{16} - \prod_{n=0}^{\infty}(1+x^{2n+1})^{24}$$

$$= -2^4\ x \prod_{n=1}^{\infty}(1+x^{2n})^8 \left[\prod_{n=0}^{\infty}(1-x^{2n+1})^{16} + \prod_{n=0}^{\infty}(1+x^{2n+1})^{16}\right].$$

A systematic exploration of the consequences of (5.39) can be based on the equations

$$F_{2,-2N} = \sum_{i=0}^{N} c_i f_2^{-i}, \quad N \in \mathbb{Z}^+.$$

Direct calculation of undetermined constants show that $F_{2,-2} = 2^7 + f_2^{-1}$. We can base our calculations on the modular equation (5.30). They immediately imply (using our calculations for $F_{2,0}$ and $F_{2,1}$) that $F_{2,-1} = -2^3$ and as a second application, the formula for $F_{2,-2}$.

10.2. $k = 3$. The basic identity $F_{3,1} = 3^2 f_3$ tells us that

$$\frac{1}{3}\sum_{l=0}^{2}\left(\frac{\eta(3(\tau+l))}{\eta(\frac{\tau+l}{3})}\right)^3 = 3^2\ \left(\frac{\eta(3\tau)}{\eta(\tau)}\right)^{12}.$$

Its power series version is

$$\sum_{m=1}^{\infty} P_3(3m-1)x^m = 3^2\ x \frac{\prod_{n=1}^{\infty}(1-x^{3n})^9}{\prod_{n=1}^{\infty}(1-x^n)^{12}}, \tag{5.41}$$

which implies immediately $P_3(3m-1)$ is congruent to zero mod 3^2.

That $P_3(3m \pm 1) \equiv 0 \mod 3$, for all $m \in \mathbb{Z}^+ \cup \{0\}$, is trivial. It is a consequence of the following general

Proposition 10.3. *If $N \in \mathbb{Z}$ is prime, then for all $n \in \mathbb{Z}^+ \cup \{0\}$,*

$$P_N(n) \equiv P_{\frac{N}{|N|}}\left(\frac{n}{|N|}\right) \mod |N|;$$

in particular,

$$P_N(n) \equiv 0 \mod |N| \text{ if } n \not\equiv 0 \mod |N|$$

and

$$P_N(n|N|) \equiv P_{\frac{N}{|N|}}(n) \mod |N|.$$

Proof. From the definitions, using the Frobenius map,

$$\sum_{n=0}^{\infty} P_N(n)x^n = \prod_{n=1}^{\infty}(1-x^n)^{-N} = \prod_{n=1}^{\infty}\left((1-x^n)^{|N|}\right)^{-\frac{N}{|N|}} \equiv \prod_{n=1}^{\infty}(1-x^{n|N|})^{-\frac{N}{|N|}}$$

$$= \sum_{n=0}^{\infty} P_{\frac{N}{|N|}}(n)x^{n|N|} \mod |N|.$$

Remark 10.4. The same argument establishes that for all (positive) primes k and all integers N,

$$P_{kN}(n) \equiv P_N\left(\frac{n}{k}\right) \mod k, \ n \in \mathbb{Z}^+ \cup \{0\}.$$

It is perhaps interesting to observe that the congruence $P_3(3m-1) \equiv 0 \mod 9$ also has another proof. It is clear from Jacobi's identity that

$$\sum_{n=0}^{\infty} P_3(n)x^n \sum_{n=0}^{\infty}(-1)^n(2n+1)x^{\frac{n(n+1)}{2}} = 1,$$

and as a consequence that for each $n \geq 1$ we have[125]

$$\sum_{m=0}^{\infty}(-1)^m(2m+1)P_3\left(n - \frac{m(m+1)}{2}\right) = 0.$$

We therefore have the following recursion relation:

$$P_3(n) = -\sum_{m=1}^{\infty}(-1)^m(2m+1)P_3\left(n - \frac{m(m+1)}{2}\right).$$

[125] All the sums are finite

It is obvious that $\frac{m(m+1)}{2}$ is always congruent to either 0 (if $m \not\equiv 1 \mod 3$) or 1 (if $m \equiv 1 \mod 3$) mod 3 so that $\left(3n - 1 - \frac{m(m+1)}{2}\right)$ is always congruent to either 1 or 2 mod 3. Therefore

$$P_3\left(3n - 1 - \frac{m(m+1)}{2}\right) \equiv 0 \mod 3.$$

From the last recursion formula,

$$P_3(3n-1) = -\sum_{m=1}^{\infty}(-1)^m(2m+1)P_3\left(3n - 1 - \frac{m(m+1)}{2}\right). \tag{5.42}$$

We use this last identity to prove by induction that $P_3(3n-1) \equiv 0 \mod 9$. This claim is certainly valid for $n = 1$ (because $P_3(2) = 9$). So assume that the assertion holds for all $n \leq l$, $l \in \mathbb{Z}^+$. Therefore from (5.42),

$$\begin{aligned} P_3\left(3(l+1) - \frac{m(m+1)}{2}\right) \\ = -\sum_{m=1}^{\infty}(-1)^m(2m+1)P_3\left(3l + 2 - \frac{m(m+1)}{2}\right). \end{aligned} \tag{5.43}$$

However, $\left(3l + 2 - \frac{m(m+1)}{2}\right)$ is congruent to either 2 or 0 mod 3. In the first case,

$$P_3\left(3l + 2 - \frac{m(m+1)}{2}\right) \equiv 0 \mod 9$$

by the induction hypothesis (since $\left(3l + 2 - \frac{m(m+1)}{2}\right) \leq 3l$). In the second case,

$$P_3\left(3l + 2 - \frac{m(m+1)}{2}\right) \equiv 0 \mod 3$$

only. But this case occurs if and only if m is congruent to 1 mod 3, that is, when $2m + 1$ is congruent to 0 mod 3. Hence the induction argument is completed by (5.43).

The formula $F_{3,-8} = 3^2[2^7 f_3^{-2} - 3^9]$ is equivalent to the case $k = 3$ of Mordell's theorem.

From the observation

$$\operatorname{ord}_{P_\infty} F_{3,-N} = -\left\lfloor \frac{N}{3} \right\rfloor,$$

we conclude (for example) that

$$F_{3,-2} = cF_{3,-1} \text{ and } F_{3,-5} = c_0 + c_1 F_{3,-4}.$$

These two identities translate to

$$\sum_{m=0}^{\infty} P_{-6}(3m+2)x^m = -3\sum_{m=0}^{\infty} P_{-3}(m)x^{3m} \sum_{m=0}^{\infty} P_{-3}(3m+1)x^m$$

and

$$2\sum_{m=0}^{\infty} P_{-15}(3m+2)x^m$$

$$= 3^6 x \sum_{m=0}^{\infty} P_{-15}(m)x^{3m} - 3\cdot 5 \sum_{m=0}^{\infty} P_{-3}(m)x^{3m} \sum_{m=0}^{\infty} P_{-12}(3m+1)x^m,$$

respectively. The first of these has

$$P_{-6}(3m+2) = -3 \sum_{j=0}^{\lfloor \frac{1}{9}+\frac{m}{3} \rfloor} P_{-3}(j)P_{-3}(m-j)$$

as a consequence, a formula that should be contrasted with

$$P_{-6}(m) = \sum_{j=0}^{m} P_{-3}(j)P_{-3}(m-j)$$

obtained from the usual Cauchy product.

A formula involving P_{-24} is obtained from the relation

$$F_{3,-8} = c_0 + c_1 F_{3,-4} + c_2 F_{3,-7}.$$

10.3. $k = 5$. The formula $F_{5,1} = 5f_5$ has the formal power series identity given by the elegant equation (5.2) as a consequence. That equation tells us that $P(5m-1)$ is congruent to zero mod 5, yielding perhaps the most illuminating proof of the level one congruence for the prime five.

The power series version of the identity for $F_{5,2}$ is

$$\sum_{m=1}^{\infty} P_2(5m-2)x^m = 5x\frac{\prod_{n=1}^{\infty}(1-x^{5n})^4}{\prod_{n=1}^{\infty}(1-x^n)^6}\left[5^2 x\frac{\prod_{n=1}^{\infty}(1-x^{5n})^6}{\prod_{n=1}^{\infty}(1-x^n)^6}+2\right].$$

It implies the previously established fact in Theorem 6.8 that $P_2(5m+3)$ is congruent to zero mod 5 and the recursion relation

$$P_2(5m+3) = 2\cdot 5\sum_{j=0}^{\lfloor \frac{m}{5} \rfloor} P_{-4}(j)P_6(m-5j) + 5^3 \sum_{j=0}^{\lfloor \frac{m-1}{5} \rfloor} P_{-10}(j)P_{12}(m-1-5j).$$

When combined with the calculations of Theorem 9.1 (using only the equality $d(5,1,6) = 5$) this yields the Ramanujan level two congruences for the prime 5 for the function P_2. The formula for $F_{5,3}$ is

$$\prod_{n=1}^{\infty}(1-x^{5n})^3 \sum_{m=1}^{\infty} P_3(5m-3)x^m = \sum_{i=1}^{3} c_i x^i \frac{\prod_{n=1}^{\infty}(1-x^{5n})^{6i}}{\prod_{n=1}^{\infty}(1-x^n)^{6i}},$$

with

$$c_1 = 3^2,\ c_2 = 3\cdot 5^3 \text{ and } c_3 = 5^5.$$

It implies

Corollary 10.5. *We have the formal relations*

$$\sum_{m=1}^{\infty} P_3(5m-3)x^m \equiv 3^2 x \prod_{n=1}^{\infty}(1-x^n)^9 = 3^2 \sum_{m=0}^{\infty} P_{-9}(m)x^{m+1} \mod 5$$

and hence for all $m \in \mathbb{Z}^+ \cup \{0\}$,

$$P_3(5m+2) \equiv 3^2 P_{-9}(m) \mod 5.$$

The investigation of the situation for $F_{5,-5}$ leads to the interesting identity

$$\sum_{m=-1}^{\infty} P_{-5}(5m+5)x^m = x^{-1}\prod_{n=1}^{\infty}\frac{(1-x^n)^6}{1-x^{5n}}$$

and hence (using Frobenius) that

$$P_{-5}(5m) \equiv P_{-1}(m) \mod 5, \text{ for all } m \in \mathbb{Z}^+ \cup \{0\},$$

a special case of Proposition 10.3. A study of $F_{5,-24}$ concludes that the partition coefficient $P_{-24}(5m+4)$ is congruent to zero mod 5, a fact established in Theorem 6.8 that also follows from the multiplicative properties of T.

We now turn to the material of §4.4. We consider the function $G_{5,1}$. We know that for some constant c,

$$G_{5,1} = c f_5^{-1}.$$

We evaluate c to be 1, which leads us to the power series identity

$$\sum_{j=0}^{\infty} P(j)x^{5^2 j} - \sqrt{5}x \sum_{m=1}^{4} c_m \left(\sum_{j=0}^{\infty}\left(\exp\left\{\frac{2\pi \imath m j}{5}\right\}\right) P(j)x^j\right)$$
$$= \frac{\prod_{n=1}^{\infty}(1-x^n)^5}{\prod_{n=1}^{\infty}(1-x^{5n})^6},$$

where[126]

$$c_m = \left(\frac{-m}{5}\right)\left(\exp 2\pi\imath\left\{\frac{m}{5}\right\}\right) = \begin{cases} \exp\frac{2\pi\imath}{5} & \text{for } m=1 \\ \exp\frac{-\pi\imath}{5} & \text{for } m=2 \\ \exp\frac{\pi\imath}{5} & \text{for } m=3 \\ \exp\frac{-2\pi\imath}{5} & \text{for } m=4 \end{cases}.$$

We hence have a new type of identity:

$$\sum_{j=0}^{\infty} P(j)x^{5^2 j} - 5x\left(\sum_{j\in\mathbb{Z}^+\cup\{0\},\ j\equiv 0,3 \mod 5} P(j)x^j - \sum_{j\in\mathbb{Z}^+,\ j\equiv 1,2 \mod 5} P(j)x^j\right)$$
$$= \sum_{n=0}^{\infty} P_{-5}(j)x^j \sum_{n=0}^{\infty} P_6(j)x^{5j}.$$

---

[126]Recall that $(\frac{\cdot}{\cdot})$ is the Legendre symbol.

In particular,

$$\sum_{j=0}^{\lfloor \frac{n}{5} \rfloor} P_6(j)P_{-5}(n-5j) = \begin{cases} P\left(\frac{n}{5^2}\right) \text{ for } n \equiv 0 \mod 5 \\ P\left(\frac{n}{5^2}\right) - 5P(n-1) \text{ for } n \equiv 1,\ 4 \mod 5 \\ P\left(\frac{n}{5^2}\right) + 5P(n-1) \text{ for } n \equiv 2,\ 3 \mod 5 \end{cases}.$$

The fact that $F_{5,1}$ is a constant multiple of $G_{5,-5}$ (both functions vanish at P_∞) has the more interesting consequence

$$\prod_{n=1}^{\infty}(1-x^{5n}) \sum_{m=0}^{\infty} P(5m+4)x^{m+1} = \prod_{n=1}^{\infty}(1-x^n)^{-5}$$

$$\times \begin{bmatrix} 5^2\ x^5\ \sum_{j=0}^{\infty} P_{-5}(j)x^{5^2 j} \\ +\sum_{j\in\mathbb{Z}^+,\ j\equiv 2,3 \mod 5} P_{-5}(j)x^j \\ -\sum_{j\in\mathbb{Z}^+,\ j\equiv 1,4 \mod 5} P_{-5}(j)x^j \end{bmatrix}.$$

When combined with Proposition 10.3, this last identity produces once again the level one congruence for the prime 5.

10.4. $k = 7$. This is the first prime k for which $\beta(k) > 1$. As a result the formulae look different. The theta identity for $F_{7,1}$ is

$$\frac{1}{7}\sum_{l=0}^{6} \frac{\eta(7(\tau+l))}{\eta(\frac{\tau+l}{7})} = 7\left(\frac{\eta(7\tau)}{\eta(\tau)}\right)^4 \left(7\left(\frac{\eta(7\tau)}{\eta(\tau)}\right)^4 + 1\right)$$

whose power series version is equation (5.3). That equation implies that the partition coefficient $P(7m-2)$ is congruent to zero mod 7, the Ramanujan level one congruence for prime 7.

The calculations based on Theorem 9.1 suggest that the first interesting formula for $F_{7,N}$, $N > 1$ occurs for $N = 4$. Calculations lead to

$$\begin{aligned} \sum_{m=0}^{\infty} P_4(7m+6)x^m &= 2\cdot 7\cdot 41 \sum_{m=0}^{\infty} P_{-4}(m)x^{7m} \sum_{m=0}^{\infty} P_8(m)x^m \\ &+2^4 7^3 11x \sum_{m=0}^{\infty} P_{-8}(m)x^{7m} \sum_{m=0}^{\infty} P_{12}(m)x^m \\ &+5\cdot 7^4 13^2 x^2 \sum_{m=0}^{\infty} P_{-12}(m)x^{7m} \sum_{m=0}^{\infty} P_{16}(m)x^m \\ &+2^4 7^6 17x^3 \sum_{m=0}^{\infty} P_{-16}(m)x^{7m} \sum_{m=0}^{\infty} P_{20}(m)x^m \\ &+2\cdot 7^8 23x^4 \sum_{m=0}^{\infty} P_{-20}(m)x^{7m} \sum_{m=0}^{\infty} P_{24}(m)x^m \\ &+2^2 7^{10} x^5 \sum_{m=0}^{\infty} P_{-24}(m)x^{7m} \sum_{m=0}^{\infty} P_{28}(m)x^m \\ &+7^{11} x^6 \sum_{m=0}^{\infty} P_{-28}(m)x^{7m} \sum_{m=0}^{\infty} P_{32}(m)x^m \end{aligned},$$

which contains the congruences

$$P_4(7m+6) \equiv 0 \mod 7 \text{ and } P_4(7^2 m + 41) \equiv 0 \mod 7^2,$$

the Ramanujan level one and two congruences for the prime 7 for the function P_4. The level one congruence is an immediate consequence of the last complicated equality since each term of the right hand side of the equation is divisible by 7. To obtain the level two congruence we use the consequence of the Frobenius operator: $P_8(7j+5) \equiv 0 \mod 7$.

N	$\mathrm{ord}_\infty F_{7,-N}$	c
1	0	$P_{-1}(2) = -1$
2	0	$P_{-2}(4) = 1$
3	0	$P_{-3}(6) = -7$
4	-1	$P_{-4}(8) = 3^2$
5	-1	$P_{-5}(3) = 2 \cdot 5$
6	0	$P_{-6}(12) = 7^2$
7	-2	$P_{-7}(0) = 1$
8	-2	$P_{-8}(2) = 2^2 \cdot 5$
9	-2	$P_{-9}(4) = -2 \cdot 3^2 \cdot 5$
10	0	$P_{-10}(20) = 2^2 \cdot 5 \cdot 7 \cdot 13$
11	-3	$P_{-11}(1) = -11$
12	-3	$11P_{-12}(4) = -3^2 11$
24	-6	$P_{-24}(6) = -2^3 \cdot 7 \cdot 13 \cdot 23$

Table 21. CALCULATIONS OF $\alpha = \mathrm{ord}_\infty F_{7,-N}$ AND THE COEFFICIENT c OF LEADING TERM x^α OF LAURENT SERIES EXPANSION OF $F_{7,-N}$ AT P_∞.

Based on the calculations summarized Table 21showing the orders of the poles of some functions $F_{7,-N}$ (with N restricted to positive integers), it appears useful to exploit the equation

$$F_{7,1} = d_2 G_{7,-7} + d_1 G_{7,-5} + d_0,$$

which leads to the partition identity

$$\prod_{n=1}^{\infty}(1 - x^{7n}) \sum_{m=0}^{\infty} P(7m+5)x^{m+1} = \frac{7}{2^2}$$

$$-\frac{7}{2^2}\prod_{n=1}^{\infty}(1-x^n)^{-5}\left[\begin{array}{c} 7^2 x^{10}\sum_{j=0}^{\infty} P_{-5}(j)x^{49j} + \\ \sum_{j\in\mathbb{Z}^+\cup\{0\};j\equiv 0,4,5 \mod 7} P_{-5}(j)x^j - \\ \sum_{j\in\mathbb{Z}^+;j\equiv 1,2,6 \mod 7} P_{-5}(j)x^j \end{array}\right]$$

$$+\frac{7^3}{2}\prod_{n=1}^{\infty}(1-x^n)^{-7}\left[\begin{array}{c} 7^3 x^{14}\sum_{j=0}^{\infty} P_{-7}(j)x^{49j} + \\ \sum_{j\in\mathbb{Z}^+,j\equiv 3,5,6 \mod 7} P_{-7}(j)x^j - \\ \sum_{j\in\mathbb{Z}^+,j\equiv 1,2,4 \mod 7} P_{-7}(j)x^j \end{array}\right]$$

and contains (thus giving another proof of) the level one congruence for the prime 7.

10.5. $k = 11$. We have seen that Theorem 6.8 implies the Ramanujan level one congruence for the prime 11

$$P(11m+6) \equiv 0 \mod 11, \text{ for all } m \in \mathbb{Z}^+.$$

N	$\alpha = \text{ord}_\infty F_{11,-N}$	c
1	0	$P_{-1}(5) = 1$
2	0	$P_{-2}(10) = 1$
3	0	$P_{-3}(15) = -11$
4	0	$P_{-4}(20) = -11$
5	-2	$P_{-5}(3) = 2 \cdot 5$
6	0	$P_{-6}(30) = 11^2$
7	-3	$P_{-7}(2) = 2 \cdot 7$
8	0	$P_{-8}(40) = -11^3$
9	-4	$P_{-9}(1) = -3^2$
10	0	$P_{-10}(50) = 11^4$
11	-5	$P_{-11}(0) = 1$
12	-5	$P_{-12}(5) = 2^2 \cdot 3^3 \cdot 5$
13	-5	$P_{-13}(10) = 2 \cdot 3 \cdot 13$
14	0	$P_{-14}(70) = 11^6$
15	-6	$P_{15}(6) = -5 \cdot 7 \cdot 71$
24	-10	$P_{-24}(10) = 2^2 \cdot 3 \cdot 23 \cdot 1937$
26	0	$P_{-26}(130) = 11^{12}$

Table 22. CALCULATIONS OF $\alpha = \text{ord}_\infty F_{11,-N}$ AND THE COEFFICIENT c OF LEADING TERM x^α OF LAURENT SERIES EXPANSION OF $F_{11,-N}$ AT P_∞.

The general theory tells us that for negative N, the function $F_{11,N}$ is regular except at P_∞ and the order of the pole at this point can be easily evaluated. In fact (see Table 22), $F = F_{11,-5}$ has a double pole at P_∞ (reflecting the fact that $\overline{\mathbb{H}^2/\Gamma_o(11)}$ is a torus) and the function $G = F_{11,-7}$ has a triple pole there. We have hence produced generators for $\mathcal{K}(\overline{\mathbb{H}^2/\Gamma_o(11)})$, the field of meromorphic functions on the torus $\overline{\mathbb{H}^2/\Gamma_o(11)}$. The defining equation for the function field for this torus is then of the form

$$G^2 - c_0F^3 - c_1FG - c_2F^2 - c_3G - c_4F = c_5.$$

It is a tedious but routine procedure to compute the six constants.

We easily see that in general for all $N \in \mathbb{Z}^+ \cup \{0\}$,

$$-\text{ord}_\infty F_{11,-(11N+r)} \leq 5N + \begin{cases} 0 \text{ for } r = 0,\ 1,\ 2, \\ 1 \text{ for } r = 3,\ 4, \\ 2 \text{ for } r = 5,\ 6, \\ 3 \text{ for } r = 7,\ 8, \\ 4 \text{ for } r = 9,\ 10, \end{cases}$$

with equality except for

$$N = 0,\ r = 1,\ 2,\ 3,\ 4,\ 8,\ 10;\quad N = 1,\ r = 3 \text{ and } N = 2,\ r = 2,$$

where $F_{11,-(11N+r)}$ is constant. In particular, by choosing the indices m as

$$5,\ 7,\ 9,\ 11,\ 15,\ 16,\ 18,\ 20,\ 22,\ 25,\ 27,\ 29,\ 31,\ \ldots,$$

$$11N,\ 11N+3,\ 11N+5,\ 11N+7,\ 11N+9, 11(N+1),\ \ldots\ (N \geq 2)\ ,$$

we obtain a basis $\{F_{11,-m} - F_{11,-m}(0)\}$ for $\mathcal{K}(\Gamma_o(11))_\infty$.

Definition 10.6. For $j \geq 2$, $\in \mathbb{Z}^+$, we let $N(j)$ be that choice of m from the above list so that $\operatorname{ord}_{P_\infty} F_{11,-N(j)} = -j$.

It follows that the functions $F_{11,-N(j)} - F_{11,-N(j)}(0)$, $j = 2, 3, 4, \ldots$ form a basis for $\mathcal{K}(\Gamma_o(11))_\infty$ and the corresponding images under precomposition by A_{11}, $G_{11,-N(j)} - G_{11,-N(j)}(\infty)$ form a basis for $\mathcal{K}(\Gamma_o(11))_0$. Alternately, we can let

$$F_2 = F_{11,-N(2)} - F_{11,-N(2)}(0),\ F_3 = F_{11,-N(3)} - F_{11,-N(3)}(0),$$

$$G_2 = G_{11,-N(2)} - G_{11,-N(2)}(\infty) \text{ and } G_3 = G_{11,-N(3)} - G_{11,-N(3)}(\infty)$$

and use

$$F_2,\ F_3,\ F_2^2,\ F_2F_3,\ \ldots,\ F_2^j,\ F_2^{j-1}F_3,\ F_2^{j+1},\ \ldots$$

and

$$G_2,\ G_3,\ G_2^2,\ G_2G_3,\ \ldots,\ G_2^j,\ G_2^{j-1}G_3,\ G_2^{j+1},\ \ldots$$

as bases for $\mathcal{K}(\Gamma_o(11))_\infty$ and $\mathcal{K}(\Gamma_o(11))_0$, respectively.

An evaluation of the constants in

$$F_{11,1} = d_5 G_{11,-11} + d_4 G_{11,-9} + d_3 G_{11,-7} + d_2 G_{11,-5} + d_0 \tag{5.44}$$

leads to a complicated looking partition identity:

$$\begin{aligned}
&3^2 5 \cdot 7 \prod_{n=1}^{\infty} (1 - x^{11n}) \sum_{m=0}^{\infty} P(11m+6) x^{m+1} \\
&= 2 \cdot 3^3 7 \cdot 11 \cdot 43 \cdot 109 \left(-1 + \prod_{n=1}^{\infty} (1 - x^n)^{-5} \right) \\
&\times \left[\begin{array}{r} 11^2 x^{25} \sum_{j=0}^{\infty} P_{-5}(j) x^{121j} + \\ \sum_{j \in \mathbb{Z}^+ \cup \{0\}; j \equiv 0,2,5,9,10 \mod 11} P_{-5}(j) x^j - \\ \sum_{j \in \mathbb{Z}^+; j \equiv 1,4,6,7,8 \mod 11} P_{-5}(j) x^j \end{array} \right] \\
&-2 \cdot 3^2 5 \cdot 11^2 \cdot 17 \cdot 43 \left(-1 + \prod_{n=1}^{\infty} (1 - x^n)^{-7} \right) \\
&\times \left[\begin{array}{r} 11^3 x^{35} \sum_{j=0}^{\infty} P_{-7}(j) x^{121j} + \\ \sum_{j \in \mathbb{Z}^+ \cup \{0\}; j \equiv 0,3,5,6,7 \mod 11} P_{-7}(j) x^j - \\ \sum_{j \in \mathbb{Z}^+; j \equiv 1,4,8,9,10 \mod 11} P_{-7}(j) x^j \end{array} \right]
\end{aligned}$$

$$-2^2 5\cdot 7\cdot 11^3 47\left(-1+\prod_{n=1}^{\infty}(1-x^n)^{-9}\right)$$
$$\times\left[\begin{array}{c} 11^4 x^{45}\sum_{j=0}^{\infty}P_{-9}(j)x^{121j}+\\ \sum_{j\in\mathbb{Z}^+\cup\{0\};j\equiv 0,3,7,8,9 \mod 11}P_{-9}(j)x^j-\\ \sum_{j\in\mathbb{Z}^+;j\equiv 2,4,5,6,10 \mod 11}P_{-9}(j)x^j\end{array}\right]$$
$$+3^2 5\cdot 7\cdot 11^3\prod_{n=1}^{\infty}(1-x^n)^{-11}\left[\begin{array}{c} 11^5 x^{55}\sum_{j=0}^{\infty}P_{-11}(j)x^{121j}+\\ \sum_{j\in\mathbb{Z}^+;j\equiv 1,3,4,5,9 \mod 11}P_{-11}(j)x^j-\\ \sum_{j\in\mathbb{Z}^+;j\equiv 2,6,7,8,10 \mod 11}P_{-11}(j)x^j\end{array}\right].$$

This last displayed equation, despite its complexity, yields our second proof of the level one congruence for the prime 11 . We can replace $G_{11,-11}$ by $G_{11,-12}$, $G_{11,-13}$ (each of these functions has a pole of order 5 at P_0) or $f_{11,1}^{12}$ (a function with divisor $\frac{P_\infty^5}{P_0^5}$) and obtain different partition identities. The equation we have produced gives an identity for the prime 11. If we only want to obtain the level one congruence for the prime eleven, we are able to dispense with the evaluation of some constants. To proceed, we consider a slight variation of equation (5.44):

$$\begin{aligned}(5.45)\qquad F_{11,1} &= d_5(G_{11,-11}-G_{11,-11}(\infty))+d_4(G_{11,-9}-G_{11,-9}(\infty))\\ &+d_3(G_{11,-7}-G_{11,-7}(\infty))+d_2(G_{11,-5}-G_{11,-5}(\infty)).\end{aligned}$$

We apply the involution A_{11} to the above equation and obtain

$$\begin{aligned}G_{11,1} &= d_5(F_{11,-11}-F_{11,-11}(0))+d_4(F_{11,-9}-F_{11,-9}(0))\\ &+d_3(F_{11,-7}-F_{11,-7}(0))+d_2(F_{11,-5}-F_{11,-5}(0)).\end{aligned}$$

Calculations using the expansions of functions in terms of the local coordinate $x=\exp(2\pi\imath\tau)$ lead to (in these equations c stands for a generic integer, different in each appearance, whose exact value is not needed)

$$d_5=\frac{c}{11P_{-11}(0)},$$

$$d_4=\frac{c}{11P_{-11}(0)P_{-9}(1)},$$

$$d_3=\frac{c}{11P_{-11}(0)P_{-9}(1)P_{-7}(2)}$$

and

$$d_2=\frac{c}{11P_{-11}(0)P_{-9}(1)P_{-7}(2)P_{-5}(3)}.$$

Use

$$P_{-11}(0)=1,\ P_{-9}(1)=-3^2,\ P_{-7}(2)=2\cdot 7 \text{ and } P_{-5}(3)=2\cdot 5$$

to conclude that the $d_i\in\mathbb{Z}\left[\frac{1}{2\cdot 3^2 5\cdot 7\cdot 11}\right]$. Now we use the fact that the Laurent series expansion $\sum_{m=0}^{\infty}P(11m+6)x^{m+1}$, essentially the left hand side of (5.45)), has integer coefficients. Hence so does the right hand side. By

Theorem 4.19, each of the Taylor series coefficients of each of the functions $G_{11,j} - G_{11,j}(\infty)$ appearing in equation (5.45) is an integer divisible by 11^2. Thus the Taylor series coefficients of the function described by the right hand side of equation (5.45) are integers divisible by 11.

The computation of "the" abelian differential of the first kind on the torus $\overline{\mathbb{H}^2/\Gamma_o(11)}$ from the corresponding objects on the surface $\overline{\mathbb{H}^2/\Gamma(11,11)}$ of genus six has consequences that should be explored. For the moment we record only that

$$(\eta(11\tau)\eta(\tau))^2 = x\prod_{n=1}^{\infty}(1-x^{11n})^2(1-x^n)^2,\ x = \exp(2\pi\imath\tau),$$

is an abelian differential of the first kind on $\overline{\mathbb{H}^2/\Gamma_o(11)}$.

10.6. $k = 13$. Using MATHEMATICA to evaluate the constants for the case $N = 1$ of part (b) of Theorem 4.7 yields

$$c_1 = 11,\ c_2 = 2^2 3^2 13,\ c_3 = 2\cdot 13^2 19,\ c_4 = 2^2 5\cdot 13^3,\ c_5 = 2\cdot 3\cdot 13^4$$

and

$$c_6 = 13^5 = c_7.$$

The gcd of the seven constants is 1; however, the gcd of the last six (excluding c_1) is 13. Hence we have obtained from Corollary 4.8

Corollary 10.7. *We have the formal power series identity*

$$\sum_{m=1}^{\infty} P(13m-7)x^m \equiv 11\ x\prod_{n=1}^{\infty}\frac{1-x^{13n}}{(1-x^n)^2} \mod 13.$$

Since by the Frobenius automorphism

$$(1-x^n)^{13} \equiv 1-x^{13n} \mod 13,$$

this last result implies

Corollary 10.8. *For all $m \in \mathbb{Z}^+ \cup \{0\}$,*

$$P(13m+6) \equiv 11\ P_{-11}(m) \mod 13.$$

Remark 10.9. (a) We have on several occasions remarked (or implied) that when a certain gcd is of a given form there are no congruence implications. We have meant, of course, no obvious immediate congruence implications. In the case $k = 13$ there were also no immediate implications; however, there are implications as stated. Note that Corollary 10.5 implies $P_3(5m-3) \equiv 9P_{-9}(m-1) \mod 5$. We also saw in §10.3 that

N	$\text{ord}_\infty F_{13,-N}$	c
1	0	$P_{-1}(7) = 1$
2	-1	$P_{-2}(1) = -2$
3	0	$P_{-3}(21) = 13$
4	-2	$P_{-4}(2) = 2$
5	-2	$P_{-5}(9) = 2^2 5$
6	-3	$P_{-6}(3) = 2 \cdot 5$
7	-3	$P_{-7}(10) = 2 \cdot 7^2$
8	-4	$P_{-8}(4) = -2 \cdot 5 \cdot 7$
9	-4	$P_{-9}(11) = -2 \cdot 3^4$
10	-5	$P_{-10}(5) = 2 \cdot 7 \cdot 17$
11	-5	$P_{-11}(12) = -2 \cdot 11 \cdot 41$
12	-6	$P_{-12}(6) = -2 \cdot 11 \cdot 19$
13	-7	$P_{-13}(0) = 1$

Table 23. CALCULATIONS OF $\alpha = \text{ord}_\infty F_{13,-N}$ AND THE COEFFICIENT c OF LEADING TERM x^α OF LAURENT SERIES EXPANSION OF $F_{13,-N}$ AT P_∞.

$P_{-5}(5m) \equiv P_{-1}(m) \bmod 5$.
(b) Our calculations for $F_{13,1}$ showed that

$$c_1 = 11 \text{ and } \gcd\{c_2, \ldots, c_7\} = 13.$$

Thus a lack of a mod 13 congruence is caused by the presence of a simple zero at infinity for $F_{13,1}$, and the Taylor coefficients of $F_{13,1} - 11f_{13}$ certainly are congruent to 0 mod 13. Calculations of the constants $c_2, \ldots, c_{14}$ appearing in the formula for $F_{13\,2}$ yield

$$\gcd\{c_2, \ldots, c_{14}\} = 1 \text{ and } \gcd\{c_3, \ldots, c_{14}\} = 13.$$

To help organize our data we construct Table 23, which can be used by the reader to extract consequences from relations such as

$$F_{13,-5} = c_1 F_{13,-4} + c_2 F_{13,-2} + c_3.$$

10.7. $k = 4$. Nonprimes seem to offer new avenues for exploration. For arbitrary k, $\left(\frac{\eta(k\tau)}{\eta(\tau)}\right)^{24}$ and $\left(\frac{\eta(k\tau)}{\eta(\frac{\tau}{k})}\right)^{24}$ still respectively define $\Gamma_o(k)$ and $\Gamma(k,k)$-invariant functions, but it is more difficult to compute their divisors.

Consider $k = 4$. The surface $\mathbb{H}^2/\Gamma(4)$ has type $(0,6)$; the punctures are determined by the cusps ∞, 0, 1, 2, 3 and $\frac{1}{2}$. The type of $\mathbb{H}^2/G(4)$ is $(0,3)$. The punctures P_∞ and $P_{\frac{1}{2}}$ on $\mathbb{H}^2/\Gamma(4)$ are fixed by[127] $\tilde{B}$; the

[127] Recall that $\tilde{B}$ is the automorphism of $\mathbb{H}^2/\Gamma(4)$ induced by $B \in \Gamma$.

other four punctures are permuted cyclically by $<\tilde{B}>$. We also note that $G(4)=\Gamma_o(4)$ is torsion free. The divisor of the

$$\Gamma_o(4)\text{-invariant function } g(\tau)=\left(\frac{\eta(4\tau)}{\eta(\tau)}\right)^8 \text{ is } \frac{P_\infty}{P_0}$$

(the numerator has divisor $P_\infty^{\frac{4}{3}}\ P_0^{\frac{1}{3}}\ P_{\frac{1}{2}}^{\frac{1}{3}}$; the denominator $P_\infty^{\frac{1}{3}}\ P_0^{\frac{4}{3}}\ P_{\frac{1}{2}}^{\frac{1}{3}}$). The computation of the divisors is facilitated by the fact that η is a $\frac{1}{4}$-form and that the order at P_∞ is more or less given. We also use the transformation properties of η. The divisor of the

$$\Gamma(4,4)=\Gamma(4)\text{-invariant function } f_o(\tau)=\left(\frac{\eta(4\tau)}{\eta(\frac{\tau}{4})}\right)^8 \text{ is } \frac{P_\infty^5\ P_{\frac{1}{2}}}{P_0^5\ P_2}$$

because the numerator has divisor

$$P_\infty^{\frac{16}{3}}\ P_0^{\frac{1}{3}}\ P_1^{\frac{1}{3}}\ P_2^{\frac{1}{3}}\ P_3^{\frac{1}{3}}\ P_{\frac{1}{2}}^{\frac{4}{3}},$$

while the denominator has divisor

$$P_\infty^{\frac{1}{3}}\ P_0^{\frac{16}{3}}\ P_1^{\frac{1}{3}}\ P_2^{\frac{4}{3}}\ P_3^{\frac{1}{3}}\ P_{\frac{1}{2}}^{\frac{1}{3}}.$$

Summing over the group generated by B, we are adding four functions with divisors

$$\frac{P_\infty^5\ P_{\frac{1}{2}}}{P_0^5\ P_2},\ \frac{P_\infty^5\ P_{\frac{1}{2}}}{P_1^5\ P_3},\ \frac{P_\infty^5\ P_{\frac{1}{2}}}{P_2^5\ P_0},\ \frac{P_\infty^5\ P_{\frac{1}{2}}}{P_3^5\ P_1}.$$

The sum (on $\mathbb{H}^2/\Gamma_o(4)$) has a pole at only one point: P_0 (it is obviously of order five). We have already accounted for at least three zeros (two at P_∞ and one at $P_{\frac{1}{2}}$). Hence there can be at most three interior zeros. The sum has a zero of order eight at infinity (as a $\Gamma(4)$-invariant function). Hence the divisor of the

$$\Gamma_o(4)\text{-invariant function } f(\tau)=\sum_{l=0}^{3}\left(\frac{\eta(4(\tau+l))}{\eta(\frac{\tau+l}{4})}\right)^8 \text{ is } \frac{P_\infty^2\ P_{\frac{1}{2}}\ Q_1\ Q_2}{P_0^5},$$

with $Q_i\neq P_\infty$ and $Q_i\neq P_0$. It follows that

$$\frac{f}{g^2} \text{ has divisor } \frac{P_{\frac{1}{2}}\ Q_1\ Q_2}{P_0^3},$$

and is hence a polynomial of degree three in g. We have thus established

Theorem 10.10. *There are constants c_i, $i=0,\ 1,\ 2,\ 3$ such that*

$$\frac{1}{4}\sum_{l=0}^{3}\left(\frac{\eta(4(\tau+l))}{\eta(\frac{\tau+l}{4})}\right)^8=\sum_{i=0}^{3}c_i\ \left(\frac{\eta(4\tau)}{\eta(\tau)}\right)^{8(i+2)}.$$

This theorem translates to the identity in

N	$c_{N,0}$	$c_{N,1}$	$c_{N,2}$	$c_{N,3}$
1	0	-2^3		
2	-2^{15}	-2^{11}	$2^3 13$	
3	$2^{18}3$	$2^{14}3$	$-2^6 23$	0

Table 24. THE CONSTANTS $c_{N,i}$ APPEARING IN THEOREM 10.14.

Corollary 10.11. *For all $x \in \mathbb{C}$, $|x| < 1$,*

$$\sum_{m=2}^{\infty} P_8(4m-5)x^{m-2} = \sum_{i=0}^{3} c_i x^i \frac{\prod_{n=1}^{\infty}(1-x^{4n})^8(i+1)}{\prod_{n=1}^{\infty}(1-x^n)^8(i-2)}.$$

Calculations (using MATHEMATICA) show that

$$c_0 = 2^6 3,\ c_1 = 2^{10}19,\ c_2 = 2^{19} \text{ and } c_3 = 2^{22}.$$

Since the gcd of these four integers is 2^6 we have established

Corollary 10.12. *For each $m \in \mathbb{Z}$, $m \geq 2$, $P_8(4m-5)$ is congruent to zero* mod 2^6.

Remark 10.13. Corollary 11.3, below, provides an alternate way to establish the last congruence.

We now turn to the reciprocal of g. Averaging the N-th power ($N \in \mathbb{Z}^+$) of the function $\frac{1}{f_o}$ over the group $< \tilde{B} >$, we get a $\Gamma_o(4)$ -invariant function with a pole of order $\lfloor \frac{5N}{4} \rfloor$ at P_∞ and a pole of order at most $\lfloor \frac{N}{4} \rfloor$ at $P_{\frac{1}{2}}$. Thus for $N = 1,\ 2$ or 3, this function is holomorphic at $P_{\frac{1}{2}}$. We have hence shown that

Theorem 10.14. *For fixed $N = 1,\ 2,\ 3$ there are constants $c_{N,i}$ such that*

$$\frac{1}{4}\sum_{l=0}^{3}\left(\frac{\eta(\frac{\tau+l}{4})}{\eta(4(\tau+l))}\right)^{8N} = \sum_{i=0}^{N} c_{N,i}\left(\frac{\eta(\tau)}{\eta(4\tau)}\right)^{8i}.$$

The calculations of these constants are summarized in Table 24.

The corresponding power series identities read

Corollary 10.15. *For all $x \in \mathbb{C}$, $|x| < 1$. ($N = 1$):*

$$\sum_{m=0}^{\infty}\left(P_{-8}(4m+1) + 2^3 P_{-8}(m)\right)x^m = 0,$$

($N = 2$):

$$\sum_{m=0}^{\infty}\left(P_{-16}(4m-2) - 2^3 13 P_{-16}(m)\right)x^m = -2^{15}x^2\prod_{n=1}^{\infty}(1-x^{4n})^{16}$$

$$-2^{11}x\prod_{n=1}^{\infty}(1-x^{4n})^8\sum_{m=0}^{\infty}P_{-8}(m)x^m$$

and $(N=3)$:

$$\sum_{m=0}^{\infty}P_{-24}(4m+3)x^m = 2^6 3x^3\prod_{n=1}^{\infty}(1-x^{4n})^{24}$$

$$+2^{14}3x^2\prod_{n=1}^{\infty}(1-x^{4n})^{16}\sum_{m=0}^{\infty}P_{-8}(m)x^m - 2^6 23x\prod_{n=1}^{\infty}(1-x^{4n})^8\sum_{m=0}^{\infty}P_{-16}(m)x^m.$$

As a consequence of the equation for $N=1$, we see that

$$P_{-8}(4m+1) = -2^3P_{-8}(m) \text{ for all } m\in\mathbb{Z}^+\cup\{0\}.$$

Since $P_{-8}(3)=0=P_{-8}(7)$, we conclude that for all $n\in\mathbb{Z}^+\cup\{0\}$,

$$P_{-8}\left(4^n 3+\frac{4^n-1}{3}\right) = 0 = P_{-8}\left(4^n 7+\frac{4^n-1}{3}\right).$$

We see, for example, that $P_{-8}(13)=0=P_{-8}(29)$. These conclusions should be contrasted with Theorem 6.7 which implies that

$$P_{-8}(m) = 0 \text{ for all } m\in\mathbb{Z}^+,\ m\equiv 3 \mod 4.$$

The equation for $N=2$ tells us that $P_{-16}(4m+2)\equiv 0 \mod 2^3$. The reader is invited to investigate the case $N=2$ of equation (5.38) and compare the result obtained with this last congruence. The equation for $N=3$ allows us to conclude that $P_{-24}(4m+3)\equiv 0 \mod 2^6$. As before, we can use the multiplicativity of the function T to conclude that

$$T(4(m+1)) = T(4)T(m+1) = -2^6 23T(m+1) \text{ provided } m\not\equiv 3 \mod 4.$$

We have produced, above, a holomorphic universal of $\mathbb{H}^2/\Gamma_o(4)$, $\left(\frac{\eta(\tau)}{\eta(4\tau)}\right)^8$, with divisor $\frac{P_0}{P_\infty}$ and in §3.4 of Chapter 3, the holomorphic cover

$$\tau\mapsto\left(\frac{\theta^4\begin{bmatrix}0\\0\end{bmatrix}(0,2\tau)}{\theta^4\begin{bmatrix}1\\0\end{bmatrix}(0,2\tau)}\right)$$

with divisor $\frac{P_{\frac{1}{2}}}{P_\infty}$. These functions are affinely related. The resulting affine relation between the meromorphic functions is equivalent to the infinite product identity given by equation (5.40).

10.8. $k = 6$. The compact surface $\overline{\mathbb{H}^2/\Gamma_o(6)}$ is a four (respectively, three) sheeted cover of $\overline{\mathbb{H}^2/\Gamma_o(2)}$ ($\overline{\mathbb{H}^2/\Gamma_o(3)}$). It has genus 0 and the punctures on this surface are P_∞, $P_{\frac{1}{2}}$, P_0 and $P_{\frac{1}{3}}$. We can compare three functions: $\left(\frac{\eta(2\tau)}{\eta(\tau)}\right)^{24}$, $\left(\frac{\eta(3\tau)}{\eta(\tau)}\right)^{12}$ and $\left(\frac{\eta(6\tau)}{\eta(\tau)}\right)^{24}$. To calculate their divisors, we note that $\Gamma_o(2)$ is generated over $\Gamma_o(6)$ by (we are using left cosets $\Gamma_o(6)\gamma$ with γ from the list below)

$$\begin{bmatrix} 1 & 0 \\ 0 & 1 \end{bmatrix}, \begin{bmatrix} 1 & 0 \\ 2 & 1 \end{bmatrix}, \begin{bmatrix} 1 & 0 \\ 4 & 1 \end{bmatrix} \text{ and } \begin{bmatrix} -1 & -2 \\ 2 & 3 \end{bmatrix},$$

while $\Gamma_o(3)$ is generated over $\Gamma_o(6)$ by

$$\begin{bmatrix} 1 & 0 \\ 0 & 1 \end{bmatrix}, \begin{bmatrix} 1 & 0 \\ 3 & 1 \end{bmatrix} \text{ and } \begin{bmatrix} -1 & -1 \\ 3 & 2 \end{bmatrix}.$$

We also note the canonical projections (these covers are not Galois)

$$\pi_i : \overline{\mathbb{H}^2/\Gamma_o(6)} \to \overline{\mathbb{H}^2/\Gamma_o(i)},\ i = 2,\ 3,$$

$$\deg \pi_2 = 4,\ \deg \pi_3 = 3,$$

$$\pi_2(P_\infty) = P_\infty = \pi_2(P_{\frac{1}{2}}),\ \pi_2(P_0) = P_0 = \pi_2(P_{\frac{1}{3}}),$$

$$\pi_3(P_\infty) = P_\infty = \pi_3(P_{\frac{1}{3}}),\ \pi_3(P_0) = P_0 = \pi_3(P_{\frac{1}{2}}),$$

$$b_{\pi_2}(P_\infty) = 0 = b_{\pi_2}(P_{\frac{1}{3}}),\ b_{\pi_2}(P_{\frac{1}{2}}) = 2 = b_{\pi_2}(P_0),$$

and

$$b_{\pi_3}(P_\infty) = 0 = b_{\pi_3}(P_{\frac{1}{2}}),\ b_{\pi_3}(P_{\frac{1}{3}}) = 1 = b_{\pi_3}(P_0).$$

Note that

$$\left(\frac{\eta(2\cdot)}{\eta}\right) = P_\infty^{\frac{1}{24}}\, P_0^{-\frac{1}{24}} \text{ on } \overline{\mathbb{H}^2/\Gamma_o(2)} \text{ and } \left(\frac{\eta(3\cdot)}{\eta}\right) = P_\infty^{\frac{1}{12}}\, P_0^{-\frac{1}{12}} \text{ on } \overline{\mathbb{H}^2/\Gamma_o(3)}.$$

Hence on $\overline{\mathbb{H}^2/\Gamma_o(6)}$,

$$\left(\frac{\eta(2\cdot)}{\eta}\right) = P_\infty^{\frac{1}{24}}\, P_{\frac{1}{2}}^{\frac{1}{8}}\, P_0^{-\frac{1}{8}}\, P_{\frac{1}{3}}^{-\frac{1}{24}} \text{ and } \left(\frac{\eta(3\cdot)}{\eta}\right) = P_\infty^{\frac{1}{12}}\, P_{\frac{1}{2}}^{-\frac{1}{12}}\, P_0^{-\frac{1}{6}}\, P_{\frac{1}{3}}^{\frac{1}{6}}.$$

Slightly more complicated calculations tell us that

$$\left(\frac{\eta(6\cdot)}{\eta}\right) = P_\infty^{\frac{5}{24}}\, P_0^{-\frac{5}{24}} \text{ on } \overline{\mathbb{H}^2/\Gamma_o(6)}.$$

Thus $\left(\frac{\eta(6\cdot)}{\eta}\right)^{\frac{24}{5}}$ and $\frac{\eta(2\tau)^{24}\eta(3\tau)^{36}}{\eta(\tau)^{\frac{204}{5}}\eta(6\tau)^{\frac{96}{5}}}$ are meromorphic functions on the sphere $\overline{\mathbb{H}^2/\Gamma_o(6)}$ with divisors $\frac{P_\infty}{P_0}$ and $\frac{P_{\frac{1}{3}}^5}{P_0^5}$, respectively. Thus

$$\frac{\eta(2\tau)^{24}\eta(3\tau)^{36}}{\eta(\tau)^{\frac{204}{5}}\eta(6\tau)^{\frac{96}{5}}} = \sum_{i=0}^{5} c_i \left(\frac{\eta(6\cdot)}{\eta}\right)^{\frac{24}{5}i}.$$

The undetermined constants can be computed; these lead to (perhaps interesting) relations between power series.

11. The higher level Ramanujan congruences

The Ramanujan level n congruences are obtained by averaging $\Gamma(k^n, k)$-invariant functions to obtain $\Gamma_o(k)$-invariant functions. They are consequences of the expansions of the functions $F_{k,n,N}$ in terms of simpler functions (at times linear combinations of f_k^i). For the prime k, the positive integers N and n, we will discuss in this section congruences for the coefficients

$$P_N\left(k^n m - l_{k,n,N}\right),$$

where $l_{k,n,N} \in \mathbb{Z}^+$. Our discussion is limited to the cases $k = 2, 3, 5, 7, 13$ (those primes for which $p(\overline{\mathbb{H}^2/\Gamma_o(k)}) = 0$) and 11 ($p(\overline{\mathbb{H}^2/\Gamma_o(11)}) = 1$). We have seen (Theorem 6.7) that for negative N we have analogous equalities rather than congruences.

This section is organized as follows. In §11.1, we derive or rederive those level 2 and 3 congruences that are easy consequences of the analytic machinery established so far. As samples,[128] we develop (at least outline the development of) all the level $n \in \mathbb{Z}^+$ congruences for the primes 2 (11.2), 5 (11.3) and 11 (11.4).

11.1. The level two and three congruences for small primes. We have already established the Ramanujan level two congruences for the primes 2, 3, 5 and 7. Other methods seem to be required to produce the level $n \geq 3$ congruences for these primes as well as all the higher level congruences for the prime 11.

We start a detailed discussion of the case $k = 5$. We define for each $n \in \mathbb{Z}^+$ a positive integer $l_n = l_{5,n,1}$ as the smallest positive integer such that $24 l_n \equiv 1 \mod 5^n$. The level n congruence for the prime 5 is equivalent to the statement

$$P(5^n m + l_n) \equiv 0 \mod 5^n \text{ for all } m \in \mathbb{Z}^+ \cup \{0\}.$$

As we saw, the level one congruence is a consequence of equation (5.2) which also implied the recurrence relation given by equation (5.25). The calculations following Theorem 9.1 tell us that $5 | P_6(5m+4)$. Hence

$$P(5(5m+4)+4) \equiv 0 \mod 5^2.$$

This is the level two congruence for the prime 5.

[128] In part, because we do not have a general theory.

Lemma 11.1. *We have the formula*

$$24l_n = \begin{cases} 19 \cdot 5^n + 1 \text{ for } n \text{ odd} \\ 23 \cdot 5^n + 1 \text{ for } n \text{ even} \end{cases},$$

and the recursion relation

$$l_{n+1} = l_n + \begin{cases} 4 \cdot 5^n \text{ for } n \text{ odd} \\ 3 \cdot 5^n \text{ for } n \text{ even} \end{cases}.$$

Further, we have

$$5^n = l_n + \begin{cases} \frac{5^{n+1}-1}{24} \text{ for } n \text{ odd} \\ \\ \frac{5^n-1}{24} \text{ for } n \text{ even} \end{cases}. \tag{5.46}$$

Proof. Obviously, l_n is the multiplicative inverse (with $1 < l_n < 5^n$) of 24 in the ring $\mathbb{Z}_{5^n}$; it exists since gcd(24,5^n) $= 1$. The last assertion gives a formula for l_n. The first assertion follows from the last one, and the second, from the first.

We examine another proof of the level two congruence for the prime five. The level one congruence is a consequence of the relation $F_{5,1,1} = 5f_5$. To obtain the level two congruence we can start with

$$F_{5,2,1} = V_{5,0}(F_{5,1,1}) = V_{5,0}(5f_5) = 5f_5^{-1}F_{5,1,6}, \tag{5.47}$$

and use the modular equation to conclude that

$$F_{5,2,1} = 5(5^2F_{5,1,5} + 5^2F_{5,1,4} + 3 \cdot 5F_{5,1,3} + 5F_{5,1,2} + F_{5,1,1}).$$

We hence see that the level two congruence follows from the observation that $F_{5,1,1} = 5f_5$. Alternatively, we can conclude that $d(5,2,1) = 5^2$ from equation (5.47) and the calculation $d(5,1,6) = 5$. We try to obtain by a similar method the level three congruence for the prime five. Our starting point is the equation $F_{5,3,1} = V_{5,1}(F_{5,2,1})$. We know from the level two result that $F_{5,2,1} = \sum_{i=1}^{5} c_i f_5^i$ with $c_i \in \mathbb{Z}$ and $5^2 | c_i$. The properties of the operator V tell us that $V_{5,1}(f_5^i) = f_5^{-i}F_{5,1,1+6i}$. Hence

$$F_{5,3,1} = \sum_{i=1}^{5} c_i f_5^{-i} F_{5,1,1+6i}.$$

We could easily obtain the level three congruence for the prime five if we knew that $5 | d(5,1,1+6i)$ for $i = 1, \ldots, 5$. However, our calculations show that $d(5,1,13) = 1$ (among others). Hence we must use the properties of the constants c_i and the integers $d(5,1,1+6i)$ to obtain a level three congruence. In the notation of §8.3, $c_i = 5c_{6,i+1}$ and as a result of the calculations in that subsection,

$$\gcd\{c_i; i = 1, \ldots, 5\} = 5^2.$$

Since

$$\gcd\{c_i; i = 2, \ldots, 5\} = 5^5,$$

our calculation that $d(5,1,7) = 5$ gives us that

$$P(5^3 m + l_3) \equiv 0 \mod 5^3 \text{ for all } m \in \mathbb{Z}^+\{0\},$$

the level three congruence for the prime five. Similar arguments yield the level two and three congruences for the prime seven. To obtain more general results, we must examine more carefully the consequences of the modular equations. We turn to the first case in the next subsection. We will also discuss in detail the case $k = 5$ and very briefly the case $k = 11$. As will be seen, different tools are needed for the last case. The cases $k = 3$ and 7 (this case is discussed in [**16**, Ch. 8]) are left to the reader.

11.2. The level n congruences for the prime 2. We study the case $N = 1$. We have by straightforward calculations,

$$F_{2,1,1} = 2^3 f_2,\ F_{2,2,1} = 2^6(1+2)f_2 + 2^{14} f_2^2$$

and

$$F_{2,3,1} = 2^7(1+2^3)f_2 + 2^9(1+2+2^2+2^7+2^{10})f_2^2$$
$$+2^{10}(1+2^3+2^8+2^9+2^{11})f_2^3 + 2^{25}(1+2+2^2)f_2^4 + 2^{33} f_2^5.$$

Lemma 11.2. *We have for $n \in \mathbb{Z}^+$,*

$$F_{2,n,1} = \begin{cases} \sum_{i=1}^{\frac{1}{3}(2^{n+1}-1)} c_i f_2^i, & n \text{ odd} \\ \sum_{i=1}^{\frac{2}{3}(2^n-1)} c_i f_2^i, & n \text{ even} \end{cases},$$

where $c_i \in \mathbb{Z}$ and $\pi_2(c_i)$

$$\begin{array}{r} = 3\left(\frac{n+1}{2}\right) \text{ if } n \text{ is odd and } i = 1 \\ \geq 3\left(\frac{n+1}{2}\right) + 3(i-1) \text{ if } n \text{ is odd and } i > 1 \\ = 3\left(\frac{n+2}{2}\right) \text{ if } n \text{ is even and } i = 1 \\ \geq 3\left(\frac{n+2}{2}\right) + 6i - 7 \text{ if } n \text{ is even and } i > 1 \end{array}.$$

Proof. The calculations preceding the statement of the lemma have established the conclusion for $n = 1$ and 2. So let $n > 2$ and assume the results for indices below n. If n is even, then

$$F_{2,n,1} = V_2(F_{2,n-1,1}) = V_2\left(\sum_{i=1}^{\frac{1}{3}(2^n-1)} c_i f_2^i\right) = \sum_{i=1}^{\frac{1}{3}(2^n-1)} c_i f_2^{-i} F_{2,3i}$$

$$= \sum_{i=1}^{\frac{1}{3}(2^n-1)} c_i f_2^{-i} \sum_{j=\lceil \frac{3i}{2} \rceil}^{3i} c_{3i,j} f_2^j = \sum_{i=1}^{\frac{2}{3}(2^n-1)} d_i f_2^i,$$

where, by induction, $\pi_2(c_i) \geq 3\left(\frac{n}{2} + (i-1)\right)$ and, by Lemma 8.3, $\pi_2(c_{3i,j}) \geq 3 + 8\left(j - \frac{3i+1}{2}\right)$. Since

$$d_i = \sum_{j=1}^{\frac{1}{3}(2^n-1)} c_j c_{3j,i+j} = \sum_{j=\lceil \frac{i}{2} \rceil}^{2i} c_j c_{3j,i+j},$$

we conclude that

$$\pi_2(d_i) \geq \frac{3n}{2} + \min\left\{8i - 4 - j;\ j = \left\lceil \frac{i}{2} \right\rceil, \ldots, 2i\right\} = \frac{3n}{2} + 6i - 4$$
$$= \frac{3(n+2)}{2} + 6i - 7$$

and

$$\pi_2(d_1) \geq \min\{\pi_2(c_1) + \pi_2(c_{3,2}),\ \pi_2(c_2) + \pi_2(c_{6,3})\} = 3\frac{n}{2} + 3.$$

If n is odd, then

$$F_{2,n,1} = V_2(f_{2^2,1}^8 F_{2,n-1,1}) = V_2\left(f_{2^2,1}^8 \sum_{i=1}^{\frac{2}{3}(2^{n-1}-1)} c_i f_2^i\right)$$

$$= \sum_{i=1}^{\frac{2}{3}(2^{n-1}-1)} c_i f_2^{-i} F_{2,1+3i} = \sum_{i=1}^{\frac{2}{3}(2^{n-1}-1)} c_i f_2^{-i} \sum_{j=\lceil \frac{1+3i}{2} \rceil}^{1+3i} c_{1+3i,j} f_2^j = \sum_{i=1}^{\frac{1}{3}(2^{n+1}-1)} d_i f_2^i$$

and

$$d_i = \sum_{j=1}^{\frac{2}{3}(2^n-1)} c_j c_{1+3j,i+j} = \sum_{j=\lceil \frac{i-1}{2} \rceil}^{2i-1} c_j c_{1+3j,i+j}.$$

The induction assumption tells us that

$$\pi_2(d_i) \geq \frac{3(n+1)}{2} + \min\left\{8i - 12 - 2j;\ j = \left\lceil \frac{i-1}{2} \right\rceil, \ldots, 2i-1\right\}$$
$$= \frac{3(n+1)}{2} + 4i - 8 \geq \frac{3(n+1)}{2} + 3(i-1) \text{ if } i \geq 5.$$

Direct calculations verify the last conclusion for $i = 1, 2, 3$ and 4.

We have hence proven

Corollary 11.3. *For each n and $m \in \mathbb{Z}^+$,*

$$P_8\left(2^n m - \frac{2^{2\lfloor \frac{n+1}{2} \rfloor} - 1}{3}\right) \equiv 0 \mod 2^{3\lceil \frac{n+1}{2} \rceil}.$$

11.3. The level n congruences for the prime 5. The most important special case in the study of $F_{5,1,n}$, $n \in \mathbb{Z}^+$ odd, yields the identity

$$\prod_{m=1}^{\infty}(1-x^{5m})\sum_{m=1}^{\infty} P\left(5^n m - \frac{5^{n+1}-1}{24}\right)x^m$$

$$= \sum_{i=1}^{\frac{5^{n+1}-1}{24}} c_i \left[x\frac{\prod_{m=1}^{\infty}(1-x^{5m})^6}{\prod_{m=1}^{\infty}(1-x^m)^6}\right]^i.$$

It is extremely cumbersome to use the last displayed formula since even the case $n=3$ involves the evaluation of 26 constants c_i. For even $n \in \mathbb{Z}^+$, we have

$$\prod_{m=1}^{\infty}(1-x^m)\sum_{m=1}^{\infty} P\left(5^n m - \frac{5^n-1}{24}\right)x^m = \sum_{i=1}^{\frac{5(5^n-1)}{24}} c_i \left[x\frac{\prod_{m=1}^{\infty}(1-x^{5m})^6}{\prod_{m=1}^{\infty}(1-x^m)^6}\right]^i.$$

As before, equation (5.46) holds. We rewrite the "last two" formulae as ($n \in \mathbb{Z}^+$)

$$\sum_{m=0}^{\infty} P(5^n m + l_n)x^{m+1}$$

$$= \begin{cases} \sum_{i=1}^{\frac{5^{n+1}-1}{24}} c_i x^i \sum_{m=0}^{\infty} P_{1-6i}(m)x^{5m}\sum_{m=0}^{\infty} P_{6i}(m)x^m, & n \text{ odd} \\ \sum_{i=1}^{\frac{5^{n+1}-5}{24}} c_i x^i \sum_{m=0}^{\infty} P_{-6i}(m)x^{5m}\sum_{m=0}^{\infty} P_{1+6i}(m)x^m, & n \text{ even} \end{cases}.$$

These formulae imply the recursion relations

$$P(5^n m + l_n)$$

$$= \begin{cases} \sum_{i=1}^{\frac{5^{n+1}-1}{24}} c_i \sum_{j=0}^{\left\lfloor\frac{m+1-i}{5}\right\rfloor} P_{1-6i}(j)P_{6i}(m+1-i-5j), & n \text{ odd} \\ \sum_{i=1}^{\frac{5^{n+1}-5}{24}} c_i \sum_{j=0}^{\left\lfloor\frac{m+1-i}{5}\right\rfloor} P_{-6i}(j)P_{1+6i}(m+1-i-5j), & n \text{ even} \end{cases}.$$

We start to prove the higher level congruences for the prime 5 with the calculations

$$F_{5,1,1} = F_{5,1} = 5f_5 \equiv 0 \mod 5,$$

$$F_{5,2,1} = 5V_5(f_5) = 5f_5^{-1}F_{5,6}$$

$$= 5(5^2F_{5,5} + 5^2F_{5,4} + 3\cdot 5F_{5,3} + 5F_{5,2} + 5f_5) \equiv 0 \mod 5^2.$$

Lemma 11.4. *We have for all $n \in \mathbb{Z}^+$,*

$$F_{5,n,1} = \begin{cases} \sum_{i=1}^{\frac{1}{24}(5^{n+1}-1)} c_i f_5^i, \ c_i \in \mathbb{Z}, \ 5^{n-2+\left\lfloor\frac{5i-1}{2}\right\rfloor} | c_i, & n \text{ odd} \\ \sum_{i=1}^{\frac{1}{24}(5^{m+1}-5)} c_i f_5^i, \ c_i \in \mathbb{Z}, \ 5^{n-2+\left\lfloor\frac{5i}{2}\right\rfloor} | c_i, & n \text{ even} \end{cases}.$$

Proof. The two calculations preceding the statement of the lemma have established the conclusion for $n = 1$ and 2. So let $n > 2$ and assume the results for indices below n. In the equations that follow, the c_i are integers. We show that the claim for $F_{5,n,1}$ for a given n proves the claim for $F_{5,n+1,1}$. We treat the case of even and odd indices separately. If n is even, then

$$F_{5,n,1} = V_5(F_{5,n-1,1}) = V_5\left(\sum_{i=1}^{\frac{1}{24}(5^n-1)} c_i f_5^i\right) = \sum_{i=1}^{\frac{1}{24}(5^m-1)} c_i f_5^{-i} F_{5,6i}$$

$$= \sum_{i=1}^{\frac{1}{24}(5^n-1)} c_i f_5^{-i} \sum_{j=\lceil \frac{6i}{5}\rceil}^{6i} c_{i,j} f_5^j = \sum_{i=1}^{\frac{1}{24}(5^{m+1}-5)} d_i f_5^i,$$

where, by induction, $5^{n-3+\lfloor \frac{5i-1}{2}\rfloor}|c_i$ and, by Lemma 8.5, $5^{\lfloor \frac{5j-6i-1}{2}\rfloor}|c_{i,j}$. Since

$$d_i = \sum_{j=\lceil \frac{i}{5}\rceil}^{5i} c_j c_{j,i+j}, \quad 5^{n-3+\min\{\lfloor \frac{5j-1}{2}\rfloor + \lfloor \frac{5i-j-1}{2}\rfloor ; j=\lceil \frac{6i}{5}\rceil,\ldots,6i\}}|d_i.$$

The fact that the minimum occurs at $j = 1$ (even if not in range) completes the induction argument (the exponent is then $n - 2 + \lfloor \frac{5i}{2}\rfloor$). If n is odd, then the proof proceeds along similar lines using

$$F_{5,n,1} = V_5(f_{5^2,1} F_{5,n-1,1}) = V_5\left(f_{5^2,1} \sum_{i=1}^{\frac{1}{24}(5^n-5)} c_i f_5^i\right) = \sum_{i=1}^{\frac{1}{24}(5^n-5)} c_i F_{5,1+6i}.$$

We have hence proven

Corollary 11.5. *For each n and $m \in \mathbb{Z}^+$,*

$$P\left(5^n m - \frac{5^{2\lfloor \frac{n+1}{2}\rfloor} - 1}{24}\right) \equiv 0 \mod 5^n.$$

Alternately, for each $n \in \mathbb{Z}^+$ and each $m \in \mathbb{Z}^+ \cup \{0\}$,

$$P(5^n m + l_n) \equiv 0 \mod 5^n.$$

The case $N = 2$ involves the formulae

$$\prod_{m=1}^{\infty}(1-x^{5m})^2 \sum_{m=1}^{\infty} P_2\left(5^n m - \frac{5^{n+1}-1}{12}\right) x^m$$

$$= \sum_{i=1}^{\frac{5^{n+1}-1}{12}} c_i \left[x \frac{\prod_{m=1}^{\infty}(1-x^{5m})^6}{\prod_{n=1}^{\infty}(1-x^m)^6}\right]^i, \ n \text{ odd},$$

and

$$\prod_{m=1}^{\infty}(1-x^m)^2 \sum_{m=1}^{\infty} P_2\left(5^n m - \frac{5^n-1}{12}\right) x^m$$

$$= \sum_{i=1}^{\frac{5(5^n-1)}{12}} c_i \left[x \frac{\prod_{m=1}^{\infty}(1-x^{5m})^6}{\prod_{m=1}^{\infty}(1-x^m)^6}\right]^i, \quad n \text{ even}.$$

It leads to: For each n and $m \in \mathbb{Z}^+$,

$$P_2\left(5^n m - \frac{5^{2\lfloor \frac{n+1}{2} \rfloor}-1}{12}\right) \equiv 0 \mod 5^{1+\lfloor \frac{n}{2} \rfloor}.$$

11.4. The level two congruences for the prime 11. A study of the level n congruences for the prime 11 can be based on an exploration of the equation (here $n \in \mathbb{Z}^+$)

$$F_{11,n,1} = \sum_{j=2}^{J} c_j (G_{11,-N(j)} - G_{11,-N(j)}(\infty)),$$

where $J = \begin{cases} \frac{11^{n+1}-1}{24}, & n \text{ odd} \\ 11\frac{11^n-1}{24}, & n \text{ even} \end{cases}$, or the equation

$$F_{11,n,1} = \sum_{j=1}^{\lfloor \frac{J}{2} \rfloor} c_j G_2^j + \sum_{i=1}^{\lfloor \frac{J-3}{2} \rfloor} G_2^i G_3,$$

where the functions G_2 and G_3 are defined in §10.5. The study involves techniques not developed in this book.

12. Taylor series expansions for infinite products

In our work we have established many formulae giving the Taylor series expansions for analytic functions defined as infinite products. We summarize these results below. In each case, we give, if known to us, the commonly assigned name (usually its discoverer) to the formula, the function involved in the proof or the formula, the formula relating a product with a sum and the location of its proof (if found in this book). In general, τ and x are related by either $x = \exp(2\pi\imath\tau)$ or $x = \exp(\pi\imath\tau)$. All products and series converge (absolutely and uniformly on compact subsets of the appropriate region) on $|x| < 1$, $z \neq 0$.

1. Euler[129]
   $\eta(\tau)$
   $\prod_{n=1}^{\infty}(1-x^n) = 1+\sum_{n=1}^{\infty}(-1)^n\left[x^{\frac{n(3n-1)}{2}}+x^{\frac{n(3n+1)}{2}}\right]$
   Chapter 2, §8.

2. $F_{7,1}(\tau), F_{13,-2}, \eta^2(\tau)$
   $2\prod_{n=1}^{\infty}(1-x^n)^2 = -x\sum_{n=0}^{\infty}P_{-2}(n)x^{13n}-\sum_{n=0}^{\infty}P_{-2}(13n+1)x^n$
   §5, 6.5.
   The formula displayed above should, of course, be contrasted with the definition: $\prod_{n=1}^{\infty}(1-x^n)^2 = \sum_{n=0}^{\infty}P_{-2}(n)x^n$.

3. Jacobi
   $\eta^3(8\tau)$
   $\prod_{n=1}^{\infty}(1-x^n)^3 = \sum_{n=0}^{\infty}(-1)^n(2n+1)x^{\frac{n(n+1)}{2}}$
   Chapter 2, §8.
   The above formula reinterprets slightly the identity (3.41) of Chapter 3 as

$$\theta'\begin{bmatrix}1\\1\end{bmatrix}(0,\tau) = 2\pi\imath\theta^3\begin{bmatrix}\frac{1}{3}\\1\end{bmatrix}(0,3\tau) = -2\pi\theta^3\begin{bmatrix}1\\\frac{1}{3}\end{bmatrix}\left(0,\frac{\tau}{3}\right) = -\pi\eta^3(\tau).$$

4. Jacobi
   $\eta^{-1}(2\tau)\eta^2(\tau), \theta\begin{bmatrix}0\\1\end{bmatrix}$
   $\prod_{n=1}^{\infty}(1-x^n)^2(1-x^{2n})^{-1} = 1+2\sum_{n=1}^{\infty}(-1)^n x^{n^2}$
   Chapter 2, §8.1.

5. $F_{2,1}(\tau),\ \eta^{16}(2\tau)\eta^{-24}(\tau)$
   $2^3x\prod_{n=1}^{\infty}(1-x^{2n})^{16}\prod_{n=1}^{\infty}(1-x^n)^{-24} = \sum_{m=1}^{\infty}P_8(2m-1)x^m$

[129]In relating the formula to the function, we have ignored the factor $x^{\frac{1}{24}}$ that usually multiplies each side of the formula. We have used similar cancellations in all the formulae on the list in this section.

§10.1.
This identity is algebraically equivalent to the Jacobi quartic.

6. $F_{2,1}(\tau),\ \eta^{24}(2\tau)\eta^{-8}(\tau)$
$$2^4\prod_{n=1}^{\infty}(1-x^n)^{-8}(1-x^{2n})^{24}=2^7x\sum_{n=0}^{\infty}P_{-16}(n)x^{4n}-\sum_{n=0}^{\infty}P_{-16}(2n+1)x^{2n}$$
§5.

7. $F_{2,1}(\tau),\eta^{24}(2\tau)$
$$2^3 3\prod_{n=1}^{\infty}(1-x^n)^{24}=-2^{11}x\sum_{n=0}^{\infty}P_{-24}(n)x^{2n}-\sum_{n=0}^{\infty}P_{-24}(2n+1)x^n$$
§5.
The above formula is a special case of Mordell's theorem: For every prime k,
$$P_{-24}(k-1)\prod_{n=1}^{\infty}(1-x^n)^{24}$$
$$=k^{11}x^{k-1}\sum_{n=0}^{\infty}P_{-24}(n)x^{kn}+\sum_{n=0}^{\infty}P_{-24}(kn+k-1)x^n.$$
Chapter 4, §5.

8. $F_{3,1}(\tau),\eta^9(3\tau)\eta^{-12}(\tau)$
$$3^2x\prod_{n=1}^{\infty}(1-x^{3n})^9(1-x^n)^{-12}=\sum_{n=1}^{\infty}P_3(3n-1)x^n$$
§10.2.

9. $F_{3,1}(\tau),\eta^{12}(3\tau)\eta^{-3}(\tau)$
$$3^2x\prod_{n=1}^{\infty}(1-x^{3n})^{12}(1-x^n)^{-3}$$
$$=3^4x^3\sum_{n=0}^{\infty}P_{-9}(n)x^{9n}+\sum_{n=0}^{\infty}x^{3n+1}\left(P_{-9}(3n+2)x-P_{-9}(3n+1)\right)$$
§5.

10. $F_{3,1}(\tau),\eta^{12}(3\tau)$
$$2^2 3x\prod_{n=1}^{\infty}(1-x^{3n})^{12}=-3^5x^4\sum_{n=0}^{\infty}P_{-12}(n)x^{9n}-\sum_{n=0}^{\infty}P_{-12}(3n+1)x^{3n+1}$$
§5.

11. $F_{3,1}(\tau), \eta^{12}(3\tau)\eta^3(\tau)$

$$2 \cdot 3 \cdot 5 \prod_{n=1}^{\infty} (1 - x^{3n})^{12}(1 - x^n)^3$$

$$= 3^7 x^4 \sum_{n=0}^{\infty} P_{-15}(n)x^{9n} - \sum_{n=0}^{\infty} x^{3n} \left(2P_{-15}(3n+1) + P_{-15}(3n+2)x\right)$$

§5.

12. Jacobi

$\eta^{-2}(4\tau)\eta^5(2\tau)\eta^{-2}(\tau), \theta \begin{bmatrix} 0 \\ 0 \end{bmatrix}$

$$\prod_{n=1}^{\infty} (1 - x^{2n})^5 (1 - x^n)^{-2} (1 - x^{4n})^{-2} = 1 + 2 \sum_{n=1}^{\infty} x^{n^2}.$$

13. $F_{5,-5}, \eta^{-1}(5\tau)\eta^6(\tau)$

$$x^{-1} \prod_{n=1}^{\infty} (1 - x^{5n})^{-1}(1 - x^n)^6 = \sum_{m=-1}^{\infty} P_{-5}(5m+5)x^m$$

§10.3.

14. Ramanujan

$F_{5,1}(\tau), \eta^5(5\tau)\eta^{-6}(\tau)$

$$5x \prod_{n=1}^{\infty} (1 - x^{5n})^5 (1 - x^n)^{-6} = \sum_{n=1}^{\infty} P(5n-1)x^n$$

§10.3.

15. $F_{5,1}(\tau), \eta^6(5\tau)\eta^{-1}(\tau)$

$$5 \prod_{n=1}^{\infty} (1 - x^{5n})^6 (1 - x^n)^{-1} = 5^2 x^4 \sum_{n=0}^{\infty} P_{-5}(n)x^{25n}$$

$$+ \sum_{n \in \mathbb{Z}^+, n \equiv 2,3 \mod 5} P_{-5}(n)x^{n-1} - \sum_{n \in \mathbb{Z}^+, n \equiv 1,4 \mod 5} P_{-5}(n)x^{n-1}$$

§5.

16. $F_{5,1}(\tau), \eta^6(5\tau)$

$$2 \cdot 3 \prod_{n=1}^{\infty} (1 - x^n)^6 = 5^2 x \sum_{n=0}^{\infty} P_{-6}(n)x^{5n} + \sum_{n=0}^{\infty} P_{-6}(5n+1)x^n$$

§5.

17. $F_{5,1}(\tau), \eta^6(5\tau)\eta(\tau)$

$$2 \cdot 7x \prod_{n=1}^{\infty}(1-x^{5n})^6(1-x^n) = 5^3x^7 \sum_{m=0}^{\infty} P_{-7}(n)x^{25n}$$
$$- \sum_{n\in\mathbb{Z}^+, n\equiv 2 \mod 5} P_{-7}(n)x^n - 2 \sum_{n\in\mathbb{Z}^+, m\equiv 1,3 \mod 5} P_{-7}(n)x^n$$

§5.

18. $F_{5,1}(\tau), \eta^6(5\tau)\eta^3(\tau)$

$$2 \cdot 3^2x \prod_{n=1}^{\infty}(1-x^{5n})^6(1-x^n)^3 = -5^4x^9 \sum_{n=0}^{\infty} P_{-9}(n)x^{25n}$$
$$- 2 \sum_{n\in\mathbb{Z}^+, n\equiv 1,2 \mod 5} P_{-9}(n)x^n - \sum_{n\in\mathbb{Z}^+, n\equiv 4 \mod 5} P_{-9}(n)x^n$$

§5.

19. $F_{5,-4}(\tau)$

$$\prod_{m=1}^{\infty}(1-x^{5m})^4 = (-5)^{-(n+1)} \sum_{m=0}^{\infty} P_{-4}\left(5^{2n+1}m + \frac{5^{2(n+1)}-1}{6}\right) x^m,$$

for all $n \in \mathbb{Z}^+ \cup \{0\}$

§6.4.

20. $F_{5,-4}(\tau)$

$$\prod_{m=1}^{\infty}(1-x^m)^4 = (-5)^{-n} \sum_{m=0}^{\infty} P_{-4}\left(5^{2n}m + \frac{5^{2n}-1}{6}\right) x^m, \text{ for all } n \in \mathbb{Z}^+$$

§6.4.

The last two formulae are sample illustrations of the many similar formulae that are consequences of the fact that the functions $F_{k,N}$ are constant for a number of primes k and negative integers N.

21. $G_{5,1}(\tau)$

$$\prod_{n=1}^{\infty}(1-x^n)^5 \prod_{n=1}^{\infty}(1-x^{5n})^{-6} = \sum_{j=0}^{\infty} P(j)x^{5^2 j}$$
$$- 5x\left(\sum_{j\in\mathbb{Z}^+\cup\{0\},\ j\equiv 0,3 \mod 5} P(j)x^j - \sum_{j\in\mathbb{Z}^+,\ j\equiv 1,2 \mod 5} P(j)x^j\right)$$

§10.3.

The above formula followed from the fact that f_5^{-1} is a constant multiple of $G_{5,1}$. Using the fact that f_5 is a constant multiple of $G_{5,-N} - G_{5,-N}(\infty)$ for $N = 5$, 6, 7, and 9, we obtain the Taylor series expansion for

$$\prod_{n=1}^{\infty}(1-x^{5n})^5\prod_{n=1}^{\infty}(1-x^n)^{-6+N}.$$

Note that $G_{5,-N}(\infty)=0$.

22. Ramanujan
$\eta^{-1}(5\tau)\eta^5(\tau)$
$$\prod_{n=1}^{\infty}\frac{(1-x^n)^5}{(1-x^{5n})}=1-5\sum_{n=1}^{\infty}f(n)x^n,\ f(n)=\sum_{d|n}\chi(d)d,$$
$$\chi(n)=\begin{cases}+1 \text{ if } n\equiv\pm1 \mod 5\\ -1 \text{ if } n\equiv\pm2 \mod 5\\ 0 \text{ if } n\equiv 0 \mod 5\end{cases}$$
Chapter 4, §9.

23. $\eta^5(5\tau)\eta^{-1}(\tau)$
$$x\prod_{n=1}^{\infty}\frac{(1-x^{5n})^5}{(1-x^n)}=\sum_{n=1}^{\infty}g(n)x^n,\ g(n)=\textstyle\sum_{d|n}\chi\left(\frac{n}{d}\right)d$$
Chapter 4, §9. See also 15.

24. Köhler-Macdonald
$\eta^5(6\tau)\eta^{-2}(3\tau)$
$$x\prod_{n=1}^{\infty}(1-x^{6n})^5(1-x^{3n})^{-2}=\sum_{n=1}^{\infty}(-1)^{n-1}\left(\frac{n}{3}\right)nx^{n^2}.$$

25. $F_{7,1}(\tau),\eta^4(7\tau)$
$$-2^2\prod_{n=1}^{\infty}(1-x^n)^4=7x\sum_{n=0}^{\infty}P_{-4}(n)x^{7n}+\sum_{n=0}^{\infty}P_{-4}(7n+1)x^n$$
§5.

26. $F_{7,1}(\tau),\eta^4(7\tau)\eta(\tau)$
$$2\cdot 5x\prod_{n=1}^{\infty}(1-x^{7n})^4(1-x^n)=7^2x^{10}\sum_{n=0}^{\infty}P_{-5}(n)x^{49n}$$
$$-\sum_{n\in\mathbb{Z}^+,n\equiv3 \mod 7}P_{-5}(n)x^n-2\sum_{n\in\mathbb{Z}^+,n\equiv1,2,6 \mod 7}P_{-5}(n)x^n$$
§5.

27. $\eta^3(9\tau)\eta^{-2}(3\tau)\eta^3(\tau)$
$$x\prod_{n=1}^{\infty}\frac{(1-x^{9n})^3(1-x^n)^3}{(1-x^{3n})^2}=\sum_{k=0}^{\infty}\sigma(3k+1)x^{3k+1}-\sigma(3k+2)x^{3k+2}$$
Chapter 7, §3.2.

28. $\theta\begin{bmatrix}0\\0\end{bmatrix}$
$$\sum_{n=-\infty}^{\infty}x^{n^2}=\prod_{n=1}^{\infty}(1+x^{2n})(1+x^{2n-1})^2$$
Chapter 2, §8.1.

29. $\theta\begin{bmatrix}0\\1\end{bmatrix}$
$$\sum_{n=-\infty}^{\infty}(-1)^n x^{n^2}=\prod_{n=1}^{\infty}(1-x^{2n})(1-x^{2n-1})^2$$
Chapter 2, §8.1.

30. $\theta\begin{bmatrix}1\\0\end{bmatrix}$
$$\sum_{n=0}^{\infty}x^{n(n+1)}=\prod_{n=1}^{\infty}(1-x^{2n})(1+x^{2n})^2=\prod_{n=1}^{\infty}(1+x^{2n})(1-x^{4n})$$
Chapter 2, §8.1.

31. Jacobi triple product
$\theta(\zeta,\tau)$
$$\prod_{n=1}^{n=\infty}(1-x^n)(1+x^n z)\left(1+\frac{x^{n-1}}{z}\right)=\sum_{n=-\infty}^{n=\infty}x^{\frac{n(n+1)}{2}}z^n$$
Chapter 2, §8.1.

32. Quintuple product
$$(1+z)\prod_{n=1}^{\infty}(1-x^n)(1+x^n z)\left(1+\frac{x^n}{z}\right)(1-x^{2n-1}z^2)\left(1-\frac{x^{2n-1}}{z^2}\right)$$
$$=\sum_{n=-\infty}^{\infty}(-1)^n x^{\frac{n(3n-1)}{2}}z^{3n}+z\sum_{n=-\infty}^{\infty}(-1)^n x^{\frac{n(3n+1)}{2}}z^{3n}$$
Chapter 2, §8.2.

33. Septuple product

$$(1+z)(1-z)^2 \prod_{n=1}^{\infty} (1-x^n)^2(1-x^n z)\left(1-\frac{x^n}{z}\right)(1-x^n z^2)\left(1-\frac{x^n}{z^2}\right)$$

$$= \sum_{n=-\infty}^{\infty} (-1)^n x^{\frac{n(5n+1)}{2}}$$

$$\times \left(\sum_{n=-\infty}^{\infty} (-1)^n x^{\frac{n(5n+3)}{2}} z^{5n+3} + \sum_{n=-\infty}^{\infty} (-1)^n x^{\frac{n(5n-3)}{2}} z^{5n} \right)$$

$$- \sum_{n=-\infty}^{\infty} (-1)^n x^{\frac{n(5n+3)}{2}}$$

$$\times \left(\sum_{n=-\infty}^{\infty} (-1)^n x^{\frac{n(5n+1)}{2}} z^{5n+2} + \sum_{n=-\infty}^{\infty} (-1)^n x^{\frac{n(5n-1)}{2}} z^{5n+1} \right)$$

Chapter 4, §1.1.

34. $\prod_{n=1}^{\infty} (1-x^{2n-1})^2(1-x^n)^3 = \sum_{n=-\infty}^{\infty} (6n+1)x^{\frac{n(3n+1)}{2}}$

Chapter 2, §8.2.

Remark 12.1. 1. Only two of the above formulae were not established in the manuscript. The first of these (Jacobi) is an immediate consequence of the Jacobi triple product. The second (Köhler-Macdonald) is discussed in the bibliographical notes.

2. Not all the formulae derived are independent. We have listed in item 7 of the above list two formulae for an infinite product, derived by different methods. The first of these is obviously a special case of the second. The last three formulae listed (that can be represented by the symbols JTP, QPI and SPI, respectively) demand a unifying generalization.

Exercise 12.2. The formulae for the Jacobi triple product identity in Item 31 of the above list is not the same as the one given earlier in the text. Prove that the two forms are equivalent.

Chapter 6

Identities related to partition functions

In this chapter we begin with some simple observations relating fundamental groups and coverings and once again derive some identities among theta constants. We point out that some of the identities we obtain are quite elementary and may be viewed as tautologies, but others seem to be quite deep. We then begin a study of the roots of the discriminant function Δ, "the" nontrivial 6-cusp form for Γ; and the $\jmath$-function, "the" uniformizer for $\mathbb{H}^2/\Gamma$. We show that the congruences obtained for the partition function in the previous chapter are part of a general theory for coefficients of modular forms. We do not go into this theory, but do just enough to convince the reader that there is a lot more to learn.

1. Some more identities related to covering maps

For all $k \in \mathbb{Z}^+$ we have the following commutative diagram of group inclusions (represented by arrows $\longrightarrow$)

$$(6.1)\qquad \begin{array}{ccc} \Gamma(k) & \longrightarrow & \Gamma(k,k) \\ \downarrow & & \downarrow \\ G(k) & \longrightarrow & \Gamma_o(k) \end{array} .$$

For primes k, the vertical inclusions are of index k and the horizontal of index $\lfloor \frac{k-1}{2} \rfloor$; in particular,

$$\Gamma(k) = \Gamma(k,k) \text{ and } G(k) = \Gamma_o(k) \text{ iff } k = 2 \text{ or } 3.$$

1.1. $k = 2$. Based on the work in §3.2 of Chapter 3 and §4.2 of Chapter 5, we see that $\dfrac{\theta^4\begin{bmatrix}0\\1\end{bmatrix}}{\theta^4\begin{bmatrix}1\\0\end{bmatrix}}$ is a $\Gamma(2)$-invariant meromorphic function with divisor $\frac{P_0}{P_\infty}$, while $\frac{\eta^8(\frac{\tau}{2})}{\eta^8(2\tau)}$ is a $\Gamma(2,2) = \Gamma(2)$-invariant meromorphic function with the same divisor. We thus derive the θ-identity

$$\frac{\theta^4\begin{bmatrix}0\\1\end{bmatrix}(0,\tau)}{\theta^4\begin{bmatrix}1\\0\end{bmatrix}(0,\tau)} = \frac{1}{2^4}\frac{\theta'^{\frac{8}{3}}\begin{bmatrix}1\\1\end{bmatrix}(0,\frac{\tau}{2})}{\theta'^{\frac{8}{3}}\begin{bmatrix}1\\1\end{bmatrix}(0,2\tau)}.$$

The above is transparent in terms of the infinite product expansions of the respective sides of the identity in the local coordinate $x = \exp(\pi\imath\tau)$. It can hence be viewed as a trivial identity. Similarly, $\dfrac{\theta^4\begin{bmatrix}0\\0\end{bmatrix}\theta^4\begin{bmatrix}0\\1\end{bmatrix}}{\theta^8\begin{bmatrix}1\\0\end{bmatrix}}$ is a $G(2)$-invariant meromorphic function with divisor $\frac{P_0}{P_\infty}$, while $\frac{\eta^{24}(\tau)}{\eta^{24}(2\tau)}$ is a $\Gamma_o(2) = G(2)$-invariant meromorphic function with the same divisor. We thus establish the identity

$$\frac{\theta^4\begin{bmatrix}0\\0\end{bmatrix}(0,\tau)\theta^4\begin{bmatrix}0\\1\end{bmatrix}(0,\tau)}{\theta^8\begin{bmatrix}1\\0\end{bmatrix}(0,\tau)} = \frac{1}{2^8}\frac{\theta'^8\begin{bmatrix}1\\1\end{bmatrix}(0,\tau)}{\theta'^8\begin{bmatrix}1\\1\end{bmatrix}(0,2\tau)}.$$

The resulting identity is again trivial.

1.2. $k = 3$. The quotient $u = \dfrac{\theta^3\begin{bmatrix}\frac{1}{3}\\1\end{bmatrix}}{\theta^3\begin{bmatrix}1\\\frac{1}{3}\end{bmatrix}}$ is a $\Gamma(3)$-invariant meromorphic function with divisor $\frac{P_0}{P_\infty}$, while $v = \frac{\eta^3(\frac{\tau}{3})}{\eta^3(3\tau)}$ is a $\Gamma(3,3) = \Gamma(3)$-invariant meromorphic function with the same divisor. The equality among functions

$$\frac{\theta^3\begin{bmatrix}\frac{1}{3}\\1\end{bmatrix}(0,\tau)}{\theta^3\begin{bmatrix}1\\\frac{1}{3}\end{bmatrix}(0,\tau)} = \frac{\sqrt{3}\imath}{3^2}\frac{\theta'\begin{bmatrix}1\\1\end{bmatrix}(0,\frac{\tau}{3})}{\theta'\begin{bmatrix}1\\1\end{bmatrix}(0,3\tau)}$$

is, in view of (4.8) of Chapter 4 (see also (3.41) of Chapter 3), equivalent to the easily derived identity

$$\theta\begin{bmatrix}\frac{1}{3}\\ 1\end{bmatrix}(0,3^2\tau) = \left(\frac{\sqrt{3}}{2}+\frac{1}{2}\imath\right)\theta\begin{bmatrix}1\\ \frac{1}{3}\end{bmatrix}(0,\tau).$$

The $G(3)$-invariant function $U = \dfrac{\theta^3\begin{bmatrix}\frac{1}{3}\\ \frac{1}{3}\end{bmatrix}\theta^3\begin{bmatrix}\frac{1}{3}\\ 1\end{bmatrix}\theta^3\begin{bmatrix}\frac{1}{3}\\ \frac{5}{3}\end{bmatrix}}{\theta^9\begin{bmatrix}1\\ \frac{1}{3}\end{bmatrix}}$ with divisor $\frac{P_0}{P_\infty}$ is a nonzero constant multiple of the $\Gamma_o(3) = G(3)$-invariant meromorphic function $V = \frac{\eta^{12}(\tau)}{\eta^{12}(3\tau)}$. We hence have the identity

$$\frac{\theta^3\begin{bmatrix}\frac{1}{3}\\ \frac{1}{3}\end{bmatrix}(0,\tau)\theta^3\begin{bmatrix}\frac{1}{3}\\ 1\end{bmatrix}(0,\tau)\theta^3\begin{bmatrix}\frac{1}{3}\\ \frac{5}{3}\end{bmatrix}(0,\tau)}{\theta^9\begin{bmatrix}1\\ \frac{1}{3}\end{bmatrix}(0,\tau)} = -\frac{\sqrt{3}}{3^5}\imath\frac{\theta'^4\begin{bmatrix}1\\ 1\end{bmatrix}(0,\tau)}{\theta'^4\begin{bmatrix}1\\ 1\end{bmatrix}(0,3\tau)}.$$

The identity is once again rather transparent in its infinite product form. In this case, we have another way to produce a $\Gamma(3)$-invariant function; the methods of §8.1 of Chapter 3 tell us that $\dfrac{\theta'\begin{bmatrix}\frac{1}{3}\\ 1\end{bmatrix}(0,3\tau)}{\theta'\begin{bmatrix}1\\ 1\end{bmatrix}(0,3\tau)}$ is such a function with divisor $\frac{P_a}{P_\infty}$ for some $a \in \mathbb{H}^2$. Comparing this function with v leads to the θ-identity (3.40) and its power series analogue (3.42).

In the following commutative diagram (a consequence of (6.1)) of maps between compact Riemann surfaces:

$$\begin{array}{ccc}\overline{\mathbb{H}^2/\Gamma(3)} & \longrightarrow & \overline{\mathbb{H}^2/\Gamma(3,3)}\\ \downarrow & & \downarrow\\ \overline{\mathbb{H}^2/G(3)} & \longrightarrow & \overline{\mathbb{H}^2/\Gamma_o(3)}\end{array},$$

where the vertical maps are of degree 3 and the horizontal maps are conformal equivalences. The field of meromorphic functions on each of these surfaces is isomorphic to the rational functions. Using the functions u, v, U and V as generators for the respective fields, we can express the above maps by the equations

$$v = -3\sqrt{3}\imath u,\quad V = 3^4\sqrt{3}\imath U,\quad U = u^3+\sqrt{3}\imath u^2 - u,$$

and

$$V = v^3 + 3^2 v^2 + 3^3 v = 3^4\sqrt{3}\imath u^3 + 3^5 \imath u^2 - 3^4\sqrt{3}\imath u.$$

1.3. $k = 5$**.** The function $\frac{\eta(\frac{\tau}{5})}{\eta(5\tau)}$ is $\Gamma(5,5)$-invariant with divisor $\frac{P_0}{P_\infty}$; hence certainly $\Gamma(5)$-invariant with divisor $\frac{P_0 P_{\frac{5}{2}}}{P_\infty P_{\frac{2}{5}}}$. We also know that $\frac{\theta\begin{bmatrix}\frac{1}{5}\\1\end{bmatrix}(0,5\tau)}{\theta\begin{bmatrix}\frac{3}{5}\\1\end{bmatrix}(0,5\tau)}$ is $\Gamma(5)$-invariant with divisor $\frac{P_{\frac{2}{5}}}{P_\infty}$. Hence

$$\frac{\theta'^{\frac{1}{3}}\begin{bmatrix}1\\1\end{bmatrix}(0,\frac{\tau}{5})}{\theta'^{\frac{1}{3}}\begin{bmatrix}1\\1\end{bmatrix}(0,5\tau)}$$

$$= -1 + \exp\left(\frac{\pi\imath}{5}\right)\frac{\theta\begin{bmatrix}\frac{1}{5}\\1\end{bmatrix}(0,5\tau)}{\theta\begin{bmatrix}\frac{3}{5}\\1\end{bmatrix}(0,5\tau)} - \exp\left(\frac{-\pi\imath}{5}\right)\frac{\theta\begin{bmatrix}\frac{3}{5}\\1\end{bmatrix}(0,5\tau)}{\theta\begin{bmatrix}\frac{1}{5}\\1\end{bmatrix}(0,5\tau)}.$$

The translation to infinite products is

$$\prod_{n=1}^{\infty}\frac{1-x^n}{1-x^{25n}} + x$$

$$= \prod_{n=0}^{\infty}\frac{(1-x^{25n+10})(1-x^{25n+15})}{(1-x^{25n+5})(1-x^{25n+20})} - x^2\prod_{n=0}^{\infty}\frac{(1-x^{25n+5})(1-x^{25n+20})}{(1-x^{25n+10})(1-x^{25n+15})}.$$

It is quite easy to derive many similar identities. We give one more example. The function $\frac{\theta\begin{bmatrix}\frac{1}{5}\\1\end{bmatrix}(0,\tau)\theta\begin{bmatrix}\frac{3}{5}\\1\end{bmatrix}(0,\tau)}{\theta\begin{bmatrix}\frac{1}{5}\\1\end{bmatrix}(0,5\tau)\theta\begin{bmatrix}\frac{3}{5}\\1\end{bmatrix}(0,5\tau)}$ is $\Gamma(5)$-invariant with divisor $\frac{P_0 P_{\frac{5}{2}}}{P_\infty P_{\frac{2}{5}}}$. One sees that it is indeed a function by computing its divisor. Hence

$$\frac{\theta\begin{bmatrix}\frac{1}{5}\\1\end{bmatrix}(0,\tau)\theta\begin{bmatrix}\frac{3}{5}\\1\end{bmatrix}(0,\tau)}{\theta\begin{bmatrix}\frac{1}{5}\\1\end{bmatrix}(0,5\tau)\theta\begin{bmatrix}\frac{3}{5}\\1\end{bmatrix}(0,5\tau)}$$

$$= -1 + \exp\left(\frac{\pi\imath}{5}\right)\frac{\theta\begin{bmatrix}\frac{1}{5}\\1\end{bmatrix}(0,5\tau)}{\theta\begin{bmatrix}\frac{3}{5}\\1\end{bmatrix}(0,5\tau)} + \imath\frac{\theta\begin{bmatrix}\frac{3}{5}\\1\end{bmatrix}(0,5\tau)}{\theta\begin{bmatrix}\frac{1}{5}\\1\end{bmatrix}(0,5\tau)}.$$

The translation to infinite products is

$$\prod_{n=1}^{\infty}\frac{(1-x^n)(1-x^{5n})}{(1-x^{25n})^2}=\prod_{n=0}^{\infty}(1-x^{25n+10})^2(1-x^{25n+15})^2$$

$$-x\prod_{n=0}^{\infty}(1-x^{25n+5})(1-x^{25n+10})(1-x^{25n+15})(1-x^{25n+20})$$

$$-x^2\prod_{n=0}^{\infty}(1-x^{25n+5})^2(1-x^{25n+20})^2.$$

It is more interesting to work with modified theta constants (§4 of Chapter 3). The function $\frac{\varphi_0}{\varphi_1}$ is $\Gamma(5)$-invariant with divisor $\frac{P_{\frac{2}{5}}}{P_{\frac{1}{5}}}$, and $\frac{\varphi_0^5}{\varphi_1^5}$ is $G(5)$-invariant with the same divisor (on a different surface). Hence averaging over $G(5)\backslash\Gamma_o(5)$, we get a function with a simple pole at P_∞. Comparing this function with f_5^{-1} (§4.1 of Chapter 5), we get the identity

$$\frac{\varphi_0^5(\tau)}{\varphi_1^5(\tau)}+c_1\frac{\varphi_1^5(\tau)}{\varphi_0^5(\tau)}=c_2\left(\frac{\eta(\tau)}{\eta(5\tau)}\right)^6+c_3,$$

with constants c_1 (of absolute value 1), c_2 and c_3 to be determined. The resulting identity in $x=\exp(2\pi\imath\tau)$ (and a number of subsequent identities) is a consequence of (2.52), which in the current notation for the prime k reads

$$\theta\left[\begin{array}{c}\frac{2l+1}{k}\\1\end{array}\right](0,k\tau)$$

$$=x^{\frac{(2l+1)^2}{8k}}\exp\left(\frac{\pi\imath(2l+1)}{2k}\right)\prod_{n=0}^{\infty}(1-x^{k(n+1)})(1-x^{\frac{k+1}{2}+kn+l})(1-x^{\frac{k-1}{2}+kn-l}),$$

and infinite product expansions for the η-function. In the present case, we have

$$\prod_{n=0}^{\infty}\frac{(1-x^{5n+2})^5(1-x^{5n+3})^5}{(1-x^{5n+1})^5(1-x^{5n+4})^5}-x^2\prod_{n=0}^{\infty}\frac{(1-x^{5n+1})^5(1-x^{5n+4})^5}{(1-x^{5n+2})^5(1-x^{5n+3})^5}$$

$$=\prod_{n=1}^{\infty}\frac{(1-x^n)^6}{(1-x^{5n})^6}+11x.$$

1.4. $k=7$. We start the reader on a route to exploit further the commutative diagram (6.1). Our first observation is that the infinite product version of equation (3.44) of Chapter 3 is

$$\prod_{n=0}^{\infty}(1-x^{7n+1})(1-x^{7n+3})^3(1-x^{7n+4})^3(1-x^{7n+6})$$

$$-\prod_{n=0}^{\infty}(1-x^{7n+2})^3(1-x^{7n+3})(1-x^{7n+4})(1-x^{7n+5})^3$$
$$+x\prod_{n=0}^{\infty}(1-x^{7n+1})^3(1-x^{7n+2})(1-x^{7n+5})(1-x^{7n+6})^3=0.$$

The divisors of the functions $\left(\frac{\eta(\tau)}{\eta(7\tau)}\right)^4$, $\frac{\varphi_0^3}{\varphi_1\varphi_2^2}$, $\frac{\varphi_1^3}{\varphi_2\varphi_0^2}$ and $\frac{\varphi_2^3}{\varphi_0\varphi_1^2}$ on $\overline{\mathbb{H}^2/G(7)}$ are $\frac{P_0P_{\frac{7}{2}}P_{\frac{7}{3}}}{P_\infty P_{\frac{2}{7}}P_{\frac{3}{7}}}$, $\frac{P_{\frac{2}{7}}}{P_\infty}$, $\frac{P_{\frac{3}{7}}}{P_{\frac{2}{7}}}$ and $\frac{P_\infty}{P_{\frac{3}{7}}}$, respectively. Hence

$$\left(\frac{\eta(\tau)}{\eta(7\tau)}\right)^4=\exp\left(\frac{5\pi\imath}{7}\right)\frac{\varphi_0^3}{\varphi_1\varphi_2^2}-\exp\left(-\frac{2\pi\imath}{7}\right)\frac{\varphi_1^3}{\varphi_2\varphi_0^2}-\exp\left(-\frac{4\pi\imath}{7}\right)\frac{\varphi_2^3}{\varphi_0\varphi_1^2}-5;$$

alternately we have the infinite product identity

$$\prod_{n=1}^{\infty}\frac{(1-x^n)^4}{(1-x^{7n})^4}=\prod_{n=0}^{\infty}\frac{(1-x^{7n+3})^3(1-x^{7n+4})^3}{(1-x^{7n+1})^2(1-x^{7n+2})(1-x^{7n+5})(1-x^{7n+6})^2}$$
$$-x\prod_{n=0}^{\infty}\frac{(1-x^{7n+2})^3(1-x^{7n+5})^3}{(1-x^{7n+1})(1-x^{7n+3})^2(1-x^{7n+4})^2(1-x^{7n+6})}$$
$$-x^2\prod_{n=0}^{\infty}\frac{(1-x^{7n+1})^3(1-x^{7n+6})^3}{(1-x^{7n+2})^2(1-x^{7n+3})(1-x^{7n+4})(1-x^{7n+5})^2}-5x.$$

Exercise 1.1. Verify the identities for the primes $k=2$, 3, 5 and 7 described in the previous parts of this chapter and provide complete details for their derivations.

1.5. $k=11$. (See §8.5 of Chapter 3.) Let us start on a more systematic exploration of the function theory for the prime 11 based on much of our previous work; particularly, on our work on the space $V(11)$ introduced in Chapter 3. To compute the orders of the modified theta constants, we introduce a generator $\gamma=\begin{bmatrix}2&-1\\11&-5\end{bmatrix}$ of $\Gamma_o(11)/G(11)$. The first 4 powers of the automorphism induced by γ on $\mathbb{H}^2/\Gamma(11)$ send $P_{\frac{1}{11}}$ to $P_{\frac{2}{11}}$, $P_{\frac{4}{11}}$, $P_{\frac{3}{11}}$, $P_{\frac{5}{11}}$, respectively. To compute the induced permutation on the modified theta constants, we consider the action of $\tilde{\gamma}=\begin{bmatrix}2&-11\\1&-5\end{bmatrix}$ on characteristics. The first 4 powers of this motion send the equivalence class of the characteristic $\begin{bmatrix}\frac{1}{11}\\1\end{bmatrix}$ to the equivalence classes of the characteristics $\begin{bmatrix}\frac{9}{11}\\1\end{bmatrix}$, $\begin{bmatrix}\frac{7}{11}\\1\end{bmatrix}$, $\begin{bmatrix}\frac{3}{11}\\1\end{bmatrix}$, $\begin{bmatrix}\frac{5}{11}\\1\end{bmatrix}$, respectively. Hence the first 4 powers of $\gamma_{\frac{1}{4}}^*$ send φ_0 to a constant multiple of φ_4, φ_3, φ_1, φ_2, respectively. The above information is more than enough to compute the entries in Table 25. We saw in §6.3 of

	$P_{\frac{1}{11}}$	$P_{\frac{2}{11}}$	$P_{\frac{4}{11}}$	$P_{\frac{3}{11}}$	$P_{\frac{5}{11}}$
φ_0	$\frac{1}{8}$	$\frac{81}{8}$	$\frac{49}{8}$	$\frac{9}{8}$	$\frac{25}{8}$
φ_4	$\frac{81}{8}$	$\frac{49}{8}$	$\frac{9}{8}$	$\frac{25}{8}$	$\frac{1}{8}$
φ_3	$\frac{49}{8}$	$\frac{9}{8}$	$\frac{25}{8}$	$\frac{1}{8}$	$\frac{81}{8}$
φ_1	$\frac{9}{8}$	$\frac{25}{8}$	$\frac{1}{8}$	$\frac{81}{8}$	$\frac{49}{8}$
φ_2	$\frac{25}{8}$	$\frac{1}{8}$	$\frac{81}{8}$	$\frac{49}{8}$	$\frac{9}{8}$

Table 25. THE ORDERS OF THE MODIFIED THETA CONSTANTS AT THE DISTINGUISHED PUNCTURES OF $\mathbb{H}^2/\Gamma(11)$.

Chapter 3 that the functions $\frac{\varphi_1}{\varphi_0}$ and $\frac{\varphi_2}{\varphi_1}$ generate the function field $\mathcal{K}(\Gamma(11))$. The functions $\phi = \frac{\varphi_0^2}{\varphi_1\varphi_4}$ and $\frac{\varphi_0\varphi_3}{\varphi_2^2}$ are thus in $\mathcal{K}(G(11))$ and the discussion in §6.3 easily shows, because we computed their divisors, that these functions generate the function field.

As an exercise, the reader should determine the algebraic relations satisfied by the respective generators of $\mathcal{K}(\Gamma(11))$ and $\mathcal{K}(G(11))$ and the period matrices of the tori $\overline{\mathbb{H}^2/G(11)}$ and $\overline{\mathbb{H}^2/\Gamma_o(11)}$. We are particularly interested in producing functions on $\overline{\mathbb{H}^2/\Gamma_o(11)}$ and relating them to previous constructions. For each $N \in \mathbb{Z}$, the function $\Phi_N = \sum_{i=0}^{4} \phi^N \circ \gamma^i \in \mathcal{K}(\Gamma_o(11))$. The function Φ_1 has at most a simple pole at P_∞ and hence must be constant (so is Φ_{-1}). The fact that Φ_1 is constant translates to

(6.2)
$$\varphi_0^3\varphi_2\varphi_3 + c_1\varphi_0\varphi_1\varphi_4^3 + c_2\varphi_2\varphi_3^3\varphi_4 + c_3\varphi_0\varphi_1^3\varphi_3 + c_4\varphi_1\varphi_2^3\varphi_4 = c_5\varphi_0\varphi_1\varphi_2\varphi_3\varphi_4,$$

with constants c_i that can, with perseverance, be determined. It is more convenient (using the consequences of the Jacobi triple product) to translate the identity to one involving infinite products and to evaluate the constants at that stage. The resulting identity in $x = \exp(2\pi\imath\tau)$ (after computing constants) reads

$$\frac{1}{x}\prod_{n=0}^{\infty}\frac{(1-x^{11n+5})^2(1-x^{11n+6})^2}{(1-x^{11n+1})(1-x^{11n+4})(1-x^{11n+7})(1-x^{11n+10})}$$

$$-x\prod_{n=0}^{\infty}\frac{(1-x^{11n+1})^2(1-x^{11n+10})^2}{(1-x^{11n+2})(1-x^{11n+3})(1-x^{11n+8})(1-x^{11n+9})}$$

$$+x\prod_{n=0}^{\infty}\frac{(1-x^{11n+2})^2(1-x^{11n+9})^2}{(1-x^{11n+4})(1-x^{11n+5})(1-x^{11n+6})(1-x^{11n+7})}$$

$$-\frac{1}{x}\prod_{n=0}^{\infty}\frac{(1-x^{11n+4})^2(1-x^{11n+7})^2}{(1-x^{11n+1})(1-x^{11n+3})(1-x^{11n+8})(1-x^{11n+10})}$$

$$-\prod_{n=0}^{\infty}\frac{(1-x^{11n+3})^2(1-x^{11n+8})^2}{(1-x^{11n+2})(1-x^{11n+5})(1-x^{11n+6})(1-x^{11n+9})}=-1.$$

Remark 1.2. Our procedure for the computation of the five constants appearing in (6.2) is indirect and requires some comments. The first four constants can be computed from the transformation theory for the modified theta constants. The last constant uses the fact that Φ_1 is constant. Instead of working with modified theta constants, we chose to use undetermined coefficients in the infinite product (power series) analogue of equation (6.2) (the last displayed equation, before determining the constants). For each $n \in \mathbb{Z}^+$, we can use only the first $n+1$ terms of the power series expansion of this last identity, obtaining $o(|x|^n)$. We expected to be able to determine the five constants, using $n=4$. However, the resulting set of linear equations has a one parameter family of solutions. To get a unique solution, we need to use $n \geq 5$. An explanation of these observations requires a digression, that we will not make, to Weierstrass points of spaces of meromorphic functions. Out of respect for the family of identities we are producing, using MATHEMATICA, we verified the last one numerically up to $O(|x|^{1001})$. Equation (6.2) describes the algebraic relation satisfied by $\Phi(\mathbb{H}^2/\Gamma(11)) \subset \mathbf{P}\mathbb{C}^4$. (See §7 of Chapter 3.)

The function Φ_2 has at most a double pole at P_∞. Hence

$$\frac{\varphi_0^4}{\varphi_1^2\varphi_4^2}+c_1^2\frac{\varphi_4^4}{\varphi_2^2\varphi_3^2}+c_2^2\frac{\varphi_3^4}{\varphi_0^2\varphi_1^2}+c_3^2\frac{\varphi_1^4}{\varphi_2^2\varphi_4^2}+c_4^2\frac{\varphi_2^4}{\varphi_0^2\varphi_3^2}=d_1F_{11,-5}+d_2,$$

where the constants c_1, ..., c_4 are the same as those appearing in (6.2) and d_1 and d_2 are to be determined. The infinite product (power series) identity becomes

$$\frac{1}{x^2}\prod_{n=0}^{\infty}\frac{(1-x^{11n+5})^4(1-x^{11n+6})^4}{(1-x^{11n+1})^2(1-x^{11n+4})^2(1-x^{11n+7})^2(1-x^{11n+10})^2}$$

$$+x^2\prod_{n=0}^{\infty}\frac{(1-x^{11n+1})^4(1-x^{11n+10})^4}{(1-x^{11n+2})^2(1-x^{11n+3})^2(1-x^{11n+8})^2(1-x^{11n+9})^2}$$

$$+x^2\prod_{n=0}^{\infty}\frac{(1-x^{11n+2})^4(1-x^{11n+9})^4}{(1-x^{11n+4})^2(1-x^{11n+5})^2(1-x^{11n+6})^2(1-x^{11n+7})^2}$$

$$+\frac{1}{x^2}\prod_{n=0}^{\infty}\frac{(1-x^{11n+4})^4(1-x^{11n+7})^4}{(1-x^{11n+1})^2(1-x^{11n+3})^2(1-x^{11n+8})^2(1-x^{11n+10})^2}$$

$$+\prod_{n=0}^{\infty}\frac{(1-x^{11n+3})^4(1-x^{11n+8})^4}{(1-x^{11n+2})^2(1-x^{11n+5})^2(1-x^{11n+6})^2(1-x^{11n+9})^2}$$

$$=\frac{1}{5}\prod_{n=1}^{\infty}(1-x^{11n})^{-5}\sum_{m=-2}^{\infty}P_{-5}(11m+25)x^m+\frac{34}{5}.$$

2. The $\jmath$-function and generalizations of the discriminant Δ

Let $k = 2$, 3, 5, 7 or 13. In these cases $\left(\frac{\eta(k\tau)}{\eta(\tau)}\right)^{\frac{24}{k-1}}$ is a $\Gamma_o(k)$-invariant function with divisor $\frac{P_\infty}{P_0}$. Hence averaging over $\Gamma_o(k)\backslash\Gamma$ produces a Γ-invariant function (on $\overline{\mathbb{H}^2/\Gamma}$, $P_0 = P_\infty$) with at most a simple pole at P_∞; thus there exist constants $c(k)$ and $d(k)$, not both zero, such that

$$k^{\frac{12}{k-1}}\left(\frac{\eta(k\tau)}{\eta(\tau)}\right)^{\frac{24}{k-1}}+\sum_{l=0}^{k-1}\left(\frac{\eta\left(\frac{\tau+l}{k}\right)}{\eta(\tau+l)}\right)^{\frac{24}{k-1}}=c(k)\jmath(\tau)+d(k).$$

(Recall that the $\jmath$-function was described in §3.1 of Chapter 3.) The translation to the local coordinate $x=\exp(2\pi\imath\tau)$ tells us that the left hand side of the last equation equals

$$k^{\frac{12}{k-1}}x\frac{\sum_{m=0}^{\infty}P_{\frac{24}{1-k}}(m)x^{km}}{\prod_{n=1}^{\infty}(1-x^n)^{\frac{24}{k-1}}}+k\frac{\sum_{m=0}^{\infty}P_{\frac{24}{1-k}}(mk+1)x^m}{\prod_{n=1}^{\infty}(1-x^n)^{\frac{24}{k-1}}}.$$

Hence the averaging process killed the possible pole at P_∞, and we conclude (after evaluating $d(k)$) that

$$k^{\frac{13-k}{k-1}}x\sum_{m=0}^{\infty}P_{\frac{24}{1-k}}(m)x^{km}+\sum_{m=0}^{\infty}P_{\frac{24}{1-k}}(mk+1)x^m=\frac{24}{1-k}\sum_{m=0}^{\infty}P_{\frac{24}{1-k}}(m)x^m.$$

The resulting three term recursions agree with those obtained previously by Corollary 6.20 of Chapter 5. Persevering with the methods suggested by the above considerations, we let k be an arbitrary prime and consider those $N\in\mathbb{Z}$ that satisfy $(k-1)|12N$. Under these assumptions, $\left(\frac{\eta(k\tau)}{\eta(\tau)}\right)^{\frac{24N}{k-1}}$ is a $\Gamma_o(k)$-invariant function with divisor $\frac{P_\infty^N}{P_0^N}$. The averaging process leads us to

$$k^{\frac{12N}{k-1}}\left(\frac{\eta(k\tau)}{\eta(\tau)}\right)^{\frac{24N}{k-1}}+\sum_{l=0}^{k-1}\left(\frac{\eta\left(\frac{\tau+l}{k}\right)}{\eta(\tau+l)}\right)^{\frac{24N}{k-1}},\tag{6.3}$$

whose expansion in terms of the local coordinate x is

$$k^{\frac{12N}{k-1}} x^N \frac{\sum_{m=0}^{\infty} P_{\frac{24N}{1-k}}(m) x^{km}}{\sum_{m=0}^{\infty} P_{\frac{24N}{1-k}}(m) x^m} + k \frac{\sum_{m=-\lfloor \frac{N}{k} \rfloor}^{\infty} P_{\frac{24N}{1-k}}(mk+N) x^m}{\sum_{m=0}^{\infty} P_{\frac{24N}{1-k}}(m) x^m}.$$

It is thus easy to analyze the results of our averaging process since only one of the above two terms contributes to the singularity of our new function. If $N > 0$, the function may have a pole at P_∞ of order at most $\lfloor \frac{N}{k} \rfloor$, whereas for $N < 0$ the pole is precisely of order $|N|$. For $1 \le N \le (k-1)$, we have obtained a constant function. Thus

Theorem 2.1. *Let k be a prime. For all $N \in \mathbb{Z}^+$, $N < k$, such that $(k-1)|12N$ and for all $n \in \mathbb{Z}$,*

$$k^{\frac{12N+1-k}{k-1}} P_{\frac{24N}{1-k}}\left(\frac{n-N}{k}\right) + P_{\frac{24N}{1-k}}(nk+N) = P_{\frac{24N}{1-k}}(N) P_{\frac{24N}{1-k}}(n).$$

In the special case $N = k-1$, the identity we obtain,

$$k^{11} P_{-24}\left(\frac{n-k+1}{k}\right) + P_{-24}(nk+k-1) = P_{-24}(k-1) P_{-24}(n),$$

is easily seen to be equivalent to Mordell's theorem (5.4 of Chapter 4).

Corollary 2.2. *Let ν be a positive factor of 12. For all primes k, $k \equiv 1 \mod \nu$, and all $n \in \mathbb{Z}^+ \cup \{0\}$,*

$$\begin{aligned} k^{\frac{12}{\nu}-1} P_{-\frac{24}{\nu}}\left(\frac{n\nu-(k-1)}{k\nu}\right) &+ P_{-\frac{24}{\nu}}\left(\frac{nk\nu+(k-1)}{\nu}\right) \\ &= P_{-\frac{24}{\nu}}\left(\frac{k-1}{\nu}\right) P_{-\frac{24}{\nu}}(n). \end{aligned}$$

Proof. We use $N = \frac{k-1}{\nu}$.

To obtain a more function theoretic interpretation of the last result, it is convenient to introduce ($\Delta_1 = \Delta$)

$$\Delta_\nu(\tau) = x \prod_{n=1}^{\infty} (1-x^{\nu n})^{\frac{24}{\nu}} = \sum_{n=1}^{\infty} T_\nu(n) x^n = x \sum_{n=0}^{\infty} P_{-\frac{24}{\nu}}(n) x^{\nu n}. \tag{6.4}$$

In the above $x = \exp\left(\frac{2\pi\imath\tau}{\nu}\right)$. The second of the above equalities defines the function

$$T_\nu : \mathbb{Z}^+ \to \mathbb{Z} \quad (T_1 = T).$$

Note that

$$T_\nu(n) = P_{-\frac{24}{\nu}}\left(\frac{n-1}{\nu}\right) \text{ and thus } T_\nu(n) = 0 \text{ if } n \not\equiv 1 \mod \nu.$$

We have established a restricted multiplicativity property for the family of functions T_ν.

Theorem 2.3. *For all $\nu \in \mathbb{Z}^+$ such that $\nu|12$, and for all primes k with $k \equiv 1$ mod ν and for all $n \in \mathbb{Z}^+ \cup \{0\}$,*

$$k^{\frac{12}{\nu}-1}T_\nu\left(\frac{n}{k}\right) + T_\nu(nk) = T_\nu(k)T_\nu(n). \tag{6.5}$$

Proof. The result for $n \equiv 1$ mod ν is a consequence of the previous corollary. The result is true for all n because whenever n does not satisfy the congruence the result is trivial in the sense that each term vanishes separately.

Remark 2.4. Several remarks should be made concerning the last result and the choice of the relation between x and τ in (6.4).
(a) The functions T_ν, $\nu > 2$, have nontrivial (that is, $T_\nu(n) = 0$ for some $n \equiv 1 \mod \nu$) zeros; for example, the values of $T_{12}(85)$, $T_{12}(133)$, ..., $T_6(55)$, $T_6(85)$, ..., $T_4(21)$, $T_4(33)$, ..., $T_3(10)$, $T_3(21)$, ... are zero. We do not know whether T_2 and T have nontrivial zeros.
(b) Note that $T_\nu(1) = 1$. The recursion formula (6.5) does not hold for arbitrary primes for $\nu > 1$; for example, for $\nu = 2$ or 3, $k = \nu = n$,

$$T_\nu(\nu) = 0 = T_\nu(\nu^2) \text{ but } T_\nu(1) = 1.$$

Similarly, for $\nu = 4$,

$$T_4(7) = 0 \text{ but } T_4(7^2) = 7^2.$$

(c) We show below that the relation $x = \exp\left(\frac{2\pi\imath\tau}{\nu}\right)$ is in a sense dictated by the function theory involved.

Note that

$$\Delta_\nu(\tau) = \exp\left(\frac{2\pi\imath\tau}{\nu}\right)\prod_{n=1}^{\infty}(1-\exp(2\pi\imath n\tau))^{\frac{24}{\nu}} = \Delta(\tau)^{\frac{1}{\nu}} = \eta(\tau)^{\frac{24}{\nu}}.$$

Since Δ is a cusp 6-form for Γ, Δ_ν is a e-automorphic function for this group that satisfies the cusp condition for a factor of automorphy of weight $\frac{6}{\nu}$. Assume now that $\nu|6$. In this case $e = c\kappa^{\frac{6}{\nu}}$, where κ, as usual, is the canonical factor of automorphy (for Γ) and c is a normalized character on Γ. Let Γ_ν be the kernel of c. Then $\Delta_\nu \in \mathbb{A}_{\frac{6}{\nu}}(\mathbb{H}^2, \Gamma_\nu)$. The fact that $c^\nu = 1$ tells us that $c(\Gamma) \subset \mathbb{Z}_\nu$ (viewed as ν-th roots of unity). From equation (3.43), we conclude that $B^\mu \notin \Gamma_\nu$ for $1 \le \mu < \nu$ and $B^\nu \in \Gamma_\nu$. We have thus justified the choice of local coordinate in our definitions of Δ_ν and T_ν and shown that $c(\Gamma) = \mathbb{Z}_\nu$. Thus $\chi(\Gamma_\nu) = \frac{\nu}{6}$. The invariance properties of η also tell us that

$$c(A) = \exp\left(-\frac{6\pi\imath}{\nu}\right) \text{ and } c(B) = \exp\left(-\frac{2\pi\imath}{\nu}\right)$$

and

$$A^*_{\frac{6}{\nu}}(\Delta_\nu) = \exp\left(\frac{6\pi\imath}{\nu}\right)\Delta_\nu \text{ and } B^*_{\frac{6}{\nu}}(\Delta_\nu) = \exp\left(\frac{2\pi\imath}{\nu}\right)\Delta_\nu.$$

ν	Signature
1	$(0;\ 2,3,\infty)$
2	$(0;\ 3,3,\infty)$
3	$(0;\ 2,2,2,\infty)$
6	$(1;\ \infty)$

Table 26. THE SIGNATURE OF Γ_ν.

Let the branch schema of Γ_ν be $(\nu;\nu_2,\nu_3,\nu_\infty)$. Short calculations using (that the case $\nu = 1$ has been settled and) the facts that Γ_ν contains parabolics, that $\chi(\Gamma_\nu) = \frac{\nu}{6}$ and that $\dim \mathbb{A}_\nu(\mathbb{H}^2,\Gamma_\nu) \geq 1$ reduces us to a short list of possibilities as well as the conclusion that each possibility is realized.

The information we have is sufficient to determine descriptions of the groups Γ_ν, generators P of their parabolic subgroups representing the puncture and representatives $\mathcal{R}$ for the left cosets $\Gamma_\nu\backslash\Gamma$. We proceed with the description of these groups:

$$\Gamma_1 = \Gamma =< A, AB;\ A^2 = (AB)^3 = 1 >,\ P = A(AB) = B,\ \mathcal{R} = \{1\},$$

$$\Gamma_2 =< AB, BA;\ (AB)^3 = (BA)^3 = 1 >,$$
$$P = (BA)(AB) = B^2,\ \mathcal{R} = \{1,\ A\},$$

$$\Gamma_3 =< A, BAB^{-1}, B^2AB^{-2};\ A^2 = (BAB^{-1})^2 = (B^2AB^{-2})^2 = 1 >,$$
$$P = (B^2AB^{-2})(BAB^{-1})A = B^3,\ \mathcal{R} = \{1,\ B,\ B^2\}$$

and

$$\Gamma_6 =< B^3A, B^4AB^{-1} >,$$
$$P = (B^4AB^{-1})(B^3A)^{-1}(B^4AB^{-1})^{-1}(B^3A) = B^6, \mathcal{R} = \{1,\ B,\ B^2,\ ...,\ B^5\}.$$

We proceed to interpret Theorem 2.3 in terms of automorphic forms. That theorem leads us to a power series identity. For $\nu \in \mathbb{Z}^+$, $\nu|12$ and primes k, $k \equiv 1 \mod \nu$,

$$k^{\frac{12}{\nu}} \sum_{n=1}^{\infty} T_\nu(n)x^{kn} + \sum_{l=0}^{k-1}\sum_{n=1}^{\infty} T_\nu(n)\left(\epsilon^l x^{\frac{l}{k}}\right)^n$$

$$= kT_\nu(k)\sum_{n=1}^{\infty} T_\nu(n)x^n,\ \epsilon = \exp\left(\frac{2\pi\imath}{k}\right);$$

the second term in the above equality equals, of course, $k\sum_{n=1}^{\infty} T_\nu(kn)x^n$. The last identity is equivalent to

$$k^{\frac{12}{\nu}}\Delta_\nu(k\tau) + \sum_{l=0}^{k-1}\Delta_\nu\left(\frac{\tau+l}{k}\right) = kT_\nu(k)\Delta_\nu(\tau).$$

This identity needs an explanation! At the moment all we can say is that it is a consequence of the restricted multiplicativity property of the functions T_ν. In another direction $\Theta_{\Gamma_\nu\backslash\Gamma}(\Delta_\nu) = 0$, leading to trivial identities.

Before proceeding, we derive a more convenient formula for $\jmath$. We note that

$$\theta^{24}\begin{bmatrix}0\\0\end{bmatrix} + \theta^{24}\begin{bmatrix}0\\1\end{bmatrix} + \theta^{24}\begin{bmatrix}1\\0\end{bmatrix} \in \mathbb{A}_6^+(\mathbb{H}^2,\Gamma) - \mathbb{A}_6(\mathbb{H}^2,\Gamma)$$

and

$$\eta^{24} \in \mathbb{A}_6(\mathbb{H}^2,\Gamma);$$

as a matter of fact the Fourier series expansions of these functions are of the form

$$2 + 4\cdot 23\cdot 24x + O\left(|x|^2\right) \text{ and } x - 24x^2 + O\left(|x|^3\right),\ x \to 0,$$

respectively. It follows that

$$\jmath = \frac{1}{2}\frac{\theta^{24}\begin{bmatrix}0\\0\end{bmatrix} + \theta^{24}\begin{bmatrix}0\\1\end{bmatrix} + \theta^{24}\begin{bmatrix}1\\0\end{bmatrix}}{\eta^{24}} - 2^3\cdot 3\cdot 47$$

and that the Fourier series coefficients of $\jmath$ are integers. We can get more precise information on the Laurent expansion of $\jmath$ in terms of the local coordinate $x = \exp(2\pi\imath\tau)$ at P_∞ on $\mathbb{H}^2/\Gamma$. This Laurent series is derived from

$$\jmath(\tau) =$$

$$\frac{1}{2}\left[\left(1+2\sum_{n=1}^{\infty}x^{\frac{n^2}{2}}\right)^{24} + \left(1+2\sum_{n=1}^{\infty}(-1)^n x^{\frac{n^2}{2}}\right)^{24} + 2^{24}x^3\left(\sum_{n=0}^{\infty}x^{\frac{n(n+1)}{2}}\right)^{24}\right]$$

$$\times x^{-1}\prod_{n=1}^{\infty}(1-x^n)^{-24} - 2^3\cdot 3\cdot 47.$$

Proposition 2.5. *The $\jmath$-function has the Laurent expansion*

$$\frac{1}{x} + \sum_{n=1}^{\infty} c_n x^n,\ c_n \in \mathbb{Z}^+ \cup \{0\},$$

in terms of the local coordinate $x = \exp(2\pi\imath\tau)$ at P_∞ on $\mathbb{H}^2/\Gamma$.

Proof. The only claim that needs verification is that each coefficient c_n is nonnegative. Clearly the Taylor series expansion of $\prod_{n=1}^{\infty}(1-x^n)^{-24}$ has only nonnegative coefficients. A negative coefficient could only arise from a contribution from the Taylor series expansion of $1+2\sum_{n=1}^{\infty}(-1)^n x^{\frac{n^2}{2}}$, but such a negative contribution is canceled by the corresponding term from $1+2\sum_{n=1}^{\infty} x^{\frac{n^2}{2}}$. We can easily compute the value of any particular c_n for reasonably small parameter n. For example,

$$c_1 = 196884 = 2^2\cdot 3^3\cdot 1823,\ \ c_2 = 21493760 = 2^{11}\cdot 5\cdot 2099,$$

$$c_3 = 864299970 = 2\cdot 3^5\cdot 5\cdot 355679,\ \ c_4 = 20245856256 = 2^{14}\cdot 3^3\cdot 45767,$$

$$c_5 = 333202640600 = 2^3\cdot 5^2\cdot 2143\cdot 777421,$$

$$c_6 = 4252023300096 = 2^{13}\cdot 3^6\cdot 11\cdot 13^2\cdot 383$$

and

$$c_7 = 44656994071935 = 3^3\cdot 5\cdot 7\cdot 271\cdot 174376673.$$

It is obvious that an examination of these raw numbers to see patterns among them is rather difficult. We have established that for all $N\in\mathbb{Z}^+$ such that $(k-1)|12N$, there exist constants $e_0,\ \ldots,\ e_{\lfloor\frac{N}{k}\rfloor}$ such that

$$k^{\frac{12N}{k-1}}\left(\frac{\eta(k\tau)}{\eta(\tau)}\right)^{\frac{24N}{k-1}}+\sum_{l=0}^{k-1}\left(\frac{\eta\left(\frac{\tau+l}{k}\right)}{\eta(\tau+l)}\right)^{\frac{24N}{k-1}}=\sum_{n=0}^{\lfloor\frac{N}{k}\rfloor}e_n j^n(\tau).$$

We can certainly set $N=2(k-1)$ giving us, after evaluation of constants,

Theorem 2.6. *For all primes k,*

$$k^{23}x^{2(k-1)}\frac{\sum_{m=0}^{\infty}P_{-48}(m)x^{km}}{\sum_{m=0}^{\infty}P_{-48}(m)x^m}+\frac{\sum_{m=-1}^{\infty}P_{-48}((m+2)k-2)x^m}{\sum_{m=0}^{\infty}P_{-48}(m)x^m}$$

$$=(48P_{-48}(k-2)+P_{-48}(2k-2))+P_{-48}(k-2)\left(\frac{1}{x}+\sum_{n=1}^{\infty}c_n x^n\right).$$

The results for $N<0$ are of a different kind. The basic equation, a consequence of (6.3), is

$$k^{\frac{12N}{k-1}}\left(\frac{\eta(k\tau)}{\eta(\tau)}\right)^{\frac{24N}{k-1}}+\sum_{l=0}^{k-1}\left(\frac{\eta\left(\frac{\tau+l}{k}\right)}{\eta(\tau+l)}\right)^{\frac{24N}{k-1}}=\sum_{n=0}^{-N}e_n j^n(\tau),$$

with $e_{-N}\neq 0$.

3. Congruences for the Laurent coefficients of the $\jmath$-function

For all $N \in \mathbb{Z}$, the function $\jmath^N(\tau)$ is Γ-automorphic. Thus for every $k \in \mathbb{Z}^+$, $\jmath^N\left(\frac{\tau}{k}\right)$ is $\Gamma^o(k)$-automorphic, hence certainly $\Gamma(k,k)$-automorphic. [130] Assume now that k is a prime. Averaging over $\Gamma(k,k)\backslash\Gamma_o(k)$, we obtain ($J_k = J_{k,1}$)

$$J_{k,N}(\tau) = \frac{1}{k}\sum_{l=0}^{k-1} \jmath^N\left(\frac{\tau+l}{k}\right),$$

a $\Gamma_o(k)$-automorphic function. For positive N, $\jmath^N(\tau)$ has a pole only at P_∞ and Laurent series expansion

$$\jmath^N(\tau) = \sum_{n=-N}^{\infty} c_{n,N}x^n,$$

in terms of $x = \exp(2\pi\imath\tau)$. Further,

$$c_{n,N} \in \mathbb{Z}^+ \cup \{0\},\ c_{-N,N} = 1,\ c_{n,1} = c_n \text{ for } n \in \mathbb{Z}^+,\ c_{0,1} = 0.$$

Hence in these cases, $J_{k,N}$ is holomorphic except possibly at P_∞ and P_0; its Laurent series expansion at P_∞ is

$$J_{k,N}(\tau) = \sum_{n=-\left\lfloor \frac{N}{k}\right\rfloor}^{\infty} c_{kn,N}x^n, \tag{6.6}$$

and we conclude that

$$\operatorname{ord}_\infty J_{k,N} \geq -\left\lfloor \frac{N}{k}\right\rfloor \text{ and } \operatorname{ord}_\infty J_k = 1.$$

We need to investigate the behavior of $J_{k,N}$ at P_0. We use the automorphism of $\Gamma_o(k)$ induced by A_k and the invariance of $\jmath$ under A:

(6.7)

$$\begin{aligned} J_{k,N}\left(-\frac{1}{k\tau}\right) &= \frac{1}{k}\sum_{l=0}^{k-1}\jmath^N\left(-\frac{1}{k^2\tau}+\frac{l}{k}\right) = \frac{1}{k}\jmath^N(k^2\tau) + \frac{1}{k}\sum_{m=1}^{k-1}\jmath^N\left(\tau+\frac{m}{k}\right) \\ &= \frac{1}{k}\left(\jmath^N(k^2\tau) - \jmath^N(\tau)\right) + J_{k,N}(k\tau) \end{aligned}$$

---

[130] We will need to iterate the operator that sends $\jmath$ to $J_{k,N}$. Thus we want to view both $\jmath$ and $J_{k,N}$ as $\Gamma_o(k)$-invariant functions. The first of these functions is Γ-invariant, hence certainly $\Gamma_o(k)$-invariant. We see below that the averaging process we are using leads to a $\Gamma_o(k)$-invariant function $J_{k,N}$. Now in general, if $f(\tau)$ is $\Gamma_o(k)$-invariant, it certainly is $\Gamma_o(k^2) = \Gamma(1,k^2)$-invariant. Hence $f\left(\frac{\tau}{k}\right)$ is $\Gamma(k,k)$-invariant. Thus, for k prime,

$$F(\tau) = \frac{1}{k}\sum_{l=0}^{k-1} f\left(\frac{\tau+l}{k}\right)$$

is $\Gamma_o(k)$-invariant.

$$= \frac{1}{k}\sum_{n=-N}^{\infty} c_{n,N}(x^{k^2n} - x^n) + \sum_{n=-\lfloor \frac{N}{k} \rfloor}^{\infty} c_{kn,N}x^{kn}.$$

(In the second equality, we have used the fact that the map $l \mapsto m = -l^{-1}$ is an automorphism of the multiplicative subgroup of the field $\mathbb{Z}_k$, as was already done on several occasions. See also the proof of Theorem 4.6 in Chapter 4.) Hence

$$\mathrm{ord}_0 J_{k,N} = -k^2 N.$$

We have established most of

Theorem 3.1. *Let $k = 2,\ 3,\ 5,\ 7$ or 13 and let $N \in \mathbb{Z}^+$. There exists a unique set of integers $\alpha_1,\ \ldots,\ \alpha_{k^2N},\ \beta_0,\ \ldots,\ \beta_{\lfloor \frac{N}{k} \rfloor}$ such that*

$$k^{\frac{12i}{1-k}+1}\alpha_i \in \mathbb{Z} \tag{6.8}$$

and

$$J_{k,N} = \sum_{i=1}^{k^2N} \alpha_i f_k^i + \sum_{j=0}^{\lfloor \frac{N}{k} \rfloor} \beta_j f_k^{-j}. \tag{6.9}$$

Proof. The function f_k is $\Gamma_o(k)$-automorphic with divisor $\frac{P_\infty}{P_0}$. Hence we can choose $\alpha_i \in \mathbb{C}$ such that $J_{k,N}(\tau) - \sum_{i=1}^{k^2N} \alpha_i f_k^i(\tau)$ is holomorphic at $\mathbb{C}$. The equations we must solve in order to determine the α_i are linear equations with coefficients in the integers (the integers $c_{n,N}$). The calculations become easier if we switch our attention from P_0 to P_∞. Using the involution A_k, we see that the constants to be chosen are specified by the condition that

$$\frac{1}{k}\left(\jmath^N(k^2\tau) - \jmath^N(\tau)\right) + J_{k,N}(k\tau) - \sum_{i=1}^{k^2N} \alpha_i k^{\frac{12i}{1-k}} f_k^{-i}(\tau)$$

is holomorphic at ∞. So the equations to be solved are (equating to 0 the coefficient of x^{-m}, for $1 \le m \le k^2N$, in the Laurent series of the last displayed function in terms of the local coordinate $x = \exp(2\pi\imath\tau)$)

$$k^{-1}\left(c_{-\frac{m}{k^2},N} - c_{-m,N}\right) + \epsilon_m c_{-m,N} - \sum_{j=m}^{k^2N} \alpha_j k^{\frac{12j}{1-k}} d_{-m,j} = 0,$$

where the constants $c_{n,N}$ are understood to be zero where not previously defined,

$$\epsilon_m = \begin{cases} 1 \text{ if } m \equiv 0 \mod k \text{ and } m \ge -k\left\lfloor \frac{N}{k} \right\rfloor \\ 0 \text{ if } m \not\equiv 0 \mod k \text{ or } m < -k\left\lfloor \frac{N}{k} \right\rfloor \end{cases},$$

and

$$f_k^{-i}(\tau) = \sum_{m=-i}^{\infty} d_{m,i}x^m \quad (d_{m,i} \in \mathbb{Z},\ d_{-i,i} = 1).$$

Thus we conclude inductively starting from the equation for $m = k^2N$ that

$$\alpha_{k^2N} = k^{\frac{12k^2N}{k-1}-1}$$

and, in general, for each i,

$$\alpha_i = k^{\frac{12i}{k-1}-1}c,$$

where c is an integer (that depends on i). After the determination of the α_i, we see that the β_j are determined by the condition that

$$J_{k,N} - \sum_{i=1}^{k^2N} \alpha_i f_k^i - \sum_{j=0}^{\lfloor \frac{N}{k} \rfloor} \beta_j f_k^{-j}$$

be holomorphic at ∞. The equations to be solved are

$$c_{k\alpha,N} - \sum_{l=\alpha}^{\lfloor \frac{N}{k} \rfloor} \beta_l d_{-\alpha,l} = 0, \ \alpha = 0, \ldots, \left\lfloor \frac{N}{k} \right\rfloor.$$

Again, solving the last equation first, we see that

$$\beta_{\lfloor \frac{N}{k} \rfloor} = c_{k\lfloor \frac{N}{k} \rfloor, N}$$

and, in general, for each j,

$$\beta_j = c_{kj,N} - \sum_{l=j+1}^{\lfloor \frac{N}{k} \rfloor} \beta_l d_{-j,l}$$

are integers. Note that a key observation in the proof is that

$$f_k^{-1}(\tau) = \frac{1}{x} + R(x),$$

where $R(x)$ is holomorphic at ∞ and has good expansion.

Corollary 3.2. *For $k = 2,\ 3,\ 5,\ 7$ or 13 and $1 \leq N < k$,*

$$c_{m,N} \equiv 0 \mod k^{\frac{12}{k-1}-1} \ \textit{for}\ m \in \mathbb{Z}^+ \equiv 0 \mod k.$$

Proof. We use (6.6) and (6.9). We note of course that the second sum in (6.9) only consists of the one term for the index $j = 0$. The proof thus follows from the fact that the greatest common divisor of the α_i is $k^{\frac{12}{k-1}-1}$.

Remark 3.3. It is rather remarkable that the corollary (despite its elementary nature) is sharp for $k = 2, 3, 5$ and 7, $N = 1$, as the calculations of the coefficients c_2, c_3, c_5 and c_7 showed.

Before we proceed further, we point out that it is possible to obtain the coefficients of the $\jmath$-function in a different fashion, using the infinite product expansions of the theta functions.

The formula for the $\jmath$-function (in the variable $x = \exp(2\pi\imath\tau)$) is

$$\jmath(\tau) = \frac{\prod_{n=1}^{\infty}(1+x^{\frac{2n-1}{2}})^{48} + \prod_{n=1}^{\infty}(1-x^{\frac{2n-1}{2}})^{48} + 2^{24}x^3\prod_{n=1}^{\infty}(1+x^n)^{48}}{2x} + \text{const.}$$

Define

$$f(y) = \prod_{n=1}^{\infty}(1+y^{2n-1})^{48} = 1 + \sum_{n=1}^{\infty} a_n y^n$$

and

$$g(y) = \prod_{n=1}^{\infty}(1+y^{2n})^{48} = 1 + \sum_{n=1}^{\infty} b_{2n} y^{2n}.$$

Note that $f(-y) = \prod_{n=1}^{\infty}(1-y^{2n-1})^{48} = 1 + \sum_{n=1}^{\infty} a_n(-y)^n$. Hence

$$\jmath(\tau) = \frac{1}{x} + a_4 x + \sum_{n=2}^{\infty}(a_{2n+2} + 2^{23}b_{2n-4})x^n, \ b_0 = 1.$$

We have therefore shown that the coefficients of the $\jmath$-function are given as

$$c_1 = a_4 \text{ and } c_n = a_{2n+2} + 2^{23}b_{2n-4}, \ n \geq 2.$$

The integer $a_2 = 1128$ is the constant term in some normalizations of the the $\jmath$-function.

The last corollary should be considered as the level one congruences for the $\jmath$ function (see below). It can be improved considerably to obtain higher level correspondences. We consider only the case $N = 1$. Before proceeding to the congruences, we use the trivial estimates

$$a_{2n} \geq 24 \cdot 47 \text{ and } b_{2n} \geq 1, n \in \mathbb{Z}^+,$$

to obtain

Corollary 3.4. *For all $n \in \mathbb{Z}^+$, $c_n \in \mathbb{Z}^+$.*

Exercise 3.5. Verify the alternate form of the $\jmath$-function given above in terms of infinite products.

Theorem 3.6. *For all $m \in \mathbb{Z}^+$,*

$$c_m \equiv \begin{cases} 0 \mod 2^{8+3n} \ \textit{for } m \equiv 0 \mod 2^n \\ 0 \mod 3^{3+2n} \ \textit{for } m \equiv 0 \mod 3^n \\ 0 \mod 5^{n+1} \ \textit{for } m \equiv 0 \mod 5^n \\ 0 \mod 7^n \ \textit{for } m \equiv 0 \mod 7^n \\ 0 \mod 11^n \ \textit{for } m \equiv 0 \mod 11^n \end{cases}.$$

We call the above results the *level n congruences for the function $\jmath$ for the primes* 2, 3, 5, 7 *and* 11, respectively. We will complete the proof of Theorem 3.6 for the prime 5, prove it for the prime 11 for level 1 only, and leave the rest of the cases to the reader. We have already established the level one congruences for the primes 2, 3, 5 and 7. Assume now that the prime $k \leq 7$.

Our basic equation in these cases for $N = 1$, equation (6.9), can be rewritten in operator notation as

$$J_k = \tilde{U}_k(\jmath) = \sum_{i=1}^{k^2} \alpha_i f_k^i, \ \alpha_i \in \mathbb{Z}, \ k^{\frac{12i}{k-1}-1} | \alpha_i. \tag{6.10}$$

We shall see that (this equation is also valid for $k = 11$)

$$\tilde{U}_k^n(\jmath)(\tau) = \sum_{m=1}^{\infty} c_{k^n m} x^m.$$

So the proof of the last theorem involves a study of the operators $\tilde{U}_k^n$. Since

$$\tilde{U}_k^n(\jmath) = \tilde{U}_k^{n-1}\left(\tilde{U}_k(\jmath)\right) = \tilde{U}_k^{n-1}\left(\sum_{i=1}^{k^2} \alpha_i f_k^i\right),$$

we see that, as in the case of partition functions, we need to study how the operators $\tilde{U}_k^n$ act on the functions f_k^i. The case $k = 11$ has additional complications and is handled in §3.3.

3.1. Averaging f_k^i. We need to compute $\tilde{U}_k(f_k^i)$. Let k be a prime and $N \in \mathbb{Z}$ such that $(k-1)|12N$. We start with the $\Gamma^o(k)$-invariant function $g_k^N = \left(\frac{\eta(\tau)}{\eta(\frac{\tau}{k})}\right)^{\frac{24N}{k-1}}$ with divisor $\frac{P_\infty^N}{P_0^N}$. Since $\Gamma^o(k) \supset \Gamma(k,k)$, this function is certainly $\Gamma(k,k)$-invariant, and hence it makes sense to average it over $\Gamma(k,k)\backslash\Gamma_o(k)$. We obtain a $\Gamma_o(k)$-function previously denoted by the symbol $Y_{k,N}$. We have shown that for $N > 0$,

$$\text{ord}_\infty Y_{k,N} \geq \left\lceil \frac{N}{k} \right\rceil \ \text{ and } \ \text{ord}_0 Y_{k,N} = -kN, \tag{6.11}$$

while for $N < 0$,

$$\text{ord}_\infty Y_{k,N} \geq \left\lfloor N - \frac{N}{k} \right\rfloor \ \text{ and } \ \text{ord}_0 Y_{k,N} \geq (k-1)N.$$

The Laurent series of $Y_{k,N}$ at ∞ can be easily described as a result of our work in the previous chapter. To describe the corresponding series at 0, we can use the automorphism A_k of $\Gamma_o(k)$ and invariance properties of the η-function to compute $g_k \circ A_k$. However, we can avoid reliance on this computation by use of previously established properties.

We have shown that

$$\tilde{U}_k(f_k^i) = Y_{k,i} = f_k^{-i} F_{k,\frac{(k+1)i}{\beta(k)}}. \tag{6.12}$$

We now assume that the prime k is restricted to the small values 2, 3, 5 or 7 so that $\frac{12}{k-1} \geq 2$ and equation (6.10) holds. It is convenient to rewrite that equation as

$$\tilde{U}_k(j) = k^{\frac{r}{2}-1}\left(b_1 f_k + k^2 b_2 f_k^2 + \ldots + k^{k^2} b_{k^2} f_k^{k^2}\right),\ r = \frac{24}{k-1},\ b_i \in \mathbb{Z}.$$

3.2. Completion of the proof of Theorem 3.6 for $k=5$. The reproducing equation in this case reads

$$\tilde{U}_5(j) = 5^2\left(b_1 f_5 + 5^2 b_2 f_5^2 + \ldots + 5^{5^2} b_{5^2} f_5^{5^2}\right). \tag{6.13}$$

We need to establish a key lemma.

Lemma 3.7. *For all $n \in \mathbb{Z}^+$,*

$$\tilde{U}_5^n(j) = 5^{1+n}\left(b_1 f_5 + 5^2 b_2 f_5^2 + \ldots + 5^{5^{1+n}} b_{5^{1+n}} f_5^{5^{1+n}}\right), b_i \in \mathbb{Z}, \tag{6.14}$$

and hence

$$\mathrm{ord}_\infty \tilde{U}_5^n(j) \geq 1 \text{ and } \mathrm{ord}_0 \tilde{U}_5^n(j) \geq -5^{1+n}.$$

Proof. The claim about the orders of vanishing of $\tilde{U}_5^n(j)$ follows at once from equation (6.14) since $(f) = \frac{P_\infty}{P_0}$. We have already established equation (6.14) for $n=1$. By induction, using equations (6.13) and (6.12),

$$\tilde{U}_5^{n+1}(j) = \tilde{U}_5\left(\tilde{U}_5^n(j)\right)$$

$$= 5^{1+n}\left(b_1 f_5^{-1} F_{5,6} + 5^2 b_2 f_5^{-2} F_{5,2\cdot 6} + \ldots + 5^{5^{1+n}} b_{5^{1+n}} f_5^{-5^{1+n}} F_{5,6\cdot 5^{1+n}}\right).$$

From (6.12), $\tilde{U}_5(f_5^i) = Y_{5,i} = f_5^{-i} F_{5,6i}$, we see that (using Lemma 8.5 of Chapter 5 or (6.11)) for each $i \in \mathbb{Z}^+$,

$$\mathrm{ord}_\infty \tilde{U}_5^n(f_5^i) \geq \left\lceil \frac{i}{5} \right\rceil \text{ and } \mathrm{ord}_0 \tilde{U}_5^n(f_5^i) = -5i.$$

We conclude that

$$\tilde{U}_5^{1+n}(j) = 5^{1+n}\left(b_1 f_5^{-1} F_{5,6} + 5^2 b_2 f_5^{-2} F_{5,12} + \ldots + 5^{5^{1+n}} b_{5^{1+n}} f_5^{-5^{1+n}} F_{5,6\cdot 5^{1+n}}\right)$$

$$= 5^{1+n}\left(\tilde{b}_1 f_5 + 5^2 \tilde{b}_2 f_5^2 + \ldots + 5^{5^{1+n}} \tilde{b}_{5^{2+n}} f_5^{5^{2+n}}\right).$$

We must show that $5|\tilde{b}_i$. From Lemma 8.5 of Chapter 5 we see that

$$f_5^{-i} F_{5,6i} = \epsilon d_1 f_5^{i_o} + 5\left(d_2 f_5^{i_o+1} + \ldots\right), d_j \in \mathbb{Z},$$

where

$$i_o = \left\lceil \frac{i}{5} \right\rceil \text{ and } \epsilon = \begin{cases} 1 \text{ for } i \equiv 0,3,4 \mod 5 \\ 5 \text{ for } i \equiv 1,2 \mod 5 \end{cases}.$$

Each $\tilde{b}_i$ is a sum of terms consisting of a product of a b_i with power of 5. This power is at least 1 unless $i \equiv 0, 3$ or $4 \mod 5$ and the index of the corresponding d_j is 1 (that is, $i = \lceil \frac{i}{5} \rceil$). Since this case never occurs, we have completed the proof of the lemma.

3.3. Proof of Theorem 3.6 for $k = 11$, $n = 1$. The function $J_{11} \in \mathcal{K}(\Gamma_o(11))_0$ has a pole of order at most 121 at P_0 and vanishes at P_∞. Hence

$$J_{11} = \tilde{U}_{11}(\jmath) = \sum_{j=2}^{121} \alpha_j (G_{11,-N(j)} - G_{11,-N(j)}(\infty)).$$

We need to determine the constants α_j. We know that the Taylor series coefficients (at P_∞ in terms of the local coordinate $x = \exp(2\pi\imath\tau)$) of each of the functions $G_{11,-N(j)} - G_{11,-N(j)}(\infty)$ are integers divisible by 11^j. It hence suffices for our purposes to show that the α_j are rationals whose denominators are not divisible by 11^2. Using the involution A_{11}, the last equation is seen to be equivalent to

$$\frac{1}{11}(\jmath(11^2\tau) - \jmath(\tau)) + J_{11}(11\tau) = \sum_{j=2}^{121} \alpha_j (F_{11,-N(j)}(\tau) - F_{11,-N(j)}(0)).$$

Translating to the local coordinate $x = \exp(2\pi\imath\tau)$, we see that the left hand side is given by equation (6.7) with $k = 11$ and $N = 1$, that is, by

$$\frac{1}{11} \sum_{n=-1}^{\infty} c_n (x^{11^2 n} - x^n) + \sum_{n=1}^{\infty} c_{11n} x^{11n}.$$

We expand

$$F_{11,-N(j)}(\tau) - F_{11,-N(j)}(0) = \sum_{m=-j}^{\infty} d_{m,j} x^m, \quad d_{m,j} \in \mathbb{Z}.$$

Our equations to be solved, for $2 \leq m \leq 11^2$, are (equating coefficients of x^{-m})

$$\frac{1}{11}\left(c_{-\frac{m}{121}} - c_{-m}\right) = \sum_{j=m}^{121} \alpha_j d_{-m,j}.$$

Thus (for $m = 11^2$)

$$\alpha_{121} d_{-121,121} = \frac{1}{11}$$

and in general

$$\alpha_m d_{-m,m} = -\sum_{j=m+1}^{121} \alpha_j d_{-m,j} + \frac{1}{11}\left(c_{-\frac{m}{121}} - c_{-m}\right).$$

In an effort to obtain better formulae we switch to a basis adapted to P_0. In this case our basic equations become

$$J_{11} = \sum_{j=2}^{121} \alpha_j F_j \circ A_{11} \text{ and } \frac{1}{11}(\jmath(11^2\tau) - \jmath(\tau)) + J_{11}(11\tau) = \sum_{j=2}^{121} \alpha_j F_j(\tau)$$

(see §10.5 of Chapter 5 for the definition of the functions F_j). It follows that

$$\frac{1}{11}\left(c_{-\frac{m}{121}} - c_{-m}\right) = \alpha_m.$$

Our equations can be easily solved since for the range of the indices m under consideration, $c_{-m} = 0$ and $c_{-\frac{m}{121}} = 0$ except for $m = 121$. The solutions are

$$\alpha_2 = \alpha_3 = \ldots = \alpha_{120} = 0, \ \alpha_{121} = \frac{1}{11}.$$

We are led to the remarkable identity (because it relates two seemingly different processes)

$$11J_{11} \circ A_{11} = F_{121}.$$

It remains for us to study the Laurent series expansions of the functions $F_j \circ A_{11}$. We shall use the uniqueness of adapted bases and the fact that precomposition with A_{11} is a field isomorphism on functions to conclude that we may take the function $F_j \circ A_{11}$ as

$$F_j \circ A_{11} =$$

$$\begin{cases} \frac{1}{c}(G_{11,-N(2)} - G_{11,-N(2)}(\infty))^{\frac{j}{2}} \text{ for even } j \\ \frac{1}{c}(G_{11,-N(2)} - G_{11,-N(2)}(\infty))^{\frac{j-3}{2}}(G_{11,-N(3)} - G_{11,-N(3)}(\infty)) \text{ for odd } j \end{cases},$$

where c is an integer that is a product of powers of 2, 5 and 7 subject to an adjustment that consists of subtracting a linear combination with rational coefficients of functions of the same type with lower indices. It is important to determine the denominators of these rationals. We proceed by induction. All variables appearing in the rest of this paragraph are integers (whose exact values are irrelevant). The Laurent coefficients of $F_2 \circ A_{11}$ are certainly of the form $\frac{11^2}{2^\mu 5^\nu 7^\sigma}e$ (and no adjustment is needed in this case). The Laurent coefficients of $\frac{1}{c}(G_{11,-N(3)} - G_{11,-N(3)}(\infty))$ are of the form $\frac{11^3}{2^\mu 5^\nu 7^\sigma}e$, and the needed adjustment to obtain $F_3 \circ A_{11}$ changes the coefficients to be of the form $\frac{11^2}{2^\mu 5^\nu 7^\sigma}e$. By induction so also are the coefficients of $F_j \circ A_{11}$ for all $j > 3$.

This establishes the level 1 congruence for the prime 11 for the $\jmath$-function. To study the level n congruences, we need to establish

Lemma 3.8. *For each prime $k \in \mathbb{Z}^+$ and all $n \in \mathbb{Z}^+$,*

$$\tilde{U}_k^n(j)(\tau) = \sum_{m=1}^{\infty} c_{k^n m} x^m = \frac{1}{k^n} \sum_{l=0}^{k^n-1} j\left(\frac{\tau + l}{k^n}\right). \tag{6.15}$$

Furthermore, this function has a simple zero at P_∞ and a pole at P_0 of order k^{n+1}.

Proof. The proof of the first equality is contained in the footnote at the beginning of this section. For the second equality, note that $j\left(\frac{\tau}{k^n}\right)$ is a $\Gamma^o(k^n) = \Gamma(k^n, 1)$-invariant function. Hence certainly $\Gamma(k^n, k)$-invariant. Averaging over $\Gamma(k^n, k)\backslash\Gamma_o(k)$, we obtain a $\Gamma_c(k)$-function given by the last term in equation (6.15) whose power series at P_∞ agrees with the middle term of that equation. We leave it to the reader to verify the statement about the orders of the averaged function at the two punctures of $\overline{\mathbb{H}^2/\Gamma_o(k)}$.

The higher level congruences for the prime 11 should now be obtained by investigating the properties of the constants α_j that solve

$$\tilde{U}_{11}^n(j) = \sum_{j=2}^{k^{n+1}} \alpha_j F_j \circ A_{11}.$$

These constants α_j satisfy the equations

$$c_{11^n m} = \sum_{j=2}^{k^{n+1}} \alpha_j d_{m,j}, \ m \in \mathbb{Z}^+,$$

where the $d_{m,j}$ are rationals, coefficients of the Taylor series expansions of the functions $F_j \circ A_{11}$.

3.4. A further analysis of the $k = 2$ case. To better understand what is going on, we examine a simpler case: $k = 2$. Our basic equation here is equation (6.10) for $k = 2$. If we apply the involution A_2 to this equation, it transforms to

$$\frac{1}{2}(j(2^2\tau) - j(\tau)) + J_2(2\tau) = \sum_{i=1}^{4} \alpha_i 2^{-12i} f_2^{-i}(\tau). \tag{6.16}$$

Using power series expansions, we evaluate the constants α_i:

$$\alpha_1 2^{-12} = 2^{-1} \cdot 5 \cdot 2099, \ \alpha_2 2^{-24} = 2^3 \cdot 3 \cdot 7^2, \ \alpha_3 2^{-36} = 2^4 \cdot 3, \ \alpha_4 2^{-48} = 2^{-1}.$$

Thus the last displayed equation, even after evaluation of the constants[131] $(\alpha_i 2^{-12i})$, is not strong enough to give us the level one congruence for the prime two. We need to use instead the equation obtained by precomposing

[131] See more below.

the functions in the last displayed equality by A_2, that is, the equation (a special case of equation (6.9) or (6.10))

$$J_2(\tau) = \sum_{i=1}^{4} \alpha_2 f_2^i(\tau) \tag{6.17}$$

(the additional information about the transformation of the function f_2 under A_2). Our computations show that $2^{11}|\alpha_i$ for each i. The level 1 congruence for the prime 2 follows. Our previous proof of this fact did not require the computations.

We can replace (with new definitions of the constants α_i) equation (6.16) by

$$\frac{1}{2}(\jmath(2^2\tau) - \jmath(\tau)) + J_2(\tau) = \alpha + \sum_{j=1}^{4} \alpha_j F_{2,-2j}(\tau).$$

We know the Laurent series expansion of the function on the left hand side of the last equation. The Laurent series for the functions $F_{2,2j}$ have been used many times:

$$F_{2,-2j}(\tau) = \prod_{n=1}^{\infty}(1 - x^{2n})^{-16j} \sum_{m=-j}^{\infty} P_{-16j}(2m + 2j)x^m.$$

We evaluate the undetermined constants:

$$\alpha = 2^6 \cdot 3^2 5 \cdot 7 \cdot 59 \cdot 67 \cdot 163,$$

$$\alpha_1 = -\frac{5 \cdot 139 \cdot 191 \cdot 3769}{2}, \ \alpha_2 = 2^3 \cdot 19 \cdot 29 \cdot 181, \ \alpha_3 = -2^4 61, \ \alpha_4 = \frac{1}{2}.$$

We try one more approach. We replace equation (6.17) by

$$J_2(\tau) = \sum_{i=1}^{4} \alpha_2 F_{2,i}(\tau).$$

Computations show that

$$\alpha_1 = 2^8 \cdot 13 \cdot 59, \ \alpha_2 = 2^{20} \cdot 19, \ \alpha_3 = 2^{29}, \ \alpha_4 = 2^{32}.$$

Again, from this calculation alone, we do not have sufficient information to conclude the level one congruence for the prime 2 for the function $\jmath$. However, if we also use the level one congruence for the prime 2 for the partition function, we conclude that each coefficient in the Taylor series expansion of $F_{2,1}$ is an integer divisible by 2^3; the level one congruence for the prime 2 for the $\jmath$-function follows readily.

The key seems to be to study the Taylor series expansions at P_∞ of functions with singularities only at P_0 and to use the "correct" functions to produce an identity.

Chapter 7

Combinatorial and number theoretic applications

The first four chapters of this book deal essentially with the function theory on the Riemann surfaces which are the compactifications of the ones produced by the action of a subgroup of $\mathrm{PSL}(2,\mathbb{Z})$ on the upper half plane $\mathbb{H}^2$. Chapter 1 was a general introduction. Chapter 2 introduced the theta functions which we used to construct meromorphic functions and differentials on the Riemann surfaces. Chapter 3 used the ideas of the previous chapter for constructions in the particular cases of the modular curves, and Chapter 4 dealt with the notion of theta identities. In Chapters 5 and 6 we studied the partition function, the $\jmath$-function, Ramanujan congruences and their generalizations. This may have alerted the reader to the fact that there is combinatorial content to identities. In this chapter we show that the identities derived have interesting combinatorial consequences. It is not our intention to be exhaustive, but rather to point out some of the beautiful interpretations of some of the identities and leave the reader with the pleasant task of constructing other identities and interpreting them.

The combinatorial number theory in this context involves three things: representations of integers as sums of squares, triangular, and in general n-gonal numbers; partitions of positive integers; and the σ-function defined earlier in Chapter 4.

We also hint at the relation of topics such as Lambert series, Euler series and continued fractions to theta functions.

1. Generalities on partitions

We start with a definition that generalizes an object studied (the functions P_N with $N \in \mathbb{Z}$) in Chapter 5.

Definition 1.1. A *partition* of the positive integer N from a set S of positive integersis a choice of positive integers n_i from S with $\sum_{i=1}^{p} n_i = N$, for some $p \in \mathbb{Z}^+$. The summands n_i are called the *parts* of the partition and unless otherwise noted a partition is called *even* or *odd* depending on whether it has an even or odd number of parts.[132] We denote the number of even and odd partitions of n by $E(n)$ and $O(n)$, respectively, when the meaning of parity and the set S are clear from the context.

Remark 1.2. 1. In general the order of the parts n_i in the partition $\sum_{i=1}^{p} n_i$ is immaterial, and we can thus assume for convenience that $n_i \leq n_{i-1}$.
2. We shall have to distinguish partitions that allow repetitions from those that do not allow repetitions. A partition from S that allows repetitions can of course be viewed as a partition from $\mathcal{S}$ that does not allow repetitions, where $\mathcal{S}$ consists of infinitely many copies of S.
3. Unless otherwise noted, for all partitions of integers, $S = \mathbb{Z}^+$ and repetitions are allowed (the case $P = P_1$ of Chapter 5).

According to the above definitions and conventions, the integer 4 has exactly five different partitions:

$$4, \quad 3+1, \quad 2+2, \quad 2+1+1, \quad 1+1+1+1.$$

There are thus 3 even partitions of 4 and 2 odd partitions of 4. The last partition would not appear if we took for S two copies of the positive integers and did not allow repetitions.

There is a very simple graphical representation of partitions. If a partition is given by $\sum_{i=1}^{p} n_i$, then its *graphical representation* is a figure with p rows of dots with the r-th row having n_r dots. The partition of 4 given by 2+1+1 has the graphical representation

$$\begin{array}{ll} \bullet & \bullet \\ \bullet & \\ \bullet & \end{array}$$

The above partition of 4 is an example of a partition which has no part exceeding 2. By turning columns into rows we obtain

$$\begin{array}{lll} \bullet & \bullet & \bullet \\ \bullet & & \end{array}$$

[132] We will have *some* exceptions to these conventions. The *parity* of a partition is thus in general the same as the parity of its number of parts.

another partition of 4, namely 3+1, a partition with 2 parts. This geometric artifice gives the following elementary result concerning partitions. The number of partitions of N with no part exceeding m (see also the next subsection) is the same as the number of partitions of N with at most m parts.

We denote (as before) the number of (unrestricted) partitions of n by $P(n)$ and easily compute the first few values:

$$P(1) = 1, \quad P(2) = 2, \quad P(3) = 3, \quad P(4) = 5, \quad P(5) = 7, \ \dots \ .$$

A little thought shows that there is a simple generating function for $P(n)$, a function that was the principal object of study of Chapter 5, defined by $\mathbf{P}(x) = \prod_{n=1}^{\infty} \frac{1}{1-x^n} = \sum_{n=0}^{\infty} P(n)x^n$ with $P(0) = 1$.

The function P describes the simplest kind of partitions, where we have placed no restrictions on the parts. We can restrict the parts to disallow repetitions. A different type of restriction may be to only allow parts which are odd integers. The above two cases are intimately related. This is due to the identity $\prod_{n=1}^{\infty}(1 + x^n) = \prod_{n=1}^{\infty} \frac{1}{1-x^{2n-1}}$ we have already used. The function on the left is the generating function for partitions disallowing repetitions, and the function on the right is the generating function for partitions whose parts are odd integers.

Let us now consider partitions which disallow repetitions (P_{-1} of Chapter 5), hence with generating function $\prod_{n=1}^{\infty}(1+x^n)$. The generating function for $E(n) - O(n)$ (of this case) is given by

$$\mathbf{P}^{-1}(x) = \prod_{n=1}^{\infty}(1 - x^n) = 1 + \sum_{n=1}^{\infty}(E(n) - O(n))x^n. \tag{7.1}$$

A well-known theorem of Euler asserts that in the case we are discussing $E(n) - O(n) = 0$, unless $n = \frac{3m^2+m}{2}$ for some $m \in \mathbb{Z}$ in which case $E(n) - O(n) = (-1)^m$.

Euler's theorem follows immediately from Euler's identity (2.2), which we have already obtained as a consequence of the Jacobi triple product identity. We now give an independent, combinatorial, proof. We establish, in part, a bijection between the even and odd partitions of n when n is not of the form $\frac{3m^2+m}{2}$, for any $m \in \mathbb{Z}$.

To this end, let $n \in \mathbb{Z}^+$ not be of the form $\frac{3m^2+m}{2}$ and consider the graphical representation of a partition λ of n with p parts. Let $\beta(\lambda) = \beta \geq 1$ be the least part in the partition. Since we are disallowing repetitions the number of dots is a strictly decreasing function of the rows. Let $\sigma(\lambda) = \sigma \geq 1$ be one plus the number of consecutive initial rows that contain one more dot than the next row.

There are now two cases to consider: $\beta \leq \sigma$ and $\beta > \sigma$. In the first case, we remove from λ its last row (of length β) and add one dot to each of the first β rows to produce a new partition λ' of the same integer n of opposite parity provided $\beta < \sigma$ or $\beta = \sigma$ and $p > \sigma$. If $\beta = \sigma = p$, then this process will not change the parity. The new partition λ' will have the same number of parts as λ with $\beta' = \beta(\lambda') = 1$. In this case, however, we can describe λ completely. Its graphical representation consists of σ rows; the last row has σ dots, the next to last row has $\sigma + 1$ dots, ..., the first has $\sigma + (\sigma - 1)$ dots. Hence

$$n = \sum_{j=0}^{\beta-1} (\beta + j) = \frac{\beta(3\beta - 1)}{2} = \frac{-\beta(-3\beta + 1)}{2}.$$

In the second case we use the inverse map and remove one dot from each of the first σ rows and add an additional row of σ dots as a last row. We get a new partition λ'. It represents a partition without repetitions if $\beta \geq \sigma + 2$ or $\beta = \sigma + 1$ and $p > \sigma$. In the excluded case, $\beta = \sigma + 1$ and $p = \sigma$, the last two rows of the new partition λ' are equal. Once again we are fortunate enough to be able to describe the partition λ completely; it consists of σ rows, the last row has $\sigma + 1$ dots, the next to the last has $\sigma + 2$ dots, ..., the first has $\sigma + \sigma$ dots. Thus

$$n = \sum_{j=0}^{\sigma} (\sigma + j) = \frac{\sigma(3\sigma + 1)}{2}.$$

The above shows that when n is not of the form $\frac{3m^2+m}{2}$, there is a bijection of the set of even partitions of n onto the set of odd partitions of n. When n is of the form $\frac{3m^2+m}{2}$, exactly one partition of n does not have an image in the set of partitions of opposite parity; its parity depends on the parity of m. This finishes the combinatorial proof of Euler's identity.

There are other types of restrictions we can impose on partitions. For example, we can restrict the number of repetitions in some way. One way of doing this is to set up a set in advance from which we can take the parts. An example of such a set could be three copies of the positive integers (P_{-3} of Chapter 5). We wish to also distinguish between copies of the same integer, so let us consider as our set S all the positive integers in three colors. To fix ideas, let us take a specific case. If we partition the integer 3 using this set, we have

$$3, 2 + 1, 1 + 1 + 1.$$

Notice that we have three different ways of getting 3, nine different ways of getting 2+1 and only one way of getting 1+1+1. Thus the total number of partitions of 3 is 13. Of these, 4 are odd and 9 are even; hence $E(3) - O(3) =$

5. In fact there is a general theorem here (also due to Jacobi) which asserts:

$$E(n)-O(n) = \begin{cases} 0 \text{ if } n \neq \frac{m(m+1)}{2}, & \text{for all } m \in \mathbb{Z}^+ \cup \{0\} \\ (-1)^m(2m+1) \text{ if } n = \frac{m(m+1)}{2}, & \text{for some } m \in \mathbb{Z}^+ \cup \{0\} \end{cases}.$$

The analytic statement of the above theorem is the identity of Jacobi (2.1) previously encountered.

Chapter 5 contains a study of the partition coefficients $P_N(n)$, $n \in \mathbb{Z}^+$ ($P_N(0) = 1$), $N \in \mathbb{Z}$. For $N \in \mathbb{Z}^+$, $P_N(n)$ represents the total number of partitions of n using positive integers of N colors and allowing repetitions. Of course, $P_0(n) = 0$. For $N \in -\mathbb{Z}^+$, $P_N(n)$ represents the difference between the number of even and odd partitions of n using positive integers of $-N$ colors and not allowing repetitions. As we mentioned several times already, the only nontrivial cases for which we have a formula for $P_N(n)$ are $N = -1$ and -3.

1.1. Euler series and some old identities. This section is devoted to establishing some classical identities of Euler and some mild generalizations. We begin with a simple example to illustrate some ideas.

Euler proved the following identity:

$$1 + \sum_{k=1}^{\infty} \frac{x^k}{\prod_{l=1}^{k}(1-x^l)} = \prod_{n=1}^{\infty} \frac{1}{(1-x^n)} = \sum_{n=0}^{\infty} P(n)x^n. \tag{7.2}$$

We shall give several proofs[133] of this identity. We begin with some definitions and a lemma.

Definition 1.3. For $l \in \mathbb{Z}^+$, let $P_{\leq l}(n)$ denote the number of partitions of $n \in \mathbb{Z}^+$ with no part exceeding l, thus also the number of partitions of n with at most l parts.

Lemma 1.4. *We have the recursion formula for n, $l \in \mathbb{Z}^+$ with $2 \leq l \leq n-1$,*

$$P_{\leq l}(n) - P_{\leq (l-1)}(n) = P_{\leq l}(n-l). \tag{7.3}$$

[133] We are interested in exploring what each proof reveals. The simplest proof is a consequence of an induction argument that shows that for all $n \in \mathbb{Z}^+$,

$$1 + \sum_{k=1}^{n} \frac{x^k}{\prod_{l=1}^{k}(1-x^l)} = \prod_{k=1}^{n} \frac{1}{(1-x^k)}.$$

Proof. It is clear that for all $l \in \mathbb{Z}^+$, the generating function of $P_{\leq l} : \mathbb{Z}^+ \to \mathbb{Z}^+$ is[134]

$$\mathbf{P}_{\leq 1}(x) = \frac{1}{\prod_{m=1}^{l}(1 - x^m)} = \sum_{n=0}^{\infty} P_{\leq l}(n)x^n. \tag{7.4}$$

Thus

$$\mathbf{P}_{\leq l}(x) - \mathbf{P}_{\leq (l-1)}(x) = x^l \mathbf{P}_{\leq l}(x).$$

The lemma follows by comparing power series coefficients.

Remark 1.5. We have chosen to give the above analytic proof of the lemma. One can also give a purely combinatorial proof as follows. The power series coefficient of x^n of the left hand side of the equation in the statement of the lemma is simply the number of partitions of n with exactly l parts. This is clearly the same as the number of partitions of $n - l$ with at most l parts.

Euler's identity is now a simple consequence since

$$1 + \sum_{k=1}^{\infty} \frac{x^k}{\prod_{l=1}^{k}(1 - x^l)} = 1 + \sum_{k=1}^{\infty} x^k \mathbf{P}_{\leq k}(x) = 1 + \sum_{k=1}^{\infty} (\mathbf{P}_{\leq k}(x) - \mathbf{P}_{\leq (k-1)}(x)),$$

which in fact gives the whole story since the partial sums of the last of the above series are just the $\mathbf{P}_{\leq k}(x)$ (that is

$$1 + \sum_{l=1}^{k} (\mathbf{P}_{\leq l}(x) - \mathbf{P}_{\leq (l-1)}(x)) = \mathbf{P}_{\leq k}(x)$$

for all $k \in \mathbb{Z}^+$) which converge to $\mathbf{P}(x)$ as k tends to infinity. This, of course, suffices to prove the identity given by equation (7.2).

It is however worthwhile to observe that

$$1 + \sum_{k=1}^{\infty} \frac{x^k}{\prod_{l=1}^{k}(1 - x^l)} = 1 + \sum_{n=1}^{\infty} P(n)x^n, \; P(n) = \sum_{l=1}^{n} P_{\leq l}(n - l). \tag{7.5}$$

Our last lemma and the remark following it tell us that $P_{\leq l}(n-l)$ equals the number of partitions of n with exactly l parts. Hence equation (7.5) gives us $P(n)$ as a sum of those partitions with 1, 2, ..., n parts. Precisely the same reasoning can be applied to the series

$$1 + \sum_{k=1}^{\infty} (-1)^k \frac{x^k}{\prod_{l=1}^{k}(1 - x^l)}$$

to obtain that its power series coefficients are $E(n) - O(n)$, where (we recall our convention that) $E(n)$ and $O(n)$ are the numbers of even and odd partitions of n with no restrictions on the parts (thus very different from the

---

[134] By our conventions $P_{\leq l}(0) = 1$ for all $l \in \mathbb{Z}^+$ and $P_{\leq 0}(n) = 0$ for all integers $n \neq 0$.

same symbols appearing in (7.1)). The argument goes as follows:

$$1+\sum_{k=1}^{\infty}(-1)^k\frac{x^k}{\prod_{l=1}^{k}(1-x^l)} = 1+\sum_{k=1}^{\infty}(-1)^k x^k \mathbf{P}_{\le k}(x)$$

$$= 1+\sum_{k=1}^{\infty}(-1)^k(\mathbf{P}_{\le k}(x)-\mathbf{P}_{\le(k-1)}(x))$$

$$= 1+\sum_{k=1}^{\infty}(-1)^k\left(\sum_{n=1}^{\infty}(P_{\le k}(n)-P_{\le(k-1)}(n))\right)x^n$$

$$= 1+\sum_{k=1}^{\infty}\sum_{n=1}^{\infty}(-1)^k\left((P_{\le k}(n)-P_{\le(k-)}(n))x^n\right)$$

$$= 1+\sum_{n=1}^{\infty}\sum_{k=1}^{\infty}(-1)^k\left((P_{\le k}(n)-P_{\le(k-1)}(n))x^n\right) = 1+\sum_{n=1}^{\infty}(E(n)-O(n))x^n.$$

Series of the form considered above are sometimes called Euler series; we shall use this terminology. In order to see the consequences of our last equality we turn to yet another proof of Euler's identity. We begin with

Definition 1.6. For m a nonnegative integer, define the *Euler series*

$$E_m(x,z) = 1+\sum_{k=1}^{\infty}\frac{x^{mk}}{\prod_{l=1}^{k}(1-x^l)}z^k.$$

In the above as in all our Euler series x is a complex number (parameter) with $|x|<1$. It is clear for example that for fixed x the series can be thought of as a power series in the complex variable z and as such it has a positive radius of convergence. Moreover, as m tends to ∞, the function of z tends to the constant function 1. It is immediate from the definition that

$$E_m(x,z)-E_{m+1}(x,z) = x^m z E_m(x,z),$$

and hence by induction

$$E_0(x,z) = \frac{E_{m+1}(x,z)}{\prod_{l=0}^{m}(1-x^l z)}.$$

Letting m approach infinity we obtain

$$E_0(x,z) = 1+\sum_{k=1}^{\infty}\frac{1}{\prod_{l=1}^{k}(1-x^l)}z^k = \frac{1}{\prod_{n=0}^{\infty}(1-x^n z)}. \tag{7.6}$$

If we set $z=x$ in the last formula, we obtain Euler's identity (7.2). We also obtain

$$E_m(x,z) = \frac{1}{\prod_{l=m}^{\infty}(1-zx^l)}. \tag{7.7}$$

More important for us however is

Proposition 1.7. *For all x of modulus less than one*

$$1+\sum_{k=1}^{\infty}\frac{(-1)^k x^k}{\prod_{l=1}^{k}(1-x^l)}=\frac{1}{\prod_{n=1}^{\infty}(1+x^n)}.$$

Proof. Set $z=-x$ in the expression for $E_0(x,z)$.

The remarks we have made thus far indicate that there is a big difference between partitions without any restrictions on the parts and partitions which disallow repetitions. Euler's identity $\prod_{n=1}^{\infty}(1-x^n)=\sum_{n=-\infty}^{\infty}(-1)^n x^{\frac{3n^2+n}{2}}$ has told us the "entire story" in the latter case. In the former case we have

Theorem 1.8. *Let $n\in\mathbb{Z}^+$. For n odd, there are always more odd partitions than even partitions. If $n=2$, then there is exactly one even and one odd partition. If n is even and $n\neq 2$, then there are more even than odd partitions, and*

$$\lim_{n\to\infty}|E(n)-O(n)|=\infty,$$

with E and O referring to unrestricted (allowing repetitions) partitions from $\mathbb{Z}^+$.

Proof. The last proposition has shown us that

$$\frac{1}{\prod_{n=1}^{\infty}(1+x^n)}=1+\sum_{n=1}^{\infty}(E(n)-O(n))x^n.$$

The elementary identity $\prod_{n=1}^{\infty}(1+x^n)=\prod_{n=1}^{\infty}\frac{1}{(1-x^{2n-1})}$ therefore gives

$$\prod_{n=1}^{\infty}(1-x^{2n-1})=1+\sum_{n=1}^{\infty}(E(n)-O(n))x^n. \tag{7.8}$$

We need the partition theoretic interpretation of the left hand side of the above equation. The coefficient of x^n, $n>0$, in the Taylor series (with center 0) expansion of the product is the difference between the number of even and odd partitions of the integer n with the parts restricted to being odd integers and not allowing repetitions. This implies that when n is odd, one obtains only odd partitions, and when n is even and not 2, only even partitions. When $n=2$ there are no such partitions. This suffices to complete the proof.

One can now go a bit further and determine the number of even and odd unrestricted partitions of an integer n. We use a simple consequence of the Jacobi triple product identity. That formula of Chapter 2 with $z=-1$ is the identity $\sum_{n=-\infty}^{\infty}(-1)^n x^{n^2}=\prod_{n=1}^{\infty}(1-x^{2n})(1-x^{2n-1})^2$ whose right hand side can be rewritten as $\prod_{n=1}^{\infty}(1-x^n)(1-x^{2n-1})$, so that we obtain the identity

$$\sum_{n=0}^{\infty} P(n)x^n \sum_{n=-\infty}^{\infty} (-1)^n x^{n^2} = \frac{\sum_{n=-\infty}^{\infty}(-1)^n x^{n^2}}{\prod_{n=1}^{\infty}(1-x^n)} = \prod_{n=1}^{\infty}(1-x^{2n-1}).$$

Taking the Cauchy product of the two power series on the left and replacing the infinite product on the right with (7.8), we find that

$$E(n) - O(n) = \sum_{l=-\infty}^{\infty} (-1)^l P(n-l^2),$$

where the sums are finite since $P(m) = 0$ when m is not a positive integer. Since $E(n) + O(n) = P(n)$, we have proven

Proposition 1.9. *For all $n \in \mathbb{Z}^+$,*

$$E(n) = \sum_{l=0}^{\infty}(-1)^l P(n-l^2) \text{ and } O(n) = \sum_{l=1}^{\infty}(-1)^{l+1} P(n-l^2).$$

The reader cannot have failed to notice that the restricted partitions where repetitions are disallowed and the general partitions are related in a very simple fashion through their generating functions. The generating function for general partitions is $\prod_{n=1}^{\infty} \frac{1}{(1-x^n)}$, and the generating function for the difference between the even and odd partitions with no repetitions allowed is $\prod_{n=1}^{\infty}(1-x^n)$. It therefore follows that

$$\sum_{n=0}^{\infty} P(n)x^n \sum_{n=-\infty}^{\infty} (-1)^n x^{\frac{3n^2+n}{2}} = 1.$$

Hence for $n \geq 1$ we have the following formula which expresses $P(n)$ in terms of $P(m)$ with $m < n$:

$$\sum_{l=-\infty}^{\infty} (-1)^l P\left(n - \frac{3l^2+l}{2}\right) = 0. \tag{7.9}$$

Remark 1.10. 1. Before the advent of fast computers, the last formula was probably the most efficient way for (recursively) computing the partition coefficients $P(n)$.

2. Euler series appear in the right hand side of the famous Rogers-Ramanujan identities:

$$\frac{1}{\prod_{n=0}^{\infty}(1-x^{5n+1})(1-x^{5n+4})} = 1 + \sum_{n=1}^{\infty} \frac{x^{n^2}}{\prod_{l=1}^{n}(1-x^l)}$$

and

$$\frac{1}{\prod_{n=0}^{\infty}(1-x^{5n-2})(1-x^{5n+3})} = 1 + \sum_{n=1}^{\infty} \frac{x^{n^2+n}}{\prod_{l=1}^{n}(1-x^l)}.$$

Fairly straightforward proofs of these identities are available ([**13**, §19.13], among others) that do not shed any light on any theory that might exist behind the identities. The reader is invited (this is probably a very hard problem) to supply proofs that are either purely combinatorial (as most of the arguments of this section) or based on θ-functions (as most of the arguments in this book).

1.2. Partitions and sums of divisors. In this section we show how the σ-function enters naturally into the theory of partitions. We begin with some computations. For

$$\mathbf{P}(x)=\frac{1}{\prod_{n=1}^{\infty}(1-x^n)}=\sum_{n=0}^{\infty}P(n)x^n,\ \frac{x\mathbf{P}'(x)}{\mathbf{P}(x)}=\sum_{n=1}^{\infty}\frac{nx^n}{(1-x^n)}=\sum_{n=1}^{\infty}\sigma(n)x^n.$$

We have shown that

$$\sum_{n=1}^{\infty}nP(n)x^n=x\mathbf{P}'(x)=\mathbf{P}(x)\sum_{n=1}^{\infty}\sigma(n)x^n=\sum_{n=0}^{\infty}P(n)x^n\sum_{n=1}^{\infty}\sigma(n)x^n.$$

Once again taking the Cauchy product, we find

Proposition 1.11. *For all $n\in\mathbb{Z}^+$, we have*

$$nP(n)=\sum_{j=0}^{n-1}P(j)\sigma(n-j).$$

The above identity indicates the very tight relation between the partition function P and the divisor function σ. We have seen in equation (7.9) that the partition coefficient $P(n)$ is determined by the values of P at the numbers $n-\frac{3l^2+l}{2}$ (we need to use only the finite set of integers l for which $n-\frac{3l^2+l}{2}\geq 0$). This is not surprising since the partition function depends on the additive properties of the integers. It is more surprising that a similar formula is valid for the σ-function.

Proposition 1.12. *Let $n\in\mathbb{Z}^+$. Then*

$$\sum_{j=-\infty}^{\infty}(-1)^j\sigma\left(n-\frac{3j^2+j}{2}\right) \tag{7.10}$$

$$=\begin{cases} 0 \ \textit{if}\ n\neq\frac{3m^2+m}{2}\ \textit{for all}\ m\in\mathbb{Z} \\ (-1)^{m+1}\frac{3m^2+m}{2}\ \textit{if}\ n=\frac{3m^2+m}{2}\ \textit{for some}\ m\in\mathbb{Z}\end{cases}.$$

Proof. Apply the argument in the previous proposition to

$$\mathbf{P}^{-1}(x)=\prod_{n=1}^{\infty}(1-x^n)=\sum_{n=-\infty}^{\infty}(-1)^n x^{\frac{n(3n+1)}{2}}.$$

The reader is invited to apply the same reasoning to the more general identity (2.51) which we write for k and l of the same parity and $l \leq k$ in slightly modified form as

$$\prod_{n=0}^{\infty}(1-x^{kn+\frac{k-l}{2}})(1-x^{kn+\frac{k+l}{2}})(1-x^{k(n+1)}) = \sum_{n=-\infty}^{\infty}(-1)^n x^{\frac{kn^2+ln}{2}}.$$

If we let $f(x)$ denote the infinite sum in the last equation, then

$$\frac{-xf'(x)}{f(x)} = \sum_{n=1}^{\infty}\widehat{\sigma_k(n)}\,x^n, \text{ where } \widehat{\sigma_k(n)} = \sum_{d|n,d\equiv\frac{k-l}{2},\frac{k+l}{2},\,0 \mod k} d.$$

By completing an argument similar to the one used to establish the last two propositions, one obtains more identities. Some of them should be of interest.

Exercise 1.13. Use the last suggestion to obtain new identities of the type discussed above.

1.3. Lambert series. The connections discussed above between the partition function and the σ-function lie at the surface of the theory and depend for the most part on simple power series identities which arise from, for example, the Jacobi triple product. We continue this line of thought a bit more.

Our starting points have been infinite product expansions of series of the form $\sum_{n=1}^{\infty} a_n \frac{x^n}{1-x^n}$. A series of this type is called a *Lambert series*. For example, we saw that the product expansion of $\prod_{n=1}^{\infty}(1-x^n)$ led us via logarithmic differentiation to the Lambert series $\sum_{n=1}^{\infty} n\frac{x^n}{1-x^n}$, and this we recognized as the power series $\sum_{n=1}^{\infty}\sigma(n)x^n$.

It is easy to see that in the general case of a Lambert series

$$\sum_{n=1}^{\infty} a_n \frac{x^n}{1-x^n} = \sum_{n=1}^{\infty} b_n x^n, \text{ where } b_n = \sum_{d|n} a_d. \tag{7.11}$$

There are some simple and interesting observations one can make in this connection that are related to the material under study.

For example, if we denote the number of divisors of a positive integer n by $d(n)$ (we include the trivial divisors 1 and n among the divisors of n), then

$$\sum_{n=1}^{\infty} d(n)x^n = \sum_{n=1}^{\infty}\frac{x^n}{1-x^n}.$$

In the same vein

$$\sum_{n=1}^{\infty}(-1)^n\frac{x^n}{1-x^n} = \sum_{n=1}^{\infty}(d_E(n)-d_O(n))x^n,$$

where d_E and d_O represent the number of even and odd divisors, respectively.

A series similar to a Lambert series leads to

$$\sum_{n=1}^{\infty} \frac{x^n}{1-x^{2n}} = \sum_{n=1}^{\infty} d_O(n) x^n.$$

Perhaps the easiest way of obtaining this last identity is based on the observation that

$$\frac{x^n}{1-x^n} - \frac{x^n}{1-x^{2n}} = \frac{x^{2n}}{1-x^{2n}}, \tag{7.12}$$

so that we have the identity

$$\sum_{n=1}^{\infty} d(n) x^n - \sum_{n=1}^{\infty} \delta(n) x^n = \sum_{n=1}^{\infty} d(n) x^{2n},$$

where $\sum_{n=1}^{\infty} \frac{x^n}{1-x^{2n}} = \sum_{n=1}^{\infty} \delta(n) x^n$. Thus

$$\delta(n) = \begin{cases} d(n) = d_O(n) \text{ for } n \in \mathbb{Z}^+ \text{ odd} \\ d(n) - d\left(\frac{n}{2}\right) = d_O(n) \text{ for } n \in \mathbb{Z}^+ \text{ even} \end{cases}.$$

To verify the last equality, write $n = 2^\alpha m$ with α and $m \in \mathbb{Z}^+$, and $(2, m) = 1$. Using the multiplicativity of the function d,

$$\begin{aligned} d(n) - d\left(\frac{n}{2}\right) &= d(2^\alpha m) - d(2^{\alpha-1} m) = \left(d(2^\alpha) - d(2^{\alpha-1})\right) d(m) \\ &= (\alpha + 1 - \alpha) d_O(m) = d_O(2^\alpha m). \end{aligned}$$

Proposition 1.14. *For all $x \in \mathbb{C}$, $|x| < 1$,*

$$\sum_{n=1}^{\infty} \frac{x^n}{1-x^{2n}} = \sum_{n=1}^{\infty} \frac{x^{2n-1}}{1-x^{2n-1}}.$$

Proof. From (7.12) we see that

$$\sum_{n=1}^{\infty} \frac{x^n}{1-x^{2n}} = \sum_{n=1}^{\infty} \frac{x^n}{1-x^n} - \sum_{n=1}^{\infty} \frac{x^{2n}}{1-x^{2n}} = \sum_{n=1}^{\infty} \frac{x^{2n-1}}{1-x^{2n-1}}.$$

Definition 1.15. Since each positive integer has a unique prime factorization[135] $n = \prod_{i=1}^{k} p_i^{\alpha_i}$, with p_i prime and $\alpha_i \in \mathbb{Z}^+$, we define the *degree of* the positive integer n to be $\deg(n) = \sum_{i=1}^{k} \alpha_i$ if $n > 1$ and $\deg(1) = 0$.

It is clear that for $n \in \mathbb{Z}$, $n > 1$, $d(n) = \prod_{i=1}^{k} (\alpha_i + 1)$. Thus a positive integer n is a square if and only if $d(n)$ is odd.

Theorem 1.16. *Let N be a positive integer. If N is not a square, N has as many divisors of even degree as odd degree. If N is a square, then the excess of divisors of even degree over divisors of odd degree is precisely* 1.

[135]The indexing set is empty for $n = 1$.

Proof. We begin by defining the Lambert series

$$f(x) = \sum_{n=1}^{\infty} (-1)^{\deg(n)} \frac{x^n}{1-x^n}.$$

The power series representation of the function f is given by

$$f(x) = \sum_{n=1}^{\infty} b_n x^n, \text{ where } b_n = \sum_{d|n} (-1)^{\deg(d)}.$$

It is a well known fact [**13**, Th. 265] that if $g(n)$ is a multiplicative function, then so is $h(n) = \sum_{d|n} g(d)$. Since $(-1)^{\deg(n)}$ is a multiplicative function so is b_n. Thus if $n = \prod_{i=1}^{k} p_i^{\alpha_i}$, with the p_i distinct primes and $\alpha_i \in \mathbb{Z}^+$, then $b_n = \prod_{i=1}^{k} b_{p_i^{\alpha_i}}$. Since the divisors of $p_i^{\alpha_i}$ are $1,\ p_i,\ p_i^2,\ \ldots,\ p_i^{\alpha_i}$, it is clear that $b_n = 0$ unless n is a square (so that all the α_i are even). It thus follows that the powers series representation of f is $f(x) = \sum_{n=1}^{\infty} x^{n^2}$. This suffices to prove the theorem.

One of the first results we proved in our discussion of theta functions and theta constants in Chapter 2 was Lemma 1.6, which we now use. As a consequence of the lemma we get

$$\theta^2 \begin{bmatrix} 0 \\ 0 \end{bmatrix} (0,\tau) = \theta^2 \begin{bmatrix} 0 \\ 0 \end{bmatrix} (0,2\tau) + \theta^2 \begin{bmatrix} 1 \\ 0 \end{bmatrix} (0,2\tau) \tag{7.13}$$

and

$$\theta^2 \begin{bmatrix} 1 \\ 0 \end{bmatrix} (0,\tau) = 2\theta \begin{bmatrix} 0 \\ 0 \end{bmatrix} (0,2\tau) \theta \begin{bmatrix} 1 \\ 0 \end{bmatrix} (0,2\tau).$$

One iterate of the lemma (use of the last two equations), starting from the identity (7.13), yields

$$\theta \begin{bmatrix} 0 \\ 0 \end{bmatrix} (0,\tau) = \theta \begin{bmatrix} 0 \\ 0 \end{bmatrix} (0,4\tau) + \theta \begin{bmatrix} 1 \\ 0 \end{bmatrix} (0,4\tau), \tag{7.14}$$

which in the local coordinate with $x = \exp(\pi\imath\tau)$ becomes

$$1 + 2\sum_{n=1}^{\infty} x^{n^2} = 1 + 2\sum_{n=1}^{\infty} x^{(2n)^2} + 2\sum_{n=1}^{\infty} x^{(2n-1)^2}. \tag{7.15}$$

The last equation gives us a decomposition of the power series appearing on the left side of the equality into its even and odd parts. We conclude that (7.14) is a trivial identity. In the language of infinite products, we get from the last power series decomposition (as a consequence of the Jacobi triple product)

$$\prod_{n=1}^{\infty} (1-x^{2n})(1+x^{2n-1})^2 = \prod_{n=1}^{\infty} (1-x^{8n})(1+x^{8n-4})^2 + 2x \prod_{n=1}^{\infty} (1-x^{8n})(1+x^{8n})^2.$$

In particular we obtain

Proposition 1.17. *For all $x \in \mathbb{C}$, with $|x| < 1$,*

$$\sum_{n=1}^{\infty} x^{(2n-1)^2} = x \prod_{n=1}^{\infty} (1 - x^{8n})(1 + x^{8n})^2.$$

The above proposition can be given a combinatorial interpretation in terms of partitions, but we leave this to the reader.

In terms of Lambert series the identity (7.15) is

$$\frac{1}{2} + \sum_{n=1}^{\infty} (-1)^{\deg(n)} \frac{x^n}{1 - x^n} = \frac{1}{2} + \sum_{n=1}^{\infty} (-1)^{\deg(4n)} \frac{x^{4n}}{1 - x^{4n}}$$
$$+ \sum_{n=0}^{\infty} (-1)^{\deg(4n+1)} \frac{x^{4n+1}}{1 - x^{4n+1}} + \sum_{n=0}^{\infty} (-1)^{\deg(4n+2)} \frac{x^{4n+2}}{1 - x^{4n+2}}$$
$$+ \sum_{n=0}^{\infty} (-1)^{\deg(4n+3)} \frac{x^{4n+3}}{1 - x^{4n+3}},$$

so that we also have

Proposition 1.18. *For all $x \in \mathbb{C}$, with $|x| < 1$,*

$$\sum_{n=1}^{\infty} x^{(2n-1)^2} = \sum_{n=0}^{\infty} (-1)^{\deg(4n+1)} \frac{x^{4n+1}}{1 - x^{4n+1}}$$
$$+ \sum_{n=0}^{\infty} (-1)^{\deg(4n+2)} \frac{x^{4n+2}}{1 - x^{4n+2}} + \sum_{n=0}^{\infty} (-1)^{\deg(4n+3)} \frac{x^{4n+3}}{1 - x^{4n+3}}.$$

The power series version (using $x = \exp(\pi \imath \tau)$) of (7.13) is

$$\left(\sum_{n=-\infty}^{\infty} x^{n^2} \right)^2 = \left(\sum_{n=-\infty}^{\infty} x^{2n^2} \right)^2 + x \left(\sum_{n=-\infty}^{\infty} x^{2(n^2+n)} \right)^2. \tag{7.16}$$

Fix two positive integers m and k. We are interested in writing k as a sum of m integers of a given form, for example, squares or triangular numbers. We also want to count the number of ways to do so. Towards this end, we denote by $S_m(k)$ the number of representations of k as a sum of m squares, by $T_m(k)$ the number of representations of k as a sum of m triangular numbers, and by $\tilde{T}_m(k)$ the number of representations of k as a sum of m double triangular numbers.[136] We must make clear what we mean by different representations. In this context, each representation of $n \in \mathbb{Z}^+$ is determined by a vector $(n_1, \ldots, n_m) \in \mathbb{Z}^m$, where each n_i is a square, triangular number, or double triangular number and $n = \sum_{i=1}^m n_i$; distinct vectors correspond to distinct representations. We shall denote the

[136] Here *double triangular* means of the form $n(n+1)$ and *triangular* means $\frac{n(n+1)}{2}$ with $n \in \mathbb{Z}$.

number of representations of n so obtained by $\mathcal{S}_m(n)$, $\mathcal{T}_m(n)$, and $\tilde{\mathcal{T}}_m(n)$, respectively. A triangular number is of the form $\frac{m(m+1)}{2}$, with $m \in \mathbb{Z}$, and each triangular number n has two representations: $n = \frac{k(k+1)}{2} = \frac{(-k-1)(-k)}{2}$. Thus we may rewrite our representation of n by triangular numbers as $n = \sum_{i=1}^{m} n_i = \sum_{i=1}^{m} \frac{k_i(k_i+1)}{2}$ with $k_i \in \mathbb{Z}^+ \cup \{0\}$, and to obtain $\mathcal{T}_m(n)$ we count vectors $(k_1, \ldots, k_m)$ with $k_i \in \mathbb{Z}^+ \cup \{0\}$ that represent a given $n \in \mathbb{Z}^+$. If we allow $k_i \in \mathbb{Z}$ and we count vectors that represent a given n, we obtain $T_m(n)$. Similar statements hold for sums of double triangular numbers. The situation for squares is slightly different (since all square integers except zero have two representations); however, there exists, in this case, an obvious distinction between $\mathcal{S}_m(n)$ and $S_m(n)$. It is clear that the two counting methods are related:

$$\mathcal{S}_m(n) \leq S_m(n) \leq 2^m \mathcal{S}_m(n),\ 2^m \mathcal{T}_m(n) = T_m(n),\ 2^m \tilde{\mathcal{T}}_m(n) = \tilde{T}_m(n),$$
$$\tilde{T}_m(2n) = T_m(n).$$

We will state results in terms of $S_m(n)$ and $T_m(n)$ whenever possible. The symbols and methods of counting discussed here are consistent with those used in Corollary 7.11 of Chapter 4.

The identity (7.16) can be rewritten as

$$1 + \sum_{n=1}^{\infty} S_2(n)x^n = 1 + \sum_{n=1}^{\infty} S_2(n)x^{2n} + \sum_{n=0}^{\infty} T_2(n)x^{4n+1}. \tag{7.17}$$

This identity has the following combinatorial content:

$$S_2(2k) = S_2(k),\ T_2(k) = S_2(4k+1),\ S_2(4k+3) = 0,\ k \in \mathbb{Z}^+.$$

It is not difficult to prove the last three equalities without the identity (7.17) yielding an alternate proof of the latter.

There is a lovely number theoretic result [**13**, Theorems 278 & 311] which says that the number of representations of N as a sum of two squares is given by $4(d_1(N) - d_3(N))$, where $d_i(N)$ is the number of divisors of N congruent to i mod 4. We derive a corollary of this result. We start with equation (7.13). This gives rise to (we are using the local coordinate $x = \exp(\pi\imath\tau)$)

$$\theta^2\begin{bmatrix}1\\0\end{bmatrix}(0,2\tau) = \theta^2\begin{bmatrix}0\\0\end{bmatrix}(0,\tau) - \theta^2\begin{bmatrix}0\\0\end{bmatrix}(0,2\tau)$$
$$= 4\sum_{k=0}^{\infty}(-1)^k\left(\frac{x^{2k+1}}{1-x^{2k+1}} - \frac{x^{4k+2}}{1-x^{4k+2}}\right) = 4\sum_{k=0}^{\infty}(-1)^k\frac{x^{2k+1}}{1-x^{4k+2}}.$$

We need to explain the second of the above equalities:

$$\theta^2\begin{bmatrix}0\\0\end{bmatrix} = \left(\sum_{n=-\infty}^{\infty} x^{n^2}\right)^2 = \sum_{n=0}^{\infty} S_2(n)x^n = 4\sum_{n=0}^{\infty}(d_1(n) - d_3(n))x^n$$

$$= 4\sum_{k=0}^{\infty}(-1)^k \frac{x^{2k+1}}{1-x^{2k+1}}.$$

We still owe the reader an explanation of the last equality. It is a consequence of (7.11). Our conclusion is hence

Proposition 1.19. *For all $x \in \mathbb{C}$, with $|x| < 1$,*

$$\frac{1}{4}\sum_{k=0}^{\infty} T_2(k)x^{4k+1} = \sum_{k=0}^{\infty}(-1)^k \frac{x^{2k+1}}{1-x^{4k+2}} = \sum_{k=0}^{\infty}(d_1(2k+1) - d_3(2k+1))x^{2k+1}.$$

As a last example of ideas of this sort, we derive a general recursion formula for $S_m(n)$ which involves the σ function.

We define

$$f(x) = 2\sum_{k=1}^{\infty} \frac{x^{2k-1}}{(2k-1)(1+x^{2k-1})} = \log \prod_{n=1}^{\infty}(1-x^{2n})(1+x^{2n-1})^2.$$

The last equality follows from the sequence of identities:

$$\sum_{k=1}^{\infty} \frac{x^{2k-1}}{(2k-1)(1+x^{2k-1})} = \sum_{k=1}^{\infty} \frac{x^{2k-1}}{2k-1} \sum_{n=0}^{\infty}(-1)^n x^{n(2k-1)}$$

$$= \sum_{n=0}^{\infty}(-1)^n \sum_{k=1}^{\infty} \frac{x^{(n+1)(2k-1)}}{2k-1} = \frac{1}{2}\sum_{n=0}^{\infty}(-1)^n \log\left(\frac{1+x^{n+1}}{1-x^{n+1}}\right)$$

$$= \frac{1}{2}\log\left[\prod_{n=1}^{\infty} \frac{(1+x^{2n-1})(1-x^{2n})}{(1-x^{2n-1})(1+x^{2n})}\right]$$

$$= \frac{1}{2}\log\left[\prod_{n=1}^{\infty}(1+x^{2n-1})(1-x^{2n})(1+x^n)(1-x^{4n-2})\right]$$

$$= \frac{1}{2}\log\left[\prod_{n=1}^{\infty}(1+x^{2n-1})(1-x^{2n})(1+x^{2n})(1+x^{2n-1})(1-x^{4n-2})\right]$$

$$= \frac{1}{2}\log \prod_{n=1}^{\infty}(1+x^{2n-1})^2(1-x^{2n}).$$

Hence for any $m \in \mathbb{R}$,

$$2m\sum_{k=1}^{\infty} \frac{x^{2k-1}}{(2k-1)(1+x^{2k-1})} = \log \prod_{n=1}^{\infty}(1+x^{2n-1})^{2m}(1-x^{2n})^m.$$

By the Jacobi triple product identity,

$$\prod_{n=1}^{\infty}(1+x^{2n-1})^2(1-x^{2n}) = \sum_{n=-\infty}^{\infty} x^{n^2}.$$

Hence

$$e^{f(x)} = \sum_{n=-\infty}^{\infty} x^{n^2} = 1 + 2\sum_{n=1}^{\infty} x^{n^2}.$$

Thus

$$f'(x)e^{f(x)} = 2\sum_{n=1}^{\infty} n^2 x^{n^2-1},$$

or equivalently

$$xf'(x) \sum_{n=-\infty}^{\infty} x^{n^2} = 2\sum_{n=1}^{\infty} n^2 x^{n^2}.$$

More generally, for any $m \in \mathbb{Z}^+$,

$$e^{mf(x)} = \left(\sum_{n=-\infty}^{\infty} x^{n^2}\right)^m = 1 + \sum_{n=1}^{\infty} S_m(n)x^n,$$

and as a consequence

$$xmf'(x)\left(1 + \sum_{n=1}^{\infty} S_m(n)x^n\right) = \sum_{n=1}^{\infty} nS_m(n)x^n. \tag{7.18}$$

In order for the above to be useful we require a bit more information concerning the function $f(x)$ we have been using. This is provided in the next lemma.

Lemma 1.20. *Let $n \in \mathbb{Z}^+$ and $n = 2^\alpha p_1^{\alpha_1}...p_k^{\alpha_k}$ be its primary decomposition. Denote by n_o the positive integer $p_1^{\alpha_1}...p_k^{\alpha_k}$, the odd part of n. Then*

$$f(x) = 2\sum_{n=1}^{\infty}(-1)^{n-1}\frac{\sigma(n_o)}{n_o}x^n. \tag{7.19}$$

Proof. Observe that

$$f(-x) = -2\sum_{n=1}^{\infty} \frac{x^{2k-1}}{(2k-1)(1-x^{2k-1})}$$

is a Lambert series. It follows from our remarks at the beginning of the discussion that

$$f(-x) = -2\sum_{n=1}^{\infty} \sum_{d \text{ odd}, d|n} \frac{1}{d}x^n = -2\sum_{n=1}^{\infty} \frac{\sigma(n_o)}{n_o}x^n.$$

Therefore (7.19) and (for use in the next theorem)

$$xf'(x) = 2\sum_{n=1}^{\infty}(-1)^{n+1}\frac{n}{n_o}\sigma(n_o)x^n.$$

Theorem 1.21. *The following power series identity holds for all positive integers m.*

$$\left(2m\sum_{k=1}^{\infty}(-1)^{k+1}\frac{k}{k_o}\sigma(k_o)x^k\right)\left(1+\sum_{n=1}^{\infty}S_m(n)x^n\right)=\sum_{n=1}^{\infty}nS_m(n)x^n.$$

In particular, we have the recursion relation

$$\frac{N}{2m}S_m(N)=(-1)^{N+1}\frac{N}{N_o}\sigma(N_o)+\sum_{l=1}^{N-1}(-1)^{l+1}\frac{l}{l_o}\sigma(l_o)S_m(N-l).$$

Proof. The proof is by substitution into (7.18).

The identities we have derived in Chapter 4 are deeper than the ones we have just now demonstrated; these require almost no function theory, although ultimately we used the Jacobi triple product identity in the translation to infinite product expansions. In the next section we explore the combinatorial content of some of the Chapter 4 identities.

Exercise 1.22. Define a multiplicative function ϕ_2 on $\mathbb{Z}^+$ as follows. For a prime p and a positive integer α let

$$\phi_2(p^\alpha)=\begin{cases}-1 & \alpha \quad \text{if } \alpha \text{ is odd}\\ 1 & \alpha \quad \text{if } \alpha \text{ is even}\end{cases}.$$

Prove the Lambert series identity

$$\sum_{n=1}^{\infty}\frac{\phi_2(n)x^n}{1-x^n}=\sum_{n=1}^{\infty}x^{n^2}.$$

Exercise 1.23. Generalize the function ϕ_2 of the previous exercise to a function ϕ_k defined for an arbitrary integer $k\geq 3$ so that

$$\sum_{n=1}^{\infty}\frac{\phi_k(n)x^n}{1-x^n}=\sum_{n=1}^{\infty}x^{n^k}.$$

2. Identities among partitions

In this section we establish relations between partitions taken from different sets. The problems we solve here are not motivated by a burning desire to know the answer to the question raised, but rather to show the combinatorial content of identities derived previously. We however have no doubt that these are questions which should be addressed by combinatorialists.

2.1. A curious property of 8. Let Z denote 8 copies of the positive integers. Decompose this set into the two subsets E and O consisting of even and odd integers, respectively. Let n be an arbitrary positive integer. One can now ask the question, how many partitions of the number n can one obtain from the set O? One can clearly ask the same question regarding the set E, but here the immediate answer is none, if the number n is chosen to be odd. With this in mind let us turn to the question: Is the number of partitions of the odd number $2k+1$ from the set O obtainable from knowledge of the number of partitions of (even) integers from the set E? In general, for a set S of positive integers and $n \in \mathbb{Z}^-$, we let $L_S(n)$ denote the number of partitions of the n with parts from S (without "additional" repetitions). We set $L_S(0) = 0$.

Theorem 2.1. *For all* $k \in \mathbb{Z}^+ \cup \{0\}$,

$$L_O(2k+1) = 8L_E(2k).$$

Proof. Use (4.20). The partition theoretic interpretation of the terms of this identity is the statement of the theorem. The left hand side is the generating function of $L_O(n)$, the last product on the right hand side is the generating function for $L_E(n)$, and the middle product is the generating function for the difference between the number of even and odd partitions from O. The important point to keep in mind is that if the positive integer n is odd, there are no even partitions of n (from O); and if n is even, there are no odd partitions.

Remark 2.2. The theorem should have a purely combinatorial proof since in this case the basic identity used is the Jacobi quartic identity which has a purely combinatorial proof. We have not tried too hard to find one, but suspect that it exists.

2.2. A curious property of 3. Let O denote the set consisting of 4 copies of the positive odd integers which are congruent to 3 mod 6 and 2 copies of the remaining positive odd integers. Let E denote the set consisting of 4 copies of the positive even integers which are congruent to 0 mod 6 and 2 copies of the remaining positive even integers. Once again we ask the question of the previous section. Can we get information about partitions from O in terms of information about partitions from E?

Theorem 2.3. *For all* $k \in \mathbb{Z}^+ \cup \{0\}$,

$$L_O(2k+1) = 2L_E(2k).$$

Proof. The theorem is the partition theoretic interpretation of the identity (4.21).

2.3. A curious property of 7. In this subsection, where we present our last result of this type, we let O be the set consisting of 2 copies of the set of positive odd integers congruent to 7 mod 14 and one copy of the remaining positive odd integers, and the set E be the set consisting of 2 copies of the positive even integers congruent to 0 mod 14 and one copy of the remaining positive even integers.

Theorem 2.4. *For all* $k \in \mathbb{Z}^+ \cup \{0\}$,

$$L_O(2k+1) = L_E(2k).$$

Proof. The proof is the interpretation of the identity (4.22).

Remark 2.5. The curious property of 7 is in a certain sense the simplest among the examples given. It is however the most difficult one to prove since the identity on which it is based is more difficult to prove than the identities on which the other curious properties were based. It would be nice to find a simpler proof of the identity of equation (4.22).

3. Partitions, divisors, and sums of triangular numbers

In this section we derive a beautiful result which relates three objects. We begin with a set S consisting of 4 copies of the positive integers which are congruent to 0 mod 4 and 4 copies of the set of positive odd integers. We will construct partitions from this set. A partition so constructed is called *even* provided it has an even number of even parts in it. Otherwise it is called an *odd* partition. Let us denote the difference between the number of even partitions and odd partitions of the positive integer n by $D(n)$. The generating function for D is given by

$$\prod_{n=1}^{\infty}(1-x^{4n})^4(1+x^{2n-1})^4 = 1+\sum_{n=1}^{\infty} D(n)x^n.$$

This is clear since each partition of the number n from the set S contributes either ± 1 to the coefficient of x^n in the power series of the infinite product depending on whether the number of integers congruent to 0 mod 4 is even or odd.

We now ask the following question. Given a positive integer n, in how many different ways can it be represented as a sum of 4 triangular numbers?

It is classically known from the time of Lagrange that each positive number is representable as the sum of 4 squares[137] and that for triangular numbers 3 suffice. One of the interpretations that one can give to the Jacobi

[137]There is a necessary and sufficient condition for a positive integer to be representable as a sum of two squares [**13**, Th. 366]. It is easy to see that three will not do [**13**, §20.2]. The four squares result due to Lagrange is established in [**13**, Th. 369].

quartic identity is a relation between the number of representations of the positive integer n as a sum of squares and as a sum of triangular numbers.

Our first result is

Proposition 3.1. *For all $n \in \mathbb{Z}^+$,*

$$T_4(n) = 16D(n).$$

Proof. The proof is rather elementary and direct. Start with the remark that

$$\prod_{n=1}^{\infty}(1-x^{4n})^4(1+x^{2n-1})^4 = \prod_{n=1}^{\infty}(1-x^{2n})^4(1+x^{2n})^4(1+x^{2n-1})^4$$
$$= \prod_{n=1}^{\infty}(1-x^n)^4(1+x^n)^8.$$

The beautiful result we promised is

Theorem 3.2. *For all $n \in \mathbb{Z}^+$,*

$$T_4(n) = 16\sigma(2n+1).$$

Proof. This theorem could have been proven in Chapter 2 and follows from Theorem 5.3 and the Jacobi triple product. The proof we give here (which is basically the same as the argument we could have given earlier) uses (4.30) and Proposition 7.3, which gave us the identity (4.26). The right hand side of that identity can be rewritten as the power series $\sum_{n=1}^{\infty}\sigma(2n-1)x^{2n-1}$. A simple way to see this is to start with

$$\sum_{n=1}^{\infty}\sigma(n)x^n = \sum_{n=1}^{\infty}\frac{nx^n}{1-x^n} = \sum_{n=1}^{\infty}\left(\frac{(2n-1)x^{2n-1}}{1-x^{2n-1}} + \frac{2nx^{2n}}{1-x^{2n}}\right).$$

The right hand side of (4.26) is just the odd part of the last series. We therefore have (after replacing x^2 by x)

$$\prod_{n=1}^{\infty}(1-x^n)^4(1+x^n)^8 = \sum_{n=1}^{\infty}\sigma(2n-1)x^{n-1}.$$

In the proof of the last proposition we saw that the left hand side of the last equation is the power series $1+\sum_{n=1}^{\infty}T_4(n)x^n$.

Theorem 3.2 has some interesting corollaries. Perhaps the most interesting is a primality condition.

Corollary 3.3. *For all $l \in \mathbb{Z}^+$, the integer $2l+1$ is prime if and only if $T_4(l) = 2^5(l+1)$.*

Proof. Since $T_4(l) = 16\sigma(2l+1)$ the result follows.

Another immediate corollary is

Corollary 3.4. *For all $l \in \mathbb{Z}^+$, $\frac{\mathcal{T}_4(l)}{16}$ is odd if and only if $2l+1$ is a square.*

Proof. Since $\mathcal{T}_4(l) = \sigma(2l+1)$, the result follows from facts concerning the divisor function. We claim that $\sigma(2l+1)$ is even unless $2l+1$ is a square. If $2l+1$ is not a square, then it has at least one odd factor p^k with p a prime and k an odd number (chosen as large as possible). Since $\sigma(n)$ is multiplicative, it follows that $\sigma(n)$ is a multiple of

$$\sigma(p^k) = \frac{p^{k+1}-1}{p-1} = 1 + p + p^2 + ... + p^k$$

which is even since there are an even number of odd integers in the last sum.

The above results are simple consequences of the easily derived properties of the σ function. This function satisfies many congruences. For example, it is easy to establish

Proposition 3.5. *For all $k \in \{0\} \cup \mathbb{Z}^+$,*

$$\sigma(3k+2) \equiv 0 \mod 3.$$

Proof. In the prime decomposition of $3k+2$ there must be at least one prime p congruent to 2 mod 3 which appears with odd multiplicity k. We have (the first step is as before) $\sigma(p^k) = \frac{p^{k+1}-1}{p-1} \equiv 0 \mod 3$ since p^{k+1} is congruent to 1 mod 3.

We leave it for the reader to show that the following congruences hold:

$$\sigma(4k+3) \equiv 0 \mod 4,\ \sigma(6k+5) \equiv 0 \mod 6,\ \sigma(8k+7) \equiv 0 \mod 8,$$
$$\sigma(12k+11) \equiv 0 \mod 12 \text{ and } \sigma(24k+23) \equiv 0 \mod 24.$$

As a consequence of the relation between $\mathcal{T}_4$ and σ we conclude

Proposition 3.6. *For all $k \in \{0\} \cup \mathbb{Z}^+$,*

$$T_4(2k+1) \equiv 0 \mod 2^6,\ T_4(3k+2) \equiv 0 \mod 2^5 3,\ T_4(4k+3) \equiv 0 \mod 2^7,$$
$$T_4(6k+5) \equiv 0 \mod 2^6 3 \text{ and } T_4(12k+11) \equiv 0 \mod 2^7 3.$$

Remark 3.7. It is not necessary to use the relation between $\mathcal{T}_4$ and σ in order to prove all the above congruences. Some follow easily from the definition of $\mathcal{T}_4$.

We now prove a general result about $\mathcal{T}_4$ which does follow from its relation to σ.

Theorem 3.8. *If $2m+1$ and $2n+1$ are relatively prime positive odd integers, then*

$$2^4 T_4(2mn+m+n) = T_4(m)T_4(n).$$

Proof. Under the given hypothesis we have

$$\begin{aligned}\mathcal{T}_4(2mn+m+n) &= \sigma(2(2mn+m+n)+1) = \sigma((2m+1)(2n+1)) \\ &= \sigma(2m+1)\sigma(2n+1) = \mathcal{T}_4(m)\mathcal{T}_4(n).\end{aligned}$$

As an application of the above theorem let us take $m = 4$. In order to apply the theorem, we require that $2n+1$ be relatively prime to 9. To accomplish this it is sufficient to require that n be congruent to either 0 or 2 mod 3.

If we take $n \equiv 0 \mod 3$, we obtain $\mathcal{T}_4(4)\mathcal{T}_4(3k) = \mathcal{T}_4(27k+4)$, and therefore

Corollary 3.9. *For all* $k \in \{0\} \cup \mathbb{Z}^+$,

$$\mathcal{T}_4(27k+4) \equiv 0 \mod 2^4 13.$$

Proof. We use the fact that $\mathcal{T}_4(4) = 13$.

If we take $n \equiv 2 \mod 3$, we obtain $\mathcal{T}_4(4)\mathcal{T}_4(3k+2) = \mathcal{T}_4(27k+22)$, and therefore

Corollary 3.10. *For all* $k \in \{0\} \cup \mathbb{Z}^+$,

$$\mathcal{T}_4(27k+22) \equiv 0 \mod 2^5 3 \cdot 13.$$

Proof. In addition to $\mathcal{T}_4(4) = 13$, we use the part of the last proposition that tells us that $\mathcal{T}_4(3k+2) \equiv 0 \mod 6$.

We close this section with some more results related to primality. We have seen that the equation $\mathcal{T}_4(l) = \sigma(2l+1)$ gives a primality condition on $2l+1$ in terms of $\mathcal{T}_4(l)$. Recall that $\mathcal{T}_4(l)$ is the number of ways l can be represented as a sum of exactly 4 triangular numbers, if we think of 0 as a triangular number. In this setting, a representation of l as a sum of 4 triangular numbers is a vector $(\Delta_1, \Delta_2, \Delta_3, \Delta_4) \in \mathbb{Z}^4$, where each Δ_i is a triangular number and $\sum_{i=1}^4 \Delta_i = l$. Two representations are considered to be the same if and only if the vectors are equal. There are 24 different representations of a positive integer which use the same 4 distinct triangular numbers. We shall call the number of different representations of l obtained from the four triangular numbers $\Delta_i, i = 1, 2, 3, 4$ the *weight* of the representation. The possible weights of representations are 1, 4, 6, 12 and 24. The first occurs when all the Δ_i are equal, and the last occurs when all the Δ_i are distinct. If there are only two distinct triangular numbers in the representation and one occurs with multiplicity three, the weight is 4. If there are two distinct triangular numbers in the representation and each occurs with multiplicity 2, the weight is 6. If there are three distinct triangular numbers in the representation (so that exactly one of them occurs with multiplicity 2), the weight is 12.

The above considerations allow us to prove the following

Corollary 3.11. *Let $p = 4k+1$ be a prime number $\equiv 1 \mod 4$. The number of representations of k as a sum of two triangular numbers has odd parity. In particular, there is always such a representation.*

Proof. By Theorem 3.2, $\mathcal{T}_4(2k) = \sigma(4k+1)$. The hypothesis that $4k+1$ is prime yields $\sigma(4k+1) = 2(2k+1)$. Thus $\mathcal{T}_4(2k) \equiv 2 \mod 4$. Note that $2k$ cannot possibly be a number of the form 4Δ, with Δ a triangular number. If it were, it would follow that

$$4k+1 = 8\Delta + 1 = 4n(n+1) + 1 = (2n+1)^2,$$

contradicting the fact that $4k+1$ is a prime.

It follows that there are no representations of $2k$ of weight 1. The number of representations of l of each of the other weights with the exception of weight 6 is equivalent to 0 mod 4. In order that $\mathcal{T}_4(2k) \equiv 2 \mod 4$, it is necessary that $2k$ have an odd number of representations of weight 6. Hence $2k = 2\Delta_1 + 2\Delta_2$, Δ_i triangular with $\Delta_1 \neq \Delta_2$ and therefore $k = \Delta_1 + \Delta_2$. Moreover, there must be an odd number of such representations.

Remark 3.12. Suppose that for $n \in \mathbb{Z}^+$, the number n^2+1 is an odd prime. Then n must be even so that $n^2+1 = 4k^2+1$, and by the previous corollary, it must be the case that k^2 is representable as a sum of two triangular numbers. This is true whether n^2+1 is prime or not since $k^2 = \frac{k(k+1)}{2} + \frac{k(k-1)}{2}$. The last corollary is not new and can be derived from Gauss' result that each prime of the form $4k+1$ is representable as a sum of two squares. The reader is invited to provide the details.

3.1. Sums of 4 squares. Theorem 3.2 gave a formula for the number of representations, $T_4(n)$, of the positive integer n as a sum of 4 triangular numbers. We have chosen theorems on representations as sums of triangular numbers rather than sums of squares as an area for application because the statements of results seemed cleaner in this setting. However, the Jacobi quartic identity gives a precise relation between the number of representations of n as a sum of 4 triangular numbers and the number, $S_4(n)$, of its representations as a sum of 4 squares. The Jacobi quartic, which in terms of infinite series can be written as

$$\left(\sum_{n=-\infty}^{\infty} x^{n^2}\right)^4 - \left(\sum_{n=-\infty}^{\infty} (-1)^n x^{n^2}\right)^4 = x\left(\sum_{n=-\infty}^{\infty} x^{n(n+1)}\right)^4,$$

asserts that

$$2S_4(2n+1) = T_4(n).$$

To verify this assertion, note that the left hand side of the equation is twice the odd part of $\left(\sum_{n=-\infty}^{\infty} x^{n^2}\right)^4$; hence $2S_4(2n+1)$ is the coefficient of the

x^{2n+1} term in its Taylor series expansion (about 0). The coefficient of the x^{2n+1} term of the Taylor series expansion of the right hand side is $\tilde{T}_4(2n) = T_4(n)$. It follows that any theorem proven about representations of integers by 4 triangular numbers gives a theorem about representations by 4 squares and vice-versa. Thus we conclude that $S_4(2n+1) = 8\sigma(2n+1)$. In particular we recover part of Lagrange's theorem which asserts that every positive integer is expressible as a sum of four squares.

In Chapter 4, §8, we studied the function $\left(\frac{(\eta(\tau))^{\frac{(k^2+1)}{k}}}{\eta(k\tau)\eta(\frac{\tau}{k})}\right)^{\frac{24}{k^2-1}}$ and showed that when $k = 2$ or 3, it projects to a meromorphic differential on $\overline{\mathbb{H}^2/\Gamma(k,k)}$ whose only singularities are simple poles at P_0 and P_∞. For $k = 2$, where $\Gamma(2,2) = \Gamma(2)$, we can write the above in terms of the local coordinate $x = \exp(\pi\imath\tau)$ as the infinite product $\prod_{n=1}^{\infty} \frac{(1-x^{2n})^{20}}{(1-x^{4n})^8(1-x^n)^8}$. A sequence of algebraic manipulations transforms the last expression to

$$\prod_{n=1}^{\infty} \frac{(1-x^{2n})^{12}}{(1+x^{2n})^8(1-x^n)^8} = \prod_{n=1}^{\infty} \frac{(1-x^{2n})^{12}}{(1+x^{2n})^8(1-x^{2n-1})^8(1-x^{2n})^8}$$

$$= \prod_{n=1}^{\infty} \frac{(1-x^{2n})^4}{(1+x^{2n})^8(1-x^{2n-1})^8} = \prod_{n=1}^{\infty} \frac{(1-x^{2n})^4(1+x^n)^8}{(1+x^{2n})^8}$$

$$= \prod_{n=1}^{\infty} (1-x^{2n})^4(1+x^{2n-1})^8 = \theta^4\begin{bmatrix} 0 \\ 0 \end{bmatrix}(0,\tau).$$

We have seen that $\left(\frac{\eta(2\tau)}{\eta(\frac{\tau}{2})}\right)^8$ is a meromorphic function on $\overline{\mathbb{H}^2/\Gamma(2)}$ with a simple zero at P_∞ and a simple pole at P_0. Its infinite product expansion is given by $x\prod_{n=1}^{\infty} \frac{(1-x^{4n})^8}{1-x^n)^8}$.

It follows that

Theorem 3.13. *There exists a constant c so that for all $\tau \in \mathbb{H}^2$,*

$$\frac{d}{d\tau}\log\frac{\eta(2\tau)}{\eta(\frac{\tau}{2})} = c\frac{(\eta(\tau))^{20}}{\eta(2\tau)^8\eta(\frac{\tau}{2})^8} = c\theta^4\begin{bmatrix} 0 \\ 0 \end{bmatrix}(0,\tau). \tag{7.20}$$

Proposition 3.14. *We have the identity*

$$\sum_{n=1}^{\infty} S_4(n)x^n = 8\sum_{n=1}^{\infty} \sigma(n)(x^n - 4x^{4n}).$$

Proof. The identity is the translation of the equality of the extreme terms in Theorem 3.13 to infinite products.

The power series on the right hand side of the equality in the last equation can be rewritten as $\sum_{n=1}^{\infty} \left(\sigma(n) - 4\sigma(\frac{n}{4})\right) x^n$, where it is understood that (according to our usual conventions) $\sigma(m) = 0$ when m is not a positive integer. It follows that if n is not congruent to 0 mod 4, then the coefficient of x^n in the last series is is $\sigma(n)$. What happens when n is congruent to 0 mod 4? To this end, let $n = 4k$ and $k = 2^l N$ with N an odd positive integer. Then

$$\sigma(n) - 4\sigma\left(\frac{n}{4}\right) = \sigma(4k) - \sigma(k) = \sigma(2^{l+2}N) - 4\sigma(2^l N)$$

$$= (2^{l+3} - 1)\sigma(N) - 4(2^{l+1} - 1)\sigma(N) = 3\sigma(N) = \sigma(2N).$$

Hence the coefficient of x^n in the power series on the right hand side of the equality in Proposition 3.14 can be written as $8\tilde{\sigma}(n)$, where $\tilde{\sigma}(n) = \sum_{d|n} d$, where d is a positive integer not congruent to 0 mod 4. This has given us

Corollary 3.15. *For all $n \in \mathbb{Z}^+$,*

$$S_4(n) = 8\tilde{\sigma}(n).$$

It is worth noting that the above also follows from Corollary 7.11 of Chapter 4 as a consequence of the observation that

$$\sum_{n=1}^{\infty} \left(\frac{(2n-1)x^{2n-1}}{1 - x^{2n-1}} + \frac{2nx^{2n}}{1 + x^{2n}} \right) = \sum_{n=1}^{\infty} \tilde{\sigma}(n)x^n.$$

We leave the proof of this formula as an exercise for the reader.

Our last result in this section relates the function $\tilde{\sigma}$ to partitions. We have seen that

$$\prod_{n=1}^{\infty} (1 - x^{2n})^4 (1 + x^{2n-1})^8 = 1 + \sum_{n=1}^{\infty} S_4(n)x^n = 1 + 8\sum_{n=1}^{\infty} \tilde{\sigma}(n)x^n.$$

The left hand side of the above equation has a simple partition theoretic interpretation. Consider the set S consisting of 4 copies of the positive even integers and 8 copies of the positive odd integers. Call a partition of n with parts from S *even* if it has an even number of even parts and *odd* otherwise. Hence

Corollary 3.16. *We have*

$$E(n) - O(n) = 8\tilde{\sigma}(n), \ n \in \mathbb{Z}^+.$$

Therefore, among other things, $E(n) - O(n) \geq 8$ and for all $k \in \mathbb{Z}^+$, $E(2^k) - O(2^k) = 24$.

3.2. A remarkable formula. In the previous section we showed that there is a connection between partitions and the σ-function. In this section we derive a formula for σ which follows from Proposition 7.7 in Chapter 4. The identity (4.29) of that proposition translates to

$$(7.21)\quad x\prod_{n=1}^{\infty}\frac{(1-x^{9n})^3(1-x^n)^3}{(1-x^{3n})^2}=\sum_{k=0}^{\infty}(\sigma(3k+1)x^{3k+1}-\sigma(3k+2)x^{3k+2}).$$

The above identity gives a remarkable formula.

Let S, our set from which partitions are formed, consist of four copies of the positive integers which are congruent to 0 mod 9; three copies of the positive integers which are congruent to 1,2,4,5,7,8 mod 9; and one copy of the integers congruent to 3, 6 mod 9. We have

Theorem 3.17. *For all integers $n \geq 2$,*

$$E(n-1)-O(n-1)=\begin{cases}0 \text{ for } n\equiv 0 \mod 3\\ \sigma(n) \text{ for } n\equiv 1 \mod 3\\ -\sigma(n) \text{ for } n\equiv 2 \mod 3\end{cases}.$$

We once again have a primality condition.

Corollary 3.18. *If $n \geq 5$ is a prime, then*

$$E(n-1)-O(n-1)=\pm(n+1)$$

depending on whether n is congruent to 1 or 2 $\mod 3$. Conversely, if for $n \in \mathbb{Z}^+$, $E(n)-O(n)=\pm(n+2)$, then $n+1$ is a prime.

The above formula came from the theory for $k=3$, and the integer 9 seems to play a special role in it. We return to the meromorphic differential $\left(\frac{(\eta(\tau))^{\frac{(k^2+1)}{k}}}{\eta(k\tau)\eta(\frac{\tau}{k})}\right)^{\frac{24}{k^2-1}}$ on $\mathbb{H}^2/\Gamma(k,k)$. For $k=3$, this meromorphic differential $\frac{\eta^{10}(\tau)}{\eta^3(3\tau)\eta^3(\frac{\tau}{3})}$ on $\overline{\mathbb{H}^2/\Gamma(3,3)}=\overline{\mathbb{H}^2/\Gamma(3)}$ has simple poles at P_0 and P_∞. The ratio $\frac{\eta^3(3\tau)}{\eta^3(\frac{\tau}{3})}$ is a function on the same surface with a simple zero at P_∞ and a simple pole at P_0. It follows that there is a constant c such that

$$c\frac{\eta^{10}(\tau)}{\eta^3(3\tau)\eta^3(\frac{\tau}{3})}=\frac{d}{d\tau}\log\frac{\eta^3(3\tau)}{\eta^3(\frac{\tau}{3})}.$$

The last equation translates to the infinite product/series expansion

(7.22)
$$\prod_{n=1}^{\infty}\frac{(1-x^{3n})^{10}}{(1-x^{9n})^3(1-x^n)^3}=1+3\sum_{n=1}^{\infty}\sigma(n)(x^n-9x^{9n})=1+3\sum_{n=1}^{\infty}\overline{\sigma}(n)x^n,$$

where

$$\overline{\sigma}(n) = \sum_{d|n,\ d\not\equiv 0 \mod 9} d.$$

The above equation has an interesting consequence which we now explore. We rewrite equation (7.22) as

$$\prod_{n=1}^{\infty} \frac{(1-x^{3n})^{10}}{(1-x^{9n})^3} = \prod_{n=1}^{\infty}(1-x^n)^3 \left(1 + 3\sum_{n=1}^{\infty} \overline{\sigma}(n) x^n\right).$$

Using Jacobi's formula for the product on the right and partition notation, we obtain

(7.23)

$$\prod_{n=1}^{\infty} \frac{(1-x^{3n})^{10}}{(1-x^{9n})^3} = 1 + \sum_{n=1}^{\infty} a_n x^n \; a_n = 3\sum_{l=0}^{n-1} \overline{\sigma}(n-l) P_{-3}(l) + P_{-3}(n).$$

Since $a_n = 0$ for $n \not\equiv 0 \mod 3$, $P_{-3}(l) = 0$ whenever l is not a triangular number, and $P_{-3}(\frac{l^2+l}{2}) = (-1)^l(2l+1)$, we have

Theorem 3.19. *If n is congruent to 1 or 2 $\mod 3$, then*

$$3\sum_{l=0}^{\infty} (-1)^l (2l+1) \overline{\sigma}\left(n - \frac{l^2+l}{2}\right) + P_{-3}(n) = 0.$$

Corollary 3.20. *If n is congruent to 2 $\mod 3$, then*

$$\sigma(n) = \sum_{l=1}^{\infty} (-1)^{l+1} (2l+1) \sigma\left(n - \frac{l^2+l}{2}\right).$$

Proof. Since triangular numbers are congruent to 0 or 1 mod 3, n cannot be triangular if it is congruent to 2 mod 3. Thus, in this case, $a_n = 0 = P_{-3}(n)$ in equation (7.23). When n is congruent to 2 mod 3, $n - \frac{l^2+l}{2}$ is never a multiple of 3, so surely, not a multiple of 9. Hence

$$\overline{\sigma}\left(n - \frac{l^2+l}{2}\right) = \sigma\left(n - \frac{l^2+l}{2}\right).$$

The above corollary gives us a way of computing $\sigma(n)$ given sufficiently many $\sigma(n-k)$ with k triangular numbers; but only when n is congruent to 2 mod 3. It turns out that one can get this formula and more in a simpler way. We define a function that has two apparently different and useful forms as a result of Jacobi's formula,

$$f(x) = \prod_{n=1}^{\infty}(1-x^n)^3 = \sum_{n=0}^{\infty}(-1)^n(2n+1)x^{\frac{n^2+n}{2}}.$$

It follows that

$$-x\frac{f'(x)}{f(x)} = \sum_{n=1}^{\infty} \frac{3x^n}{1-x^n} = 3\sum_{n=1}^{\infty} \sigma(n)x^n,$$

and hence

$$-\sum_{n=1}^{\infty}(-1)^n(2n+1)\frac{n(n+1)}{2}x^{\frac{n(n+1)}{2}}$$
$$= \left(\sum_{n=0}^{\infty}(-1)^n(2n+1)x^{\frac{n(n+1)}{2}}\right)\left(3\sum_{n=1}^{\infty}\sigma(n)x^n\right).$$

Thus we have

$$-\sum_{n=1}^{\infty}(-1)^n(2n+1)\frac{n(n+1)}{2}x^{\frac{n(n-1)}{2}} = 3\sum_{n=1}^{\infty} a_n x^n,$$

with

$$a_n = \sum_{j=0}^{\infty}(-1)^j(2j+1)\sigma\left(n - \frac{j(j+1)}{2}\right).$$

We conclude that if n is not a triangular number, then

$$\sum_{j=0}^{\infty}(-1)^j(2j+1)\sigma\left(n - \frac{j(j+1)}{2}\right) = 0;$$

while if n is the triangular number $n = \frac{m(m+1)}{2}$, we have

$$\sum_{j=0}^{\infty}(-1)^j(2j+1)\sigma\left(n - \frac{j(j+1)}{2}\right) = (-1)^{m+1}\frac{m(m+1)(2m+1)}{6}.$$

We therefore see that the formula of the corollary is also correct for numbers that are congruent to 0 or 1 mod 3 provided they are not triangular.

The above observations together with the last theorem therefore give

Theorem 3.21. *Assume n is congruent to 1 modulo 3. If n is not a triangular number, then*

$$\sum_{j=0}^{\infty}(-1)^j(2j+1)\sigma\left(n - \frac{j(j+1)}{2}\right) = \sum_{j=0}^{\infty}(-1)^j(2j+1)\overline{\sigma}\left(n - \frac{j(j+1)}{2}\right),$$

and in particular

$$\sum_{j\in\mathbb{Z}^+,\ j\equiv 1 \mod 3}(-1)^j(2j+1)\sigma\left(n - \frac{j(j+1)}{2}\right)$$
$$= \sum_{j\in\mathbb{Z}^+,\ j\equiv 1 \mod 3}(-1)^j(2j+1)\overline{\sigma}\left(n - \frac{j(j+1)}{2}\right).$$

If n is a triangular number, then

$$\sum_{j=0}^{\infty}(-1)^j(2j+1)\sigma\left(n-\frac{j(j-1)}{2}\right)=n\sum_{j=0}^{\infty}(-1)^j(2j+1)\overline{\sigma}\left(n-\frac{j(j+1)}{2}\right).$$

In the theorem, we interpret (as usual) both $\sigma(n)$ and $\overline{\sigma}(n)$ to be zero if $n \notin \mathbb{Z}^+$.

We continue with another curious formula involving triangular numbers. At the beginning of this chapter we showed the combinatorial content of Euler's identity. It said that unless an integer[138] n is pentagonal ($n = \frac{3n^2+n}{2}$), the number of even partitions of n is the same as the number of odd partitions of n when dealing with partitions with no repetitions. There is a curious analogue. Denote by $\sigma_o(n)$ and $\sigma_e(n)$, respectively, the sum of the odd and even positive divisors of the positive integer n, and as usual let the value of both be zero if $n \notin \mathbb{Z}^+$.

Theorem 3.22. *If n is not a triangular number, then*

$$\sum_{j=0}^{\infty}\sigma_o\left(n-\frac{j^2+j}{2}\right)-\sigma_e\left(n-\frac{j^2+j}{2}\right)=0.$$

If n is a triangular number, $n=\frac{k^2+k}{2}$, then

$$\sum_{j=0}^{\infty}\sigma_o\left(n-\frac{j^2+j}{2}\right)-\sigma_e\left(n-\frac{j^2+j}{2}\right)=\frac{k^2+k}{2}.$$

Proof. The Jacobi triple product identity implies that

$$f(x)=\sum_{n=0}^{\infty}x^{\frac{n^2+n}{2}}$$

$$=\prod_{n=1}^{\infty}(1-x^n)(1+x^n)^2=\prod_{n=1}^{\infty}(1-x^{2n})(1+x^n)=\prod_{n=1}^{\infty}\frac{(1-x^{2n})}{(1-x^{2n-1})}.$$

Hence

$$x\frac{f'(x)}{f(x)}=\sum_{n=1}^{\infty}\left(\frac{(2n-1)x^{2n-1}}{1-x^{2n-1}}-\frac{2nx^{2n}}{1-x^{2n}}\right)=\sum_{n=1}^{\infty}(\sigma_o(n)-\sigma_e(n))x^n.$$

Thus

$$\sum_{n=1}^{\infty}\frac{n^2+n}{2}x^{\frac{n^2+n}{2}}=\left(\sum_{n=1}^{\infty}(\sigma_o(n)-\sigma_e(n))x^n\right)\left(\sum_{n=0}^{\infty}x^{\frac{n^2+n}{2}}\right)=\sum_{n=1}^{\infty}a_nx^n,$$

[138] Actually pentagonal means that $n=\frac{3n^2-n}{2}$, and the above should be called generalized pentagonal

where

$$a_n = \sum_{j=0}^{\infty} \left(\sigma_o \left(n - \frac{j^2+j}{2} \right) - \sigma_e \left(n - \frac{j^2+j}{2} \right) \right).$$

Remark 3.23. One can remove all reference to even and odd divisors in the previous theorem. For all $n \in \mathbb{Z}^+$, $\sigma(n) = \sigma_o(n) - \sigma_e(n)$. Write $n = 2^k \Delta$ with $(2, \Delta) = 1$. Observe that for $k \in \mathbb{Z}^+$,

$$\sigma_e(2^k \Delta) = \sigma_e(2^k)\sigma(\Delta)$$

$$= 2(2^k - 1)\sigma(\Delta) = 2\sigma(2^{k-1})\sigma(\Delta) = 2\sigma(2^{k-1}\Delta),$$

while for (for $k = 0$) $\sigma_e(\Delta) = 0$. Hence for all $n \in \mathbb{Z}^+$, $\sigma_e(n) = 2\sigma\left(\frac{n}{2}\right)$. Hence $\sigma_o(n) - \sigma_e(n) = \sigma(n) - 2\sigma_e(n) = \sigma(n) - 4\sigma\left(\frac{n}{2}\right)$.

As a result of the remark we have

Corollary 3.24. *If n is not a triangular number, then*

$$\sum_{j=0}^{\infty} \sigma\left(n - \frac{j^2+j}{2}\right) = 4\sum_{j=0}^{\infty} \sigma\left(\frac{n - \frac{j^2+j}{2}}{2}\right).$$

In particular, the left hand side is congruent to 0 modulo 4. If n is a triangular number, $n = \frac{k^2+k}{2}$, then

$$\sum_{j=0}^{\infty} \sigma\left(n - \frac{j^2+j}{2}\right) - 4\sum_{j=0}^{\infty} \sigma\left(\frac{n - \frac{j^2+j}{2}}{2}\right) = \frac{k^2+k}{2}.$$

3.3. Weighted sums of triangular numbers.

Let N be a positive integer. If N is not a triangular number, its *weight* is defined to be 0. If N is a triangular number, $N = \frac{m^2+m}{2}$ $(m \in \mathbb{Z}^+ \cup \{0\})$, its *weight*[139] is defined to be $\frac{\cos\left(\frac{(2m+1)\pi}{6}\right)}{\cos\frac{\pi}{6}}$.

If N is triangular, its weight is either $0, +1$ or -1 depending on the residue class of m mod 6. If $m \equiv 1, 4 \bmod 6$, the weight is zero; if $m \equiv 2, 3 \bmod 6$, the weight is -1; while if $m \equiv 0, 5 \bmod 6$, the weight is $+1$. It is clear that for $m \equiv 1, 4 \bmod 6$, $\frac{m^2+m}{2} \equiv 1 \bmod 3$, while in the other cases $\frac{m^2+m}{2} \equiv 0 \bmod 3$. In fact, weight -1 corresponds to those triangular numbers which are of the form $3\left(\frac{3m^2+m}{2}\right)$ with m odd, and weight 1 corresponds to those of the same form with m even. In other words, the only triangular numbers

[139] Each triangular number N has two representations: $\frac{m(m+1)}{2}$ and $\frac{(-m-1)(-m)}{2}$, with $m \in \mathbb{Z}$; because the cosine is an even function, the weight of the number N is independent of the representation.

with nonzero weight correspond to multiples of (generalized) pentagonal numbers by 3. These facts have the following interpretation:

$$\sum_{m=0}^{\infty} \frac{\cos \frac{(2m+1)\pi}{6}}{\cos \frac{\pi}{6}} x^m = \sum_{n=-\infty}^{\infty} (-1)^n x^{\frac{3(3n^2+n)}{2}} = \prod_{n=1}^{\infty} (1-x^{3n}).$$

This example is the case $k=3$ of the general situation described next.

Let k be an odd integer ≥ 3. Let N be a positive integer and let the vector $\left(m_1, \ldots, m_{\frac{k-1}{2}}\right)$ correspond to the representation of N as a sum of $\frac{k-1}{2}$ triangular numbers, $N = \sum_{i=1}^{\frac{k-1}{2}} \frac{m_i^2+m_i}{2}$.

Define the *weight* of the representation of N corresponding to the vector $\left(m_1, \ldots, m_{\frac{k-1}{2}}\right)$ to be

$$\prod_{i=1}^{\frac{k-1}{2}} \frac{\cos\left(\frac{(2m_i+1)(2i-1)\pi}{2k}\right)}{\cos\left(\frac{(2i-1)\pi}{2k}\right)}.$$

Note that the weight of a representation is not necessarily an integer, but is independent of the choice of which of the two integers is used to represent a fixed triangular number. If we take $k=5$ and $N=1$, the representations are $(1,0)$ and $(0,1)$. The weights are $\frac{\cos \frac{3\pi}{10}}{\cos \frac{\pi}{10}}$ and $\frac{-\cos \frac{\pi}{10}}{\cos \frac{3\pi}{10}}$. These are respectively $\frac{\sqrt{5}-1}{2}$ and $-\left(\frac{\sqrt{5}+1}{2}\right)$; their sum is -1, an integer.

Theorem 3.25. *Let $k \geq 3$ be an odd integer and let N be an arbitrary positive integer. Let*

$$S_N = \left\{ \left(m_1, \ldots, m_{\frac{k-1}{2}}\right) \in (\mathbb{Z}^+ \cup \{0\})^{\frac{k-1}{2}};\ \sum_{i=1}^{\frac{k-1}{2}} \frac{m_i^2+m_i}{2} = N \right\}.$$

Then

$$\sum_{\left(m_1,\ldots,m_{\frac{k-1}{2}}\right) \in S_N} \prod_{i=1}^{\frac{k-1}{2}} \frac{\cos\left(\frac{(2m_i+1)(2i-1)\pi}{2k}\right)}{\cos\left(\frac{(2i-1)\pi}{2k}\right)} = A_N, \tag{7.24}$$

with A_N an integer defined by the following series:

$$\prod_{n=1}^{\infty} (1-x^n)^{\frac{k-3}{2}} (1-x^{kn}) = 1 + \sum_{N=1}^{\infty} A_N x^N. \tag{7.25}$$

Proof. Fix an odd integer $k > 1$, and let m be a positive odd integer $< k$. Let $x = e^{2\pi i\tau}$. Then we have

$$\theta\left[\begin{array}{c}1\\ \frac{m}{k}\end{array}\right](0,\tau) = \sum_{n=-\infty}^{\infty} \exp 2\pi\imath\left[\frac{1}{2}\left(n+\frac{1}{2}\right)^2\tau + \left(n+\frac{1}{2}\right)\frac{m}{2k}\right]$$

$$= \sum_{n=-\infty}^{\infty} \exp 2\pi\imath\left[(n+\frac{1}{2})\frac{m}{2k}\right] x^{\frac{1}{2}(n+\frac{1}{2})^2}$$

$$= x^{\frac{1}{8}} \sum_{n=-\infty}^{\infty} \exp 2\pi\imath\left[(n+\frac{1}{2})\frac{m}{2k}\right] x^{\frac{n^2+n}{2}}$$

$$= x^{\frac{1}{8}} \sum_{n=0}^{\infty}\left(\exp 2\pi\imath\left[(n+\frac{1}{2})\frac{m}{2k}\right] + \exp 2\pi\imath\left[(-(n+1)+\frac{1}{2})\frac{m}{2k}\right]\right) x^{\frac{n^2+n}{2}}$$

$$= x^{\frac{1}{8}} \sum_{n=0}^{\infty}\left[\exp\left(\pi\imath(2n+1)\frac{m}{2k}\right) + \exp\left(-\pi\imath(2n+1)\frac{m}{2k}\right)\right] x^{\frac{n^2+n}{2}}$$

$$= x^{\frac{1}{8}} \sum_{n=0}^{\infty} 2\cos\left(\frac{(2n+1)m\pi}{2k}\right) x^{\frac{n^2+n}{2}}.$$

Thus using (7.24)

$$\prod_{l=0}^{\frac{k-3}{2}} \theta\left[\begin{array}{c}1\\ \frac{1+2l}{k}\end{array}\right](0,\tau) = x^{\frac{k-1}{16}} \prod_{l=0}^{\frac{k-3}{2}} \sum_{n=0}^{\infty} 2\cos\left(\tfrac{(2n+1)(1+2l)\pi}{2k}\right) x^{\frac{n^2+n}{2}}$$

$$= 2^{\frac{k-1}{2}} x^{\frac{k-1}{16}} \left(\prod_{l=0}^{\frac{k-3}{2}} \cos\tfrac{(1+2l)\pi}{2k}\right)\left(1+\sum_{N=1}^{\infty} A_N x^N\right).$$

It remains to prove that A_N is an integer given by (7.25). This follows from the Jacobi triple product formula, which implies the formula

$$\prod_{l=0}^{\frac{k-3}{2}} \theta\left[\begin{array}{c}1\\ \frac{1+2l}{k}\end{array}\right](0,\tau) = 2^{\frac{k-1}{2}} x^{\frac{k-1}{16}} \prod_{l=0}^{\frac{k-3}{2}} \cos\frac{(1+2l)\pi}{2k} \prod_{n=1}^{\infty}(1-x^n)^{\frac{k-3}{2}}(1-x^{kn}).$$

Remark 3.26. The integers A_N have a combinatorial interpretation. Let Y be the set of $\frac{k-3}{2}$ copies of the nonnegative integers not multiples of k and $\frac{k-1}{2}$ copies of the multiples of k. The integer A_N is the difference between the number of even and odd partitions of N from the set Y.

4. Counting points on conic sections

Almost all of the combinatorics dealt with till now involved some form of partitions. We did encounter in our discussion the question: In how many different ways can one represent a positive integer as the sum of 4 squares or 4 triangular numbers? This can be viewed as a geometric question.

Given a positive integer n, we ask for the number of solutions of the diophantine equations (that is, we are looking only for integral solutions)

$$x_1^2 + x_2^2 + x_3^2 + x_4^2 = n \text{ or } x_1^2 + x_2^2 + x_3^2 + x_4^2 + x_1 + x_2 + x_3 + x_4 = 2n.$$

In the first case we are asking for the number of lattice points in $\mathbb{R}^4$ which lie on the sphere

$$x_1^2 + x_2^2 + x_3^2 + x_4^2 = n,$$

and in the second case we are asking for the number of lattice points in $\mathbb{R}^4$ which lie on the sphere

$$\left(x_1 + \frac{1}{2}\right)^2 + \left(x_2 + \frac{1}{2}\right)^2 + \left(x_3 + \frac{1}{2}\right)^2 + \left(x_4 + \frac{1}{2}\right)^2 = 2n + 1.$$

The number of solutions of the first problem is $S_4(n)$, while the number of solutions of the second problem is $T_4(n)$. The Jacobi quartic identity gives the relation $T_4(n) = 2S_4(2n+1)$ between the solutions to the two questions.

We continue to study the geometric content of some of the identities we have derived in Chapter 4.

Consider the sphere S in $\mathbb{R}^3$ defined by the equation

$$x^2 + y^2 + \left(z + \frac{1}{2}\right)^2 = \frac{8n+1}{4}, \quad n \in \mathbb{Z}^+ \cup \{0\}.$$

Thus, for example, when $n = 0$, there are precisely 2 lattice points on the sphere. They are the points $(0,0,0)$ and $(0,0,-1)$. We shall call a lattice point $(x, y, z) \in \mathbb{Z}^3$ *even* provided y is an even integer and call it *odd* otherwise. Denote by $E(n)$ and $O(n)$ respectively the cardinality of the set of even and odd lattice points on the sphere S.

Theorem 4.1. *Let $n \in \mathbb{Z}^+$. Then*

$$E(n) - O(n) = \begin{cases} 0 \text{ for } n \text{ not a triangular number} \\ (-1)^k(4k+2) \text{ for } n \text{ the triangular number } \frac{k^2+k}{2} \end{cases}.$$

Proof. The theorem is a translation of the Jacobi derivative formula, Corollary 5.8 of Chapter 2, which in the variable $x = \exp(2\pi\imath\tau)$ is written as

$$\sum_{n\in\mathbb{Z}} x^{\frac{n^2}{2}} \sum_{n\in\mathbb{Z}} (-1)^n x^{\frac{n^2}{2}} \sum_{n\in\mathbb{Z}} x^{\frac{n^2+n}{2}} = \sum_{n\in\mathbb{Z}} (-1)^n (2n+1) x^{\frac{n^2+n}{2}}.$$

One can ask a similar question concerning the lattice points on the sphere

$$\left(x + \frac{1}{6}\right)^2 + \left(y + \frac{1}{6}\right)^2 - \left(z + \frac{1}{6}\right)^2 = \frac{8n+1}{12}, \quad n \in \mathbb{Z}^+ \cup \{0\}.$$

In this case, the parity of the lattice point (x, y, z) is the parity of the integer $x + y + z$. The result now is

$$E(n) - O(n) = \begin{cases} 0 \text{ for } n \text{ not a triangular number} \\ (-1)^k(2k+1) \text{ for } n \text{ the triangular number } \frac{k^2+k}{2} \end{cases}.$$

Its proof depends on the identity (4.8).

We can of course continue in this fashion going through our identities and interpreting them. Some will be more interesting than others. We shall be satisfied with just one more example of this type. The identity we shall use is the content of Theorem 4.4 of Chapter 4.

Theorem 4.2. *Let k be an odd prime. For $n \in \mathbb{Z}^+$, denote by $A(n)$ the set of lattice points in $\mathbb{R}^{\frac{k-1}{2}}$ which lie on the ellipsoid*

$$3x_0^2 + x_0 + k \sum_{l=1}^{\frac{k-3}{2}} (3x_l^2 + x_l) = 2n,$$

and by $B(n)$ the set of lattice points in $\mathbb{R}^{\frac{k-1}{2}}$ which lie on the ellipsoid

$$\sum_{l=0}^{\frac{k-3}{2}} (kx_l^2 + (2l+1)x_l) = 2n.$$

Then

$$\sum_{\left(x_0,..,x_{\frac{k-3}{2}}\right) \in A(n)} (-1)^{\sum_{l=0}^{\frac{k-3}{2}} x_l} = \sum_{\left(x_0,..,x_{\frac{k-3}{2}}\right) \in B(n)} (-1)^{\sum_{l=0}^{\frac{k-3}{2}} x_l}.$$

Proof. The proof is the translation of Theorem 4.4 to the language of power series and the fact that for some constant c, $\theta' \begin{bmatrix} 1 \\ 1 \end{bmatrix} (0, \tau) = c\theta^3 \begin{bmatrix} \frac{1}{3} \\ 1 \end{bmatrix} (0, 3\tau)$.

Our next example of a problem of similar type is a consequence of the septuple product identity derived in Chapter 4. If we differentiate that identity with respect to the variable z and then set $z = -1$, we obtain

$$\sum_{n=-\infty}^{\infty} (-1)^n x^{\frac{5n^2+n}{2}} \left(\sum_{n=-\infty}^{\infty} (5n+3)x^{\frac{5n^2+3n}{2}} - \sum_{n=-\infty}^{\infty} 5nx^{\frac{5n^2-3n}{2}} \right)$$
$$- \sum_{n=-\infty}^{\infty} (-1)^n x^{\frac{5n^2+3n}{2}} \left(\sum_{n=-\infty}^{\infty} (5n+1)x^{\frac{5n^2-n}{2}} - \sum_{n=-\infty}^{\infty} (5n+2)x^{\frac{5n^2+n}{2}} \right)$$
$$= 4 \prod_{n=1}^{\infty} (1 - x^{2n-1})^2 (1 - x^{2n})^4.$$

The right hand side of this identity has a clear combinatorial interpretation. Consider the set S consisting of two copies of the positive odd integers and four copies of the even positive integers. The coefficient of x^N in the power series defined by the product is $4(E(N)-O(N))$. The reader is invited to formulate an associated lattice point problem using the left hand side.

After deriving the septuple product identity we showed how one can obtain an identity, (4.2), of equal complexity using the odd functions in $\mathcal{O}_3\begin{bmatrix}1\\1\end{bmatrix}$. The particular case $z=-1$ of that identity is

$$
\begin{aligned}
(7.26)\qquad & 3\sum_{n=0}^{\infty}(-1)^n(2n+1)x^{\frac{3(n^2+n)}{2}}\sum_{n=-\infty}^{\infty}x^{\frac{3n^2+n}{2}}\\
&+\sum_{n=0}^{\infty}x^{\frac{3(n^2+n)}{2}}\sum_{n=-\infty}^{\infty}(-1)^n(6n+1)x^{\frac{3n^2+n}{2}}\\
&=4\left(\sum_{n=0}^{\infty}x^{\frac{n^2+n}{2}}\right)^3\sum_{n=-\infty}^{\infty}(-1)^n x^{\frac{3n^2+n}{2}}.
\end{aligned}
$$

An infinite product/power series version of the last identity is

$$
\begin{aligned}
&3\prod_{n=1}^{\infty}(1-x^{3n})^4(1+x^{3n-2})(1+x^{3n-1})\\
&+\sum_{n=-\infty}^{\infty}(-1)^n(6n+1)x^{\frac{3n^2+n}{2}}\prod_{n=1}^{\infty}(1-x^{3n})(1+x^{3n})^2\\
&=4\prod_{n=1}^{\infty}(1-x^n)^4(1+x^n)^6.
\end{aligned}
$$

To give a geometric (combinatorial) interpretation of equation (7.26), define for $N\in\mathbb{Z}^+\cup\{0\}$,

$$A_N=\left\{(m,n);\ \frac{3(m^2+m)}{2}+\frac{3n^2+n}{2}=N\right\},$$

where $m\in\mathbb{Z}^+\cup\{0\}$, $n\in\mathbb{Z}$ and

$$C_N=\left\{(p,q,r,s);\ \frac{p^2+p+q^2+q+r^2+r}{2}+\frac{3s^2+s}{2}=N\right\},$$

where $p\in\mathbb{Z}^+\cup\{0\}$, $q\in\mathbb{Z}^+\cup\{0\}$, $r\in\mathbb{Z}^+\cup\{0\}$, $s\in\mathbb{Z}$. Our identity says that

$$3\sum_{(m,n)\in A_N}(-1)^m(2m+1)+\sum_{(m,n)\in A_N}(-1)^n(6n+1)=4\sum_{(p,q,r,s)\in C_N}(-1)^s.$$

This identity however also has a more geometric version. Consider the semi-circle in $\mathbb{R}^2$

$$\left(x+\frac{1}{2}\right)^2+\left(y+\frac{1}{6}\right)^2=\frac{12N+5}{18}, \quad x>0,\ N\in\mathbb{Z}^+\cup\{0\}.$$

We count the number of lattice points on this circle weighted by

$$\rho(x,y)=(-1)^x(6x+3)+(-1)^y(6y+1).$$

A consequence of the last identity is

$$\sum_{\text{lattice points}}\rho(x,y)\equiv 0 \mod 4.$$

It is interesting to evaluate the sum appearing above.

Consider the ellipsoid in $\mathbb{R}^4$

$$\frac{1}{3}\left(\left(x+\frac{1}{2}\right)^2+\left(y+\frac{1}{2}\right)^2+\left(z+\frac{1}{2}\right)^2\right)+\left(w+\frac{1}{6}\right)^2=\frac{12N+5}{18},$$

where $x\geq 0,\ y\geq 0,\ z\geq 0$ and $N\in\mathbb{Z}^+\cup\{0\}$. Weigh a lattice point on the ellipsoid by $R(x,y,z,w)=4(-1)^w$. Our identity thus asserts that

$$\sum_{\text{lattice points}}\rho(x,y)=\sum_{\text{lattice points}}R(x,y,z,w)=\sum_{\text{lattice points}}4(-1)^w.$$

5. Continued fractions and partitions

At the beginning of this chapter we derived some identities of Euler. The main tool was a two term recursion relation among a sequence of functions. It allowed us to derive the Euler identities relating infinite products and what we called Euler series.

We return to this theme but for problems with an added degree of difficulty. The recursion relations we now establish involve three terms. In general, these are more difficult to handle. The reason we treat this case here is that three term recursion relations are intimately related with continued fraction expansions and it seemed appropriate to close this book with what appears to us as a rather striking identity which resembles the continued fraction discovered by Ramanujan that is connected to the Rogers-Ramanujan identity.

We begin with the sequence of functions

$$F_n(x)=1+\sum_{k=1}^{\infty}\frac{x^{k^2+nk}}{\prod_{l=1}^{k}(1-x^l)^2}, \quad k\in\mathbb{Z}^+,\ n\in\mathbb{Z}^+\cup\{0\}.$$

It is a short calculation to obtain the following three term recursion relations for this sequence:

$$F_n(x) = 2F_{n+1}(x) + (x^{n+1} - 1)F_{n+2}(x), \tag{7.27}$$

and, in particular

$$F_0(x) = 2F_1(x) + (x - 1)F_2(x). \tag{7.28}$$

Lemma 5.1. *There are polynomials* $P_n(x)$ *and* $Q_n(x)$, $n \in \mathbb{Z}^+ \cup \{0\}$, *such that*

$$F_0(x) = P_n(x)F_n(x) + Q_n(x)F_{n+1}(x). \tag{7.29}$$

Proof. We set $P_0 = 1$, $Q_0(x) = 0$, $P_1(x) = 2$ and $Q_1(x) = x-1$, establishing the lemma for the cases $n = 0$ and 1. Assuming the result (7.29) for a given n, we have

$$F_0(x) = P_n(x)F_n(x) + Q_n(x)F_{n+1}(x),$$

which implies by the recursion relations that

$$F_0(x) = P_n(x)(2F_{n+1}(x) + (x^{n+1} - 1)F_{n+2}(x)) + Q_n(x)F_{n+1}(x).$$

We may thus define

$$P_{n+1}(x) = 2P_n(x) + Q_n(x), \qquad Q_{n+1}(x) = P_n(x)(x^{n+1} - 1).$$

In particular, we have obtained a sequence of polynomials which satisfy the recursions

$$P_{n+1}(x) = 2P_n(x) + (x^n - 1)P_{n-1}(x). \tag{7.30}$$

The sequence of polynomials, $\{P_n\}$, we have obtained satisfies $P_n(0) = n + 1$ and is not very interesting in the sense that it is not even formally convergent to a power series. In order to remedy this defect we construct a new sequence from it by defining for $n \in \mathbb{Z}^+$,

$$R_n(x) = P_n(x) - P_{n-1}(x).$$

This new sequence has a lot better chance of converging since $R_n(0) = 1$ for all n. It also satisfies a recursion relation since (7.30) can be rewritten as

$$P_{n+1}(x) - P_n(x) = P_n(x) - P_{n-1}(x) + x^n P_{n-1}(x)$$

or

$$R_{n+1}(x) = R_n(x) + x^n P_{n-1}(x).$$

Obviously $P_{n-2}(x) =$

$$(P_{n-2}(x) - P_{n-3}(x)) + (P_{n-3}(x) - P_{n-4}(x)) + ... + (P_1(x) - P_0(x)) + P_0(x).$$

It follows that

$$R_n(x) = R_{n-1}(x) + x^{n-1}(R_{n-2}(x) + R_{n-3}(x) + ... + R_1(x) + P_0(x)).$$

The recursion for R_n shows us that the sequence R_n stabilizes at the n-th stage (in the sense that $R_{n+1}(x) - R_n(x) = o(x^n)$, $\to 0$) and therefore that the sequence converges formally to a power series.

Lemma 5.2. *The sequence R_n converges (formally) to the Euler series F_0.*

Proof. We start with

$$\begin{aligned}F_0(x) &= P_n(x)(1 + \epsilon_n(x)) + (x^n - 1)P_{n-1}(x)(1 + \epsilon_{n+1}(x)) \\ &= R_n(x) + P_n(x)\epsilon_n(x) + x^n P_{n-1}(x)(1 + \epsilon_{n+1}(x)) - P_{n-1}(x)\epsilon_{n+1}(x),\end{aligned}$$

where $F_n = 1 + \epsilon_n I$ and $\lim_{n\to\infty} \epsilon_n = 0$ (since formally $\epsilon_n(x) = O(x^n)$, $x \to 0$). Thus $P_n(x)\epsilon_n(x) = O(x^n)$, $P_{n-1}(x)\epsilon_{n+1}(x) = O(x^{n+1})$ and hence $R_n \to F_0$ formally as $n \to \infty$.

It is a classical result that $F_0(x)$ is Ramanujan's partition function $\frac{1}{\prod_{n=1}^{\infty}(1-x^n)}$. We only outline the main ideas of the proof since we shall re-prove this result and the theorems following it in the next section.

Define the sequence $q_l(n)$ by

$$\frac{1}{\prod_{r=1}^{l}(1 - x^r)^2} = 1 + \sum_{n=1}^{\infty} q_l(n)x^n, \quad q_l(n) = \sum_{k=0}^{n} P_{\leq l}(n-k)P_{\leq l}(k),$$

and recall the sequence $P_{\leq l}(n)$ (defined by (7.4)). Now

$$F_0(x) = 1 + \sum_{n=1}^{\infty} a_n x^n, a_n = \sum_{l=1}^{\infty} q_l(n - l^2).$$

We know that

$$q_l(n - l^2) = \sum_{k=0}^{n-l^2} P_{\leq l}(n - l^2 - k)P_{\leq l}(k).$$

We need to suitably interpret the above quantity. Recall that each partition has a graphical representation. The sum defining $q_l(n - l^2)$ is counting the number of partitions of n which have a square with a side of length l in the upper left hand corner of their graphical representation. The term $P_{\leq l}(n - l^2 - k)$ represents the number of partitions of $n - k$ with l parts with each part at least l. We are multiplying this by $P_{\leq l}(k)$. Geometrically, we are therefore bordering the square of side l with geometric partitions of $n - l^2 - k$ on the right with at most l parts and bordering underneath with geometric partitions of k with at most l parts. This gives rise to all the geometric partitions of n after we sum over l showing that $F_0(x)$ is Ramanujan's partition function.

An argument similar to the one used in the analysis of F_0 yields

$$F_1(x) = 1 + \sum_{n=1}^{\infty} a_n x^n, \quad a_n = \sum_{l=1}^{\infty} q_l(n - (l^2 + l)).$$

As before,

$$q_l(n-(l^2+l)) = \sum_{k=0}^{\infty} P_{\le l}(k)P_{\le l}(n-(l^2+l)-k)$$

$$= \sum_{k=0}^{\infty} P_{\le l}(k)\left[P_{\le l+1}(n-(l^2+l)-k) - P_{\le l+1}(n-1-(l^2+2l)-k)\right].$$

(The last equality uses (7.3).)

To analyze the last equality, we study a closely related partition problem. Given three positive integers n, l and k, consider those partitions of n with graphical representation containing a rectangle of height $l+k$ and width l in the upper left hand corner. The union over l of all such partitions is the set of partitions with more than k parts thus containing $P(n) - P_{\le k}(n)$ partitions. The last displayed equation is a difference of two sums. The first sum counts the number of partitions of n with graphical representation containing a rectangle of dimension $(l+1)$ by l in the left hand corner. Summing over l therefore gives $P(n) - P_{\le 1}(n)$. What does the second sum represent? We reuse the simple trick of replacing $P_{\le l+1}(m)$ by $P_{\le l+2}(m) - P_{\le l+2}(m-(l+2))$ obtaining

$$\sum_{k=0}^{\infty} P_{\le l}(k)P_{\le l+1}(n-1-(l^2+2l)-k)$$

$$= \sum_{k=0}^{\infty} P_{\le l}(k)\left[P_{\le l+2}(n-1-(l^2+2l)-k) - P_{\le l+2}(n-3-(l^2+3l)-k)\right]$$

$$= \sum_{k=0}^{\infty} P_{\le l}(k)P_{\le l+2}(n-1-(l^2+2l)-k)$$

$$- \sum_{k=0}^{\infty} P_{\le l}(k)P_{\le l+2}(n-3-(l^2+3l)-k).$$

The first sum in the last line counts the partitions of $n-1$ with an $(l+2)$ by l rectangle in the left hand corner of its graphical representation, and therefore summing over l gives $P(n-1) - P_{\le 2}(n-1)$. Continuing in this fashion leads to

$$a_n = \sum_{l=0}^{\infty} (-1)^l \left(P\left(n - \frac{l^2+l}{2}\right) - P_{\le l+1}\left(n - \frac{l^2+l}{2}\right)\right).$$

We are, of course, using our convention that $P(\cdot) = 0$ whenever the argument is not a nonnegative integer. This has led us to

Theorem 5.3. *The power series coefficients a_n of the function $F_1(x)$ are given by*

$$a_n = \sum_{l=0}^{\infty} (-1)^l P\left(n - \frac{l^2+l}{2}\right).$$

Proof. We need only show that $\sum_{l=0}^{\infty}(-1)^l P_{\leq l+1}\left(n - \frac{l^2+l}{2}\right) = 0$. This follows from the identity

$$1 + \sum_{l=0}^{\infty} (-1)^{l+1} \frac{x^{\frac{l^2+l}{2}}}{\prod_{k=1}^{l+1}(1-x^k)} = 0,$$

which we now prove. It suffices to show that

$$\frac{x}{1-x} + \sum_{l=1}^{\infty} (-1)^l \frac{x^{\frac{l^2+l}{2}}}{\prod_{k=1}^{l+1}(1-x^k)} = 0.$$

It is easy to see by induction that the partial sums S_n of the above series satisfy

$$S_n(x) = \frac{(-1)^n x^{\frac{(n+1)(n+2)}{2}}}{\prod_{k=1}^{n+1}(1-x^k)},$$

and $\lim_{n\to\infty} S_n(x) = 0$ (since $|x| < 1$).

We now come to the continued fraction which is the title of this section. Let

$$f(x) = 2 + \cfrac{(x-1)}{2 + \cfrac{(x^2-1)}{2 - \cfrac{(x^3-1)}{2 + \cfrac{(x^4-1)}{\cdots}}}}.$$

We prove the following striking

Theorem 5.4. *We have*

$$\frac{1}{f(x)} = \sum_{n=0}^{\infty} (-1)^n x^{\frac{n^2+n}{2}}.$$

Proof. We show first that $f(x) = \frac{F_0(x)}{F_1(x)}$. As a consequence of the recursion relations (7.28) and (7.27), we see that

$$\frac{F_0(x)}{F_1(x)} = 2 + \frac{(x-1)}{\frac{F_1(x)}{F_2(x)}} \text{ and } \frac{F_n(x)}{F_{n+1}(x)} = 2 + \frac{(x^{n+1}-1)}{\frac{F_{n+1}(x)}{F_{n+2}(x)}}.$$

This together with the fact that the sequence $\{F_n\}$ converges to the constant 1 gives the formula $\frac{1}{f(x)} F_0(x) = F_1(x)$. We have already computed the power series coefficients of the functions F_0 and F_1. The last equality in terms of power series is

$$\sum_{n=0}^{\infty} a_n x^n \sum_{n=0}^{\infty} P(n) x^n = \sum_{n=0}^{\infty} b_n x^n,$$

where the a_n are the power series coefficients of $\frac{1}{f(x)}$ and the b_n are the power series coefficients of $F_1(x)$, which we have seen to be

$$b_n = \sum_{k=0}^{\infty} (-1)^k P\left(n - \frac{k^2+k}{2}\right).$$

Thus

$$\sum_{n=0}^{\infty} a_n x^n = \sum_{n=0}^{\infty} (-1)^n x^{\frac{n^2+n}{2}},$$

the statement of the theorem.

6. Return to theta functions

In the previous section we defined a sequence of functions $\{F_n\}$ and showed that F_0 was Ramanujan's partition function in disguise. We gave the standard (combinatorial) proof of this classical fact. We now give a second proof of this theorem using theta functions.

At the beginning of this chapter we defined a sequence of functions $\{E_m\}$ and obtained their infinite product expansions (7.7) and (7.6). More generally we define for each $l \in \mathbb{Z}^+$ and each $m \in \mathbb{Z}^+ \cup \{0\}$, the series

$$E_{m,l}(x, z) = 1 + \sum_{k=1}^{\infty} \frac{x^{mk}}{\prod_{r=1}^{k} (1 - x^{rl})} z^k,$$

and derive the formula

$$E_{m,l}(x, z) = \frac{1}{\prod_{n=0}^{\infty} (1 - x^{ln+m} z)}.$$

The case $m = 1$, $l = 2$ yields

$$\frac{1}{\prod_{n=0}^{\infty} (1 - x^{2n+1} z)} = 1 + \sum_{k=1}^{\infty} \frac{x^k}{\prod_{r=1}^{k} (1 - x^{2r})} z^k.$$

In the same fashion, define a different sequence of functions

$$H_{m,l}(x, z) = 1 + \sum_{k=1}^{\infty} \frac{x^{k^2+mk}}{\prod_{r=1}^{k} (1 - x^{rl})} z^k$$

and conclude that

$$H_{m,l}(x, z) - H_{m+l,l}(x, z) = x^{m+1} z H_{m+l,l}(x, z).$$

In the cases $m = 0$ and 1, $l = 2$ we find that

$$H_{0,2}(x,z) = \prod_{n=0}^{\infty}(1 + x^{2n+1}z) \text{ and } H_{1,2}(x,z) = \prod_{n=0}^{\infty}(1 + x^{2n+2}z),$$

so that

$$\prod_{n=0}^{\infty}(1 + x^{2n+1}z) = 1 + \sum_{k=1}^{\infty} \frac{x^{k^2}}{\prod_{l=1}^{k}(1 - x^{2l})} z^k$$

and

$$\prod_{n=0}^{\infty}(1 + x^{2n+2}z) = 1 + \sum_{k=1}^{\infty} \frac{x^{k^2+k}}{\prod_{l=1}^{k}(1 - x^{2l})} z^k. \tag{7.31}$$

The above calculations are routine, and the reader who has come this far in the text should have no difficulty with them. We now prove again that $F_0 = \mathbf{P}$.

From the Jacobi triple product formula, we see that

$$\frac{\sum_{n\in\mathbb{Z}} x^{n^2} z^n}{\prod_{n=0}^{\infty}(1 - x^{2n+2})}$$

$$= \left(1 + \sum_{k=1}^{\infty} \frac{x^{k^2}}{\prod_{r=1}^{k}(1 - x^{2r})} z^k\right)\left(1 + \sum_{k=1}^{\infty} \frac{x^{k^2}}{\prod_{r=1}^{k}(1 - x^{2r})} \frac{1}{z^k}\right).$$

We compute the constant term coefficient of the Laurent expansion about the origin of each side. We find (after a trivial change of variables) that

$$\frac{1}{\prod_{n=1}^{\infty}(1 - x^n)} = 1 + \sum_{k=1}^{\infty} \frac{x^{k^2}}{\prod_{l=1}^{k}(1 - x^l)^2}.$$

The above argument shows once again how theta functions (in conjunction with the Jacobi triple product identity) can be used in this sort of combinatorial problem. We give an additional example which will be related to a prominent result in this area, the Rogers-Ramanujan identities, and will culminate with a beautiful identity relating Euler series and infinite products and a beautiful identity between two continued fractions.

We begin by defining two Euler series:

$$f_0(x) = 1 + \sum_{n=1}^{\infty} \frac{(-1)^n x^{n^2}}{\prod_{l=1}^{l=n}(1 - x^l)} \text{ and } f_1(x) = 1 + \sum_{n=1}^{\infty} \frac{(-1)^n x^{n^2+n}}{\prod_{l=1}^{l=n}(1 - x^l)}.$$

The above series are closely related to the series

$$g_0(x) = 1 + \sum_{n=1}^{\infty} \frac{x^{n^2}}{\prod_{l=1}^{l=n}(1 - x^l)} \text{ and } g_1(x) = 1 + \sum_{n=1}^{\infty} \frac{x^{n^2+n}}{\prod_{l=1}^{l=n}(1 - x^l)}.$$

The Euler series $g_0(x)$, $g_1(x)$ are each one of the two sides of the celebrated Rogers-Ramanujan identities and thus have well known partition theoretic interpretations. The power series coefficients of g_0 count the number of ways to partition a positive integer under the restriction that the difference between successive parts be at least 2. The power series coefficients of g_1 count the number of ways to partition an integer under the restriction that the difference between successive parts be at least 2 and that no part be less than 2. The coefficients for the Taylor series about 0 of f_0 and f_1 count the difference between the number of even and odd partitions, respectively.

We once again use the Jacobi triple product identity and the Euler identities for $\prod_{n=1}^{\infty}(1+\frac{x^{2n-1}}{z})$ and $\frac{1}{\prod_{n=1}^{\infty}(1+x^{2n}z)}$ to conclude

$$\frac{\sum_{-\infty}^{\infty} x^{n^2} z^n}{\prod_{n=0}^{\infty}(1+x^n z)} = \frac{\prod_{n=1}^{\infty}(1-x^{2n})\left(1+\frac{x^{2n-1}}{z}\right)}{\prod_{n=0}^{\infty}(1+x^{2n}z)}$$

$$= \prod_{n=1}^{\infty}(1-x^{2n})\left(1+\sum_{n=1}^{\infty}\frac{x^{n^2}z^{-n}}{\prod_{l=1}^{n}(1-x^{2l})}\right)\left(1+\sum_{n=1}^{\infty}(-1)^n\frac{z^n}{\prod_{l=1}^{n}(1-x^{2l})}\right).$$

The first term on the left of the above identity can be rewritten as

$$\left(\sum_{-\infty}^{\infty} x^{n^2} z^n\right)\left(1+\sum_{n=1}^{\infty}(-1)^n\frac{z^n}{\prod_{l=1}^{n}(1-x^l)}\right).$$

We compute the constant term of the Laurent expansion of the above function and set it equal to the constant term of the Laurent expansion of the term on the extreme right hand side of the identity and obtain

$$1+\sum_{n=1}^{\infty}(-1)^n\frac{x^{n^2}}{\prod_{l=1}^{n}(1-x^l)} = \prod_{n=1}^{\infty}(1-x^{2n})\left(1+\sum_{n=1}^{\infty}(-1)^n\frac{x^{n^2}}{\prod_{l=1}^{n}(1-x^{2l})^2}\right).$$

The left hand side of this identity is our function $f_0(x)$.

In order to obtain a formula for the function $f_1(x)$, we adopt the previous argument to begin with the quotient

$$\frac{\sum_{n=-\infty}^{\infty} x^{n^2} z^n}{\prod_{n=1}^{\infty}\left(1+\frac{x^n}{z}\right)}.$$

Computation of the constant term of the Laurent series expansions yields

$$f_1(x) = \prod_{n=1}^{\infty}(1-x^{2n})\left(1+\sum_{n=1}^{\infty}(-1)^n\frac{x^{n^2+2n}}{\prod_{l=1}^{n}(1-x^{2l})^2}\right).$$

If we define

$$F_k(x) = 1+\sum_{n=1}^{\infty}(-1)^n\frac{x^{n^2+kn}}{\prod_{l=1}^{n}(1-x^{2l})^2},$$

we have proven

Theorem 6.1. *For $n = 0, 1$ we have*

$$f_n(x) = F_{2n}(x)\prod_{k=1}^{\infty}(1 - x^{2k}).$$

The above theorem, which should be viewed as a relation between Euler series and infinite products, yields

Corollary 6.2. *We have*

$$\frac{f_0(x)}{f_1(x)} = \frac{F_0(x)}{F_2(x)}.$$

The above leads to an identity between continued fractions. We only sketch the idea of the proof.

The family of functions f_m satisfies the recursion relations

$$f_n(x) = f_{n+1}(x) - x^{n+1}f_{n+2}(x).$$

Just as in the section discussing continued fractions this gives rise to a sequence of polynomials which satisfy the relations

$$P_{n+1}(x) = P_n(x) - x^n P_{n-1}(x).$$

Assuming the initial conditions $P_0(x) = 1 = P_1(x)$ gives us the sequence

$$1,\ 1,\ 1 - x,\ 1 - x - x^2,\ 1 - x - x^2 - x^3 + x^4,$$
$$1 - x - x^2 - x^3 + x^5 - x^6,\ \ldots\ .$$

This sequence of polynomials converges to the power series expansion of $f_0(x)$. More important for us though is the construction of the continued fraction which follows from the recursions satisfied by the functions $f_n(x)$.

We begin with

$$f_0(x) = f_1(x) - xf_2(x).$$

Dividing by $f_1(x)$ gives

$$\frac{f_0(x)}{f_1(x)} = 1 - \frac{x}{\frac{f_1(x)}{f_2(x)}}.$$

Writing now the relation connecting $f_1(x)$ with $f_2(x)$ and $f_3(x)$ and continuing in this fashion gives us the continued fraction expansion

$$\frac{f_0(x)}{f_1(x)} = 1 - \frac{x}{1 - \frac{x^2}{1-x^3\cdots}}.$$

This is not surprising if one knows the Rogers-Ramanujan continued fraction, which is almost the same except for the signs. The signs in Rogers-Ramanujan are all positive. This is because the recursion relations in the Rogers-Ramanujan situation are

$$g_m(x) = g_{m+1}(x) + x^{m+1}g_{m+2}(x),$$

and the sequence of polynomials which converge to $g_0(x)$ is defined by

$$P_{n+1}(x) = P_n(x) + x^n P_{n-1}(x),$$

sometimes called the generalized Fibonacci sequence.

The reason we obtain something new here is that the family of functions F_k also satisfies recursion relations. It is easy to check that

$$F_{2m}(x) = (2 - x^{2m+1})F_{2m+2}(x) - F_{2m+4}(x).$$

If we begin with

$$\frac{F_0(x)}{F_2(x)} = 2 - x + \frac{1}{\frac{F_2(x)}{F_4(x)}}$$

and use the recursions, we find that

$$\frac{F_0(x)}{F_2(x)} = 2 - x - \frac{1}{2 - x^3 - \frac{1}{2 - x^5 - ..}}.$$

The equality given by the corollary of this section now gives an identity between two continued fraction expansions.

We now present a more function theoretic proof of Theorem 5.4. We begin with the Jacobi triple product (where we replace x by $x^{\frac{1}{2}}$ and z by $x^{\frac{1}{2}}z$) which tells us that

$$\sum_{n=-\infty}^{\infty} x^{\frac{n^2+n}{2}} z^n = \left(1 + \frac{1}{z}\right) \prod_{n=1}^{\infty} (1 - x^n)(1 + x^n z) \left(1 + \frac{x^n}{z}\right)$$

or equivalently

$$\prod_{n=1}^{\infty} (1 - x^n)(1 + x^n z) \left(1 + \frac{x^n}{z}\right) = \left(\sum_{n=0}^{\infty} (-1)^n z^n\right) \left(\sum_{n=-\infty}^{\infty} x^{\frac{n^2+n}{2}} z^{n+1}\right).$$

We view the last identity as an equality of two functions of the variable z. We now use (7.31) to translate the equation to

$$\prod_{n=1}^{\infty} (1 - x^n) \left(1 + \sum_{n=1}^{\infty} \frac{x^{\frac{n^2+n}{2}}}{\prod_{l=1}^{n} (1 - x^l)} z^n\right) \left(1 + \sum_{n=1}^{\infty} \frac{x^{\frac{n^2+n}{2}}}{\prod_{l=1}^{n} (1 - x^l)} \frac{1}{z^n}\right)$$
$$= \left(\sum_{n=0}^{\infty} (-1)^n z^n\right) \left(\sum_{n=-\infty}^{\infty} x^{\frac{n^2+n}{2}} z^{n+1}\right).$$

If we compute the coefficient of the constant term in the Laurent series of each side, we get

$$\prod_{n=1}^{\infty} (1 - x^n) F_1(x) = \sum_{n=0}^{\infty} (-1)^n x^{\frac{n^2+n}{2}},$$

the result we want.

Exercise 6.3. Define three sequences of polynomials by the recursions

$$T_0(x) = 1, \qquad T_n(x) = (1 - x^n)^2\, T_{n-1}(x) + x^{n^2},$$

$$T_0(x) = 1, \qquad T_n(x) = (1 - x^n)\, T_{n-1}(x) + (-1)^n x^n$$

and

$$T_0(x) = 1, \qquad T_n(x) = (1 - x^n)\, T_{n-1}(x) + x^{\frac{n^2+n}{2}}.$$

Prove that the respective sequences $T_n(x)$ converge to $\prod_{n=1}^{\infty}(1 - x^n)$, $1 + 2\sum_{n=1}^{\infty}(-1)^n x^{n^2}$ and $\prod_{n=1}^{\infty}(1 - x^{2n})$.

Bibliography

[1] L.V. Ahlfors. *Complex Analysis (third edition)*. McGraw-Hill, 1979.

[2] G.E. Andrews. *The Theory of Partitions*, volume 2 of *Encyclopedia of Mathematics and its Applications*. Addison-Wesley, 1976.

[3] J. Bak and D.J. Newman. *Complex Analysis*. Springer-Verlag, 1982.

[4] A.F. Beardon. *The Geometry of Discrete Groups*, volume 91 of *Graduate Texts in Mathematics*. Springer-Verlag, 1983.

[5] J.H. Conway. *The Atlas of Finite Simple Groups*. Oxford Clarendon Press, 1985.

[6] H.M. Farkas and I. Kra. *Riemann Surfaces (second edition)*, volume 71 of *Graduate Texts in Mathematics*. Springer-Verlag, 1992.

[7] N.J. Fine. *Basic Hypergeometric Series and Applications*, volume 27 of *Mathematical Surveys and Monographs*. American Mathematical Society, 1989.

[8] L.R. Ford. *Automorphic Functions*. Chelsea, 1951.

[9] J.B. Fraleigh. *A First Course in Abstract Algebra*. Addison-Wesley, 1969.

[10] G. Gasper and M. Rahman. *Basic Hypergeometric Series*, volume 35 of *Encyclopedia of Mathematics and its Applications*. Cambridge University Press, 1990.

[11] P. Griffiths and J. Harris. *Principles of Algebraic Geometry*. John Wiley & Sons, 1978.

[12] R.C. Gunning. *Lectures on Modular Forms*, volume 48 of *Annals of Mathematics Studies*. Princeton University Press, 1962.

[13] G.H. Hardy and E.M. Wright. *An Introduction to the Theory of Numbers (fifth edition)*. Clarendon Press, 1995.

[14] Y. Imayoshi and M. Taniguchi. *An Introduction to Teichmüller Theory*. Springer-Verlag, 1992.

[15] A.W. Knapp. *Elliptic Curves*, volume 40 of *Mathematical Notes*. Princeton University Press, 1992.

[16] M.I. Knopp. *Modular Functions in Analytic Number Theory*. Chelsea Publishing Company, 1993.

[17] N. Koblitz. *Introduction to Elliptic Curves and Modular Forms (second edition)*, volume 97 of *Graduate Texts in Mathematics*. Springer-Verlag, 1993.

[18] I. Kra. *Automorphic Forms and Kleinian Groups.* Benjamin, 1972.

[19] S. Lang. *Algebra.* Addison-Wesley, 1971.

[20] S. Lang. *Albebraic Number Theory (second edition).* Springer-Verlag, 1994.

[21] B. Maskit. *Kleinian Groups*, volume 287 of *Grundlehren der mathematischen Wissenschaften.* Springer-Verlag, 1988.

[22] T. Miyake. *Modular Forms.* Springer-Verlag, 1989.

[23] D. Mumford. *Tata Lectures on Theta I.* Birkhäuser, 1983.

[24] H.E. Rauch and H.M. Farkas. *Theta Functions with Applications to Riemann Surfaces.* Williams and Wilkins, 1974.

[25] H.E. Rauch and A. Lebowitz. *Elliptic Functions, Theta Functions and Riemann Surfaces.* Williams and Wilkins, 1973.

[26] J.-P. Serre. *A Course in Arithmetic*, volume 7 of *Graduate Texts in Mathematics.* Springer-Verlag, 1973.

[27] G. Shimura. *Introduction to the Arithmetic Theory of Automorphic Functions*, volume 11 of *Publications of the Mathematical Society of Japan, Kanô memorial lectures 1.* Iwanami Shoten and Princeton University Press, 1971.

[28] H. Weber. *Lehrbuch der Algebra (dritter band).* Chelsea, 1961.

Bibliographical Notes

The bibliographical entries in the main body of the text contained only references to books that filled in gaps in our presentation or described prerequisites. The main purpose of these notes is to give the reader an idea of the many relevant books and scientific manuscripts in the literature on one variable theta functions as related to to the topics of the book. It is incomplete and will be maintained and updated periodically on the web at

`http://www.math.sunysb.edu/~irwin/bkbiblio.dvi`

We (our respective e-mail addresses are

`farkas@sunset.huji.ac.il`

and

`irwin@math.sunysb.edu`)

would appreciate receiving corrections and additions from readers, especially from experts. The authors of the current volume are not experts in the field of combinatorial number theory and have attempted in their work to open this area to a wider audience. We have emphasized the function theoretic aspects and may have missed many points known to number theorists and combinatorialists. Edited communications to us will be posted on the web and acknowledged appropriately.

1. Books – background material

1.1. Analytic and other prerequisites. The basic facts on complex analysis come from [**1**]. We have assumed the reader is familiar with the material of a first year graduate course on the subject.

We refer the reader to [**1**, Ch. 7] for elementary facts concerning elliptic functions and curves. Many more complete treatments of function theory on tori are available in the literature including, for example, [**60**].

The necessary preliminaries about Riemann surfaces were taken from [**35**]. There are of course many other excellent books on this subject; among them we list only three: [**100**] (an English translation of an earlier classic), [**76**] and [**45**].

The prerequisites on automorphic forms and Fuchsian groups may be found in portions of [**17**], [**74**], [**52**] and [**65**]. The computations of the topological invariants associated to a subgroup of the modular group are based on [**96**].

In Chapter 3, we discussed automorphic forms. Although not extremely relevant for our work, we refer to two papers on the subject (in part because they deal with averaging processes, a topic central to this book): the first by one of the coauthors of this volume [**66**]; the second by McMullen [**75**]. The reader interested in number theory should consult [**58**].

1.2. Other approaches. There is some overlap between our treatment of theta functions and resulting identities (especially involving the partition function) and that of the number theoretic work [**61**].

[**7**] is a thorough treatment of the theory of partitions, particularly the interplay between combinatorial and analytic methods.

The book [**48**] on hypergeometric functions has many points of contact with Ramanujan identities and partition theory.

The encyclopedic book [**9**] contains chapters on q-series, partitions and the Rogers-Ramanujan identities.

Each of the books mentioned in this subsection contains an extensive bibliography.

1.3. Recent developments. Although it is not our purpose to record recent progress in Teichmüller theory or other fields encountered in our presentation but not directly related to the main topics under study, we mention that the problem of describing quasi-Fuchsian space for once punctured tori as a subset of $\mathbb{C}^2$ is treated in [**95**] in the smooth, not complex analytic, setting.

2. Papers by other authors

2.1. Curves represented by subgroups of Γ. In §8.7 of Chapter 3, we described the curve $\overline{\mathbb{H}^2/\Gamma(9)}$. A refinement of the description can be found in [**64**].

2.2. The $\jmath$-function. Divisibility properties of the Fourier coefficients of the $\jmath$-function for the primes 5, 7 and 11 (the level 1 and 2 only for the prime 11) are found in [**70**]. The complete picture for the prime 11 is established in [**11**]. Multiplicative properties (see §2.4) of these Fourier coefficients are studied in [**68**].

2.3. The partition functions P_N. For $N = 1$, Ramanujan [**86**], [**87**] conjectured partition congruences for the primes 5, 7 and 11 and established the level 1 and 2 results. Watson [**99**] proved the conjecture for all levels n for the prime 5 and a suitably modified level n conjecture for the prime 7. More work is required for the prime 11 since the surface involved is no longer of genus 0. The level 1 congruence was proven by Winquist [**101**]; Atkin [**11**] established the general conjecture (see also the earlier manuscript [**69**]). The level 1 congruence for the prime 5 has probably received more attention than any other case; among the shortest proofs are [**28**] and [**55**]. Partition congruences (as well as properties of the Fourier coefficients of the $\jmath$-function) are studied in [**14**].

Among the papers studying the function P_N for arbitrary N are [**13**], [**50**] and [**56**]; the computer connection is explored in [**12**].

For negative N, one obtains recursion identities rather than only congruences for the partition coefficients $P_N(n)$. This case has been thoroughly investigated by M. Newman, whose papers on this and related topics include [**78**], [**79**], [**80**], [**82**], [**81**], [**83**], [**84**] and [**85**].

2.4. The Hecke connection. A series $\sum_{n=0}^{\infty} a_n x^n$ that converges on the unit disc is *multiplicative* if $a_{nm} = a_n a_m$ whenever $(n, m) = 1$. The multiplicativity of Ramanujan's τ-function (η^{24}), a theorem first proved by Mordell [**77**], was established in Chapter 4. In [**29**], the authors determine all the multiplicative products of η-functions. Two further papers should be mentioned here which connect the η-function to the representation theory of Lie algebras. One is the 1972 paper of Macdonald [**71**]; the second is the 1978 paper of V. Kac [**59**]. See also [**72**]. Hecke operators are discussed in portions of [**60**, Chs. VIII,IX and X].

One of the motivations of the authors in writing this volume was to show that η is in fact just one example of a theta function with rational characteristics and that introduction of these more general functions can be extremely useful in the theory of uniformizations of the Riemann surfaces which arise from congruence subgroups of the modular group and in the study of theta constant identities. While this was presumably known to Klein, it is only with the development of the theory of theta characteristics that it becomes an efficient procedure.

2.5. The η-function. Papers dealing principally or exclusively with the eta-function and having at least tangential connection to our work include [**51**], [**29**], [**73**] and [**72**].

2.6. Theta identities. This rich subject is studied in many papers. Among the papers dealing with short and elementary proofs of the Jacobi triple product formula and the quintuple product identity are [**3**] and [**27**]. Proofs of the latter also appear in [**98**], [**94**] and [**49**]. [**47**] is a follow up of some of the work in our book.

2.7. Combinatorial proofs and interpretations of some identities. Among the many papers on the subject are [**4**] and [**46**].

3. Papers by the authors

Our interest in this subject may be traced to an early paper by one of the authors [**31**]. His interest in the subject reemerged as a consequence of the doctoral studies of one of his students resulting in [**32**] and [**33**].

The collaboration that led to this book started with [**36**] and [**37**], where the use of theta constants with rational characteristics was applied to the study of uniformization of surfaces represented by subgroups of the modular group. In [**34**] modified theta constants were introduced as a tool in the study of uniformization of surfaces represented by the prime level principal congruence subgroups; related topics are investigated in part of [**43**]. Our papers on partitions and the Ramanujan τ-function include [**38**], [**67**], [**41**], [**39**] and [**42**]. Our work that relates to Hecke operators includes [**44**]. The quintuple product identity is reproven in [**40**].

4. Original sources, conference proceedings

The primary sources for the mathematics of S. Ramanujan, the main subject of much of this book, are [**54**] and [**89**]; the Hardy lectures [**53**] are extremely useful in any attempt to evaluate Ramanujan's genius. So are his notebooks [**88**], [**90**] and [**18**]. The last of these contains extensive scholarly and bibliographic information. Among the many conference proceedings devoted to Ramanujan's legacy is [**10**]; [**57**] is among the latest. At least two journals, The Ramanujan Journal and The Hardy-Ramanujan Journal are named in his honor. Among the many books and papers that include the name Ramanujan in their title are [**19**], [**20**], [**21**], [**22**], [**23**], [**24**], [**5**], [**6**], [**8**], [**25**] and [**26**].

5. Related questions

Among the many issues related but not treated in the book are the higher level congruences for $\jmath$ and P_N and the Rogers-Ramanujan identities. These identities and their generalizations appear in many books and in the following papers: [**91**], [**92**], [**98**], [**15**], [**16**], [**97**] and [**30**], among others.

The Rogers-Ramanujan identities quoted in Chapter 7 read

$$\frac{1}{\prod_{n=0}^{\infty}(1-x^{5n+1})(1-x^{5n+4})} = 1+\sum_{n=1}^{\infty}\frac{x^{n^2}}{\prod_{l=1}^{n}(1-x^l)}$$

and

$$\frac{1}{\prod_{n=0}^{\infty}(1-x^{5n+2})(1-x^{5n+3})} = 1+\sum_{n=1}^{\infty}\frac{x^{n^2+n}}{\prod_{l=1}^{n}(1-x^l)}.$$

These identities are related via the Jacobi triple product formula to the theta constants which appeared in Chapter 3, namely

$$\theta\begin{bmatrix}\frac{1}{5}\\1\end{bmatrix}(0,5\tau) \text{ and } \theta\begin{bmatrix}\frac{3}{5}\\1\end{bmatrix}(0,5\tau).$$

The Jacobi triple product yields

$$\prod_{n=0}^{\infty}(1-x^{5n+1})(1-x^{5n+4})(1-x^{5n+5}) = \sum_{n=-\infty}^{\infty}(-1)^n x^{\frac{5n^2+3n}{2}}$$

and

$$\prod_{n=0}^{\infty}(1-x^{5n+2})(1-x^{5n+3})(1-x^{5n+5}) = \sum_{n=-\infty}^{\infty}(-1)^n x^{\frac{5n^2+n}{2}},$$

while straightforward calculation from the definitions shows that in terms of the local coordinate $x=\exp(2\pi\imath\tau)$,

$$\theta\begin{bmatrix}\frac{1}{5}\\1\end{bmatrix}(0,5\tau) = \exp\left(\frac{\pi\imath}{10}\right)x^{\frac{1}{40}}\sum_{n\in\mathbb{Z}}(-1)^n x^{\frac{n(5n+1)}{2}}$$

and

$$\theta\begin{bmatrix}\frac{3}{5}\\1\end{bmatrix}(0,5\tau) = \exp\left(\frac{3\pi\imath}{10}\right)x^{\frac{9}{40}}\sum_{n\in\mathbb{Z}}(-1)^n x^{\frac{n(5n+3)}{2}}.$$

Since

$$\eta(5\tau) = x^{\frac{5}{24}}\prod_{n=1}^{\infty}(1-x^{5n}),$$

the left hand side (the infinite products) of the Rogers-Ramanujan identities can be interpreted as multiplicative automorphic forms for $\Gamma(5)$. Part of the fascination with these identities is that the right hand side (the infinite sum) does not seem to have such an interpretation.

The fact that the Rogers-Ramanujan identities are connected to the above theta constants was known to Schur. In [**93**], this connection is made and Schur derives the identities in two different ways. One way uses the theta constants but he also gives a second proof using combinatorial methods. The second proof uses an identity between the binomial coefficients extended to the q-binomial coefficients and some cosine identities.

There is also a generalization of the Rogers-Ramanujan identities known as the Gollnitz-Gordon identities where the modulus 8 replaces 5. While this topic is also not treated in the text, it seems interesting to point out that some of the techniques used to study these identities seem to be related to material in this text. It is pointed out in a paper by K. Alladi [**2**] that proofs of these identities can be based on splitting the respective functions (defined as infinite sums) into their even and odd parts and that these parts have interesting product representations. The variable x (in most of the literature q) is related to the parameter τ by $x = \exp(2\pi\imath\tau)$. Our proof of the quintuple product identity also used a decomposition of a function into even and odd parts in the variable $z = \exp(2\pi\imath\zeta)$. These ideas arise from certain continued fraction identities which were also considered by us in the last chapter of the book. The reader is encouraged to look at Alladi's paper and the references given there to the work of Andrews, Gordon and others.

Theta functions with characteristics enrich the set of tools available to investigate diverse questions. We believe that the material in the text opens areas for further investigations (in particular for doctoral work) and can be used to derive many identities not considered in the book. We illustrate both points by providing two different proofs of the Köhler-Macdonald (the name comes from [**73**]) identity quoted in Chapter 5. It reads

$$x\prod_{n=1}^{\infty}(1-x^{6n})^5(1-x^{3n})^{-2} = \sum_{n=1}^{\infty}(-1)^{n-1}\left(\frac{n}{3}\right)nx^{n^2}. \tag{1}$$

The right hand side is easily seen to be the same as

$$x\sum_{m=0}^{\infty}(-1)^m(3m+1)x^{9m^2+6m} + x\sum_{m=0}^{\infty}(-1)^m(3m+2)x^{9m^2+12m+3}.$$

It now makes sense to replace x by $x^{\frac{1}{3}}$ and rewrite the identity as

$$\prod_{n=1}^{\infty}(1-x^{2n})^5(1-x^n)^{-2} = \sum_{m=-\infty}^{\infty}(-1)^m(3m+1)x^{m(3m+2)}.$$

To translate this identity to the language of theta functions, we use the change of variable $x = \exp(2\pi\imath\tau)$. The left hand side of the Köhler-Macdonald identity is thus

$$\exp\left(-\frac{2\pi\imath\tau}{3}\right)\eta(2\tau)^5\eta(\tau)^{-2}.$$

We have shown that

$$\theta'\begin{bmatrix}\frac{2}{3}\\1\end{bmatrix}(0,\tau)=\frac{2\pi\imath}{3}\exp\left\{\pi\imath\frac{1}{3}\right\}x^{\frac{1}{18}}\sum_{n\in\mathbb{Z}}(3n+1)(-1)^n x^{\frac{n}{6}(3n+2)}.$$

Thus the right hand side of the Köhler-Macdonald identity is

$$\frac{3}{2\pi\imath}\exp\left\{-\pi\imath\frac{1}{3}\right\}\exp\left\{-2\pi\imath\tau\frac{1}{3}\right\}\theta'\begin{bmatrix}\frac{2}{3}\\1\end{bmatrix}(0,6\tau).$$

We conclude that the function theoretic version of Köhler-Macdonald is

$$(2)\qquad 2\pi\imath\eta(2\tau)^5\eta(\tau)^{-2}=3\exp\left\{-\pi\imath\frac{1}{3}\right\}\theta'\begin{bmatrix}\frac{2}{3}\\1\end{bmatrix}(0,6\tau).$$

Our first proof of the above identity follows from the following product formula established by Y. Godin, which will appear as part of his doctoral thesis at the Hebrew University of Jerusalem. Godin proves

$$\theta\begin{bmatrix}\epsilon\\\epsilon'\end{bmatrix}(\zeta,\tau)\theta\begin{bmatrix}\delta\\\delta'\end{bmatrix}(\zeta,2\tau)$$

$$=\sum_{\mu=0}^{2}\theta\begin{bmatrix}\frac{2\mu+\epsilon+\delta}{3}\\2\epsilon'+\delta'\end{bmatrix}(3\zeta,6\tau)\theta\begin{bmatrix}\frac{2\mu+\epsilon-2\delta}{3}\\\epsilon'-\delta'\end{bmatrix}(0,3\tau).$$

It thus follows that

$$\theta\begin{bmatrix}1\\0\end{bmatrix}(\zeta,\tau)\theta\begin{bmatrix}1\\1\end{bmatrix}(\zeta,\tau)$$

$$=\theta\begin{bmatrix}\frac{1}{3}\\1\end{bmatrix}(0,3\tau)\left(\theta\begin{bmatrix}\frac{2}{3}\\1\end{bmatrix}(3\zeta,6\tau)-\theta\begin{bmatrix}\frac{2}{3}\\1\end{bmatrix}(-3\zeta,6\tau)\right).$$

Differentiation (with respect to ζ) and then evaluation at $\zeta=0$ yields

$$\theta'\begin{bmatrix}\frac{2}{3}\\1\end{bmatrix}(0,6\tau)=\frac{\theta\begin{bmatrix}1\\0\end{bmatrix}(0,\tau)\theta'\begin{bmatrix}1\\1\end{bmatrix}(0,2\tau)}{\theta\begin{bmatrix}\frac{1}{3}\\1\end{bmatrix}(0,3\tau)}.$$

An application of the Jacobi triple product formula completes the proof of (2).

It is appropriate to illustrate the combinatorial content of the Köhler-Macdonald identity. The identity may be rewritten as

$$\sum_{n=-\infty}^{\infty}(-1)^n(3n+1)x^{3n^2+2n}=\prod_{n=1}^{\infty}(1-x^{4n})^3(1-x^{4n-2})(1+x^{2n-1})^2.$$

In this form the combinatorial content is as follows: Let S be the set containing 3 copies of the positive integers congruent to 0 modulo 4, 1 copy of the positive integers congruent to 2 modulo 4, and 2 copies of the odd positive integers. Call such a partition *even* if it has an even number of even parts and call it *odd* otherwise. Denote by $E(N)$ and $O(N)$ the number of even and odd partitions of $N \in \mathbb{Z}^+$ using parts from the set S. The Köhler-Macdonald identity says that for all $N \in \mathbb{Z}^+$,

$$E(N) - O(N) = \begin{cases} 0 & \text{for } N \neq 3m^2 + 2m \text{ with } m \in \mathbb{Z} \\ (-1)^m(3m+1) & \text{for } N = 3m^2 + 2m \text{ with } m \in \mathbb{Z} \end{cases}.$$

The second proof of the identity, more in the spirit of the book, is based on the connection of θ-constants to automorphic forms. It starts by "rewriting" (2) as

$$(3) \qquad (2\pi)^{24}\eta(2\tau)^{120}\eta(\tau)^{-48} = 3^{24}\left(\theta'\begin{bmatrix} \frac{2}{3} \\ 1 \end{bmatrix}(0, 6\tau)\right)^{24}.$$

The equivalence class of the characteristic $\begin{bmatrix} \frac{2}{3} \\ 1 \end{bmatrix}$ is fixed by the group $\Gamma^o(6)$. Using this observation one shows that

$$f(\tau) = \left(\theta'\begin{bmatrix} \frac{2}{3} \\ 1 \end{bmatrix}(0, 6\tau)\right)^{24}$$

is a multiplicative 6-form for $\Gamma_o(6)$ (perhaps with a nontrivial multiplier system); that is,

$$\left(\theta'\begin{bmatrix} \frac{2}{3} \\ 1 \end{bmatrix}(0, 6\gamma(\tau))\right)^{24}(\gamma'(\tau))^6 = c_\gamma\left(\theta'\begin{bmatrix} \frac{2}{3} \\ 1 \end{bmatrix}(0, 6\tau)\right)^{24}, \ \tau \in \Gamma_o(6),$$

where c_γ is a constant of absolute value 1. The four punctures on $\mathbb{H}^2/\Gamma_o(6)$ are P_∞, $P_{\frac{1}{2}}$, P_0 and $P_{\frac{1}{3}}$. The divisor (f) is computed to be $P_\infty^8 P_0^9 P_{\frac{1}{2}}^{18} P_{\frac{1}{3}}$. The function $\frac{\eta(2\tau)^{120}}{\eta(\tau)^{48}}$ is a multiplicative 6-form for $\Gamma_o(2)$, hence certainly for $\Gamma_o(6)$. It has the same divisor as the automorphic form f. The formulae for the divisors show that these are ordinary cusp forms. Hence, except for the constants, (3) follows. If we extract a 24-th root and evaluate the one (so far undetermined) constant, we obtain (2).

The above two proofs are based on recognizing that both sides of (1) can be reinterpreted in terms of θ-constants. A proof of Köhler-Macdonald based on theta functions (in particular on **[62]**) can be found in **[63]**. It is different in flavor from the second of our proofs because it relies more directly and more exclusively on the properties of the η-function. Our second proof has an unexpected dividend. While it is more or less obvious that $\theta'\begin{bmatrix} \frac{2}{3} \\ 1 \end{bmatrix}(0, 6\tau)$ is a multiplicative form for $\Gamma_o(6)$, we have shown that it is a form for the bigger group $\Gamma_o(2)$.

As a special case of a more general formula Godin also proves

$$\theta\begin{bmatrix}\epsilon\\ \epsilon'\end{bmatrix}(0,\tau)\theta\begin{bmatrix}\delta\\ \delta'\end{bmatrix}(0,\tau)$$

$$=\sum_{n=0}^{4}\theta\begin{bmatrix}\frac{2\mu+\epsilon+\delta}{5}\\ 4\epsilon'+\delta'\end{bmatrix}(0,20\tau)\theta\begin{bmatrix}\frac{-2\mu-\epsilon+4\delta}{5}\\ -\epsilon'+\delta'\end{bmatrix}(0,5\tau).$$

Choosing $\begin{bmatrix}\epsilon\\ \epsilon'\end{bmatrix}=\begin{bmatrix}1\\ 0\end{bmatrix}$ and $\begin{bmatrix}\delta\\ \delta'\end{bmatrix}=\begin{bmatrix}0\\ 1\end{bmatrix}$ we obtain after some algebraic manipulation

$$\prod_{n=1}^{\infty}(1+x^{2n-1})^2$$

$$=\frac{1}{\prod_{n=0}^{\infty}(1-x^{5n+1})(1-x^{5n+4})(1-x^{4(5n+1)})(1-x^{4(5n+4)})}$$

$$+\frac{x}{\prod_{n=0}^{\infty}(1-x^{5n+2})(1-x^{5n+3})(1-x^{4(5n+2)})(1-x^{4(5n+3)})},$$

a formula, related to the Rogers-Ramanujan identities, which also has a clear combinatorial interpretation which we now give.

Let S be the set consisting of two copies of the odd positive integers. Denote the number of partitions of N from S by $S(N)$. Let T be the set consisting of the positive integers congruent to ± 1 modulo 5 with an extra copy of those which are also congruent to 0 modulo 4. Denote the number of partitions of N from T by $T(N)$. Finally, let U be the set of the positive integers congruent to ± 2 modulo 5 with an extra copy of those which are also congruent to 0 modulo 4. Denote the number of partitions of N from U by $U(N)$. The above identity says that for all $N \in \mathbb{Z}$, $N \geq 2$,

$$S(N) = T(n) + U(N-1).$$

References

[1] L.V. Ahlfors. *Complex Analysis (third edition)*. McGraw-Hill, 1979.

[2] K. Alladi. Some new observations on the Gollnitz-Gordon and Rogers-Ramanujan identities. *Trans. Amer. Math. Soc.*, 347:897–914, 1995.

[3] G.E. Andrews. A simple proof of Jacobi's triple product identity. *Proc. Amer. Math. Soc.*, 16:333–334, 1965.

[4] G.E. Andrews. Enumerative proofs of certain q-identities. *Glasgow Math. J.*, 8:33–40, 1967.

[5] G.E. Andrews. *On the general Rogers-Ramanujan theorem*, volume Number 152 of *Memoirs of the Amer. Math. Soc.* Amer. Math. Soc., 1974.

[6] G.E. Andrews. On Rogers-Ramanujan type identities related to the modulus 11. *Proc. London Math. Soc.*, 30:330–346, 1975.

[7] G.E. Andrews. *The Theory of Partitions*, volume 2 of *Encyclopedia of Mathematics and its Applications*. Addison-Wesley, 1976.

[8] G.E. Andrews. An introduction to Ramanujan's "lost" notebook. *Amer. Math. Monthly*, 86:89–108, 1979.

[9] G.E. Andrews, R. Askey, and R. Roy. *Special Functions*, volume 71 of *Encyclopedia of Mathematics and its Applications*. Cambridge University Press, 1999.

[10] G.E. Andrews and R.A. Askey et al. (editors). *Ramanujan Revisited*. Academic Press, 1988.

[11] A.O.L. Atkin. Proof of a conjecture of Ramanujan. *Glasgow Math. J.*, 8:14–32, 1967.

[12] A.O.L. Atkin. Congruences and modular forms. In *Computers in Mathematical Research*, pages 8–19. North-Holland, 1968.

[13] A.O.L. Atkin. Ramanujan congruences for $p_{-k}(n)$. *Can. J. Math.*, 20:67–78, 1968.

[14] A.O.L. Atkin and J.N. O'Brien. Some properties of $p(n)$ and $c(n)$ modulo powers of 13. *Trans. Amer. Math. Soc.*, 126:442–459, 1967.

[15] W.N. Bailey. Identities of the Rogers-Ramanujan type. *Proc. London Math. Soc.*, 50:1–10, 1949.

[16] W.N. Bailey. On the simplification of some identities of the Rogers-Ramanujan type. *Proc. London Math. Soc.*, 1:217–221, 1951.

[17] A.F. Beardon. *The Geometry of Discrete Groups*, volume 91 of *Graduate Texts in Mathematics*. Springer-Verlag, 1983.

[18] B.C. Berndt. *Ramanujan's Notebooks Parts I-V*. Springer-Verlag, 1985–1998.

[19] B.C. Berndt. Ramanujan's modular equations. In *Ramanujan Revisited*, pages 313–333. Academic Press, 1988.

[20] B.C. Berndt and D.C. Bowman. Ramanujan's short unpublished manuscript on integrals and series related to euler's constant. In *Constructive, experimental, and nonlinear analysis (Limoges, 1999)*, pages 17–27. American Mathematical Society, 2000.

[21] B.C. Berndt, H.H. Chan, and S.-S Huang. Incomplete elliptic integrals in Ramanujan's lost notebook. *Contemporary Math.*, 254:79–126, 1998.

[22] B.C. Berndt and P.T. Joshi. *Chapter 9 of Ramanujan's Second Notebbok, Infinite series identities, transformations, and evaluations*, volume 23 of *Contemporary Mathematics*. American Mathematical Society, 1983.

[23] B.C. Berndt, R.L. Lamphere, and B.M. Wilson. Chapter 12 of Ramanujan's second notebook: Continued fractions. *Rocky Mt. J. Math.*, 15:235–310, 1985.

[24] B.C. Berndt and R.A. Rankin. *Ramanujan: Letters and Commentary*, volume 9 of *History of Mathematics*. American Mathematical Society, 1995.

[25] S. Bhargava and C. Adiga. On some continued fraction identities of Srinivasa Ramanujan. *Proc. Amer. Math. Soc.*, 92:13–18, 1984.

[26] S. Bhargava and C. Adiga. Two generalizations of Ramanujan's continued fraction. In *Number Theory, Lecture Notes in Mathematics 1122*, pages 56–62. Springer, Berlin, 1985.

[27] L. Carlitz and M.V. Subbarao. A simple proof of the quintuple product identity. *Proc. Amer. Math. Soc.*, 32:42–44, 1972.

[28] J.L. Drost. A shorter proof of the Ramanujan congruence modulo 5. *Amer. Math. Monthly*, 104:963–964, 1997.

[29] D. Dummit, H. Kisilevsky, and J. McKay. Multiplicative products of η-functions. *Contemporary Math.*, 45:89–98, 1985.

[30] L. Ehrenpreis. Function theory for Rogers-Ramanujan-like partition identities. *Contemporary Math.*, 143:259–320, 1993.

[31] H.M. Farkas. Elliptic functions and modular forms. In *Contributions to Analysis*, pages 133–145. Academic Press, 1974.

[32] H.M. Farkas and Y. Kopeliovich. New theta constant identities. *Israel J. Math.*, 82:133–140, 1993.

[33] H.M. Farkas and Y. Kopeliovich. New theta constant identities II. *Proc. Amer. Math. Soc.*, 123:1009–1020, 1995.

[34] H.M. Farkas, Y. Kopeliovich, and I. Kra. Uniformization of modular curves. *Comm. Anal. Geom.*, 4:207–259 and 681, 1996.

[35] H.M. Farkas and I. Kra. *Riemann Surfaces (second edition)*, volume 71 of *Graduate Texts in Mathematics*. Springer-Verlag, 1992.

[36] H.M. Farkas and I. Kra. Automorphic forms for subgroups of the modular group. *Israel J. Math.*, 82:87–131, 1993.

[37] H.M. Farkas and I. Kra. Automorphic forms for subgroups of the modular group. II: Groups containing congruence subgroups. *J. d'Analyse Math.*, 70:91–156, 1996.

[38] H.M. Farkas and I. Kra. Theta constant identities with applications to combinatorial number theory. *Contemporary Math.*, 211:227–252, 1997.

[39] H.M. Farkas and I. Kra. A function theoretic approach to the Ramanujan partition identities with applications to combinatorial number theory. *Contemporary Math.*, 240:131–157, 1999.

[40] H.M. Farkas and I. Kra. On the quintuple product identity. *Proc. Amer. Math. Soc.*, 127:771–778, 1999.

[41] H.M. Farkas and I. Kra. Ramanujan partition identities. *Contemporary Math.*, 240:111–130, 1999.

[42] H.M. Farkas and I. Kra. Partitions and theta constant identities. *Contemporary Math.*, 251:197–203, 2000.

[43] H.M. Farkas and I. Kra. Special sets of points on compact Riemann surfaces. *Contemporary Math.*, 256:75–94, 2000.

[44] H.M. Farkas and I. Kra. Three term theta identities. *Contemporary Math.*, 256:95–101, 2000.

[45] O. Forster. *Lectures on Riemann Surfaces*, volume 81 of *Graduate Texts in Mathematics*. Springer-Verlag, 1981.

[46] F.G. Garvan. Combinatorial interpretations of Ramanujan's partition identities. In *Ramanujan Revisited*, pages 29–45. Academic Press, 1988.

[47] F.G. Garvan. A combinatorial proof of the Farkas-Kra theta function identities and their generalizations. *J. Math. Anal. Appl.*, 195:354–375, 1995.

[48] G. Gasper and M. Rahman. *Basic Hypergeometric Series*, volume 35 of *Encyclopedia of Mathematics and its Applications*. Cambridge University Press, 1990.

[49] B. Gordon. Some identities in combinatorial analysis. *Quart. J. Math. Oxford*, 12:285–290, 1961.

[50] B. Gordon. Ramanujan congruences for $p_{-k}(n) \equiv 11^r$. *Glasgow Math. J.*, 24:107–123, 1983.

[51] B. Gordon and D. Sinor. Multiplicative properties of η-products. In *Number Theory, Madras*, pages 173–200. Springer Lecture Notes 1395, 1987.

[52] R.C. Gunning. *Lectures on Modular Forms*, volume 48 of *Annals of Mathematics Studies*. Princeton University Press, 1962.

[53] G.H. Hardy. *Ramanujan, Twelve Lectures on Subjects Suggested by His Life and Work*. Chelsea Publishing Company, 1959.

[54] G.H. Hardy, P.V. Seshu Aiyar, and B.M. Wilson (editors). *Collected Papers of Srinivasa Ramanujan*. Chelsea Publishing Company, 1927.

[55] M.D. Hirschhorn. Another short proof of Ramanujan's mod 5 partition congruence, and more. *Amer. Math. Monthly*, 106:580–583, 1999.

[56] K. Hughes. Ramanujan congruences for $p_{-k}(n)$ modulo powers of 17. *Can. J. Math.*, 43:506–525, 1991.

[57] M.E.H. Ismail and D.W. Stanton (editors). *q-Series from a Contemporary Perspective*, volume 254 of *Contemporary Mathematics*. American Mathematical Society, 2000.

[58] H. Iwaniec. *Topics in Classical Automorphic Forms*, volume 17 of *Graduate Studies in Mathematics*. American Mathematical Society, 1997.

[59] V.G. Kac. Infinite dimensional algebras, Dedakind's η-function, classical Möbius function and the very strange formula. *Adv. in Math.*, 30:85–136, 1978.

[60] A.W. Knapp. *Elliptic Curves*, volume 40 of *Mathematical Notes*. Princeton University Press, 1992.

[61] M.I. Knopp. *Modular Functions in Analytic Number Theory*. Chelsea Publishing Company, 1993.

[62] G. Köhler. Theta series on the theta group. *Abh. Math. Seminar Univ. Hamburg*, 58:15–45, 1988.

[63] G. Köhler. Some eta-identities arising from theta series. *Math. Scand.*, 66:147–154, 1990.

[64] Y. Kopeliovich and J.R. Quine. On the curve X(9). *The Ramanujan J.*, 2:371–378, 1998.

[65] I. Kra. *Automorphic Forms and Kleinian Groups*. Benjamin, 1972.

[66] I. Kra. On the vanishing of and spanning sets for Poincaré series for cusp forms. *Acta Math.*, 153:47–116, 1984.

[67] I. Kra. On a Ramanujan partition identity. In *Proceedings of the Ashkelon Conference on Complex Function Theory (1996)*, pages 167–172. Israel Conference Math. Proceedings, volume **11**, Bar-Ilan University, 1997.

[68] D.H. Lehmer. Properties of the coefficients of modular invariant $j(\tau)$. *Amer. J. Math.*, 64:488–502, 1942.

[69] J. Lehner. Ramanujan identities involving the partition function for the moduli 11^α. *Amer. J. Math.*, 65:495–520, 1943.

[70] J. Lehner. Divisibility properties of the Fourier coefficients of the modular invariant $j(\tau)$. *Amer. J. Math.*, 71:136–148, 1949.

[71] I.G. Macdonald. Affine root systems and Dedakind's η-function. *Invent. Math.*, 15:91–143, 1972.

[72] Y. Martin. Multiplicative η-quotients. *Trans. Amer. Math. Soc.*, 348:4825–4856, 1996.

[73] Y. Martin and K. Ono. η-quotients and elliptic curves. *Proc. Amer. Math. Soc.*, 125:3169–3176, 1997.

[74] B. Maskit. *Kleinian Groups*, volume 287 of *Grundlehren der mathematischen Wissenschaften*. Springer-Verlag, 1988.

[75] C. McMullen. Amenability, Poincaré series and quasiconformal maps. *Invent. Math.*, 97:95–127, 1989.

[76] R. Miranda. *Algebraic Curves and Riemann Surfaces*, volume 5 of *Graduate Studies in Mathematics*. American Mathematical Society, 1995.

[77] L.J. Mordell. Note on certain modular relations considered by Messrs. Ramanujan, Darling, and Rogers. *Proc. London Math. Soc.*, 20:408–416, 1922.

[78] M. Newman. Remarks on some modular identities. *Trans. Amer. Math. Soc.*, 73:313–320, 1952.

[79] M. Newman. An identity for the coefficients of certain modular forms. *J. London Math. Soc.*, 30:488–493, 1955.

[80] M. Newman. On the existence of identities for the coefficients of certain modular forms. *J. London Math. Soc.*, 31:350–359, 1956.

[81] M. Newman. Congruences for the coefficients of modular forms and some new congruences for the partition function. *Can. J Math.*, 9:549–552, 1957.

[82] M. Newman. Some theorems about $p_r(n)$. *Can. J. Math.*, 9:68–70, 1957.

[83] M. Newman. Further identities and congruences for the coefficients of modular forms. *Can. J. Math.*, 10:577–586, 1958.

[84] M. Newman. Construction and application of a class of modular functions II. *Proc. London Math. Soc.*, 9:373–387, 1959.

[85] M. Newman. Modular functions revisited. In *Analytic Number Theory*, pages 396–421. Springer Lecture Notes 899, 1980.

[86] S. Ramanujan. Some properties of $p(n)$, the number of partitions of n. *Proc. Camb. Phil. Soc.*, 19:207–210, 1919.

[87] S. Ramanujan. Congruence properties of partitions. *Math. Z.*, 9:147–153, 1921.

[88] S. Ramanujan. *Notebooks (2 volumes)*. Tata Institute of Fundamental Research, 1957.

[89] S. Ramanujan. *Collected Papers*. Chelsea, 1962.

[90] S. Ramanujan. *The Lost Notebooks and other Unpublished Papers*. Narosa, 1988.

[91] L.J. Rogers. Second memoir on the expansion of certain infinite products. *Proc. London Math. Soc.*, 25:318–343, 1894.

[92] L.J. Rogers. On two theorems of combinatory analysis and allied identities. *Proc. London Math. Soc.*, 16:315–336, 1917.

[93] I. Schur. Ein Beitrag zur additiven Zahlentheorie und zur Theorie der Kettenbrueche. *S.-B. Preuss. Akad. Wiss. Phys.-Math. Kl.*, pages 302–321, 1917.

[94] D.B. Sears. Two identities of Bailey. *J. London Math. Soc.*, 27:510–511, 1952.

[95] C. Series. Lectures on pleating coordinates on once punctured tori. *RIMS (Kyoto Univ.) Kokyuroku*, 1104:31–90, 1999.

[96] G. Shimura. *Introduction to the Arithmetic Theory of Automorphic Functions*, volume 11 of *Publications of the Mathematical Society of Japan, Kanô memorial lectures 1*. Iwanami Shoten and Princeton University Press, 1971.

[97] L.J. Slater. Further identities of the Rogers-Ramanujan type. *Proc. London Math. Soc.*, 54:147– 167, 1952.

[98] G.N. Watson. A new proof of the Rogers-Ramanujan identities. *J. London Math. Soc.*, 4:4–9, 1929.

[99] G.N. Watson. Ramanujans Vermutung über Zerfällungsanzahlen. *J. Reine Angew. Math.*, 179:97–118, 1938.

[100] H. Weyl. *The Concept of a Riemann Surface*. Addison-Wesley, 1955.

[101] L. Winquist. An elementary proof of $p(11m+6) \equiv 0 \pmod{11}$. *J. Comb. Th.*, 6:56–59, 1969.

Index